Out of the Crystal Maze

Out of the Crystal Maze

Chapters from the History of Solid-State Physics

Edited by

LILLIAN HODDESON
University of Illinois

ERNEST BRAUN
Open University

JÜRGEN TEICHMANN
Deutsches Museum

SPENCER WEART
American Institute of Physics

New York   Oxford
OXFORD UNIVERSITY PRESS
1992

Oxford University Press

Oxford   New York   Toronto
Delhi   Bombay   Calcutta   Madras   Karachi
Kuala Lumpur   Singapore   Hong Kong   Tokyo
Nairobi   Dar es Salaam   Cape Town
Melbourne   Auckland

and associated companies in
Berlin   Ibadan

Library of Congress Cataloging-in-Publication Data
Out of the crystal maze : Chapters from the history of solid-state physics
edited by Lillian Hoddeson . . . [et al.].
p. cm.   Bibliography: p.   Includes index.
ISBN 0-19-505329-X
1. Solid-state physics—History.   I. Hoddeson, Lillian.
QC176.O98 1990
530.4′1′09—dc20   89-33498

196711167

9 8 7 6 5 4 3 2 1

Printed in the United States of America
on acid-free paper

Foreword

Future generations looking back on the twentieth century, with its tremendous variety of intellectual and social developments, may find none of these more significant than the way in which solid-state physics has become of major importance. At the turn of the century, it was of interest to only a few specialists. In the meantime, the discoveries have opened a new world not only for such obvious fields as communications, computation, and entertainment, but also in fields ranging from astronomy to defense. It is evident that the story of this great development must be an important component of the history of our times. Remarkably enough, that history has remained virtually unknown not only to historians and the public at large, but even to many young investigators who work in the field.

We have, therefore, watched with pleasure and excitement the creation and progress of the International Project on the History of Solid State Physics. The historians who have joined in the work were not too late to talk directly with most of the pioneers who created the field and brought it to fruition from the 1920s through the 1950s. In addition to spending several hundred hours with these pioneers, the historians spent even more time examining private correspondence, laboratory records, journals, and textbooks.

As we reviewed the drafts of this book, we found that the project had gone far to explain how solid-state physics began and how it grew into a central scientific discipline. Our only regret is that still more remains to be done. Everyone involved in the project has become aware that not only individual contributions to knowledge but entire areas of solid-state physics have had to be omitted for lack of space and time. It is our hope that this book will stimulate others, both physicists and historians, to look more deeply into the broad and fascinating history of the field. To aid in this process, the American Institute of Physics will issue a catalog of historical source material for the benefit of those who wish to explore further.

Knowing the history of a subject adds enormously to our appreciation of its more detailed aspects today. It is our hope that, in addition to shedding new light on how solid-state physics developed, the program that has led to the production of this volume will provide a model for those in other fields of science that deserve similar attention.

E. Mollwo
Nevill Mott
Frederick Seitz

Preface

The following chapters from the history of solid-state physics are a result of what may well be the largest international research project ever done in the history of modern science. To understand the book, the reader should know why and how this project emerged and what it has attempted to accomplish. We would also like to explain its limitations. The book covers a restricted period of time, and even inside that period it does not pretend to give a complete description. We do not come as cartographers with a detailed and definitive map of a country; we come rather as explorers of an almost unknown land, with sketches of those regions we have been able to penetrate and reports of some remarkable features there.

The International Project in the History of Solid State Physics

The term *solid-state,* popularized by advertisers' tags on transistorized devices, properly refers to a field of basic research that grew rapidly to maturity between 1920 and 1960—the branch of physics dealing with the properties of solid matter. This field has enormous scope and addresses most of the materials that humans handle and use; each of these materials has a multitude of properties, from hidden internal structures to obvious external characteristics, such as cohesion and response to heat and light.

In both fundamental and applied research, solid-state physics has been overwhelmingly successful. Many simple and ancient questions about things immediately at hand, once among the first questions of philosophers, have been confidently answered. Why, for example, is ordinary glass transparent, fragile, and a poor conductor of heat, whereas metals are opaque, ductile, and good conductors? The fact that such questions about everyday appearances have now been answered may have unexpected effects on common human understanding. Within science itself, certain solid-state developments (such as the discovery of the role of "defects" in the properties of crystals) imply a fundamentally new way of thinking, an acceptance of the irreducible complexity of real objects.

Once questions about solids were answered, people could proceed to applications of the highest economic, military, and cultural importance. Devices from electric motors to jet airplanes—indeed, nearly all of modern physical technology—relies on solid-state physics. In particular, the vitally important semiconductor industry was built, far more directly than anything in prior history, on pure scientific thought as much as on chance phenomenological discoveries and craft techniques. Solid-state physics is thus the main scientific impulse behind the revolution in information and communications, with its consequent fundamental changes in society. A case could be made that no special field of scientific inquiry is as important to the history of the late twentieth century as solid-state physics.

Yet only within the past decade have historians of science begun to pay any attention to this field. Although numerous articles and books documented the rise of quantum mechanics, relativity theory, nuclear technology, and molecular biology, solid-state physics was invisible. Even general surveys of the history of physics went past the field with scarcely a mention, although for many decades many physicists, sometimes almost a majority, have worked on the properties of solids.

We can guess some of the causes for this neglect: the field is huge and varied and lacks the unifying features beloved of historians—neither a single hypothesis or set of basic equations, such as quantum mechanics and relativity theory established for their fields, nor a single spectacular and fundamental discovery, as uranium fission did for nuclear technology or the structure of DNA for molecular biology. Solid-state physics owes what unity it has simply to a common concern with solid materials, but this quality allows such a large range of theories and methods that the field resembles a confederation of interest groups rather than a single entity. And even this confederation is recent; before the Second World War, the various strands that would form solid-state physics existed as isolated and seemingly peripheral subjects. Historians have only recently begun to probe postwar science, and it would have been too much to expect them to treat prewar solid-state physics as a single field when even physicists did not perceive the unity. Finally, what identity solid-state physics has attained comes largely from a shared concern for applied questions of interest chiefly to industry, and historians who are themselves academics tend to be more interested in academic science. Except for a small minority who have personally had at least a few months' experience in an industrial establishment, we do not know of any historian of modern science who has studied physics in industry.

By the early 1970s, several eminent solid-state physicists had spoken out on the urgent need to preserve documentation and to write a well-researched history of their field. In the United States, Cyril Stanley Smith wrote letters to historians and physicists urging them to arrange a conference to open up the subject. Other progenitors of a research project include Frederick Seitz and John Bardeen in the United States, and Margaret Gowing and Sir Nevill Mott, in Britain. The need for solid-state history was also noted by the Advisory Committee of the Center for History of Physics at the American Institute of Physics (AIP), but the center was too busy with other projects to do much. By the mid-1970s, a few solid-state physicists, including two editors of this volume (Lillian Hoddeson and Ernest Braun), had begun serious historical work on solid-state developments. But it became clear that, given the great breadth and complexity of solid-state physics, progress in the traditional mode of research would be painfully slow.

A break occurred when Mott organized a small meeting on the history of solid-state physics. At the meeting, held from 30 April to 2 May 1979 in the rooms of the Royal Society in London, 15 leading solid-state physicists of the older generation presented memoirs on their contributions to the subject. Historians of science were also present: Braun and Gowing from Britain, Hoddeson from the United States, and Karl von Meyenn from West Germany. The papers presented at the meeting, or prepared for it, were published in the *Proceedings of the Royal Society* and as a separate volume edited by Mott, *The Beginnings of Solid State Physics* (London: Royal Society, 1980).

Those at the meeting unanimously felt that, useful as this exchange of memories was, it could not transcend anecdotal reflection. Serious academic work was needed on the history of this vital branch of physics. Braun undertook to organize a British project, and Hoddeson undertook the same for the United States. The idea of conducting historical work with teams of scholars in more than one country had little precedent, but it increasingly seemed necessary in order to present a sufficiently broad picture within a reasonable time. To make use of the reflections and files of pioneers in the field while they were still living, we believed it was important to conduct the research immediately, aiming for results within three to five years.

As a result of Braun's efforts, substantial financial support was obtained for a British solid-state history project. (Sources of funds are noted in the Acknowledgments.) A small research group was established in the Technology Policy Unit of the University of Aston in Birmingham; the group's members were Braun, Paul Hoch, and Stephen Keith. Meanwhile, Hoddeson worked with Spencer Weart, the director of the AIP Center for History of Physics, and Bardeen and Seitz, to organize a United States project. The AIP center, following the advice of its advisory committee, had independently decided that solid-state history should be its next major project, and, coincidentally, Seitz had become head of the Friends of the Center for History of Physics, a fund-raising group. Grants were obtained, and a research group was established in the Department of Physics of the University of Illinois at Urbana-Champaign; the group's members were Hoddeson, Gordon Baym, Paul Henriksen, Jerome Rowley, and Krzysztof Szymborski. Weart and Joan Warnow of the AIP were associated with the project throughout its existence. Steve Heims later joined as a consultant based in Cambridge, Massachusetts.

The British and American research groups were supported by separate advisory committees, which consisted largely of senior physicists but also included historians and sociologists of science. (The members are listed in the Acknowledgements.) The first meetings of the advisory committees were devoted to establishing which parts of solid-state physics should be included in the research of the international project and what form the publication of results should take. Particularly influential meetings were held at the Institute for Theoretical Physics at the University of California, Santa Barbara, in February 1980, at the AIP Center for History of Physics in New York City in November 1980, and at the University of Aston in Birmingham, England, in December 1980.

It was decided to publish the project's results in three ways. The main results would be collected in a single volume that would attempt to relate as full and as accurate a history of solid-state physics as could be assembled with the time and funds available. It would be written chiefly for people with some education in physics, including basic quantum mechanics, although some sections would be accessible to others. Subsequently, the AIP center would publish a guide to source materials for the history of solid-state physics; this guide would contain details of relevant archives of correspondence, laboratory notebooks, oral history interviews, and other materials to make it easier for future historians to find the vast amount of raw material uncovered during the course of our work. Finally, members of the research teams were permitted to publish additional material in whatever form they chose. Oxford University Press agreed to publish the main volume—the present work.

One further meeting shaped the method of work and the forms of cooperation. This meeting, held in Stuttgart in April 1980, was hosted by Armin Hermann and Karl von Meyenn and was supported by the Stiftung Volkswagenwerk. At this meeting, several members of the American and British research teams, supplemented by other invited historians, discussed the novel prospect of writing history through a collective international effort.

It was becoming clear that the subject was too large and too poorly explored to be tackled in a few years by historians acting as isolated individuals. The subject was also too recent to be reduced with adequate historical perspective to a tiny number of truly vital aspects. There seemed no option but to create a team with a scope unprecedented in work on the history of science, perhaps on any field of history. The team had to span geographical borders, for solid-state physics had roots in many countries and it would not have been practical for a single institution to approach all the national sources without burdening the project with exorbitant expenses and, most probably, national jealousies. It was therefore decided to construct a team consisting of at least three national groups.

We would tackle each separate portion of the subject through the personal cooperation of a few authors, but information and commentary would generally be supplied by more than one national group. Each author and co-author would carry full responsibility for his or her own work, and all help by suppliers of information and commentary would be suitably acknowledged. The method was designed to foster teamwork without detriment to individual effort, scholarship, and responsibility.

As a further outcome of the Stuttgart meeting, efforts to establish a German research group were redoubled. Braun and Jürgen Teichmann of the Deutsches Museum in Munich obtained funds for such a group, to be based at the Deutsches Museum; the members were Teichmann, Michael Eckert, Helmut Schubert, and Gisela Torkar. This group started a little later than the others, in the spring of 1981, and, like the other groups, was supported by its own advisory committee.

A fourth national group was formed in the latter half of 1981. Based at the Parc de la Villette science museum in Paris, it consisted mainly of Christine Blondel and Pierre Quédec. André Guinier was instrumental in facilitating the establishment of the French group and its cooperation with the other, already established groups. In Italy, two historians of physics (S. Galdabini and G. Giuliani, University of Pavia) started their own intensive work at the end of 1983, and in Japan Atsushi Katsuki and others gave individual assistance. It would, of course, have been extremely desirable to establish a substantial group in the Soviet Union as well, but we were able to obtain no more than rather loose contacts, made possible chiefly through Viktor Ya. Frenkel in Leningrad. We have not been able to describe the important contributions of Russian scientists as fully as we would have liked. Indeed, more generally the authors naturally tended to concentrate on work within their own geographical region; further research in many places will be needed before the complete story of developments can be written.

The efficiency of the research efforts by the national groups was much enhanced by annual meetings of the entire international team, at which the state of the art and details of collaboration were thrashed out. Three such meetings were held: at the University of Illinois in Urbana in October 1981, at the Deutsches Museum in Munich in September 1982, and at the Parc de la Villette in Paris in September 1983. Each meeting filled several busy days and ironed out many problems.

Authors often needed to use sources in other nations, and it was here that cooperation met its most severe tests. Sometimes authors were able to travel in person, a welcome opportunity to meet with other team members, who gave much assistance in using their local sources. More normal, if more difficult, was collaboration by mail.

More detailed notices of selected source materials, along with progress reports from each group and general information, were issued in three newsletters published by the AIP Center for History of Physics. The first was distributed in May 1982; the second, in February 1983; and the third, in February 1984. It was hoped that the newsletters would be another mechanism to keep team members aware of each group's work, but in practice the newsletter turned out to be too slow for this purpose, and it served mainly to keep outside historians and physicists informed and interested in the project's progress.

In retrospect, given the delays in communication, the collaboration would probably have been much closer if the project could have lasted for at least six years. However, the financial support lasted for only three years, and the work was scheduled to be completed by the spring of 1984. The editorial work was originally planned to be completed by the end of 1984, and the manuscript was to have been delivered to Oxford University Press early in 1985. As might have been expected, everything took longer than scheduled, except, of course, the financial support.

This Book and Its Method of Research

This book tells the story of the emergence of modern solid-state physics out of a collection of unrelated industrial, scientific, and even artistic efforts. These ranged from observations on the symmetry of natural crystals, through technological concerns like the manufacture of vacuum-tube filaments and the development of pliable alloys, to academic theories of the electron and the quantum. Almost the only thing the efforts had in common was an interest in understanding and controlling the properties of real solids.

Determining just when the emergence of solid-state physics took place would depend on the answer to important philosophical questions, such as how one defines a field of science. Some would date the formation of a field from the first key experiments; for solid-state physics, one might choose the classic x-ray diffraction experiments in 1912 that first revealed the crystal lattice. Others would say a field is created along with the first successful fundamental theory; here one could begin with the quantum theory of solids, developed from 1926 on. A very different criterion is sociological: the formation of the first specialized departments, institutions, research programs, textbooks, and courses; for solid-state physics, that took place between 1935 and 1950. Or one could start with the earliest discoveries of relevant phenomena and the first speculations about their nature; that would take us all the way back to primitive chippers of flint. Despite such ambiguity, it is still meaningful to search for a starting date, for this points to explanations of why and how the field emerged.

Scientific and technical endeavors that hindsight recognizes as solid-state physics can be found in antiquity, from philosophers' speculations about the nature of solids to glassmakers' craft secrets, and such efforts redoubled as modern science grew from the sixteenth century on. We have not taken on the task of examining this immense and fascinating prehistory, for it did not lead directly and uniquely into the modern field. Rather, the early theories, experiments, and techniques were an inseparable part of the entire rise of academic and industrial research. That rise was related just as closely to many other factors: on the one hand, social factors such as European nationalism and imperialism, the industrial revolution, and the advent of institutionalized education in science and technology; on the other hand, intellectual factors, including all the great breakthroughs in physical science—each of which concerned solids in part, but only in part. The historical perspective of this book therefore opens with the nineteenth century, when all these factors became plain, and the chief narrative begins in the first decades of the twentieth century as the strands of solid-state physics began to separate from the tangle.

Our terminal point is more definite. During the 1950s, many physicists around the world recognized that they were members of a distinct field of science—that they were in fact "solid-state physicists." By the end of the decade, meanwhile, some longstanding puzzles around which the field had originally coelesced had been solved, from the source of colors in crystals to the explanation of superconductivity. In parallel with the intellectual development, physicists established a full range of sturdy institutions for their field, and solid-state technology attained tremendous practical importance. From these roots, the field would grow luxuriously in many directions, and it would take a hundred historians to trace all the branches. We therefore halt around 1960.

This book begins with two introductory chapters examining the chief strands that would later come together as solid-state physics. Chapter 1 describes major scientific and social developments that preceded the discovery of quantum mechanics. Among these developments were new microscopic conceptions of matter introduced around the turn of the century, the development of new experimental means for studying solids, and the rise of industrial research. Chapter 2 deals with the emergence from 1926 to 1933 of the quantum theory of solids. This theory provided the essential theoretical nucleation center for the field, setting the terms for explaining almost every subsequent discovery.

The main period of our history is roughly 1930 to 1960, although individual chapters set their own periods by the main events that shape their subjects. Five major themes are examined in parallel through this time period. Chapter 3 continues the story of Chapter 2, discussing how understanding grew of the band structure of electron energies in solids and of the Fermi surface, which is the key feature of this structure. We then turn from idealized theory to the phenomenology of real crystals. Chapter 4 tells how physicists, beginning with color centers and using them as a model for later research, discovered that point defects were the key to numerous crystal properties. Chapter 5 focuses on the ancient puzzle of the plastic properties of solids, which was solved with the discovery of dislocations and the proof of their existence. Chapter 6 addresses magnetism, studying cases such as the important

French school in order to give some insight into this complex and sprawling field of inquiry. Chapter 5 takes up another enormous field, semiconductor physics, and looks into topics such as the role of impurities and the lifetimes of injected electrons, which are so important for the industrial development of semiconductors. Chapter 8 returns to the quantum-theoretical approach of Chapters 2 and 3. It describes how this line of inquiry culminated in the theory of many-particle systems and collective phenomena, thereby explaining old puzzles such as superconductivity and superfluidity. In treating each of these themes, we tried to start at a point where modern ideas began to emerge and to finish at a point in the 1950s where an interim plateau was reached.

Finally, Chapter 9 discusses the internal social structure of the field. It relates how a coherent solid-state-physics community emerged from the dispersed social system described in Chapter 1. However, all the authors agreed that social factors could not be ignored in any chapter. Each chapter includes much information about particular research schools (for example, their organization and sources of funds) and about the influence of historical events beyond the confines of science.

For the benefit of readers who are not historians, we wish to say a few words about our research methods. The first sources of information were not only the scientific journals in which the major contributions were first published, but also textbooks, which represent a more consolidated form of consensus knowledge; monographs, an in-between category; and the limited secondary literature, including the work of earlier historians of science and the published reminiscences and biographies of scientists. We also searched intensively for archival materials, such as letters, notebooks, photographs, apparatus, and diaries. In every category, we found more material than we could possibly use; yet we also became painfully aware that much archival material that would have been of immense help has been irretrievably lost. Historians and archivists have become interested in the subject too late, and scientists themselves usually attached too little value to historical source materials, particularly apparatus and instruments. The Second World War also destroyed valuable materials, especially in Germany. In certain cases, however, a great wealth of material is available—for example, correspondence and notes of John Bardeen and of Walter Schottky.

This brings up a fundamental problem: unpublished materials greatly enrich and enliven the history of a particular area, but they may distort the relationship to other areas where correction through unpublished materials is not possible. A case in point is the relative inaccessibility of many business archives, for this strengthens the historian's tendency to focus on academic rather than industrial science. We have tried to find and make balanced use of every category of material, and in the companion guide to source materials we hope to pave the way for future historians to correct our omissions.

Besides reading published and unpublished written sources, we used a third, well-established approach in the investigation of recent history: we interviewed many of the major participants. This process mainly involved speaking to older members of the solid-state-physics community who had reached some prominence through their work in shaping the development of the field. This method proved extremely valuable. Not that all "facts" related in interviews are necessarily correct—memory fades and distorts—although some memories do seem able to produce a remarkably faithful record of events long past. Whenever possible, factual information was cross-checked against other interviews and written information. In every case, the notes identify reminiscences as such, so that the reader will not mistake these for confirmed historical fact. Even in uncertain cases, the interviews are a unique source for suggesting connections among people and for pointing to the importance of small events never set down on paper. Above all, the interviews describe personal motivations and inspirations; even if told only in hindsight, the testimony says much about the personalities of the scientists. There is nothing like oral history for giving a dead body of facts something akin to a soul.

Whenever possible, the interviews were recorded on tape, and permission was obtained for future scholars to use the information. Many of the tapes were transcribed, edited, and provided with indexes by the AIP center. Most tapes, whether transcribed or not, are available at the Niels Bohr Library of the American Institute of Physics in New York City. We are extremely grateful to all those we interviewed for their time and effort, which in a very real sense made the writing of the history possible. Often we would scarcely have known where to begin without some guidance from leading older figures of the field. This is their history in a double sense, the sense of doing and the sense of relating.

The Choice and Treatment of Topics

The principal aim of the solid-state-history project has been to open a new area of study in the history of science. From the beginning, we knew that we could not possibly write a complete history, in any sense of the word *complete*, for ours is the first extended reconnaissance of a highly complex and still rapidly developing field. We do not hope or even desire to say the final word, but will judge our success above all by the extent to which our work stimulates others to carry on the study of the subject.

All writing of history implies distillation from large amounts of raw material, a valuation of what is important and what is less so. Writing history also requires drawing boundaries around one's topic. Unfortunately, in our case the application of these common constraints would still have left too large a task: we had no small region but a continent to explore, and almost everywhere we had to make our own maps from scratch. We had to face the reality that we could do justice to only parts of the subject.

The final choice of topics was made through a combination of selection or elimination following deliberate criteria, together with the choices of the team members' personal interests and available time. We considered that importance was imparted to a topic either by fundamental scientific significance or by the role the topic played in technological innovations. The electron theory of metals is an example of a topic with strong claims for inclusion on the grounds of scientific significance, and the physics of semiconductors, so vital to an entire industry, found an equally assured place in our history. On the other hand, subjects that are currently extremely active and turbulent do not allow the historical perspective of 20 years or more that is commonly recommended; therefore, any topic that had not reached a plateau, albeit a temporary one, by the early 1960s seemed unapproachable at present. For example, the physics of amorphous solids fell a victim to such discrimination. Some topics that all agreed it would be well to explore in detail—low-temperature research and surface physics are but two examples—simply had to fall by the wayside because we lacked the resources to venture in every direction. The choice or elimination of the various topics was endorsed by our scientific advisers, and we feel this book presents an enlightening, if far from complete, first look at our subject.

Within the selected topics, further choices of detail were necessary; this followed the authors' judgment, often in close contact with scientist advisers. Other historians would probably disagree on details of choice, for this is where history becomes a personal art. Additional decisive factors were, again, the limited time and staff, as well as some gaps in the available source materials.

Cutting up the material into discrete topics naturally meant that internal relationships and cross-fertilization would not be fully displayed. For example, the close relationship between dislocations and point defects was obscured by dividing the topic into two separate chapters. Many concerns that run through all of solid-state physics had to be divided or treated superficially. Among these we would like to note the concern with real rather than ideal materials, the effects of "dirt" or impurities, the use of approximation, and the vital role of measuring and instrumental techniques. Social factors too, such as the formation and influence of scientific schools and the crucial influences of politics, could be touched on only intermittently. Readers will encounter these factors again and again in the following chapters and should draw their own conclusions.

The way to approach such overarching questions has long been debated among historians of science. The two extreme approaches are known as "internalist" and "externalist." A strictly internalist historian feels that the development of science follows its own internal logic, one discovery leading to another, as scientists choose their path according to scientific judgment and personal research interests. An externalist feels that science is buffeted by social and economic forces that drive it in externally determined directions, thus leaving scientists with mere remnants of free choice.

Few historians remain at either extreme. Externalists admit that "nuggets" of science may develop according to internal logic while only the surrounding field falls under social influences; internalists admit that a research effort will advance more rapidly if its practical possibilities attract generous funding. Some reject the externalist–internalist division altogether. It is our view that the relationships of "real" history, in contrast to written and interpreted history, are fundamentally more complicated than any historian could depict. Therefore, different methods of historical treatment may be appropriate for different historical subjects.

For the history of solid-state physics we have found that no distinct barriers existed between fundamental and applied research or between pure science and the needs of society. This lack of distinct barriers, perhaps seen most clearly in technical developments such as instrumentation, was very closely related to both fundamental and applied research, and vigorously stimulated them both.

Techniques were sometimes developed in the academic laboratory and for the laboratory, but they quickly entered a common pool of knowledge. Such laboratory techniques often grew into technologies with broad applications. One example is the art of growing single crystals of metals and then, infinitely more important, of producing single crystals of semiconductors. These became the raw material for an entire new industry. Other examples may be found among scientific instruments, which usually arise out of disinterested research and the discovery of new phenomena, but which engineers may take up and develop into something of great commercial or military value. The solid-state infrared detector is a case in point. The flow of technique in the opposite direction is just as important; throughout the history of solid-state physics, we find academic researchers dependent on pure materials and instruments that they can get only from their industrial colleagues. Technique and instrumentation are only the most tangible of many links between science and technology. The deliberate application of science to the search for and development of technological innovations is one of the hallmarks of our time and has affected the course of science as much as the course of technology. To disregard the role of technological innovation in the development of solid-state physics would be as foolish as disregarding the role of solid-state physics in technological innovation.

At a subtler level are influences from the dominant beliefs and values of the time. Compartmentalized though humans may be, the compartments are permeable and scientists cannot escape widespread demands and fashions. Sometimes these constraints are openly expressed in job opportunities and support for research, which are determined not only by scientific timeliness, but also by economic and political factors—that is, by perceived social needs. Examples of these factors are business opportunities, "defense" requirements, and less material goals, such as the current concern for environmental problems. At the crudest level, extreme political regimes coerce scientists to do their bidding, and armed conflicts engulf science as much as other spheres of life.

The Interrelation of Industry, Politics, and Science as Solid-State Physics Emerged

The 1930s and 1940s were a time of overwhelming political change, turbulence, and tragedy. These tumultuous events influenced solid-state physics too profoundly for us to ignore. We give an overview of their influence here so that readers can understand the examples dispersed throughout the book.

We see three developments as most important for the history of our field. First, by 1950, physics, particularly solid-state physics, had become an established part of industrial science, which, in turn, had gained a major place in economic life—what we might call the "scientific–technical complex." From an unobtrusive occupation of academic gentlemen, science became a central economic activity, part of a concerted effort to produce ever new goods and satisfy ever new perceived needs. The second important influence was the emigration of scientists from German-dominated countries, primarily to the United States and Britain. The Fascist seizure of power launched a wave of emigration by people threatened in their livelihood and their lives because of their origins or their beliefs, and by people sickened beyond endurance by the cruelty of the new regimes. Finally, faced with the appalling carnage of the Second World War, all sides mobilized physics as an essential part of their military effort. This mobilization did not entirely end in 1945: ever since then, a major part of the funding for solid-state physics has flowed more or less directly from national military establishments.

Of these developments, the most pervasive influence on solid-state physics has been its connection with industry. Before the First World War, the world knew only a handful of industrial research laboratories, concentrated mostly in the chemical, pharmaceutical, rubber, electrical, optical, and photographic industries. The First World War accelerated the rise of such laboratories in several ways. It cut off the belligerents from normal supplies of raw materials and industrial products, so domestic substitutes had to be developed, and it introduced novel technical means, such as poison gases, submarines, and airplanes. Scientists, industrialists, and the public all came more than ever before to see science as a key element in national strength. Many felt that the end of the war only turned national rivalries back to the sphere of commerce. The incorporation of physics into efforts to increase competitiveness by technical excellence and innovation was accelerated and became firmly established during the interwar years. Many new industrial laboratories were founded, cooperation between universities and industry was strengthened, the number of physicists engaged in industry became an increasing proportion of a rising total, and universities ceased to be the sole suppliers of scholarly papers to journals.

The ravages of the Great Depression interrupted the movement for only a few years, after which industrial science, particularly solid-state physics, returned to its steady upward curve. The Second World War did nothing to slow this, and afterward, just as after the First World War, the proportion of physicists going into industry moved still higher. By this time, the interaction between science and industry had become essential to both.

The nature of these interactions is clear in the famous example of semiconductors, the first economically central technical development to grow almost entirely out of systematic scientific investigations, in the style if not always on the premises of academe. In return, important scientific advances were stimulated by the intense commercial and military interest in semiconductors and by the resulting powerful new techniques. Scientists have never been "free" in their research, least of all in the areas of solid-state physics that are close to technical developments.

Our second major influence, the emigration of Central European scientists, was less important for its effect on the world as a whole than for changing the balance between nations. Viewed from the comfortable position of hindsight, the migration of many talented and trained refugees undoubtedly greatly strengthened American and British science, while it destroyed a large part of German science. The process was not, however, painless to anyone. Apart from the hardship suffered by flight from a homeland that had become monstrous, the problems of absorption and reestablishment were formidable. The stagnant economies of the early 1930s lacked the absorptive capacity of growing systems, and many refugee scientists had difficulty finding appropriate, or indeed any, jobs. Eventually, a good number of them became established in and integrated into their host countries, with undoubtedly beneficial effects to the latter. The center of gravity of solid-state-physics research shifted decisively to the United States and, to a lesser extent, to Britain, while the remnant of German science took a generation to reach respectability again and never resumed its earlier preeminence.

The emigration had already begun in the 1920s, when Americans recognized that they lagged far behind Europe in theoretical physics and set out to catch up. Students went to Europe for postgraduate training, while their elders brought foreign scholars on visits to the United States. Some of the visitors came to stay, finding better economic opportunities than in the overcrowded European universities. However, the subsequent pressure of Nazi anti-Semitism and other forms of persecution greatly accelerated the process, so that physics in the United States not only caught up with but wholly surpassed the enfeebled German science.

During the Second World War, the United States enjoyed the further advantage of distance from battle. In every other major scientific nation, many laboratories were disrupted by actual bombardment. The political and economic devastation of Europe and East Asia left the United States predominant in solid-state physics, as in most other areas of culture.

In some other respects, however, the Second World War brought increased effort and increased importance to solid-state physics in every nation. This was the third major influence of the period. Although some aspects of physics were totally neglected during the war because they did not show promise of immediate military utility (for example, dislocation theory), physics as a whole gained greatly in power and public esteem. The fact that prowess in war brings esteem may be a sad commentary on human values, but such developments as radar (including solid-state detectors) and atomic bombs did impress everyone, although in the latter case not without bringing permanent distress as well.

Deliberately or unwittingly, the Second World War engendered many techniques that became useful to the postwar generation of solid-state physicists. For example, the wartime atomic bomb projects intensively studied how neutrons interacted with solids, and afterward neutrons were found to be effective probes for exploring magnetic and other properties of solids. Further examples originated in the rapid development of microwave radar, which produced a variety of useful new techniques. One such technique was microwave spectroscopy, in which radiation is tuned to coincide with certain natural vibrational or rotational frequencies of atoms and molecules. The wartime radar workers also developed methods for producing perfect crystals with accurately controlled impurity content; the methods of producing pure crystals of the semiconductors silicon and germanium, which had been used as microwave detectors and frequency converters, became so highly developed that an enormous number of careful studies in the postwar period employed silicon and germanium as prototypes for the study of solid-state phenomena in general. In such matters of technique, the distinctions among military, industrial, and purely scientific influences tend to disappear. For example, digital-computing methods had been rapidly developed for wartime applications, but in the postwar period they were used by physicists to attain a new level of accuracy in solving problems of quantum mechanics, and a little later they became an industry of their own.

Entirely apart from its utility, physics had retained its fascination with inquiry into the most fundamental structures and forces of nature. In the immediate postwar period, many talented young people wished to take part in what seemed to be not only the most dynamic and most relevant, but also the most idealistic enterprise of their time. After the unprecedented destruction, they could help turn the world into a more beautiful place for human life—a hope that far outweighed the depressing effect of the war. When industrial and government support for physics increased dramatically in the postwar period, that too was not only for strict utilitarian reasons. Finally, as the following chapters show, despite all the influences that society exerted, on the level of day-to-day research things usually went in a given direction because scientific theory and experiment themselves pointed that way.

Acknowledgments

It remains for us to thank the many people and organizations that helped our project generously with their time, advice, knowledge, hospitality, facilities, and money. Our first thanks, of course, are to the authors of the chapters, who often worked far beyond the call of duty, and to those who helped them, as listed in the chapters' individual acknowledgments.

The advisory committees gave indispensable help, and we thank all the members. *Great Britain:* Nevill Mott, Margaret Gowing, Cyril Hilsum, Rudolf Peierls, Alan Wilson, and Hendrik B. G. Casimir. *United States:* John Bardeen, Felix Bloch, Erwin Hiebert, Conyers Herring, Martin Klein, Arthur Norberg, Gerald Pearson, Frederick Seitz, Cyril S. Smith, and David Turnbull. *Germany:* Heinrich Welker, Friedrich Hund, Erich Mollwo, Heinz Pick, Walter Rollwagen, and Günther Küppers. *France:* André Guinier, Pierre Aigrain, Jacques Friedel, André Lebeau, and G. Delacote.

We thank all those who graciously consented to be interviewed. They are acknowledged in individual footnotes, and the tape-recorded interviews will be described in the guide to historical source materials that the American Institute of Physics will publish.

We also owe thanks to many physicists and other scholars who helped with information and advice, and to the archivists and private individuals who allowed us access to their precious holdings. For making source material available to us, we wish particularly to thank: David Cassidy, New York; Günther Glaser, Stuttgart; Karl Hecht, Kiel; Annemarie Hilsch, Göttingen; Eduard Justi, Brauschweig; Holger Meissner, Munich; Karl von Meyenn, Barcelona; Physikalisches Institut der Universität Göttingen; the family of R. W. Pohl, Göttingen; Helmut Rechenberg, Max-Planck-Institut für Physik und Astrophysik—Werner Heisenberg-Institut, Munich; Rijksmuseum voor de Geschiedenis de Naturwetenschappen, Leiden; Martin Schottky, Pretzfeld; and the Siemens-Museum, Munich. The staffs of the AIP Niels Bohr Library and the Deutsches Museum, in particular, responded to hundreds of requests from team members for materials and for aid in processing the oral-history interviews.

Our sincere thanks also go to the industrial and other institutions and charitable foundations that gave financial support to our project, sensitively, generously, and without stipulating any form of control over the final products. In Great Britain, the principal support came from the Leverhulme Trust in London, with further help from the British Academy. In Germany, the principal supporter was the Stiftung Volkswagenwerk. In the United States, principal donors were the Department of Energy Division of Material Sciences, the National Science Foundation Program in History and Philosophy of Science, the Alfred P. Sloan Foundation, Bell Laboratories, David Packard, IBM, General Electric, Motorola, Texas Instruments Foundation, and Xerox; additional help came from Exxon Research & Engineering, Hewlett-Packard Company Foundation, Intel, Raytheon, Rockwell International Corporate Trust, Schlumberger-Doll, Scientific American, the Southwest Research Institute, TRW, and the Friends of the AIP Center for History of Physics.

Finally, we thank the institutions that offered homes and indirect financial aid to our research teams, or that hosted our meetings: the American Institute of Physics in New York, the University of Aston in Birmingham, the Deutsches Museum in Munich, the physics department of the University of Illinois at Urbana, the Institute of Theoretical Physics of the University of California at Santa Barbara, the Parc de la Villette in Paris, and the Royal Society of London. And we thank Oxford University Press for waiting patiently for the manuscript and publishing it in exemplary fashion.

Our greatest appreciation is reserved for those future historians who will take up the task where we have faltered. We know better than anyone how incomplete our work has been; the principal aim of the International Project in the History of Solid State Physics, in preserving and cataloging documents, conducting oral-history interviews, and presenting our research, has been to get the task under way. This book is meant to open doors into an astonishingly rich and complex subject that until now has scarcely touched the consciousness of historians and even of many physicists. We trust the very inadequacies of our volume will work as a stimulus to further research into the history of this grand field of knowledge.

Vienna	E. B.
Urbana	L. H.
Munich	J. T.
New York	S. W.
June 1990	

Contents

Abbreviations

Used in citations throughout the book

AHQP	Archive for History of Quantum Physics, microfilm copies located in AIP, DM, and elsewhere.
AIP	Niels Bohr Library, American Institute of Physics, New York
Ann. Phys.	*Annalen der Physik*
Beginnings	*The Beginnings of Solid State Physics* (London: Royal Society, 1980)
C.R.	*Comptes rendus hebdomadaires des séances de l'Académie des Sciences*
DM	Deutsches Museum, Munich
Nachr Gött.	*Nachrichten von der Gesellschaft der Wissenschaften Göttingen, Mathematisch–Physikalische Klasse*
Phil. Mag.	*Philosophical Magazine*
Phys. Rev.	*Physical Review*
Phys. Zs.	*Physikalische Zeitschrift*
Proc. Roy. Soc.	*Proceedings of the Royal Society of London*
Sitz. Preuss. Akad.	*Sitzungsberichte der Königlich-Preussischen Akademie der Wissenschaften Berlin*
Verh. Ges.	*Verhandlungen der Deutschen Physikalischen Gesellschaft*
Zs. Phys.	*Zeitschrift für Physik*

Out of the Crystal Maze

1 | *The Roots of Solid-State Physics Before Quantum Mechanics*

MICHAEL ECKERT, HELMUT SCHUBERT,
AND GISELA TORKAR
WITH CHRISTINE BLONDEL AND PIERRE QUÉDEC

The plain goal of science is to understand and use the objects around us, and of these objects the most obvious are solids. They were among the first things to be mastered and among the last things to be understood. The growing ability to manipulate stone, then ceramics, then glass, metals, jewels, and so forth was central to the history of the human race. If there was no corresponding progress in finding concepts that could unify the understanding of solids, it was not for lack of trying; the works of Greek philosophers, medieval alchemists, and seventeenth-century Cartesian natural philosophers abounded in fanciful models of the solid state. Then, as

3

Cyril Stanley Smith has pointed out, the breakthrough of the new quantitative science "interrupted the age-old tradition of qualitative thought about the properties of solids." Properties such as color and hardness, even after people learned to measure them, could not be calculated with the concepts at hand. Those who strove to do physics on the model of Galileo and Newton had to set these questions aside and abandon everything that depended on the detailed interaction of many parts. With reference to metallurgy, Smith noted that the era of Descartes "was the last time for three centuries that respectable thinkers concered themselves with the properties of real solids."[1]

Unifying concepts that offered a genuine ability to calculate the properties of solids had to await the coming of quantum mechanics.[2] But even after the first applications of quantum mechanics to solid-state problems in the 1920s, over a decade would pass before anything emerged that physicists would recognize as an independent discipline devoted to the solid state.[3] The history of early solid-state physics is thus a subject that does not exist; what we have instead are histories of a bewildering variety of separate topics.[4]

It is impossible to answer every question about every discovery that fed into what eventually became solid-state physics. Hence, we will indicate the general scientific, institutional, and technical background from which such studies of solids emerged before they were combined to form an independent area of physics. In particular, we indicate the points of departure for the developments treated in this book. Therefore, most of this chapter covers the first decades of the twentieth century, along with some brief remarks on the nineteenth-century institutional framework. We shall see how deep and varied are the roots of solid-state physics, but we shall also see how the different studies of solids themselves contributed to the unified understanding on which the new field would eventually be built.

The differentiation of scientific knowledge into the spectrum of separate disciplines recognized today began in the nineteenth century. Solid-state themes were isolated from one another in this process, and research was recorded in the context of disciplines that had diverse problems and goals. Themes such as crystal chemistry and the properties of metals became topical in newly institutionalized—and for the most part isolated—areas of chemistry, mineralogy, and engineering, which flourished in mining schools, the revived universities, and the recently established "polytechnic" schools.

The institutional base—that is, the source of the money that made research possible—had much to do with practical affairs; this was quite important for research on solids, as we shall see throughout this chapter. The industries that blossomed in the latter nineteenth century, particularly the electrical industry, placed new demands on science. State and industrial research institutes occupied chemists, engineers, and physicists with problems of cables, the fabrication of filaments for incandescent lamps, and the investigation of alloys for standards of electrical resistance, among others. In the face of intense international competition, it became more and more important for certain branches of industry to take academic research into account. Conversely, academic physics was influenced by new questions that arose in industrial contexts. For example, the engineering question of the temperature dependence of the electrical and thermal conductivities of a metal or an alloy played an important role in theories about electrons, and thus made its way into academic discussions.

Meanwhile, the world view of academic physics changed radically. Since the early nineteenth century, crystals, for example, had been studied primarily in the abstract, as idealized systems in which the interplay and symmetry of the forces of nature could be examined. The existence of microscopic elements such as atoms could be inferred, but almost nothing could be learned about them. The great nineteenth-century advances were in macroscopic concepts, such as the laws of thermodynamics and electromagnetism. But at the end of the century, new concepts of statistical mechanics, electrons, and ions, combined with the discovery of x rays, of radioactivity, and of Planck's quantum, revolutionized physics. These concepts were a breakthrough in microscopic models, which would eventually explain a multitude of properties that had previously been handled only phenomenologically.

As a rule, the revolutionary new thought was not aimed primarily at solids in their diverse manifestations and properties. Rather, many effects in solids, most of them known for a long time, served as objects upon which to test and develop the new theories. The treatments of the specific heat of solids by Albert Einstein, Peter Debye, Max Born, and others provides an example of a solid-state theme that was also a forerunner of quantum physics.

The breakthrough in microscopic concepts, which together with quantum mechanics achieved a unified formulation in the mid-1920s, was not purely a matter of theoretical physics, for new experimental techniques were equally important. Thus the efforts of theorists to grasp the prominent effects of the quantum at low temperatures advanced in parallel with a drive toward ever lower temperatures, which led to the discovery of unexpected phenomena such as superconductivity. Other important examples may be found in spectroscopy, where discussions of the thermal vibration of the atoms in solids were intimately connected with infrared techniques, which could be practically applied.

That science was more than just an intellectual endeavor became evident to all during the First World War. Those states that had underestimated science as a tool of power now began to take institutional measures along the lines already taken, for example, in Germany, with the establishment of the Kaiser Wilhelm Foundation in 1911. National research councils, cooperation between academic research and industry, and state and private research support began to be organized. Typical groups were the Department of Scientific and Industrial Research in England and the Research Institute of Iron and Steel (RIIS) as well as the Institute for Physical and Chemical Research in Japan. Many of these institutions were concerned with specialized solid-state research. For example, Kaiser Wilhelm Institutes in Berlin-Dahlem studied metals, the chemistry of fibrous materials, and silicates; the Japanese RIIS supported Kotaro Honda in his magnetic and metallurgical research.

After the First World War, these organizational efforts were systematically continued, so that solid-state research was institutionalized and financially supported in many places in the 1920s. Review articles, textbooks, handbooks, and conferences of the period provide evidence that themes like "the electrical conductivity of metals" and "the dynamics of crystal lattices" were considered as identifiable areas of current research and were already integrated into the main body of physics instruction. However, there was still little evidence, institutional or intellectual, for the existence of a fully formed and independent discipline of solid-state physics.

This chapter spans the period from the beginning of the nineteenth century to approximately 1925. Using representative cases, we sketch the background on which solid-state physics would eventually appear. We will not discuss some specific developments that are covered in their full context in later chapters (e.g., point defects). In addition, many other important developments are omitted. The events that we do discuss usually appear as unconnected incidents because they were not connected except by the broadest economic and intellectual themes, and it is only hindsight that singles out these examples as parts of an eventual solid-state physics. A mosaic representation is almost inevitable in any case when several authors with varying backgrounds and interests write a single chapter. But we hope the very heterogeneity of the mosaic will show how solid-state ideas emerged from a multitude of contexts and were combined into subtle patterns that can be comprehended only from a distance.

1.1   Institutional Settings for the Scientific Investigation of Solids in the Nineteenth Century

The industrial revolution and the rise of the bourgeoisie as the dominant class during the nineteenth century imposed new conditions on the organization of society, giving rise to many new institutions; some of them were devoted to scientific investigation of the properties of solids.

There were forerunners of such institutions even before the nineteenth century. In some states, the extraction and processing of raw materials were supported by absolutist rulers in order to enhance their financial power and military strength. In particular, the extraction of precious metals had been a source of feudal wealth for centuries, and this wealth was increased by the raw materials of industry, whether imported from colonies or mined at home. But many sites could be mined only at the price of increasing effort. Governments therefore founded specialized academies, such as the Austro-Hungarian Bergakademie Schemnitz (1770) and the Ecole des Mines (1783) in France. Technological questions attracted scientific interest especially in the new French schools, where methods were developed to investigate such matters as the strength of materials.[5] In the Bergakademie Schemnitz, mineralogy was taught with the aid of "magnificent specimens of gold and iron ore, calcspar . . . geognostic pieces, especially of granite, syenite [and] crystal models made out of gypsum," while a laboratory equipped "with regal splendor" fostered clear empirical instruction.[6]

In 1794, in the midst of revolution, the French state founded a new kind of institution, the Ecole Polytechnique, which would become far more than an engineering school.[7] It was meant to serve as a preparatory school for the reformed or newly established engineering schools for mining, artillery, civil engineering, surveying, and shipbuilding, providing students in these specialties with a broad scientific foundation. Under Napoleon, the school was assigned to the War Ministry, and students could look forward to careers as officers in the imperial army. Professors as well as students enjoyed great social prestige. The publication of a journal as well as the obligatory publication of the professors' lectures provided stimulation, a specialized body of knowledge, and a high standard. Teaching came to be seen as linked to research, which included both theoretical and experimental investigations. The overriding goals were to reduce empirical observations to mathematics and to attain the greatest possible precision in experiments, carried out in the special laboratories that had been set up for teaching.

There are many examples of the role played by the Ecole Polytechnique and other engineering schools in the development of physical science.[8] In the area of solids, professors conducted important studies in the theory of elasticity, strength of materials, thermal conductivity, and optics. Many such investigations were closely tied to applied problems. Thus Louis-Marie-Henri Navier, a Polytechnique student employed in the 1820s as a professor at the Ecole des Ponts et Chaussées, the civil-engineering school, was interested in the construction of iron bridges which were just coming into use. In his *Lois de l' équilibre et du mouvement des corps solides,* he used molecular conceptions to develop the differential equations for the elasticity of three-dimensional isotropic bodies. This approach to solids, combining a concrete model with abstract mathematics, was so fundamental that physicists eventually came to take it almost for granted.[9]

The example of the Ecole Polytechnique influenced other countries. The United States Military Academy at West Point was organized in 1802 along similar lines, and through the nineteenth century this was one of the most progressive American institutions for scientific instruction and research. In Europe, especially in the German states, the model was imitated in polytechnic schools from which many technical universities later sprang.

The next great impulse toward educational reform came in the early nineteenth century in Prussia and spread throughout Germany. Associated with the reformers was an idealistic, neohumanist ethos: scholarship should be free from ulterior motive and carried out for its own sake. The circle of existing German universities was imprinted with the new ideals, summarized in the motto "Freedom to teach and freedom to study." Thus teaching and research presented a unified front. The reform movement proclaimed education for its own sake to be a duty of the state, and the educational system became the responsibility of a state cultural ministry. When these ministries came to choose men for university positions, they began to use recognized achievements in research as an essential prerequisite. This "research imperative" for professorial careers made research at the universities a regular professional activity.[10]

The new supremacy of research opened the way for new characteristics of science to evolve. A reform that strove toward humanistic education paradoxically ended up encouraging the dissociation of science from philosophy and a progressive specialization into independent, professionalized disciplines: the era of the amateur was ending. At the same time, a "'polytechnicism,' or whatever one wants to call the materialistic aims" that some researchers adopted, increasingly conflicted with the reformers' ideals.[11] Toward the end of the nineteenth century, the conflict would erupt in intense struggle as the engineering schools demanded a status approaching that of the universities. Nevertheless, the coexistence of "training and education," or of "soul [*geist*] and industry" (as slogans described the division), broadened the range of possibilities for both scientific and applied activity.

The strength of the German universities was much admired in England, and helped lobbyists for science to argue for increased national support of both research and teaching. Here, where the industrial revolution was born and where a prosperous middle class surpassed the old feudal and clerical leaders of society earlier than on the Continent, new institutions and societies were organized on a private basis. Not Oxford and Cambridge but "dissenting academies" and "mechanics institutes," most often founded in the aspiring new industrial cities, were breeding grounds for research oriented toward the needs of industrial society. For the first time, a significant number of people earned their living explicitly by doing research as well as by teaching.[12]

Another indicator of the establishment of a new class of professional scientists and engineers was the founding of associations for the support of science. In the mid-nineteenth century, scientists in many countries banded together in such associations,[13] which brought new pressure on governments to increase scientific instruction.

The new associations also reflected the differentiation of science, as seen particularly in the case of the Gesellschaft Deutscher Naturforscher und Ärzte. Founded without subdivisions in 1822, the organization by 1894 had 41 specialized sections, including geognosy, mineralogy, chemistry, physics, technology, and mechanics, to name only those relating to the investigation of solids.[14] In the second half of the nineteenth century, some scientific disciplines achieved complete independence. Metallurgy, mineralogy, chemistry, and physics became well-organized areas, with their own specific interests and methods, gathering and disseminating information in their own separate research institutions, societies, and organs of publication. Now more than ever when solids were studied, they were studied within the boundaries of one or another of these new professional disciplines.

Engineering, Chemistry, and Physics as Specialties

It may be quite true, that many of the practical and technical details of civil engineering may be best learned in the offices of engineers engaged in the execution of important works; but the knowledge of mathematics and of mechanical principles, as involved in the estimate of the strength and distribution of materials, the effects of elasticity, and generally of the operations of forces and pressures, is so necessary in all the more important and difficult applications of this science, that no amount of practical skill and experience can ever replace the want of this theoretical knowledge.[15]

In such terms a need was formulated in mid-nineteenth-century England that, there and in other countries, caused the study of solids to be institutionalized as a part of engineering. In particular, knowledge of elasticity and the strength of materials was crucial for such mechanical-engineering tasks as constructing railways and erecting large buildings of iron and reinforced concrete.

In technical universities like the Eidgenössische Technische Hochschule (ETH), founded in 1855 in Zurich, and the Munich Polytechnikum, regular engineering instruction came to include subjects such as construction materials, metallurgy, mineralogy, and crystallography.[16] In the 1870s, many of these engineering schools set up laboratories for materials testing, where properties of iron and steel became the principal object of intensive research. The practical goals were explicitly promoted by new social organizations in which engineers were coming together. For example, in 1877 a "technische Commission des Vereines Deutscher Eisenbahn-Verwaltungen," a railroad group working toward a national standard for iron and steel classification, insisted on the need for better knowledge of the properties of materials. "For the investigation of these laws we propose the establishment of a research institute," said the technical commission, and it formally moved "that the Association [of German Railroad Administrations] bring its considerable influence to bear in order that this conviction also takes hold within the government and in order to bring such institutes to life as soon as possible."[17]

Other striving industrial nations, particularly the United States, also had a "blossoming of technical teaching institutes," as a German observer reported in 1894: "The number of American technical teaching institutes is enormous, there are now already over 200 on record, with new ones still to come."[18] The curricula of new technical unversities, such as the Massachusetts Institute of Technology (founded in 1861), show that here too the essential components of engineering began to include materials testing, the theory of elasticity, and similar problems of the mechanical behavior of solids.

Chemistry provides a second example of the incorporation of solid-state themes into an emergent specialty. Subjects like mineralogy and crystallography had long been of practical importance for mining, but they became increasingly useful for chemistry as well. The usefulness hinged on the recognition that similar chemical compounds correspond extensively in their crystal forms.[19] Thus the tremendous upsurge of industrial chemistry, one of the central transformations of the nineteenth century, aided mineralogy and crystallography. For example, it was a combination of chemical and crystallographic research, inspired by interests connected with the French wine industry, that led young Louis Pasteur to discover that tartaric acid can have two forms that are chemically identical but physically different. A crystal of one type rotates polarized light to the left; the other turns it to the right.[20] Later, in 1874, Jacobus van t'Hoff and Achille Le Bel learned that this sort of optical activity is caused by the asymmetrical positioning of carbon atoms in organic molecules.[21] Studying such interactions between molecules and light became a main concern of late-nineteenth-century scientists. Such work would eventually be at the center of solid-state physics, but at the outset it was a chemistry technique.

Organic chemistry, busily giving rise to new industries such as the production of artifical dyes, urgently needed crystallography to identify and classify the rapidly growing number of chemical compounds. Long before the discovery of x-ray diffraction in crystals, chemical crystallography, carried out in new institutes founded expressly for the purpose, brought an abundance of crystallographic investigations, chiefly of organic compounds.[22]

Physicists were meanwhile forming their own professional organizations and establishing their own characteristic problems, which naturally included problems of solids. A series of phenomena, such as thermoelectricity and the Hall effect, were discovered and studied as part of the general program of physics.[23] But such phenomena were not grouped together under a general heading as questions of the solid state; rather, they were placed as seemed best among the classical subdivisions of mechanics, heat, electricity, magnetism, and optics.

Where special areas like crystal physics were touched on (as discussed in what follows), the physicists were chiefly interested in overriding fundamental ideas, things that could be taught to students as examples of generalized, mathematical, and instructive principles. The archetypical physicist's attention was directed toward ideal crystals, in hopes of progressing inward toward the molecules and atoms themselves. This direction was opposite to that followed by the chemists, metallurgists, and engineers, who concerned themselves with real solids such as steel.

This does not mean, however, that physical research and instruction were everywhere removed from practical needs. In the 1870s, "practical physics" became a catchphrase for a movement under which new laboratories were founded at universities and engineering schools, along with corresponding laboratory courses and textbooks.[24] The basic subject of physics proved well suited to help out industry, particularly as physics-based industries spread. These industries required an international effort in measurement, calibration of instruments, and standards, all based on proven scientific principles; thus practical investigation of materials moved into the realm of physics. For example, physics principles and experiments were required by those who calibrated thermometers by using the temperature dependence of the electrical resistance of certain materials. Other problems were found, according to the foreword of Friedrich Kohlrausch's textbook of practical physics, published at the end of the century, "in the subtler parts of optics, in photometry and thermal radiation; but especially in the theory of electricity, where . . . the determination of magnetic and dielectric constants . . . have been given a wider place than they had before."[25] The academic physicist still did not see studies of solids as gathered together in a fundamental research program; what was studied depended on the accidents of interest in particular applications. Yet this new interest in problems of solids helped physics to get increasing attention from industry and the state.

The Influence of Industry: Cryogenics, Steel, and Electrical Engineering

As scientific laws, or at least scientific methods, were increasingly applied to the production process, industry became a main source of employment for graduates trained in science. This meant a corresponding increase in the number of academic professors of science. The professors were naturally interested in the problems of the industries where their students were headed, and this interest sometimes led to new results of a strictly scientific nature.

The study of matter at very low temperatures, which would eventually have particular significance for understanding the properties of solids, is an example of a field that began among highly practical interests. For one thing, increasing trade and the growth of cities raised a need to cool food on long transport routes and in warehouses. For another, chemists were interested in using low temperatures to liquefy gases, an interest connected with the search for easy extraction of nitrogen and oxygen, gases that played an essential role in the chemical industry and steel production.[26]

Many engineers took up the construction of different sorts of refrigerators.[27] Carl von Linde was the first to apply himself to the question of the efficiency of refrigerators. At the ETH in Zurich, he studied mechanical engineering and attended the physics lectures of Rudolf E. Clausius. Employed in 1868, at the age of 26, at the Technical University in Munich, Linde advocated this practice of giving instruction in theoretical mechanics along with practical machine construction. Along with theory, he was given responsibility for applications, since he had worked in several industrial concerns. Making a systematic comparison of refrigeration processes, he found that "existing ice machines reached at most 10% of the theoretical maximum output."[28] With the financial support of a Munich brewery, he succeeded in his first attempt to construct a machine with doubled output. Encouraged by this success, Linde founded his own firm, which took up so much of his time that, in 1879, he resigned his professorship.[29] Linde's combination of basic research and practical results emerged again in 1892, when he took up his university career once more and investigated methods for liquefying air. Within a few years, he constructed the first economical air-liquefaction equipment. The new method was put to work by James Dewar in the Royal Institution, who carried systematic investigations of the properties of solids down to a temperature only 20 K above absolute zero.[30]

As in cryogenics development, so in the steel industry the application of scientific methods to traditional processes led to marked improvements. The tremendous increase in demand for iron in the 1850s, mainly for railroad tracks, showed up the limitations in the hearth refining and puddling processes for making iron. In London, Henry Bessemer found through laboratory experiments a more efficient way to make steel, but he lacked exact knowledge of the chemical reaction involved and encountered many difficulties in putting his ideas into practice. A little later, the chemist Robert Wilhelm Bunsen and the physicist Gustav Kirchhoff developed spectral analysis; a pupil of Bunsen used it on the light from the flames in an English ironworks and found the conditions for the optimum air supply in the Bessemer converter.[31] The new discoveries helped bring a sharp rise in steel production; production in the Krupp firm, for example, increased from 5000 tons in 1861 to 50,000 tons in 1865.[32] The importance of science to steel production was now evident, and all the larger producers changed over to the new methods and hired chemists.

The increase in steel production also boosted the importance of metallurgy as an engineering science. In Germany, France, England, the United States, and Russia, important works on the structure of solids appeared between 1860 and 1890 in engineering journals. These studies used iron and steel, not simple solids, as examples. For example, Dimitri K. Tschernoff, a Russian metallurgist, reported on the fracture grain size in steel in relation to heat treatment and carbon content at various temperatures; he also investigated dendritic crystal growth. His point of departure was observations made during the forging of guns. Tschernoff was one of the first to notice that the properties of materials are directly connected with their crystalline fine structure, and that casting, cold working, or heat treatment determines the quality of the products by influencing this structure.[33]

In 1878 the *Zeitschrift des Vereins deutscher Ingenieure* published an article on the microstructure of metals[34] written by a bridge-construction engineer of the Prussian railroad. The article showed that metals could be analyzed by etching thin sheets with acid and observing them through a microscope. This method, which beautifully reveals the grain structure, was developed into an independent branch of science by Henry C. Sorby in Sheffield, a center for the English metal industry.

In contrast to such traditional industries, the nascent electrical industries were closely linked to scientific research from the start. No practical use could be made of electricity until disinterested research had revealed some of its basic phenomena. One early application was electric telegraphy. The electric telegraph's utility for trade, railways, and military maneuvers soon caused it to grow into a profitable industry. By the middle of the century, the European nations were linked with one another by copper wire. The connection between Continental Europe and America and between Europe and its colonies became an international enterprise inspired by the needs of trade. The completion in 1866 of a permanent telegraphic connection between the stock-exchange centers on Wall Street and in London was celebrated as one of the greatest technical accomplishments of the nineteenth century.

The problems that arose during this undertaking, such as design of insulation and cable strength, were overcome by the collaboration of scientists from various disciplines.[35] For example, in 1864 the chemist August Matthiessen developed a method to control the purity of copper, still one of the standard methods today. He discovered that the specific resistance of metals is composed of the sum of two components; one depends on the temperature, and the other measures the purity of the metal. Matthiessen learned how to separate these components through measurements at varying temperatures, and thus could compare the purity of every type of copper found in trade.[36]

The explosive growth of the electrical industry from 1880 to 1900—for example, the number of people employed in the industry in Germany rose from 1292 in 1875 to 18,704 in 1895—gave still greater impetus for investigating properties of solids.[37] The development of electric illumination now took the lead. The carbon-arc lamp, which was turning night into day on streets, in public buildings, and in factories, was the point of departure for the discovery of the anomalous temperature dependence of electrical resistance and of the anomalous specific heat of carbon.

When incandescent lamps eclipsed gas illumination in private households toward the end of the century, thus giving rise to a huge branch of production, these lamps also became a lucrative object of research. For example, in 1888 H. F. Weber, professor of theoretical and applied physics at the Zurich ETH, published a relationship between the change in the wavelength of light emitted by a glowing body and its temperature, from which "all properties of the electric incandescent light—except for the lifetime of the lamp—[could] be deduced."[38] Such work paved the way for Max Planck's radiation formula and the quantum theory, but at the time it had a distinctly practical flavor.

Practical electrical research that would lead to fundamental understanding was embodied in towering brick laboratories. In addition to universities, corporations such as General Electric and Philips created these laboratories with the original intent of studying lamps, but scientific results of broader interest flowed out. The telephone was another electrical device that furthered the tendency to establish basic research in industrial contexts. In the United States at the turn of the century, the Bell Telephone Company was among the first corporations to set up a research

laboratory, which eventually became one of the greatest centers of solid-state physics.[39]

Many such institutions were founded to bring academic and industrial science
together. We will look in more detail at two that illustrate the range of possibilities.

The Physikalisch-Technische Reichsanstalt and the Göttinger Vereinigung

The more closely new technologies such as the telegraph brought nations together,
the stronger was industry's call for standardized, precisely defined electrical units.
In 1881 the International Commission on the Definition of Electrical Units agreed
on a resistance standard corresponding largely with a suggestion of the German
industrialist Werner von Siemens. But no German laboratory was in position to
undertake the costly precision work on the exact representation of the "ohm," so
this was accomplished in England.[40] Although Germany was second only to the
United States as an exporting country, it had to import precision measuring equipment from England and France.[41] Not only the German precision-tool and optical
industries, but leaders of scientific institutes as well deplored the quality of optical
and mechanical instruments and demanded an institute to further scientific instrumentation.[42]

Siemens and the physicist Herman von Helmholtz argued that in such an institute applied as well as general scientific problems should be investigated. In a letter
to the Prussian Ministry of Education, Siemens emphasized the national advantages of a place where scientists could be freed from teaching duties to concentrate
on research. "The Reich would gain important material as well as idealistic advantages from a scientific workplace," he wrote. "In the competition between nations,
presently waged so actively, the country that first sets foot on new paths and first
develops them into established branches of industry has a decisive upper hand."
Not only power but prestige were at stake: "It is not scientific education, but scientific accomplishment that assigns a nation to the position of honor among civi-
lized people."[43] Siemens lent further weight to his demands by donating a plot of
land in Berlin for the financial foundation of the institute. In 1887, the Physikalisch-Technische Reichsanstalt (PTR) was founded with two departments: one tech-
nical-mechanical, and the other physical-scientific.

Research and testing duties for industry and the military dominated PTR's early
years. But Helmholtz, its first president, placed a high value on fundamental
research as the basis for technical innovation, and almost the entire scientific staff
was made up of his university students.[44] The PTR scientists could solve their
assigned problems within a wider scientific context than could their colleagues in
industry, and the PTR became one of the most important centers for the investigation of the properties of matter. It was the forerunner of national research establishments where the costs of innovation were socialized while the profitability of
private firms was enhanced.

The PTR's physical-scientific department consisted of three groups: thermal,
electrical, and optical. In the first years, the main work of the thermal group was to
improve thermometers. For example, it investigated the thermal expansion of var-

ious types of glass, manufactured chiefly by Zeiss and Schott. It also carried out pyrometric investigations in order to get a grasp on temperatures up to the melting point of platinum, a task that originated in the steel industry.[45]

Meanwhile, the electrical group looked for ways to determine more precisely the values of the fundamental electrical units. It also carried out magnetic measurements of iron and steel on commission for the German navy. The navy, finding that its compasses deviated up to 50% from the correct values because of the iron used in its ships and torpedoes, asked the PTR to investigate which types of steel exhibited the least residual magnetism.[46]

Likewise, the optical group existed to serve practical needs. Its first investigations stemmed from an 1889 petition of the German Association for Gas and Water Experts, calling for the development of a reliable measure of illumination. The search focused on the emission from platinum foil at constant temperature as a standard of comparison. This search opened a series of radiation investigations, leading to the famous examination of the spectral energy distribution of blackbody radiation that became a root of quantum mechanics. Besides this photometric work, the group carried out polarimetric investigations, prompted in 1891 by the German Association for the Beet Sugar Industry. The rotation of polarization by solutions could be used to determine the concentration of sugar and other organic substances. To find a standard of comparison, investigations were carried out on quartz, which furthered understanding of the optical properties of solids.

The PTR's importance for Germany's economic and scientific hegemony was recognized abroad by the turn of the century. Similar establishments were founded in all the major industrial nations: the National Physical Laboratory in England in 1900, the National Bureau of Standards in the United States, the Laboratoire d'Essai in France, and, in 1917, the Japanese physical-chemical institute Riken.

The Göttingen Association for the Support of Applied Physics and Mathematics is an example of a different form of collaboration between university science and industry at the end of the nineteenth century. The impetus came from Felix Klein, professor of mathematics at Göttingen University. He was an *eminence grise* for the organization of science in Germany around 1900 because of his connections with Friedrich Althoff, the overseer of the Prussian Ministry of Education, and with influential industrialists such as Henry Theodor Böttinger, a director of the Bayer dye factory and a representative in the Prussian Diet.[47] Klein realized that German industrialists might be willing to support applied science financially if he spurred them with the threat of foreign competition.

The chemical industry took the lead in 1896 when, with Böttinger's support, the Institute for Physical Chemistry and Electrochemistry was founded at Göttingen and put under the leadership of Walther Nernst.[48] A year later, donations from industry enabled an Applied Physics Department to be established at the university's physics institute.[49] Meanwhile, in 1896 industry's support of science at the university was institutionalized in the Göttinger Vereinigung. The founding assembly announced that its aim was "extending instruction and research in the area of applied physics."[50] The association at once raised money for an institute for applied electricity.[51]

Investigations at the new applied institutes typically bordered purely scientific and purely applied disciplines: electrochemistry (Nernst), electrometallurgy (Hans

A cartoon from 1908 (the tenth anniversary of the Göttinger Vereinigung) showing the cooperation between science and industry. Industry brings money and gains knowledge. The mathematician Felix Klein, who supported such collaboration, is shown as the sun. (From A. Hermann, *Weltreich der Physik* [Esslingen: Bechtle, 1980], 193)

Lorenz), theory of solids along with hydrodynamics and aerodynamics (Ludwig Prandtl), and the theory and technology of magnetism (Hermann Simon, Erwin Madelung). In the Göttinger Vereinigung, the common interests of science, industry, and the state brought diverging disciplines together again.[52] In sum, industry donated money, the state created prestige and positions, and science contributed to practical ends. This was a pattern that in the twentieth century would create many other organizations that pursued large-scale, interdisciplinary solid-state research.

1.2 Conceptions of Solids in Classical Physics

The Mechanical Crystal

Franz Ernst Neumann's move in the fall of 1826 from the capital city of Berlin to the remote city of Königsberg, "a trip by post-chaise lasting eight days and eight

Franz Ernst Neumann (1798–1895). (AIP Niels Bohr Library, See Collection)

nights,"[53] was the midpoint of a journey for classical physics, a journey during which crystals came to be understood in terms of the laws of mechanics. The development had begun in France with the mathematization of specialized physical knowledge, an approach Neumann carried with him in his head. Neumann had accepted his appointment to a provincial university reluctantly: he was to take over the teaching of mineralogy, an area in which Neumann had already won distinction, from a professor who had been teaching nearly all the scientific disciplines. After only a few semesters, Neumann began to intersperse his mineralogy lectures with physical insights.

Neumann had become interested in physics in Berlin, where he particularly admired French work in the field. To master Fourier's book on heat theory, he copied it by hand, absorbing its mathematical approach.[54] Neumann's first experimental studies were, accordingly, in the theory of heat. He depended on works by Alexis-Thérèse Petit and Pierre Louis Dulong that had recently appeared at the Ecole Polytechnique, works that culminated in the law named after the pair.[55] The Dulong–Petit law states that for simple bodies the product of the specific heat and atomic weight is a constant. The French tended to place experiment in the foreground, and Neumann kept it there. He created no new theory, but extended known laws through further experiment. For example, he extended the Dulong–Petit law to compound materials; the extended law became known as the Kopp–Neumann rule. This work was published three decades later, by one of his students. Neumann's habit of scarcely publishing anything makes it difficult to evaluate his contributions with precision.[56]

Turning from thermal measurements, Neumann next applied himself to two topics then of much interest: optics and the theory of elasticity. Interest in the connection between the two had recently revived when the wave theory of light triumphed over Newton's theory of light as a stream of particles. Through work by

Thomas Young and Jean Fresnel, even the skeptics became convinced that light is a wave in a light-ether that interpenetrates all bodies. The key was the idea that the wave is transverse; this explained not only diffraction but polarization phenomena, including the rotation of the plane of polarization by certain substances, a phenomenon discovered at the beginning of the century.[57]

Fresnel's ideas led to a momentous conclusion: since transverse vibrations are possible only in solids, the ether had to be something like a solid. Thus optics and mechanics were linked, and the theory of light would depend on a satisfactory mechanical theory of elasticity in solids.

There was already a long tradition of observations of elastic phenomena and attempts to describe them in a theory.[58] However, despite much progress in solving individual problems, the sum of elasticity theory at the beginning of the nineteenth century has been described as "an unsatisfactory theory of bending; an incorrect theory of torsion; an unproven theory of rod and plate vibrations; and the definition of Young's Modulus."[59]

Along with the new hope of providing a theory for the light-ether, elasticity had always held out hopes for practical applications, and this became more pressing as the nineteenth century opened. In the wake of Napoleon's military expeditions, analysis of the construction of roads, bridges, and even fortresses received enormous impetus through numerous prize competitions. For example, in 1808 the question of virbrating plates was offered, drawing the attention of scholars to problems of surface elasticity.

Louis Navier, a road and bridge engineer, responded with the equations of equilibrium and motion for a two-dimensional isotropic medium.[60] He began with the Newtonian concept of molecules as material points connected by forces, and supposed the forces to be proportional to the change in distance between the molecules, thereby deriving Hooke's law for the extension of a spring. But Navier's efforts to move up to three-dimensional equations failed. Today we know that one needs more than the single elasticity constant that he defined with reference to Young's spring modulus.

The problem found a surprising solution. Auguste Cauchy, after resigning his professorship in road and bridge engineering at the Ecole Polytechnique, concerned himself with the elasticity of solids.[61] To describe the behavior of an isotropic body, he introduced a quantity that was a collection of 21 independent coefficients—a symmetric tension and deformation tensor in modern terminology. In 1827, Cauchy formulated general equations on the basis of a model of the solids as a continuum; this was the foundation of classical linear elasticity theory. However, when he extended his ideas to anisotropic substances, such as typical crystals, he was forced to abandon the continuum and imagine the substance as a lattice of material points linked by attractive and repulsive forces. Taking the known geometric systems under which such three-dimensional space lattices could be grouped, he found relationships among the various elasticity coefficients, reducing the number of coefficients from 21 to 15 in the general case and to 2 for the simplest lattice, the cubic system.

However, experiments failed to demonstrate these "Cauchy relations." The large number of coefficients became the subject of a decades-long debate among physicists. The community split into supporters of the multiple-constants theory and the reduced-constants theory—a question not settled, as we shall see, until the twenti-eth century.

When Neumann took up the entire complex of questions in 1832, he agreed with Navier's theory based on Newtonian point molecules.[62] Neumann managed to rig-orously derive a theory for the interaction of light and matter, making the mechan-ical theory of elasticity the basis for optics. One result was an understanding of bire-fringence, the splitting of an image by certain anisotropic crystals; this phenomenon depends on the differing velocities of propagation of light in different directions, as Fresnel had determined through observation. Neumann went further; for example, he described in mechanical terms the birefringence that appears when an isotropic body is subjected to pressure or uneven heating.

Such matters had to be considered from a new angle later in the century as James Maxwell's electromagnetic theory of light gradually supplanted the simple mechan-ical theory. The new ideas gave new motives for work on crystals. For example, Heinrich Rubens, professor at the Technical University in Berlin-Charlottenburg, was solidly convinced of the validity of Maxwell's theory; he did not use radiation to understand crystals, but he used crystals to understand the spectrum of electro-magnetic radiation.[63] He pursued the invisible heat radiation that had already occu-pied numerous researchers. Measuring how the refractive index of quartz depends on wavelength, Rubens found indications of absorption in particular infrared wavelength ranges. By 1896, he had hunted down and described what he named the "residual radiation bands." He saw these as a filter that would allow him to separate out an almost monochromatic beam from the broad band of infrared radiation that his sources produced.[64] But as we shall note later, the values that he and his students measured were not just useful for studying radiation; they gave a decisive impetus to work on the dynamics of crystal lattices.

Neumann's contribution to this later work would come through his students. In 1876 he retired because of age. Following his wishes, his teaching duties were taken up by Woldemar Voigt, who had taken his degree, on the elasticity of rock salt, under Neumann.[65] Voigt's research, devoted in the following years to elasticity the-ory and its application to optics, was in the spirit of Neumann's work.[66] When in 1883 he accepted a call to Göttingen, Voigt brought the Königsberg traditions with him; while lecturing on mathematical physics, he devoted himself to experimental work.[67] To Voigt, measurements of elastic and piezoelectric constants, of pyroelec-tricity, of magnetization, and of optical constants of transparent, absorbing, or opti-cally active bodies all fell under crystal physics, that area "in which the music of physical laws sounds in full and rich chords as in no other area of physics."[68] At first he held to the elastic theory of light, and when he moved in the direction of Max-well's theory he was criticized for having "brought the entire Neumann school into disrepute."[69] So he returned to a purely phenomenological way of thinking, seldom asking about the mechanism of a process. This style of work in crystal optics brought great success to him and his favorite students, Paul Drude and Friedrich Pockels. In 1910 he published his monumental book on crystal physics, dedicated to Neumann's memory. Here Voigt summed up his life's work, "the great, glorious area whose cultivation I have returned to again and again for the last 36 years."[70]

Crystallographers before Voigt had connected physical properties with crystal structure in their books, yet he was the first to bring all the crystallographers' theories about structure together in a physics textbook. He could rightly claim to be pioneering the path to "the promised land of crystal physics" through the "desert of crystallography."[71] He described in detail the effects of heat, electricity, magnetism, and pressure, using experimental results including error analysis, and he tried to provide a theoretical explanation based strictly on classical mechanical and thermodynamic laws. Not once did his textbook mention Maxwell's equations or even the emerging microscopic notions of matter. Despite what was already an old-fashioned standpoint, Voigt's book remained current as a basic source for unified mechanical crystal theory because of its completeness. It was an "inexhaustible mine for all physical questions on crystals,"[72] "the bible for workers in this field."[73]

As one of the last supporters of a school with outdated views, Voigt would be overshadowed by the growing acceptance of Maxwell's theory, by the revolutionary quantum physics, and by the x-ray experiments that would make the crystal lattice visible. In his personal life, he would not recover from the defeat of Germany in the First World War, with the loss of his international ties and friendships. But something of his way of thinking, passed on from Neumann and the French before him, survives. Voigt saw his research "not only as mental work and the fulfillment of his desire for knowledge, but also as an aesthetic duty, a devotion and an expression of his personality."[74] He would certainly not have wanted to be called a father of solid-state physics, insofar as that became a practical-minded science. In fact, the central message of his textbook was a specific method of looking at things that could scarcely have been more abstract and ideal. This method seems so inevitable today that it is hard to imagine that physics ever lacked it, but it arose only late in the nineteenth century out of crystal studies, and is perhaps the greatest contribution such studies have made to scientific thought—analysis through symmetry.

Symmetry

In ancient cultures, symmetry, as it appears in natural objects, had made its way into art and mathematics.[75] The idea was taken up during the Renaissance and left its mark, for example, in Kepler's planetary orbits.[76] But symmetry was not used as a well-defined scientific approach until René-Just Haüy's studies of crystals.[77] He found general rules for the growth and regeneration of crystals, deriving, for example, the numerous secondary forms of calcite as expressions of a single form of nucleus, a parallelepiped, under various symmetry operations.[78] The work Haüy published in 1801 and 1802, summing up his previous publications on the theory of structure, marked the point where crystallography split from mineralogy to become a science of its own.

Haüy's work was immediately translated into German, and one of the young scholars who helped in the translating was the mineralogist Christian Samuel Weiss, later a respected professor at the University of Berlin. Weiss made a name for himself by opposing Haüy's static "atomistic" picture of crystals with a representation based on the "dynamic" approach that characterized many German natural philosophers of the early nineteenth century.[79] Weiss saw crystal forms as a necessary consequence of attractions and repulsions. This led him to a systematic characterization of crystals in terms of crystallographic axes, a system that even today is the first thing one encounters in a typical solid-state-physics textbook.[80] To

Weiss, crystal forms were an expression of generative forces, and the axes had a geometric as well as a physical significance. This led to the concept of a vector, a quantity characterized by a numerical value and a direction. With some help from mineralogical works on crystal geometry published by Franz Neumann (who had been his student), Weiss developed Haüy's laws of symmetry into a more useful "law of rational indices."

Haüy's work was also followed up by Johann Hessel, Auguste Bravais, and Axel Gadolin, who each independently derived limits on the set of possible symmetry classes of crystals.[81] Meanwhile, mathematicians developed the theory of groups, and in 1884 Ludwig Bernhard Minnigerode, a student of Neumann, recognized the relationship with crystallography and analyzed the 32 possible crystal classes in terms of group theory. It was not recognized until later that it is precisely the mathematical group properties that make symmetry significant for crystals.[82]

A theory of lattices developed in parallel with the theory of crystal classes. The classes were based on symmetry considerations, but ideas about lattices were built partly on ideas from crystallographers and partly on the geometric methods of number theory. Christiaan Huygens had already used a lattice construction based on ellipsoids of rotation in an attempt to explain the birefringence and the remarkably easy cleavability of calcite.[83] The crowning achievement in this line of development was the work of Auguste Bravais.[84] Proceeding from geometric considerations, he showed that there are only a few ways to order a set of similar points so that the neighborhood of any given individual lattice point looks the same. In his classification, which followed from symmetry considerations, he found the 14 space lattice types that are still used today. Bravais was able to order them into seven groups corresponding exactly to the crystal systems then known; this was the first convincing demonstration that the observed shapes of crystals reflect the properties of invisibly small lattices.[85]

Leonhard Sohncke, another student of Neumann, took up the Bravais theory in Germany. He noticed that Bravais had started with the lattice molecules in parallel. If one gave up this restriction and allowed other stable equilibrium positions, adding translations, point rotations, and screw rotations, one found a total of 65 regular point systems.[86] Further development of the Sohncke theory added reflections and their combination with rotations and eventually gave the complete set of 230 possible space groups.

This basically mathematical line of work went along with attempts to connect the abstractions to physical properties. In 1690, Huygens had suggested that the regular external form of crystals could be related to their physics. He then discovered that not only calcite (Iceland spar) but also quartz showed birefringence. Through the investigations of David Brewster and others, the number of substances recognized as birefringent exceeded 150 by the beginning of the nineteenth century. Around that time, Jean-Baptiste Biot discovered crystals with two optical axes— that is, ones with different degrees of birefringence along different axes. It was becoming clear that the symmetries shown by crystals were central to their properties. Brewster not only could classify crystals through their optical behavior, but also succeeded in associating this behavior with particular crystal classes. Only regular crystals were simply refracting. Trigonal, hexagonal, and tetragonal crystals had single optical axes, and crystals of the three other systems were biaxial.[87]

Shortly after, the chemist Eilhard Mitscherlich related an additional property of crystals—their thermal expansion—to their external form.[88] Neumann took the next step, working to understand the mechanical elasticity of crystals. It was a short step for him, since he had already undertaken to connect Fresnel's law of birefringence with fundamental equations of mechanics and had proposed a theory describing the birefringence that is produced in isotropic crystals by mechanical pressure or a thermal gradient. Neumann went on to relate a crystal's elastic response to stress to its external form.

All this raised the question of what the shape of a crystal really signified. Haüy had conceived of molecules as polyhedra, the crystals being constructed as stacks of odd-shaped bricks, but this simple view was becoming less and less tenable. One blow was Mitscherlich's discovery that chemically different materials can have similar crystal forms and that chemically similar materials can have distinguishable crystal forms. Using the newly developed optical goniometer, scientists were able to make exact measurements of crystal angles, which showed numerous deviations between Haüy's theory and fact. Ideas about atoms in crystals increasingly approached the chemist's atom, with its spherical form. The matter was settled by Pasteur's discovery that a single chemical, tartaric acid, can have two distinguishable crystal forms, one the mirror image of the other, with one form rotating polarized light to the right and the other rotating it to the left.[89] Plainly it was not the atoms themselves but the way they were linked that made the difference. The basic structures of crystallography were not atoms at all but "unit cells," specific combinations of atoms endlessly repeated in a lattice.

It took a long struggle to come to this realization, for it meant a new approach to physics—one that would eventually become central to solid-state physics. According to this approach, to study properties of interest one need not always seek to reduce them to ever more fundamental properties of ever more elementary objects. Rather, it may be the relationships among a collection of things that count.

The web that was binding the symmetries and the physical properties of crystals still lacked an important strand: electrical phenomena. In 1880 the brothers Pierre and Jacques Curie began to look for a way to generate electricity in crystals. They had in mind the phenomenon of pyroelectricity, known since antiquity and made famous by Haüy, in which heating certain types of crystals produced an uneven distribution of charge. While investigating hemihedral crystals, they discovered almost at once piezoelectricity: imposing dissymmetry by stressing the crystal produced a dissymmetric electrical phenomenon.[90] They noted that both pyroelectricity and piezoelectricity were present only in crystals with polar axes.

Pierre Curie followed the clue to a far more general conclusion. As he remarked in 1894, "I think it would be useful to introduce into the study of physical phenomena the ideas about symmetry that are familiar to crystallographers."[91] He was developing ideas he had learned from the works of Pasteur, who had proceeded from crystallographic studies to philosophical statements about the importance of dissymmetry in life and the universe. Curie went beyond these ideas to a fundamental law: "The dissymmetry creates the phenomenon."[92] Stated baldly, the aphorism may seem so obvious as to be almost trivial; Curie's great advance was to recognize the power of this principle as a tool for physical analysis, which he demonstrated by applying it to experimental results. In particular, from knowledge of the symmetry elements of an external crystal form, he could make broad statements about what physical properties could be present. This was soon common knowledge among crystallographers, and this knowledge influenced mathematicians (e.g., Schoenfliess) as well.

Voigt's textbook of crystal physics[93] won its lasting fame by disseminating these ideas. Investigating the functional dependence between any two given physical quantities, he began with their transformation properties. The rank of the measurable quantity, the observable (e.g., is it an axial or a polar quantity?), was given by the rank of the corresponding physical quantity. Whether it was pyroelectricity and the electrocaloric effect, corresponding to the interaction between scalar temperature and the vectorial electric field, or the suspected analogous phenomenon that brought about a magnetic rather than an electric field; or a result of the interaction between two vectors, the electric field and thermal conduction, in thermoelectricity; or the interaction between two tensor triplets, whose numerous effects expressed themselves in the various phenomena of elasticity; or the correlation between vector and tensor that caused piezoelectricity and piezomagnetism, Voigt described them all in terms of symmetry and through experimental results.

In his book, Voigt gave a lasting memorial to his teacher Neumann. Mindful of Neumann's lectures, which had pointed to a connection between the geometric symmetries of crystals and the symmetries of their physical behavior, Voigt introduced the term "Neumann's principle."[94] This embedded the point groups in crystal physics, but the space groups still had no correspondingly obvious value. Voigt was horrified by the "crushingly high number of 230 possible orderings," yet he admitted that "to judge the value of that type of consideration would be premature."[95] Only two years later x-ray analysis would arrive, and the great diversity of crystal structures that geometry allows would begin to appear as visible constellations of spots on photographic plates.

The experience with symmetry that emerged from the study of crystalline solids turned out to be useful in chemistry, biology, and, especially, physics. The principles have found application in molecular and atomic physics, in quantum mechanics, and in the study of fundamental conservation laws. Modern particle physics in particular has so refined Pierre Curie's tool that it is hard to imagine the field proceeding without it.[96]

Magnetism

Despite all efforts to use the microscopic nature of matter to understand the macroscopic properties of crystals, and vice versa, as the end of the nineteenth century approached, atoms remained mysterious points; whatever had been explained about crystals had not come through any understanding of the nature of atoms but through more abstract considerations. The difficulties of this situation were revealed with exceptional clarity in researchers' constantly frustrated attempts to understand magnetism.

It had long been known that a broken magnet yields as many new magnets as there are fragments. Combined with the notion of atoms, this fact had long since inspired the idea that a magnetic substance resembles an assemblage of miniature magnetized needles; each needle, bearing a magnetic moment, would represent an atom or a molecule. But no attempt to explain magnetism solely on the basis of a property of atoms or molecules had proved adequate. Rather, observations on paramagnetic and ferromagnetic substances revealed the existence of collective phenomena. It was evident, for example, that the magnetization of a bar of iron is modified by temperature variations or mechanical constraints and thus does not depend solely on the characteristics of an atom isolated from its neighbors. It was therefore necessary to approach the observed macroscopic phenomena by way of interactions involving more than one atom or molecule.

The electron, playing the dual role of a source of electromagnetic field and a receiver of electromagnetic perturbations from elsewhere, seemed the likely agent for these interactions. But the calculation of electronic interactions presented such great mathematical difficulties that researchers preferred to approach the problem in terms of a simple magenetic dipole interactions.

Thus there were two viewpoints from which magnetism was taken up. One, directed at the isolated molecule, was helpless to approach a major part of the experiments; the other, with a broader perspective, embraced collective phenomena but was frequently hampered by an inability to imagine any type of interaction except that between two magnetic dipoles. Progress would be made whenever a researcher could extend his gaze beyond the individual molecule without being blinded by the difficulties of the problem of interactions.

This was a type of challenge that faced not only studies of magnetism but, as will be seen in later chapters, many other areas of the emerging field of solid-state physics. Learning how to deal with mutual interactions, how to understand collective phenomena without losing sight of the individual elements, would be one of the keys to the field and eventually to much else in physics and other areas of modern science.

In this section, we restrict ourselves to the period between 1850 and the early twentieth century. During these years, metallurgists, chemists, and physicists considerably extended the range of experimental knowledge. Since this resulted only in augmenting the already abundant descriptive literature, we will not linger on that aspect of the field. One outstanding result appeared in 1895, with Pierre Curie's publication of the first experimental law for magnetism; in Chapter 6, we take this law as the starting point for the recent history of magnetism.[97] As for theory, our period is marked by the work of Wilhelm Eduard Weber and Alfred Ewing on ferromagnetism, Joseph John Thomson and Woldemar Voigt on diamagnetism and paramagnetism, and Curie, who, as we noted, cast a new light on the entire problem of magnetism in showing a relationship between the symmetries of phenomena and their causes.

The story of magnetism from the mid-nineteenth century to the early twentieth is chiefly a story of mechanical hypotheses. Throughout this period, physicists were accustomed to picturing the elementary magnetic moments as something like tiny magnetized needles. Each of the various models contained ancillary hypotheses that tried to account for as many experimental facts as possible.

The first significant model was proposed by Weber in 1852. Molecules were seen as magnets free to pivot around their own axes.[98] Applying an external magnetic field caused them to align with the field; perfect alignment explained the saturation that takes place when a substance cannot be further magnetized. To account for the fact that magnetization increases only gradually toward saturation as the applied field increases, Weber had to assume that some undefined cause opposed the free movement of the magnets. Without attempting to understand the origin of this opposition, he compared its effect with that of a restoring torque that pushes each molecule back toward the position it originally occupied. Such a torque could be produced by a magnetic field acting on the elementary magnets, and he therefore postulated a "local magnetic field" at the site of each molecule, oriented in the direction that the molecular moment would point in the absence of any external influence. Because matter is usually unmagnetized when there is no external magnetic field, he concluded that the molecular moments were distributed uniformly in all directions, and likewise for the local field.

The local field thus had nothing to say about anisotropy in the material, and it was only as an artificial aid to calculation that Weber called it a magnetic field at all. The local field was fundamentally no more than a supplementary attribute of the molecule—its preferred orientation, maintained without change in the presence of any external disturbance. Weber kept his eyes on the individual molecule; his model was not a collective one. James Clerk Maxwell added a supplementary hypothesis in 1873, and it too concerned the individual molecule.[99] His concern was residual magnetization—the phenomenon that makes it possible to create magnetic compass needles. He suggested that if a molecule shifted too far from its preferred position, it would fail to return after the disturbance was removed, so some magnetization would remain.

About 1885, Alfred Ewing attempted to adapt these models to his pioneering experimental results, which led him to attribute differing coefficients of friction to various molecules while leaving other molecules entirely free. Such additional hypotheses, although making clear the complexity of the problem, tended to obscure Weber's original idea. Ewing began from the beginning with an entirely new perception of magnetism: he would take into account a collective phenomenon.

Ewing described his new approach in a paper published in 1890.[100] He began by quoting Maxwell, who had written that molecules do not take up a position in perfect alignment with the applied field, either because "each molecule is acted on by a force tending to preserve it in its original direction, or because an equivalent effect is produced by the *mutual action* of the entire system of molecules." Weber had selected the first possibility, Maxwell had followed his example, and Ewing had originally done the same, but now he took an interest in the second possibility. He was inspired by several physicists who had recently put forth ideas about mutual interactions between molecules. In particular, it had been suggested that molecules might take up their privileged positions under the influence of their neighbors, and arrange themselves in such a way as to minimize the interaction energy.[101] This hypothesis did not exclude the existence of mechanical friction, which continued to be assumed. Arthur Edwin Kennelly subsequently showed that the rapid rise observed in a part of the magnetization curve might be due to a sudden break in the positioning of the set of molecules: for a sufficiently strong applied field, the arrangement of the moments would become unstable and the moments would swing into the direction of the field.[102] The stable arrangements were assumed to resemble closed loops of chain.

Ewing undertook to verify the existence of such arrangements. He constructed a large-scale version of Weber's model, a two-dimensional analogue made up of magnetized needles, balanced on nails hammered into a table and free to swing horizontally. He placed this model in a large solenoid, which created a uniform field that could orient the needles without changing the internal magnetic state of iron from which they were made. Ewing particularly noted the positions taken by the needles in the absence of the solenoid when they were subject to only the weak action of neighboring needles.[103] They did not arrange themselves into closed chains, but nevertheless they fell into a certain order. He saw groups of needles all pointing in the same direction; if he destroyed the configuration, another replaced it. There were many possible configurations, all of them giving a zero net moment.

When the solenoid was added, the needles would turn as the magnetic field was gradually applied. Ewing found that the interest lay not in any given needle, but in the group to which it belonged. As the field increased, a group would became unstable and suddenly swing into the direction of the externally applied field. The same level of applied field would not make every group unstable, but all would swing as the field crossed a rather narrow range—the range where there was a rapid rise in magnetization.

Ewing's model thus reproduced all the essential characteristics of the iron magnetization curve. He had restored clarity to Weber's model while retaining only the hypothesis of mutual action between elementary magnets. He had achieved still more: through observing his model, he had recognized that magnetism is a collective phenomenon. He had also explained why the net magnetization was zero in his mosaic of groups. It would be anachronistic to call these groups domains (the term was later applied to microscopic regions of uniform magnetic field within a magnetic metal), but when he described the multiple equivalent configurations of his system of needles, Ewing was ahead of his time in using the language of statistical physics.

In pointing to a large-scale phenomenon, Ewing was opening up a direction for research that much later would lead to an understanding of the magnetic force within crystals. At the time, however, another problem remained in the fore. Each element within Ewing's groups represented a complex object, a molecule whose magnetic character remained to be demonstrated.

Pierre Curie set about to resolve this problem by using the symmetry viewpoint, as noted earlier. Although no effect analogous to pyroelectricity or piezoelectricity was known for magnetism, Curie predicted that spontaneous magnetic polarization of matter was indeed possible. He had noted in particular that the electrical polarization that naturally occurs along one axis of a tourmaline crystal is compatible with the structure of the crystal, whose symmetry falls within the mathematical group that describes the symmetry of an electrical field. Since other crystals have structures whose symmetries are similarly related to the magnetic field's symmetry group, Curie suggested that these crystals might be naturally magnetized as a result of their internal structure, just as tourmaline is naturally polarized. Taking these considerations down to the level of the molecule would allow one to say that under certain conditions of asymmetry, the molecule could be magnetic.

However, that was about as far as one could go. Although great discoveries were made in atomic physics around the turn of the century, their application to magnetism did not lead to any remarkable results. A striking example of the obstacles confronting a microscopic theory of magnetism may be seen in the attempts made by the dean of atomic physicists, Joseph John Thomson.

In 1903, Thomson addressed himself to the magnetic element in isolation from its surroundings and attempted to determine the conditions under which the element could give rise to diamagnetic and paramagnetic effects in matter.[104] He viewed the atom, on the one hand, as a source of magnetic field and, on the other hand, as a system responding to the action of an external field. Using one of his characteristically detailed mental models, he pictured this atom as a positive charge with electrons clustered around it, the electrons being located at regular intervals along a uniformly rotating ring. Applying the equations of electromagnetism to these currents, he calculated that in the presence of an applied field each electron would make a positive contribution to the coefficient of magnetization, provided the radius of the ring decreased as time passed. Paramagnetism was therefore an atomic property that came into being only if, by some unknown process, an electron slowly spiraled toward the center instead of remaining in a stable orbit. Thus in Thomson's model, the origin of paramagnetism was the damped movement of electrons.

Thomson compared this with a radioactive atom emitting charged particles that had insufficient energy to escape the atom's attraction—that is, satellite particles that would dissipate their energy and fall back into the atom. He suggested that a permanent energy flux would emanate from such an atom; it was therefore not impossible that the temperature "in the middle of a mass of a magnetic substance like iron, whose surface is kept at a constant temperature, [differs] from the temperature inside the mass of a non-magnetic substance like brass whose surface is kept at the same temperature."[105]

Thomson had also come up with a hypothetical explanation for diamagnetism. He attributed it to the fact that a wholly free electron would assume a circular or helical orbit when a magnetic field was applied, producing a magnetic moment opposing the field; this induction effect would make the substance diamagnetic. Permanent continuous current loops, bound in some manner, would supposedly make the substance paramagnetic.

For Thomson, magnetic properties were "inherent in the atom," aside from ferromagnetic properties, for which he was willing to go beyond the isolated atom. He looked for a collective phenomenon in aggregates of molecules and proposed as "a rough analogy" that the orbit of charged bodies might resemble "a tube bored through the aggregate so that the orbit and aggregate move like a rigid body."

As of 1903, Thomson's ideas about magnetism were obviously little more than ingenious guesswork. He was not alone; for example, around the same time Voigt asserted, independently of Thomson, that permanent magnetic phenomena could only be explained by the damped motion of electrons.[106] Theorists would generally have more immediate success in areas other than magnetism. For the time being, the subject, despite Ewing's clarification, remained bogged down in unreliable metaphors such as Voigt and Thomson had offered; some of these ideas were entirely incompatible with the observations that Curie reported from 1895 on.

Paul Langevin took a decisive step toward understanding paramagnetic and diamagnetic phenomena in 1905, when he calculated the magnetic field of an orbiting electron giving more attention to precise assumptions and conclusions than Thomson had.[107] Another step came in 1907, when Pierre Weiss transformed the vague notion of a local field into a precisely defined collective quantity that could help explain ferromagnetism.[108] Thanks to the work of these researchers, theories of the magnetism of matter finally acquired some independent authority rather than being constructed solely on ideas borrowed from other fields of physics. The inquiry opened up by Langevin and Weiss—difficult, long in reaching fruition, and closely bound up with the emerging theory of the quantum—is described in Chapter 6.

1.3 Breakthrough of Microscopic Conceptions

The Free Electron Gas. I: Classical Electron Theory of Metals from Drude to Bohr (1900–1911)

The discovery of the electron at the end of the nineteenth century marked the beginning of a new era for theoretical physics. After the bound electron in the atom and the free particles of cathode rays were recognized as identical (since the Zeeman effect and cathode ray experiments produced the same ratio of charge to mass, e/m), an electron theory appeared to be the key to a microscopic understanding of matter. Physicists soon derived phenomena in electricity, magnetism, and optics from equations of motion for electrons. Themes like the electrodynamics of moving bodies, which led to relativity theory, and attempts to explain radiation laws, which opened the door for quantum theory, likewise belonged within the electron theory's range of interest, and so did the electrical properties of metals.

Atomistic concepts of the electrical properties of metals could already be found in a number of nineteenth-century theories,[109] but the revolution in microphysics in the first decade of the twentieth century made such concepts concrete. The emerging picture imagined a metal as permeated by a gas of free electrons, a gas interacting with the thermal motion of the metal's atoms. Corresponding theories developed in which the fundamental constant of statistical mechanics (Boltzmann's constant, k) played a role alongside quantities connected with the electron (e and m). The first milestones on the way to such an electron theory of metals were the "classical" theories of Paul Drude,[110] Joseph John Thomson,[111] and Hendrik Antoon Lorentz.[112]

Drude's work began with problems of optics. In his 1887 dissertation under Woldemar Voigt, Drude had based his arguments on a mechanical concept of light propagating through an elastic medium, an ether such as Fresnel, Neumann, and Voigt had taught. In the 1890s, he reluctantly adopted a theory of light based on Maxwell's theory; with optics now a branch of electrodynamics, new problems came to the fore. Among them was the question of the connection between optical absorption coefficients and electrical conductivity, a connection that under the old theory of light had no reason to exist at all.[113] In numerous experiments, Drude determined the optical constants of metals. In December 1899, in his paper "On the Ionic Theory of Metals," he tried to explain such optical properties with the help of equations of motion for "insulating" (i.e., bound) and "conducting" (i.e., free) ions, either positively or negatively charged, to which he assigned different masses.[114]

Joseph John Thomson (1856–1940). (Burndy Library. Courtesy of AIP Niels Bohr Library)

Paul Drude (1863–1906). (AIP Niels Bohr Library, Physics Today Collection)

Two months later, Drude published his work "On the Electron Theory of Metals."[115] Here he dealt exclusively with the "conducting ions," which he now called "particles," "electrons," or "electrical cores." He assumed particles of both positive and negative charges and used "for the elementary quantum e of electricity the number found by J. J. Thomson . . . , which also closely agrees with the data obtained from electrolysis." He added, "Whether or not an electron carries with it a very small ponderable mass, we leave for the present undecided." The dependence of the electron mass on its state of motion and on concepts of the ether remained an unsettled problem; Drude avoided it by using the gas kinetic expression $mv^2/2 = \alpha T$ to connect the velocity v and the "apparent mass" m of each "free-flying particle" with the absolute temperature T. The proportionality α was introduced here as a "universal constant" in his theory.

This borrowing from the kinetic theory of gases pointed to the next step: derivation of the conductivity. Drude calculated the heat flow of free electrons in a temperature gradient by using Boltzmann's gas theory, with Boltzmann's formula for the mean free path L applied to the case of gases having two types of particle. Also, he calculated the electric current for the electron gas in an applied electric field in the same way as it had been calculated for ions in an electrolyte.[116] He found that both the thermal conductivity κ and the electrical conductivity σ were proportional to the product vLN, where N is the concentration of electrons. Thus the quotient κ/σ became independent of N, a most useful result, since N could not then be reached through experiment. Drude found that

$$\frac{\kappa}{\sigma} = \frac{4}{3}\frac{\alpha^2}{e}T$$

This was the Wiedemann–Franz law, derived for the first time on the basis of a microscopic theory.

Drude's formula was consistent with the latest experimental values from the Physikalisch-Technische Reichsanstalt, and was widely regarded as an outstanding indication of the validity of the electron gas concept.[117] Drude's plain derivation of the proportionality factor $\kappa/\sigma T$ also called attention to this natural constant in physics (named the Lorenz constant after Ludwig Lorenz, who, in 1881, had determined the temperature dependence in the Wiedemann–Franz law). When Max Thiessen of the PTR summed up the situation regarding the natural constants for a meeting of the Berlin Academy of Science, he cited not only the latest work by Planck on the radiation law and by Thomson on the fundamental electrical charge, but also Drude's derivation of the Lorenz constant.[118] In appreciation of Drude's work, Max Planck named α the Boltzmann–Drude constant.[119]

Almost simultaneously, and apparently independently of Drude, Thomson also published a free-electron gas theory of metals. Unlike Drude, Thomson assumed only one type of "corpuscle," which he identified with the particles of the cathode rays, hardly a surprising move given the long-standing tradition of cathode ray experiments in Thomson's Cavendish laboratory.

Hendrik Antoon Lorentz also worked with a single particle, the electron. In his general electron theory, problems such as normal and anomalous dispersion of light, rotation of polarization, and the Zeeman effect were shown to follow from the laws of motion of electrons; refining Drude's model for metals by assuming a pure electron gas was only a small portion of Lorentz's grand scheme.[120] In 1903 he tried to derive a formula for the energy distribution of blackbody radiation from Drude's theory. He found a solution for only long wavelengths, which agreed with Kirchhoff's law and with the long-wavelength limit of Planck's formula.[121]

In a talk in Berlin a year later, Lorentz reported that he had repeated Drude's calculations, using only one type of moving particle, and that he had "tried to penetrate somewhat more deeply into the details of the motion of individual particles of the electron swarm."[122] Drude's calculations had used only an average velocity of the electrons, but Lorentz also took into account the statistical distribution of the velocities. He inserted a Maxwell distribution function, modified by an additional term that represented an external influence such as an electrical field or a temperature gradient. He considered collisions of electrons, not with one another but only with the atomic cores, which he treated as positively charged hard spheres fixed in place, and from the Boltzmann equation for the number of collisions he derived expressions for the current of electrons driven by the external influence. With this move, Lorentz brought the theory of the free-electron gas in metals into a self-contained and logically satisfying form. According to a contemporary, Lorentz had thus accomplished "for electron statistics the same as Boltzmann and Maxwell had accomplished in their time by their extension of the original equations of Clausius."[123] To be sure, the results changed little in Drude's results beyond modifying a few numerical factors. In the case of the Wiedemann–Franz law, Lorentz found a constant approximately two-thirds of Drude's, which considerably worsened the beautiful numerical agreement with experimental values.

The conceptual clarity of the electron gas model stood in contrast to a bewildering array of experimental findings that seemed to contradict the concept. One dilemma concerned the specific heat of metals. Specific heat was understood to have something to do with the motion of all the free particles in the material. If each metal atom contains one free electron, then, according to the Dulong–Petit law and the classical theorem of equipartition of energy, the specific heat should take on 1½ times the value that one would get without taking the electrons into account. But this is well above what is actually observed. In short, in Lorentz's theory the electrons were numerous and freely moving, but somehow they failed to take up their share of the energy when the metal was heated. This dilemma had already been referred to by Max Reinganum immediately after Drude's first work in 1900.[124] Drude and Lorentz quoted the work of Reinganum, but they did not mention the specific heat dilemma in their articles. To others, like Thomson and Willy Wien, Drude's theory appeared so convincing that they postulated mechanisms according to which only a small fraction of the metal electrons would be free at a given time.[125] Around 1911 it was probably the majority opinion among physicists competent in the area that one had "to either declare the gas equation for the free electrons as invalid, or regard their number as vanishingly small compared with the metal atoms."[126]

The Hall effect posed another dilemma. With only one type of electron, the negative one, the appearance of positive Hall coefficients for metals seemed completely inexplicable. As for the quantitative values, in 1906 Richard Gans applied the Lorentz theory to galvano- and thermomagnetic effects for metals like bismuth with negative Hall effect and obtained results nowhere near the experimental findings.[127] Karl Bädeker's monograph of 1911 on "the electrical phenomena in metallic conductors," which described not only Hall coefficient measurements but also special properties of ferromagnetic materials like bismuth and antimony, changes of resistance in magnetic fields, and other phenomena, summed up from the point of view of the experimenter: "Galvanomagnetic phenomena are an area where the electron theory of metals is for the time being still insufficient."[128]

In his doctoral dissertation in 1911, Niels Bohr produced what was at that time the most comprehensive analysis of the electron gas model. Along the way, he discovered a further dilemma: A classical free-electron gas exhibits no diamagnetism.[129] We have noted that for diamagnetism in metals (which had been experimentally observed), Thomson's theory appealed to an induction effect stemming from electrons that orbited freely in the applied field. Bohr now demonstrated that a magnetic field performs no work on free electrons in thermal and electrical equilibrium and leaves the distribution of electrons unchanged; consequently, a piece of metal in thermal and electrical equilibrium can possess no magnetic properties attributed to free electrons.

The example of Bohr's dissertation makes it clear that the electron gas theories did not arise simply as applications of new microscopic concepts, but were harbingers of the quantum revolution. The dilemma of diamagnetism in metals, a property Bohr could not derive as a characteristic of the free-electron gas, was one reason for him to turn his attention to bound electrons.

Langevin's 1905 theory had invoked changes in the velocities of bound atomic electrons as the source of diamagnetism, and changes in the orientation of their orbits as the source of paramagnetism (for details, see Chapter 6).[130] Bohr's dissertation presented an argument against Langevin's diamagnetism theory similar to

Hendrik Antoon Lorentz (1853–1928), around 1900. (Algemeen Rijksarchief, The Hague. Courtesy of AIP Niels Bohr Library)

Walther Nernst (1864–1941). (Rijksmuseum voor de Geschiedenis der Natuurwetenschappen, Leiden. Courtesy of AIP Niels Bohr Library)

the argument in the free electron case. According to the "general laws of mechanics," when a magnetic field was applied to change the velocity of atomic electrons, and so to build up a magnetic dipole moment, if the magnetic field reached a constant value the energy had to be distributed rapidly and evenly among all atomic electrons, which would destroy the dipole moment. To save Langevin's diamagnetic theory, one had to call on some sort of nonmechanical "freezing" of the changes in velocity, some mechanism hindering the equipartition of energy inside the atom that classical mechanical theory of heat demanded. This problem is one of several factors that influenced Bohr on his path from the electron theory of metals in 1911 to his atomic model of 1913, based on quantized electron orbits.[131]

The Specific Heat of Solids

"If Planck's theory of radiation has hit upon the heart of the matter, then we must also expect to find contradictions between the present kinetic molecular theory and practical experience in other areas of heat theory, contradictions which can be removed in the same way."[132] With this statement, Albert Einstein took up the specific heat of solids, his attention drawn by the increasingly numerous deviations between experiment and current theory. This was in the fall of 1906, a year after he explained the photoelectric effect with the help of Planck's quantum. As a student, Einstein had heard in a lecture about the notable deviations of specific heat from theory in the case of diamond, where not only was the value at room temperature wrong, but the change of the value with temperature could not be explained. In the meantime, especially through James Dewar's measurements at low temperatures, numerous other substances had been found whose specific heat deviated from the value of the Dulong–Petit law, which prescribed for all solids a value for the specific heat c_v of 5.96 calories per degree Celsius. Rather than staying constant, the specific heat decreased with temperature. Once again, somehow the particles of the solid were failing to take up energy as expected.

With his "improbable ability to empathize with the workings of nature,"[133] Einstein freed himself from the classical equipartition principle and pressed ahead on his own. He considered the energy content of a solid to be the sum of the individual energies of its charged atoms. This was the energy whereby each atom oscillates sinusoidally with a fixed frequency, the "residual ray" frequency, about its equilibrium position, as determined by Planck. By requiring this fixed frequency, Einstein was, in effect, saying that the particles could take up energy only one quantum at a time; eventually, this would not only explain the low specific heat, but also open the way to a clearer understanding of the elusive quantum concept. Meanwhile, the monochromatic "Einstein formula" described the temperature variation of the specific heat of diamond and also enabled Einstein to calculate the residual ray frequencies of numerous crystals from the known values of the specific heats. The good agreement of the frequencies calculated from his heat formula with the optically determined data of Rubens and his students convinced Einstein of the validity of his approach: "We have to accept, therefore, that the elementary carriers of heat in diamond are nearly monochromatic structures."[134] But although this work was published in the 1907 *Annalen der Physik,* it attracted little attention for several years, and Einstein laid the topic aside for the time being.[135]

The way for the next step was prepared in 1905 when Walther Nernst was appointed professor at the University of Berlin.[136] Through his versatile education, his collaboration with distinguished physicists and chemists, and his successful activity as leader of the Institute for Physical Chemistry and Electrochemistry in Göttingen, Nernst seemed especially suitable for the position. As a younger man, Nernst became well known for his work on galvanic current production; his efforts to provide a theoretical foundation for chemical empirical laws through physical chemistry had also won respect from his colleagues. Meanwhile, his business skills, especially his financial success with the Nernst lamp, made him prosperous.

Nernst emerged from his work on galvanic current production with a major idea: as temperatures approach absolute zero, certain curves not only intersect but become tangential to one another. This was subsequently codified as the universal law that entropy becomes zero at absolute zero (the Third Law of Thermodynamics).[137] From the new law, which Nernst named "my heat theorem," he drew the conclusion that as temperature approaches absolute zero, the specific heat must approach a limiting value independent of the nature of the body.

To create an experimental basis for his heat theorem, Nernst set about building a costly low-temperature facility, with all the financial means his position in Berlin made available to him. This was a difficult assignment, which for years required much of him and his students. Nernst also occupied himself, at the same time as Einstein, with the theoretical problem of specific heat.

Einstein's motive was to use the quantum concept on radiation and matter as well as on the interaction between the two; Nernst's motive was to prove the validity of his heat theorem. Exactly when each became aware of the other's work is not known. Although by 1906 Einstein had corresponded with and met Laue, who was then Planck's assistant and living in Berlin, there is no evidence of an exchange between Nernst and Einstein on the subject of specific heat.[138] The discussion began in earnest at the eighty-first annual meeting of the Gesellschaft Deutscher Natur-forscher und Ärzte, which took place in Salzburg in September 1909. Einstein, still a little-known outsider, gave a stirring speech to the great experts, speaking not only on his special theory of relativity, but much more on the quantum problem.[139] Soon after, Nernst set out for Zurich to talk with Einstein.

Einstein reported to a friend: "For me, the quantum theory is certain. My suppositions regarding specific heat appear to be splendidly confirmed. Nernst, who just visited me, and Rubens are hard at work with the experimental test, so we will soon be put on the right track."[140] For Nernst, the pursuit of Einstein's work on specific heat had two consequences. First, it gave him a plausible description of the results of the measurements that were being made in his cryogenic laboratory. Second, he recognized that his heat theorem had further significance: it represented a necessary consequence of the quantum theory. Nernst was open to everything new and became one of the earliest and most important supporters of the quantum. The great esteem that he commanded in the scientific world, his numerous contacts, and his skillful tactics helped the revolutionary quantum concept take hold faster than could otherwise have occurred.

Although Nernst's first low-temperature results agreed with the Einstein formula, systematic discrepancies were soon found. Nernst and his student Frederick Alexander Lindemann (later famous as Churchill's adviser Lord Cherwell) remarked that they did not want "to completely lose contact with Planck's radiation formula" or stray too far from "the quantum theory, otherwise so eminently useful." So in 1911 they came up with a formula constructed in a way analogous to Einstein's, but composed of two additive terms. Found "by trial," the modified Einstein formula accounted for the experimental results. But attempts by Rubens and others to sustain a physical interpretation fell flat.[141]

Einstein was nevertheless stimulated by Nernst's measurements, which he believed gave the theory of specific heat a "real triumph."[142] He confronted specific heat again in work on elasticity, seeking a connection between infrared absorption and mechanical theory.[143] He was convinced that molecular oscillators were complex systems, far from monochromatic.[144] Proceeding from the concept of the lattice in crystals, he considered the effect on each atom of the many neighboring atoms. But he could not get beyond an order-of-magnitude estimate.

When Einstein accepted a position in Prague in 1911, Arnold Sommerfeld saw to it that his assistant Peter Debye became Einstein's successor in Zurich; Debye promptly took up the problem of the specific heat of solids.[145] His lectures on thermodynamics in the winter of 1911/1912 forced him to take a new look at the theory of radiation and specific heat.[146] He began with Einstein's recognition that it is not just a question of monochromatically oscillating structures. As we will see later, just at this time the idea of a lattice of bound atoms was much discussed, for such lattices had just then been observed for the first time with x rays, and it was even becoming possible to measure the lattice spacing directly. Debye calculated the frequency of the body from elasticity theory, ignoring the electrons and concentrating on the lattice structure; he limited the number of degrees of freedom in the case of a body with N atoms to $6N$. By associating Planck's energy with each pair of degrees of freedom, he obtained a formula in which specific heat varied as T^3 at low temperatures; this convincingly described the experimental results.[147] Since Debye's formula has a very simple form, it is still used today in calculating the heat content of a solid. Many years later, a student of Max Born proved that the formula breaks down at very low temperatures.[148] This at last confirmed Nernst's suspicion that "it is not permissible to neglect the atomic structure of matter at wavelengths of hydrodynamic oscillation commensurable with the distance between atoms."[149]

In Göttingen in the spring of 1912, two friends, Max Born and Theodore von Kármán, began to rework Einstein's monochromatic formula by considering the coupling between the oscillating lattice elements.[150] Starting from the vibration spectrum of a one-dimensional, mechanically coupled chain of mass particles, they found an approximate solution for the spectrum of the three-dimensional crystalline lattice.[151] They found the energy distribution according to Planck's formula, by using the principle vibrations (normal modes) rather than the individual atoms as the independent resonators. The normal modes of lowest frequency were simply the usual elastic vibrations, and the highest mode numbers fell in the frequency region of infrared light. These considerations yielded an expression for the specific heat in excellent agreement with experimental results, even at low temperatures. However, because the Born theory is so complicated in comparison with the Debye formula, it is little used in practice.

Von Kármán soon turned his attention to hydro- and aerodynamics, but Born continued to work on crystals. With the help of the theory of lattice vibrations in space, he and his students succeeded in describing many properties of certain types of solids.

Lattice Dynamics

When Max Born died in January 1970, his student Maria Goeppert-Mayer chose as the title of his obituary notice, "Pioneer of Quantum Mechanics Max Born Dies in Göttingen."[152] Even she, who had composed the *Handbuch der Physik* article "Dynamic Lattice Theory of Crystals"[153] under Born's guidance, ranked his contribution to quantum mechanics considerably higher than the numerous works that helped create a unified crystal theory. Yet it was precisely his concern about the physical properties of crystals that led Born first to quantum mechanics, later to atomic physics, and finally to the famous paper that enters the history of quantum mechanics as the *Dreimännerarbeit* (three-man work).[154]

Born entered the field of lattice dynamics in April 1912 with the publication of the paper "On Vibrations in the Space Lattice."[155] The heart of the work was not its result—the exact formulation of the Einstein formula for the specific heat of solid bodies—but a derivation of the vibration spectrum of a cubic crystalline lattice.

Solving the equations of motion for the lattice elements under the approximation that only closest neighbors exert forces on one another, Born obtained a type of dispersion equation, a dependence of the frequency v on a "wavelength" λ. The solution was noteworthy for the case in which the lattice possesses a "basis"—that

James Franck (1882–1964) and Max Born (1882–1970). (AIP Niels Bohr Library, W. F. Meggers Collection)

is, alternating mass points with distinguishable masses and charges. There will then exist two types of distortion (Fig. 1.1). One is a slow vibration, like a sound wave, in which a lattice cell moves as a whole; this is called the acoustic branch of the dispersion curve. The other is a more rapid vibration, in which the distortion takes place within the basis group, as a "hidden distortion"; this is called the optical branch of the dispersion curve, and its threshold frequency was found to have a revealing connection with the frequency of the infrared "residual radiation" bands that Rubens had discovered and measured. Having determined that each distortion state of the lattice could be represented by a superposition of plane waves, Born defined these as the "normal" and "characteristic" vibrations, respectively. Using Planck's formula on normal vibrations then brought forth the specific heat formula.

When Voigt retired from lecturing in the spring of 1914, Peter Debye took over part of his teaching duties. In his autobiography, Born reflected on Debye: "It was great luck for me to have this brilliant man as a colleague. We held a colloquium together and had many interesting discussions."[156] One of these discussions stimulated Born to take up again the elasticity of crystals, that unbeloved theme to which he had already sacrificed much of his time.

In 1905, Born had found himself responsible for giving a talk for a sick colleague in Felix Klein and Carl Runge's elasticity seminar.[157] Born's method of solution using the calculus of variations had pleased the "great Felix" so much that he nominated the theme as the prize-winning work of the year. But at that time, Born was

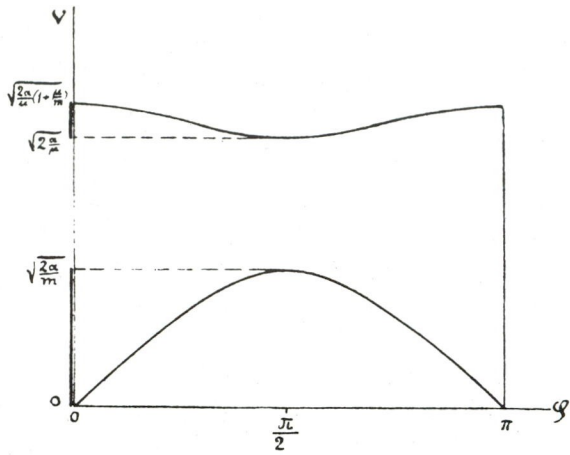

Fig. 1.1   The vibration spectrum of a cubic ionic crystal. The possible frequencies are divided into two distinct parts. (From M. Born and T. von Kármán, "Über Schwingungen in Raumgittern," *Phys. Zs.* 13 [1912]: 304)

more interested in the electrodynamics of moving bodies, and at first he put off working on the problem. Only after a confrontation with Klein did he hasten to finish the task, winning the prize while taking his degree under Voigt with this work. But the endless calculations and extensive experiments left him thoroughly disgusted not only with the subject but also with Göttingen. He resolved never to return.[158]

But Born did return to Göttingen in 1908, and in 1914 voluntarily took up the theme of elasticity—this time with a crucial question. The dispute over the number of independent elasticity coefficients had dragged on for almost 100 years. This confrontation had begun with Cauchy's equations and had been given new force by Voigt's measurements.[159] Proceeding from a determination of the structure of diamond done with novel x-ray techniques (discussed in a later section), Born showed that one arrives at the many-parameter theory, with 21 parameters. The experts must have been thunderstruck by this refutation of the six Cauchy equations that had attempted to reduce the number of parameters in the general lattice. Born not only had settled the long-standing debate, but also had cleared away the last remaining obstacle for Cauchy's fundamental picture of a crystal as an atomic lattice, a picture whose acceptance had been held back by the differing number of parameters in the discontinuous and continuum models.

Born moved on to a chair in Berlin and proceeded to write *The Dynamics of Crystal Lattices,* a book that Voigt called "the most important advance in crystal physics in a long time."[160] Many years later, Ewald addressed Born with special mention of the book: alongside "Laue's and Bragg's related discoveries," he said, it was "the most important landmark in the history of solid-state physics. Your theory, so simple in its foundation and so perfect in its execution, broke through the thorn hedge and rescued the beauty who had been asleep since Cauchy's contribution."[161]

Born's little book is pure theory; there is scarcely any attempt at clarification and no reference to experiment. Born transposed the mechanical model of a chain to three dimensions, where the most general lattice consists of an arbitrary number of basis elements. He assumed that the linear force law between the lattice elements is not known, but noted his suspicion that "the chemical forces, which bind atoms in liquids and gases into molecules, are identical with the forces in the crystalline state that hold atoms together in a lattice formation."[162]

Born put off the step from the mechanics of crystal lattices (part A of the book) to electrodynamics (part B) until he came to treat optical phenomena. He connected the equations for the mechanical vibration of the lattice elements with Maxwell's equations, which were based on a continuum model, thus producing a mixture of molecular and continuum theory, and so obtained Fresnel's law for the propagation of light in crystals. Born showed the invalidity of Cauchy's equations, and remarked that this "clarification is the first lovely result of carrying out the calculation on the compound lattice."[163] He also showed that "our theory takes piezoelectric phenomena into account without any new assumptions."[164] Further, without additional ad hoc hypotheses he could describe the rotation of the plane of polarization during the passage of light through optically active crystals, explaining a phenomenon that during the past century had fascinated chemists and biologists as well as physicists.[165]

The book's summary focused on the unsolved problem of the nature of the chemical forces between the particles that made up the crystal. This already hinted at Born's future work, including his turn away from solids toward fundamental questions of atomic physics. Although (according to von Kármán) Born had begun to work on lattice dynamics in order to show, among other things, that the quantum hypothesis is not necessary, he now spoke of crystals as an area for applying quantum theory.[166]

But the First World War had begun, and once the manuscript for his book was sent to the printer Born did not hold back, despite his unstable health, from "doing my duty to my fatherland."[167] He allowed himself to be transferred to the Artillery Testing Commission, where "a small army of physicists" developed a method of locating enemy artillery by measuring the sound of its firing.[168] Later he was joined by acquaintances, among them Erwin Madelung and, at Born's intervention, his previous student Alfred Landé. As the pace of the war slackened, Born resolved to continue scientific work on crystals at his new location.

He was convinced of the validity of the new Bohr–Sommerfeld model, which pictured a free atom as surrounded by quantized rings or shells of electrons. Born now tried to apply this to the ions of the alkali halides, whose crystal structure had become known in the meantime thanks to x-ray analysis. For the calculation of the potential energy of the lattice, he added two terms, which depended on a theory of binding of ionic crystals just published by Walter Kossel. The first term described the Coulomb attraction between oppositely charged ions; the second characterized the electromagnetic repulsion of the electron shells, which kept the crystal from collapsing on itself.

After the war, Born began to work out the chemical ramifications of his theory. "I calculated the lattice energy, that is, the work that must be done by the atomic forces in order to assemble the crystal out of ions. However, there was no available experimental material for comparison."[169] As Born was discussing this one day with his friend James Franck, Fritz Haber came in and took part in the argument.[170] "He was immediately enthusiastic about my ideas. . . . I learned from him that apart from some crude estimates . . . chemical energy had never before been calculated, and that my attempt was the first which in his opinion looked very promising."[171] The two men now found a graphical calculation technique, the Born–Haber cycle for the representation of the chemical energies.[172] As Haber had predicted, the work caused a small sensation in chemical circles. For it finally proved that the cohesive forces in ionic crystals are identical with the chemical forces that act between molecules and with the electrical forces that act between atomic nuclei and the electron shells.

In 1920, Born accepted an offer to be Debye's successor in Göttingen. Influenced by direct collaboration with Franck, who was appointed to a Göttingen experimental physics chair at the same time, Born turned his interests more and more toward quantum theory and its application to the construction of individual atoms. But he also kept pursuing the further development of the theory of the solid state in the context of revised editions of his book, although essentially he left direct work to his students.[173] Some of this work, such as Friedrich Hund's theory of the formation of crystal lattice types and the polarizability of crystal ions, was significant for the further development of crystal physics; other work, like Carel Jan Brester's group-theoretical classification of lattice vibrations, had an influence that spread to other areas of physics.[174]

In the spring of 1933, Born, who was of Jewish descent, took a leave of absence and, like a great many of his friends and students, decided to emigrate. His theory of the dynamics of crystal lattices did not become of widespread interest until lattice vibrations could be observed by means of neutron and Raman spectroscopy. In 1962, when Born was invited to the first conference on the subject, he was overwhelmed: "In my time, the interest in lattice dynamics was limited to a small number of specialists. Now I have the impression that it has spread over the entire world and has become a blossoming branch of physics."[175] Nevertheless, one can agree with Ewald, who said, "I have often found that your accomplishment didn't get the recognition that it deserved."[176]

The Free Electron Gas. II: The Impact of Quantum Theory (1911–1926)

The full development of the electron theory of metals required quantum mechanics and Fermi statistics. Neither was available until 1926. However, it had already become clear by 1911, after more than a decade of classical electron gas concepts, that there had to be fundamental reasons for the utter failure of these concepts to explain most phenomena in metals. New effects, like the onset of superconductivity at extremely low temperatures, followed on the heels of older unexplained phenomena. The electron theories put forth between 1911 and 1926 provide interesting examples of how physics progressed during a phase when old concepts were recognized as unreliable while new ones were not yet in clear view.

The quantum theory and the electron theory of metals did not develop independently of each other during this period: the electron gas model served time and again as a test case of a gas in which quantum effects were supposed to occur. In this way, the physics of electrons in metals remained of current interest, catching the attention of physicists who in these years of the quantum mechanics revolution had problems in mind more fundamental than this or that property of solids. Work to apply the quantum theory to gases provided an important impetus toward the formulation of wave mechanics and quantum statistics. The work thus led in the end of Sommerfeld's electron theory, based on Fermi statistics, which brought a renaissance of the concept of the free electron gas in metals, as will be related in Chapter 2.

Theories of "degeneracy"—that is, the divergence of the ideal gas from classical gas laws—marked the path toward quantum mechanics ever since the first Solvay Conference in 1911, when in a sense the quantum concept was officially allowed into physics.[177] For example, Nernst's law of thermodynamics required the specific heat of an ideal gas to vanish as temperature goes to absolute zero, whereas the classical gas laws predicted a value independent of the temperature. We have seen how Debye showed in 1912 that the experimentally observed vanishing specific heat of solids could be explained with the help of the quantum theory. Could not an analogous procedure be applied to gases?

In 1913 two studies attempted to use Debye's method to explain Nernst's degenerate gas.[178] The role of lattice vibrations was transferred to the acoustic vibrations in the gas, for it was "merely the singular behavior of the velocity of sound which gives rise to the difference between gases and solids."[179] In the limiting case of temperature approaching zero, these theories succeeded in predicting a vanishing specific heat and entropy for the degenerate ideal gas. According to the formulas, the temperature threshold where degenerate behavior appears is inversely proportional to the mass of a gas particle. It followed that "in the case of electrons in metallic conductors, with their small mass, the quantum deviations investigated above should become apparent at low temperatures."[180]

Willem Hendrik Keesom varied these degenerate gas theories by introducing a "zero-point energy"; this brought the theoretically calculated deviations into good agreement with experimental values. The introduction of the zero-point energy did not interfere with the disappearance of entropy and specific heat as temperature approaches zero. Once again, the metal electrons served as a test case: Keesom argued that "those consequences which do not arise for the ideal gas until extremely low temperatures are reached, can occur for free electrons at experimentally accessible temperatures, even at fairly high ones."[181] For the case of 1.7×10^{20} electrons/cm^3, Keesom's theory established a "characteristic temperature" of 5500°C. At 0°C the electron gas would already be so degenerate that its specific heat would contribute only one-fiftieth of the classically expected value.[182]

Written 13 years before the discovery of Fermi statistics and their application to metal electrons, Keesom's statement appears to be an unparalleled prophetic anticipation of the methods that would ultimately overcome the specific heat dilemma posed by the classical electron gas theory. But Keesom's theory did not remain an isolated example. In 1924, for example, Erwin Schrödinger declared in one of his own works on the theory of gas degeneracy that "it has already often been mentioned that the application of a correct theory of gas degeneracy to free electrons might well help cure the known difficulties in the theory of metals." He conceived of the metal electrons as a two-phase system; the relative proportion of these phases would be "similar to that between a condensate, which has already lost its specific heat at low temperatures, and its vapor, which has not yet become degenerate." For electrons in the condensate, a rough calculation suggested degeneration temperatures "between 4,500 and 18,000°K, that is, the metal electrons would already be completely degenerate at average temperatures . . . and the lack of a contribution to the specific heat would be understandable."[183]

Einstein also thought of metal electrons in 1924, when he applied the new statistics developed by Satyendranath Bose to extend Planck's radiation formula to the ideal gas. In section 10 of his "Quantum Theory of the Monatomic Ideal Gas," Einstein faced the old problem "that the free electrons produce no appreciable contribution to the specific heat of metals. This difficulty disappears, however, if we base our work on the present theory of gases." However, Einstein immediately repudiated the idea:

> During consideration of this theoretical possibility one encounters the difficulty that one has to assume a very large mean free path (order of magnitude 10^{-3} cm) to explain the measured conductivity of metals for heat and electricity, due to the very minute density per unit volume of electrons which, according to our results, take part in the thermal agitation. On the basis of this theory, it also appears impossible to explain the behavior of metals exposed to infrared radiation (reflection, emission).[184]

The Bose–Einstein and Fermi–Dirac quantum statistics that would emerge in the mid-1920s were merely the end point of a long tradition of gas degeneracy theories that repeatedly called on the electron gas as an example.[185] Not all the new electron theories were by-products of attempts to apply the quantum theory to gases. Review articles of the 1920s show evidence of continuing interest in the electron theory of metals itself.[186] "The electrical conductivity of metals and problems related to it" were thought to be so important that the Fourth Solvay Congress in 1924 was organized around this theme.[187]

One can scarcely summarize the multitude of electron theories put forth between 1911 and 1926, and only with difficulty can they even be classified under a few general headings. Some workers made only negligible modifications to older, classical gas theories. Others deviated radically from the notion of an electron gas; for example, electron chain and lattice theories assumed that within the lattice of fixed atomic cores, the electrons form their own rigid chains or lattices, which are more or less movable as a whole. Other theories may be labeled "dipole theories," "conduction electrons in quantum orbits," or simply "intermediate theories."[188]

A theory of Willy Wien[189] provides an example of how the uncertain state of the theory of electrons in metals gave rise to bold speculations, in part remarkably apt and in part false. From the outset, Wien rejected the gas kinetic equation for the electron gas, for "the specific heats at higher temperatures behave as if only the molecules took part in the thermal motion. . . . Therefore u [electron velocity] must have another significance than in the previous theory; there must be a velocity which has nothing to do with the temperature, and which therefore also exists unchanged at T = 0." The only temperature-dependent quantity in Wien's theory was the mean free path L of the electrons, which he took to be a measure of the number of collisions of the electrons with the metal atoms. He calculated L from the number and amplitude of the lattice vibrations, borrowing a formula from Debye's theory of specific heats. Wien thereby found the correct temperature dependence of the electrical conductivity: σ is proportional to L, which in turn is proportional to $1/T$, in the limiting case of high temperatures (that is, $kT \gg h\nu_m$, where ν_m is the Debye cutoff frequency).[190] Yet the Wiedemann–Franz law, celebrated for more than a decade as a crowning achievement of the classical electron theory, could not be explained in this way. Wien consoled himself with the argument that "the question of thermal conduction and its connection with quantum theory does not appear to be developed to the point where a theory can be formulated."

For practical applications, the electron theory of metals did not yet play any role at all.

> To those immediately concerned, it has become increasingly apparent that Physical Metallurgy is at present in an unsatisfactory position . . . little or no attempt [has been made] to investigate the properties of metals and alloys systematically from the point of view of the Periodic Table, in order to discover the general principles involved . . . the Metallurgist has, in the past, been profoundly suspicious of the value of theoretical work.[191]

William Hume-Rothery wrote these words in the foreword of his textbook on the electrical properties of metals, which appeared in 1931 when the first quantum mechanical works on the electron theory of metals were already in existence. People who actually worked with metals still got no help from theory and had to be content with their combination of craft traditions and strictly empirical research. However, these were now being increasingly helped by important new developments in experimental technique and instrumentation.

1.4 New Experimental Techniques

At the beginning of the twentieth century, the growing understanding of the properties of solids gave rise to research methods that could be used only at great material and organizational cost. Thus arose the first precursors of big science, such as the low-temperature laboratory of Heike Kamerlingh Onnes in Leiden, and Aimé Cotton's giant electromagnet near Paris, and later Peter Kapitza's Mond Laboratory in Cambridge, which reached for the highest magnetic fields. But as important as large-scale research in solid-state physics was, most of this research drew on a small-scale approach, characterized not by large teams and heavy expenditures but by ingenious laboratory methods.

Extending the range of traditional methods of investigation was one approach that brought to light unexpected properties of solids. About 1900, Percy William Bridgman was able to increase the attainable pressure from a few thousand to 100,000 kg/cm² by using the high pressure itself rather than the conventional external screw, to tighten the packing. He thereby enlarged the range of atomic compressions and correspondingly altered the spatial ordering of atoms in solids, affecting not only mechanical but also electrical, magnetic, thermal, and optical properties.[192] Not very much later in the Leiden laboratory, helium liquefaction made extremely low temperatures accessible, which increased the body of data on the properties of solids and opened the door to unexpected qualitative insights.

Traditional experimental methods also had to be adapted to meet new demands; for example, instruments for measuring the specific heat of gases had to be changed to be of use for studying solids.[193] Familiar phenomena like the Hall effect and thermoelectric effects were pressed into service to determine magnetic fields and the density or mobility of charge carriers, or to measure temperature and voltage.[194] And new discoveries eventually led to entirely new experimental techniques. X-ray diffraction in crystals was put to use in determining structures; the photoelectric effect made it possible to determine the work function of electrons in metals; and radioactive rays could be used to produce defects in the crystal lattice.[195]

Since materials in their natural form were seldom suitable for research, the production of samples with high purity or controlled impurities became another important task for experimenters. In the 1920s, methods were developed for producing large single crystals, whose dimensions greatly exceeded those of naturally occurring crystals.[196] Single crystals would become a strict prerequisite for research into many phenomena, such as the behavior of semiconductors.

The significance of new methods in physical research was expressed in the demand for specialized scientific journals. The *Zeitschrift für Instrumentenkunde,* published by the end of the nineteenth century under the auspices of the Physikalisch-Technische Reichsanstalt, served as a model. For example, an article in *Science* in 1916 called for a journal of scientific instruments in English, and also for an encyclopedia of instruments and methods of research.[197] In meetings and physics symposia, much space was given to discussion of instrumental techniques.[198]

A description of only the most important new measurement techniques and their effects on subsequent solid-state physics would lead far beyond the boundaries of this chapter. We therefore treat in the following sections only three important examples: low-temperature research, the discovery of x-ray diffraction and its consequences for structure analysis, and the powder method for the determination of the structures of solids.

Low-Temperature Methods

The path to increasingly lower temperatures offers an early example of the transition from little to big science. In 1883, after the Frenchmen Louis Cailletet and Raoul Pictet had independently noted a fog created during the sudden expansion of compressed and cooled oxygen gas (showing that this was a way to further cooling), three laboratories (Cracow, London, and Leiden) set out to liquefy gases with this expansion method of refrigeration. In Cracow, the chemist Karol Stanislaw Olszewsky and the physicist Zygmunt Florenty von Wroblewsky succeeded in liquefying a larger amount of some cubic centimeters of oxygen. The poor financial situation at the Austrian-Galician university prevented the two researchers from maintaining their lead on the path toward absolute zero.[199] But at first, the other competing laboratories also had no success, since the expansion method failed for the liquefaction of hydrogen because of this element's low inversion point.

Compared with the Cracow laboratory, the chemical laboratory of the Royal

Heike Kamerlingh Onnes (1853–1926). (Rijksmuseum voor de Geschiedenis der Natuurwetenschappen, Leiden. Courtesy of AIP Niels Bohr Library)

Institution in London was lavishly equipped. There, James Dewar could use the majority of the budget, donated by private individuals and industrial firms, for his low-temperature experiments. In addition, much of his measurement apparatus was made available by loan from industry.[200] Several collaborators, among them two permanently employed engineers, assisted him in his work. The compressors, pumps, and liquefaction equipment gave his workplace the appearance of a machine shop rather than a chemical laboratory.[201]

In this laboratory, Dewar took up the Joule–Thomson effect and the counter-current method that Carl von Linde had used in Germany to liquefy air; in 1898 Dewar managed to collect hydrogen in liquid form at −252°C in a glass container. He immediately informed his competitors of this latest success. Yet a telegram of 18 May 1898 to Heike Kamerlingh Onnes in Leiden proves that Dewar misinterpreted his results. "Liquefied Hydrogen and Helium," he wrote, 10 years before helium was actually liquefied.[202]

In Leiden, work on the liquefaction of gases was viewed as experimental fulfillment of the theoretical work on gases by the Dutch physicist Johannes Diderik van der Waals. Van der Waals's theory would allow one to predict, from measurements of isotherms, the critical temperature and boiling point of the as-yet unliquefied gases. In 1882, in his inaugural address as occupant of the first Dutch chair for experimental physics, Kamerlingh Onnes had already set himself the task of studying deviations from van der Waals's theory.

Kamerlingh Onnes was not concerned with producing spectacular effects; the motto he wanted to inscribe above each entrance to a physics laboratory was "Door meten tot weten" (Knowledge through measurement). In his inaugural address he declared, "It is measurement finally, that in the deviation [from theory] provides the natural material for new hypotheses regarding the properties of molecules. . . .

While constantly referring the deviations to ever new approximate laws, the extension of the limits of research and the increase of accuracy must keep pace with one another."[203]

As the son of a factory owner, Kamerlingh Onnes knew the importance of detailed organization for a large enterprise. In his physics laboratory, he accordingly applied himself to planning and organization to a degree far beyond anyone before him. Recognizing that his laboratory, with its many modern pieces of equipment, would require a staff of skilled and specially trained assistants, he founded a school for instrument makers and glass blowers, financed by the Dutch government.[204] Later the students of this school would flock to laboratories all over the world, with the Dutch electrical industry particularly profiting from their education. Another organizational step was the founding in 1885 of the laboratory's own journal, *Communications from the Physical Laboratory of the University of Leiden.* In addition to publishing the laboratory's scientific investigations, the journal contained exact descriptions of the apparatus used, and thus became the bible for low-temperature physicists for several decades.[205]

The most telling sign of the arrival of large-scale organization in physics was specialization in the Leiden cryogenic laboratory. For example, three chemists were occupied mainly with extracting purified helium gas; after three years, they had a supply of 360 liters at their disposal. The division of labor even extended to the series of physicists who assisted Kamerlingh Onnes. Gilles Holst was responsible for the electrical parts of the experiments and pursued measurement of the smallest resistances, and Dorsman took care of the cryogenic equipment itself. The first great result of all this effort was the liquefaction of helium in 1908, which won Kamerlingh Onnes the 1913 Nobel Prize. For such a liquefaction, the gases had to be compressed to about 100 atmospheres, which at that time was a dangerous procedure: two of Dewar's assistants each lost an eye and von Wroblewsky was killed during explosions of low-temperature apparatus. Karol S. Olzewski was hit by parts of an exploding apparatus, and only the protection provided by his thick coat protected him from serious injury.[206]

After several structural extensions, the Leiden low-temperature institute even looked like a factory; Paul Ehrenfest jokingly called it a brewery because of its copper kettles and tubes.[207] Reaching temperatures down to 1.2 K, the Leiden cryostat remained the coldest point on earth for 12 years. The laboratory became the center for the investigation of properties at temperatures that in some cases were effectively equivalent to absolute zero. Kamerlingh Onnes enjoyed the prestige of his institute in the scientific world, as he elaborated in his Nobel Prize acceptance speech in 1913:

> Thanks to specialization, I was able to become the collaborator of various men of learning, who accepted the hospitality of the laboratory. . . . J. Becquerel came to investigate the absorption spectra of rare earths with and without a magnetic field (Zeeman phenomena); the Becquerels, father and son, the phosphorescence of uranium compounds; Lenard and Pauli, the long-lasting phosphorescence of earth alkali sulfides; Madame Curie, the penetrating rays of radium . . . ; Bengt Beckmann and his wife came to investigate the Hall effect of alloys and the increase in resistance of lead due to pressure . . . ; Weiss came to determine the magnetic moment of the elementary magnets of ferromagnetic materials.[208]

The laboratory for low-temperature research of Heike Kamerlingh Onnes in Leiden. Kamerlingh Onnes is sitting in the middle, and W. H. Keeson is sitting on the right. In this laboratory, helium was liquefied in 1908, and superconductivity was discovered in 1911. (Kamerlingh Onnes Laboratorium, Leiden)

This was characteristic of physics at the time—an interest in miscellaneous solid-state properties mixed together almost at random with atomic physics problems.

The first and simplest experiments in all low-temperature laboratories were measurements of the electrical resistance of metals. The curve, linear at higher temperatures, promised to be useful as a simple method for measuring low temperatures. In 1885 Wroblewsky concluded, on the basis of measurements of copper at −200°C (the temperature of liquid air), that the resistance would decrease faster than the absolute temperature, and upon further cooling would completely disappear even before absolute zero was reached:

> Such a conductor would have quite curious properties. The electricity would continue moving in it without the generation of heat, and in such a conductor the efficiency for the transport of electrical power would approach unity. Although at this time this could lead to no practical results, it is important to state that the above suggestions do not belong to the realm of fantasy and are realizable within the means which the liquefaction of gas has made available to us.[209]

This was a plain challenge to extend measurements to still lower temperatures. Dewar and J. Andrew Fleming, who from 1892 had measured the resistance of metals in liquid oxygen[210] and from 1903 in liquid hydrogen, showed that Wroblewsky's conclusions were false: the resistance appeared to take on a constant value at sufficiently low temperature. But for certain metals, this result would have to be corrected.

In 1911 Kamerlingh Onnes and his collaborators, particularly Gilles Holst, investigated mercury in liquid helium and found the surprising result that the resistance below 4.15 K suddenly fell several orders of magnitude. It became too small to measure with the means available: mercury became superconducting. A current once set in motion could circulate indefinitely in a loop of this metal. Other metallic elements, such as tin and lead, and many alloys also showed the same astonishing behavior. Almost all eminent physicists, beginning with Kamerlingh Onnes himself, would occupy themselves at one time or another with attempts to explain this unforeseen property in terms of the quantum, yet almost half a century would pass before a useful theory would be put forth (see Section 2.5 and Chapter 8).[211]

It was different for specific heat. This subject became popular because experimental results deviated from the classical theory at low temperatures, leading physicists to extended series of measurements. But as we have seen, here they had more success in establishing theories that made use of the quantum concept. By 1911, measurements of specific heat had led to the conclusion that in a solid body in the neighborhood of absolute zero the great majority of the atoms stop their thermal motion. This suggested that transfer of heat would be barred.[212] However, measurements that year showed that the thermal conductivity, which approaches a constant at room temperature, does not fall but rises with decreasing temperature in almost all crystals.[213] Many attempts to explain this behavior were attempted, but success was not achieved until Rudolf Peierls's work at the end of the 1920s.[214]

The investigation of properties of solids at low temperature was gaining practical significance as industry used low-temperature methods more and more often; the number of cryogenic laboratories at universities also grew. Since materials could behave very differently at low temperatures than at room temperature, it was first necessary to find materials with the most advantageous properties for low-temperature technology. Extensive measurements of these properties were carried out in many locations, but particularly at the Physikalisch-Technische Reichsanstalt, whose leadership was quick to recognize the significance of low-temperature physics. In 1914 they commissioned Walther Meissner to establish the first helium liquefaction plant in Germany (after Toronto, the third in the world).[215] Thus low-temperature apparatus moved from its university origins to a role straddling pure science and practice, producing some results of interest to quantum theorists and others of interest to industrialists.

Our next example of apparatus was a matter of delicate vacuum tubes and tiny samples suspended on slender rods, in physical appearance at the opposite extreme from the Leiden "brewery." Yet it would play much the same sort of role as low-temperature apparatus did in intellectual and practical life.

The Laue Experiment and Its Interpretation

In 1910, Sommerfeld offered his student Peter Paul Ewald an unusually stimulating dissertation topic.[216] Ewald was to show that optical birefringence could be calculated by assuming an orthorhombic lattice (i.e., an anisotropic arrangement) of iso-

Peter Paul Ewald (1888–1985) and Lise Meitner (1878–1968) in Tübingen in front of the Institute of Physics, around 1925. (AIP Niels Bohr Library)

tropic resonators. Ewald got positive results, but as he was writing them up he realized that he had not seriously followed through a problematic point.[217] His calculations had shown that the incident waves can be neglected in determining the refractive index and the characteristic frequency of the mechanical system. For these latter depend on only the free vibrations of the entire arrangement: a dipole is excited by the field of the neighboring dipoles, and that lattice oscillation is selected that can exist without external excitation. The incident wave itself no longer exists in the interior of the crystal.

This "rather radical departure from the traditional theory"—the dispersion theory of Lorentz and Planck—brought Ewald to seek advice from an optics expert.[218] He consulted a former student of Planck, Max von Laue, who had been working for several years in Sommerfeld's group and had just composed, at Sommerfeld's request, a contribution on wave optics for the *Encyklopädie der mathematischen Wissenschaften*.[219] Ewald did not find a solution to his problem. But it appears that the three-dimensional space lattice structure of crystals mentioned in this discussion was a point of departure for an idea of Laue. This space lattice structure seemed new to Laue.[220] His new idea was that x rays could be diffracted by a three-dimensional crystal lattice, just as light waves are diffracted by an optical grating.[221]

Laue's idea was discussed in the Café Lutz, one of the regular meeting places of the students of Sommerfeld and W. C. Röntgen. At this time, the pursuit of the puzzling x rays was a core problem for scientists in Munich. Röntgen, occupant of the chair for experimental physics there since 1900, had personally secured Sommerfeld's appointment to the theoretical physics position, which had been unoccupied for years after Boltzmann's death. Shortly before that, Sommerfeld had made a new estimate of the wavelengths of x rays from diffraction experiments on a slit, "the most reliable determination at the time of the wavelength of x rays," as Ewald later noted. Like many of his contemporaries, Sommerfeld was convinced

Arnold Sommerfeld (1868–1951) lecturing on x-ray diffraction. (DM. Courtesy of AIP Niels Bohr Library)

that x rays were electromagnetic in nature, ever since Charles G. Barkla had proved that they could be polarized. A close connection existed between the students of Röntgen and of Sommerfeld, and also with the crystallographers, whose chair was held by "the world's most famous authority on crystallography," Paul von Groth. As a colleague of Leonhard Sohncke (who had extended the theory of Bravais lattice symmetries), Groth held the "rock-solid" conviction "that the crystal's space lattice structure possessed definite dimensions." In sum, the intellectual atmosphere of Munich was uniquely suited for the birth and growth of Laue's idea.[222]

A former student of Röntgen, Walther Friedrich, offered to carry out the experiment. As Sommerfeld's assistant, Friedrich was just then busy with experiments on the distribution of x rays. When Friedrich hesitated, Laue also engaged the help of Paul Knipping, one of Röntgen's doctoral candidates.

The work began on 21 April 1912. After an initial failure, the photographic plate was placed behind the crystal. The experiment, carried out on a copper sulfate crystal in a jury-rigged apparatus, was successful. Friedrich reported, "It was an unforgettable experience for me, as late in the evening I stood all alone at the developing tray in my workroom in the institute, and saw the traces of the deflected rays emerge on the plate."[223] The photographs of substances investigated in the first weeks (zinc blende, copper, galena, rock salt, diamond), with their regular arrangements of dots, drove home the analogy with the suspected spatial lattice. "This fact, that a complete fourfold [symmetry] is present on the plate, is probably one of the most beautiful proofs for the space lattice of the crystal."[224]

The significance of the photographs was immediately clear to those directly participating in the experiment (Laue, Sommerfeld, Friedrich, and Knipping). Yet

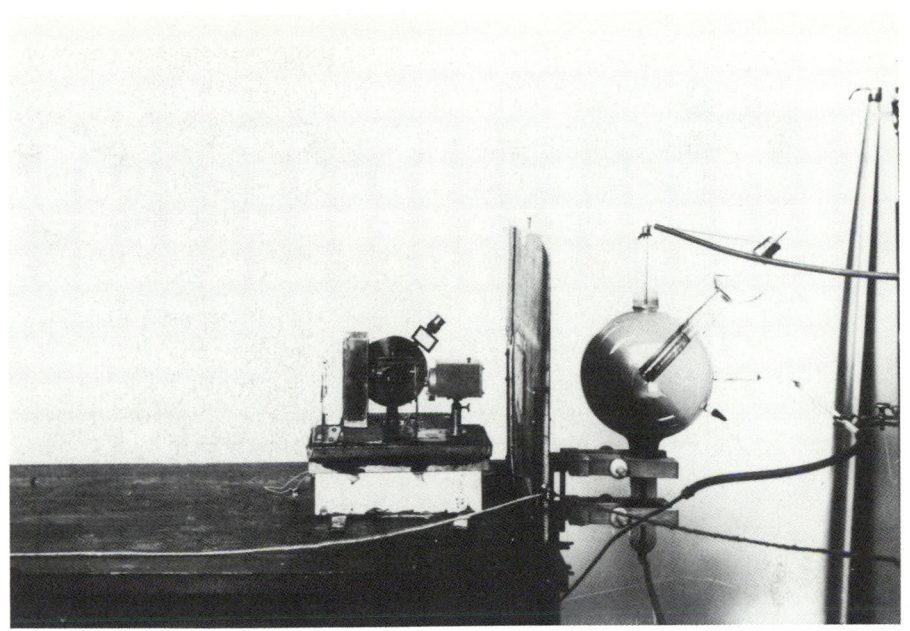

(above) Original device used for the discovery of x-ray diffraction in Sommerfeld's institute in 1912. (DM. Courtesy of AIP Niels Bohr Library). *(below)* Apparatus with which W. Friedrich, P. Knipping, and Max von Laue discovered the diffraction of x rays by crystals in 1912, exhibited at the Deutsches Museum. *(left to right)* x-ray tube, goniometer with crystal, and photo plate. (Courtesy of DM)

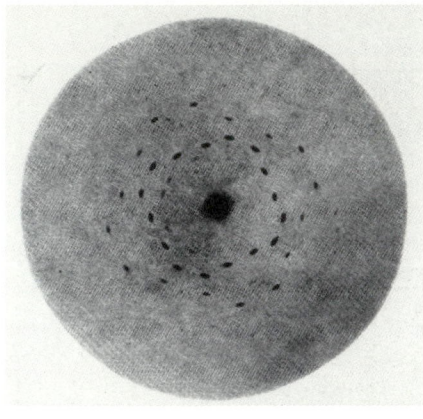

Fig. 1.2   The famous "Fig. 5" from 1912, where for the first time perfect symmetry in a crystal (ZnS) was proved by x-ray diffraction. (From W. Friedrich, P. Knipping, and M. von Laue, "Interferenzerscheinungen bei Röntgenstrahlen," in M. von Laue, *Gesammelte Schriften und Vorträge* [Braunschweig: Vieweg, 1961], 204)

there were skeptics, even in the same camp, including Röntgen himself.[225] Among Laue's foreign friends, to whom he sent photographs, uncertainty alternated with enthusiasm.[226] To secure priority, on 4 May 1912 Sommerfeld deposited a sealed envelope with the Bavarian Academy of Sciences containing a sketch of the experimental setup, a brief description of the experiment, and some photographs. On 8 June 1912, he submitted the results of the experiment along with a theory of diffraction that Laue had composed in the meantime. Laue reported on the experiment at his former workplace while on a visit to Berlin. Planck reflected on the event 25 years later: "When the first typical Laue diagram [Fig. 1.2] became visible . . . a general, slightly restrained 'ah' propagated through the gathering. Each of us felt that a great achievement had been made."[227]

The foregoing tells how the participants in that discovery represented the course of events in countless reflections on this special period of 1912, and how it was commonly accepted by physicists and historians and printed in history of physics and physics textbooks. Only the historian Paul Forman examined the sequence of events more critically and discovered some points that call this smooth sequence into question.[228] Forman's work was motivated by Ewald's contribution to the fiftieth-anniversary celebration of x-ray diffraction. Ewald, finding his recollections criticized, strongly defended himself in a countering article.[229] But this is not merely a question of who was right or wrong: it is a question of understanding the layers of complexity in any historical event.

There is no doubt that from the beginning Laue had in mind the diffraction of x rays on a three-dimensional crystal lattice. Yet how was the incident x-ray spectrum to be provided? Why did he speak of fluorescent radiation (with reference to some experiments by Charles Barkla) and then allow copper sulfate to be investigated first—a crystal that "contained metal of considerable atomic weight as a component, possibly to obtain intense and also homogeneous secondary radiation"?[230] This problem and others noted by Forman suggest that at the time and for months following there was no clear view about the origin of the rays that created the interference pattern. Moreover, Sommerfeld was opposed to letting the experiment, as conceived by Laue, be carried out. Did he consider the thermal motion of the lattice,[231] as Ewald later objected, to be so considerable that it would have washed out

Sir William Lawrence Bragg (1890–1971), Max von Laue (1879–1960), and Isidor Fankuchen (1904–1964). (AIP Niels Bohr Library, Fankuchen Collection)

the interference patterns? Or were Sommerfeld's own experiments so important to him that he did not recognize Laue's ingenious idea? Or was Laue's idea of interference patterns not at all as clear as it appears to us today, so perhaps for this reason Sommerfeld withheld his support? Yet whatever the discovery scenario may have been, it was the results that affected further understanding of the crystal lattice.

The documents show that despite the geometric theory advanced by Laue, and the good agreement between the measured photograph of zinc sulfide (Fig. 1.2) and an interference pattern calculated with Laue's theory, the geometry and specific mechanisms of interference remained unclear.[232] Not until 1913, when the first measurements and theories of William Henry Bragg and William Lawrence Bragg were put forth and physicists and crystallographers in many other places were confronted with the Laue experiment, did the situation clear up.

In the fall of 1913, the interaction of x rays and crystals was discussed at three conferences, held shortly one after another. The first was a meeting of the British Association for the Advancement of Science at Birmingham, where Ewald went on his honeymoon to hear a talk by W. H. Bragg on crystals and x rays and to report on it in the Munich colloquium. The second conference was in Vienna, where students of Sommerfeld and Röntgen reported on new experiments proving that x rays that produce one interference spot, regardless of the continuous incident spectrum, contain only one wavelength. Not until this confirmation were the last doubts laid aside: Laue had been right to interpret this as an interference phenomenon. The third meeting, the most important for international dissemination, took place a month later in Brussels—the Second Solvay Conference, whose theme, "The Structure of Matter," was especially up to date. Along with researchers from England, Laue spoke and reported not only on his own work but also on results that Ewald had published some months before.[233]

The First World War broke out the following year. This led to tensions and mutual condemnation even between physicists who had been friends but were of opposing nationalities. The confident community life came to a halt.[234] In 1915, Ewald was called for military duty and escorted an x-ray device, in the service of medicine, to the eastern front. At that time, he began to produce a general microscopic crystal optics based on his dissertation, which became topical due to x-ray diffraction. He sent the manuscript to Sommerfeld for his judgment. Sommerfeld then wrote to Wien, to whom he had transmitted the thesis for publication, that "it is a somewhat painstaking work. Whether much new will come of it for experiment remains to be seen."[235] His doubts did not at first appear unjustified. The experimental proof of the correctness and necessity of Ewald's "dynamic theory" would be years in coming.[236]

The Beginnings of X-ray Structure Analysis of Single Crystals

It had apparently been Voigt who brought the news of Laue's sensational experiment to England, in July 1912 when he took part in a Royal Society anniversary celebration.[237] W. H. Bragg, a member of the Royal Society and Cavendish Professor at the University of Leeds, had discussed the results with his son William Lawrence by the time of their summer vacation in Yorkshire.[238] At first, the elder Bragg strove to interpret the results by supposing x rays to be streams of corpuscles—a hypothesis of which he was the chief protagonist—using notions of channels in the crystal.[239] Yet he soon came to agree with those who had championed the wave nature of x rays; Laue's photographs and their explanation as an interference phenomenon were too convincing. His son, at that time a research student, also kept after the problem: "When I came back to Cambridge and thought about Laue's papers, though I was convinced of the validity of his conclusion that it was a matter of a diffraction phenomenon of waves, I was just as solidly convinced that his explanation of the process was not right."[240]

Lawrence Bragg's investigations, together with study of the elliptical form of the interference spots, brought him to an entirely new idea of the geometry of the diffraction pattern. In analogy with optical concepts, there was "a much simpler interpretation of the phenomena, by considering the reflection of waves from parallel layers of atoms or diffraction points, each typical set of parallel crystal planes acting as a reflecting surface of radiation." He derived a simple law, the Bragg law, for the relation among the wavelength of the x ray, the distance between the crystal planes, and the incident angle of the radiation.[241] Thus at one stroke he found a precise way to measure two quantities, x-ray wavelengths and the spacing between layers of atoms in crystals, whose magnitudes and whose very existence had been debatable.

While Laue still spoke of an "unexplained question,"[242] Bragg recognized that from the incoming beam containing a continuous spectrum, the reflecting planes select certain wavelengths. Lawrence had known of the lattice planes because his chemistry professor William Pope and the mineralogist William Barlow had advanced a theory of crystal binding, on which Lawrence had made a seminar report.[243] He could explain the lack of some spots that Laue expected on the zinc blende photographs by assuming that the lattice was not only cubic but face-centered cubic (i.e., the unit cell whose repetitions made up the lattice was a cube with atoms at each corner and in the center of each face). This was an important insight for the determination of structure. During the next month, he investigated crystals with simple structures, such as rock salt and sylvite (KCl). He found that the space lattice of alkali halides is composed, in effect, of two ion lattices interwoven with each other, corresponding to a model that Barlow and the Göttingen physicist Erwin Madelung had proposed for rock salt. Now at last one could state what kinds of atoms a salt was made of, and how those atoms were arranged within the crystal, which is information of the most fundamental importance for any investigation.

Lawrence Bragg had also tried to verify experimentally his basic picture of x rays from a monochromatic incident ray reflected by the lattice planes. But it was his father who made the breakthrough after constructing a new x-ray spectrometer with which he could vary the incident angle, an arrangement that proved to be superior to Laue's.[244]

In April 1913 father and son published their first joint paper, which has been called the foundation of the science of x-ray crystal analysis.[245] Their joint experimental work did not begin until the summer vacation of 1913, which Lawrence spent in Leeds. "The capital" that the son brought into the "family firm" was his knowledge of optical diffraction and crystallography as well as his success in analyzing the first crystal structures.[246] His father's contribution was an extensive experience with experimental methods, including a thorough knowledge of x rays. Having determined that the radiation emitted in the x-ray tube contained not only a continuous spectrum but also a characteristic line spectrum dependent on the target metal, the elder Bragg was soon able to make important predictions about the spectroscopy of x rays and the relation of the lines to atomic weight.[247] Next, while testing his spectrometer, he analyzed the structure of diamond. In the following years, both Braggs determined the structure of many minerals: fluorspar, potassium iodide, potassium bromide, and the more complicated pyrite.[248] Only a qualitative analysis was possible at the beginning, but a complete structural determination of NaCl was undertaken in June 1913, yielding the first absolute data on lattice constants.[249]

When the First World War broke out in 1914, the Braggs were drawn into different war projects and their joint research came to an end. In the middle of these projects, the news came that they were awarded the 1915 Nobel Prize in Physics for their structural determination of crystals by x rays. A year earlier, Laue had won a Nobel Prize for his work. This showed that the physics community had immediately recognized the central importance of x-ray crystal diffraction. Yet until the end of the war, only a few researchers were concerned with the subject: 10 in Germany, 10 in England, 5 in France, 5 in the United States, 4 in Japan, 2 in Russia, and 5 in the Netherlands and Scandinavia combined.[250]

The most fertile soil for the Laue experiment was in England. The establishment of the new discipline required combining two different branches of science, for experience in the theory and experimental practice of x rays was just as important as crystallographic and theoretical knowledge of the structure of solids. Except in Munich, both these prerequisites were immediately fulfilled only in England. The lead that the two Braggs gained in their first year of structural analysis was so large that others could not catch up. Although there were collaborative groups in other countries—for example, in the Netherlands, where the x-ray expert Hermann Haga joined with the chemistry and symmetry expert Frans Maurits Jaeger—few noteworthy successes were achieved outside England during the next 10 years.

In Germany the development of structural analysis went astray as the Munich group of x-ray experts fell apart. Laue was not much interested in experiments or in the application of his method to particular cases, and he accepted an appointment in Zurich in the fall of 1912. Friedrich went to Freiburg in the winter of 1912 and from then on devoted himself to medical use of x rays. Knipping took a position with the Siemens company after completing his dissertation, and at the beginning of the war transferred to the Kaiser Wilhelm Institute under Haber. During the war, Röntgen's assistants, as well as Ewald, were sent with x-ray equipment to the eastern and western fronts. Aside from a small group in Leipzig and Würzburg, only the Dutchman Peter Debye and the Swiss Paul Scherrer, in Göttingen, kept up relevant research, as described in the next section.

Elsewhere, the movement toward structural analysis with x rays was entirely halted by the First World War. Even in countries that did not participate directly in the war, such as the Netherlands and Sweden, university activities were substantially disrupted. The international cooperation as well as the necessary exchange of knowledge among groups had been broken.

After the end of the war, scientific research slowly returned to normal. Both Braggs, now highly respected for their Nobel Prize, moved into new positions. Lawrence obtained Rutherford's old chair in Manchester. His father, who in 1920 became Sir Henry (as much for his war work on acoustic methods for finding submarines as for his scientific eminence), took up a position granted him in 1915 as Quain Professor for Physics at University College in London. The chairs of the two Braggs remained for many years the leading centers for the specialty of x-ray analysis. Both places were of critical importance for the education of those who would study the structure of materials and for the exchange of knowledge with researchers from other countries. The many researchers who were trained by, or otherwise were in contact with, the Braggs formed one of the first of the various schools of research that would eventually become the solid-state physics community.

Researchers in other countries followed this healthy start to some extent. The x-ray experts Maurice de Broglie in France and Manne Siegbahn in Sweden had each mastered and improved on the Laue experiment, and in these countries collaboration with structural researchers soon developed. Another center was established in Japan, where Torahiko Tereda had independently derived the Bragg equation shortly after Laue's experiment,[251] and Shoji Nishikawa had even applied group-theoretical considerations. Additional groups set to work in the United States, where researchers showed less interest in abstract knowledge of the solid state than in specific applications.

Structural research received a further impulse through its practical applications, and the investigations shifted toward nonacademic institutions such as the Kaiser

Wilhelm Institutes in Germany, the Metallographic Institute in Stockholm, and industrial laboratories. The number of centers and researchers, the number of structures investigated, and the effect on physics in general increased rapidly. The most important facts and episodes of this development fill a thick volume written for the fiftieth anniversary of Laue's discovery.[252]

The Powder Method

The methods used by the Braggs and others to determine the first crystal structures required crystals of adequate size. This prevented scientists from determining the structure of materials that were difficult to obtain in the form of single crystals, which meant most materials of practical interest. The problem was overcome during the First World War by Debye and Scherrer in 1916 at Göttingen and, independently, by Albert W. Hull in 1917 at the General Electric Research Laboratory in the United States.[253]

Debye, a theorist, had found opportunity to experiment when he took over Voigt's teaching post. He began to build an apparatus for x-ray diffraction to analyze fundamental phenomena such as the distribution of electrons in the atom.[254] Scherrer, who had been carrying out measurements of the Zeeman effect for the now retired Voigt, took great pleasure in Debye's investigations and changed fields.[255]

In their attempt to determine the distribution of electrons on their atomic orbits, Debye and Scherrer projected monochromatic x rays onto powdered substances such as carbon and boron, both amorphous and crystalline.[256] Only the crystalline powders showed clear interference patterns—a set of lines. This phenomenon was quite unexpected, since in his article announcing the discovery of x-ray diffraction by crystals Laue had argued that a powdered crystal would not produce interference phenomena. However, Debye and Scherrer now recognized that the effect could easily be explained within the context of the Braggs' theory, and they demonstrated how the lines in the powder diagram could be used to calculate the spacing between crystal planes. In particular, they showed that the powder diagram of lithium fluoride could be explained by a cubic structure and that silicon has the same structure as diamond.

Any communication between Hull at General Electric and the Göttingen pair would have been greatly hindered during the First World War, and in any case Hull had been working with Irving Langmuir on different matters. However, stimulated by the Braggs' failure to determine the structure of iron and hoping to cast light on the magnetic properties of iron, Hull tackled this new problem. Faced with the impossible task of obtaining large single crystals of iron, Hull used iron filings and showed that the position of the lines obtained in the diffraction diagram confirmed the hypothesis of a body-centered cubic lattice. He and Wheeler P. Davey drew charts enabling them to find the lattice structure of a particular sample from the powder diagrams. In the case of mixtures, the relative intensity of the diagrams of each constituent told them the composition.

The powder method opened the way for the determination of a vast number of atomic structures for materials not available as single crystals, most notably metals. In addition, the diagram of a crystalline powder proved to be characteristic of the substance, providing an easy and sure method of identification. However, that required an atlas of standard diagrams for comparison with the experimental results. Such atlases were built up gradually in industrial laboratories toward the end of the 1930s.[257] Most crystallography laboratories equipped themselves with Debye–Scherrer chambers for powders, used from 1919 on alongside the Bragg rotating crystal chambers for single crystals. With these tools, Hull and then Davey, during the 1920s identified the structure of most of the common metals (the transition metals, Cd, Zn, Pt, Cr, Mb, etc.) and showed that they could be divided into three structures: body-centered cubic, face-centered cubic, and close-packed hexagonal.[258] The method was always the same: a comparison with the standard diagram corresponding to a given structure.

Since the end of the nineteenth century, it has been known that some metals are allotropic (i.e., they have more than one crystal form), with the "phase" depending in particular on the temperature. From the 1920s on, some of the metallurgists' macroscopic phase changes were linked with specific microscopic structural changes. For instance, although Hull studied ordinary (alpha) iron and found the body-centered cubic type, the structures of the other allotropic varieties of iron still had to be determined: beta-iron (in the range 670–910°C) and gamma-iron (910–1390°C). Arne Westgren, a metallographer in the Swedish ball-bearing company S.K.F., demonstrated that the magnetic phase change from alpha-iron to beta-iron did not correspond to any change in the crystal structure. Westgren had studied physics at Göttingen, learned to use x-ray equipment in Siegbahn's laboratory, and went on to study the structures of iron and steel at high temperatures. Using the first Debye–Scherrer high-temperature chamber, he demonstrated that gamma-iron is face-centered cubic, like austenite. In 1921, when the Institute for Metal Research of Stockholm was founded, Westgren was appointed there and continued his research with Gösta Phragmen. Using vacuum chambers, Phragmen resolved the technical problems occurring at high temperatures. Together Westgren and Phragmen determined the structure of delta-iron and extended the high-temperature study of steel and alloys such as cementite and chrome steels.[259]

Using equipment developed by Phragmen, Albert J. Bradley determined the structure of additional metals.[260] A student of Lawrence Bragg in Manchester, Bradley was at the time spending a study year in Stockholm; this was then the largest center for determining the structures of metals and alloys, although Manchester was a close second. Bradley went on to construct many diagrams for gamma-brass, and after returning to Manchester he determined its structure. In 1926 he unraveled the structure of alpha-manganese. An element with an extraordinarily complex structure, whose unit cell consists of no fewer than 58 atoms, alpha-manganese was the last word for metallography at that time.

The practical importance of metals such as copper, aluminum, and, particularly, iron and its alloys for industries ranging from electrical transmission to aircraft made them the first choice for the application of x-ray diffraction. The metallurgists' craft had consisted above all in mixing one type of atom with another; the prime example is carbon mixed with iron to produce steel. After crystals, the simplest case is that of solid solutions: the atoms in the solute either substitute for certain atoms in the solvent's lattice (substitutional solutions) or position themselves at interstices within the lattice (interstitial solutions). The distinction between these two types of solid solution was made in 1923 by Edgar C. Bain in the United States and by Edwin A. Owen and George D. Preston in Great Britain, using an accurate measurement of the lattice parameters together with the density of the alloy.[261] The lattice of a solid solution is slightly deformed in comparison with the original solvent lattice, and this warp shows up in the diffraction pattern; thus accurate measurements of the parameters could open the door to the study of solid solutions. The development of new forms of x-ray chambers made such measurements possible.

The study of steels and iron carbides was a specialty of Swedish physicists. For example, Westgren and Phragmen, followed by Gunnar Hägg, demonstrated that carbon forms a solid interstitial solution with gamma-iron at high temperatures, that slow cooling brings about the formation of cementite (Fe_3C) mixed with alpha-iron, and that a simple operation will produce an interstitial solution of carbon in the alpha-iron lattice.[262] In such a fashion, between the two world wars x-ray crystallography built up a body of detailed knowledge about alloys. But the connection with industrial practice remained tenuous, and the new knowledge was not fully adopted by metallurgists until after the Second World War. In any case, the knowledge itself was far from complete; for example, the highly important structure of cementite was not completely determined until 1940.

The metallurgists' mixtures can become still more varied when different metals are combined. It was well known at the beginning of the century that the solidification of a mixture of two or more metals leads to numerous possible solids, stable or metastable, homogeneous or heterogeneous. Many phase diagrams were constructed to show the phases as a function of the temperature and composition of the alloy. A distinction was then made between solid solutions of continuously variable composition and compounds with a definite ratio of metal atoms—for example, Mg_2Sn. Westgren and Phragmen were the first to distinguish clearly, in 1925, between the two extreme cases: "In an ideal compound, structurally equivalent atoms are chemically identical. In an ideal solid solution all atoms are structurally equivalent."[263] Little by little, they incorporated all cases into the phase concept. Thus they observed that in various zinc alloys ($CuZn$, $AgZn$, $AuZn$), phases having the same structure corresponded to similar atomic concentrations of zinc. For copper alloys they also observed that the same phase (body-centered cubic, or beta) appeared for different atomic concentrations, corresponding to simple formulas ($CuZn$, Cu_3Al, Cu_5Si, etc.).

Only a month before Westgren and Phragmen published these results, the Oxford chemist and metallurgist William Hume-Rothery, using a microscope, had announced similarities between the phases of various alloys. He had also found a criterion common to all the phases that exhibited these similarities: the ratio of the number of valence electrons to the number of atoms in the lattice was the same. For instance, for the body-centered cubic beta phases it was in the neighborhood of 3/2. Hume-Rothery believed that the existence of integral ratios such as 3/2 and 21/13 was a consequence of a model devised by F. A. Lindemann, in which the electrons in metals form a lattice in the same way that atoms do. This model was roundly criticized by most physicists, but Hume-Rothery, as well as Lindemann, remained attached to it until around 1930.[264]

Westgren and Bradley took up Hume-Rothery's hypothesis on the influence of the electron concentration and demonstrated that in metal alloys, phases having the same structure do indeed appear for clearly defined electron concentrations (e.g., 3/2 for the beta phase). In 1936, Hume-Rothery set down two empirical conditions for the appearance of solid solutions: a size condition, that the two kinds of atoms must not differ in diameter by more than 15%; and a valence condition, that the atoms must not have two different electrochemical characteristics, since otherwise they would form a chemical compound.

In that same year, two fundamental works appeared covering the theoretical and empirical progress in the field of metals and alloys during the preceding 15 years: *The Theory of the Properties of Metals and Alloys* by Nevill Mott and Harry Jones and *The Structure of Metals and Alloys* by Hume-Rothery. Jones went on to connect the influence of electron concentration and the formation of solid solutions with Brillouin's theory of bands. The concrete results of metallurgical and of x-ray analysis, which had gone their own way for decades, were beginning to link back up with fundamental physics and the quantum. However, there was still a long way to go before physicists would arrive at a theory that could explain any important property of an alloy in terms of its composition.

1.5   Solid-State Research Outside the Academic Establishments

We now turn to industrial research and the more applications-oriented institutions of the early decades of the twentieth century, institutions that sustained much of the work described here and in following chapters. We will say no more about the academic institutions, which by this time were standing firmly on the foundations laid in the nineteenth century and playing roles that will be familiar to our readers. We must, however, say something about the industrial and government laboratories, which turned out to be scarcely less important for solid-state physics. These institutions had entered the era of their most explosive growth and change, taking up roles that are still not often clearly seen from the viewpoints traditional in the history of physics.

Despite the increasing connection between science and technology, the number of physicists employed in industry by 1900 remained small. To satisfy new demands for the application of science, industrial concerns began to establish their own research laboratories. Within a few decades, these facilities had so increased in number and size that they employed a majority of all physicists, including people with no more than an undergraduate education in the subject. Much of the work in these laboratories was to develop materials with properties planned in advance and to use these materials in new products, subject to increasingly precise methods of materials testing.

The Emergence of Industrial Research

The first significant industrial research establishments arose not long before 1900 out of small chemical or physical-chemical laboratories, in which nature's laws were applied mainly by trial and error—a far cry from systematic investigation of the properties of matter. In 1876, Thomas Alva Edison had established the largest of these laboratories in Menlo Park, New Jersey. Two years later, some 20 employees were working there on the development of the light bulb; only a few had scientific training.[265] The origin of the research laboratory of the Bell Telephone Company, which grew into the world's largest industrial laboratory and a leading center for solid-state research, lay in "many problems daily arising in the broad subject of telephony which require solution but are not studied as they will not lead to any direct advantage to ourselves," as the head of the technical staff, Hammond V. Hayes, wrote to his superior in 1887. Hayes continued: "All these questions should be answered and I write to ask you to allow me to broaden the field of our work to embrace such problems."[266] His request was met, and in the following years the company recruited several physicists with Ph.D. degrees. A separate research department was organized in 1911. At General Electric, too, several scientists were occupied with the solution of technical problems before a research laboratory devoted to basic research was established in 1900. The development and influence of these two laboratories has been described in detail elsewhere.[267]

In Europe, the research laboratory of the Siemens firm played a leading role. Its development goes back to a physical-chemical laboratory run separately from the industrial plant and, since 1890, housed in a makeshift building ("doctor's kennel" in the vernacular) in Siemens's city of Berlin. From 1896, Werner von Bolton, a chemist who had graduated from Leipzig in 1895 and had been hired by Siemens in 1891 as a laboratory assistant, was in charge of the firm's first studies of the properties of solids. His task was to find new materials for incandescent filaments.[268] The extension of this laboratory and other small testing stations into a research center was first called for in a memorandum written in 1903 by Ludwig Fischer, the leader of the patent department. At first, the Siemens leadership gave little support to the proposal, but in 1905 the physical-chemical laboratory was resettled in a new building, and by 1913 the laboratory had grown so large that the building had to be expanded. Fischer repeated his demand for a central research department so that the entire Siemens–Halske group could avoid duplication of effort, pursue more basic research, and "attract suitable additional assistants" more easily. Furthermore, he insisted on regular conferences of all members of the laboratory to discuss any problems that arose.[269]

Hans Gerdien, a student of Voigt in Göttingen, entered the physical-chemical laboratory in 1908 and assumed its leadership in 1912. When he supported Fischer's suggestion, he was commissioned to work out plans for a new research laboratory. In the spring of 1914, he took a tour of physics institutes in the Netherlands and Germany to learn from their example.[270] At the request of Wilhelm von Siemens, a son of Werner, he was also permitted to inspect the laboratories of the firms of Krupp and I. G. Farben, although company secrets usually were carefully guarded; it was not until the early 1920s that industrial groups issued journals containing their scientific works.[271] Gerdien submitted a comprehensive plan according to which, among other things, physical-technical, metallurgical, physical-chemical, and strength-of-materials laboratories, as well as the most important workshops, were to be united under one roof.[272] The task of establishing the new laboratory began in the summer of 1914, but was interrupted by the outbreak of war.

In 1918, Siemens took stock of its workplaces, laboratories, and personnel; about 70 people were working in physical-technical laboratories, and about 30 in chemical laboratories.[273] The new central laboratory was ready for occupancy in 1920. Its research would be overseen by a governing board, which also made decisions about appointments. About half of the researchers surveyed in 1918 had been recruited from academia, so it was important for the laboratory leadership to maintain close contacts with university colleagues. For example, Gerdien placed great emphasis on good "contacts with the Göttingen institutes. . . . So I had the possibility to get to know early on the young talent being educated there and constantly to acquire the best workers for the research laboratory."[274] But recruitment did not always go smoothly. Gerdien's proposal for hiring a physicist was turned down in August 1928 with the following argument:

> The research laboratory has now grown to 90 employees; in July of 1914 there were 19. The number of personnel grows from year to year. Prof. Gerdien will come after a time and complain about the lack of space. In my opinion the number of personnel in the research laboratory should be greatly curtailed.[275]

Such decisions were reinforced by the world economic crisis that began in 1929. A confidential letter to Sommerfeld from Georg Gehlhoff, co-editor of the *Zeitschrift für technische Physik* and director of the Osram Society, gives evidence of this. In 1930, Gehlhoff let the influential professor know that "in the present economic situation the receptivity of industry to young physicists can no longer be relied upon, and accordingly the number of physicists to be trained should first of all slacken somewhat."[276] There is evidence that the number of physicists employed in many industrial laboratories dropped in the early 1930s.

But even the Great Depression caused only a temporary pause in the rise of industrial laboratories as they grew to become research establishments equal in rank to the universities. In 1927 a visitor from the Leningrad Institute for Electrical Engineering had been impressed after an inspection tour through the laboratories of the largest European firms; he especially noted the Siemens laboratory,

> which is in a separate five-story building. This central laboratory is magnificently equipped as a scientific research institute and is occupied with all sorts of questions, even those of a purely scientific character, such as chemistry, physics and electrical engineering. As soon as . . . it appears that one or another scientific achievement can also be used for industrial purposes, it is very thoroughly worked through. . . . Scientists of international reputation work in all laboratories, especially those of Siemens and Halske. Industry guarantees these savants a good living.[277]

Electrical-engineering firms were not the only important branches of industry that turned toward science early in the twentieth century. For example, in 1909 the Krupp steel firm restructured its research department by subdividing the physics research institute, now housed in a new building, into four departments: metallurgical, metallographical, mechanical, and physical. The firm also found positions for several scientists.[278]

It became commonplace for industry to employ good students of noted scientists. They were visible carriers in the transfer of science from universities to industry.[279] An additional flow of information stemmed from personal contacts and projects shared between industrial physicists and their colleagues at national research establishments; frequently, only industry was equipped with the tools to produce, for example, samples of single crystals, special alloys, or very high-purity materials. Scientists in industry were keenly aware of which professors offered the most fruitful contacts. Marcello von Pirani, leader of the laboratory in Siemens' incandescent-lamp factory, wrote to Werner von Siemens in 1913:

> [I]t has come to my attention that the Auer company and the AEG [Allgemeine Elektrizitätsgesellschaft] support Prof. Lummer in his work by constructing and delivering (as a gift) apparatus, etc., obviously to keep up-to-date themselves and eventually to utilize technical ideas. I don't think it is impossible that new ideas for arc lamp technology may come about or have already come about. Wouldn't it be possible for Siemens and Halske too to connect up with Prof. Lummer in some way, possibly by sending over one of the gentlemen of the physical-chemical laboratory?[280]

University professors, especially in the United States, often took a semester off to work in an industrial research laboratory, such as those of the Bell System or General Electric. Conversely, industrial physicists would conduct experiments at national research establishments, such as the low-temperature laboratory in the Physikalisch-Technische Reichsanstalt. Meanwhile, industrial laboratories held colloquia in which the newest developments and methods were presented. Outside speakers invited to Bell Labs included Robert A. Millikan, Frederick W. Aston, Ernest Rutherford, and Arnold Sommerfeld; General Electric invited Percy W. Bridgman, Arthur H. Compton, and Niels Bohr. Paul Ehrenfest gave talks regularly at Philips, and Richard Becker at Siemens.[281] Further, the industrial laboratories sent their representatives to scientific conferences. Thus Gerdien of the Siemens laboratory took part in the annual sessions of the Bunsen Society, the Society for Metallurgy, the German Physical Society, and the Society for Applied Physics.[282] Industrial laboratories also established libraries; in 1914 the General Electric Company already had available over 1400 technical books and 64 different journals.[283]

The interpenetration of academic and industrial research was embodied in institutions during the early twentieth century. For example, in 1914 the American Committee of One Hundred on Scientific Research of the American Association for the Advancement of Science established the Subcommittee on Research in Industrial Laboratories. It drew its members from both industry and academe, and they discussed such themes as the selection and training of students and the removal of trade secrecy.[284] In the same year, the Canadian government appointed the Advisory Council on Scientific and Industrial Research, to which belonged, among others, John C. McLennan, the director of the physics laboratory of the University of Toronto, and Robert Hobson, president of the Steel Company of Canada.[285]

The First World War redoubled the contacts between industry and academia, with many university scientists suddenly thrust for the first time into industrial settings. In all the warring nations, industrial research was accelerated by the search for substitutes for raw materials that were no longer available, the employment of scientists in the development of new military technologies, and the urgent need to increase production of war materials. In peacetime, of course, the cost of research was kept as low as possible, but wartime disregard for cost accounting led to a great surge of technological investigation in certain directions. Research tasks were adapted to the new market. At the Siemens laboratory, Gerdien reported,

It soon turned out that, in view of the expected long duration of hostilities, a complete readjustment of all work toward direct and indirect war problems must be the only thing imaginable and—in a patriotic sense, as well as in the interest of our firm—the only thing required.[286]

In 1916 a professor at the University of Nevada wrote in *Science:* "Whatever else may be the results of the European war, one thing is certain and this is the inevitable stimulus to research in the industries."[287] The science most stimulated was chemistry, which supplied governments with the newly developed poison gases and giant quantities of explosives that could be manufactured with the Haber–Bosch process. As noted in the following section, another part of war-related studies was applied solid-state research.

New Products and Materials

One of the chief intellectual pursuits that occupied the proliferating industrial researchers, along with many academics who mingled with them, was the old preoccupation with the practical properties of materials.

Well before the twentieth century, the problem had come to the fore in connection with error and disaster—exploding steam boilers, cracking rails, collapsing bridges. Researchers of the nineteenth century had clarified the connections between crucial properties of steels, such as hardness and brittleness, and the varying content of carbon and trace metals. Subsequently, with the help of phase diagrams, they arrived at a great variety of types of steel suitable for various purposes. For example, an especially hard steel for metalworking tools, which enabled the cutting velocity to be increased by a factor of about 6 above earlier steel tools, was celebrated as a milestone at the 1900 World Exhibition in Paris.[288] The triumph was credited to engineer Frederick W. Taylor and metallurgist M. White of the Bethlehem Steel Corporation of America, who recognized that carbides increase the resistance against erosion, so they added carbide-containing material to harden the steel. Further investigations in subsequent decades led to the fabrication of many specialized alloys. One example was Widia (from the German *Wie Diamant,* "like diamond"), used to make dies for drawing the tungsten wires needed in incandescent-lamp filaments; this alloy was developed in the laboratories of the Osram company and was later used by the Krupp firm in metalworking tools.[289]

The more widely used a specialized material was, the more important for profitability were even small improvements, and none was so specialized yet widely used as lamp filaments. The carbon-filament lamp became the outstanding research problem for electrical-engineering firms. Although business flourished, the market was at first limited to the wealthier classes. After the Siemens firm lowered its price in July 1903, "based on the fact that our competitors by producing greater quantities have lowered the prices of incandescent lamps," the selling price still came to 25 marks for a low-voltage lamp, which exceeded the weekly earnings of an ordinary worker.[290] A cheaper lamp would promise a vastly expanded market. Edison's trial-and-error methods (he had scoured the world for usable materials) would no longer suffice to win a share of the market. Therefore, manufacturers of incandescent lamps everywhere employed researchers to solve problems on scientific grounds in well-equipped laboratories.

The research competition surged back and forth with all the intensity, the crushing setbacks, and startling victories of Napoleonic warfare. Even before the turn of the century, the carbon-filament lamps were challenged by two new light sources. The first was an osmium filament developed by the Auer company in 1898, which had a longer lifetime and gave 60% more light. However, this lamp was expensive and delicate. The second light source was developed by Nernst, then a Göttingen chemistry professor; by 1897 he had doubled the output of light with the ceric oxide lamp named after him. Nernst sold his patent to the AEG for more than 1 million marks.[291]

Bolton took a different tack in the chemical laboratory of the Siemens firm. He looked for a material that would be malleable yet could be heated to very high temperatures without rapidly vaporizing. He found that tantalum fulfilled these conflicting requirements, and after tedious experiments produced the first incandescent filament from this brittle material in 1902. Two years later, the lamp was put on the market and was also produced abroad under license; it gave the Siemens firm a competitive advantage for many years. The number of lamps the firm sold rose steadily to almost 10 million in 1911, but then in the next several years fell to 500,000 per year.[292] The drop was due to a process developed at the General Electric research laboratory in which a ductile material could be produced from tungsten, an element even more brittle than tantalum.[293]

General Electric had recruited William D. Coolidge, a prominent scientist from the Massachusetts Institute of Technology, by guaranteeing him double pay and the prospect of spending half his time on research of his own choosing.[294] The tungsten lamps built under his direction and distributed from 1911 on were cheaper and more versatile than the tantalum lamps. Further difficulties had to be overcome in the production of light bulbs—for example, in developing leads to carry the current into the interior. The development of chromium alloys whose coefficient of expansion matched that of the glass allowed a stable construction, where the leads would not melt or the glass shatter when current was applied.

The production of materials with specified properties became an equally essential research activity in the munitions industry. For example, a central research bureau established in 1901 by the leading weapons factories of Germany attempted to develop a light metal alloy that could replace brass in the production of cartridge shells. Measurements of a tempered aluminum–magnesium alloy made on different days showed that the metal had become harder during aging.[295] The alloy turned out to be unsuitable for cartridge shells, but the method of age-hardening opened up the use of Duralumin in lightweight construction. The new alloy was especially useful for supports in rigid airships (98 were built during the First World War) and other aircraft fuselages. Duralumin also replaced wood in propellers and was used as a material for pistons, first in airplane motors and then, in the 1920s, in automobiles.

Systematic empirical investigation had equally striking results when applied to magnetic alloys, a matter of deep concern to the electrical and telephone industries. For example, theory showed that the troublesome attenuation of the signal in an underwater cable could be defeated by increasing the cable's inductance, and it was suggested that the copper wire be wrapped in a band of some material with high magnetic permeability and negligible hysteresis loss.[296] At Bell Laboratories, the problem was investigated under the leadership of Gustav W. Elmen, beginning in 1906. Elmen studied variations in magnetic properties of alloys as the composition was varied, and he found that these properties strongly depend on the speed of cooling during production. A Bell patent of 1917 introduced an iron–nickel alloy called Permalloy, whose treatment brought a remarkably high permeability. In their first scientific publication on this theme, Elmen and Bell Labs research director Harold D. Arnold reported in 1923 that the permeability of the new alloy was 30 times above that of soft iron, but the area of the hysteresis curve amounted to only one-sixteenth that of iron.[297] Permalloy found numerous applications in underwater cables, transfer circuits, transformers, filters, and (in the form of sintered powder) magnetic cores in coils.

The protection of such discoveries by patents was very advantageous for a firm, provided it held the patents and did not have to acquire the process by license from a competitor. The establishment of patent rights had great influence on the creation of research laboratories in industry. It was in the late nineteenth century (for example, beginning in 1877 for the German Reich as a whole) that effective protection began, but it was some years before it prevailed against the free-trade opponents of the patent system.[298] In his memoirs, Werner von Siemens called the inadequate pre-1877 German patent law "one of the greatest obstacles to the free and independent development of the German industry." Patents had lasted for three years at most, and "even for this short time they provided insufficient protection against violations . . . as a rule patent applications were made only in order to get a witness for the discovery."[299] It was only with the guarantee of exclusive use of patented processes and materials that in-house research became profitable, as illustrated in a statement by an executive engineer at General Electric: "Were it not for the patent system, the industrial research laboratory would be non-existent, for no corporation however wealthy could afford the great expense of research, if its results, as soon as they were commercially developed, could be copied by competitors at the bare cost of reproductions."[300]

Once protection of invention became widespread, the industrial laboratories took on an additional function: where patent rights of competitors could be obtained only at high cost or not at all, firms tried to get around the patents through research. Thus the scientists at Siemens developed the alloy Perminvar, and in Japan Sendüst came on the market, both materials having properties resembling Permalloy's.[301]

Of still greater urgency, and with similar goals, was research on substitute materials during the First World War. For example, since Germany had to make do without imported copper, brass in fuses was replaced by sintered or alloyed zinc. To reduce the use of copper for guide rings around bullets, Siemens laboratory physicists studied the electrolytic production of iron, since very pure iron was as soft as copper; finger-thick rods could be manipulated with bare hands. After the war, this research led to further investigations of alloys.[302]

Materials Testing

All research along the lines we have described, although now conducted in the systematic and thoughtful manner of the scientist, still went on without any satisfactory theory of underlying causes; at heart it remained empirical, a matter of trial and error. Thus there was no clear distinction between research and materials testing. Meanwhile, the importance of testing was enhanced by the widespread use of new materials in applications that relied increasingly on precisely specified properties.

The origins of the materials-testing bureaus of the twentieth century may be found in nineteenth-century institutions largely concerned with testing steel and iron. Railroads and other commercial users of steel required reproducible quality, but the most stringent requirements arose in the construction of weapons such as pistols and artillery. For as nitrogen-based explosives supplanted black powder, only the hardest steel alloys could withstand the extreme bore pressures. Standardized manufacturing, introduced first with rifles, also raised new demands for uniformity and high precision in metalworking. It was no wonder, then, that materials-testing establishments were set up in the neighborhood of weapons manufacturers.[303]

Another significant trend was the availability of ever greater power, both from improved steam turbines and from the extension of the electrical network. This meant greater pressures, higher temperatures, and an increase in the running speed of machines. The forces now at hand allowed alloys to be sintered under high temperature and pressure or cold-worked by rolling, pulling, and pressing. In the new high-speed factories, the dynamic properties of solids gained importance, and static methods of materials testing had to give way to dynamic ones. In particular, materials subject to continually changing stress were destroyed far below the static pressure limit, so vibration strength as well as tensile strength had to be investigated. It was found, for example, that after several million stresses the static pressure limit of light metals was reduced by one-third.[304] Inspection of the break points under the microscope revealed steps, which suggested that the break proceeded along *Gleitlinien* (slip lines), a hint that lattice properties had great significance for plastic deformation.[305]

An indication of the pervasive importance of such work was the formation of institutions, indeed an entire community. The International Group for Material Testing in Technology was founded in 1885 in Zurich. Quality control of products was a major theme at its first congress, where the leaders of materials-research institutes acted as middlemen between producers and consumers. Talks at such meetings had a plainly scientific character; an example was a talk in 1900 in Paris entitled "Experiments for the Determination of Hardness, the Flow Limit, the Limit of Elasticity, by Means of a Stamp Made of Hardened Steel," and at other congresses, talks such as "Inception, Development and Determination of the Non-metallic Inclusions in Iron and Steel" and "Use of Magnetic and Electrical Measurement Methods for the Continuation of Processes during Mechanical Testing" marked the beginning of modern nondestructive testing methods.[306]

Theodore von Kármán referred to the extensive body of observations made at the testing institutes in a comprehensive article on the physical basis for knowledge of strength of materials, published in the *Encyklopädie der mathematischen Wissenschaften* in 1913. He treated mechanical hysteresis and aftereffect phenomena in relation to the danger of breakage.[307] A model developed that year by Ludwig Prandtl had led to the conclusion that such hysteresis phenomena simply "did not occur" in the case of a "defect-free crystal." The explanation was rather to be found somewhere in the concepts of "grain boundaries, structural defects, or molecules which were not part of the ideal lattice." But this was barely a start on a still intractable problem, and Prandtl did not publish a mathematical elaboration of his model until 1928, by which time "thanks to the study of crystal structures, and also atomic physics, physical research on strength has become modern again (among pure physicists it is for the first time really modern)."[308]

Materials testing first entered the submicroscopic world in earnest during the 1920s with the help of x-ray diffraction in crystals. It was now possible to study the distortion of the atomic lattice by elastic and plastic deformation. The transformation during transitions among allotropic forms, the construction of mixed crystals, and the processes of refining and age-hardening became observable.[309]

All this represented only one side of the use of x rays in the analysis of materials. Röntgen's first work on x rays in 1895 (such as a silhouette of his hunting rifle) already hinted that when a photographic plate is illuminated through a piece of metal, impurities and defects would show up.[310] It was soon realized that air pockets, cavities, and cracks could be revealed as shadows, and welded joints could be checked without destroying the material. The technique spread rapidly, until by the end of the 1920s improvements in x-ray tubes and apparatus made it possible for such testing methods to be used in bridge and road construction.[311]

A second method of nondestructive materials testing, also commonly employed today, dates to the First World War. At the beginning of 1915 after the success of the German U-boats, the French government assigned Paul Langevin the problem of locating submarines by reflection of ultrasonic waves.[312] He found a way to generate these waves through use of the piezoelectric effect in crystals. Developed by the Allies as the Asdic or Sonar system, the discovery came too late to make a difference in the First World War but played a major role in the next war. During the Second World War, means were also developed to send ultrasonic waves through metals in order to discover defects in objects too large to be penetrated by x rays.

1.6   The Organization of Solid-State Research Before 1920

Let us now step back from the specific examples we have described, a colorful but diverse mosaic in which each tile seems scarcely related to the next. Will a general picture emerge if we view the mosaic from a greater distance, or will each piece remain only a part of its own local matrix? The problem may best be attacked by using tools that, on other occasions, have proved useful for revealing underlying social patterns: we will follow where the money came from and where it went, and we will trace the lines of intellectual communication.

Financial Support

It is difficult to see any emphasis on any particular field within the support of physics at the turn of the century. It was simply academic physics as a whole that profited from the foundation of new institutes and the expansion of older ones in all the industrialized nations.[313] In an era characterized by international military and commercial competition and unprecedented industrial growth, the lobbyists of science demanded their share: "Like big government and big industry, so is big science . . . an important element of our cultural development. . . . Big science requires working capital just as big industry does."[314] State and private financial backers shared in the financing of the "big science," and which group of backers predominated depended strongly on national characteristics. In the United States, for example, much of the support of science came from private universities and private foundations like those of Carnegie and Rockefeller, whereas in Germany academic research and teaching were traditionally a concern chiefly of the Ministries of Culture of the various regional governments. But in Germany, too, around 1900, big industry began to show an interest in science and to participate in its financing, as we discussed.

In every nation, the level of research support available before the First World War barely sufficed for well-focused research into solid-state subjects. Even when funds came directly from industry, interest tended to focus on projects that were expected to enhance prestige. Thus the Carnegie Institution of Washington, founded in 1902 with a grant of $10 million from steel magnate Andrew Carnegie (about 0.1% of his estimated fortune), had no less a goal than to overcome America's "national poverty in science."[315] In the first years, Carnegie's money went primarily toward proven and established areas like astrophysics, geophysics, and geology, where exceptional men appeared most easily available and certain of a result, and not for the new atomic physics or systematic solid-state investigations, where results seemed uncertain or, in the case of solids, not likely to be sensational enough. From 1900 on, the Rockefeller Foundation mainly supported medical projects, and it was not until the 1920s that the Rockefeller International Education Board became an important funding source for the transfer of theoretical physics from Europe to America.[316]

The Kaiser Wilhelm Society for the Support of Science (Kaiser Wilhelm Gesellschaft [KWG]), established in Germany in 1911, was similarly intended to help diminish the "emergency situation of science in the fatherland."[317] A call for donations raised an endowment of 12.6 million marks by 1914, primarily from the chemical industry, the mining and steel industries, banks, and trades. This endowment financed the construction of the Kaiser Wilhelm Institutes where, as we have seen, considerable solid-state research would be done. But the first two institutes, built in 1912, were for the exceptional men of chemistry and physical chemistry, a reflection of both the prestige of chemistry and the commitment of the chemical industry in supporting the movement.

The First World War brought increased support for research. For example, the British formed an advisory council to suggest research problems and to fund them.[318] In the council's first fiscal year of 1915/1916, 20 projects were supported from a government budget of £12,241. Most of them were for research into such subjects as the optical properties of various materials, the corrosion of nonferrous metals, and the rigidity and melting of crystalline substances. These projects were usually entrusted to professional institutions like the National Physical Laboratory, the Faraday Society, the Iron and Steel Institute, and the Institution of Mining and Metallurgy.

But the British, who attributed Germany's initial success in the war to the large-scale application of scientific knowledge in industry, demanded much more in the way of a nationally supported organization for industrial research. In 1917 the government accordingly established a Department of Scientific and Industrial Research (DSIR); its tasks included support of scientists by research stipends and the promotion of technological innovation in industry. The DSIR eventually took responsibility for the National Physical Laboratory and established additional separate research institutions, such as a chemical research laboratory and a low-temperature station.[319]

The impetus of the war gave other countries a similar spur to follow Germany's example in scientific organization. In the United States, the National Research Council was founded in 1916 to support research for "the national security and welfare."[320] Similar efforts led the Japanese in 1916 and 1917 to found two of the most important Japanese institutions for solid-state studies, the Research Institute for Iron and Steel, where Kotaro Honda was commissioned by the Japanese army to develop magnetic steels,[321] and the Institute for Physical and Chemical Research, which was oriented along the lines of Germany's Physikalisch-Technische Reichsanstalt and took charge of similar tasks.[322]

These and many other new institutions for the support of applied science were continued after the end of the war, and usually they expanded further.[323] A striking case is offered by Germany, where revolutionary disturbances, inflation, high unemployment, and scarcity of goods and raw materials seemed to offer an inauspicious climate for the support of science. Yet new organizations, such as the Notgemeinschaft der Deutschen Wissenschaft and the Helmholtz-Gesellschaft, along with newly founded Kaiser Wilhelm Institutes for iron research (1921), metal research (1921), fiber chemistry (1922), and so forth, worked to ensure that, scientifically at least, despite the loss of the war Germany would remain "sitting on the world's board of directors."[324] Such motives benefited not only applied research, but also fundamental research, which was aided in particular by the Notgemeinschaft.[325]

The funds of the Notgemeinschaft and the Helmholtz-Gesellschaft were distributed by specialized committees of scientists, an autonomous mechanism that promoted work on scientific criteria. A wide distribution of research funds resulted, furthering, among other things, a number of solid-state studies. The Notgemeinschaft supported, for example, Walther Meissner's low-temperature research; work by Werner Heisenberg, Hans Bethe, Erich Hückel, and others on quantum mechanics; and research on x-ray structural analysis by Hermann Mark, Michael Polanyi, and others.[326] The Helmholtz-Gesellschaft supported almost exclusively projects with technological relevance, and this included many studies of solids: "investigations of photoelectric properties of metals" by Peter Pringsheim; Ludwig Prandtl's "experiments on the changes in structure of stressed bodies, especially metals"; the "production of synthetic crystals for the investigations of photoelectric conductivity" by Bernhard Gudden and Robert W. Pohl; Max Born's theoretical investigations "into whether solid hydrogen halogens form a molecular or ionic lattice"; and "x-ray photographs of crystal powders" in Sommerfeld's Institute in Munich.[327] Overall, Paul Forman estimates that the amount of support for physical research from both organizations resulted in at least a doubling of real expenditures compared with the prewar period.[328]

A concentrated support of solid-state themes is most clearly visible in the Kaiser

Wilhelm Institutes for fiber chemistry, metal research, and silicate research, created in Berlin-Dahlem in the 1920s. At these institutes, x-ray structural analysis became an important diagnostic tool for crystallographers, chemists, physicists, and metallurgists. The close proximity of the Dahlem institutes fostered intensive interaction among the various disciplines. Thus the crystallographer Arthur L. Patterson, who came to Dahlem as a guest, recalled that

> the Pulp and Paper Industry Research Laboratory at McGill had become interested in the work of Herzog and Jancke on the x-ray diffraction from cellulose, and suggested that I go to Dahlem to join the group headed by Hermann Mark, which had been built up in Herzog's Institute [for fiber chemistry] to study such problems. . . . Down the street was Haber's Institut für physikalische Chemie, where I made many personal friends as I did in Eitel's Institut für Silikatforschung, next door.[329]

Such personal contacts across the barriers between disciplines facilitated the appearance and spread of new concepts; for example, an analogy between the x-ray diffraction pictures of threads and the pictures of work-deformed metals led Polanyi to investigate the deformation of single metallic crystals.[330]

We could almost call the Berlin-Dahlem complex as a whole the first of the world's great solid-state physics laboratories, but it certainly did not recognize itself as such. If in all this funding and institution building anything was seen as an intellectual unit, it was not solid-state physics but more narrowly defined subjects, notably metal physics. In such cases, the definition was provided less by the physics itself than by the hope for applications.

Industry quickly profited from its donations to such fields of research. For example, the Kaiser Wilhelm Institute for metal research in Berlin-Dahlem and the one for iron research in Düsseldorf proved their worth soon after their founding with research on solidity, alloys, and magnetic properties, which became important for the entire metal industry.[331] The lessons of the war were thus reinforced by peacetime experience. It became accepted everywhere that funds must be found for large-scale scientific research with an orientation, not too specific, toward practical problems.

By the end of the 1920s, support of science was widely understood as a sort of capital investment made by science-based industries and national governments. In most cases, the details of the distribution of funds were organized on the principle of self-management by scientists, which turned out to meet practical goals while still allowing much diversity in research. The result was that solid-state research found more favorable financial conditions than ever before. In the form of specific projects and entire institutes, certain applied branches of the field were beginning to emerge from under the shadow of the older patronage of science based on generalized cultural advancement and prestige.

Communication Within the Scientific Community

Conferences, review articles, and books are the usual means for establishing and propagating the intellectual structure of science. They are the obvious places to look for the first signs of a physics of solids, clearly differentiated as a field of its own. By the early 1910s, with the existence of the Drude–Lorentz electron theory of metals, Born's lattice dynamics, Ewald's dynamical theory of crystal optics, Laue's proof of the space lattice structure, and the Braggs' x-ray structure analysis, was not everything present that was needed for the establishment of a field? Were not the themes of the first Solvay conferences,[332] such as specific heat of solids (1911) and the experiments and theories relating to x-ray diffraction by crystals (1913), signs that the study of solids was being established on an independent footing?[333]

They were not. There was still no independent solid-state physics, for studies of solids still were seen as falling within the numerous other subfields of physical research. The Versammlungen der Gesellschaft Deutscher Naturforscher und Ärzte, annual meetings in whose second section German-speaking physicists reported their work, give evidence of the limited role then played by research on solids. To be sure, in 1903 Rubens reported on the connection between optical and electrical properties of metals; in 1905, Weiss discussed the ferromagnetism of crystals; the same year, Voigt spoke on optically biaxial crystals, and in 1906 Sommerfeld, too, addressed this subject; in 1907, Leo Königsberger reported on the conduction of electricity in solids; in 1911, Nernst talked about a "general law relating to the behavior of solid materials at very low temperatures"; and in 1913, Laue and some of Röntgen's students discussed x-ray diffraction by crystals, while Hans Thirring presented the theory of space lattice oscillators. But these topics represented a small fraction of the total number of talks and were not differentiated from the rest in any way. The principal discussions were on such subjects as the nature of x rays and radioactivity, wireless telegraphy and electrical waves in wires, the liquefaction of air, liquid crystals, and physical aspects of geology and meteorology. A large fraction of the talks dealt simply with apparatus, which is not surprising, considering that physicists had to build nearly all their experimental tools themselves.

No central theme had yet emerged when the meetings resumed after the First World War.[334] Each year, there was at least one talk about solids on the program, but these addressed a variety of questions and were scattered among many other subjects. Then suddenly in 1926 an entire morning of the conference was devoted to the theme of research on metals. The first talk was on metal research in industry, followed by discussion of topics such as the retention and deformation of shape, the effects of crystal structure, the molecular theory of solidity, and single crystals. Here, as in funding and institution building, we note that the first signs of an independent field did not cover solid-state physics as a whole, but the particular subject of metal physics.

Textbooks are even better than congresses for suggesting how a field is organized. When in the 1920s the standard German university physics textbook *Müller–Pouillet* was published in its eleventh edition (the previous edition was already two decades old), it revealed how physics had developed.[335] The established division into fields was taken up unchanged from earlier editions: mechanics, optics, thermodynamics, magnetism and electricity, geophysics, and cosmology. Where did solids find their place? In each volume, their properties and theories were treated, but not as an independent theme; the solid state was lined up on an equal footing with the liquid and gaseous states.

Shortly afterward, an attempt was made to compile the greatly increased knowledge of physics for the use of research physicists themselves: the *Handbuch der Physik* and the *Handbuch der Experimentalphysik*.[336] The prototype for these handbooks was the *Encyklopädie der mathematischen Wissenschaften,* conceived by Felix Klein at the turn of the century, whose fourth and fifth volumes were reserved for physics.[337] The publication of individual volumes of this encyclopedia had begun in 1903, but it progressed so slowly that modern solid-state themes eventually entered, if only in piecemeal fashion: interference of x rays by crystals (1912), the electron theory of metals (1921), and in the 1921 supplement, a reprint of the second edition of Born's book on lattice dynamics.[338]

A somewhat different picture emerged in the *Handbuch der Physik,* whose 24 volumes appeared between 1926 and 1929. Through Volume 18, these maintained the traditional scheme: mechanics, optics, thermodynamics, and electricity and magnetism. From Volume 19 on, this principle of organization no longer held, as the work went on to treat light, including its interaction with matter, and modern atomic physics. A final volume appeared with the title *Cohesive (Zusammenhängende) Matter*—a physics of solids, so to speak.

The trend became even clearer in the new edition of 1933, whose final three volumes, now grown to six, were rewritten to accommodate a great outburst of new results. The revised *Cohesive Matter* volume contained Born's lattice dynamics and Sommerfeld's electron theory of metals, an article by Adolf Smekal on the structure-sensitive properties of metals, and one by Ralph Kronig on the connection between the molecular structure and crystalline structure. Along with Ewald's contribution on the construction of solid matter and its investigation by x rays in Volume 23 of the second edition, and with the several contributions on various themes within crystal optics, the *Handbuch* seemed to contain a quite comprehensive solid-state physics. Yet the appearance is illusory. The catchphrase "solid bodies" *(Festkörper),* already used in Born's first book, is to be found only once in the index. Even though many of the elements of the future field of solid-state physics were present and gathered for convenience in one place, there is no evidence that they were thought to be related as parts of a single coherent field.

The *Handbuch der Experimentalphysik,* begun almost simultaneously, also recognized the traditional subfields (mechanics, thermodynamics, optics, electricity and magnetism, geophysics, and astrophysics), although new themes of interest were placed within these categories, particularly themes oriented toward applications.[339] Thus, noting that "technology has an interest in research on structure" and pointing to the "rapid progress of knowledge" in areas such as crystal growth methods and materials testing, the *Handbuch* devoted a volume to "plastic deformation, especially of metals," until recently an unfamiliar group of problems for physicists.[340] The foreword to the entire edition had noted explicitly that the structural properties of solids would be included in detail.[341] In this volume "basic crystallographic and theoretical concepts of structure" were clarified, as one of Ewald's students wrote in the context of x-ray diffraction analysis.[342] Further, the "lattice theory of solids," a comprehensive revision of Born's article, was taken up by one of Born's students.[343] Even Darwin's and Ewald's dynamical theories were included, although only where they were useful to explain the results of the ever more refined structural investigations of crystals.[344]

Here, too, the collection in one place of review articles dealing with solids shows that some sort of intellectual reorientation had gotten under way by the 1930s; this reorientation had just barely begun and received scant recognition. Atomic physics and quantum mechanics were the topics that held center stage, with nuclear and elementary particle physics waiting in the wings.

After all, most phenomena that characterized solids remained not merely at odds with theory, but divorced from any reasonable explanation whatsoever. There was no trace of any understanding why, for example, some metals become superconducting near absolute zero (Chapters 2 and 8), nor why certain other metals seemed to lie in a peculiar condition intermediate between conductors and insulators (Chapter 7). For that matter, until the full development of quantum mechanics, nobody could get close to saying why some substances conduct at all and others do not (Chapters 2 and 8). Nor could the experts cast much light on still more basic phenomena: Why are some crystals clear (diamond), whereas others are colored (ruby) (Chapter 4)? Or why are some materials ductile, whereas others are brittle (Chapter 5)? Any field that stood helpless in the face of its most obvious and important phenomena could scarcely be called a field at all.

Yet off in the background, more and more concepts were developing that could explain that "most willful of all aggregate states, the solid."[345] Many pieces of the puzzle of solid-state physics were on hand, and some were already connected to one another, although the full picture could not yet be seen.

Notes

We wish to thank Jerome Rowley and Spencer Weart for their translation of this chapter, and Stephen Keith for his critical reading of the material and helpful suggestions. Michael Eckert is the author of the introduction and the following sections: "Institutional Settings for the Scientific Investigations of Solids in the Nineteenth Century" (with Helmut Schubert), "The Free Electron Gas. I: Classical Electron Theory of Metals from Drude to Bohr (1900–1911)," "The Free Electron Gas. II: The Impact of Quantum Theory (1911–1926)," and "Financial Support." Helmut Schubert is the author of "Institutional Settings for the Scientific Investigations of Solids in the Nineteenth Century" (with Michael Eckert), the introduction to "New Experimental Techniques," "Low-Temperature Methods," and "Solid-State Research Outside the Academic Establishments." Gisela Torkar is the author of "The Mechanical Crystal," "Symmetry ," "The Specific Heat of Solids," "Lattice Dynamics," "The Laue Experiment and Its Interpretation," "The Beginnings of X-ray Structure Analysis of Single Crystals," and "Communication Within the Scientific Community." Christine Blondel is the author of "The Powder Method." And Pierre Quédec wrote the section "Magnetism."

1. C. S. Smith, "The Prehistory of Solid-State Physics," *Physics Today* 17, no. 12 (1964): 18–30, p. 19.

2. See Chapter 2.

3. See Chapter 9.

4. For historiographic remarks, see the general introduction.

5. Examples of such research are in S. P. Timoshenko, *History of Strength of Materials* (New York: McGraw-Hill, 1953), 41–60.

6. J. Mihalovits, *Die Entstehung der Bergakademie in Selmecbánya (Schemnitz) und ihre Entwicklung bis 1846* (Sopron: Facultas Rerum Metallicarum et Saltuariarum Universitatis Regiae Hungaricae Scientiarum Technicarum et Oeconomicarum de Palatino Josepho nominatae, 1938) fascicle 2, p. 48.

7. Ecole Polytechnique, *Livre du centenaire 1794–1894), 3 vols. (Paris: Gauthier-Villars, 1895–

1897); T. Shinn, *Savoir scientifique et pouvoir social. L'Ecole Polytechnique, 1794–1914* (Paris: Presses de la Fondation Nationale der Sciences Politiques, 1980); C. C. Gillispie, *Science and Policy in France at the End of the Old Regime* (Princeton, N.J.: Princeton University Press, 1980).

8. E. Frankel, "J. B. Biot and the Mathematization of Experimental Physics in Napoleonic France," *Historical Studies in the Physical Sciences* 8 (1977): 33–72; R. M. Friedmann, "The Creation of a New Science: Joseph Fourier's Analytical Theory of Heat," *Historical Studies in the Physical Sciences* 8 (1977): 73–99.

9. Ecole Polytechnique (n. 7), 1: 159.

10. R. S. Turner, "The Growth of Professorial Research in Prussia, 1818–1848—Causes and Context," *Historical Studies in the Physical Sciences* 3 (1971): 137–182; R. Stichweh, *Zur Entstehung des modernen Systems wissenschaftlicher Disziplinen. Physik in Deutschland 1740–1890* (Frankfurt am Main: Suhrkamp 1984).

11. According to a philologist (Creuzer) to C. F. Nebenius, the organizer of the Karlsruhe Polytechnikum, in 1830. Cited in K.-H. Manegold, *Universität, Technische Hochschule und Industrie. Ein Beitrag zur Emanzipation der Technik im 19. Jahrhundert unter besonderer Berücksichtigung der Bestrebungen Felix Kleins* (Berlin: Duncker und Humblot, 1970), 33.

12. D.S.L. Cardwell, *The Organisation of Science in England,* rev. ed. (London: Heinemann, 1972).

13. In the first half of the nineteenth century there was first in Germany the Gesellschaft Deutscher Naturforscher und Ärzte (GDNÄ [Society of German Naturalists and Doctors]); then in Great Britain the British Association for Advancement of Science (BAAS); and in the United States, the American Association for Advancement of Science (AAAS). See F. R. Pfetsch, *Zur Entwicklung der Wissenschaftspolitik in Deutschland 1750–1914* (Berlin: Duncker and Humblot, 1974); Cardwell (n. 12); S. G. Kohlstedt, *The Formation of the American Scientific Community* (Urbana: University of Illinois Press, 1976).

14. Pfetsch (n. 13), 272, 280, 309.

15. Quoted in E. Ashby, "Education for an Age of Technology," in *A History of Technology,* ed. C. Singer, E. J. Holmyard, A. R. Hall, and J. Williams (Oxford: Clarendon Press, 1958), 5: 776–798, p. 785.

16. *Festschrift zum 75-jährigen Bestehen der Eidgenössischen Technischen Hochschule in Zürich* (Zurich: Kommissionsverlag Orell Füssli, 1930), 388 (here the curriculum for mechanical engineering of 1858/1859); *Bericht über die Königl. Polytechnische Schule zu München für das Studienjahr 1869–1870* (Munich: Straub, 1870), 17–19; see also the subsequent annual reports.

17. "Denkschrift über die Einführung einer staatlich anerkannten Classification von Eisen und Stahl," *Zeitschrift des Vereins Deutscher Ingenieure* 21, issue 11 (1877). For the institutionalization of material testing in Germany, see, for example, W. Ruske, *100 Jahre Materialprüfung in Berlin* (Berlin: Bundesanstalt für Materialprüfung, 1971).

18. A. Riedler, "Amerikanische technische Lehranstalten," *Zeitschrift des Vereins Deutscher Ingenieure* 38 (1894): 405–409, 507–514, 608–614, 629–636.

19. The first systematic research on the dependence of crystal forms on their chemical nature was by Eilhard Mitscherlich during the 1820s. See P. Groth, *Entwicklungsgeschichte der mineralogischen Wissenschaften* (Berlin: Springer, 1926).

20. J. D. Bernal, *Science and Industry in the Nineteenth Century* (London: Routledge and Kegan Paul, 1953), 181–219.

21. H.A.M. Snelders, "The Birth of Stereochemistry. An Analysis of the 1874 Papers of J. H. van't Hoff and J. A. Le Bel," *Janus* 60 (1973): 261–278.

22. Groth (n. 19), 126–146.

23. For the history of individual areas of physics related to the solid state in the nineteenth century, there are detailed reviews, such as B. S. Finn, "Thermoelectricity," *Advances in Electronics and Electron Physics* 50 (1980): 175–240, and L. L. Campbell, *Galvanomagnetic and Thermomagnetic Effects. The Hall and Allied Phenomena* (New York: Longmans, Green, 1923, 1960).

24. See, for example, D. Cahan, "The Institutional Revolution in German Physics, 1865–1914," *Historical Studies in the Physical Sciences* 15, no. 2 (1985): 1–66; R. Sviedrys, "The Rise of Physics Laboratories in Britain," *Historical Studies in the Physical Sciences* 7 (1976): 405–436.

Examples of textbooks for practical physics are those composed by Lord Rayleigh and by Friedrich Kohlrausch, *Lehrbuch der praktischen Physik,* 9th rev. ed. (Berlin: Teubner, 1901).

25. Kohlrausch (n. 24), viii.

26. Oxygen for Bessemer methods, and nitrogen for the dye and fertilizer industries.

27. R. Planck, "Zur Geschichte der Kältemaschinen und Kältetechnik," in *Handbuch der Kältetechnik* (Berlin: Springer, 1954), 1: 42–68.

28. C. von Linde, "Über die Wärmeentziehung bei niedrigen Temperaturen durch mechanische Mittel," *Bayerisches Industrie- und Gewerbeblatt* 2 (1870): 205–210, 321–326, 363–367.

29. C. von Linde, *Aus meinem Leben und von meiner Arbeit* (Munich: Oldenbourg, 1979) (reprint of notes that appeared in 1916).

30. J. Dewar, *Collected Papers* (Cambridge: Cambridge University Press, 1927).

31. U. Troitzsch, "Wissenschaft und industrielle Praxis am Beispiel des Bessemer-Verfahrens," in *Wissenschaftsreport der Universität Bielefeld* (Bielefeld: Universität Bielefeld, 1976), 161–175.

32. For the increase in steel production due to the Bessemer process, see H. Gummert, "Die Entwicklung neuer technischer Methoden unter Anwendung wissenschaftlicher Erkenntnisse im Bereich der Deutschen Schwerindustrie, gezeigt am Beispiel der Firma Krupp in Essen," in *Studien zur Naturwissenschaft, Technik und Wirtschaft im 19. Jahrhundert* (Göttingen: Vandenhoek and Rupprecht, 1976), 351ff.

33. C. S. Smith, *A History of Metallography* (Chicago: University of Chicago Press, 1960), 116–119, 140. There are also details on other investigations along the same lines in France and the United States.

34. A. Martens, "Über die mikroscopische Untersuchung des Eisens," *Zeitschrift des Vereins Deutscher Ingenieure* 22 (1878): 11–18; Smith (n. 33), 270, 276.

35. For the problems connected with the construction of the Atlantic cable, see H. I. Sharlin, *Lord Kelvin, The Dynamic Victorian* (University Park: Pennsylvania State University Press, 1979), 124–147.

36. A. Matthiessen and C. Vogt, "Über den Einfluss der Temperatur und die elektrische Leitfähigkeit der Legierungen," *Ann. Phys.* 122 (1864): 19–78.

37. P. Czada, "Die Berliner Elektroindustrie in der Weimarer Zeit; Eine regionalstatistisch-wirtschaftshistorische Untersuchung," *Einzelveröffentlichungen der Historischen Kommission zu Berlin beim Friedrich Meinecke-Institut der Freien Universität Berlin* 4 (1969): 53.

38. H. F. Weber, "Untersuchungen über die Strahlung fester Körper," *Sitz. Preuss. Akad.* 2 (1888): 933–957, p. 957.

39. M. D. Fagen, ed., *A History of Engineering and Science in the Bell System* (Murray Hill, N.J.: Bell Telephone Laboratories, 1975); L. Hoddeson, "The Emergence of Basic Research in the Bell Telephone System, 1875–1915," *Technology and Culture* 22 (1981): 512–544; Hoddeson, "The Entry of the Quantum Theory of Solids into Bell Telephone Laboratories, 1925–40: A Case-Study of the Industrial Application of Fundamental Science," *Minerva* 18, no. 3 (1980): 422–447; Hoddeson, "The Discovery of the Point-Contact Transistor," *Historical Studies in the Physical Sciences* 12, no. 1 (1981): 41–76.

40. Siemens used this argument to emphasize the importance of the foundation of the PTR in a letter to the Prussian government. Cited in H. von Helmholtz, "Vorrede," *Abhandlungen der Physikalisch-Technischen Reichsanstalt* 1 (1894): 1–8.

41. Ibid., 1.

42. Pfetsch (n. 13), 109–123.

43. Siemens to the Prussian government (n. 40), 3.

44. D. L. Cahan, "The Physikalisch-Technische Reichsanstalt, A Study in the Relations of Science, Technology and Industry in Imperial Germany" (Ph.D. diss., Johns Hopkins University, 1980), 37.

45. "Die Tätigkeit der PTR in den Jahren 1891–92," *Zeitschrift für Instrumentenkunde* 13 (1893): 113–140.

46. Hydrographic Office of the Naval Office of the Reich to H. von Helmoltz, 13 November 1891, cited in Cahan (n. 44), 216.

47. Manegold (n. 11), 157–188.

48. W. Nernst, "Das Institut für physikalische Chemie und besonders Elektrochemie an der Universität Göttingen," *Zeitschrift für Elektrochemie* 2 (1896): 629–636.

49. Besides Böttinger, the scientist and industrialist Carl Linde and the locomotive manufacturer Krauss contributed to it. The technical department was shortly converted into an independent institute for applied mathematics and mechanics. See *Die physikalischen Institute der Universität Göttingen, Festschrift* (Berlin: Teubner, 1906).

50. In addition to industrialists, professors of Göttingen University were also members (for example, Riecke, Voigt, Nernst). See ibid., 191.

51. F. Klein, "Über die Neueinrichtungen für Elektrotechnik und allgemeine technische Physik an der Universität Göttingen," *Phys. Zs.* 1 (1899): 143–145.

52. A decade after its foundation, the Göttinger Vereinigung already included all leading industrialists in Germany and all of Göttingen's full professors of the chairs of mathematics, physics, chemistry, agriculture and mineralogy. See *Die physikalischen Institute* (n. 49).

53. A. Wangerin, *Franz Neumann und sein Wirken als Forscher und Lehrer* (Braunschweig: Vieweg, 1907), 14.

54. P. Volkmann, *Franz Neumann* (Leipzig: Teubner, 1896), 7.

55. Ecole Polytechnique (n. 7), 1: 269–284, 312–313.

56. His pupils later not only reprinted his publications, *Franz Neumanns Gesammelte Werke, herausgegeben von seinen Schülern* (Leipzig: Teubner, 1906, 1912), but also published his cycle of lectures, *Vorlesungen über mathematische Physik,* 7 vols. (Leipzig: Teubner, 1885).

57. See A. Hermann, *Lexikon Geschichte der Physik A–Z* (Cologne: Aulis Verlag Deubner, 1972): "Wellentheorie (des Lichts)," 403–405; "Polarisation," 304–305; "Äther," 16–17. For Fresnel, see also A. Wangerin, in *Encyklopädie der mathematischen Wissenschaften,* vol. 5 (Leipzig: Teubner, 1903–1926); Ecole Polytechnique (n. 7), 1: 291–311.

58. K. Stiegler, "Elastizitätstheorie im 17. Jahrhundert," *Janus* 56 (1969): 107–122; I. Szabó, "Die Geschichte der Materialkonstanten der linearen Elastizitätstheorie homogener isotroper Stoffe," *Die Bautechnik* 51, no. 1 (1974): 1–8; Szabó, "Die Grundlegung der linearen Elastizitätstheorie für homogene und isotrope Körper," *Technikgeschichte* 40 (1973): 301–336; Szabó, "Die Entwicklung der Elastizitätstheorie nach Cauchy," *Die Bautechnik* 53, no. 4 (1976): 109–116.

59. A.E.H. Love, *Lehrbuch der Elastizitätstheorie* (Berlin: Teubner, 1907), 7.

60. Ecole Polytechnique (n. 7), 1: 157–161.

61. Ibid., 104–110.

62. Wangerin (n. 53), 69–106.

63. H. Kangro, "Ultrarotstrahlung bis zur Grenze elektrisch erzeugter Wellen, das Lebenswerk von Heinrich Rubens," *Annals of Science* 26 (1970): 235–259 and *Annals of Science* 27 (1971): 165–200; see also "Dem Andenken an Heinrich Rubens," *Die Naturwissenschaften* 10 (1922): 1017–1040.

64. The first work stemmed from his American student, E. F. Nichols, "On The Behavior of Quartz in the Presence of Rays of Large Wavelength Investigated by the Radiometric Method," *Ann. Phys.* 60 (1896): 401–417. See also H. Rubens and E. F. Nichols, "Wärmestrahlen von grosser Wellenlänge," *Naturwissenschaftliche Rundschau* 11 (1896): 545–549. The name dated from 1898: H. Rubens and E. Aschkinass, "Die Reststrahlen von Steinsalz and Sylvin," *Ann. Phys.* 65 (1898): 241–256.

65. Notes of the wife of W. Voigt in possession of his grandson, E. Mollwo, Erlangen; microfilm in DM.

66. A. Sommerfeld, "Woldemar Voigt," *Jahrbuch der Königlich Bayerischen Akademie der Wissenschaften* (1919): 83–84.

67. C. Runge, "Woldemar Voigt," *Nachr. Gött.* (1920): 46–52, p. 49.

68. W. Voigt, *Lehrbuch der Kristallphysik* (Berlin: Teubner, 1910).

69. K. Försterling, "Woldemar Voigt zum einhundertsten Geburtstage," *Die Naturwissenschaften* 38 (1951): 217–221.

70. Voigt (n. 68), viii.

71. Ibid., 10.

72. M. von Laue, *Geschichte der Physik* (Bonn: Universitätsverlag, 1946), 140.

73. W. G. Cady, *Piezoelectricity* (New York: McGraw-Hill, 1946), 9.

74. Notes of Voigt's wife (n. 65), 221.

75. F. M. Jaeger, *Lectures on the Principle of Symmetry* (Amsterdam: Elsevier, 1920); A. Speiser, *Die Theorie der Gruppen von endlicher Ordnung* (Berlin: Springer, 1927).

76. A. Speiser, *Die mathematische Denkweise* (Zurich: Rascher, 1932), 110–135; E. H. Lockwood and R. H. Macmillan, *Geometric Symmetry* (Cambridge: Cambridge University Press, 1978), 9–10.

77. Regarding the early history of crystal studies, see, for example, Groth (n. 19), 3–4.

78. K. H. Wiederkehr, "Von frühen Ideen über eine regelmässige Gestalt kleinster Materieteilchen bis zu Delisles und Bergmanns Vorarbeiten für Haüys Kristallstrukturtheorie," *Centaurus* 21 (1977): 27–43; Wiederkehr "René-Just Haüys Strukturtheorie der Kristalle," *Centaurus* 21 (1977): 278–299; Wiederkehr, "René-Just Haüys Konzeption vom individuellen integrierenden Molekül, ihre Widerlegung, und seine Ansichten über kristallbindende Kräfte," *Centaurus* 22 (1978): 131–156; Wiederkehr, "Das Weiterwirken der Haüyschen Idee von der Polyedergestalt der Moleküle in der Chemie, die Umgestaltung der Haüyschen Strukturtheorie durch Seeber und Delafosse, und Bravais' Entdeckung der Gittertypen," *Centaurus* 22 (1978): 177–186. J. G. Burke, *Origins of the Science of Crystals* (Berkeley: University of California Press, 1966), 147–175.

79. Burke (n. 78); Groth (n. 19), 59–62.

80. Groth (n. 19), 65–70.

81. K. H. Wiederkehr, "Über die Entdeckung der Röntgenstrahlinterferenzen durch Laue und die Bestätigung der Kristallgittertheorie," *Gesnerus* 3 (1981): 351–369, pp. 356, 357. See also Wiederkehr (1977, 1978) (n. 78). H. Burzlaff and H. Zimmermann, *Symmetrielehre* (Stuttgart: Thieme, 1977), 214–217. It was half a century before Sohncke made reference to the forgotten work of Hessel. L. Sohncke, "Die Entdeckung des Eintheilungsprincips der Krystalle durch J.F.C. Hessel," *Zeitschrift für Kristallographie und Mineralogie* 18 (1891): 486–498. Ecole Polytechnique (n. 7), 1: 332–337.

82. Speiser (n. 76), 21.

83. L. Sohncke, *Entwicklung einer Theorie der Kristallstruktur* (Leipzig: Teubner, 1879), 5–27.

84. C. J. Bradley and A. P. Cracknell, *The Mathematical Theory of Symmetry in Solids* (Oxford: Clarendon Press, 1972), 1–2.

85. Sohncke (n. 83), 18–23.

86. He published his knowledge in ibid.

87. Groth (n. 19), 141.

88. Both phenomena were used in the following years to order the corresponding crystal groups.

89. H. Weyl, *Symmetrie* (Basel: Birkhäuser, 1955); J. Nicolle, *Die Symmetrie und ihre Anwendungen* (Berlin: Deutscher Verlag der Wissenschaften, 1954).

90. Cady (n. 73), 2–3.

91. P. Curie, "Sur la symétrie dans les phenomenes physiques, symétrie d'un champ électrique et d'un champ magnétique," *Journal de physique,* 3d ser., vol. 3 (1894): 393–415, p. 393. Prior to Curie, Groth had come to a similar conclusion: "The symmetry of crystals, that is the existence or lack of symmetry planes, is what determines their variety" *Physikalische Kristallographie* [Leipzig: Engelmann, 1876], 174).

92. Curie (n. 91), 400.

93. Voigt (n. 68).

94. Ibid., 19.

95. Ibid., 119.

96. H. Schopper, *Die Symmetrieprinzipien in der Physik* (Karlsruhe: Müller, 1968).

97. P. Curie, "Propriétés magnétiques des corps à diverses températures," *Annales de chimie et de physique,* 7th ser., vol. 5 (1895).

98. W. E. Weber, *W. Weber's Werke* (Berlin: Springer, 1893), 3: 570.

99. J. C. Maxwell, *Treatise on Electricity and Magnetism* (Oxford: Oxford University Press, 1873).

100. J. A. Ewing, "Contributions to the Molecular Theory of Induced Magnetism," *Proc. Roy. Soc.* 48 (1890): 342–359.

101. D. E. Hughes, "Theory of Magnetism Based upon New Experimental Researches," *Proc. Roy. Soc.* 35 (1883): 178–202.

102. A. E. Kennelly, *The Electrician,* 7 and 13 June 1890, cited in Ewing (n. 100), 344.

103. Ewing (n. 100), 345.

104. J. J. Thomson, "On the Magnetic Properties of Systems of Corpuscles Describing Circular Orbits," *Phil. Mag.,* 6th ser., vol. 6 (1903): 673–692.

105. Ibid., 689.

106. W. Voigt, "Elektronenhypothese und Theorie des Magnetismus," *Ann. Phys.* 9 (1902): 115–146.

107. P. Langevin, "Sur la théorie du magnétisme," *Journal de physique,* 4th ser., vol. 4 (1905): 678.

108. P. Weiss, "L'hypothèse du champ moléculaire et la propriété ferromagnétique," *Journal de physique,* 4th ser., vol. 6 (1907): 661–690.

109. For example, those by W. E. Weber, F. Kohlrausch, W. Giese, and E. Riecke. For references, see R. Seeliger, "Elektronentheorie der Metalle," in *Encyklopädie der mathematischen Wissenschaften,* (1922), 777–872, pp. 778–781. A recent historical study is W. Kaiser, "Early Theories of the Electron Gas," *Historical Studies in the Physical and Biological Sciences* 17 (1987): 271–297.

110. P. Drude, "Zur Electronentheorie der Metalle," *Ann. Phys.* 1 (1900): 566–613; *Ann. Phys.* 3 (1900): 369–402; Drude, "Optische Eigenschaften und Elektronentheorie," *Ann. Phys.* 14 (1904): 677–725; *Ann. Phys.* 14 (1904): 936–961.

111. J. J. Thomson, "Indications relatives à la constitution de la matière fournies par les recherches récentes sur le passage de l'électricité à travers les gaz," *Rapports présentés au Congrès International de Physique* (Paris: Gauthier-Villars, 1900), 3: 138–151, Thomson, *The Corpuscular Theory of Matter* (London: Methuen, 1907).

112. H. A. Lorentz, "De beweging der electronen in de metalen," *Verslagen der Koniglichen Akademie der Wetenschapen Amsterdam* 13 (1905): 493–508, 565–573, 710–719; Lorentz, *Ergebnisse und Probleme der Elektronentheorie* (Berlin: Springer, 1906); Lorentz, *The Theory of Electrons and Its Applications to the Phenomena of Light and Radiant Heat* (Leipzig: Teubner, 1909).

113. In 1892 and 1893 Drude gave lectures in Göttingen on the "properties of the electromagnetic field and their use in the explanation of optical properties," which he published in 1894 as a book: *Physik des Aethers auf elektromagnetischer Grundlage* (Stuttgart: Enke, 1894).

114. P. Drude, "Zur Ionentheorie der Metalle," *Phys. Zs.* 1 (1900): 161–165.

115. Drude (1900, vol. 1) (n. 110).

116. Electrolysis was one of the themes belonging to physical chemistry that became popular at the end of the nineteenth century. Drude quoted Walter Nernst's *Theoretische Chemie* many times; it first appeared in 1893 and by 1926 had gone through 15 editions. Nernst was the first director of the Institute for Physical Chemistry, which was supported by the Göttinger Vereinigung.

117. W. Jaeger and H. Diesselhorst, "Wärmeleitung, Elektricitätsleitung, Wärmecapacität und Thermokraft einiger Metalle," *Sitz. Preuss. Akad.* (1899): 719–726. Jaeger and Diesselhorst belonged to the department for electrical measurement techniques at the PTR.

118. M. Thiessen, "Über allgemeine Naturconstanten," *Verh. Ges.* 2 (1900), 116–121. Also see M. Reinganum, "Theoretische Bestimmung des Verhältnisses von Wärme- und Elektricitätsleitung der Metalle aus der Drude'schen Elektronentheorie," *Ann. Phys.* 2 (1900): 398–403.

119. M. Planck, "Über die Elementarquanta der Materie und der Elektrizität," *Ann. Phys.* 4 (1901): 514–566; in this work he set α equal to $3k/2$, where k is Boltzmann's constant, which he had determined in 1900, by comparing his famous radiation formula with measurements, to be $k = 1.346 \times 10^{-10}$ erg/K.

120. For the significance of Lorentz's electron theory for physics around 1900, see R. McCormmach, "H. A. Lorentz and the Electromagnetic View of Nature," *Isis* 61 (1970): 459–497.

121. H. A. Lorentz, "On the Emission and Absorption by Metals of Rays of Heat of Great Wave-Lengths," *Proceedings of the Royal Academy Amsterdam* 5 (1903): 666–687.

122. Lorentz (1906) (n. 112), 33.

123. Seelliger (n. 109), 789.

124. ". . . if one assumes free electrons, one would—at least, if their number is comparable to that of the metallic atoms—have to replace the theory of the Dulong-Petit law of specific heat . . . by a completely different theory of this regularity" (Reinganum [n. 118], 401).

125. Thomson (1907) (n. 111). See W. Wien, "Elektromagnetische Lichttheorie," in *Encyklopädie der mathematischen Wissenschaften* (1908), 5: 157.

126. W. Nernst, "Der Energieinhalt fester Stoffe," *Ann. Phys.* 36 (1911): 395–439, p. 439.

127. R. Gans, "Zur Elektronenbewegung in Metallen," *Ann. Phys.* 20 (1906): 293–326.

128. K. Bädeker, *Die elektrischen Erscheinungen in metallischen Leitern* (Braunschweig: Vieweg, 1911), 111.

129. N. Bohr, "Studier over Metallernes Elektrontheori" (Ph.D. diss., Copenhagen University, 1911), reprinted in Danish and in English translation in L. Rosenfeld, ed., *Niels Bohr Collected Works,* vol. 1: *Early Work (1905–1911),* ed. J. R. Nielsen (Amsterdam: North-Holland, 1972).

130. P. Langevin, "Magnétisme et la théorie des électrons," *Annales de chimie et de physique* 5 (1905): 70–127.

131. J. L. Heilbron and T. S. Kuhn, "The Genesis of the Bohr Atom," *Historical Studies in the Physical Sciences* 1 (1969): 211–290.

132. A. Einstein, "Die Plancksche Theorie der Strahlung und die Theorie der spezifischen Wärme, *Ann. Phys.* 22 (1907): 180–190, 800.

133. M. Born, *Mein Leben* (Munich: Nymphenburger Verlagshandlung, 1975), 234.

134. Einstein (n. 132).

135. See also M. J. Klein, "Einstein, Specific Heat and Early Quantum Theory," *Science* 148 (1965): 173–180.

136. K. Mendelssohn, *Walter Nernst und seine Zeit* (Weinheim: Physik-Verlag, 1976); F. Hofmann, "Walter Nernst zum Gedächtnis," *Phys. Zs.* 43 (1942): 110–116.

137. M. von Laue, "Walther Nernst zum Gedächtnis," *Zeitschrift medizinischen Klinik* (1942): 7–8 (reprinted in Laue, *Gesammelte Schriften und Vorträge* [Braunschweig: Vieweg, 1961], 3: 180–181).

138. J. Wickert, *Einstein* (Hamburg: Rowohlt, 1972), 19. See also letters from Laue to Einstein, copies in DM.

139. *Verh. Ges.* 11 (1909): 415–417. See also A. Hermann, *Frühgeschichte der Quantentheorie* (Mosbach: Physik-Verlag, 1969), 77–81.

140. A. Hermann, "Von Planck zu Bohr. Die ersten 15 Jahre in der Entwicklung der Quantentheorie," *Angewandte Chemie* 82 (1970): 1–7.

141. W. Nernst and F. A. Lindemann, "Untersuchungen über die spezifische Wärme bei tiefen Temperaturen V," *Sitz. Preuss. Akad.* (1911): 494–501; A. Eucken, *Die Theorie der Quanten und der Strahlung* (Halle: Knapp, 1914).

142. A. Einstein to M. Besso, 13 May 1911, in *Albert Einstein–Michele Besso Correspondence, 1903–1955,* ed. P. Speziali (Paris: Hermann, 1972), 20.

143. This concerns the work of the English physicist W. Sutherland, "The Mechanical Vibration of Atoms," *Phil. Mag.* 20 (1910): 657–660, which nearly simultaneously and independently arose from related work by Madelung in Göttingen (see also n. 151).

144. A. Einstein, "Eine Beziehung zwischen dem elastischen Verhalten und der spezifischen Wärme bei festen Körpern mit einatomigem Molekül," *Ann. Phys.* 34 (1911): 170–174; Einstein "Elementare Betrachtungen über die thermische Molekular bewegung im festen Körpern," *Ann. Phys.* 35 (1911): 679–694.

145. I. Waller, "Early History of Lattice Dynamics," in *Proceedings of the International Conference on Lattice Dynamics,* ed. M. Balkanski (Paris: Flammarion Sciences, 1978), 3–5.

146. P. Debye, "The Solid State Around 1910," in *Ferroelectricity,* ed. E. F. Weller (Amsterdam: Elsevier, 1967), 297–303. See also P. Debye, interview with D. M. Kerr and L. P. Williams 22 December 1965, transcript in Kerr's possession.

147. P. Debye to A. Sommerfeld, 29 March 1912, DM. At the Wolfskehl Congress in 1914, Sommerfeld noted that his assistant Debye had grappled with the Rayleigh–Jeans normal modes in connection with Planck's radiation formula before turning to the specific heat of solids. See A. Sommerfeld, "Probleme der freien Weglänge," in *Gesammelte Schriften,* ed. F. Sauter (Braunschweig: Vieweg, 1968), 2: 287–327.

148. See, for example, M. Blackman, "Heat Capacity of Crystals and the Vibrational Spectrum," in *Beginnings,* 116–119. Much earlier, Planck and others were already aware that the Debye

law failed to work at low temperatures. See K. F. Herzfeld, ed., *Müller–Pouillets Lehrbuch der Physik,* 11th ed. (Braunschweig: Vieweg, 1925), 3, pt. 2: 366–371.

149. W. Nernst at Wolfskehl Congress, "Vorträge über die kinetische Theorie der Materie und der Elektrizität, gehalten in Göttingen auf Einladung der Kommission der Wolfskehlstiftung" (Berlin: Teubner, 1914), 73.

150. M. Born, "Reminiscences of My Work on the Dynamics of Crystal Lattices," in *Proceedings of the International Conference on Lattice Dynamics,* ed. R. F. Wallis (Oxford: Pergamon, 1965), 1–7. See also P. Debye, "The Early Days of Lattice Dynamics," in Balkanski (n. 145).

151. The first joint publication of the two followed in April 1912. See M. Born and T. von Kármán, "Über Schwingungen im Raumgitter," *Phys. Zs.* 13 (1912): 297–309. It appeared before Laue's confirmation of the spatial lattice hypothesis: "We consider the existence of atomic lattices to be obvious, not only because we know of the group theory of crystal structure established by Schoenflies and Fedorow, which explains the geometric properties of crystals, but also on other grounds. A short while ago, Erwin Madelung in Göttingen had derived the first dynamic consequences of the lattice hypothesis, namely, the connection between the characteristic vibration of a crystal observable in the infrared and its elastic properties" (Born [n. 150], 2).

152. M. Goeppert-Mayer, "Pioneer of Quantum Mechanics Max Born Dies in Göttingen," *Physics Today* 23, no. 3 (1970): 97–99.

153. M. Born and M. Goeppert-Mayer, "Der Aufbau der festen Materie," in *Handbuch der Physik,* 2nd ed., vol. 24, pt. 2 (Berlin: Springer, 1933).

154. M. Born, W. Heisenberg, and P. Jordan, "Zur Quantenmechanik II," *Zs. Phys.* 35 (1926): 557–615.

155. Born and von Kármán (n. 151).

156. Born (n. 133), 220. His student F. Hund did not always consider Born's representations in this volume to be reliable. See Hund to the DM, 4 May 1984. For Born, see also J. Lemmerich, "Der Luxus des Gewissens," in *Katalog zur Ausstellung anlässlich des 100. Geburtstages von Born und Franck* (Berlin: Staatsbibliothek Preussischer Kulturbesitz, 1982).

157. M. Born, "Über meine Arbeiten," in *Ausgewählte Abhandlungen* (Göttingen: van den Hoeck and Ruprecht, 1963), 1: xiv.

158. Born (n. 133), 159.

159. Born gives a historical overview in *Dynamik der Kristallgitter* (Berlin: Teubner, 1915).

160. W. Voigt, "Fragen der Pyro- und Piezoelektrizität der Kristalle I," *Phys. Zs.* 17 (1916): 287–293, p. 293. See also Born (n. 159).

161. P. P. Ewald on Born's eightieth birthday, New Milford, Conn., 2 December 1962, in *Born Papers* (Berlin: Staatsbibliothek Preussischer Kulturbesitz).

162. Born (n. 159), 1. The idea was followed up by W. Nernst.

163. P. P. Ewald, "Besprechung von M. Born: Dynamik der Kristallgitter," *Phys. Zs.* 17 (1916): 212–213.

164. Born (n. 159), 5.

165. Ibid.

166. T. von Kármán, interview with John Heilbron, 1962, AHQP: "Now it was Born's idea that we should calculate the spectrum by calculating the lattice exactly. But he wanted to calculate these to show that the quantum theory is not necessary." This assertion is also supported by some statements of Born's as well as his remarks in the first publication. See Born and von Kármán (n. 151).

167. Born (n. 133), 235.

168. P. P. Ewald to A. Sommerfeld, 5 September 1915, DM.

169. Born (n. 133), 261–262.

170. The chemist Fritz Haber repreatedly dealt with problems of crystal physics, such as the connection between infrared and ultraviolet characteristic vibrations and their dependence on ionic or electronic mass. These considerations led him to advance an electron lattice hypothesis for metals that seemed sound until 1919, when it was refuted by Debye's structural investigations.

171. Born (n. 133), 261–262.

172. Fajans also took part in the discussion of the solution of this problem. See seven letters from M. Born to K. Fajans, 1919–1920, DM.

173. The first recasting of "dynamics of the crystal lattice" *(Dynamik der Kristallgitter)* into an

"atomic theory of the solid state" *(Atomtheorie des festen Zustandes)* came in the early 1920s (Berlin: Teubner, 1923). Some years later, there was a revision (done twice) for the *Handbook of Physics.* See M. Born and O. F. Bollnow, "Der Aufbau der festen Materie," *Handbuch der Physik,* vol. 24 (Berlin: Springer, 1927, 1933).

174.   F. Hund, interview with M. Eckert, G. Torkar, H. Rechenberg, H. Schubert, and J. Teichmann, 18 May 1982, DM and AIP. Born (n. 157).

175.   Born (n. 150).

176.   Ewald (n. 161).

177.   A summary is J. Mehra, *The Solvay Conferences on Physics. Aspects of the Development of Physics since 1911* (Boston: Reidel, 1975), xv, 3–11. The themes of the first Solvay conferences show that the development of quantum theory by no means followed exclusively from problems of atomic physics à la Bohr and Sommerfeld. See, in particular, A. Eucken, "Die Entwicklung der Quantentheorie vom Herbst 1911 bis Sommer 1913," *Abhandlungen der Deutschen Bunsen-Gesellschaft für angewandte physikalische Chemie* 7 (1914): 371–405.

178.   H. Tetrode, "Bemerkungen über den Energieinhalt einatomiger Gase und über die Quantentheorie für Flüssigkeiten," *Phys. Zs.* 14 (1913): 212–215; Sommerfeld (n. 147). D. Hilbert, ed., *Vorträge über die kinetische Theorie der Materie und der Elektrizitat,* Mathematische Vorlesungen an der Universität Göttingen, vol. 6 (Berlin: Teubner, 1914).

179.   Sommerfeld, in Hilbert (n. 178), 136. The sole author of the work was W. Lenz, who earned his habilitation under Sommerfeld, and whose work Sommerfeld presented at a conference in Göttingen in April 1913 ("I concur completely with an as yet unconcluded investigation of my collaborator, W. Lenz.")

180.   Tetrode (n. 178), 214.

181.   W. H. Keesom, "Über die Anwendung der Quantentheorie auf die Theorie der freien Elektronen in Metallen," in Hilbert (n. 178), 195.

182.   W. H. Keesom, "Zur Theorie der freien Elektronen in Metallen," *Phys. Zs.* 14 (1913): 670–675.

183.   E. Schrödinger, "Gasentartung und freie Weglänge," *Phys. Zs.* 25 (1924): 41–45.

184.   A. Einstein, "Quantentheorie des einatomigen idealen Gases," *Sitz. Preuss. Akad, Physikalisch–Mathematische Klasse* (1924): 261–267; (1925): 3–14, 18–25.

185.   P. A. Hanle, "The Coming of Age of Erwin Schrödinger: His Quantum Statistics of Ideal Gases," *Archive for History of Exact Sciences* 17 (1977): 165–192; L. Belloni, "A Note on Fermi's Route to Fermi–Dirac Statistics," *Scientia* 113 (1978): 421–430; M. J. Klein, "Einstein and the Wave–Particle Duality," *The Natural Philosopher* 3 (1964): 3–49.

186.   Seeliger (n. 109); E. Kretschmann, "Kritischer Bericht über neue Elektronentheorien der Elektrizitäts- und Wärmeleitung in Metallen," *Phys. Zs.* 28 (1927): 565–592; W. Hume-Rothery, *The Metallic State: Electrical Properties and Theories* (Oxford: Clarendon, 1931).

187.   Institut Solvay, "Conductibilité électrique des métaux et problèmes connexes," *Rapports et discussions du Quatrième Conseil de Physique* (Paris: Gauthier-Villars, 1927). The congress was held on 24–29 April 1924 in Brussels. See also Mehra (n. 177), 115–130.

188.   Seeliger (n. 109); Kretschmann (n. 186); Hume-Rothery (n. 186).

189.   W. Wien, "Zur Theorie der elektrischen Leitung in Metallen," *Sitz. Preuss. Akad.* (1913): 184–200. Hume-Rothery classified this theory as one of the "intermediate theories" (n. 186).

190.   Fifteen years later, W. V. Houston, then Sommerfeld's research fellow, used Wien's work and Fermi statistics to make "the first decent treatment of the resistance law," according to Sommerfeld, at least for high temperatures (conductivity is indeed proportional to $1/T$ according to modern knowledge). For lower temperatures the correct result (conductivity proportional to $1/T^5$) was not produced until 1929, by Bloch, with the help of quantum mechanics: see Chapter 2.

191.   Hume-Rothery (n. 186), v.

192.   P. W. Bridgman, *The Physics of High Pressure* (London: Belland, 1949).

193.   For a description of apparatus for measuring specific heat, see W. Nernst, "Der Energieeinhalt fester Stoffe," *Ann. Phys.* 36 (1911): 395–439.

194.   For the use of the Hall effect and the thermoelectrical effect, see F. Kohlrausch, *Praktische Physik* (Berlin: Teubner, 1935), 157–158, 604.

195. On the use of x-ray diffraction, photoelectric effects, electron diffraction, and radioactive rays in measurement techniques for the investigation of the properties of solids, see K. Lark-Horowitz and V. A. Johnson eds., *Methods of Experimental Physics,* vol. 6: *Solid State Physics* (New York: Academic Press, 1959).

196. For methods of growing crystals, see G. Tammann, *Kristallisieren und Schmelzen* (Leipzig: Barth, 1903); J. Czochralsky, "Ein neues Verfahren zur Messung der Kristallisationsgeschwindigkeit der Metalle," *Zeitschrift für physikalische Chemie* 92 (1918): 219–221; S. Kyropoulos, "Ein Verfahren zur Herstellung grosser Kristalle," *Zeitschrift für anorganische und allgemeine Chemie* 154 (1926): 308–313.

197. A. Zeleny, "The Dependence of Progress in Science on the Development of Instruments," *Science* 180 (1916): 185–193, pp. 192–193.

198. From 1898 on, there was a separate division for scientific instruments within the Gesellschaft Deutscher Naturforscher und Ärzte, which held meetings in common with the physics division. See Pfetsch (n. 13), 309. After 1920, there were two divisions: physics and technical physics.

199. For the conditions in Polish laboratories, see H. Kamerlingh Onnes, "Karol Olszewsky," *Chemiker-Zeitung* 82 (1915): 517–519.

200. D. Chilton and H. G. Coley, "The Laboratories of the Royal Institution in the 19th Century," *Ambix* 27 (1980): 175–203.

201. R. Cory, "Fifty Years at the Royal Institution," *Nature* 166 (1949): 1049–1053.

202. J. Dewar to H. Kamerlingh Onnes, 15 May 1898, Archive 8, Museum Boerhaave, Leiden.

203. H. Kammerlingh Onnes, "Inaugural Address" (1882), cited in W. H. Keesom, "Prof. Dr. H. Kamerlingh Onnes. His Life-work, the Founding of Cryogenic Laboratory," *Communications from the Physical Laboratory of the University of Leiden,* suppl. 57 (1926): 3–21, p. 4.

204. H.B.G. Casimir, "Superconductivity," in *Proceedings of the International School of Physics,* ed. Charles Weiner (New York: Academic Press, 1977), 172.

205. Much material about Leiden can be found in K. Mendelssohn, *Der Weg zum absoluten Nullpunkt* (Munich: Kindler, 1966).

206. J. Dewar to H. Kamerlingh Onnes, 20 July 1895, Archive 8, Museum Boerhaave, Leiden; K. Olszewski, "Die Verflüssigung der Gase," *Zeitschrift für komprimierte und flüssige Gase* 8 (1968); 103–126, p. 122.

207. Casimir (n. 204), 173.

208. H. Kamerlingh Onnes, "Untersuchungen über die Eigenschaften der Körper bei niedrigen Temperaturen, welche Untersuchungen unter anderem auch zur Herstellung von flüssigem Helium geführt haben" (Nobel Prize acceptance speech), in *Les Prix Nobel* (Stockholm: Norstedt, 1914), 1–25.

209. V. Wroblewsky, "Über den elektrischen Widerstand des Kupfers bei den höchsten Kältegraden," *Ann. Phys.* 52 (1885): 27–31, p. 31.

210. J. Dewar and J. A. Fleming, "On the Electrical Resistance of Pure Metals, Alloys and Non-Metals at the Boiling-Point of Oxygen," *Phil. Mag.* 34 (1892): 326–336; Dewar and Fleming, "The Electrical Resistance of Metals and Alloys at Temperatures Approaching Absolute Zero," *Phil. Mag.* 36 (1893): 271–299.

211. H. Kamerlingh Onnes, "The Resistance of Platinum at Helium Temperature," *Communications from the Physical Laboratory of the University of Leiden* 119b (1911): 19–26.

212. A. Eucken, "Die Wärmeleitfähigkeit einiger Kristalle bei tiefen Temperaturen," *Phys. Zs.* 12 (1911): 1005–1008, p. 1005.

213. A. Eucken, "Über die Temperaturabhängigkeit der Wärmeleitfähigkeit fester Nichtmetalle," *Ann. Phys.* 34 (1911): 185–221.

214. R. Peierls, "Zur kinetischen Theorie der Wärmeleitung in Kristallen," *Ann. Phys.* 3 (1929): 1055–1101.

215. J. Stark, ed., *Forschung und Prüfung, 50 Jahre Physikalisch-Technische Reichsanstalt* (Leipzig: Hirzel, 1937), 112.

216. P. P. Ewald, *Dispersion und Doppelbrechung von Elektronengittern (Kristallen)* (Göttingen: Dieterichsche Universitäts-Buchdruckerei, 1912).

217. P. P. Ewald, "The Setting for the Discovery of X-ray Diffraction by Crystals" (Speech given

at the banquet held at the First General Assembly of the International Union of Crystallography at Harvard University, 2 August 1948; manuscript in DM).

218. P. P. Ewald, ed., *Fifty Years of X-ray Diffraction* (Utrecht: Oosthoek's Uitgeversmaatschappij, 1962), 40.

219. Max Laue (the hereditary preposition "von" was not bestowed on his father, a lawyer in the military service, until 1914) had taken his degree with Planck with work on interference phenomena on parallel plates; his subsequent work on relativity brought him into early contact with Einstein. In 1909 he went to Munich for personal reasons.

220. Ewald (n. 218), 41.

221. A. Sommerfeld to W. Wien, 10 November 1915, DM; P. P. Ewald to Sommerfeld, 28 April 1924, DM.

222. P. P. Ewald, *Kristalle und Röntgenstrahlen* (Berlin: Springer, 1924), 52. "Most famous authority": W. L. Bragg, *The Crystalline State* (London: Reue, 1966), 269; "Rock-Solid": P. Niggli, "Die Bedeutung des Lauediagramms für die Kristallographie," *Die Naturwissenschaften* 16 (1922): 391–399, p. 392.

223. W. Friedrich, "Erinnerungen an die Entdeckung der Interferenzerscheinungen bei Röntgenstrahlen," *Die Naturwissenschaften* 36 (1949): 354–356, p. 355.

224. W. Friedrich, P. Knipping, and M. Laue, "Interferenzerscheinungen bei Röntgenstrahlen," *Sitzungsberichte der Bayerischen Akademie der Wissenschaften* (1912): 303–322, p. 317 (reprinted in *Die Naturwissenschaften* 39 [1952]: 361–372; this refers to Figure 1.14).

225. Friedrich (n. 223), 354. See also M. Laue to P. P. Ewald, 1 May 1924, DM.

226. W. Gerlach, "Münchner Erinnerungen," *Physikalische Blätter* 19 (1963): 97–109, p. 99. "Laue has sent me a photograph of his x-ray interference pictures almost without explanation. If I understand correctly, he uses the crystal structure as a diffraction lattice for x-rays. Then it is something truly wonderful, and tremendously juicy, momentous, and something one can truly be envious of" (P. Ehrenfest to A. Sommerfeld, 23 June 1912, DM). "It is the most wonderful thing I've ever seen! Diffraction on single molecules, whose arrangement is thus revealed. The photograph is so sharp that one would scarcely have suspected it, due to the thermal agitation" (A. Einstein, 21 June 1912, in C. Seelig, *Albert Einstein, Eine dokumentarische Biographie* [Zurich: Europa, 1954]).

227. M. Planck, "Zum 25 jährigen Jubiläum der Entdeckung von W. Friedrich, P. Knipping und M. v. Laue," *Verh. Ges.* 18 (1937): 77–80, p. 77.

228. P. Forman, "The Discovery of the Diffraction of X-rays by Crystals. A Critique of the Myths," *Archive for History of Exact Sciences* 6 (1969): 38–71. This comprehensive work lists the primary and secondary literature up to the time of publication. Later published articles are G.E.Z. Schulze, "M. v. Laue und die Geschichte der Röntgenfeinstrukturuntersuchung," *Kristall und Technik* 8 (1973): 527–543; Wiederkehr (n. 81).

229. P. P. Ewald, "The Myth of Myths," *Archive for History of Exact Sciences* 6 (1969): 72–81.

230. Friedrich, et al. (n. 224), 303.

231. Debye calculated the influence of thermal motion on the intensity, the Debye effect, between 1912 and 1914. Waller corrected this theory some years later. See I. Waller, "Memories of My Early Work on Lattice Dynamics and X-ray Diffraction," in *Beginnings*, 120–124.

232. In his December 1912 inaugural speech in Zurich, where he had become Debye's successor, Laue said, "The direction in which such a ray (secondary ray) occurs is so sharply limited that it can only be explained by the appearance of monochromatic oscillations. And it is at this time a still unanswered question as to where the monochromatic x-ray radiation comes from" (Laue [n. 137], 1:219–224, p. 224).

233. P. P. Ewald, "A Review of My Papers on Crystal Optics 1912 to 1968," *Acta Crystallographica* A35 (1979): 1–9.

234. M. Planck, "Gedächtnisrede auf Heinrich Rubens," *Sitz. Preuss. Akad.* (1923): cviii–cxiii.

235. A. Sommerfeld to W. Wien, 24 October 1917, DM.

236. For explanation, see G. Hildebrandt on the occasion of the award of the Max Planck Medal to Ewald: "62 Jahre Kristalloptik der Röntgenstrahlen," *Physikalische Blätter* 35 (1979): 55–64, 103–118.

237. Ewald (n. 218), 58. P. von Groth and W. C. Röntgen too had taken part in this celebration.

Voigt's wife verifies his trip in her notes (in possession of his grandson E. Mollwo, copy in DM). On the other hand, Laue stated that the offprints that he sent around had made his experiment known in England, and this is not unlikely; in 1911, Sommerfeld had carried on a brief correspondence with W. H. Bragg concerning his x-ray work (DM).

238. W. L. Bragg and G. M. Caroe, "Sir William Bragg, F.R.S.," *Notes and Records of the Royal Society of London* 16 (1961): 169–182.

239. R. Stuewer, "William Henry Bragg's Corpuscular Theory of X-rays and Gamma Rays," *British Journal for the History of Science* 5 (1971): 258–281; W. L. Bragg, *The Crystalline State, A General Survey* (London: Reue, 1966), 72; Bragg, "The Growing Power of X-ray Analysis," in Ewald (n. 218), 120–135, p. 122.

240. W. L. Bragg, "Personal Reminiscences," in Bragg (n. 239), 531–539; Bragg, *Geschichte der Röntgenspektralanalyse* (Berlin: Archiv und Kartei, 1947).

241. E. N. da C. Andrade, "William Henry Bragg," in Ewald (n. 218), 308–327, p. 315. From this basic equation of x-ray crystal optics, both the Laue case and the Bragg case can be calculated.

242. Laue (n. 137), 1: 191, 224.

243. Schulze (n. 228), 536.

244. Bragg (n. 240), 131; Ewald (n. 218), 61; M. von Laue, *Röntgenstrahlinterferenzen* (Leipzig: Akademische Verlagsgesellschaft, 1948); M. Siegbahn, *Spektroskopie der Röntgenstrahlen* (Berlin: Springer, 1924); P. P. Ewald, *Kristalle und Röntgenstrahlen* (Berlin: Springer, 1923).

245. Andrade (n. 241), 315.

246. Bragg (n. 240), 14.

247. Ibid. Independently from Bragg, the young physicists H.G.J. Moseley and C. G. Darwin started similar work in Rutherford's laboratory in Manchester. The law that Moseley discovered was a key confirmation of the Bohr theory; Darwin's work of 1914, resulting from Bragg's measurements, was the first formulation of a "dynamic theory," which took into consideration not only the interaction between incident waves and atoms, but also the effect of the waves scattered from the atoms.

248. Bragg discussed the most important results in detail in his book *X-rays and Crystal Structure* (London: Bell, 1915).

249. The main difficulty with both methods was that only the ratio of wavelength to lattice spacing (λ/d) could be determined. But if one adds the convincing arguments that the Braggs used about the atomic arrangement in rock salt, along with the numerical values for the density of rock salt and Avogadro's number, one can obtain a numerical value for the d and for the fixed λ palladium rays used in the experiment. Once λ is determined, one can find the d values of other crystals and, in the same crystal, the d values of various lattice planes: "The crystal structure can be felt out, so to speak, with the known wavelengths" (P. P. Ewald, "Kristalloptik der Röntgenstrahlen," in *Müller–Pouillets Lehrbuch der Physik,* ed. K. W. Meissner, 11th ed., pt. 1, [Braunschweig: Vieweg, 1929], 2: 1061–1103).

250. Ewald (n. 218), 696.

251. J. Nitta, "Shoji Nishikawa, 1884–1952," in Ewald (n. 218), 328–333, pp. 328–329, and "Japan," ibid., 484–492, p. 484; ibid., 76–77. Three months after W. Bragg's publication the Russian crystallographer Georg V. Wulff had also derived this equation and recognized the relationship between the Laue and Bragg methods. See A. V. Shubnikov, "Schools of X-ray Structural Analysis in the Soviet Union" in Ewald (n. 218), 493–497, p. 495, and "Autobiographical Data and Personal Reminiscences," in ibid., 647–653, p. 648.

252. Ewald (n. 218).

253. See W. Hume-Rothery, "Applications of X-ray Diffraction to Metallurgic Science," in Ewald (n. 218), 190–211; J. M. Bijvoet, W. G. Burgers, and G. Hägg, eds., *Early Papers on Diffraction of X-rays by Crystals,* vol. 2 (Utrecht: Oosthoek's Uitgeversmaatschappij, 1972); P. Debye and P. Scherrer, "Interferenz an regellos orientierten Teilchen in Röntgenlicht I," *Phys. Zs.* 17 (1916): 277–283. See also *The Collected Papers of Peter J. W. Debye* (New York: Wiley Interscience, 1954); A. W. Hull, "The Crystal Structure of Iron," *Phys. Rev.* 9 (1917): 84–87; Hull, "A New Method of X-ray Crystal Analysis," *Phys. Rev.* 10 (1917): 661–696.

254. P. Scherrer, "Personal Reminiscences," in Ewald (n. 218), 642–645, p. 642.

255. Debye (1965) (n. 146).

256. Scherrer (n. 254), 642–645.

257. J. D. Hanawalt, H. W. Rinn, and L. K. Frevel, "Chemical Analysis by X-ray Diffraction," *Industrial and Engineering Chemistry, Analytical Edition* 10 (1938): 457. This work was achieved in the Dow Chemical Company Laboratory.

258. A. W. Hull, "The X-ray Crystal Analysis of Thirteen Common Metals," *Phys. Rev.* 17 (1929): 571–588.

259. G. Hägg, *Arne Westgren, Minnesteckning av Gunnar Hägg* (Stockholm: Kungl. Vetenkap-sakademien, 1978), 182. See also Hägg, interview with A. Guinier, 26 May 1983, Cité des Sciences et de l'Industrie, Paris.

260. A. J. Bradley and J. Thewlis, "Crystal Structure of α-Manganese," *Proc. Roy. Soc.* 115 (1927): 456–471.

261. E. A. Owen and G. D. Preston, "X-ray Analysis of Solid Solutions," *Proceedings of the Physical Society of London* 36 (1923); 49.

262. A. Westgren and G. Phragmen, "X-ray Studies on the Crystal Structure of Steel," *Journal of the Iron and Steel Institute* 109 (1924): 159–172.

263. A. Westgren and G. Phragmen, "X-ray Analysis of the Cu–Zn, Ag–Zn, and Au–Zn alloys," *Phil. Mag.* 50 (1925): 311–341 (partly reprinted in Bijvoet et al. [n. 253]).

264. S. Keith and P. Hoch, "Formation of a Research School: Theoretical Solid State Physics at Bristol 1930–54," *British Journal for the History of Science* 19 (1986): 19–44.

265. K. Birr, "Industrial Research Laboratories," in *The Sciences in the American Context: New Perspectives,* ed. N. Reingold (Washington, D.C.: Smithsonian Institution Press, 1979), 193–207.

266. Hoddeson (1981) (n. 39), 512–544, p. 522.

267. M. D. Fagen, ed., *A History of Engineering and Science in the Bell System,* 2 vols. (New York: Bell Telephone Laboratories, 1975, 1978); K. Birr, *Pioneering Industrial Research* (Washington, D.C.: Public Affairs Press, 1957); G. Wise, "A New Role for Professional Scientists in Industry: Industrial Research at General Electric 1900–1916," *Technology and Culture* 21 (1980): 408–429.

268. F. Trendelenburg, "Aus der Geschichte der Forschung im Hause Siemens," *Technikgeschichte in Einzeldarstellungen* 31 (1975): 1–279.

269. L. Fischer to D. Spiecker, 26 July 1913, SAA 68/Li 185, Siemens Museum, Munich.

270. G. Gerdien to W. Siemens, 13 May and 4 June 1914, SAA 4/Lk 166, Siemens Museum.

271. W. Siemens to Dr. Hugensberg, 18 December 1913, SAA 4/Lk 166, Siemens Museum.

272. H. Gerdien, "Physikalisch-Chemisches Labor, Denkschrift betreffend Aufgaben und Vorschläge zu ihrer Lösung," 1914 and 1918, SAA 68/Li 185, Siemens Museum.

273. "Umfrage betreffend Organisation der Siemens Schuckert Werke," March 1918, SAA 11/Lb 31, Siemens Museum.

274. H. Gerdien, "Geschichte des Forschungslaboratoriums," 1944, p. 1, SAA 68/Li 185, Siemens Museum.

275. A. Stauch to M. Haller, 24 August 1929, SAA 11/Lb 63, Siemens Museum.

276. W. Gehlhoff to A. Sommerfeld, 3 December 1930, DM.

277. "Verbindung der Industrie mit der Wissenschaft in West Europa," in "Moskauer Industrie- und Handelszeitung vom Sept. 10, 1927," SAA 68/Li 185, Siemens Museum.

278. *Die Forschungsanstalten der Firma Krupp* (Essen: Krupp, 1934), 19–20.

279. H. Gerdien, Siemens (a student of W. Voigt); W. Schottky, Siemens (M. Wien), I. Langmuir, General Electric (W. Nernst); G. Holst, Philipps (H. Kamerlingh Onnes); A. Arnold, Bell (R. A. Millikan); H. Riegger, Siemens (F. Braun, M. Wien); W. R. Whitney, General Electric (W. Ostwald, Ch. Fridel); G. Campbell, Bell (F. Klein, Boltzmann), and many others.

280. M. von Pirani to W. von Siemens, 27 November, 1913, SAA 4/Lk 166, Siemens Museum.

281. For Bell Labs, see Fagen (n. 267), 987; for General Electric, see Birr (n. 267), 78; for Siemens, see F. Hund, interview with M. Eckert, H. Rechenberg, H. Schubert, G. Torkar, and J. Teichmann, 18 May 1982, DM and AIP.

282. When H. Gerdien went to the meetings, he informed the director of the Forschungsabteilung. SAA 11/Lb 63, Siemens Museum.

283. Birr (n. 267), 78.

284. "Report of the Subcommittee Meeting [26 December 1916]," *Science* 45 (1917): 34.

285. "Industrial Research in Canada," *Science* 44 (1916): 811–812.

286. Gerdien (n. 274), 46.

287. C. A. Jacobson, "The Importance of Scientific Research to Industries," *Science* 44 (1916), 456–459, p. 457.

288. F. Koenigsberger, "Production Engineering," in *A History of Technology,* ed. T. I. Williams (Oxford: Clarendon Press, 1978), 6: 1046–1047.

289. This is the same F. W. Taylor who is famous for applying scientific criteria to the production process to increase productivity. See F. W. Taylor, *Die Grundsätze wissenschaftlicher Betriebsführung,* trans. R. Roeseler (Munich: Oldenbourg, 1919), 123–137.

290. W. Ruske, *100 Jahre Materialprüfung* (Berlin: Bundesanstalt für Materialprüfung, 1971), 210. For the fall in carbon-filament-lamp prices, see Peschky (Siemens) to Abteilung für Beleuchtung und Kraft, 24 July 1903, SAA 68/Li 188, Siemens Museum.

291. Nernst had at first offered his finding to Siemens, who declined it because of its high price. A. F. Joffe, *Begegnungen mit Physikern* (Basel: Pfalz Verlag, 1967), 82–83.

292. "Zusammenstellung der seit Bestehen des Werkes eingegangenen Bestellungen auf Kohlefaden-, Tantal- und Wotanlampen," SAA 68/Li 188. Siemens Museum.

293. Birr (n. 267), 38.

294. Ibid., 37.

295. H. Y. Hunsicker and H. C. Stumpf, "History of Precipitation Hardening," in *The Sorby Centennial Symposium on the History of Metallurgy,* ed. C. S. Smith (New York: Gordon and Breach, 1965), 271–311.

296. C. E. Krarap, "Unterseeische Fernsprechkabel mit erhöhter Selbstinduktion," *Journal télégraphique* (1905): 187.

297. For the history of Permalloy, see Fagen (n. 267), 977–985.

298. F. K. Beier and J. Straus, "Das Patentwesen und seine Informationsfunktion- gestern und heute," *Gewerblicher Rechtsschutz und Urheberrecht* 79 (1977): 282–289.

299. W. von Siemens, *Lebenserinnerungen,* 16th ed. (Munich: Prestel Verlag, 1956), 234.

300. L. A. Hawkins, "Engineering Development and Research," *General Electric Review* 32 (1929): 92.

301. Trendelenburg (n. 268), 94–95; K. H. von Klitzing, private communication to author.

302. M. Schwarte, *Die Technik im Weltkriege* (Berlin: Mittler und Sohn, 1920), 90–91; Gerdien (n. 272), 49–50.

303. Ruske (n. 290), 182–183.

304. A. v. Zeerleder, *Die Entwicklung der Ermüdungsprüfung und ihre besondere Anwendung bei Aluminiumlegierungen* (Zurich: Eidgenössische Materialprüfanstalt, 1930).

305. J. A. Ewing and W. Rosenhain, "The Crystalline Structure of Metals," *Philosophical Transactions of the Royal Society of London A* 193 (1900): 353–376, p. 353.

306. H. Schulz, "100 Jahre Werkstoffprüfung," *Zeitschrift des Vereins Deutscher Ingenieure* 91 (1949): 141–147.

307. T. von Kármán, "Physikalische Grundlagen der Festigkeitslehre," in *Encyklopädie der mathematischen Wissenschaften* (Leipzig: Teubner, 1913), 4: 695–770.

308. L. Prandtl, "Ein Gedankenmodell zur kinetischen Theorie der festen Körper," *Zeitschrift für angewandte Mathematik und Mechanik* 8 (1928): 85–106, p. 87.

309. Hume-Rothery (n. 253), 190–211.

310. W. Röntgen, "Über eine neue Art von Strahlen," *Sitzungsberichte der Physikalisch Medizinischen Gesellschaft Würzburg* (1895): 132–141.

311. J. Eggert and E. Schiebold, eds., *Die Röntgentechnik in der Materialprüfung* (Leipzig: Akademie Verlag, 1930), 1.

312. F. V. Hunt, *Electroacoustics* (Cambridge, Mass.: Harvard University Press, 1954), 46–47.

313. P. Forman, J. L. Heilbron, and S. Weart, "Physics circa 1900: Personnel, Funding, and Productivity of the Academic Establishments," *Historical Studies in the Physical Sciences* 5 (1975): 1–185, pp. 92–95.

314. A. von Harnack, "Antrittsrede in der Preussischen Akademie der Wissenschaften am 3. July 1890," cited in J. Lemmerich, comp., *Dokumente zur Gründung der Kaiser-Wilhelm-Gesellschaft* (catalog of an exhibition) (Berlin: Max-Planck-Gesellschaft, 1981), 14–15.

315. A. Carnegie to S. Newcomb, 3 January 1902, cited in R. W. Seidel, "The Evolution of Science Policy in the Foundation" (Paper delivered at the International Conference on the Recasting of Science between the Two World Wars, Florence and Rome, 23 June–3 July 1980).

316. Seidel (n. 315); D. J. Kevles, *The Physicists* (New York: Vintage, 1979), 69–70.

317. L. Burchardt, "Wissenschaftspolitik im Wilhelmininischen Deutschland. Vorgeschichte, Gründung und Aufbau der Kaiser-Wilhelm-Gesellschaft zur Förderung der Wissenschaften," *Studien zu Naturwissenschaft, Technik und Wirtschaft im 19. Jahrhundert* 1 (1975): 7–158; Denkschrift of A. von Harnack, in Lemmerich (n. 314), 48.

318. I. Varcoe, "Scientists, Government and Organized Research in Great Britain, 1914–16: The Early History of the DSIR," *Minerva* 8 (1970): 192–216.

319. E. Hutchinson, "Scientists as an Inferior Class: The Early Years of the DSIR," *Minerva* 8 (1970): 396–411; H. Rose and S. Rose, *Science and Society* (Harmondsworth: Penguin, 1969), 37–46.

320. Kevles (n. 316), 103–116.

321. N. Kawamiya, "Kotaro Honda: Founder of the Science of Metals in Japan," *Japanese Studies in the History of Science* 15 (1976): 145–158.

322. I. Kiyonobu and E. Yagi, "The Japanese Research System and the Establishment of the Institute of Physical and Chemical Research," in *Science and Society in Modern Japan,* ed. S. Nakayama et al. (Cambridge, Mass.: MIT Press, 1974), 158–201.

323. A somewhat different situation was found in France. See S. Weart, *Scientists in Power* (Cambridge, Mass.: Harvard University Press, 1979).

324. F. Haber to H. R. Kruyt, 7 July 1926, AHQP, quoted in P. Forman, "Scientific Internationalism and Weimar Physicists: The Ideology and Its Manipulation in Germany After World War I," *Isis* 64 (1973): 150–180, p. 163.

325. P. Forman, "The Environment and Practice of Atomic Physics in Weimar Germany: A Study in the History of Science," (Ph.D. diss., University of California, Berkeley, 1967).

326. S. Richter, "Forschungsförderung in Deutschland 1920–1936. Dargestellt am Beispiel der Notgemeinschaft der Deutschen Wissenschaft und ihrem Wirken für das Fach Physik," *Technikgeschichte in Einzeldarstellungen* 23 (1972): 7–69.

327. Survey of grant distribution by Helmholtz-Gesellschaft undated (ca. 1922), in Wien Papers, DM.

328. P. Forman, "The Financial Support and Political Alignment of Physicists in Weimar Germany," *Minerva* 12 (1974): 39–66, pp. 42ff.

329. A. L. Patterson, "Experiences in Crystallography—1924 to Date," in Ewald (n. 218), 612–622, p. 616.

330. For mechanical properties of solids, see Chapter 5.

331. M. Hartmann, *25 Jahre Kaiser-Wilhelm-Gesellschaft* (Berlin: Springer, 1936), 182–183.

332. Besides Laue, who explained his experiment, his theory, and Ewald's method of interpretation, W. Bragg spoke on the reflection of x rays from crystals; the English chemists Barlow and Pope, on crystal structure and chemical composition; the French x-ray specialist Maurice de Broglie, on the structure of molecules in crystals; and E. Grüneisen, an assistant at the PTR, developed the equation of state of solids and the thermal expansion from the theory of Debye and Born–von Kármán; finally, Voigt gave a talk on pyroelectricity. See Mehra (n. 177).

333. Ibid.

334. The congresses were not held between 1914 and 1919 because of the war. From 1921 on, the physicists came and held their own physicists' conferences, at first every two years. The conference program was printed in the concurrent volumes of the *Verh. Ges.* (Leipzig: Barth; subsequently, Braunschweig: Vieweg; and subsequently, Stuttgart: Teubner).

335. L. Pfaundler, ed., *Müller–Pouillets Lehrbuch der Physik und Metereologie,* 10th ed. 4 vols. (Braunschweig: Vieweg, 1906–1914); A. Eucken et al., eds., *Müller–Pouillets Lehrbuch der Physik,* 11th ed., 5 vols. (Braunschweig: Vieweg, 1925–1934).

336. H. Geiger and K. Scheel, eds., *Handbuch der Physik,* 24 vols. (Berlin: Springer, 1926–1929); 2nd ed., 27 vols. (1933). W. Wien and F. Harms, eds., *Handbuch der Experimentalphysik,* 26 vols. (Leipzig: Akademische Verlagsgesellschaft, 1926–1937); 2 suppl. vols., ed. M. Wien and G. Joos (1931–1935).

337. *Encyklopädie der mathematischen Wissenschaften,* 6 vols. (Leipzig: Teubner, 1898–1935): vol. 4, ed. F. Klein (1901–1914); vol. 5, ed. A. Sommerfeld (1903–1926).

338. "When volume 5 was sketched out, physics stood at the beginning of the development which in the last decades we have experienced with wonder" (Sommerfeld, [n. 337], 5: pt. 1, foreword).

339. Wien and Harms (n. 336), 1: foreword.

340. Ibid., 5: foreword, 5; 7: pt. 2, 2.

341. Ibid., 1: foreword.

342. Ibid., 7: pt. 1.

343. Ibid., 7: pt. 2, 323–422.

344. At the beginning of the 1930s, the crystallographers conceived of the "structure reports" (International Tables on the Determination of Crystal Structure) in which all known crystal structures were to be published. Carl Hermann, a former student of Born, then a collaborator of Ewald's in Stuttgart, was made editor. See also Ewald (n. 218), 698; Geiger and Scheel (n. 336), 23: pt. 2, 207.

345. P. P. Ewald to M. Born, 17 February 1962, in *Born Papers* (n. 161).

2 | *The Development of the Quantum Mechanical Electron Theory of Metals, 1926–1933*

LILLIAN HODDESON, GORDON BAYM, AND MICHAEL ECKERT

The electron theory of metals underwent dramatic development between the turn of the twentieth century, when Paul Drude and H. A. Lorentz first proposed a free-electron theory of metals,[1] and 1933, when Arnold Sommerfeld and Hans Bethe published a monumental review in the *Handbuch der Physik*.[2] From crude classical conceptions, the field had reached a point where physicists could undertake realistic calculations of some of the chief properties of particular solids.

This remarkable advance may be divided into four distinct phases: the classical (1900–1926), the semiclassical (1926–1927), the laying of the quantum-mechanical foundations by Felix Bloch (1928), and the creation of the modern theory (1928–1933).

The classical period was dominated by the model of Drude and Lorentz, in which a metal was supposed to contain an ideal classical gas of conduction electrons, governed by kinetic theory. Although the failures and contradictions of the model were strikingly apparent by the time of the First World War, few useful new concepts were added until 1926, when Wolfgang Pauli's application of Fermi–Dirac statistics to metals opened up the semiclassical period.[3] In the following year and a half, Sommerfeld and others in his circle, by further application of the new statistics within the framework of the classical Drude–Lorentz theory, were able to resolve most of that theory's outstanding difficulties. But it was not until Bloch's 1928 paper on the quantum mechanics of electrons in a crystal lattice[4] that the full machinery of quantum mechanics, developed in 1925 to 1926, was brought to bear on solids, thus spearheading the creation in the next five years of the modern quantum theory of metals.

This culminating period saw a growing confidence that the new quantum theory could explain, at least qualitatively and occasionally even quantitatively, all the varied properties of solids. A rush of interrelated successes, each following in the wake of previous ones, by many theoretical physicists, including Bethe, Bloch, Rudolf Peierls, Lev Landau, Léon Brillouin, Alan Wilson, Werner Heisenberg, and John Slater, built up the framework of concepts that structured the modern theory of solids. From a small number of problems worked on at relatively few institutions, solid-state theory grew into a substantial field of research based at centers in a number of countries.[5]

In the late 1920s, the solid state came temporarily to the fore because it provided a large and ready set of concrete problems treatable by quantum theory. Solid-state topics functioned initially as an extensive proving ground, and then as a target of opportunity, for the new mechanics. Unlike other subfields of physics that developed rapidly in the 1930s (e.g., nuclear physics and quantum electrodynamics), where theory depended strongly on new experimental findings, in these years theorists approached the solid state primarily to explain phenomena that had been observed for decades, if not centuries (Chapter 1). As soon as the tools of quantum mechanics and quantum statistics were used to create the fundamental building blocks of the modern theory of solids, remarkably many experimentally observed phenomena—for example, paramagnetism, diamagnetism, magnetoresistance, the Hall effect (the transverse voltage generated by a current flow in a transverse magnetic field), and the behavior of semiconductors—were explained in microscopic terms. As later chapters will show, by the end of the Second World War the subject would mature to the point where theory could advance only by joining hand-in-hand with experimental research. But for a few years, theorists ranged through old data without need of further experiment: an interesting and rare case of theory advancing, or in some cases failing to advance, all on its own.

Up to 1933, the story revolved around three research departments: the Institute for Theoretical Physics at the University of Munich, the similar institute at Leipzig University, and the Eidgenössische Technische Hochschule (ETH) in Zurich. The physics directors at these institutions, respectively *Geheimrat* Sommerfeld, Heisenberg, and Pauli, having recently been at the center of the development of quantum mechanics, were eager to explore and test their new capability on a variety of physical problems beyond atoms and molecules. These three institutes stand out among the other important centers for research in theoretical physics in those years—including the universities of Copenhagen, Göttingen, Cambridge, Leiden, and Utrecht—as places where quantum mechanics was applied to solids.

The year 1933 marked both an intellectual and an institutional break in the development, heralded by the appearance of reviews of the quantum electron theory of metals (most notably Bethe and Sommerfeld's, but also reviews by Peierls, Bloch, Brillouin, Lothar Nordheim, Slater, and others).[6] These reviews served as texts in new graduate programs that trained the first generation of specialists in quantum solid-state physics. At the same time, for political as well as intellectual reasons, many of the earlier workers in quantum solid-state theory left the field, new workers joined, and the center of research in solid-state theory shifted from Germany to the United States and England. At this juncture, as expressed optimistically and proudly in most of the reviews, all the observed phenomena—even superconductivity—appeared to be, if not solved, then at least solvable in terms of the existing quantum theory. Attention now shifted from qualitative and conceptually oriented problems toward more quantitative comparison of theory with experiment. As shown in Chapter 3, the principal focus turned to the development of approximate methods for describing real rather than ideal solids, with Eugene Wigner and Frederick Seitz's pivotal 1933 paper[7] on sodium serving as a prototype.

Our aim in this chapter is not to present a comprehensive history of the quantum theory of solids, but to delineate the period by tracing in detail several key developments that gave this era of solid-state history its character. Thus we do not cover optical phenomena, where important foundations were laid in this period by Jacov Frenkel and others. We do not touch on the independent tradition of the study of the mechanical properties of solids, a tradition whose roots lay in metallurgy. (Only decades later would that line of research make contact with the quantum theory of solids [Chapter 5].) Our treatment of this central phase of the development focuses on three principal examples: band theory, the most immediate triumph of the new theory of solids; magnetism, a modest success; and superconductivity, a persistent failure. This was a period above all of theoretical breakthrough, and we concentrate on theoretical rather than experimental contributions.

2.1   1900–1926: The Classical Theory of Metals

To provide a background for the events in the years 1926 to 1933, we begin with a brief review of the electron theory of metals in the period 1900 to 1926 (for more detail, see Section 1.3). The first solid-state theory capable of computing observed quantities from microscopic concepts was that of Drude, who pictured metals as an overall electrically neutral gas of positive and negative mobile particles. The mean velocities of these particles were, following Boltzmann, proportional to the square root of the absolute temperature T. The major triumph of Drude's theory was a derivation, by simple kinetic theory, of the empirical Wiedemann–Franz law for the ratio of the thermal and electrical conductivities of metals, resulting in remarkably good agreement with the experiments.

Within the next five years, Lorentz refined the Drude model by assuming three things:

1. The mobile negative carriers were a single species of electron, the same in all metals, an assumption first introduced in 1900 by J. J. Thomson.[8]
2. The electrons were described by a Maxwellian velocity distribution in equilibrium.
3. The positively charged particles remained fixed in the solid.

The presence of the atoms was neglected, except insofar as they caused a finite electron mean free path. Lorentz's result for the Wiedemann–Franz ratio, two-thirds of Drude's, although derived more rigorously, was "somewhat less satisfactory"; Lorentz nevertheless considered his results "a fair start . . . toward the understanding of the electric and thermal properties of metals." Furthermore, because he used a thermal distribution of electron velocities, his results for thermoelectric phenomena and for the emissivity and absorptivity of low-frequency "heat rays" (blackbody radiation) agreed with Kelvin's and Planck's thermodynamic theories.[9] These successes strongly confirmed the free-electron picture of metals.

Despite its successes, however, the Drude–Lorentz theory could not determine the electrical and thermal conductivities separately, since unlike their ratio, these individually involved the unknown electron density and mean free path. The most reasonable assumptions, that there was about one conduction electron per atom and that the mean free path was of the order of the interatomic spacing, could not be reconciled with either the temperature dependence or the magnitudes of the observed conductivities.

But the most glaring failure of the classical theory was its inability to explain why the electrons, while participating in thermal motion according to the Lorentz theory, did not appear to contribute to the measured specific heats.[10] Rather, the value observed at high temperatures for both metals and insulators was the contribution of the bound oscillating atoms alone. One could not simply assume that the number of electrons was much smaller than the number of atoms without contradicting optical results; through an undeciphered mechanism, either arising from interactions or perhaps analogous to Planck's quantization of radiation energy, the electrons appeared to be violating classical equipartition of energy.[11] Resolution of these inconsistencies awaited the application of the Pauli exclusion principle and quantum mechanics itself.

The thermal properties of the electrons were further clouded by the observation that specific heats decreased with decreasing temperature, and by Albert Einstein's puzzling explanation of this decrease entirely in terms of quantized thermal vibrations of the atoms, with no contribution from the electrons.[12] Einstein, searching for further applications of his and Planck's quantum hypothesis, assumed in his calculation a single oscillation frequency for the atoms, thus ignoring the effects of the lattice on the vibrational spectrum. By 1912 the idea of the periodic lattice of bound atoms in a solid had come to the fore when x-ray diffraction demonstrated its existence and showed that the lattice spacing was of the order of 10^{-8} cm.[13] A more rigorous explanation of the specific heat in terms of vibrations of a lattice of atoms (without consideration of electrons), which brought the theory into agreement with experiment, was given independently by Debye and by Born and von Kármán,[14] who found that the low-temperature specific heat was proportional to T^3.

Solids were not the only, or even the main, problem. The classical theory of gases was known to fail at low temperatures; in particular, the classical ideal gas specific heat, $3kT/2$ per particle, where k is Boltzmann's constant, did not vanish as the temperature vanished, as required by the Nernst theorem. Beginning in the 1910s, a number of "gas degeneracy" theories, concerned with understanding deviations appearing at low temperatures from the classical behavior, attempted to apply quantum ideas to the ideal gas. These theories, which were important stages in the development of quantum statistics,[15] were tried out on the free-electron specific heat problem. The existence of an electron gas in metals was then still open to question, and several alternatives (e.g., dipole theories, electron chain theories, and electron lattice theories) were also considered.[16] None of these could resolve the specific heat dilemma without creating new contradictions in areas where the classical model gave good agreement with experiment. For example, some of these theories led to the Wiedemann–Franz law, but could not explain the thermal and electrical conductivities separately without additional unreasonable assumptions, such as that the mean free path is much longer at room temperature than the interatomic distance (and even longer at low temperatures); even then, agreement with experiment was not obtained.[17]

While Einstein, Debye, and Born and von Kármán were undermining the Drude–Lorentz theory by explaining the specific heats in terms of the ions alone, Niels Bohr was uncovering other fundamental failures. His 1911 doctoral dissertation[18] aimed to develop an electron theory of metals allowing more general assumptions than Lorentz had made. Retaining the free-electron model, he replaced Lorentz's "hard elastic sphere" scattering law with the assumption that the atoms attract or repel electrons with a force varying as the inverse nth power of distance. Within this more general framework, he was able to improve on Lorentz's calculations of transport phenomena and confirm the derivation from electron theory of thermoelectric properties and the law of blackbody emission of heat. Although many observed properties, such as Hall coefficients of the wrong sign, could not be explained, the most crucial failing was that since, as he showed, free electrons can exhibit neither diamagnetism nor paramagnetism (a result independently derived by Hendrika Johanna van Leeuwen[19] in 1919), the free-electron theory could not explain the magnetic properties of metals.

Questions on the magnetic properties included the following: Why is the paramagnetism of ordinary metals, such as the alkalis, weak and finite as the temperature goes to zero? Why is there a nonzero diamagnetism? If one accepted Pierre Weiss's experimentally successful "molecular field" theory of ferromagnetism, what determined the values of the theory's parameters? And how would one compute the magnitude of the observed change of resistance of metals in strong magnetic fields, the magnetoresistance? Mysteries in the theory of metals at the time of the discovery of quantum mechanics also included the most fundamental issues, such as why the properties of a metal differ from those of an insulator.[20] There was not even a clear explanation of why metals, or anything else, were solids in the first place, the atoms holding together somehow at fixed distances rather than collapsing on one another.

Among all the phenomena, a peculiarly curious and baffling one was superconductivity, first observed in 1911 by Gilles Holst, working under Heike Kamerlingh Onnes: Why should all traces of electrical resistance suddenly disappear in certain metals and many alloys when the temperature falls below a critical value close to absolute zero?[21] Furthermore, it proved impossible to reconcile the considerable empirical information gathered about semiconducting substances, such as metal oxides and selenium, with the free-electron gas or any other general theoretical model.[22]

Since no coherent basis was available for establishing the correct microscopic theory, pre-quantum-mechanical attempts to solve the problems of metals represented a groping for reasonable conceptions. Out of these emerged many valid notions that would eventually become part of the modern theory of solids—for example, that microscopic charged particles in a metal behave in many ways like free particles and are responsible for electrical transport, and that gas degeneracy is, somehow, the solution to the specific heat dilemma. But since the framework was wrong, fitting the theory to the data, particularly as more experiments were made, required adding an increasing number of ad hoc assumptions, most of which proved untenable. Ultimately the pastiche of valid as well as incorrect models and assumptions in the electron theory of metals represented a state of confusion and complexity that was a microcosm of the state of physics as a whole before the introduction of quantum mechanics.

2.2 1926–1927: Quantum Statistics and the Semiclassical Theory

Fermi–Dirac Statistics

The break came in the immediate wake of the development of quantum mechanics. Underlying that development was the exclusion principle, proposed by Pauli in early 1925 to explain the closure of electron shells in atoms.[23] The first step toward the creation of a semiclassical theory of metals was the development, independently by Enrico Fermi and Paul A. M. Dirac in 1926,[24] of a quantum statistics applicable to a gas of particles that obeyed the exclusion principle.

Fermi, then in Florence, was disturbed by the failure of the classical ideal gas theory. In his March 1926 paper he wrote, "It is necessary to assume that the motions of the molecules of an ideal gas are quantized, and that this quantization manifests itself at low temperatures through certain degeneracy phenomena." He continued, "Since this Pauli rule has shown itself to be extraordinarily fruitful in the interpretation of spectroscopic phenomena, we want to examine whether it is not also useful for the problem of the quantization of an ideal gas." Fermi was considering an ordinary gas of atoms or molecules, rather than an electron gas in a metal.

The explicit model he studied was that of an ideal gas in an external three-dimensional harmonic oscillator potential. The individual energy levels would, according to Einstein, be quantized with values $h\nu s$, where h is Planck's constant, ν is the frequency of the oscillator, and s is a nonnegative integer. The Pauli principle for this case becomes the condition that at most one molecule can be in a given quantum state. After computing the number Q_s of quantum states for each allowed energy level, he proceeded to derive, by maximizing the entropy à la Boltzmann, the mean number of particles N_s per energy level of energy E, in a system at temperature T, and found the Fermi–Dirac result

$$N_s = Q_s[\alpha^{-1}e^{hvs/kT} + 1]^{-1}$$

where α is a constant. He then went on to calculate by a semiclassical approach, which we now call the Fermi–Thomas method, the equation of state of the gas. The result was the now familiar equation of state of a free Fermi–Dirac gas at each point in space. He noted particularly the presence of a zero-point pressure and energy, and that at low temperature the specific heat falls linearly with T, thus obeying the Nernst theorem. The basic argument for the reduction of the specific heat of the gas below its classical value was that only those particles occupying a thin energy shell within kT of a critical energy—what is now called the Fermi energy—can be thermally excited to unoccupied states of higher energy.

Heisenberg learned of the new statistics directly from Fermi in the spring of 1926, shortly before the paper appeared in print,[25] when he stopped off to see Fermi on his return from a tour through Italy. He later recalled Fermi's explanation of the relation between the Bose–Einstein statistics[26] and his new statistics as a "kind of complement . . . something like plus and minus." Heisenberg developed some of his ideas about the two statistics in June 1926 in his paper "The Many-Body Problem and Resonance,"[27] the first paper to deal with the quantum mechanics of more than one particle. Applying quantum mechanics to systems composed of many identical particles, he found two groups of stationary-state solutions, one symmetric and the other antisymmetric, that neither combine with one another nor in any way transform into the other; only the antisymmetric group of solutions—the one observed empirically—obeyed the Pauli principle.

Dirac's derivation of the new statistics, in his classic 1926 paper "On the Theory of Quantum Mechanics,"[28] independently considered the connection between the Pauli principle and the antisymmetry of the many-particle wave function; he noted that noninteracting electrons can be described by a determinantal wave function. After introducing the quantization condition through boundary conditions on the single-particle wave functions, he derived the statistics and equations of state through maximizing the entropy. He also pointed out that had one begun from a completely symmetric wave function, Bose–Einstein statistics would result.

Dirac did not refer to Fermi's previous work, and in October 1926 Fermi wrote to Dirac somewhat crisply:

> In your interesting paper . . . you have put forward a theory of the Ideal Gas based on Pauli's exclusion Principle. Now a theory of the ideal gas that is practically identical to yours was published by me at the beginning of 1926. . . . Since I suppose that you have not seen my paper, I beg to attract your attention on it.[29]

Pauli recalled long after that

> Dirac was in Copenhagen in the autumn of 1926 and I wrote to him there, whether he knows how a spin of the atoms (or electrons) would modify the results. I also

Paul Adrien Maurice Dirac (1902–1984), Wolfgang Pauli (1900–1958), and Sir Rudolf Ernst Peierls (b. 1907). (AIP Niels Bohr Library)

mentioned Fermi's paper. He answered me, that he never considered this question and that Fermi's paper was entirely new to him. Immediately after that I started to work on this question myself, and I found very quickly all answers.[30]

As Pauli set to work to apply the new statistics to metals, and thus begin the semiclassical period of the theory, Ralph H. Fowler in Cambridge began to apply the statistics to the equation of state of matter in white-dwarf stars. His paper, published in December 1926,[31] proved to be the root of modern theories of dense matter in astrophysics.

Let us now turn to Pauli's pivotal contribution to the theory of metals.

Pauli's Application of Fermi–Dirac Statistics to Electrons in Metals

With the work of Fermi and Dirac, there were now two different quantum statistics: Bose–Einstein, for which the mean number of particles in a single particle state of energy E was of the form

$$n_{BE} = [\alpha^{-1}e^{E/kT} - 1]^{-1}$$

and which applied to photons for $\alpha = 1$; and Fermi–Dirac, for which the mean number was given by

$$n_{FD} = [\alpha^{-1}e^{E/kT} + 1]^{-1}$$

In the limit of very high temperatures, both reduced to the classical Maxwell–Boltzmann distribution,

$$n_{MB} = \alpha e^{-E/kT}$$

A central question was, Which of these statistics applies to matter?[32]

Dirac discussed this question at several points in his 1926 paper, stating that "the symmetrical eigenfunctions [corresponding to Bose–Einstein] alone or the antisymmetrical eigenfunctions [corresponding to Fermi–Dirac] alone give a complete solution of the problem. The theory at present is incapable of deciding which solution is the correct one." But he pointed out that since the symmetrical solution "allows any number of electrons to be in the same orbit . . . this solution cannot be the correct one for the problem of electrons in an atom." He later speculated that "the solution with antisymmetrical eigenfunctions . . . is probably the correct one for gas molecules, since it is known to be the correct one for electrons in an atom, and one would expect molecules to resemble electrons more closely than light quanta."[33]

Pauli discussed the issue of the alternative quantum statistics with Heisenberg in a detailed, almost daily, technical correspondence.[34] On 19 October 1926, Pauli wrote: "On the question of degeneracy of gases [i.e., when the probability of occupation of the low energy states is near unity], I am now thinking considerably more mildly about the Fermi–Dirac statistics, and it seems to me now that there are many arguments that speak for it." He had come to believe that "there does after all exist a difference between crystal lattices and radiation," and this difference might be attributed to the difference between Bose–Einstein and Fermi–Dirac statistics; since solids have a zero-point energy, unlike radiation, and since "such an energy can be supported by the Einstein–Bose statistics only artificially . . . this speaks right from the beginning against this theory and for the Fermi–Dirac."[35] By November, he was convinced that Fermi–Dirac and not Bose–Einstein statistics applied to the degenerate electron gas. He wrote to Erwin Schrödinger (who had earlier struggled with gas degeneracy and was then developing wave mechanics): "Recently, I have also been occupied with gas degeneracy. With a heavy heart I have become converted to the idea that Fermi . . . not Einstein–Bose, is the correct statistics. I want to write a short note about an application of it to paramagnetism."[36]

Pauli, who had earlier worked on the diamagnetism of atomic electrons,[37] saw in the weak paramagnetism of metals a clear opportunity to check whether Fermi–Dirac statistics did indeed apply. He simply needed to derive a physical consequence that could be experimentally verified. Heisenberg had recently employed the exclusion principle in showing that there is no paramagnetism in the ground state of helium,[38] and Pauli now considered making a similar argument for metals, approximating them as a degenerate electron gas obeying Fermi–Dirac statistics.[39]

Pauli's argument soon appeared in his classic article on paramagnetism, submitted to the *Zeitschrift für Physik* on 16 December 1926.[40] Most of the paper was devoted to a close examination of the thermodynamical basis and fluctuation properties of the Fermi statistics. Only in the last section did Pauli calculate, in a now familiar way, the paramagnetic susceptibility (i.e., the extent of magnetization in an applied field) of a gas of atoms obeying Fermi statistics and having a quantized spin and a magnetic moment. His crucial result was that at low temperatures—the appropriate limit, he observed, for electrons in a metal—the spin susceptibility approaches a constant value; by contrast, the Langevin theory, applicable when there is no exclusion of states, predicted a Curie law susceptibility inversely proportional to T, and about 10^2 times larger than Pauli's result at room temperature. Pauli's argument for the reduction of the susceptibility was essentially the same as Fermi's for the reduction of the specific heat: only those electrons in metals occupying a thin energy shell within kT of the Fermi surface are capable of being aligned by the magnetic field, an effect dramatically suppressing the magnetic susceptibility in metals. Not only did the Pauli result give the correct magnitude of the observed magnetic susceptibilities of metals, but it also explained its temperature independence and provided direct confirmation that the electrons in a metal form a degenerate gas.

Thus, analogously to the way in which Heisenberg showed that the Pauli principle implies that there is no magnetic moment in the normal state of a helium atom, Pauli demonstrated that Fermi–Dirac statistics, which he regarded as the generalization of his exclusion principle, implies a suppression of the magnetic susceptibility in metals. Pauli recalled 30 years later in a letter: "I was so glad that I eventually had an answer to the question, 'If it is true that the electron has a spin, why then is there not a strong paramagnetism in metals according to the Curie law?'"[41]

By carrying out this physical example, Pauli opened the development of modern solid-state theory. He would himself often criticize the field as messy and applied: "I don't like this solid state physics . . . though I initiated it."[42] Yet in this field, the Pauli principle plays an all-important role, even answering the question of how solid matter can exist in the first place rather than collapsing. As Ehrenfest wrote to his "Dear Awesome Pauli" in January 1927, "For a long time I have had the feeling that it is your prohibiting condition that above all prevents the atoms and through them the crystal from falling together."[43]

Sommerfeld's Institute in Munich

The next crucial steps were taken by Sommerfeld in Munich, for the task of applying Fermi–Dirac statistics to the specific heat dilemma and reworking the old Drude–Lorentz theory was not to Pauli's taste. Sommerfeld seems to have had little previous attachment to the electron theory of metals, and by 1927 had not yet written anything new on the electron theory of metals or quantum statistics. He did follow developments in the field, however; for example, he lectured in 1908, 1910, and 1912 on the electron theory of metals and related questions of kinetic theory of gases. Paul Debye, his first *Assistent* (the equivalent of the modern postdoctoral research assistant), wrote his *Habilitation* thesis on electrons in metals in 1910, and another Sommerfeld *Assistent*, Karl Herzfeld, lectured in Munich in 1920 to 1922 on metals, the theory of gases, and magnetism.[44]

Sommerfeld could have taken his institute in many directions. For example, he was in a position to undertake metallurgy, but he never seems to have had any interest in developing this area of his institute, and that strand of solid-state study in the 1920s and 1930s was entirely divorced from the electron theory of metals. It was the electron theory that caught Sommerfeld's attention. The problem posed by

Arnold Sommerfeld (1869–1951) and Wolfgang Pauli at the metals conference in Geneva. October 1934. (Pauli Collection, CERN Archive, Geneva)

extending Pauli's work appealed to Sommerfeld's expertise in applying sophisticated mathematical methods to a wide variety of physical problems. Unlike Heisenberg and Pauli, he appears as a rule to have been more interested in the mathematical solution of a problem than in its underlying physics.[45]

At his institute in Munich, Sommerfeld was well suited for the role he would now play in solid-state theory. In the early years of his academic career, at the turn of the century, he was strongly influenced by the great mathematician and science organizer Felix Klein. Klein, aspiring to bridge the gap between science and technology, was attracted by Sommerfeld's mathematical and pedagogical talents and helped bring him to a chair at the Aachen Technische Hochschule. In showing the engineers whom he taught the power of mathematics in solving technological problems, Sommerfeld developed skill in applied mathematics and became accustomed to working on a wide variety of problems. Klein also appointed Sommerfeld editor of the widely read physics part (volume 5) of the *Encyklopädie der mathematischen Wissenschaften,* an enterprise that occupied him between 1898 and 1926, and made him a well-known figure in the scientific community as well as a correspondent of many eminent physicists.[46]

On assuming the chair of theoretical physics at the University of Munich in 1906,[47] Sommerfeld took command of a large new institute where general lecture courses, advanced topics courses, a research seminar, and a colloquium provided a flow of students from whom to select those suited for research. Sommerfeld taught three courses: a six-semester undergraduate course to approximately 100 students (three times a week over a three-year period), a weekly seminar on current research topics to a group of about 20, and a special-topics course to about 20 advanced graduate students (twice a week in alternate semesters).[48] Sommerfeld's *Assistent* also taught a course twice a week to approximately 20 students; by the mid-1920s, the subject in alternate years was quantum mechanics, which was not included in the main Sommerfeld sequence.[49] The colloquium enabled close contact with researchers at other institutes and brought current problems, including those of solid-state physics, within the reach of Sommerfeld's growing school.[50]

Sommerfeld's undergraduate course, out of which eventually grew the famous Sommerfeld textbooks, included classical mechanics, the mechanics of continuous bodies, thermodynamics and statistical mechanics, electrodynamics, optics, and mathematical physics.[51] The seminar pursued outstanding recent developments in all areas of theoretical physics; from 1926 to 1928, it was devoted mainly to topics in quantum mechanics. For example, Peierls reported on Dirac and Pascual Jordan's recent work on transformation theory; Bethe reported on perturbation theory, as described in the galley proofs of Schrödinger's original papers on quantum mechanics, which Sommerfeld had obtained for use by those in his seminar. The seminar participants included undergraduates, graduates, professors, and visitors.[52] The advanced-topics course focused in 1927–1928 on the quantum electron theory of metals. Bethe, a research student, and Peierls, a third-year university student, were among those whose interest in solid-state physics was kindled in this course.[53] As Bethe recalled, Sommerfeld "told us what he had discovered in the last week. It was very fascinating."[54]

Particularly in his seminar, Sommerfeld transmitted his personal style and taste in research by working closely with its members, often drawing them out by his "principal technique . . . to appear dumber than any of us."[55] On occasion, Sommerfeld would, as Peierls recalls, "grab a student" to talk with him about a problem that happened to interest him,[56] which spurred them on "'to explain to the Herr Geheimrat.' He certainly was not as dumb as he pretended to be, but he had no inhibitions about appearing dumb. Sometimes it seemed that he went out of his way to misunderstand and thus force you to become clearer."[57] Max Born later compared Sommerfeld's method of personal instruction with

> the tutoring at the old British Universities, but less methodical and formal. . . .
> Often before or after the Colloquium he was seen at the Hofgarten-Cafe, discussing problems with some collaborators and covering the marble tables with formulae. It is reported that one day an integral resisted all attempts at reduction and was left unfinished on the table; the next day Sommerfeld, returning to the same table, found the solution written under the problem, obviously meanwhile worked out by another mathematician taking his cup of coffee with greater leisure. . . . A great part was played by invitations to join a ski-ing party on the "Sudelfeld" two hours by rail from Munich. There he and his mechanic Selmayr . . . were joint owners of a ski-hut. In the evenings, when the simple meal was cooked, the dishes washed, the weather and snow properly discussed, the talk invariably turned to mathematical physics, and this was the occasion for the receptive students to learn the master's inner thoughts.[58]

The physical arrangement of the institute encouraged communication among the workers. In three large connecting rooms of comparable size were Sommerfeld's

A canonical ensemble: Robert Wichard Pohl (1884–1976) and Arnold Sommerfeld sitting on a cannon during a meeting of German physicists in Braunfels, 1932. (Courtesy DM. Courtesy of AIP Niels Bohr Library)

office and library, the office and library of Sommerfeld's *Assistent,* and a workroom for approximately 20 graduate students and visitors.[59] A fourth tiny room contained a spiral staircase leading to the basement where von Laue, Friedrich, and Knipping performed the historic x-ray diffraction experiments that provided the first look inside the crystal lattice.[60] Even before the First World War, a generation of physicists who would carry out pioneering solid-state studies emerged from Sommerfeld's institute, among them von Laue, Peter Debye, Peter Paul Ewald, and Léon Brillouin.

By the 1920s, Sommerfeld's institute was one of the major international centers for theoretical physics, attracting students and traveling fellows—including Heisenberg, Pauli, Bethe, Peierls, Gregor Wentzel, Walter Heitler, and Fritz London—members of the new generation that developed quantum mechanics and its applications. These students and fellows would, in turn, build up new centers with an active exchange of ideas as well as scientists, helping to spread the Sommerfeld teaching and research tradition.

By this time, Sommerfeld was a central figure in the scientific community; his opinion was often decisive in such matters of science policy and organization as publishing, funding, and the choice of candidates to occupy physics chairs, and he used his influence to promote his school. For example, Sommerfeld sat on a committee of the Notgemeinschaft der Deutschen Wissenschaft (Emergency Fund For German Science, a principal German organization supporting research [Chapter 1]), which dealt with all areas of theoretical physics. The Notgemeinschaft gave decisive financial help to certain professors who were building up groups of young physicists, including Heisenberg in Leipzig, Robert Pohl in Göttingen (who will be discussed in Chapter 4), and Sommerfeld himself. Other funds came to Sommerfeld's institute from the Solvay Institute, the Ausschütz-Kämpfe Foundation, the

Two photographs of Arnold Sommerfeld as proof that the circle is a degenerate form of the ellipse (a reference to Sommerfeld's atomic theory), shown at the exhibition for his eightieth birthday. (Courtesy DM)

Kaiser-Wilhelm Gesellschaft, and the Helmholtzgesellschaft, not to mention the basic government support that a leading university institute could demand.[61]

Sommerfeld's Semiclassical Electron Theory of Metals

Sommerfeld's involvement with the quantum electron theory of metals began in the spring of 1927, at a meeting organized by the physics faculty of the University of Hamburg.[62] This faculty included Pauli and Wilhelm Lenz, two former Sommerfeld students, and Pauli had invited Sommerfeld to speak "on the present state of atomic physics." On this occasion, Pauli showed Sommerfeld the proofs of his paper on paramagnetism, and as Pauli recalls,

> The next day he said to me that he was very much impressed by it and that one should make further application to other parts of metal theory like the Wiedemann-Franz law, thermoelectric effects, etc. As I was not eager to do that he made then this further application himself.[63]

Sommerfeld realized that some of the worse failings of the Drude–Lorentz theory could be overcome by using the new statistics, and when he returned to Munich he set seriously to work. He soon began obtaining interesting results. For example, using Fermi's result, he showed that the specific heat of the electrons in a metal is reduced at room temperature to an immeasurably small value, about 1% of its classical value, thus resolving the specific heat dilemma of the Drude–Lorentz theory.

By Lorentz's procedure of using the Boltzmann kinetic equation with a relaxation time, but replacing the Maxwell–Boltzmann by the Fermi–Dirac statistical distribution function, Sommerfeld was able to derive an expression for the Wiedemann–Franz ratio that agreed with experiment better than either Lorentz's or Drude's expressions. Other phenomena studied in the same way included thermionic emission and thermoelectric, galvanomagnetic, and thermomagnetic effects. Although the formalism was very simple, the resulting integrations over the Fermi–Dirac distribution required Sommerfeld to develop his low-temperature expansion technique, a type of problem that particularly appealed to him.

Sommerfeld liked the electron theory of metals well enough to make it a theme at his institute. In the summer of 1927, he lectured in his special-topics course on the "structure of matter" to a small circle of advanced students, showing basic consequences of the application of Fermi–Dirac statistics to the electron gas.[64] He demonstrated there for the first time the solution to the specific heat dilemma, arguing mathematically rather than physically. Having discussed the subject—"he could never write anything or get it in writing before he talked about it"[65]—he next wrote up an overview for *Die Naturwissenschaften*[66] and presented the theory in condensed form to the International Volta Congress, held in September 1927 at Como, Italy;[67] here Lorentz, Fermi, E. Hall, and Frenkel contributed to the discussion of Sommerfeld's talk.[68] As Sommerfeld concluded in the *Naturwissenschaften* paper: "The overall impression that this work provides is, without any doubt, that through the new statistics the contradictions of the older theory are lifted and the observational facts are, in part quantitatively, in part qualitatively, reproduced."[69] Although he somewhat overstated the case, theory and experiment had for the first time come together for a wide variety of phenomena in metals.

The theory now entered a phase of refinement. Electrons in metals were the main concern in Sommerfeld's theoretical research seminar in the winter 1927/1928 semester, whose participants included Peierls, Bethe, Carl Eckart, William Houston, and Linus Pauling.[70] Peierls recollects that "Sommerfeld . . . was going around with a little book by Karl Baedeker, which was then the reference book about the properties of metals and definitions of the various coefficients, and seeing how far things could be made to agree with the theory and how far they couldn't."[71]

In this period, Sommerfeld began to write his first long publication on the subject, a two-part treatment that appeared in the *Zeitschrift für Physik*.[72] The response to this work from outside was by and large favorable. Bohr wrote to the British metallurgist William Hume-Rothery in February 1928, answering an inquiry regarding his 1911 thesis:

> Nowadays the old theories based on the classical mechanics can hardly make claim of actual physical interest. Indeed they are left quite behind by the recent fundamental work of Sommerfeld which has just been published. . . . Although not yet complete, Sommerfeld's work surely means a decisive step as regards the adequate quantum theoretical treatment of the metallic problem.[73]

Einstein wrote, "I have read your theory with great interest and gotten the impression that this is indeed the correct recovery, in principle, of what is true in the original electron theory of metals."[74] However, not all were convinced that the theory would survive; Hume-Rothery wrote to Lawrence Bragg: "I am now hard at work on a book on the Electrical and Thermal Properties of Metals . . . wrestling with the new Sommerfeld theory, which is very difficult. I think it is wrong in spite of its success in some quarters."[75] Bethe later reflected, "I believed the whole theory only when the Bloch paper appeared."[76]

Sommerfeld encouraged those at his school to work on his new theory. Houston, who came there in the spring of 1928 as a Guggenheim Foundation traveling fellow from the California Institute of Technology, intending to work on "recent developments in the quantum theory of atomic structure,"[77] was directed into the program: "When Sommerfeld somewhat discouraged me from studying electron spin, he gave me the proof of his first long paper on the application of quantum statistics to free electrons."[78] Houston was asked to examine the mean free path of electrons in metals, the weak point of the theory, and calculate the temperature dependence of the electrical conductivity.[79] Eckart, also from Caltech on a Guggenheim fellowship, was diverted from "researches concerning the new quantum theory"[80] to the problem of electron emission. Allan C. G. Mitchell and W. W. Sleator, also Americans, were similarly encouraged to examine topics in the electron theory of metals. Eckart recalls that "he got all of us involved in . . . the big project for the year . . . to rework the Lorentz theory of electrons using the Fermi statistics."[81]

By reworking the classical electron theory of metals so as to resolve many earlier failures and give better agreement with experiments, and by casting his theory into an easily understood form, Sommerfeld brought the new theory within reach of a wide international community. Nordheim, a former Sommerfeld student, became interested in the theory by reading Sommerfeld's paper during his year with R. H. Fowler in Cambridge.[82] In March 1928, Frenkel continued by letter the discussion with Sommerfeld begun at the Volta Congress, and during the next few months so did Fermi, Georg Joos, and Walter Schottky.[83] The 1928 *Zeitschrift für Physik* alone contains 14 articles relating to the electron theory. By writing review articles, for the 1931 *Reviews of Modern Physics* with Nathaniel H. Frank of MIT,[84] formerly a traveling fellow at Munich, and with Bethe in the 1933 *Handbuch der Physik,*[85] Sommerfeld drew further attention to the theory of metals.

Sommerfeld also helped to propagate the theory in his visits abroad. For example, on his trip around the world in 1928 and 1929, he included a short account in lectures at Tokyo University in December 1928.[86] He also spoke on his theory in travels through the United States in 1931. Shortly after Walter Brattain attended Sommerfeld's course on the electron theory of metals at the 1931 Michigan Summer Symposium on Theoretical Physics, which Brattain had hoped would improve his understanding of thermionic emission, Brattain gave a special series of lectures on the electron theory of metals at Bell Laboratories.[87] John Bardeen also heard Sommerfeld lecture at Wisconsin on the electron theory of metals.[88] A decade and a half later, Bardeen and Brattain would apply this theory in the discovery of the transistor.

Despite its numerous successes, serious failings were apparent in the Sommerfeld theory, even to Sommerfeld's own circle in Munich. Predictions often disagreed with experimental findings—for example, on the size and functional dependence of the resistivity, the magnetoresistance, and various thermoelectric and galvanomagnetic effects, such as the Hall effect. The theory had no explanation of the electron mean free path; Sommerfeld estimated it to be about 100 atomic separations in silver at room temperature, but did not inquire how the electrons managed to avoid the ions so successfully. More generally, Sommerfeld does not seem to have asked why the ions did not influence the electrons between collisions, or why the effects arising from motion of the ions could be neglected. As Bethe recalls, "he didn't even care terribly much why the electrons were free, which I thought was a very important thing to know."[89] Neglect of the ions also disturbed other physicists, including Heisenberg and Frenkel;[90] Schottky wrote to Sommerfeld that "to assume a field free condition inside a metal appears to me to be too specialized for the problem."[91] Sommerfeld was aware of these problems, but, as Peierls reflected recently, he was optimistic that in one way or another they would be resolved.[92]

To solve the mystery of why Sommerfeld's free-electron theory worked so well would require developing a fully quantum-mechanical treatment of electrons in metals. Sommerfeld had used just enough quantum mechanics, via the Pauli principle, to enumerate the states and their energies correctly, but he did not avail himself of the full theoretical machinery (such as wave functions and transition probabilities) already developed by Heisenberg, Schrödinger, Dirac, and others. Indeed, the Sommerfeld theory can be construed, as F. Bopp remarks, as a "successful revival of Paul Drude's theory of conductivity."[93] For while Sommerfeld had one foot in the new modern quantum school, many of whose leading practitioners he had trained at Munich, his other foot was caught in the classical theory, in which he had invested some three decades of his career. Rather than as the beginnings of a *new* theory, he saw the Pauli paper as raising the question "whether the well-known difficulties in the *old* theory of metallic conduction could be lifted by the new statistics" (emphasis added).[94]

2.3   1928: Establishing the Quantum-Mechanical Foundation

Heisenberg's Institute in Leipzig

The transition to a quantum-mechanical theory of metals took place in Leipzig. In 1926 the Leipzig Physical Institute was still a stronghold of old-fashioned classical physics, influenced by the views and habits of the *Geheimräte* Otto Wiener and Theodor Des Coudres, who in the quarter-century of their regime allowed scarcely any infiltration of the new quantum ideas into their academic life.[95] Within one year, the institute changed radically because of the deaths of both Des Coudres and Wiener and the departure of Georg Jaffé, who had held an associate professorship for mathematical physics. Sommerfeld played an essential role in filling the newly vacant posts, first with Wentzel in 1926, and then in 1927 with Debye, who had been in Zurich, and Heisenberg.[96] The discussion of these new positions was almost a family matter for Sommerfeld's circle. Pauli wrote to Wentzel:

> I heard that for the Leipzig chair [Jaffé's] I stand after Heisenberg on the list and that you are third. Heisenberg has now declined. Should fortune overtake me, it would not be altogether out of the question that I would be offered so much in Hamburg that I could also decline. So watch out![97]

Wentzel did take the post later that year. The ETH tried in vain to keep Debye, who became the choice for Wiener's experimental physics chair.[98] The next month, Debye wrote to Sommerfeld: "If the matter goes well in Leipzig, then my favorite choice [for Des Coudres' theoretical chair] would be Heisenberg."[99] As Debye had hoped, "the matter went well in Leipzig" both for himself and for Heisenberg. By the winter 1927/1928 semester, the old Leipzig guard had been replaced by the new.[100]

However, the shift to the new physics in Leipzig began slowly. Wentzel complained to his old teacher, "The sloppy routine continues. . . . Not much can be done to improve the older student generation; my experience with the state examination for physics is indescribable . . . only one doctoral candidate has applied, and he appears to be hopelessly untalented."[101] Heisenberg recalled having but one student in his first Leipzig seminar, in the winter of 1927/1928.[102] However, close to 40 came to his first regular lecture, "certainly a good beginning."[103]

The pedagogical style that the Sommerfeld team ushered into Leipzig was modeled after that at Munich and Göttingen, where Heisenberg had served as Born's assistant, with the stimulating research environment built around courses and seminars. The theory lecture course, four hours per week, dealt with classical mechanics, thermodynamics, electrodynamics, and optics in a four-semester cycle. The special three-hour weekly lecture treated subjects such as modern problems of atomic physics and quantum mechanics. Courses in 1927 to 1928 also included Debye's on experimental physics and Wentzel's on the theory of electricity and heat radiation. Advanced students were offered a seminar on the structure of materials, organized by Heisenberg and Wentzel (later by Hund, after Wentzel's departure in 1929).[104] The seminars were informal, on a "high level," usually small (consisting of approximately six students, assistants, and professors), and focused on research by the participants or on important articles in the current journals. "Then we sat together and we started to play ping-pong—it was all very informal."[105]

Although experimentalists did not usually attend the seminars and the contact between theorists and experimentalists was tenuous, their relations, aided by the weekly colloquium arranged by Debye, were substantially greater than at other German institutions in this period, and constituted a special appeal of physics in Leipzig. Quantum-mechanical formalism was balanced by Debye's insistence on sticking close to experimental facts. Recommending one of his students to Debye, Bridgman wrote, "It seemed to us that Leipzig is perhaps the best place for him. . . . Heisenberg will certainly distill enough of the general atmosphere of the theory, and you ought to exert a steadying influence, so that he will not get too far away from a firm foundation of solid fact."[106]

By the summer semester of 1928, the physics program at Leipzig was proving attractive to students: 150 in Heisenberg's lectures on mechanics, and close to 80 in his special colloquium.[107] Wentzel reported to Sommerfeld that "our seminar now has twelve participants, certainly owing in part to the delightful newcomers from Munich"[108]—likely Peierls and Houston. Peierls came as a student in the spring of 1928 (at the time Sommerfeld was preparing to set out on his world tour),[109] and Houston followed in the fall after his stay with Sommerfeld. Bloch had been there since the previous fall, having studied in Zurich with Schrödinger, and then, primarily on Debye's advice, moving to Leipzig, where he was Heisenberg's first student. Edward Teller and I. I. Rabi also came that fall. In addition, Leipzig would attract many traveling fellows in theoretical physics, such as John Slater, who visited from Harvard on a Guggenheim fellowship for half a year in 1929/1930.[110]

Peierls reflected recently on the happy environment for theoretical solid-state

(front) Rudolf Peierls, Werner Heisenberg; *(rear)* G. Gentile, George Placzek, Giancarlio Wick, Felix Bloch, Victor Weisskopf, F. Sauter in Leipzig, 1931. (AIP Niels Bohr Library, Rudolf Peierls Collection)

research arising out of Heisenberg's realization that "there was an open problem . . . electrons in metals were one proving ground for quantum mechanics." Peierls also recalled the differences between working with Heisenberg and with Sommerfeld. Although both were approachable, Sommerfeld was such a busy man that "you didn't call on him quite as easily as on Heisenberg." On the other hand, Heisenberg was about half Sommerfeld's age, very modest, and "his ambition to excel in table tennis was more obvious than his ambition to be a great physicist."[111]

Peierls and Nevill Mott reflected:

> One could not have blamed [Heisenberg] if he had started to take himself seriously and had become a little pompous, after having taken at least two decisive steps that changed the face of physics, and after reaching at so young an age the status of professor, which made many older and lesser men feel important, but he remained as he had been—informal and cheerful in manner, almost boyish, and with a modesty that verged on shyness.[112]

Bloch recalled that after he shook hands for the first time with the 26-year-old Heisenberg, "already very famous . . . I had the feeling that I was accepted."[113]

Debye strove to make Leipzig an international meeting place for theoretical and experimental physicists by organizing annual Leipzig "lecture-weeks [*Vorträge*]" on topical research themes, as he had done earlier in Zurich. Debye wrote in an invitation to one of the meetings: "It is envisioned that each invitee hold a lecture. . . . The meeting is very unofficial."[114] To cover travel and living expenses of foreign visitors, Debye raised money from the Saxon Ministry for Popular Educa-

Table 2.1 Selected Solid State Topics at the "Leipzig Weeks"

1928: Quantum Theory and Chemistry
  Molecular processes in crystal growth: W. Kossel, Kiel
  Quantum theory and chemical bonding: F. London, Berlin
  Thermal conductivity of nonmetals and metals: A. Eucken, Breslau
  The role of electrons in chemical bonding: N. V. Sidgwick, Oxford

1929: Dipole Moments and Chemical Structure
  On a simple relation between the refractive index and the dielectric
    constant of regular crystals and the repulsion exponent in the
    potential equation of Born and Landé: K. Højendahl, Copenhagen

1930: Electron Interference
  Internal potential and electrical conductivity of crystals: E. Rupp, Berlin
  On the scattering of x rays and cathode rays: H. Mark, Ludwigshafen
  Atomic form factors: N. F. Mott, Manchester
  The temperature dependence of electrical and thermal resistance of metals according to
    theory and experiment: E. Grüneisen, Marburg
  On the interaction of metallic electrons: F. Bloch, Haarlem
  The behavior of metallic conductors in strong magnetic fields: R. Peierls, Zurich

1931: Molecular Structure
  The Raman effect and the structure of molecules and crystals: F. Rasetti, Rome

1932: (no Leipzig week)

1933: Magnetism
  The change of resistance of metals in magnetic fields: P. Kapitza, Cambridge
  Relationships between magnetization and electrical resistance of ferromagnetic bodies:
    W. Gerlach, Munich
  Paramagnetic properties of rare earth crystals: H. A. Kramers, Utrecht
  Superconductors in magnetic fields: W. J. de Haas, Leiden
  Theory of ferromagnetism: H. Bethe, Tübingen
  The technical magnetization curve: R. Becker, Berlin
  On the energetics of ferromagnetic substances: Richard Gans, Königsberg

tion, which felt "that the meeting should have an international character without any restrictions."[115] Solid-state problems were a central concern of each such Leipzig week (Table 2.1).[116]

Bloch's Dissertation on Electrons in Crystals

Bloch recalls that when he arrived in Leipzig in the fall of 1927, Heisenberg was enthusiastic about problems of solids, "to which quantum mechanics could fruitfully be applied."[117] Heisenberg first suggested that Bloch take up the problem of ferromagnetism. Two decades earlier, Pierre Weiss had shown, as described in Chapter 6, that many observed characteristics of ferromagnetism could be explained in terms of a local mean "molecular field." This field was simply a successful way to summarize the effects of poorly understood interactions among the poorly understood elementary magnetic dipoles within a magnetized metal. Through his work on the helium spectrum,[118] Heisenberg realized as early as 1926 that the quantum-mechanical exchange interaction was the likely source of Weiss's field. From study of Heitler and F. London's work on homopolar bonding in the hydrogen molecule,[119] Heisenberg became convinced that he could use their approach to derive the Weiss fields. He had made such progress on the theory of ferromagnetism that Bloch felt that "Heisenberg has it already in a nutshell. . . . I don't want to just simply work it out." Furthermore, "I'm not going to compete with Heisenberg."[120] About a year later, as we shall see, he did embark on a major effort to improve Heisenberg's treatment of ferromagnetism, but for the moment Bloch chose to work instead on the quantum mechanics of electrons in metals.

Heisenberg's interest in electrons in metals dated back to his stay in Copenhagen. In October 1926, Heisenberg had found a sympathetic ear in Hund, who was then in Copenhagen as an International Education Board Fellow. Hund had applied quantum mechanics to the many-body phenomenon of molecular spectra, in his view the next logical problem after atomic spectra.[121] In his first case study, of the wave function of a particle trapped in a two-well potential, he discovered the quantum-mechanical phenomenon of tunneling (the leaking of the particle wave function through the classically forbidden barrier between the two wells).[122] Heisenberg and Hund discussed applying such "resonant wandering" to the case of many identical neighboring potential wells—the electrons in a crystal. "Heisenberg and I are dreaming up a theory of conductivity," Hund wrote in his diary on 4 November 1926.[123] This was the problem that Heisenberg offered to Bloch for his doctoral dissertation.

Bloch's dissertation work began with a question that Heisenberg posed: How are the ions in metals to be dealt with?[124] Bloch recalled,

> The main problem was to explain how the electrons could sneak by all the ions in a metal so as to avoid a mean free path of the order of atomic distances. Such a distance was much too short to explain the observed resistances, which even demanded that the mean free path become longer and longer with decreasing temperature.[125]

Extending Hund's two-potential-well approach, Bloch took a major step forward by approximating the lattice by a three-dimensional periodic potential and ignoring the mutual interaction of the electrons so as to reduce the problem to a one-body calculation. Then drawing on the idea that Heitler and London had used in their treatment of the hydrogen molecule, that of constructing electron wave functions starting from a basis of unperturbed single-atom ground-state orbitals[126] (Bloch's familiarity with the Heitler–London method dated from 1926, when Heitler and London were in Zurich and all three would enjoy walks together), he used perturbation theory to solve the single-electron problem.[127]

Bloch assumed a potential in which the electrons were bound to the lattice with an energy much larger than the kinetic energy of their motion through it—the "tight binding" method—so that most of the time any given electron revolves about the nucleus of a certain atom and rarely moves to a different atom.

> By straight Fourier analysis I found to my delight that the wave differed from the plane wave of free electrons only by a periodic modulation. This was so simple that I didn't think it could be much of a discovery, but when I showed it to Heisenberg he said right away: "That's it!"[128]

Werner Heisenberg and Felix Bloch, around 1931. (DM)

Bloch had discovered the one-dimensional version of the important theorem bearing his name, that the wave function of an electron energy eigenstate in a perfect periodic lattice has the form (now known as a Bloch state) of a product of a free wave and a periodic function $u(\mathbf{r})$ with the period of the lattice: $\Psi = e^{i\mathbf{k}\cdot\mathbf{r}}u(\mathbf{r})$, where $\mathbf{r}$ is the electron coordinate and $\mathbf{k}$ is its "crystal wave vector."[129] By implying that electrons would move freely through a perfect lattice, this theorem provided at once the conceptual basis for Sommerfeld's semiclassical model. Furthermore, it implied that the electrical conductivity of a perfect lattice of identical atoms would be infinite, and therefore that finite conductivity would be caused only by lattice imperfections or ionic motion.[130]

At Bloch's doctoral examination in July 1928, Heisenberg and Debye awarded Bloch's dissertation the highest grade, "Note 1," for Bloch had accomplished "the task . . . of developing a rigorous theory based directly on the atomic structure of metals." Heisenberg excused Bloch's neglect of electron–electron interactions because of mathematical difficulties, mentioning also that the theory could not explain superconductivity, but emphasizing the agreement with measurements that Sommerfeld's theory could not explain. Finally, he praised the practical applicability of Bloch's theory, and mentioned that he asked Bloch to give examples to the press. Debye commented that Bloch's "calculations offer a program that should be pursued further."[131]

In August 1928, Bloch submitted the work of his dissertation for publication in the *Zeitschrift für Physik*.[132] The paper laid out the foundations and many basic principles and techniques of the quantum theory of electrons in lattices. He began by deriving the "Bloch theorem," using the then fashionable group theory,[133] and also derived the expression for the electrical current carried by an electron in a Bloch state. He then went on to derive, in the tight-binding approximation, the wave functions and energies of the "ground state band." To calculate the specific heat, he introduced essentially the notion of an effective mass at the Fermi surface, and found by the Fermi–Dirac formulas that the specific heat is proportional at low temperature to T and to the effective mass.

Turning then to the dynamics of electrons, he showed first how a (Gaussian) wave packet is accelerated by a uniform electric field, and then considered the interaction of electrons with elastic waves of the lattice. To do this, he assumed the ionic motion to be described by a continuous elastic displacement vector, a sum of quantized harmonic normal modes (the same modes entering the Debye-Born-von Kármán specific heat calculation), and assumed that the displacement of the lattice modifies the regular periodic potential felt by the electrons. This modification of the potential, linear in the lattice displacement, causes scattering of the electrons, whose rate Bloch calculated by lowest-order perturbation theory.

Now that he had the rate of absorption of lattice vibrational quanta by electrons, he showed that the electrons and lattice vibrations are in equilibrium, provided that the electron Fermi distribution is one of equilibrium, at rest with respect to the lattice, and having the same temperature as that describing the lattice vibrations. Thus any lingering doubts about whether the electrons partake in the thermal motion of a solid could be laid to rest.

Finally, Bloch calculated the electric conductivity by solving the Boltzmann kinetic equation, with the full electron-lattice vibration collision term and with a simple drifting Fermi sea approximation for the electron distribution function. Unlike in previous theories, the resistivity emerged with a well-defined and experimentally verifiable temperature dependence, linear in T for temperatures large compared with the "Debye temperature" of the lattice. Bloch showed in a later paper[134] (correcting an error in solving the Boltzmann equation in the first paper) that the resistivity is proportional to T^5 at low temperatures.[135]

By carrying out the first calculation of electron wave functions in a metal that took the ions into account, and by developing many basic principles and techniques still in use today, Bloch laid the basis for the quantum-mechanical theory of metals.

2.4   1928–1933: Constructing the Building Blocks

The next phase in the emergence of the quantum theory of solids saw the theory applied to a wide range of problems, including electrical and thermal transport in metals and semiconductors, cohesion of solids, dielectric phenomena, optical phenomena, magnetism, and superconductivity. The key to understanding the electron transport and optical properties of solids was band theory; with this major conceptual building block in place, one could finally account both for fundamentals, such as the difference between metals and insulators or the nature of semiconductors, and for particular phenomena, such as the Hall effect and magnetoresistance.

Hans Bethe (b. 1906) in Ann Arbor, 1935. (AIP Niels Bohr LIbrary, Goudsmit Collection)

The band picture, in retrospect so evident once the form of the solutions of the Schrödinger equation in a periodic potential was understood, came into focus only gradually over the three years following Bloch's thesis. As Bloch showed, the electron energy–momentum relation was no longer simply quadratic; hidden within was the structure from which the concepts of bands and "holes" would later emerge. Although Bloch derived in his paper only the ground-state-band wave functions and energies, he recalls that the concept of many bands was "completely obvious" from the start: "since an atom has excited states, to each excited state there would belong a band," with gaps between.[136] However, the role of bands and band gaps in determining the properties of solids was not yet clearly recognized.

Bethe's Dissertation on Electron Scattering in Metals

Contemporaneously with Bloch in Leipzig, in 1927 and 1928 Bethe was writing his dissertation, in Munich under Sommerfeld, on the solid-state problem of electron scattering in crystals. Although Clinton J. Davisson and Lester Germer's 1926 discovery of electron diffraction[137] was generally perceived as a confirmation of quantum mechanics, some technical problems remained. The experimental diffraction maxima did not occur at the predicted energies. "And so," Bethe recalls, "Sommerfeld asked me, 'Please clear that up and tell us why that is.'"[138] In submitting Bethe's doctoral thesis to the philosophical faculty, Sommerfeld wrote, "When in the beginning of last year the epochal work of Davisson and Germer appeared, I lectured about it in my special topics course and entrusted Mr. Bethe with its theoretical evaluation."[139] Bethe's results were "so admirable that we published two preliminary communications in *Die Naturwissenschaften,*"[140] Sommerfeld continued. "Mr. Davisson himself agreed with Bethe's conclusions. . . . The work represents fundamental progress in the theory of metal electrons." Bethe passed his doctoral examination in Munich on the same day Bloch did in Leipzig, 24 July 1928, also with "Note 1."

Following closely in his thesis the methods developed at Sommerfeld's institute in 1917 by Ewald, an earlier Sommerfeld student, in his "dynamic" theory of x-ray diffraction,[141] Bethe explained how the electrons having negative potential energy in the metal have greater kinetic energy inside than outside. Consequently, their wavelengths are shortened in the crystal, thus explaining the observed discrepancy in the position of the diffraction maxima. Among other topics, Bethe dealt with the phenomenon of "selective reflection," in which electrons impinging on a metal in certain energy intervals are observed to be totally reflected. To explain this effect, he carried out, in close correspondence with Ewald, a "weak-binding" approximation for the wave function of an electron in a periodic crystal, starting from his realization (independent of Bloch) that the electron wave functions must be of the form $e^{i\mathbf{k}\cdot\mathbf{r}}u(\mathbf{r})$.

Setting up the mathematical machinery for developing band theory, which machinery would later be employed by Peierls and others, Bethe showed, as Ewald had found earlier for x rays,[142] that for certain incident directions and energy intervals one cannot construct propagating solutions for electrons in the crystal. The connection of these intervals with the forbidden gaps between bands would not, however, be made until 1930 by Philip Morse.[143] And although the concept of band gaps lurked about in Bethe's calculations (he even went so far as to write out the weak-binding secular problem that exhibits gaps), the concept was not made sufficiently explicit, and so his dissertation did not play a significant role in the further development of band theory.

Peierls's Theory of the Hall Effect: Filled Ground-State Band, Holes

During Easter 1928, as Bloch was finishing his dissertation, Peierls arrived in Leipzig. Heisenberg, having explored the Heitler–London method in ferromagnetism,[144] suggested that Peierls study the usefulness of this approach to the conductivity problem, by constructing many-electron wave functions that took into account from the start the electron–electron interactions. (Bloch's calculation, by contrast, was based on single-electron wave functions, with no account of electron–electron interactions.) But, Peierls recalls, "I struggled very hard but couldn't get away from the conclusion that . . . this model . . . would have no conductivity."[145]

Heisenberg then suggested that Peierls look at the Hall effect, the buildup of a transverse voltage as an electric current passes through a metal in a magnetic field. Sommerfeld's semiclassical theory, based on free electrons, could not essentially improve on the classical result for the Hall voltage, which, although agreeing well with observation for the alkalis and certain other metals (copper, gold, silver, lead, palladium, and manganese), could not account for the variations of the Hall voltage with temperature or magnetic field or explain why for certain metals theory gave the wrong magnitude or sometimes even the wrong sign.[146]

The clue to understanding this "anomalous" or "positive" Hall effect lay in going beyond free-electron theory and fully exploiting the nonquadratic relation Bloch found between electron energy E and crystal momentum $\mathbf{k}$. This relation implies in particular that in the upper part of the band (Peierls, like Bloch, considered only the ground-state band, in the tight-binding approximation) electrons have a group velocity decreasing with crystal momentum, opposite to the behavior of free electrons; in other words, the electrons have a negative effective mass due to the negative curvature of $E(\mathbf{k})$. Peierls recalls that in unraveling the positive Hall effect he

> first had to convince myself that the effect of the magnetic field on the wave vector of the electron was the same as for a free electron of the same velocity, but that the mean velocity of the electron was given by $dE/d\mathbf{k}$, and therefore was different from that for a free electron of the same $\mathbf{k}$, if the energy function $E(\mathbf{k})$ was different. It was obvious, in particular, that in Bloch's tight-binding model the energy would flatten off near the band edge, so that the current would there go to zero.[147]

Peierls submitted the full account of the positive Hall effect to the *Zeitschrift für Physik*[148] at the end of 1928, and described the theory at a meeting of the Deutsche Physikalischen Gesellschaft in Leipzig in January 1929.[149] In this paper, he proved that (in a one-band model) the time rate change of the electron wave vector in electric and magnetic fields is given by the Lorentz force, generalizing Bloch's early argument[150] on the behavior of Gaussian wave packets in an electric field. Then calculating the system's response to electric and magnetic fields by generalizing Bloch's integral equation (derived from the Boltzmann equation) for the change of the electron distribution function, he derived a result for the Hall constant that reduces in the limit of a slightly filled band to the classical electron theory result. However, in the limit of a nearly filled band, it reduces instead to the classical result for carriers of *positive* charge whose number equals the number of *unfilled* states in the band. Peierls almost made explicit the idea of the "hole"—that vacancies near the top of an otherwise filled band behave as positively charged particles of positive effective mass. In fact, he pointed out how his result was connected with Pauli's 1925 reciprocity principle, which had showed a correspondence between an atomic state having a certain number of electrons outside a closed shell and the state with the same number of deficiencies (*Lücken*) in the closed shell.[151] Furthermore, in pointing out that the electrical conductivity must vanish in the case of a completely occupied band, Peierls, although not mentioning it, found the basic characterization of electrical insulators.

The picture of the hole as a positively charged entity was finally delineated in full in mid-1931, when Heisenberg used Peierls's work on the Hall effect as one illustration of the "far-reaching analogy between the terms of an atomic system with n electrons and a system that is n electrons short of having a closed shell."[152] Showing that a hole is described by a complex conjugate wave function (a result perhaps more familiar now from the viewpoint of second quantization), Heisenberg concluded that for states near the top of the band

> the holes [*Löcher*] behave exactly like electrons with positive charge under the influence of a disturbing external field . . . the electrical conduction in metals having a small number of holes can in every respect be written as the conduction in metals having a small number of positive conduction electrons.

By December 1929, Dirac had already formulated the concept of the hole in quantum electrodynamics,[153] a vacancy in the sea of negative-energy electrons.

(Although he initially suggested that this hole could be the proton, he properly identified it as an "anti-electron," or positron, in mid-1931,[154] a month before Heisenberg's paper.) The analogy, so obvious today, between the solid-state hole and the Dirac hole, and the fact that both holes were invented in the same period, has led to a common belief of a historical connection between the two concepts. Bloch recently mused on such a connection, "There was [in this period] so much interplay between all the physicists . . . that as soon as somebody had an idea, another one took it up and put it in a different form and used it somewhere else."[155] However, we have found no contemporaneous evidence of a relationship in the development of the two concepts. Neither the Peierls–Heisenberg nor the Dirac paper, in presenting its picture of holes, referred to the other. Rather, both concepts appear to have had a common root in Pauli's work on almost-filled shells of many-electron atoms, a correspondence that both Heisenberg[156] and Dirac[157] explicitly drew on. Although the analogy between the two holes may have been apparent to some— "certainly when Dirac's paper on the hole theory came out, it was obvious . . . that there was an analogy with the electrons in metals and vacant places," Peierls recalls[158]—this was not mentioned in the literature of the period. Bethe recently ascribed this omission in part to the fact that the Dirac holes were not generally taken seriously until after the positron was discovered in 1932;[159] although the infinite continuum of states in the Dirac theory made it very hard to construct a finite theory of quantum electrodynamics, in a solid "it was clear there was a finite number of states, so it was very obvious that an unoccupied state was a hole and how it would behave."[160]

Pauli's Institute in Zurich

With Heisenberg's departure from Leipzig on a half-year lecture tour in the United States in the spring of 1929,[161] the center of research on the quantum theory of metals shifted to Pauli's institute at the ETH in Zurich. The appointment of Pauli on 1 April 1928 to Debye's chair in Zurich had changed the character of physics there, much as the appointment of other Sommerfeld students had changed Leipzig. Several months after his move to Leipzig, Debye, who maintained influence in the filling of his post, had received a query from Scherrer, who held the chair of experimental physics at the ETH:

> Heisenberg . . . informs me that he does not want to come to Zurich. I regret this very much, but am pleased that at least you will get him. In this way he remains, so to speak, in the family. . . . Therefore I want to ask you once again, what do you think of appointing Pauli to Zurich?[162]

Debye had replied that Pauli "is so outstanding a scientist, and I have always found it so easy to get along with him, that with this appointment, Zurich physics will be well served."[163] When ETH president Rohn traveled to Hamburg to meet Pauli, who was generally considered a difficult character, he was favorably impressed. And since Pauli apparently could not further improve his Hamburg post, he accepted the Zurich chair.[164]

The physicists in Leipzig and Zurich formed a family with close ties to Sommerfeld in Munich. Debye and Scherrer could look back on a collaboration of more than 10 years (at Göttingen as well as Zurich [Chapters 1 and 8]). Heisenberg, Pauli, and Wentzel had all studied with Sommerfeld at the beginning of the 1920s, as had Debye some 20 years earlier. Sommerfeld remained a father figure to his students. "Hopefully you will continue to run a kindergarten for physics babies like Pauli and me!" wrote Heisenberg to Sommerfeld in 1929.[165] For Sommerfeld's sixtieth birthday in December 1928 (the festivities of which Sommerfeld evaded by taking his world tour), Debye organized a Festschrift with contributions from 30 Sommerfeld students, by then almost all professors.[166]

Whereas the interactions between Leipzig and Munich and between Zurich and Munich were formed by the relationships of the pupils Hesienberg and Pauli to Sommerfeld, the interaction between Leipzig and Zurich was shaped by the less formal and more intense relationship of the two peers. Particularly during the 1920s, these two interacted deeply in science, yet their characters could scarcely have differed more: Heisenberg—openly friendly, an early riser, as ambitious in sports as in science, and a nature lover; Pauli—aggressive, often woundingly critical, moody, and preferring nightclubs over morning outings.[167] But they shared their enthusiasm for physics and continued their active exchange as heads of their own institutes.

Heisenberg's half-year absence in the United State led to important additions to Pauli's group at the ETH. Bloch, after completing his dissertation, came in the winter 1928/1929 semester to succeed Ralph Kronig as Pauli's *Assistent*. Peierls, now sufficiently advanced to begin a doctoral thesis, came that spring on Heisenberg's recommendation.[168] A number of visiting fellows, who might otherwise have gone to Heisenberg, went to Zurich; Lev Landau, J. Robert Oppenheimer, I. I. Rabi, and Léon Rosenfeld all worked around Pauli for portions of 1929.[169] "I have now a rather bustling operation here in Zurich," Pauli wrote to Sommerfeld.[170]

For Pauli, like Heisenberg, solid state was in these years a prime testing ground for quantum mechanics, and he began to form around himself a group of physicists who would make major contributions to the early development of the quantum theory of solids. A report to Sommerfeld illustrates the extent to which Pauli regarded solid state as an area to employ students and research fellows: "Mr. Bloch is at present occupied with working out a theory of superconductivity. . . . Mr. Peierls is working on a theory of thermal conductivity in solid bodies,"[171] a problem Pauli himself had struggled with several times.[172]

Peierls's Theory of Electrical and Thermal Conductivity: Band Gaps

At Pauli's suggestion, Peierls took up further problems in solid-state physics. Maintaining an interest in lattice vibrations in an anharmonic crystal, Pauli suggested that Peierls study heat conduction in nonmetallic solids.[173] Peierls's work on heat conduction in insulators, completed in late October 1929, became his doctoral thesis, submitted to Leipzig.[174] In it, Peierls carried out a critical analysis of how lattice vibrations come into thermal equilibrium at low temperatures. Introducing the important, and soon to become controversial,[175] concept of *Umklapp* processes (scattering processes accompanied by transfer of crystal momentum to the lattice), he found that in a pure material conservation of "crystal momentum" of lattice vibrations implies that as the *Umklapp* processes become frozen out with decreasing temperature, the thermal conductivity rises exponentially. Peierls immediately reacted to his discovery by asking, "Could that be superconductivity?" He recalls that Pauli encouraged the inquiry with the remark, "If this can explain superconductivity, then you can certainly have your *Habilitation*"—the advanced, postdoctoral degree of German universities.[176] However, "It didn't, of course."[177]

Peierls then turned to applying his arguments to metals. In a paper submitted to the *Annalen der Physik* six weeks after his thesis paper,[178] he pointed out the role of *Umklapp* processes in keeping the lattice vibrations in equilibrium and limiting the electrical and thermal conductivity at low temperatures. Unlike Bloch, who assumed that the lattice vibrations remain in thermal equilibrium, Peierls wrote down coupled Boltzmann equations for both the electron and lattice vibration distributions. Also, to describe the interaction of electrons with oscillating ions, he used, for the first time, the "rigid ion" approximation familiar to modern students (in this approximation, the internal structure of the ion is assumed to remain unchanged as the ion oscillates).

The first section of this rather long paper, on the two limiting cases (tight and weak binding) for electrons in solids, was to play a seminal role in the further development of band theory. Peierls's earlier explanation of the anomalous Hall effect depended on the negative curvature of the electron energy, as a function of wave number k, near the top of the band. But so far only the tight-binding approximation had been examined, and this was clearly not a good approximation for real metals.[179] For as Bloch had recently argued, the magnetic susceptibility of ordinary metals indicated a density of states at the Fermi surface (in modern language) more nearly that of free electrons.[180] The basic problem bothering Peierls was how to connect the tight-binding limit, with its novel and apparently important structure, to the free-electron limit, closer to experiment, which does not have such negative curvature. Examining the case of weakly bound electrons in one dimension, he found that (as every solid-state student now learns) whenever two free-electron states, separated by a reciprocal lattice vector, have an energy difference comparable with the potential matrix element, gaps and, hence, negative curvature appear in the spectrum, disappearing only when the electrons are exactly free. "I still remember the excitement . . . how thrilled I was to see that."[181] His starting equations were just those written down by Bethe in his thesis, although by this point Peierls, having the experience of his Hall-effect work, knew what question to ask of the model calculation.

Peierls's result, which he illustrated with a now classic diagram (Figure 2.1), established the concept of the band gap as characteristic of electrons in solids.[182] Looking back on this work, Peierls wrote recently,

> Few pieces of work have given me as much pleasure as this discovery, which required only a few lines of calculation, both because it satisfied me that the nature of the Bloch bands was now qualitatively the same all the way from tight binding to almost free electrons, and because of the neat method of approximation I had invented.[183]

Actually, the transition from tight to weak binding and the existence of gaps had been worked out in mid-1928 by M.J.O. Strutt[184] for a sinusoidal potential in one dimension (the Mathieu problem), although this work did not receive the imme-

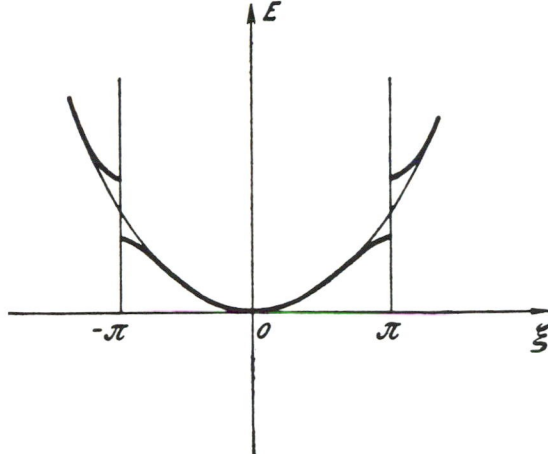

Fig. 2.1 Peierls's diagram of the formation of band gaps in a weak periodic potential. (From R. E. Peierls, "Zur Theorie der elektrischen und thermischen Leitfähigkeit von Metallen," *Ann. Phys.* 4 [1930]: 126)

Energiewerte erster Näherung

diate attention of solid-state theorists. The first application of Strutt's approach was made by Philip Morse, soon after he received his Ph.D. at Princeton in June 1929 under the supervision of Karl Compton.[185] Working during the summer at the Bell Telephone Laboratories on the interpretation of the experiments of Davisson, Germer, G. P. Thomson, and others on electron diffraction from metal surfaces,[186] Morse began a general analysis of the solutions of the Schrödinger equation for an electron in a periodic potential. The starting point, as he acknowledged in his paper on this work, were equations similar to those in Bethe's thesis. He derived the important conclusion that "the periodic variation of the potential inside the crystal creates bands of forbidden energies inside the crystal, even for electron energies greater than the maximum potential energy, a somewhat surprising result." Morse went on in his paper to work out in detail the example of the separable potential in three dimensions that is a sum of cosines, whose solution reduces to the one-dimensional Mathieu problem studied by Strutt. In applying this exactly soluble model to the Davisson–Germer experiments, he thus made the first explicit connection between the band structure of electrons in solids and the diffraction of electrons impinging on solids. Not having completed this work at Bell, he continued it at Princeton in the fall, and "after a great number of further computations, done by myself on the department's desk calculating machine," submitted his paper to the *Physical Review* in April 1930.[187] Although by this time Peierls's paper containing the weak-binding calculation had appeared, and Morse included a reference to it, Peierls's work does not appear to have had a significant influence on Morse's.

Brillouin Zones

Meanwhile Brillouin, who had worked under Sommerfeld in 1912 and 1913 as a research assistant and was now a lecturer on radio at the Ecole Supérieure d'Electricité as well as professor of science at the University of Paris, became interested in problems of electrons in solids. This interest was whetted when he prepared a book,

Paul Dirac *(front)* and Léon Brillouin *(rear)* in Ann Arbor, 1929. (AIP Niels Bohr Library)

Les Statistiques quantiques et leurs applications, published in early 1930.[188] Brillouin had extensive experience in both statistical mechanics and mathematical problems of wave propagation, including development of the WKB method in quantum theory,[189] but at the time of this book he had not yet made any independent contribution to the quantum theory of metals. The chapter on the mean free path of electrons in solids contained a description of energy bands, but only a rather mathematical one based essentially on Strutt's one-dimensional calculation. Brillouin also appears to have learned of Morse's work before the publication of the book, most likely during the previous summer when Brillouin visited the United States to lecture at the University of Michigan Summer School in Ann Arbor.[190] But Peierls's work on the Hall effect and the weak-binding calculation were not noted in the chapter, although it ended with a passing last-minute reference to Peierls's critique, in the same paper[191] as his weak-binding calculation, of Bloch's theory of electrical conductivity.[192]

The full application of Peierls's ideas on energy gaps to realistic solids was begun by Brillouin, working alone in Paris in the summer of 1930.[193] In two papers presented to the Académie des Sciences at its 28 July meeting,[194] he first generalized Peierls's result[195] to show that in three dimensions the surfaces of discontinuities in energy versus wave number for nearly free electrons form polyhedra in momentum space—the Brillouin zones. Then, counting states, he argued that each zone corresponds to a single atomic state, and showed how to transform from the extended zone scheme to the fundamental zone. He made the connection, as did Morse, between the conditions for a discontinuity in the energy and for Bragg scattering, a connection that, he remarked, Peierls did not make in his paper.

These new ideas were synthesized a week later in a paper submitted to the *Journal de physique et le radium,*[196] a paper that would become the fundamental source for anyone working out the geometry of Brillouin zones. Going beyond his recent book, with its reliance on exact and one-dimensional solutions, he realized here the generality of energy bands. He also described the curvature of the energy–wave number relation in terms of an effective mass m^*, remarking that m^* can be negative, although he did not make contact with Peierls's Hall-effect work. The paper concluded with an attempt to establish a phenomenological connection between "propagation anomalies" (in modern terms, the Fermi surface reaching a zone boundary) in polyvalent metals and lower electrical conductivity in these materials, but in the absence of any dynamical theory his arguments were inconclusive.[197]

Brillouin later described his recollections of this fertile summer:

> At first I did not realize that I was doing something that might become really important. I did it for the fun of it, following my own line of investigation by sheer curiosity and taking a great deal of pleasure in making carefully all the drawings needed to explain the properties of these Brillouin Zones.

He was also pleased by the response to his work and was "especially proud of a very affectionate letter from my old teacher, Sommerfeld, who praised warmly my contribution and said that he was so happy to be now able to understand clearly the interconnection between isolated atomic electronic levels and free electrons in metals."[198]

Brillouin took the opportunity of the translation of his book into German a year later[199] to make extensive revisions of the chapter on electrons in solids, including (nearly verbatim) his *Journal de physique* paper. He brought the discussion of electrical conductivity up to date by including Bloch's recent correction (to T^{-5}) of the low-temperature dependence found in his thesis; he also used the occasion to express deep concern about the reality of Peierls's *Umklapp* processes, arguing that since "these processes can be represented as a simple superposition of a Bragg reflection and a normal scattering . . . I have the impression that one must leave the Umklapp processes out of the theory."[200] (The validity of Peierls's arguments were not experimentally confirmed until the early 1950s, when sufficiently pure crystals could be studied.)

By the end of 1930, all the crucial pieces of band theory were waiting to be assembled. The existence of band gaps was well understood. In addition to Bloch's tight-binding and Peierls's weak-binding calculations, one had Morse's general arguments, as well as Strutt's exact example. At year's end, Kronig and William G. Penney submitted for publication their simple analytically soluble one-dimensional model of a periodic square-well potential,[201] which verified from another point of view the general features of the quantum states and energies of electrons in solids. Furthermore, the concept of holes, although not yet clearly described, was implicit in Peierls's work on the Hall effect. These ideas would be fused in 1931 by Alan Wilson in two classic papers on semiconductors.[202]

Wilson's Theory for the Difference Between Conductors, Insulators, and Semiconductors: Energy Bands in Solids

Alan Wilson's interest in solid-state physics began in 1929 when he was a research fellow at Emmanuel College in Cambridge. He attempted to explain Peter Kapitza's recent experimental discovery in Cambridge that the resistance of metals in strong magnetic fields increases linearly with the field. The problem of the influence of magnetic fields on electrical conduction, or "magnetoresistance," was outside the scope of the simple Sommerfeld theory (which provided no theory of the electron mean free path), and this problem had now become an important test for the theory of metals. It attracted the attention not only of Wilson, but of Bloch in 1928,[203] Peierls[204] and Landau[205] in 1930, and Bethe in 1931.[206] Finding Cambridge under Rutherford's influence "highly concerned with nuclear physics,"[207] Wilson obtained a Rockefeller Foundation fellowship to go to Leipzig, where he could join Heisenberg and Bloch (then an *Assistent*), who were very interested in magnetism. Also in Leipzig at that time were Hund, Teller, and Debye.

Immediately after Wilson's arrival in Leipzig during the first week of 1931, Heisenberg, sensing the significance of Peierls's work on effects of magnetic fields in metals,[208] asked Wilson to deliver an explanatory colloquium. In Wilson's words, "There were two problems, . . . one was that I had Peierls' papers and didn't really understand them, and secondly . . . to give a seminar in German at which I would be cross-questioned back and forth would be a bit of an ordeal." Wilson recalls being particularly impressed with Heisenberg's, as well as Debye's, ability to see through problems by very simple physical arguments. Contrasting the atmosphere at Leipzig with the more mathematical approach he had found at Cambridge, "particularly with Dirac and [R. H.] Fowler," he remarked, "I'd been to something like four seminars before I had to give mine, and on nearly every occasion Heisenberg would stop whoever was talking and say, 'This is all right, this is mathematics, but not physics.' And he would say, 'Wie kann mann dass physikalisch anschaulisch machen?' [How can we make that physically intuitive?]."[209]

To prepare for this ordeal, Wilson sat down to study the Peierls and Bloch papers in detail. To him the problem was that Bloch, in showing that tightly bound electrons could in fact move through the lattice, had "proved too much"—that all solids should be metals.[210] Were insulators simply very poor conductors? However, implicit in Peierls's papers on the Hall effect lay the clue, not carried further by Peierls, that a filled band would carry no current. "Suddenly one morning," Wilson realized that "I've been looking at it all wrong, of course it's perfectly simple."[211] He could make the basic Bloch–Peierls theory of electrical conductivity "intuitively more plausible if one assumed that the quasi-free electrons, like valance electrons in single atoms, could form either open or closed shells."[212] Here finally was the answer to the old question of the difference between metals and insulators: insulators have completely filled bands, whereas metals have partially filled bands, a situation nicely summed up in Wilson's remark in his first paper on semiconductors: "We have the rather curious result that not only is it possible to obtain conduction with bound electrons, but it is also possible to obtain non-conduction with free electrons."

Wilson told his idea to Heisenberg, who said, "I really must get Bloch in."[213] Wilson recalls Bloch's reply after hearing his arguments: "No, it's quite wrong, quite wrong, quite wrong, not possible at all," for Bloch, since his thesis, had assumed that the difference between metals and insulators was only quantitative, determined by the size of the electron overlap integral, which measures the ease with which an electron can hop from atom to atom. Attempting to refute Wilson, Bloch pointed out that although solids formed of monovalent elements would have only half of the uppermost band filled and would be metals, the divalent alkaline earths would

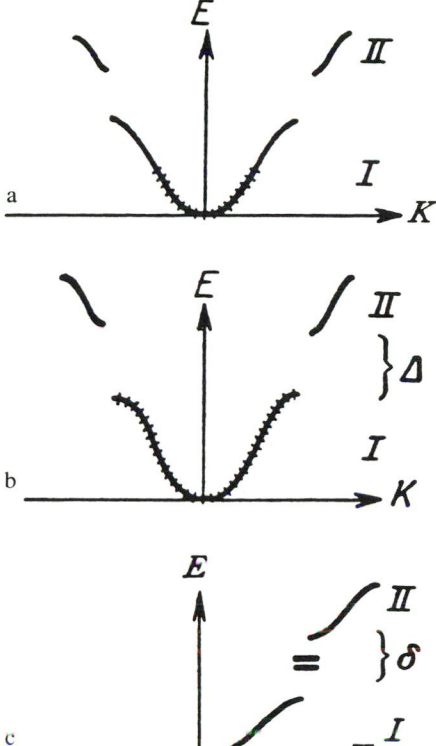

Fig. 2.2 Alan Wilson's picture of electron energy E versus wave number K, with a band gap of width Δ. (a) A metal, with all the filled electron states well below the gap; (b) an insulator, with filled states reaching up to the gap, which blocks the electron motion; (c) (after Bloch) a semiconductor with impurity states in the gap. (From F. Bloch, "Wellenmechanische Diskussion der Leitungs- und Photoeffekte," *Phys. Zs.* 32 [1931]: 883, 886)

have just enough electrons to fill the top band exactly and hence should be insulators.

But the following day, Wilson was able to point out to Bloch that, unlike in an idealized one-dimensional lattice, the bands in a three-dimensional solid can in fact overlap, so that rather than the bands being filled in the order of the corresponding atomic states, several bands in a polyvalent material could be partially filled. "It therefore followed that an elemental solid . . . with an odd valency had to be a metal, whereas elements with an even valency might produce either a metal or an insulator."[214] After a week, Bloch was convinced, and he would soon summarize Wilson's concepts in a short paper given at the seventh Deutschen Physikertages in Bad Elster, in September 1931;[215] among his illustrations are Figures 2.2(a) and (b), showing the difference between a metal and an insulator, respectively, and Figure 2.2(c), showing impurity levels in the gap in a semiconductor.

Wilson proposed to Heisenberg that he broaden his colloquium to deal more generally with bands. They decided that Wilson should give two colloquia, spaced approximately three months apart, and should also consider semiconductors, which the new concept of electrons in filled or unfilled shells might also illuminate. Heisenberg had become interested in semiconductors through the experimentalist B. Gudden, at the Theological University in Erlangen;[216] but at the time, knowledge about semiconductors was so scant that it was uncertain whether they even existed. At this time, Pauli wrote to Peierls: "One shouldn't work on semiconductors, that is a filthy mess; who knows whether any semiconductors exist."[217] While Grüneisen in his 1928 review of metallic conductivity in the *Handbuch der Physik* had distinguished semiconductors as a class of solids with a pronounced minimum in their resistance as a function of temperature, it was not clear (as Wilson noted in the introduction to his second paper on semiconductors) that this behavior might not simply be caused by oxide surface layers on otherwise metallic substances.[218]

In his first colloquium, in February 1931, Wilson described his theory of the difference between metals and insulators. He also put forward a simple picture of a semiconductor as an insulator with a gap between what we now call the valence and conduction bands, small enough so that electrons could easily be excited across the gap at finite temperature. The details were written up in a paper entitled "The Theory of Electronic Semi-Conductors" in the *Proceedings of the Royal Society.*[219] Referring in his paper to the previous work of Bloch, Peierls, Morse, Brillouin, and Kronig and Penney as indicating the general existence of energy bands, he gave a very clear review of the weak- and tight-binding approximations. He then calculated for the first time the wave functions and energies of p-state bands in tight binding. (Bloch's original calculation was for an s-state band.) The following discussion, detailing Wilson's earlier arguments to Bloch, showed how overlap of s and p bands in alkaline earths can lead to their being conductors; this section was remarkable as the first use of band theory to distinguish qualitatively the properties of realistic solids. Wilson then went on to describe the properties of realistic solids. Wilson then went on to describe his simple model of a semiconductor, showing that the chemical potential lies halfway in the band gap, and calculating the specific heat, spin paramagnetism, and electrical conductivity as limited by emission or absorption of a "sound quantum" by electrons. He concluded by remarking that even though the interpretation of the experimental results on semiconductors was still difficult, and even their very existence remained "an open question," the theory "is on the right lines."

This spring in Leipzig was also when Heisenberg wrote his paper on the Pauli exclusion principle,[220] in which—as an outgrowth of his interest in the theory of magnetic effects in metals, particularly Peierls's Hall-effect paper[221]—the concept of the solid-state "hole" first appears. In fact, this paper was received by the *Annalen der Physik* the day after Wilson's was received by the Royal Society. It is noteworthy that although he had many discussions with Heisenberg on electrons in metals, Wilson, in his paper, never treated the unoccupied states in the valence bands as hole degrees of freedom, but worked in terms of the electron states; neither did Heisenberg's paper refer to Wilson.

Between his first and second colloquia, Wilson learned through Heisenberg of Gudden's view that semiconductor behavior was always caused by impurities.[222] The second colloquium, which was attended by a group of experimentalists from Erlangen headed by Gudden and which lasted several days, addressed the role of impurities. Wilson described the detailed model in a second paper.[223] This paper began with a brief argument making an interesting connection between conductivity and optical properties—that although the experimental conductivity of cuprous oxide indicated, according to his earlier paper, an excitation energy of around 0.6 V, optical absorption implies an intrinsic band gap of around 2 V, and hence "the observed conductivity . . . must be due to the presence of impurities." The model he considered was an impure insulator in which an electron associated with an impurity has an energy in the band gap close to the conduction band, so it can be thermally excited into the conduction band. (This was the "donor" model; the concept of acceptor levels appears to have been introduced by Peierls[224] in 1932 and clarified by Schottky[225] the following year while explaining experiments of Waibel on copper oxide.) Such an impurity level is shown in Figure 2.2(c), from Bloch's talk. After determining the chemical potential, Wilson calculated the electrical conductivity and pointed out that the Hall coefficient is given by the classical formula, inversely proportional to the conduction electron density. Finally, fitting to the Hall coefficient of cuprous oxide observed by Vogt in 1930, Wilson deduced an impurity concentration of order 10^{17} per cubic centimeter, "conclusive proof that the conductivity is due to impurities and is not intrinsic."

Wilson went on later in the year to apply band concepts to develop a pioneering theory of rectification at a metal–semiconductor junction,[226] in which electrons penetrate, by quantum-mechanical tunneling, a symmetric potential barrier in the transition layer between the metal (here copper) and the oxide. The positive direction of the electron current was predicted to be from metal to semiconductor, which, unfortunately, is opposite to later experiment, as Wilson acknowledged some years after in his well-known book on semiconductors.[227] Nevertheless, this paper would provide the basis of attempts to explain experimental work on rectification during the 1930s,[228] although at the time, as Wilson recalls,[229] not much notice was taken of it by physicists.

By bringing together the elements of band theory in a simple conceptual picture, Wilson closed this chapter in the development of the fundamental quantum theory of solids. He emerges as an important figure in the transition of solid-state theory from its early conceptual to its later practical orientation, for not only did his model make it possible to begin to approach realistic solids, but because his papers were so clear, they would be widely read by subsequent generations of experimental and theoretical researchers.

2.5   1926–1932: Magnetism

Electrical transport was only one of the solid-state problems that seemed to lie open to attack with the new quantum mechanics. The same research community took up other intriguing questions, using the same tools and the same general intellectual and social style of doing theoretical physics. We will look at magnetism as a development in which this style of research was by and large successful, and also at superconductivity, as an example of an attempt where the same style failed.

Paramagnetism and Diamagnetism

As a second area of solid-state phenomena for which the quantum theory would provide the necessary theoretical building blocks, magnetism in many ways progressed in parallel with the development of electrical transport theory. The seminal papers on paramagnetism by Pauli in 1927, on diamagnetism by Landau in 1930, and on ferromagnetism by Heisenberg in 1928 were, like Sommerfeld's path-breaking 1927 work on conduction, based on drastically simplified models, shorn of all irrelevant (and realistic) detail to bring out the essential physical ideas. Detailed comparisons could be made between theory and magnetic properties of solids only later, by applying the understanding of electrons in solids that was reached through the band theory developed to account for electrical transport phenomena.

Pauli's application of Fermi–Dirac statistics to explain paramagnetism was pivotal not only to the development of the understanding of electrical transport, as discussed earlier, but also to the theory of diamagnetism as well as paramagnetism itself. He concluded his paper with a brief numerical comparison of his formula for the paramagnetic susceptibility with then available observations of the susceptibility of the alkali metals sodium, potassium, rubidium, and cesium. The difference in all cases appeared to imply the presence of a weak diamagnetism of, for Rb and Cs, comparable size to the paramagnetism. But Pauli, while not attempting to understand the residual diamagnetism in terms of contributions from the bound core and conduction electrons, opened the problem for future researchers.

The diamagnetism of ionic cores was by this time reasonably well understood. The fundamental formula for the susceptibility, given in 1920 by Pauli himself[230] (after Langevin) as a sum over the mean square radii of the electron orbits, gave results comparable with measured atomic and ionic susceptibilities.[231] On the other hand, according to the classical argument given by Bohr, Lorentz, and van Leeuwen,[232] the diamagnetic susceptibility of a free-electron gas must vanish, and thus there should be no conduction electron diamagnetism. Mathematically this result emerges from the fact that the energy of an electron in a magnetic field is proportional to its velocity squared, independent of the field. Bohr and Lorentz also explained this phenomenon physically by noting that although electrons in magnetic fields move in circles, those that bounce off the boundary slowly circulate around the edge in a direction opposite to the orbital motion of the interior freely circulating electrons, thereby producing a magnetic moment that exactly cancels that of the interior electrons.

The question we may ask at this point is whether after Pauli's paper there was any evidence for a nonzero conduction electron diamagnetism. Had Pauli taken available experimental numbers for the ionic core diamagnetic susceptibilities[233] (even by the 1930 Solvay Conference, he had not done so, in print at least), he would have concluded that the susceptibility of Rb and Cs could essentially be explained as a sum of his paramagnetic contributions plus the measured core diamagnetism. However, for Na and K he would have found the surprising result that what was needed was further paramagnetism! (The effect arises from the enhancement of the paramagnetic susceptibility above the Pauli value by electron–electron interactions, which, through the Pauli principle, favor electron spin alignment.)[234] For none of the alkalis was there any need to invoke conduction electron diamagnetism. The only hint that such diamagnetism played a role was the enormous diamagnetic susceptibility of bismuth and antimony.[235]

By sorting out the magnetic contribution of the spins in the alkalis, Pauli indirectly set out the problem that a theory of diamagnetism would have to explain. The invention of quantum mechanics provided new tools for studying this phenomenon, which Bloch recalls attracted theoreticians of the period because "it was a very clean quantum mechanics problem."[236] According to Peierls, "it was part of

Lev Landau (1908–1968). (AIP Niels Bohr Library, Margrethe Bohr Collection)

one's general interest in metals."[237] The first published attempt to deal with conduction electron diamagnetism from a quantum-mechanical point of view was by Francis Bitter[238] at Caltech in late 1929. He essentially computed the expectation value of the Pauli–Langevin atomic diamagnetism formula, using free electronic wave functions spread over a unit cell. Such an estimate must, on dimensional grounds, be reasonable, and indeed Bitter's results for the diamagnetic susceptibility were about 1.25–2 times the correct answer. Bloch's 1928 theory of electrons in solids appears not to have influenced Bitter, although Houston had returned to Caltech from Leipzig in the fall of 1928.

The full quantum-mechanical problem of electrons orbiting in a magnetic field was solved by Landau in his 1930 paper on diamagnetism.[239] In 1929, Landau, at age 21, had gone on a two-year trip, supported by the Soviet People's Commissariat of Education *(Narkompros)* and then by a Rockefeller fellowship,[240] to European research centers, including Zurich, Copenhagen, and Cambridge, with stops in Germany (notably Berlin and Leipzig), Holland, and Belgium. How Landau became involved with magnetism is unclear. Although he had not published before on solid-state physics, he was a student in Leningrad, where Paul Ehrenfest's five-year prewar stay had left an impression, and where A. F. Ioffé was a leading solid-state experimentalist. Motivation possibly came from Ehrenfest, whom Landau met in Berlin in late 1929,[241] at the time Ehrenfest, in response to a published note by Raman[242] and sundry letters, was republishing in German his 1925 paper (originally in Dutch and Russian) on the diamagnetism of bismuth.[243] According to Peierls, who became acquainted with Landau on his first visit to Zurich, Landau had the problem well under control on his arrival there in late 1929.[244] But Landau delayed submitting his paper until May 1930, when he was at the Cavendish Laboratory, during which time he had discussions about experiments with Kapitza, whom he met in Cambridge for the first time.

Landau began his paper with the remark, "It has until now been more or less quietly assumed that the magnetic properties of electrons, other than spin, originate exclusively from the binding of electrons in atoms." He proceeded to show, by a clever algebraic technique, that a quantum-mechanical electron in a uniform magnetic field H is described by a harmonic oscillator of frequency $ehH/2\pi mc$, the Larmor frequency. He then turned directly to the statistical mechanics of a degenerate free-electron gas in a field and derived the famous result that the diamagnetic susceptibility of the gas in a weak field is exactly minus one-third of the Pauli spin susceptibility, rather than zero, as in the classical theory. Landau concluded his paper with a brief qualitative attempt to understand Kapitza's recent experiments on magnetoresistance and with an acknowledgment of discussions with Kapitza.

Along the way, he noted that the diamagnetic moment should have a strong periodicity in the field. W. J. de Haas and P. M. van Alphen would report the experimental discovery of this effect in a communication to the Royal Dutch Academy of Sciences at the end of this same year, 1930.[245] Although de Haas and van Alphen gave a reference to Landau's paper in their communication, they did not make any contact with his theoretical prediction of the periodicities. And, indeed, Landau himself despaired of observing the effect, suggesting in his paper that inhomogeneities would wash it out. David Shoenberg recalls that Landau explained to him many years later how he came to make his remark about the practical limitations on seeing the oscillations: "Since he knew nothing about experimental matters he had consulted Kapitza whom he was visiting at the time, and Kapitza had told him that the required homogeneity was impracticable"[246] De Haas and van Alphen in fact carried out their experiments on single crystals of bismuth;[247] what Landau could not have realized was that lattice effects greatly increase the oscillation period and therefore the observability of the effect in bismuth. (These are the same lattice effects that would be shown to dramatically enhance bismuth's diamagnetic susceptibility.)

Landau's susceptibility results were immediately accepted. As Peierls recalls, "neither Pauli nor I had any doubt in feeling confident that Landau had got the right argument."[248] Pauli described Landau's work in his Solvay talk of October that year[249] (referring to it as being verbally communicated), and gave in addition the result for the case of nondegenerate statistics and arbitrary field strength, where again the weak field susceptibility is minus one third of the spin contribution. In the discussion of Pauli's talk, Kaptiza remarked, "Landau's new theory, in which the free electrons contribute to the magnetism of the substance, gives us great hope to see all these phenomena [such as effects of impurities and imperfections] explained by a common picture."[250]

Just how Landau's picture modified the Bohr–Lorentz argument for the vanishing of the diamagnetic susceptibility was not immediately clear. A peculiar feature of the Landau diamagnetic susceptibility of a degenerate electron gas, one that apparently was not discussed in the literature of the time, is that even though it is nonzero as a result of quantization of the orbits, it is independent of Planck's constant h. How then did one recover the classical limit? The form derived by Pauli for nondegenerate statistics, which does go over (as h^2) to the classical vanishing value as h goes to zero, showed more clearly how the nonvanishing result was a quantum phenomenon. The independence of h in the Landau result arises from the fact that degeneracy reduces the nondegenerate result by a "density of states" factor proportional to T/T_f, where T_f is the Fermi temperature; since T_f is proportional to h^2, the dependence on h curiously cancels out, a structure obscured in Landau's direct solution of the degenerate gas problem.

The relation between Landau's theory and the classical Bohr–Lorentz argument was soon addressed by C. G. Darwin in Edinburgh and by Teller, then Heisenberg's *Assistent,* in Leipzig. Landau, by calculating the partition function rather than the magnetic moments of the electron states, did not have to deal explicitly with the question of the diamagnetic contribution of the electrons near the boundary compared with those in the interior. To satisfy himself that Landau's result could be derived in terms of the electron magnetic moments, Darwin[251] studied the electron orbits in the exactly soluble model in which the container is replaced by a harmonic oscillator well. He was able to recover "the features of Bohr's argument about the creeping of the electron round the boundary wall" and to see how the cancellation between the two contributions no longer occurred in the quantum theory. Teller,[252] on the other hand, working with a more realistic confining potential, evaluated the statistically averaged magnetic moment contributions of the boundary and interior electrons to show how Landau's result emerged.

The nature of the paramagnetism and diamagnetism of conduction electrons was now understood in principle. However, the free-electron result that the diamagnetism equaled minus one-third of the paramagnetism offered no insight into the vexing anomaly of bismuth, which could not be explained in terms of core diamagnetism alone.[253] As in the problem of electron transport, meaningful comparison of theory with experiment required including effects of the lattice. Landau, in his paper, understood that even in a lattice the motion of the electrons "can in a certain sense still be considered as free . . . [and] that the principal characteristic effect in the magnetic field remains unchanged." But in a lattice, "the relation between para- and diamagnetism is altered, and it is possible that in certain cases the latter can exceed the former, so that we get a diamagnetic substance like bismuth."

The necessary extension of Landau's theory was undertaken by Peierls in Zurich and described in a paper he submitted in November 1932, a month after his arrival at Fermi's institute in Rome.[254] Peierls had a continuing interest in the behavior of electrons in solids in the presence of magnetic fields, as well as in band theory, and he soon recognized that the theory of diamagnetism of conduction electrons in real solids entailed two difficulties. One was the complexity of computing by Landau's method the exact eigenstates of electrons in a periodic potential with a magnetic field. The other was a conceptual difficulty: the broadening of levels caused by collisions with impurities and phonons in metals at most temperatures and magnetic fields exceeds the spacing between the levels, thus threatening to wipe out or modify substantially the Landau effect.[255]

Peierls easily overcame the first difficulty by examining the case of very weak magnetic fields, which could be treated as a small perturbation. The second difficulty was resolved by his realization that as long as the widths of the states induced by collisions were small compared with the temperature, although they might be large compared with the spacing between unperturbed levels, collisional broadening of the levels would have a small effect on the equilibrium thermodynamics. The important practical result that emerged from Peierls's paper was an expression for the diamagnetic susceptibility involving the electron energy–momentum relation $E(\mathbf{k})$ at the Fermi surface.[256]

Although encountering only "polite interest"[257] from Fermi, Peierls continued working on diamagnetism, and in January 1933 he submitted a second paper,[258] in which he considered the limit of strong magnetic fields, where the level spacing is now large compared with the temperature. He recalls that in examining the case of free electrons, "it suddenly dawned on me that you would in that case get a (more or less) periodic variation in the susceptibility as a function of the . . . reciprocal of the magnetic field intensity. And this of course immediately reminded me of the funny result of de Haas and van Alphen," which he had learned about during a visit to de Haas in Leiden in late 1930 or in 1931. As Peierls recounts, de Haas

> talked about this strange effect . . . which mystified him. And I remember he told me since he didn't understand what was going on, he was trying to look for the dependence of this effect on everything, including time. So he kept one particular specimen of bismuth in his cupboard, and every few months remeasured the effect to see if it was going to change. I found this phenomenon quite mystifying, but I don't think I attempted to find an explanation at that time.[259]

It was clear from the agreement of the theory that Peierls derived in his paper with measurements by de Haas and van Alphen on bismuth in strong fields that he now had the correct explanation.

The work by Peierls on strong-field diamagnetism is another of those curious situations where a scientist realizes only after making a discovery that he had earlier encountered, but had not been receptive to, the crucial ideas and facts underlying the discovery. For not only had Peierls not recognized the connection to de Haas and van Alphen's work until after he had his theory in hand, but he had not remembered that Landau had already presaged such an effect three years earlier. "Presumably I never read Landau's paper carefully, having had its main contents explained by him before publication, or if I saw the remark, I accepted Landau's assurance that it was unobservable, and promptly forgot it."[260] Peierls does not recall talking to Landau about working out the theory of the effect, but conjectures that

> "if I did, I imagine that he might have said that it was already known to him in his paper. Although as long as he thought that my result was correct, which obviously he must have done because he had obtained it himself, he might well not have bothered to point this out. This was quite within Landau's nature.[261]

The third side of this triangle, the lack of influence of Landau's work on the de Haas–van Alphen experiment, has recently been commented on by Shoenberg:

> The remarkable coincidence is that theoretical prediction and experimental observation of the oscillatory effect should have occurred almost simultaneously with neither side being aware of the other side's contribution. In fact the motive behind the Leiden experiments had nothing to do with Landau's remark, but was based on a long standing hunch of de Haas that there should be a close correlation between diamagnetic susceptibility and the change of electrical resistance in a magnetic field.[262]

Even though they proceeded along independent paths, the experimental and theoretical discoveries of the de Haas–van Alphen effect provide one of the first major examples of successful agreement between theory and contemporary experiment in quantum solid-state physics. As in electrical transport, the theory had now reached the level of sophistication where a working relation between theory and experiment could develop.[263]

Heisenberg's Theory of Ferromagnetism

In late 1926, while Pauli in Hamburg wrote his thoughts on paramagnetism to Heisenberg in Copenhagen, Heisenberg in his own letters told Pauli of his mullings on the problem of ferromagnetism.[264] The correspondence between these two friends brings out, on the one hand, the common origins of their interest in magnetism—the search for the symmetries of the wave function and statistics of a many-electron system—and, on the other, their different concerns: Pauli's with the statistics of gases, which would lead to the Sommerfeld–Bloch free-electron development, and Heisenberg's with few-electron systems, which would lead to the study of solids as extended molecular systems and to the modern theory of ferromagnetism.

Heisenberg had arrived in Copenhagen in May 1926, at the age of 24, to take up the position of lecturer.[265] Here he began to work on establishing the connection between the Pauli exclusion principle and the antisymmetry of the wave function of a several-electron system,[266] turning to "a practical problem . . . the helium atom with two electrons"[267] as a simple test case. He soon wrote two papers, submitted in June and July. The first, on the properties of systems of two and more like particles,[268] introduced the notion of the (dynamically conserved) symmetry of the wave function, the fully antisymmetric (determinantal) wave function for a many-electron system, and the concept of the exchange interaction, or "resonance" as Heisenberg termed it. The second, on the calculation of the spectrum of helium and other two-electron atoms,[269] explicitly introduced the Coulomb exchange integral.[270] The connection of these arguments to the question of statistics, a problem just then coming into focus,[271] was to be a theme in the correspondence with Pauli.

Responding on 28 October[272] to Pauli's letter of 19 October[273] on gas degeneracy that expressed Pauli's now "considerably milder" view of Fermi–Dirac statistics, Heisenberg attributed his long(!) delay in answering to the fact that Pauli's letter "constantly made the rounds here, and Bohr, Dirac and Hund are scuffling with us about it." Heisenberg continued, "With regard to the Dirac statistics we are in agreement . . . if the atoms obey your exclusion principle [*Verbot*], then so must the gas." A week later, Heisenberg revealed to Pauli:

> I myself have thought a little bit about the theory of ferromagnetism, conductivity and similar filth [*S——ereien*]. The idea is this: in order to use the Langevin theory of ferromagn[etism], one must assume a large coupling force between the spinning electrons (for *only these* turn). This force shall be obtained, as for helium, indirectly from the resonance.[274]

Here Heisenberg was taking the two crucial steps toward the theory of ferromagnetism: the first was to identify the elementary magnetic moments responsible for ferromagnetism as those of the recently discovered spinning electrons,[275] and the second was to realize that the exchange interaction could give the strong coupling required to align the moments.

By Christmas 1925, Heisenberg had accepted Samuel Goudsmit and George Uhlenbeck's hypothesis that the electron spin had associated with it an intrinsic magnetic moment,[276] and was aware that the gyromagnetic ratio g of the electron spin required by spectroscopic measurements agreed with that of ferromagnets, determined by the Einstein-de Haas-Barnett effect. The latter measurements, although from the first beset by uncertainties,[277] by 1925 had indicated with good accuracy a g factor of 2, characteristic of electrons.[278] Heisenberg's comment in his letter to Pauli, "for only these turn," in fact refers to these experiments, as is clear from his later paper on ferromagnetism,[279] in which he wrote that "it follows from the known factor $g = 2$ in the Einstein–de Haas effect (a value measured only in ferromagnetic substances), that in a ferromagnetic crystal only the intrinsic moments of the electrons are oriented, and not the atoms at all."[280]

The other basic presumption, the existence of strong internal forces between the fundamental magnetic moments in a magnetic material, underlay Weiss's phenomenological "molecular field" theory.[281] As described in Chapter 5, this theory drew on Langevin's 1905 statistical mechanical theory of the paramagnetism of magnetic dipoles,[282] as well as on earlier work by Ewing on internal forces in magnetic materials.[283] To find agreement with experiment, Weiss had to assume, ad hoc, the existence of forces orienting the atoms in a ferromagnet, forces about 10^3–10^4 times stronger than could be explained from simple magnetic interactions. Heisenberg's key contribution was to show how strong, nonmagnetic forces between electrons that favor spin alignment arise from the quantum-mechanical exchange interaction, in the same way as he had recently shown that they produce level splittings between electrons in singlet and triplet states of two-electron atoms.

He continued on 4 November,

> I believe that one can in general prove: parallel orientation of the spin vectors always give the *smallest* energy. The energy differences in question are of *electrical* order of magnitude, but fall off with increasing distance *very quickly*. I have the feeling (without knowing the material even remotely) that in principle this could be extended to give a meaning for ferromagnetism. To resolve the question of why, whereas most materials are not ferromagnetic, certain ones are, one must simply calculate quantitatively, and perhaps one can make plausible why circumstances are most favorable for Fe, Kr [*sic*] and Ni. Similarly, in conductivity the resonant wandering of the electrons à la Hund comes into play.[284]

Pauli, interestingly, made a notation on this letter: "ferromagnetism doesn't work [*geht nicht*]! (gas degeneracy),"[285] a comment probably reflecting his own fresh discovery that degeneracy of the conduction electrons leads to a striking suppression of their (para)magnetizability. When Pauli communicated his pessimism, Heisenberg responded on 15 November: "I nevertheless find very attractive the idea that it [ferromagnetism] has to do with resonance."[286]

Heisenberg's letter to Pauli confirms his admitted lack of familiarity with the problem; in particular, as his reference to Langevin indicates, he had not yet gone back to the literature, for Langevin had dealt with only paramagnetism and diamagnetism. Later, in his classic article on ferromagnetism, Heisenberg would correctly refer instead to Weiss. The origin of Heisenberg's interest in ferromagnetism is uncertain. One source was likely his friendship with physicists at the University of Hamburg, for whom magnetism was a favorite topic. Sommerfeld's student Wilhelm Lenz, and Lenz's own student Ernst Ising, both then at Hamburg, had recently developed (1920, 1924) the "Ising model" of ferromagnetism, in which spins, which can be either "up" or "down," interact with their nearest neighbors.[287] In 1926, Pauli and Wentzel held positions in Hamburg, while Otto Stern, who with Walther Gerlach had in 1920 discovered space quantization in molecular beams, directed Hamburg's program in experimental physics. Discussions about rotating electrons, with visitors such as Goudsmit in February 1926 and Frenkel in April 1926, were common at Hamburg at this time.[288] Heisenberg visited at least once, in January 1926.[289] In addition, Debye organized a magnetism week (21–26 June 1926) in Zurich, attended by Schrödinger, Pauli, Sommerfeld, Langevin, Stern, and Weiss.[290] Although Heisenberg apparently did not attend this meeting, he visited his teacher Sommerfeld in Munich the following month,[291] when Schrödinger was also present, and they may well have talked about magnetism, the topic of Sommerfeld's advanced topic lecture course at this time.[292]

Heisenberg did not submit his theory of ferromagnetism for publication until May 1928 (to the *Zeitschrift für Physik*), a year and a half after he spelled out his intuitions to Pauli. Two factors contributed to the delay. First, he did not have available in 1926 the group-theoretical machinery he would use to compute the energies of a many-spin system in terms of the exchange interaction. This would be developed in the interim by Wigner, Heitler, London, Hermann Weyl, and others, who extended the two-electron picture Heisenberg had used to study helium to many-electron systems, using the representations of the permutation group, which had been applied successfully to atomic spectrum analysis. This approach looked so promising around 1928 that not only Heisenberg but everyone working on the quantum theory of metals studied it.[293] Second, during this time Heisenberg became fully occupied with the conceptual foundations of quantum mechanics, developing the uncertainty principle the following spring. Heisenberg's thoughts appear not to have returned to ferromagnetism until after his move to Leipzig in the fall of 1927, when he suggested it as a thesis problem to Bloch. As Bloch chose instead to examine the electrons in metals, Heisenberg continued on ferromagnetism.

Once in Leipzig, Heisenberg contacted the Berlin group-theory experts, London, Wigner, and John von Neumann, to learn how to deal with calculating the exchange energy of a many-electron system. He later recalled the "close cooperation between Berlin and Leipzig. . . . I went to Berlin to discuss the matter because it was kind of an application of London's ideas on quantum chemistry."[294] London, who had studied under Sommerfeld, was at the time Schrödinger's *Assistent* at the University of Berlin. Wigner was *Assistent* to Richard Becker, another Sommerfeld student and at that time Professor of theoretical physics at the Technische Hochschule in Berlin.[295] Von Neumann, then also in Berlin, occasionally collaborated with Wigner.

Two months prior to Heisenberg, in March 1928, Frenkel at the Physical-Technical Institute in Leningrad[296] submitted a paper, published in the same volume of the *Zeitschrift für Physik* as Heisenberg's, on the magnetic and electrical properties

Paul Dirac, Jacov Frenkel (1894–1952), and Alfred Lande (1888–1975). (AIP Niels Bohr Library)

of metals at absolute zero.[297] Frenkel announced his paper to Sommerfeld in a letter that continued an intensive discussion that they had begun at the Volta conference in Como the year before (on the Sommerfeld free-electron gas model for metals).[298] In his paper, Frenkel briefly speculated that a spontaneous magnetic moment can appear as a consequence of a coupling arising from Heisenberg's resonance phenomenon between the individual spin moments of the free electrons, and between the spin vectors of the free and bound electrons,[299] "yielding, in certain circumstances, an unusually large negative value for the 'magnetic' energy." However, unlike Heisenberg, Frenkel did not develop this suggestion into a quantitative theory. No mention of Frenkel's paper appears either in Heisenberg's publications[300] or in his letters to Pauli, although hearing from Sommerfeld of solid-state studies in Leningrad might possibly have motivated Heisenberg to begin in earnest to work out his earlier ideas on ferromagnetism. These theorists kept in such close touch that they were coordinated almost like a team; for example, Sommerfeld warned another member of his circle, Georg Joos, who wanted to embark on a theory of ferromagnetism, "You had better wait. Heisenberg will presumably hit the center of the target again."[301]

Heisenberg used his scientific correspondence with Pauli, who was by now in Zurich, as a way of formulating his ideas; a series of seven letters and postcards, written between 3 and 21 May 1928, allows us to take a close look at the final stage in the development of his theory of ferromagnetism. In the first letter, of 3 May 1928, Heisenberg reveals that he considered ferromagnetism a diversion from "more important problems . . . annoying myself with Dirac['s quantum field theory]," problems that "today I have nothing new to say about." Instead, "I've dealt with . . . ferromagnetism."[302]

Of the three approximations available to compute the strong interactions between electron spins—the Pauli–Sommerfeld free-electron model, the Bloch tight-binding method, and the London–Heitler method to treat the "exchange of the valence-electrons of any two atoms in the lattice"—Heisenberg argued that when the last "is the best approximation, one can in certain circumstances obtain ferromagnetism." The condition was that the exchange term $J_{(12)}$ be positive. Using group-theoretical methods to calculate the energy levels, and neglecting their fluctuations in evaluating the partition function, he recovered "the Weiss [molecular field] formula" for the magnetization (although here for the first time the hyperbolic tangent, for spin ½, replaced Langevin's function in Weiss's theory).[303]

Four days later Heisenberg wrote,

> Today I would like to continue my epistle on ferromagnetism, and first write about the main objections, which you have naturally recognized for quite some time. . . .
> The major swindle lies in that I have inserted for all terms of the same total angular momentum j the mean energy rather than the real term value.[304]

He then proceeded to include fluctuations in the energy levels, using a Gaussian distribution, and arrived at a modified Weiss formula. Only for a lattice with the number of neighbors z at least eight did this result give a spontaneous magnetization going to zero at a critical temperature. For z at least six, he found another solution, which "has no physical meaning."[305] He concluded, "All in all, however, I find that one already understands the origin of F[erro]Magnetism substantially better." Later in the day, Heisenberg sent Pauli a short postcard in which he added that z refers to the number of "nearest neighbors," and that Fe, Co, and Ni satisfy his $z \geq 8$ condition for "ferromagnetism in the Weiss sense."[306]

Unfortunately, Pauli's letters to Heisenberg from this period are lost. Heisenberg's letters show that Pauli, playing his usual role of skeptic, expressed serious objections concerning both the physical assumptions and the mathematics; these objections served the important function of forcing Heisenberg to refine his theory. From letters to Bohr the following month, we learn that Pauli in fact considered Heisenberg's work on ferromagnetism "very beautiful"[307] and one reason for learning group theory from Weyl, although Weyl's "philosophy and life style," Pauli added, "are not to my taste."[308]

On 10 May, Heisenberg dealt with more of Pauli's objections: "I believe that I can now answer most of your questions, to a degree." First, he argued that the magnetic moments of the atomic cores could be neglected. He next turned to discuss the signs of the exchange integral J, admitting, "I have not succeeded in achieving a halfway useful evaluation of J. . . . I have the dark suspicion that J first becomes positive for p and d-states, thus not for s-states. . . . It seems not unlikely to me that in Fe-Co-Ni the d-states bear the guilt for ferromagnetism."[309] He then commented on the desirability of understanding the relationship between his theory and Pauli's theory of paramagnetism, and ended with further explanation of his calculation of the partition function. In a postcard written three days later, Heisenberg added that he had just succeeded "in gaining clarity about the sign of $J_{(12)}$. . . . J is negative for small principle quantum numbers [n], as for London and Heitl[er]; in contrast for large n it becomes positive. The boundary lies at $n = 3$, but it can just as well be 2 or 4."[310]

The next day, he responded to a comment of Pauli's concerning his own work on paramagnetism: "Actually, I also believe that your explanation of the susceptibility of the alkalies is correct; one can perhaps say that your theory is useful for metals of very large conductivity, while the third [Heitler–London] method is useful for those of smaller c[onductivity]. In reality both are but very crude approximations."[311] Finally, on 21 May, Heisenberg wrote to Pauli, "Well, I've sent Scheel [the editor of the *Zeitschrift*] a manuscript about ferromagnetism. . . . The assumption of a Gaussian distribution for the energy values, which for sufficiently low temperatures leads to false results, still seems unsatisfactory to me . . . . Perhaps I'll set one of my people here on to the calculation."[312]

Heisenberg's article[313] closely followed the lines of his correspondence with Pauli in both 1926 and 1928. After noting that neglect of electron interactions leads, according to Pauli, to paramagnetism or diamagnetism, he described the basis of his theory: "The empirical phenomenon that ferromagnetism presents is very similar to the situation we met earlier in the case of the helium atom." The clue was the splitting of the two-electron helium atom into singlet and triplet terms by the exchange interaction. He continued, "We will try to show that the Coulomb interaction together with the Pauli principle suffice to give the same result as the molecular field postulated by Weiss. Only very recently have the mathematical methods for treating such a complicated problem been developed by Wigner, Hund, Heitler and London." Recalling the Heitler–London expression for the exchange integral, and explaining how the exchange energy can tend to align spins, he then launched into a very formal calculation of the energy levels in terms of the characters of the permutation group, finally specializing to nearest neighbor interactions with a common exchange integral and introducing the Gaussian approximation.[314] His resulting version of the Weiss formula implied that a spin must have at least eight nearest neighbors for the system to become ferromagnetic,[315] a result he continued to regard as significant; it also implies that the system must become paramagnetic again at low temperature, but he did not "believe that this result has physical meaning. It arises mathematically through the assumed Gaussian distribution of the energy values."

Heisenberg concerned himself in the final section with the applicability of his theory to real ferromagnetic materials: iron, cobalt, and nickel. He found that to fit the transition temperatures of these ferromagnets required an exchange integral J_0 of order $\frac{1}{100}$ of the hydrogen atom ground-state energy, whereas due to the exponential falloff of the exchange effect, ferromagnetism should not occur in Fe or Ni solutions. However, "very much more difficult to answer is the question of the sign of J_0." Spelling out in more detail his arguments to Pauli that principal quantum number $n = 3$ is the first likely place that J_0 becomes positive, he concluded that the two conditions $n \geq 3$ and $z \geq 8$ together were "far from sufficient to distinguish Fe, Co, Ni from all other substances." Heisenberg was aware of the limitations of his theory and ended with the remark, "It was of course only to be expected that the temporary theory sketched here offers but a qualitative scheme into which ferromagnetic phenomena will perhaps later be incorporated. . . . I hope later to go into these questions as well as a thorough comparison of theory with experimental results."

Heisenberg's correspondence with Pauli on ferromagnetism included two letters in the summer of 1928,[316] the second mentioning that he had written a further paper on ferromagnetism for the *Sommerfeldfestschrift*,[317] dealing with the interaction of several valence electrons. Still mulling about his "unpleasant swindle," the Gaussian distribution, he continued,

> I'd like very much if Weyl could try this problem. I've completely given it up. The whole question seems important to me on account of the similarity between my model and Ising's. My present view is that Ising should have obtained ferromagnetism if he had assumed sufficiently many neighbors (perhaps $z \geq 8$). . . . That Ising uses this ["wild spatial"] model as an argument against ferromagnetism seems to me an indication that he did not understand in perspective his own work.

Heisenberg's immediate involvement with the foundations of ferromagnetism ends at this point; although the mathematical description of the cooperative effects in his model proved too difficult for the time, Heisenberg intuitively identified the correct physical basis of ferromagnetism as a quantum phenomenon, and thus opened the field of the quantum theory of ferromagnetism. The many problems left unsolved would in subsequent years be addressed by specialists in this field. Heisenberg's remarkable work between 1926 and 1928 further stands out as the first exploration of the physical consequences of electron–electron interactions in solids within the framework of quantum mechanics.

Spin Waves and Chains: Bloch, Slater, and Bethe

The main development of the quantum theory of ferromagnetism moved to Zurich, where Bloch, who was Pauli's *Assistent* in the 1928/1929 academic year, followed a study of the change in electrical resistance in an applied field (magnetoresistance)[318] with the first of a series of papers on ferromagnetism.[319] Recognizing Heisenberg's work as a correct insight, Bloch felt challenged to improve the "not very reliable" mathematics in which it was expressed,[320] and began with the role played by the conduction electrons. The question had been raised by J. Dorfman and co-workers[321] in Ioffé's institute in Leningrad. They argued from the observed anomaly in the thermoelectric effect at the Curie point in Ni that the specific heat discontinuity there arises from the conduction electrons; therefore, these must be the crucial actors in ferromagnetism, rather than the bound electrons considered by Heisenberg in his Heitler–London approach. By studying the free-electron model, Bloch could avoid Heisenberg's Gaussian assumption. Pauli was interested in this aspect of the problem, possibly, as Bloch later offered, because it extended Pauli's own treatment of paramagnetism, also based on a study of the conduction electrons.[322]

To determine whether conduction electrons can be the source of ferromagnetism, in his paper[323] Bloch carried out the original calculation of the now familiar exchange energy of a free-electron gas.[324] He discovered that only at low electron densities (or equivalently in narrow bands), too low for the alkalis, would the attractive exchange interaction dominate the zero-point energy of the electrons to produce a ferromagnetic state.[325] The zero-point motion of the electrons, he concluded, must be taken into account in deciding whether a metal can be ferromagnetic. While his calculation neglected the influence of the atoms on the electrons, he pointed out that these could be taken into account by using periodic (Bloch) wave functions in the exchange integral. The important contribution, he found after some calculation, "comes when the two electrons are close together, in the neighborhood of the same atom." He was led back to the region of applicability of the Heitler–London approach used by Heisenberg. But the answer was not obvious, for as he remarked, "the exchange integral can become negative, decreasing substantially the possibility for ferromagnetism. On the other hand we have shown earlier that a periodic potential can lower the zero point energy of the electrons so that in some circumstances a condition [for ferromagnetism] can be fulfilled."

In carrying out his free-electron calculation, Bloch applied the determinantal method recently developed by Slater in his theory of complex atomic spectra;[326] he thanked Slater in his paper for showing him the manuscript and for a number of friendly discussions. Slater, who visited Heisenberg's institute in Leipzig as a Guggenheim fellow during the summer and fall of 1929, recalls showing Bloch a preprint of this work in Zurich, which Bloch was "greatly taken with."[327] In this period, Bloch and Slater would have an important influence on each other's work—Bloch's earlier work on metallic structure and Slater's on complex atomic spectra converging in 1929 on the problem of ferromagnetism.

Slater had developed his theory of complex spectra in the spring of 1929 at Harvard, in an attempt to understand why Douglas Hartree's self-consistent field method[328] was so successful in analyzing atomic spectra.[329] Using a method based on Dirac's antisymmetric determinantal wave function for a many-electron system,[330] he constructed the many-electron function as a determinant of spin orbitals, and discovered that it gave a rather accurate self-consistent field.[331] This approach, which included from the outset the correct antisymmetry properties, was simpler than the group-theory method then in vogue, a point Slater emphasized at the start of his abstract: "Atomic multiplets are treated by wave mechanics, without using group theory." He recalls that when he arrived in Europe, "everyone knew of the work"; it was rumored that "Slater had slain the Gruppenpest [group plague],"[332] and so physicists could, as Bethe put it recently, "happily . . . forget all the group theory that we had learned."[333] Bloch recalls that "we were all relieved that one had a much more familiar way of expressing the content than all those general highbrow group theoretical arguments."[334]

During this stay with Heisenberg in Leipzig, Slater would contribute substantially to unraveling the origin of ferromagnetism. He entered the discussion of the merits and relationship of Heisenberg's exchange and Bloch's tight-binding methods. By comparing the two approaches in the context of the cohesion of metals, he saw that they formed different unperturbed bases for attacking metals by perturbation theory. It was, as he later put it, analogous to the relation between the "Heitler-London and the Hund-Mulliken molecular orbital approaches, respectively, to the molecular problem."[335] In his paper, he pointed out that the two methods "are essentially equivalent in their results when properly handled."[336] With this understanding, Slater pushed Bloch's recent demonstration that the nonmagnetic state of conduction electrons at metallic density has lower energy than the magnetic to "the quite general conclusion that the outer electrons, which are largely if not entirely responsible for both cohesion and conduction, cannot produce ferromagnetic effects."

He continued,

> It is a very attractive hypothesis to suppose that in the iron group the existence of the $3d$ and $4s$ electrons provides . . . the two electron groups apparently necessary for ferromagnetism; for it is only in the transition groups that we have two such sets of electrons, and this criterion would go far toward limiting ferromagnetism to the metals actually showing it.

Such inner electrons, if in well-separated orbits, could, Slater conjectured, have lower energy in a spin-aligned state. Finally, in his analysis of the cohesion of metals—a problem similar to that of the Heisenberg model of ferromagnetism, only with generally negative exchange interactions—he presented equations for the wave functions of the model. The result, basically a Schrödinger equation in difference form, would be used in subsequent studies by Bloch and Bethe. (Dirac's algebraic formulation of the Heisenberg model in terms of Pauli spin matrices[337] had not yet taken hold.)

The following spring, Slater was back at Harvard. He computed the sizes of incomplete atomic shells of various atoms in the periodic table, and showed that the $3d$ orbitals in the iron group satisfied the condition for spontaneous magnetization: their size was small enough for the energy decrease caused by the exchange effect to outweigh, à la Hund,[338] the energy increase from excitation of electrons at the top of the band.[339] He noted that if ferromagnetism "depends on the existence of incomplete shells within the atoms . . . the metals most likely to show it would be Fe, Co, Ni and alloys of Mn and Cu."

Up to this point, the work of Bloch and Slater was concerned with understanding the physical basis of the Heisenberg model, rather than with its mathematical solution. But when Bloch came to Utrecht during the fall of 1929, visiting Hans Kramers on a Lorentz Foundation fellowship (after spending the spring semester of that year as an assistant in Haarlem), he suddenly felt "quite free" after the busy time as Pauli's *Assistent,* and in this frame of mind began to "pick up old things."[340] He turned to the question of the physical predictions of the Heisenberg model itself, particularly on improving Heisenberg's treatment of ferromagnetism at low temperatures, a region in which both Weiss's molecular field theory and Heisenberg's calculations are invalid. Replacing Heisenberg's group-theoretical approach by that of Slater determinants, Bloch discovered "spin waves," the states corresponding to single or few spin flips in the fully aligned ground state. "I said, why if electrons can hop, spins can also hop."[341] Deriving a closed expression for their energy eigenvalues, Bloch went on to relate the low-lying spectrum to the thermodynamics. He made the remarkable connection that the fluctuations arising from the spin waves at low temperatures in one- and two-dimensional lattices destroy the possibility of ferromagnetism, whereas in three dimensions they give a $T^{3/2}$ falloff in the magnetization, compatible with the then-existing data.[342] In an allusion back to Heisenberg's calculation, he remarked that "not only the number of nearest neighbors, but also their arrangement plays a role."

Bloch gave a physical picture of the spin waves at the third Leipziger Vorträge,[343]

held in the summer of 1930 while he was still based in Haarlem. This meeting, one
of the earliest to specialize in solid-state topics, was organized by Debye with the
help of his *Assistent* Henri Sach, and was attended by von Laue, Bethe, Ioffé, Grüneisen, Bloch, Peierls, and Mott. The talks included Bloch on magnetism, Grüneisen on the temperature dependence of the electrical and thermal resistance of metals, and Peierls on the behavior of metallic conductors in strong magnetic fields.[344]
The degree to which the solid state was functioning as a target of opportunity for
quantum mechanics is underscored by Debye's hope, stated in the preface to the
proceedings, that the talks would not only illuminate pure electron diffraction and
interference experiments (i.e., the tools that enabled physicists to see inside the lattice) but also "verify what the wave mechanical conception can achieve in explaining the properties of metals."

Reviewing magnetism at the October 1930 Solvay Conference, Pauli summarized Heisenberg's, Bloch's, and Slater's progress in understanding ferromagnetism.[345] The open questions he noted were, first, why so few substances are ferromagnetic; Slater's relation between ferromagnetism and orbit size did not, to Pauli,
"seem solidly based." Second was the magnitude of the saturation magnetization,
which "is not compatible with the hypothesis of an electron freely circulating about
an atom, and it seems rather that several atoms must participate." And third was
the problem of including the influence of the crystal lattice on the direction of magnetization. He continued:

> Under [actual] conditions Bloch's approximation is rather bad; one should how-
> ever consider as established his general results that ferromagnetism is possible
> under conditions very different from those in which the Heitler-London method is
> applicable, and that in general it is not sufficient only to consider the signs of the
> exchange integrals.

On the difficult problem of solving the Heisenberg model, Pauli laid down what
would be a challenge to subsequent generations: "an extension of the theory of Ising
to a three-dimensional lattice might give ferromagnetism."

The following spring, Bethe, then at Fermi's institute in Rome as a Rockefeller
Foundation fellow, turned to the solution of the Heisenberg model. He decided, as
he wrote to Sommerfeld in May 1931, to "treat the problem of ferromagnetism
decently [and] . . . really calculate the eigenfunctions."[346] In this letter, he commented in detail on the limitations of Bloch's spin wave theory, which Bethe felt
did not discuss the solutions "precisely enough." In a paper, Bethe analyzed the
one-dimensional chain of spins.[347] The exchange interacton J was either positive,
corresponding to the Heisenberg ferromagnet, or negative, corresponding to the
"normal" (or now the antiferromagnetic) case, relevant for the cohesion problem.
Beginning with the fully aligned state, he determined the wave functions of states
having an arbitrary number of reversed spins, starting from his famous *Ansatz* for
the case of two interacting spin waves and then generalizing it. The calculation,
although incomplete in many respects (e.g., "for the [$J < 0$] case, the solution of
lowest energy is not yet established") is notable as the first exact solution of an interacting quantum many-body problem. Bethe expected to extend his one-dimensional analysis to the physically interesting three-dimensional case in a following
work, but did not succeed. He would argue, in the later *Handbuch* article, that when

J is positive, all three-dimensional lattices are ferromagnetic, whereas as Bloch had found, two-dimensional lattices and linear chains never are.[348]

Bethe's results did not impress Bloch. In a letter to Peierls from Copenhagen in November 1931, Bloch complained that the work did not deal adequately with the low-temperature regime, which Bloch had studied in his own spin-wave paper:

> It appears to me that Bethe's tedious algebraic manipulation [*Ixereien*] is somewhat academic in character, in particular because it does not sufficiently discuss the neighborhood of the lowest eigenvalues. I believe that in this regime, however, my calculations are reasonable, since they neglect only the exclusion of spins on the same site and this cannot play a role in a very dilute spin gas.[349]

Bloch's Dissertation on Ferromagnetism

Bloch himself had progressed on ferromagnetism during the past several months, and in the summer of 1931, at the end of his year as Heisenberg's *Assistent,* he sat down to write his postdoctoral dissertation [*Habilitationsschrift*].[350] Much of this "long and learned" paper,[351] published in 1932, was worked out during a protracted hospital stay as Bloch recovered from a broken leg resulting from a climbing accident.[352] Nominally devoted to exchange interaction problems and residual magnetization in ferromagnets, the paper presented an exceptional wealth of formalism, which has become part of the fabric of the modern theory of condensed matter physics and collective phenomena. Beyond its contribution to the theory of domain walls, this work serves as a bridge between the quantum theory of ferromagnetism in the early 1930s and present theories of many-particle systems.

Bloch began by introducing in the Heisenberg problem, as formulated by Slater, two sets of "second quantized" operators per site. These operators would create or annihilate spins pointing to the right or left (as spins in these days usually pointed, rather than up or down as they now do). By showing how the Hamiltonian remains invariant under rotation of the coordinate system, he derived a representation of quantum-mechanical angular momentum in terms of two harmonic oscillators, here the two sets of spin creation and annihilation operators (this representation was later exploited by Julian Schwinger).[353] Bloch then turned to statistical mechanical questions, and in a clear reference to Bethe's recent paper, pointed out that "in fact to answer many [statistical] questions, knowledge of the stationary states is in principle not needed." To illustrate his point, he introduced a useful connection between temperature and imaginary time, showing how finite-temperature statistical problems can be described in terms of a Schrödinger-like equation in imaginary time. Considering next dynamics of the Heisenberg ferromagnet, he derived the Heisenberg representation equations of motion of the creation and annihilation operators. Then neglecting the fluctuations of the operators (i.e., treating them as a pair of complex order parameters whose squares give the spin densities, and making a long-wavelength expansion "fully analogous to replacing the lattice by a continuum in the Debye theory of specific heat"), he derived differential equations of motion for the order parameters. In the limit of a nearly aligned system, the equations reproduced the long-wavelength spin-wave spectrum that Bloch had earlier derived.

In the final section, Bloch focused on the role of magnetic dipole interactions in leading to domain structure and residual magnetization, making the first connection between the Heisenberg model and the magnetization structure of real ferromagnets. Motivation for this work came in part from Bloch's discussions with Becker. "We met in a little village in between [Leipzig and Berlin] once, just to talk about ferromagnetism."[354] In his paper, Bloch wrote:

> We want to show here that . . . the ordinary dipole forces between the spins influence the grouping of the spins in a crystal in a decisive way. . . . The weak magnetic energies can do this since we are dealing with a very large system in which, despite the strong exchange interaction, the energies of various stationary states lie extremely close together so that even very small secular disturbances can still have a very great influence.

Including the dipole energy approximately in his long-wavelength equation for the order parameters, Bloch showed that the order parameters in the limit in which only the spin density of one spin orientation or the other is large become determined by precisely a nonlinear time-dependent Schrödinger equation. (The nonlinear terms reflect the magnetic dipole interaction.) The domain wall, which he showed in a calculation that would become familiar in all details to students of the Ginzburg–Landau equation, was then described by a "kink" solution of the nonlinear equation, proportional to sech x, in which the spin density changes in space from entirely "right" to entirely "left." The more exact equations, he pointed out, also contain a kink solution, derived by Heisenberg by a variational method. Bloch recalls discussing this work on order parameters with Landau in 1931 while visiting Kharkov,[355] and, indeed, similar ideas would underly Landau's later work, such as his theory of phase transitions in 1937 and the Ginzburg–Landau equation of superconductivity.[356]

With Bloch's dissertation, the initial development of the quantum theory of ferromagnetism reached a stage where physicists could hope, with application of realistic electronic wave functions in metals, to understand properties of physical ferromagnets. Parallel to the development of the single-electron quantum theory of metals out of Pauli's study of paramagnetism as a test case in the statistics of gases, the quantum theory of ferromagnetism evolved out of Heisenberg's concern with interacting few-electron systems—in particular, his favorite test case, the helium atom. Indeed, one may regard Heisenberg's works on helium as a second independent root of the quantum theory of solids. The one-electron theory of metals was insufficient to explain ferromagnetism, and the major works examined here, by Heisenberg, Bloch, Slater, and Bethe, indicated clearly that a satisfactory explanation required taking into account the collective interactions between electrons in the crystal lattice. These works, in their attempts to treat many-electron interactions, contain the beginnings of the modern theory of collective phenomena, a theory that would reach fruition more than two decades later.[357]

2.6   1929–1933: Superconductivity

Although superconductivity was first observed in 1911, in the decade following the invention of quantum mechanics it remained a conspicuously stubborn and insoluble problem. Between 1929 and 1933, more than a dozen theoretical physicists—including Bohr, Pauli, Heisenberg, Bloch, Landau, Brillouin, W. Elsasser, Frenkel, and Kronig—armed with the successes of the quantum theory of metals and new observations, were optimistic that the new tool would also help them to explain superconductivity. In their approaches to this phenomenon, the theorists portrayed their confidence in the power of the new mechanics, even in the face of continual frustration at the failure of their theories to agree either with experiments or with theories proposed by colleagues. As Bethe lamented in 1933, in superconductivity "only a number of hypotheses exist, which until now have in no way been worked out and whose validity cannot therefore be verified."[358]

The history here is far from a cumulative, triumphant progress, but no less typical of theoretical physics—a group inventing one odd tool after another in hopes that one of them will pry the puzzle open. To illustrate, we will examine two of the most prominent conceptions then under discussion: the spontaneous current theories of Bloch, Landau, and Frenkel centering on the notion of a current-bearing equilibrium state (1929–1933), and the electron lattice or electron chain theory of Bohr and Kronig (1932–1933). To suggest the larger picture of research on superconductivity, we also sketch three of the other theories in the air during 1932 and 1933: Elsasser's, expressed in terms of relativistic electrons; R. Schachenmeier's, based on exchange between conduction and bound electrons; and Brillouin's, in which superconductivity was associated with electrons "trapped" in metastable states.

The experimental information on superconductors available in the 1920s was quite incomplete.[359] The oustanding feature of superconductors, whose understanding was the major theoretical focus, was the loss of resistivity at very low temperatures, a loss that, as Kamerlingh Onnes and co-workers in Leiden demonstrated, was indeed total.[360] By 1913, the disappearance of superconductivity in strong magnetic fields was discovered,[361] and was realized soon after to be an effect not of local heating in "bad places," but of the field itself.

In 1923, J. C. McLennan established in Toronto the second cryogenic laboratory to engage in superconductivity research,[362] and in 1925, when experimentalists headed by Walther Meissner at the Physikalisch Technische Reichsanstalt (PTR) in Berlin liquefied helium,[363] the PTR became the third such laboratory. While in Leiden and Toronto, experiments centered on the phenomenology of superconductivity, including effects of magnetic fields and the changes of properties through the superconducting transition, the initial PTR program concerned the problem of "whether all metals become superconducting."[364] Within three years, Meissner's group had analyzed 40 metals, adding to the list of the superconducting materials tantalum, titanium, thorium, and niobium. In 1929 they also measured chemical compounds and alloys and found that even materials composed of insulating and nonsuperconducting metals (e.g., copper sulfate) could become superconducting, indicating that superconductivity was not simply an atomic property.[365] Evidence that superconductivity was not a solid state "dirt effect," like normal conductivity, was provided by Leiden measurements that showed that superconductivity does not depend essentially on the purity or crystalline order of the material.[366] Meissner, on the other hand, found that even the purest single crystals of a normal conductor like gold do not necessarily become superconducting when cooled down to temperatures as low as 1.3 K, the PTR's lowest operating temperature.[367] Adding to the mystery was the Leiden observation that heat conduction, which is primarily by electrons, remains continuous through the superconducting transition, even while the electrical conductivity becomes infinite.[368]

Two essential features of superconductors were learned too late to influence microscopic work between 1929 and 1933. The first, revealed in 1933 in the classic experiment of Meissner and Robert Ochsenfeld in Berlin (published in October 1933 in *Die Naturwissenschaften*),[369] was that superconductors expel magnetic flux.[370] This effect would suggest that the fundamental characterization of superconductors is perfect diamagnetism rather than vanishing resistivity. The second feature, which would come into focus between 1932 and 1934 through the work in Leiden of Willem Keesom, J. A. Kok, and others,[371] as well as Ehrenfest, A. J. Rutgers, Cornelius Gorter, and Hendrik G. B. Casimir,[372] was that the transition to the superconducting state is reversible and that the superconducting state—unique only if, as demonstrated by Meissner and Ochsenfeld, flux is not frozen into superconductors—can be described by thermodynamics.

The handful of theoretical attempts to understand superconductivity from microscopic principles before the development of quantum mechanics was limited by the failure to understand the behavior of electrons in normal metals.[373] For example, F. A. Lindemann in 1915[374] and J. J. Thomson in 1922[375] constructed theories of ordered electron structures that avoided the electron specific heat dilemma while providing a model for superconductivity.

Lindemann's theory of normal conductors and superconductors rested on the hypothesis "that far from forming a sort of perfect gas the electrons in a metal may be looked upon as a perfect solid"; thus, à la Born–von Kármán, the electron specific heat became greatly reduced below its classical value. Superconductivity followed; if the repulsive force between the electrons and ions were sufficiently short-ranged and the ionic motion small, as might occur at low enough temperatures, "the electron space-lattice can move unimpeded through the atom space-lattice."

Thomson's attempt to explain metallic conduction was expressed in terms of "chains of electrons lying along a line of a lattice . . . travelling along that line carrying energy and electricity from one part of the solid to another." As in Lindemann's theory, the ordered electrons would have a greatly reduced specific heat. Furthermore, "the amount of enery communicated" between the chain and lattice, Thomson wrote,

> will fall off very rapidly as the ratio of the duration of the collision to the time of [atomic] virbration increases. . . . [T]hus when the temperature gets so low that the time taken by an electron to pass [an interatomic distance] is comparable with the time of vibration of the atom, any dimunition in the temperature will produce an abnormally large increase in the conductivity, and thus the metal would show the super-conductivity discovered by Kammerlingh [*sic*] Onnes.

Unlike Lindemann, who attempted to explain perfect conductivity, Thomson predicted greatly enhanced normal conductivity. The electron space lattice and chain concepts, a foreshadowing of the need for collective ordering of the electrons in superconductors, would reappear in Bohr's and Kronig's later theories of superconductivity.

Einstein, describing problems of electron conduction in metals at the November 1922 meeting in Leiden to celebrate the fortieth anniversary of Kamerlingh Onnes's professorship,[376] shared the need to avoid free electrons: "It looks as though, according to today's state of our knowledge, free electrons do not exist in metals at all." In superconductors, "it seems unavoidable that supercurrents are carried by [electron transfer along] closed molecular conduction chains." However, he remarked with foresight, "with our far-reaching ignorance of the quantum mechanics of composite systems we are very far from being able to compose a theory out of these vague ideas."

By the end of the 1920s, quantum theory had developed to the stage where solving superconductivity seemed a more realistic goal. The turning point was again Bloch's doctoral thesis of mid-1928, which put the theory of the conductivity of normal metals on a firm foundation. He strongly suggested, through his calculation of the low-temperature resistivity, that perfect conductivity could not be obtained simply from using the single-electron approach at extremely low temperatures. But in his thesis, Bloch did not attempt to deal with superconductivity, except to remark at the very end, after illustrating "the possibility of a transition between two completely different laws of conductivity" (that between degenerate and nondegenerate electrons), that "the phenomenon of superconductivity shows that such a transition actually occurs, which . . . remains up to now not clarified."

Bloch's thesis had an immediate impact on Bohr, who, with a standing interest in the electron theory of metals since his own thesis in 1911, was apparently thinking about solving superconductivity from a single-electron picture. Bohr wrote to Heisenberg at Christmas 1928: "How have you been doing with superconductivity? Bloch's beautiful work, which you so kindly sent me, and from which I had much pleasure, taught me of course that the way out which I indicated was not possible."[377] He would make another attempt three years later, which would again fall victim to Bloch.[378]

Spontaneous Current Theories: Bloch, Landau, and Frenkel

Bloch began thinking seriously about superconductivity when he moved from Leipzig to Zurich in the fall of 1928, soon after finishing his thesis, to become Pauli's *Assistent*.[379] His theory, which he never published,[380] shared a key idea with one that Landau formulated contemporaneously and finally published in 1933.[381] This idea was that in equilibrium the thermodynamically favored superconducting ground state, corresponding to a minimum of the free energy, bears a finite spontaneous current below the critical termperature, whereas at higher temperatures, current-free equilibrium states have statistically greater probability.

Pauli's attitude toward this work, Bloch recalled, was "get on with it so as to be finally done with all these 'dirt effects.'"[382] "Pauli, after all, was a physicist, and as such he could not entirely ignore the interesting problems in solid state physics, but he didn't really have his heart in it."[383] However, Pauli himself was clearly involved in superconductivity, telling Bohr in January 1929, "On the question of superconductivity I could not come to any definite result."[384] And he proudly described Bloch's pursuit of superconductivity in numerous letters in the spring of 1929.

Writing in March to Oskar Klein, who had succeeded Heisenberg as Bohr's assistant in Copenhagen, Pauli reported that "Bloch here has made progress with a theory of superconductivity. I will not assert that he has already succeeded in finding an explanation . . . but his results bid fair hopes. In any case, I now believe that the way conjectured by Bohr last fall was completely false."[385] In a postcard to Bohr the next month: "Bloch's theory of superconductivity seems to be getting very beautiful!"[386] To Munich, he wrote in May: "Mr. Bloch is busy working out a theory of superconductivity. The job isn't finished, but it seems to be working."[387] In the same letter, he also referred to the Peierls study of heat conduction in insulators, which both Peierls and Pauli hoped might provide a model for superconductivity. And finally, to Kronig in June: "Bloch is pursuing superconductivity and is altering his theory on a daily basis (Thank God before publication!)."[388]

Both Bloch and Landau felt that ferromagnetism and superconductivity were closely joined phenomena. The currents implied by ferromagnetism and the currents of superconductivity "both persisted . . . there must be a common cause."[389] Indeed, Bloch worked on the two problems at the same time, with the thought in the back of his mind that the answer to both lay in the electron–electron interactions. As he later wrote, an

> appealing interpretation [of superconductivity] was suggested through analogy with ferromagnetism, where remanent magnetization had been explained by recognizing that parallel orientation of the magnetic moments of the atoms leads to a lower energy than random orientation. Similarly, it seemed plausible to interpret current flow in a superconductor as the result of a correlation between the velocities of the conduction electrons that is energetically favored and, therefore, manifests itself at sufficiently low temperatures."[390]

However, Bloch made far less progress in superconductivity than in ferromagnetism. In each of many calculations, he found that the minimum energy state bore no current, a result that came to be known as Bloch's first theorem on superconductivity. Brillouin later stressed the importance of this theorem at the May 1935 meeting of the Royal Society of London, since it "practically forbids any interpretation of superconductivity within the frame of classical theory."[391] The point, as Brillouin had demonstrated with a simple classical argument,[392] is that were any current to flow, one could always decrease the total energy by applying a potential difference of one sign or the other across the conductor for an instant; thus the energy could not have been at a minimum. The only way out, Brillouin realized, is for the current not to be "stable but only metastable," which seemed out of the question because "Meissner's experiment proved decisively that supra-currents were stable." (At this stage, nobody understood that superconductive states are extraordinarily metastable.) So stymied was Bloch that in exasperation he formulated a tongue-in-cheek second theorem: every theory of superconductivity can be disproved! The frequency with which this unpublished theorem was quoted indicates its appeal in this period to other physicists struggling with superconductivity.[393]

Landau had little more success. Pushing the analogy between superconducting and ferromagnetic states to the fullest, he assumed in a published article[394] that superconductors contain local "saturation currents" that flow in different directions, producing no net current unless organized by an applied field. Landau provided no microscopic justification for such a picture, and, indeed, his paper was not based on quantum mechanics; rather, he assumed a phenomenological description of the energy as a sum of the spatially varying local magnetic field energy plus a term "designated, in correspondence with the density gradient term in the theory of surface tension, as the capillary term," proportional to the curl of the local current in the superconductor (the Londons would later use the current itself).[395] Although the analogy with ferromagnetism and the theory of surface tension turns out to be false, the paper remains interesting for the seeds it contained of Landau's later work on phase transitions and the Ginzburg–Landau theory of superconductors. Particularly noteworthy was the expansion of the free energy near the transition temperature T_c in terms of an order parameter in both those theories. Here Landau expanded the free energy near T_c in the magnitude i of the saturation current, $F = F_0 + ai^2/2 + bi^4/4$, with the now familiar assumption that the coefficient a changes from positive to negative as the temperature is lowered through T_c. Although the order parameter of superconductivity was misidentified, the theory gave the correct qualitative behavior of the entropy and specific heat near T_c. Landau also ventured the possibility that the orientable internal magnetization in his model could lead to magnetic flux expulsion up to a particular external field strength. Whether he knew of the work of Meissner and Ochsenfeld, published over half a year after his paper was submitted, we do not know. The paper provides no internal evidence; Landau characteristically gave few references in his papers. He concluded with a hint that his theory was unsatisfactory, since it suggested a $(T_c - T)^{1/2}$ behavior of the critical magnetic field, compared with the linear dependence on temperature near T_c that had been recently observed.[396]

Frenkel, in Leningrad, put forth at the end of 1932 a related version of the spontaneous current theory,[397] arguing that the magnetic forces between electrons—forces "totally neglected hitherto in the electron theory of metals"—would encourage them to move in stable parallel streams rather than randomly, yielding a local current. Deducing that the magnetic interactions produce a large electron effective mass, he concluded that "so long . . . as the electrons in a metal move collectively as an organized crowd of sufficiently large size, their motion can remain unaffected by the heat motion of the crystal lattice, the quanta . . . of the heat waves being insufficient to knock out even a single electron." In analogy with ferromagnetism, he also presented the picture that there exist "in a superconducting body, regions with whirl currents whose orientations . . . vary in an irregular manner from one region to another." Frenkel also ventured that a metal in the superconducting state in an external field "must behave like a *diamagnetic* body with a large negative susceptibility . . . the interior of such a body will be screened from external magnetic fields by the system of surface currents induced by the lattice." The Meissner effect, reported in October 1933, would verify this intuition. Two months after Frenkel's theory was published, Bethe and Herbert Fröhlich severely criticized it in a paper sent to the *Zeitschrift für Physik*,[398] showing by a more precise argument that the magnetic interactions lead to only a tiny correction of the effective mass, and that "all the formulae of the usual conductivity theory remain fully in force."

Electron Lattice Theories: Bohr and Kronig

The idea of spontaneous currents as states of lowest electron energies was, according to Bethe in the 1933 *Handbuch* article, "extremely tempting; however, until now there has been no success in constructing a model with the required properties."[399] An alternative approach, adopted independently by Bohr and Kronig, was to treat supercurrents as a coherent motion of the entire ground-state electron distribution, a quantum resurrection of the earlier Lindemann–Thomson models. The lively correspondence on superconductivity among Bohr, Bloch, and Kronig between June 1932 and January 1933 provides a rare insight into this development, showing in particular how, through Bloch's persistent criticism, Bohr lost confidence in his theory, as eventually Kronig did in his.

Bohr had returned to superconductivity in the spring of 1932. In mid-June he wrote to Bloch to try out his thoughts: "I would awfully much like to talk with you on several questions concerning metallic properties, . . . namely a thought touching on superconductivity which I got and cannot get away from, even if I am far from understanding the connection between superconducting properties and ordinary electrical conduction." His idea was that "superconductivity concerns a coordinated motion of the whole electron lattice," the many-electron wave function of a supercurrent-carrying state being only a slight (effectively long wavelength) modulation of the normal ground-state wave function. Because the electron motion involves long wavelengths,

> the current will not be significantly disturbed by the thermal oscillation of the metallic lattice. . . . The transition to a state where the electrons move uncoordinated between one another should be somewhat analogous to a melting of a solid body, but it has not been possible so far for me to make my understanding of the process of this transition clear.

Expressing his faith that the solution to superconductivity lay in the new mechanics, he wrote,

> Just as quantum mechanics has first made it possible to bring the picture of "free" electrons in metals in closer correspondence with experiment, it seems also that first through quantum mechanics . . . one can understand how the two lattices can move through each other without resistance and significant deformation. I should be very happy to hear a few words from you on how you look upon all this.[400]

Bloch's reaction was rather doubting. On his way to Brussels in late June to arrange the next Solvay Conference, Bohr met with Bloch in Berlin to discuss his theory. As he related to Max Delbrück, "I had a very lively discussion with Bloch . . . and I think that I succeeded in some way to overthrow his scepticism." He also wrote that "the day before yesterday in Liege with Rosenfeld's help I wrote a little article which I have sent to *Die Naturwissenschaften*."[401] The article, entitled "Zur Frage der Supraleitung" (On the Question of Superconductivity) and dated June 1932, spelled out in more detail the basic conception he described to Bloch, and gave estimates of the magnitudes of the supercurrent associated with his modified wave function, from the limits of nearly free to tightly bound electrons. Bohr acknowledged "illuminating discussions" with Bloch and Rosenfeld. Although the

Werner Heisenberg and Niels Bohr (1885–1952) at Copenhagen conference, Bohr Institute, 1934. (AIP Niels Bohr Library, Weisskopf Collection. Photograph by Paul Ehrenfest, Jr.)

paper was accepted for the 11 July issue,[402] as a result of Bloch's continuing objections Bohr withdrew it in the proof stage.

An undated handwritten note by Bloch at this time reveals the depth of his criticism of Bohr's theory. It suggested that Bohr

> let it lie, on account of doubts: (1) experimental facts, a) magnetic fields (could possibly be understood), b) McLennan's experiments [on the superconducting transition in the presence of alternating currents] (can *not* be explained). (2) Proof of the general existence of current-carrying solutions doubtful. Difference between conductors and insulators not sufficiently considered. . . . (3) The concept could become meaningful since it deals with bringing the secured features of atomic mechanics into agreement with the empirical facts of superconductivity. (4) The concept is basically different from attempts to treat the interactions of the electrons as a rigid lattice in an intuitive [classical] picture. The latter is ruled out because the melting energy would be too high. The picture of a moving lattice is neither in agreement with the binding of the electrons to the ions nor with its motion. (5) The change of state at the critical point should not be determined by typical quantum mechanical features in an obvious way.[403]

Bloch's letters to Bohr were more deferential. In mid-July he wrote from Leipzig, telling how

> very happy I was also about the discussions in Berlin and welcome very much that you are letting a note about superconductivity appear. . . . My comments then were not meant as "negating criticism," rather I consider it entirely possible that your ideas really contain the solution of the riddle. I merely wanted to point out that one needs to be somewhat more careful in the evaluation of the order of magnitude of

the currents and believe, so far, that the discussion of the model of tightly bound electrons has *something* to do with the problem. In any case, it would interest me a great deal to see the proofs of your note![404]

Then two weeks later, Bloch sent Bohr detailed remarks on his manuscript, adding, "I'm quite in agreement with the whole 'tone' of the note and perceive only the lack of more precise conceptions still somewhat unsatisfactory." He also asked Bohr if he would hold up publication until October when he could come to Copenhagen.[405] Bohr took Bloch's comments seriously and in early September wrote back that "I have held back my article on superconductivity because I had second thoughts about it all on grounds of the paradoxes that show up when one treats the lattice field as fixed in handling the electron system."[406]

While Bohr and Bloch pursued their discussion, Kronig in Groningen developed a similar electron lattice theory, which he subsequently published in two papers in the 1932 *Zeitschrift für Physik*.[407] The first proposed that the interactions between electrons led them to form a rigid lattice intermeshed with the ionic lattice. The electron system could be superconducting, since "in analogy with Bloch's theory for a single electron, translation [by an electron lattice constant] of the whole electron lattice can experience no resistance."

On learning of Kronig's theory from McLennan in mid-October, Bohr wrote to Kronig,[408] enclosing a copy of the proofs of his paper and inviting him to Copenhagen to talk with Bloch, Rosenfeld, and himself. He remarked that he delayed returning the proofs in part because of his difficulty in bringing his theory into agreement with McLennan's experiments on the dependence of the superconducting transition temperature on the presence of high-frequency alternating currents. But after discussing McLennan's most recent results with Bloch and Rosenfeld, he felt it would be best not to wait longer but to publish his note, possibly in somewhat altered form.[409] Kronig (whose early invention of the spin of the electron had met resistance from Pauli and Heisenberg, to Bohr's later "consternation and deep regret"[410]) wrote assertively to Bohr the next day that the new physical content of Bohr's paper was in fact "covered" by his own, and commented parenthetically, "My result is somewhat more specialized but offers . . . more prospects for quantitative evaluation." He added, "Unfortunately, it was scarcely possible to mention your work, even in an added note, since I had already completed the corrections. . . . I enclose a copy of the proof, since you can best see the content of my conceptions in that."[412]

The next week Kronig visited Copenhagen,[412] and the discussion among Bohr, Bloch, and Kronig went into full swing. Bohr and Bloch raised the objection that for the electron lattice to carry a current, for N electrons it would have to be able to tunnel through the N potential hills between lattice sites, which becomes impossible as N grows large. Kronig responded after returning to Groningen,[413] invoking the zero-point motion of the electrons, and at the same time presenting Bohr with a mathematical argument (based, it would appear, on a faulty expansion of the wave function in the magnitude of the current carried) that Bohr's proposed current-carrying wave function, a modulation of the ground state, could not be a steady-state solution of the Schroödinger equation. Kronig also mentioned that he was submitting a second paper, and had sent a copy to Bloch asking him to send it on to Bohr.

Ralph Kronig (b. 1904) and Enrico Fermi (1901–1954) at a beach near Zurich, 1929. (AIP
Niels Bohr Library, Rudolf Peierls Collection)

A few days later, Bloch sent Kronig's manuscript to Bohr, as requested, with
frank comments on Kronig's idea that a large electron zero-point motion would
allow the electrons to overcome the potential hills:

> I would like straightaway to . . . make you aware of the point in Kronig's work that
> seems to me wrong. Kronig discusses the case of the linear electron lattice and finds
> that the mean square ϵ^2 of the zero-point amplitude grows as log N . . . [and] that
> the matrix elements of the interaction with the ion lattice . . . [can] become inde-
> pendent of N or even disappear with growing N. To that one can remark: 1) If the
> zero point amplitudes really become so large, then one can naturally no longer
> speak at all of a lattice. 2) The Kronig result is very specially tied to the one dimen-
> sional case.

Bloch went on to explain, using arguments familiar from his spin-wave paper, that

> the logarithmic growth of ϵ^2 with log N comes . . . mainly from the long elastic
> waves which, on account of their small frequency, have a very large zero-point
> amplitude. In the two and three dimensional case, the long wavelength waves, on
> account of their relatively small number, play practically no role, and one can also
> easily show then that for the largest values of N, ϵ^2 becomes independent of N. (It
> would also be sad if it were otherwise, for then an NaCl lattice even at absolute zero
> would fall apart.)[414]

Kronig wrote again to Bohr a week later, acquiescently:

> Mr. Bloch pointed out to me, and with full justification, that the given conception
> cannot be carried over to the three dimensional case. . . . There are then the follow-

ing possibilities: 1. The idea of the electron lattice is entirely unusable. 2. Electron exchange saves the three dimensional case . . . 3. Superconductivity must not be ascribed to a translational movement of all electrons but rather to the building of a one-dimensional electron lattice.

He went on to mention a note from Meissner, which gave the first indications of his experiments on penetration of fields in superconductors:

Mr. Meissner writes further: "It follows that in the vast majority of cases the super-conducting current in all probability is a surface-current flowing on the surface of conductors, so that only the outermost electrons can be displaced by the electron lattice. . . ." This would be in accord with the chain picture. I have asked Mr. Meissner to tell me in detail how he comes to his conception, but up to now have received no answer.[415]

The concept of a "translation of individual linear chains of electrons in the lattice" would be the basis of a revised version of Kronig's second paper on superconductivity, published at the end of the year. As he wrote to Bohr in mid-December, "On the basis of my mathematical considerations, this change of conception in fact appears to me altogether unavoidable." He remarked that "as before I would like to believe that the transition point corresponds to a phase change in which the conduction electrons go over from an unordered distribution, similar to a liquid, to an ordered state." He also inquired whether Bohr was considering publishing something of his original note, or whether he would prefer to "shroud the existence of this unpublished work in the cloak of silence."[416]

Bohr replied to Kronig just after Christmas, reiterating that

as far as I can see, the case stands just as we discussed it in Copenhagen. . . . It remains my conviction that the difference between the two phases corresponding to superconductivity and usual metallic conductivity is a purely quantum problem, which quite escapes visualization by means of basically mechanical pictures, and on grounds of the large differences in our conceptions it is difficult for me to agree with details in your second treatment.[417]

A few days later, Bloch in writing Bohr provided a more technical coup de grâce:

As far as I have understood your letter to Kronig I am completely in agreement; the more I have thought about it, the more I consider the Kronig work to be in error. The last hope that Kronig places in . . . the one-dimensional lattice is unjustified, because the condition, which according to Kronig would cause the overtaking of the potential hill, simply means that the zero-point oscillations are so big that this lattice cannot exist. I do not know whether Kronig at the time I wrote to him about it understood me completely. In any case his answer allowed the possibilitiy of giving up the idea of the electron lattice.[418]

Neither Kronig nor Bohr progressed further with electron lattice theories of superconductivity. Even while making no substantial progress himself, Bloch through his role of critic emerged from the debate as the main authority of the time in the microscopic theory of superconductivity. In his summary of superconductivity in the 1933 *Handbuch* article,[419] Bethe did not discuss Bohr's work, but suggested that despite flaws in Kronig's theory (brought out earlier by Bloch), it might, "with substantial deepening of its foundations," be the basis of a useful theory.[420]

Other Attempts to Explain Superconductivity

Solutions to superconductivity in this period were sought in many other directions. Brillouin, in 1933, hypothesized that superconductivity would occur in a metal in which the curve of single-electron energy as a function of crystal momentum contained secondary minima.[421] His agrument was the following: an electric field applied to a metal induces a current. At ordinary temperatures, when the field is removed, the electron distribution in momentum space becomes symmetric through coupling of the electrons to the lattice vibrations,[422] and the current induced by the field disappears. However, at very low temperatures this symmetry restoration mechanism, for electrons in the secondary minima, is forbidden for kinematical reasons; such electrons are metastably trapped, producing a persistent current, "which appears to me to represent the essential character of superconductivity." As the temperature is raised, the exponential onset of high-frequency lattice vibrations destroys the metastability and the persistent current disappears. But such a mechanism could not explain why the transition to superconductivity is sharp or, in fact, why the residual resistance due to impurities also disappears at the critical temperature. When Bloch, leaving Nazi Germany, passed through Paris in June 1933, he invoked his "first theorem" in pointing out to Brillouin that the hypothesized metastable states would be unstable against a small common displacement of all the electrons in momentum space. Brillouin added a short appendix to his paper, describing Bloch's objection, but could not resolve the issue. Later in the year, he published an inconclusive analysis of whether perturbations can lead to the degradation of a supercurrent, as proposed by Bloch.[423]

Elsasser, in Frankfurt, suggested in early 1932 an explanation in terms of Dirac's new relativistic electron theory.[424] He proposed that superconductivity arises from a term he extracted in the relation of an individual electron's velocity and momentum proportional to the curl of the electron spin density (not quite the analogous term in the Gordon decomposition of the Dirac current); this term, he claimed, represents an "unmechanical ... momentumless" current transport. However, Bethe pointed out, on symmetry grounds, that the expectation value of this term is zero in equilibrium. He commented more generally, "It seems certain that one cannot succeed by considering only single-electron energies," a point he illustrated with the argument that any two single-electron states with wave vectors **k** and -**k** have equal energies. Each surface of a given energy in wave-number space thus "contains as many single electron currents oriented to the right as to the left. The total current of all electronic states belonging to a certain energy, therefore, is always zero."[425]

Schachenmeier, in Berlin, around the same time tried to understand superconductivity in terms of electron–electron Coulomb interactions giving rise to "resonant" exchange between conduction electrons and electrons bound to ions.[426] For resonance frequencies above the maximum thermally excited lattice frequency, he argued, electrons would move through the vibrating lattice without being scattered by irregularities, thus forming a supercurrent. The critical temperature, calculated in terms of the lattice Debye frequency, gave order-of-magnitude agreement with experiment. Schachenmeier, too, would be taken to task by Bethe (somewhat to

Bethe's later despair and eventual amusement), who pointed out that he, as well as Kretschmann, "neglect fundamental facts of wave mechanics. Both, e.g., make a distinction between the valence electrons which are bound to individual atoms and 'free' electrons," through failure to antisymmetrize the total electron wave function.[427] Despite its execution, Schachenmeier's paper, an early attempt to employ quantum-mechanical electron–electron interactions to explain superconductivity, hints at modern ideas of mixed valence.

Bethe's evaluation had a portentous postscript. Schachenmeier would not accept Bethe's criticism, and Bethe wrote to Sommerfeld in late February 1934 from Manchester that "Schachenmeier is busy writing letters to me, which, much more briefly than he, I nevertheless respond to."[428] Two days later, Walter Henneberg, a former Sommerfeld student who was now working at the Allgemeine Elektrizitätsgesellschaft (AEG) in Berlin, reported to his old teacher: "Recently I had a turbulent debate with Professor Schachenmeier; he is very insulted by Bethe's criticism. . . . I found it very alienating that Schachenmeier reproached Dr. Bethe for being abroad." In fact, Bethe had by this time fled Nazi Germany. Henneberg continued:

> Sch's opinion is that Dr. Bethe wrote the article with the certain prospect in mind that he would go abroad and thus allow himself to be insulted. I don't believe that I succeeded in convincing Schachenmeier that Dr. Bethe had completed the article at a time when he did not yet think of going abroad. . . . Schachenmeier as an Aryan has an advantage from the beginning and my impression is that he will make use of that. Since Schachenmier is working in our research institute (nobody knows however what he is doing), I would like you to treat my communication as confidential.[429]

In his refusal to recognize Bethe's criticism, Schachenmeier went so far as to mention Bethe only in a note added in proof to a further paper on superconductivity, which he submitted at the end of March 1934, claiming in the note that the *Handbuch* article appeared after he had written the paper. Schachenmeier was also apparently quite disturbed that Bethe had referred to him only in small type in the *Handbuch* article, giving Bethe the irresistible opportunity to answer Schachenmeier a few months later in the *Zeitscrift für Physik* in an article written almost entirely in small type.[430]

Like the electron theory of metals just before Sommerfeld's work, or atomic theory preceding the invention of quantum mechanics, the microscopic understanding of superconductivity in early 1933 was but a hodgepodge of partial explanations containing ad hoc assumptions of questionable validity. None of the theories of superconductivity by this time was sufficiently developed to permit quantitative comparison with experiment. That the explanation depended on many-body interactions was recognized but not properly dealt with. Although the application of quantum mechanics to solids had culminated by 1933 in a remarkably successful theory of transport phenomena and a promising theory of ferromagnetism, superconductivity remained untreatable within the existing quantum theoretical framework.

Did the obvious failure to explain superconductivity in quantum-mechanical terms suggest that in certain circumstances the quantum theory might be inapplicable, as happens in classical mechanics when velocities approach the speed of light or when quantum-size actions are involved? There is no evidence that any leading theorist in the period drew such a conclusion. In late 1926 outstanding solid-state problems—such as spin paramagnetism, electrical conductivity, and specific heat—were used as a proving ground for the newly invented quantum mechanics, six years later, but the one remaining fundamental problem in the quantum theory of metals was not seen as such a test. Rather, as Bohr for one correctly emphasized in his correspondence with Kronig, the current models were assumed to be still too crude to deal with the full quantum problem. Bethe summed up the prevailing attitude in the 1933 *Handbuch:* "despite lack of success up until now, we may assert that superconductivity will be solved on the basis of our present-day quantum mechanical knowledge."[431]

2.7   1933: A Climax and a Transition

The year 1933 marked both a climax and a transition in the development of the quantum theory of solids. As the laying of theoretical foundations reached a temporary conclusion, attention began to shift from general formulations to computation of the properties of particular solids. Institutionally, solid-state physics began to show signs of movement toward independence, and the community of physicists working in this area underwent rapid change.

The growing recognition that certain solid-state topics had something in common was revealed from 1930 on in graduate courses and conferences in which these topics began to be grouped in categories such as solid bodies, physics of solids, and solid state. For example, at the ETH, Scherrer scheduled a two-hour-per-week lecture course—"Physics of Solid Aggregate States"—in the summer of 1930. In Leipzig, Heisenberg presented the "Quantum Theory of Solid Bodies" in his special-topics course during the winter semester of 1930/1931,[432] forewarning his students that "much is still unclear," but trusting that "by the end of the lecture course . . . some things will clear up and perhaps it will stimulate one or more of you to carry out studies." In this course, he treated subjects essentially as they became topical during the 1920s: (1) lattice theory and representations of crystals, (2) quantum theory of coherence forces, (3) magnetic properties, and (4) theory of electrical and thermal conductivity. For further reading, he recommended for subject (1) encyclopedia articles (Ewald, 1923; Born, 1923) and handbook articles (Born and Bollnow, 1927); for subject (2) textbooks on quantum mechanics (among others, Hund, Dirac, Heisenberg); and for subject (3) reports of the 1930 Solvay Conference on magnetism. For subject (4) he referred to original works, since the breakthroughs by Bloch, Peierls, and others had not yet been presented in review articles. Two years later, Hund offered a lecture course in Leipzig called "Theory of the Solid State."[433] That same year, Wigner initiated a course at Princeton also titled "Theory of the Solid State," attended by approximately 15 graduate students, including Frederick Seitz, who prepared the notes.[434] Wigner restricted the meaning of *solid* to "substances that are either crystals or aggregates of microcrystals," and covered symmetry groups in relation to crystal lattices, crystal growth, crystal structure, and the art of using "approximate methods and intuitive pictures" to explain solid-state phenomena. He particularly wanted to communicate the nature and explanation of the distinction between the various kinds of solids, based on the fundamental differences in structure.[435]

At both Leipzig and Zurich, many solid-state topics were being studied intensely by the start of the 1930s, in doctoral and *Habilitation* works as well as in seminars and conferences. At both centers, a stream of visiting research fellows provided stimulation and contributed to the rapid dissemination of new solid-state works by Slater, Van Vleck, Landau, Clarence Zener, Victor Guillemin, S. Kikuchi, and others.

While in the late 1920s, as noted in Chapter 1, individual talks on solid-state topics were increasingly included in conference programs,[436] by the early 1930s meetings devoted entirely to such topics were not uncommon—for example, the third and fifth Leipziger Vorträge, in 1930 and 1933. An international congress on experimental and theoretical x-ray physics was held in July 1929 at the ETH, in which Hartree, Ivar Waller, Kronig, Debye, and other "prominent experts of various nations" participated.[437] This congress was one of the "Zurich weeks," started initially, like the "Leipzig weeks," under Debye's entrepeneurship. Conferences elsewhere also included solid-state topics. For example, in 1934 a conference was held on metals in Geneva with the program arranged by Sommerfeld; Bethe spoke on "Quantitative Calculations of the Eigenfunctions of Metal Electrons" and Peierls on "Necessary and Unnecessary Assumptions about Disorder for the Theory of Conductivity."[438]

Around 1930, editors of the leading review journals and textbooks series (e.g., Hans Geiger and Karl Scheel of the *Handbuch der Physik* and Erich Marx of the *Handbuch der Radiologie*) solicited numerous reviews on the electron theory of metals.[439] These reviews helped to integrate the new developments into a consistent picture and made the subject accessible to students.[440] The most influential review, the bible for the electron theory of metals for more than two decades, was that by Sommerfeld and Bethe in the 1933 *Handbuch der Physik,* one volume of a 24-volume source book intended to serve as an advanced text covering the entire field of physics (Chapter 1). Scheel had asked Sommerfeld to write the article, and Sommerfeld agreed, provided that Bethe do 90% of the work (and receive 90% of the honorarium).[441] Indeed, Sommerfeld wrote only the first 35 pages out of the article's total of 289, focusing on his semiclassical treatment, and Bethe wrote the wave-mechanical part.[442]

The first textbooks on the quantum theory of solids were also conceived in this period, among them ones by Brillouin,[443] based on his articles of 1927 to 1929 on the application of quantum statistics;[444] by Wilson,[445] based on his thesis; and by Mott and Harry Jones,[446] which remains an everyday reference. Both the reviews and the texts projected the feeling, then widely held,[447] that the fundamental theoretical issues were for the most part resolved. "One gains the impression that its problem," Peierls wrote of the electron theory of metals, "to explain the typical conditions of metals from molecular properties, and to derive the quantitative laws that exist is, with exceptions . . . solved."[448] Similarly, Bloch remarked, "the electron theory of metals, insofar as it is based on the conception of independent conduction electrons, allows one to understand qualitatively or quantitatively the most important properties of metals."[449]

The electron theory of metals was not, however, a comprehensive theory of all solids. The theory had so far made only tentative contacts with crystallography in its enormous variety, even the crystallography of perfect crystals. It had made almost no contact with questions of impurities and other deviations from the ideal lattice. Therefore, the theory was still remote from any of the real properities of solids, properties that remained the preoccupation of certain experimental physicists and nearly all metallurgists (metallurgy was a strand of solid-state study still entirely divorced from quantum theory, carried out in separate institutions and by different groups of researchers). Finally, the vexing problem of collective interactions, while seeming almost within grasp, kept receding as the theorists advanced.

Although theorists remained confindent in the essential validity of the theory, this confidence was tied to a complementary belief that the subject was becoming less manageable. For example, Bethe wrote to Sommerfeld in 1931 that he doubted much more could come of certain detailed calculations until the eigenfunctions of the metallic crystal were found, an "almost hopeless" prospect: "There is no sense in calculating further, since the Bloch theory represents too coarse an approximation."[450] Peierls reflected recently:

> There you had a situation where there was a breakthrough, and you exploited that, and when it was finished, you had explained all the outstanding paradoxes and seen that things work in general; then it became less exciting. To do new things which were not just routine applications became harder, as it does in every subject.[451]

The unsolved problem of superconductivity constantly underlined a suspicion that something basic was still being overlooked. Bloch suggested that "the greatest failing still attached to . . . [the theory] is the absence of a point of view that allows one to grasp the interactions between the conduction electrons in a rational way." These "delicate pulls," he correctly surmised, "do not effect most properties of metals, but must play a prominent role in certain phenomena, notably, ferromagnetism, and superconductivity."[452] Brillouin noted that all the theories rested on simplifying assumptions and that the interactions between electrons had been almost entirely neglected: "It is not out of the question that a more complete examination (for example, taking into account exchange phenomena) will bring fundamental changes in our conception."[453] These intuitions would be confirmed 25 years later when new experimental techniques, new methods of preparing purer and more perfect samples, and a quantum field theory for treating the many-body problem would make possible another theoretical reshaping.[454]

After 1933, the central issue for the quantum theory of solids would be to explain the behavior of real solids by applying approximate mathematical models to the 1933 framework. As discussed in Chapter 3, in 1933 a simple approximate study of the band structure of sodium by Wigner and Seitz[455] marked an important step in this transition, and within a few years would give rise to an industry of band structure computations for realistic solids.[456]

Peierls reflected elsewhere on the change in solid-state research that took place over the period covered in this chapter:[457] "we were strongly influenced by the impressive successes of quantum mechanics in clearing up the basic problems of solid state physics, and this was probably responsible for a tendency to concentrate on general points of principle rather than on specific cases." As an example, Peierls noted "the attitude of Pauli, who had, in a sense, opened up the field with his paper on paramagnetism, and who maintained an interest in the field to the extent of attending conferences devoted to solid state theory (e.g., in Geneva in 1934)." However, Pauli's attitude changed: "as the work became more detailed and required more ad hoc assumptions, it seemed to him 'dirty physics.'" When the political events of 1933 gave Pauli the opportunity to hire as *Assistent* one of the German refugee physicists, in writing to Peierls he did not miss the chance to take another of his potshots at solid-state physics:

> The communication in your letter that you still stick with the physics of solids makes me feel that it would not be a tempting prospect for me to have you here again as *Assistent* . . . Fröhlich appeared here, a victim of the Third Reich and of metal physics. . . . Bethe, however, I would take only under the condition that he finish with his *Handbuch* article on the electron theory of metals by fall, and that he swear off the physics of solids.[458]

Indeed, Pauli chose as his next *Assistent* Victor Weisskopf, who had not worked on the solid state.

At just the time that the problems of the quantum theory of solids began to appear less fundamental, new discoveries were opening up the fields of nuclear physics, cosmic-ray physics, and quantum electrodynamics.[459] Many of those who had contributed to the theory of solids between 1926 and 1933—including Pauli, Heisenberg, Bloch, Landau, Peierls, and Bethe—took up the challenges of these new areas. In both Leipzig and Zurich, interest in solid-state physics declined sharply after the discovery of the neutron. Nordheim recently reflected: "In the early thirties . . . people gravitated to newer things, cosmic radiation and nuclear physics at that time . . . also the discovery of the neutron, and then of artificial radioactivity, and . . . the development of the first accelerator." However, not everyone abandoned the quantum theory of solids; Wilson, Slater, and Landau were among those who remained active in the area throughout this transition period. And there were a few who were repelled rather than attracted by the rise of nuclear physics. For example, as a student in Cambridge, Shoenberg decided to enter solid-state instead of nuclear physics, which, he recently recalled, "was getting too technical, a changeover from the sealing-wax-and-string to the big machine age, a lot of electronics." Shoenberg added, "I also felt that in nuclear physics all the big things had been done, and then students were left with the detailed jobs."[460]

For physicists of Jewish background in Germany, this tendency to change fields was reinforced by the sudden deterioration of their situation from April 1933 on, including their wholesale dismissal from positions. Bethe, who emigrated to England in 1933 and then to the United States at the end of 1934, feels that had he not left Germany at this time, he would have done more in solid-state physics: "I imagine that probably after a few more years, I would also have been captivated with nuclear physics. But it came earlier because I got into contact with people who were doing nuclear physics. England was full of nuclear physicists when I came there in '33."[461] Peierls says that his move in 1933 to Cambridge, a strong center of nuclear physics, encouraged him to enter that field.[462] Similarly, Bloch, who moved to the United States in 1934 after stays in Copenhagen, Paris, and Rome, felt that these political events had a direct influence on his change of interests.[463] America's tradition of strong experimental physics may have contributed to Bethe's choice to work on problems close to experimental interests and to do so in collaboration with other researchers. In 1936, Bethe reported to Sommerfeld that the "characteristic feature of American physics is teamwork" and that "of course nuclear physics is the subject of current interest. The result is that 90% of all work in this field is in America."[464]

The growth between 1930 and 1933 of facilities geared toward research in real physical solid-state systems would nevertheless keep the field active and attract new workers. The three prime academic examples of such facilities were the Bristol school, which grew up around John E. Lennard-Jones, Mott, and Jones; the physics department at the Massachusetts Institute of Technology under Slater's direction; and the group surrounding Wigner within the physics department at Princeton University. Bristol drew heavily on strong traditions in crystallography, low-temperature physics, and metallography, and attracted young solid-state theorists partly through a growing recognition of the potential industrial relevance of solid-state theory.[465] At MIT, Slater, who assumed the chairmanship of the Physics Department in 1930, built the department up with a strong solid-state-theory emphasis.[466] Wigner came to solid state via chemistry and crystal structure with the strong feeling that the Sommerfeld-Bloch-Peierls theory did not begin to answer fundamental questions of the structure of real crystals, such as "why the crystal exists; what is its binding energy."[467] Furthermore, major industrial laboratories in the United States, including Bell Telephone Laboratories and General Electric, recognizing that active research was essential to application of the new theoretical advances, decided to appoint many well-trained solid-state physicists. The onset of the Great Depression did not prevent these laboratories from fulfilling their hiring plans, but only delayed this until the latter 1930s.[468]

The transformation of the field, together with a shift of the center of solid-state activity from Germany to England and the United States, which was a result of the emigration from Germany and the growth of new institutions, brought to a close what physicists now look back on as an age of heroes *(Heldentage)* in the quantum theory of solids.

2.8 Conclusion

The development of the quantum-mechanical electron theory of metals, from Drude's theory in 1900 to the 1933 Sommerfeld–Bethe *Handbuch* review, replayed the stages that the theory of the atom passed through, paralleling the development of quantum mechanics itself.[469] There was a classical period, in which the inability of models to explain certain experiments created contradictions and dilemmas; then a semiclassical period, in which a few quantum conditions were included in the classical picture ad hoc, resolving certain of the dilemmas but leaving fundamental questions open; next a period during which a fully quantum-mechanical model was applied, putting the theory on a new basis; and finally a period of consolidation, during which new quantum-mechanical building blocks were constructed.

In the classical period, the basic model of a metal as formulated by Drude and Lorentz enabled computation of many thermal, electrical, and magnetic quantities, including the experimentally established Wiedemann–Franz ratio. But basic problems remained unsolved—for example, the lack of an observable electron contribution to the specific heat. With the Einstein, Debye, and Born–von Kármán explanations of the specific heat solely in terms of the ions, underlined by experimental confirmation of the crystal lattice, the classical theory entered an impasse, requiring tools not yet available. Pauli's attempts to understand the quantum statistics of a many-electron system, and Sommerfeld's extension of Pauli's method to a program reworking the classical picture using Fermi–Dirac statistics, ushered in the semiclassical period. However, although many phenomena, including the specific heat, were now accounted for, the fundamental question of why the free-electron assumption could work so well in a lattice remained a mystery. With Bloch's fully wave-mechanical theory of electrons in a solid—the analogue of the introduction of wave mechanics in the description of the atom—the electron theory of metals was put on a firm basis. Only now could more detailed developments, such as band theory and the quantum theory of magnetism, be erected. These latter development, by Peierls, Bethe, Bloch, Brillouin, Wilson, Slater, Landau, and others, made the quantum theory of metals an experimentally useful body of knowledge.

Why and how did the theory of metals, with its well-established inventory of unexplained experiments, become a problem of urgent interest starting in 1926? Quantum mechanics, having just been applied to atoms and molecules, was clearly ready for application to solids; by 1926, quantum mechanics was recognized as *the* approach to microscopic physical phenomena, and the basic concepts needed for a modern theory of metals were available. Why, then, did the development pass through a semiclassical stage at all?

The answer lay in Sommerfeld's character and particular interests. For the younger physicists of the period, the problems of metal physics, and the Drude–Lorentz theory in particular, were not a pressing issue;[470] for them, Pauli's paramagnetism paper was no more than an interesting exploration of quantum statistics. But Sommerfeld, with his long-standing familiarity with the classical theory of metals, immediately recognized the paper's significance as a fundamental advance in solid-state physics. Immersing himself in his semiclassical electron model and presenting it persuasively to his colleagues and students, he acted as a link between the old and the new theories of metals, revitalizing the classical theory and attracting to its frontier a nucleus of highly talented physicists well trained in quantum mechanics.

Sommerfeld, curiously, played similar roles in the developments of both quantum mechanics and the quantum theory of solids. Both his elliptical orbit atomic model and his electron gas model were classical with a small number of quantum concepts added. Both provided a certain level of agreement with experiment, but neither could be justified until the application of quantum mechanics or advance by the usual process of refinement and modification. Radical reformulation in nonclassical terms was needed in both cases. The glimmering of the quantum theory of metals discussed by Heisenberg and Hund in Copenhagen in 1926 was very far from being a theory capable of explaining the range of solid-state phenomena. Sommerfeld's semiclassical theory helped to bridge this gap, serving as a precursor, even indicating through its shortcomings where quantum building blocks were required.

The history of the quantum theory of metals has the striking feature of growing almost entirely out of the research program of a single network, formed by the long-standing and close relationships among Sommerfeld, Pauli, Heisenberg, and their students and other research associates. Three factors contributed. First, the groups in Zurich and Leipzig as well as in Munich functioned as a research family, sharing students, problems, seminars, celebrations, and excursions. In this family, Sommerfeld, one of the few senior theoretical physicists to whom a young theoretician in the 1920s could turn for direction, functioned as the patriarch, supplying both scientific and personal links among the members. With an open eye for current physics problems, he transmitted his interests in courses, seminars, articles, and talks. He was unusually adept at motivating students to tackle research problems, discussing them at his institute as well as in less formal environments, such as his ski hut. Second, the research at Sommerfeld's school covered an unusually broad range of physics problems, with a particular interest in the application of mathematical methods to physics problems (part of Sommerfeld's heritage from Felix Klein), a range that included the problems of solid-state physics. Finally, Sommerfeld was in the 1920s at the height of his influence in the physics community, able to place his students in professorial chairs. The resulting manner in which theoretical physics research was conducted in Munich, Leipzig, and Zurich in this period nurtured remarkable productivity in solid-state theory.

While Sommerfeld was interested in clearing up the problems in the classical theory and in mathematical description, he was less concerned with the underlying wave-mechanical basis of the solutions. To advance beyond the semiclassical theory required the talents of younger members of the Sommerfeld family, such as Heisenberg and Pauli. Having recently been at the center of the invention of quantum mechanics, they were primarily concerned, and far more familiar than Sommerfeld, with the detailed quantum-mechanics issues and their application to such problems as electrical and thermal transport, magnetism, and superconductivity.

Heisenberg and Pauli, in retrospect, played remarkably pivotal roles in the development of the quantum theory of solids (despite Pauli's remarks like "I don't like this solid state physics ... though I initiated it"[471] or "one shouldn't wallow in dirt"[472]). Heisenberg, in particular, beyond his own contribution to ferromagnetism, was responsible for Bloch's development of the basic quantum mechanics of electrons in solids, for Peierls's work on the Hall effect, and for Wilson's elucidation of band theory. The generation of students, *Assistenten,* and visiting research fellows working around these two creators of quantum mechanics between 1928 and 1933 forged the fundamental building blocks of the modern theory of solids, explaining in microscopic terms remarkably many experimentally observed phenomena, and providing, in turn, deeper insight into the workings of quantum mechanics itself.

Although the quantum theory of metals underwent its major theoretical development in Munich, Leipzig, and Zurich between 1927 and 1933, theoretical work on solids was but an episode at these schools, for political as well as scientific reasons. The work begun at these centers continued after 1933 at solid-state schools elsewhere.

Notes

The authors would like to thank the individuals and institutions that kindly and generously offered support or other assistance during the preparation of this chapter. We are particularly indebted to Friedrich Hund, Rudolf Peierls, Stephen Keith, and Paul Hoch for helpful comments based on their close reading of the manuscript; Karl von Meyenn for making available to us many Pauli letters before their publication; David Cassidy and Helmut Rechenberg for generously giving us access to other documents relating to this chapter; Mary Kay Newman for valuable research assistance; and Jerome Rowley for his considerable help, including translation from German to English. Among institutions, we would like especially to thank the Center for History of Physics of the American Institute of Physics (AIP), New York City; the University of Illinois at Urbana-Champaign; the Deutsches Museum, Munich; the Niels Bohr Institute, Copenhagen; and the Archiv der ETH, Zurich. We are also very much indebted to the many individuals who agreed to be interviewed on material relating to this chapter; their names appear in particular notes. Interview tapes or transcripts are, unless otherwise noted, available at AIP.

1. P. Drude, "Zur Elektronentheorie der Metalle," *Ann. Phys.* 1 (1900): 566–613; 43 (1900): 369–402; H. A. Lorentz, "The Motion of Electrons in Metallic Bodies. I, II, III," *Koninklitjke Akademie van Wetenschappen Amsterdam. Proceedings of the Section of Sciences* 7 (1904–1905): 438–453, 585–593, 684–691; Lorentz, *The Theory of Electrons* (Leipzig: Teubner, 1909), 63–67, 266–273. See also Chapter 1.

2. A. Sommerfeld and H. Bethe, "Elektronentheorie der Metalle," in *Hanbuch der Physik,* ser. 2, ed. H. Geiger and K. Sheel (Berlin: Springer, 1933), 24: 333–622.

3. W. Pauli, "Über Gasentartung und Paramagnetismus," *Zs. Phys.* 41 (1927): 81–102; A. Sommerfeld, "Zur Elektronen theorie der Metalle," *Die Naturwissenschaften* 15 (1927): 825–832.

4. F. Bloch, "Über die Quantenmechanik der Elektronen in Kristallgittern," *Zs. Phys.* 52 (1928): 555–600.

5. The first phase is treated in L. Hoddeson and G. Baym, "The Development of the Quantum Mechanical Electron Theory of Metals: 1900–28," in *Beginnings,* 8–23. See also M. Eckert, "Sommerfeld und die Anfange der Festkorperphysik," *Abhandlungen und Berichte des Deutschen Museums,* new series, vol. 7 (Wissenschaftliches Jahrbuch, 1990): 33–71; Eckert, "Propaganda in Science: Sommerfeld and the Spread of the Electron Theory of Metals," *Historical Studies in the Physical and Biological Sciences* 17/2 (1987): 191–233. Material has been drawn from these three articles with minor revision. Other treatments of phase one include P. B. Allen and W. H. Butler, "Electrical Conduction in Metals," *Physics Today* 31, no. 12 (1978): 44–49, and A. H. Wilson, "Solid State Physics, 1925–33: Opportunities Missed and Opportunities Seized," in *Beginnings,* 39–48. The second phase is treated in L. Hoddeson, G. Baym, and M. Eckert, "The Development of the Quantum Mechanical Electron Theory of Metals, 1928–33," *Reviews of Modern Physics* 59, no. 1 (1987): 287–327; the present chapter draws extensively, with minor revisions, from this article.

6. Sommerfeld and Bethe (n. 2); R. E. Peierls, "Elektronentheorie der Metalle," *Ergebnisse der Exakten Naturwissenschaften* (Berlin: Springer, 1932), 11: 264–322; F. Bloch, "Elektronentheorie der Metalle," in *Handbuch der Radiologie,* 2d ed., ed. Erich Marx (Leipzig: Akademische Verlagsgesellschaft, 1933), 6:226–275; G. Borelius, "Physikalische Eigenschaften der Metalle," in *Handbuch der Metallphysik,* ed. G. Masing (Leipzig: Akademische Verlagsgesellschaft, 1935), 1: 181–520; L. Brillouin, *Statistiques quantiques* (Paris: Presses Universitaires, 1930); Brillouin, *Die Quantenstatistik und ihre Anwendung auf die Elektronentheorie der Metalle,* vol. 13 of *Struktur der Materie in Einzeldarstellungen,* ed. M. Born and J. Franck (Berlin: Julius Springer, 1931); L. W. Nordheim, "Kinetische Theorie des Metallischen Zustendes," in *Elektrische Eigenschaften der Metalle und Elektrolyte; magnetische Eigenschaften der Materie,* vol. 4 of *Müller-Pouillets Lehr-*

buch der Physik, ed. C.S.M. Pouillet (Braunschweig: Vieweg, 1934), 234–389; J. C. Slater, "Electronic Structure of Metals," *Reviews of Modern Physics* 6 (1934): 208–280.

7. E. Wigner and F. Seitz, "On the Constitution of Metallic Sodium," *Phys. Rev.* 43 (1933): 804–810; 46 (1934): 509–524.

8. J. J. Thomson, in International Congress, Paris, *Rapport du Congrès de Physique* (Paris: Gauthier-Villars, 1900), 3:138.

9. Attempts to fit the Lorentz–Drude results to experiment were reviewed in N. Bohr, *Studier over Metallernes Elektronteori* (Ph.D. diss.) (Copenhagen: V. Thaning and Appel, 1911); (reprinted, with English translation, in L. Rosenfeld and J. R. Nielson, eds., *Niels Bohr Collected Works,* vol. 1 [Amsterdam: North-Holland, 1972]).

10. Lorentz pointed out this oustanding difficulty in a footnote added to his *Theory of Electrons* (n. 1) in 1915.

11. For example, J. H. Jeans, *The Dynamical Theory of Gases* (Cambridge: Cambridge University Press, 1921), 302–306, 400. For further discussuion, see Chapter 1.

12. A. Einstein, "Die Plancksche Theorie der Strahlung und die Theorie der Spezifischen Wärme," *Ann. Phys.* 22 (1907): 180–190; Einstein, "Eine Beziehung zwischen elastischen Verhalten und der spezifischen Wärme bei festen Körpern mit einatomigen Molekül," *Ann. Phys.* 34 (1911): 170–176.

13. W. Friedrich, P. Knipping, and M. Laue, "Interferenz-Erscheinungen bei Röntgenstrahlen," *Sitzungberichte bayrische Akad. Wissenschaften. Math.–Phys. Klasse* (1912): 303–322 (reprinted in *Ann. Phys.* 41 [1913]: 971–1002).

14. P. Debye, "Zur Theorie der spezifischen Wärme," *Ann. Phys.* 39 (1912): 97–100. M. Born and T. von Kármán, "Über Schwingungen in Raumgittern," *Phys. Zs.* 13 (1912): 297–309; Born and von Kármán, "Über die Verteilung der Eigenschwingungen von Punktgittern," *Phys. Zs.* 14 (1913): 15–19, 65–71.

15. W. Keesom, W. Lenz, A. Sommerfeld, H. Tetrode, O. Stern, M. Planck, P. Scherrer, E. Schrödinger, and A. Einstein, among others, contributed to the gas degeneracy problem. Schrödinger, for example, in a 1924 theory calculated that the temperature below which nonclassical behavior of the electron gas would occur was "between 4500 and 18,000 degrees K . . . this means that electrons in metals in an average temperature range would already be fully degenerate . . . and the failure to contribute to the specific heat would be understandable" ("Gasentartung und frei Weglänge," *Phys. Zs.* 25 [1924]:41–45). Two years later, Schrödinger would be an important participant in discussions of Pauli's application of Fermi-Dirac statistics to metallic electrons. See also P. A. Hanle, "The Coming of Age of Erwin Schrödinger: His Quantum Statistics of Ideal Gases," *Archive for History of Exact Sciences* 17 (1977); 165–192; L. Belloni, "A Note on Fermi's Route to Fermi-Dirac Statistics," *Scientia* 113 (1978): 421–430; Chapter 1.

16. Such theories were proposed by J. J. Thomson, J. Stark, G. Borelius, F. Lindemann, and others, and are reviewed in the literature given in n. 20.

17. E. Grüneisen, "Metallische Leitfähigkeit," in *Handbuch der Physik,* ed. H. Geiger and K. Scheel (Berlin: Springer, 1928), 13:1–75. Viewing the various theories and models in 1915 (and offering yet another), F. A. Lindemann remarked: "Each has necessitated secondary hypotheses, and none of them is very convincing. It is sufficient to point out that the main points enumerated above are in absolute contradiction with one another or with the facts" ("Note on the Theory of the Metallic State," *Phil. Mag.,* 6th ser., vol. 29 [1915]: 127–141).

18. Bohr (n. 9) Unfortunately, Bohr was unable to publish a translation of his dissertation (written in Danish), and thus its impact on the development of the theory of metals would be negligible.

19. H. J. Van Leeuwen, *Vraagstukken uit de Electronentheorie van het Magnetisme* (Ph.D. diss.) (Leiden: Eduard Ijdo, 1919), 49–51; van Leeuwen, "Problèmes de la théorie électronique du magnétisme," *Journal de physique et le radium,* 6th ser., vol. 2 (1921): 361–377. See also H. A. Lorentz, "Anwendung der kinetischen Theorie auf Elektronenbewegung," *Vorträge über die Kinetischetheorie der Materie und Elektrizität, Gehalten in Göttingen* (Leipzig: Teubner, 1914), 167–193, on p. 188; discussion in J. H. Van Vleck, *The Theory of Electric and Magnetic Susceptibilities* (London: Oxford University Press, 1932), 100–102.

20. Instituts Solvay, *Conductibilité électrique des mêtaux et problèmes connexes. Rapports et discusssions du Quatrième Conseil de Physique tenu à Bruxelles du 24 au 29 Avril 1924* (Paris: Gauthier-Villars, 1927). R. Seeliger, "Elektronentheorie der Metalle," in *Enzyklopädie der math-*

ematischen Wissenschaften, ed. A. Sommerfeld (Leipzig: Teubner, 1921), 5:777–878; W. Hume-Rothery, *The Metallic State* (Oxford: Clarendon Press, 1931).

21. H. Kamerlingh Onnes, "Further Experiments with Liquid Helium. D. On the Change of the Electrical Resistance of Pure Metals at Very Low Temperatures. . . . V. The Disappearance of the Resistance of Mercury," *Koninklitjke Akademie van Wetenschappen Amsterdam. Proceedings of the Section of Sciences* 14 (1911): 113–115 (*Communication* 122b, Physical Laboratory, Leiden). Also H. Kamerlingh Onnes, in *Nobel Lectures, Physics, 1901–1921* (Amsterdam: Elsevier, 1967), 1: 306–336; G. J. Flim, interview with student, circa 1965 ("Meester Flim"), in J. Bardeen files, University of Illinois; and H.B.G. Casimir, *Haphazard Reality, Half a Century of Science* (New York: Harper & Row, 1983), 164–166; Chapter 8.

22. For recent historical accounts of semiconductors prehistory, see C. Hempstead, "Semiconductors 1833–1914, an Historical Study of Selenium and Some Related Materials" (Ph.D. diss. Durham University, 1977); W. Kaiser, "Karl Bädeckers Beitrag zur Halbleiterforschung," *Centaurus* 22 (1978): 187–200.

23. In 1926 quantum mechanics was still in its infancy. We recall that Pauli formulated the exclusion principle in the spring of 1925: "Über die Zusammenhang des Abschlusses der Elektronengruppen im Atom mit der Komplexstruktur der Spektren," *Zs. Phys.* 31 (1925): 765–783; shortly thereafter, Heisenburg published his fundamental paper on matrix mechanics: "Über quantentheoretische Umdeutung kinematischer und mechanischer Beziehungen," *Zs. Phys.* 33 (1925): 879–893 (received 29 July). Schrödinger's wave equation was formulated in a series of papers starting in early 1926: "Quantisierung als Eigenwertproblem," *Ann. Phys.* 79 (1926): 361–376; and only in late 1926 to early 1927 did Heisenberg formulate the uncertainty relation between position and momentum: "Über den Anschaulichen Inhalt der Quantentheoretischen Kinematik und Mechanik," *Zs. Phys.* 43 (1927), 172–198 (submitted end of March). Two general sources for this development are B. L. van der Waerden, ed., *Sources of Quantum Mechanics* (Amsterdam: North-Holland, 1967), and M. Jammer, *The Conceptual Development of Quantum Mechanics* (New York: McGraw-Hill, 1966). A guide to archival study is T. S. Kuhn et al., *Sources for the History of Quantum Physics: An Inventory and Report* (Philadelphia: American Philosophical Society, 1967).

24. E. Fermi, "Zur Quantelung des Idealen Einatomigen Gases," *Zs. Phys.* 36 (1926): 902–912 (reprinted in *Atti della Accademia Nazionale dei Lincei, Rendiconti* 6, no. 3 [1926]: 145); P.A.M. Dirac, "On the Theory of Quantum Mechanics," *Proc. Roy. Soc.* A112 (1926): 661–677. For further discussion of Fermi–Dirac statistics, see Belloni (n. 15).

25. W. Heisenberg, interview with T. S. Kuhn, 27 February 1963, AHQP, 12–13.

26. A. Einstein, "Quantentheorie des Einatomigen Idealen Gases," *Sitz. Preuss. Akad.* 18–25 (1924): 261–267; 1–5 (1925): 3–14, 18–25; S. N. Bose, "Plancks Gesetz und Lichtquantenhypothese," *Zs. Phys.* 26 (1924): 178–181.

27. W. Heisenberg, "Mehrkörperproblem und Resonanz in der Quantenmechanik," *Zs. Phys.* 38 (1926): 411–427; Heisenberg, "Über die Spektra von Atomsystemen mit zwei Elektronen," *Zs. Phys.* 39 (1926): 499–518.

28. Dirac (n. 24).

29. E. Fermi to P.A.M. Dirac, 25 October 1926 AHQP, microfilm 59.

30. W. Pauli to F. Rasetti, "Recollections of Fermi Statistics . . . ," October 1956, AHQP, microfilm 66, sec. 12.

31. R. H. Fowler, "On Dense Matter," *Monthly Notices of the Royal Astronomical Society* 87 (1926): 114.

32. H.G.B. Casimir, "Pauli and the Theory of the Solid State," in *Theoretical Physics in the Twentieth Century: A Memorial Volume to Wolfgang Pauli,* ed. M. Fierz and V. F. Weisskopf (New York: Wiley Interscience, 1960), 137–139, p. 138.

33. Dirac, (n. 24).

34. Formerly in the Pauli Letter Collection, Zollikon, Switzerland. One of the authors (LH) is grateful to Mrs. Pauli for making selected letters in this collection available to her in May 1976 before their publication in *Wolfgang Pauli. Scientific Correspondence with Bohr, Einstein, Heisenberg, a.o., I: 1919–1929,* ed. A. Hermann, K. von Meyenn, and V. F. Weisskopf (Berlin: Springer, 1979), and *Wolfgang Pauli. Scientific Correspondence with Bohr, Einstein, Heisenberg, a. o., II: 1930–1939,* ed. K. von Meyenn, A. Hermann, and V. F. Weisskopf (Berlin: Springer, 1985).

35. W. Pauli to W. Heisenberg, 19 October 1926, *Scientific Correspondence I* (n. 34), 143, 340.

36. W. Pauli to E. Schrödinger, 22 November 1926, *Scientific Correspondence I* (n. 34), 147, 356. Three weeks later Pauli sent his note "Gas Degeneracy and Paramagnetism" to the *Zs. Phys.* (n. 3). See also Pauli to Heisenberg, 19 October 1926, *Scientific Correspondence I* (n. 34), 143, 340, and to G. Wentzel, 5 December 1926, *Scientific Correspondence I* (n. 34), 149, 360 and Hanle (n. 15).

37. W. Pauli, "Theoretische Bemerkungen über den Diamagnetismus einatomiger Gase," *Zs. Phys.*, 2 (1920): 201–205. See also *Scientific Correspondence I* (n. 34), 19, 53, 55–56.

38. Heisenberg (n. 27).

39. Pauli to Wentzel, 5 December 1926, *Scientific Correspondence I* (n. 34), 149, 360.

40. Pauli (n. 3).

41. Pauli to Rasetti (n. 30).

42. Quoted in Casimir (n. 32).

43. P. Ehrenfest to W. Pauli, 24 January 1927, *Scientific Correspondence I* (n. 34), 152, 371.

44. Drude, although corresponding with Sommerfeld at the time of his electron theory of metals (n. 1), did not refer to the theory in his letters to Sommerfeld (four letters between 1899 and 1901 in the Sommerfeld Papers, DM). Sommerfeld's lectures: AHQP, microfilm 21, sec. 8. Thesis: P. Debye, "Zur Theorie der Elektronen in Metallen," *Ann. Phys.* 33 (1910): 441–489.

45. W. Heisenberg, "Ausstrahlung von Sommerfelds Werk in der Gegenwart," *Physikalische Blätter* 24 (1968); 530–537, here 533.

46. Eckert (1990) (n. 5). See also U. Benz, "Arnold Sommerfeld. Eine wissenschaftliche Biographie" (Ph.D. diss, University of Stuttgart, 1973); K. H. Manegold, *Universität, Technische Hochschule und Industrie. Ein Beitrag zur Emanzipation der Technik im 19. Jahrhundert unter besonderer Bercksichtigung Felix Kleins* (Berlin: Duncker und Humblot, 1970).

47. This was one of the first full professorial chairs for theoretical physics, established in 1890 for Ludwig Boltzmann, but unoccupied from 1894 to 1906, when Röntgen reestablished it. M. Eckert and W. Pricha, "Die Besetzung der Lehrstühle für theoretische Physik in München und Wien, 1890–1917," *Mitteilungen der Österreichischen Gesellschaft für Geschichte der Naturwissenschaften* 4 (1984): 101–119.

48. R. Peierls, interview with L. Hoddeson, May 1981. Peierls's recent book, *Bird of Passage* (Princeton, N.J.: Princeton University Press, 1985), contains a wealth of reminiscences expanding on the material in this interview. See also H. Bethe, interview with L. Hoddeson, April 1981; n. 44.

49. Bethe interview (n. 48).

50. A list of themes and speakers is preserved in "Müncher Kolloquiums vorträge, 1909–1939," AHQP, microfilm 20.

51. According to Bethe, Sommerfeld, at heart a mathematical physicist, gave a particularly beautiful series of lectures in the mathematical physics section. Bethe interview (n. 48).

52. Bethe and Peierls interviews (n. 48); L. Nordheim, interview with J. Heilbron, AHQP, 17–19.

53. Peierls interview (n. 48).

54. Bethe interview (n. 48); Bethe, "My Life as a Physicist" (talk delivered at Erice, April 1981, tape at AIP).

55. Bethe ("My Life") (n. 54).

56. Peierls interview (n. 48).

57. C. Eckart, interview with J. Heilbron, AHQP, 10.

58. M. Born, "Arnold Johannes Wilhelm Sommerfeld, 1868–1951," *Obituary Notices of Fellows of the Royal Society of London* (London: Cambridge University Press, 1952–53), 275–287.

59. Bethe ("My Life") (n. 54).

60. For the role of Munich in these experiments, see Chapter 1; Friedrich et al. (n. 13); P. Forman, "The Discovery of the Diffraction of X-rays by Crystals: A Critique of the Myths," *Archives for the History of Exact Sciences* 6 (1970): 38–71; P. P. Ewald, *Fifty Years of X-ray Diffraction* (Utrecht: Oosthoeks, for the International Union of Crystallography, 1962) 6–80; interviews with P. Debye, A. Einstein, P. P. Ewald, and W. Friedrich, AHQP. In his interview (n. 48), Bethe reminisced about the creative work carried out in the institute's basement by Sommerfeld's machinist Karl Selmayr, who out of little balls and wire would construct excellent models of crystals that were sold throughout the world. F. Hund, for example, traveled to Munich several times to obtain crystal

models for the physics institutes in Göttingen, Rostock, and Leipzig. Hund, interview with M. Eckert, H. Schubert, J. Teichmann, and G. Torkar, 18 May 1982, DM and AIP.

61. For details see Eckert (1990) (n. 5); S. Richter, "Forschungsförderung in Deutschland 1920–1936," *Technikgeschichte in Einzeldarstellungen* 23 (1972); 7–69, pp. 34, 45–56. See also A. Sommerfeld, "Das Institut für Theoretische Physik," in *Die wissenschaffilichen Anstalten der Ludwig-Maximilians-Universität zu München,* ed. K. A. von Müller (Munich: Oldenbourg, 1926), 290–292. Much documentary material on Sommerfeld's role in placing students in physics positions is in the Sommerfeld Papers, DM. Sommerfeld himself was offered other chairs; for example, in 1918 he succeeded Einstein as president of the Deutsche Physikalische Gesellschaft, and in 1927 he was offered Planck's Berlin chair, but declined to leave his Munich position. Bethe and Peierls interviews (n. 48). For a list of Sommerfeld's prominent students as of 1928, see P. Debye, ed., *Probleme der modernen Physik. Arnold Sommerfeld zum 60. Geburtstage gewidmet von seinen Schülern* (Leipzig: Hirzel, 1928).

62. A. Sommerfeld to G. Wentzel, 5 December 1926, *Scientific Correspondence* (n. 34), 149, 362.

63. Pauli to Rasetti (n. 30).

64. We thank Unsöld for kindly providing notes from these lectures. This manuscript, however, does not treat transport by electrons, such as electrical and thermal conductivity, and thermoelectricity; since Sommerfeld referred to these lectures in his article in *Die Naturwissenschaften* (n. 3), stating that he attempted the work in the recent summer semester and had "obtained remarkable results," it is likely that Unsöld's notes are incomplete.

65. P. Debye, Sommerfeld's first assistant, interview with D. M. Kerr, Jr., and L. P. Williams, 22 December 1965, 7. We would like to thank Dr. Kerr for giving us access to his transcript of this interview.

66. Sommerfeld (n. 3). In the letter of 6 August 1927 that accompanied his manuscript to A. Berliner, the editor of *Die Naturwissenschaften,* Sommerfeld wrote, "Since the subject is of general interest, I have written my communication in a broadly understandable form for your journal. . . . Naturally, I will also report about the same subject extensively to the *Zeitschrift für Physik,* as well as to the Volta Congress in Como" (Sommerfeld Papers, DM).

67. Sommerfeld discussed this conference in letters to J. Franck, 20 July 1926; W. Meggers, 4 July 1927; and A. Berliner, 6 August 1927, Sommerfeld Papers, DM.

68. A. Sommerfeld, "Elektronentheorie der Metalle und des Volta-Effektes nach der Fermischen Statistik." *Atti del Congresso Internazionale dei Fisici Como-Pavia-Roma II,* September 1927, 449–473 (reprinted in A. Sommerfeld, *Collected Works,* vol. 2 [Braunschweig: Vieweg, 1968]).

69. Sommerfeld (n. 3), 831. Although it was now certain that electrons in a metal obeyed Fermi–Dirac statistics, the question of the statistics obeyed by a gas of atoms or ions remained open. Sommerfeld commented that these statistics "would hold for a possible 'proton gas' [!]. It appears doubtful that the extension suggested by Fermi of his statistics to the ordinary electrically neutral gas is legitimate . . . ," but he did point out that ^{4}He gas would become degenerate at about 5 K if it obeys Fermi statistics (826). The full experimental and theoretical resolution of this issue would await the establishment of a connection between spin and statistics, over a decade later.

70. Eckart interview (n. 57), sec. I; H. Bethe, interview with T. S. Kuhn, and R. Peierls, interview with J. L. Heilbron, 17 June 1963, AHQP.

71. K. Bädeker, *Die elektrischen Erscheinungen in metallischen Leitern* (Braunschweig: Sammlung Wissenschaft, 1911). Peierls interview (n. 48), 11–13. Peierls also recalls that Sommerfeld was in good touch with recent experiments, in particular those of Grüneisen in Berlin on the temperature dependence of the electrical resistivity. See also T. Kuhn et al. (n. 23), 147; P. Kirkpatrick, "Arnold Sommerfeld," *American Journal of Physics* 17 (1949): 312.

72. A. Sommerfeld, "Zur Elektrontheorie der Metalle auf Grund der Fermischen Statistik, I. Teil: Allgemeines, Strömungs und Austrittsvorgänge," *Zs. Phys.* 47 (1928): 1–32; Sommerfeld, "Zur Elektronentheorie der Metalle auf Grund der fermischen Statistik, II. Teil: Thermo-elektrische, galvano-magnetische und thermo-magnetische Vorgänge," *Zs. Phys.* 47 (1928): 43–60.

73. N. Bohr to W. Hume–Rothery, 28 February 1928, AHQP, Bohr Scientific Correspondence, microfilm 12.

74. A. Einstein to A. Sommerfeld, 9 November 1927; A. Hermann, *Einstein/Sommerfeld Briefwechsel. Sechzig Briefe aus dem goldenen Zeitalter der modernen Physik* (Basel: Schwabe, 1968). See also Sommerfeld to Einstein, 1 November 1927, and correspondence with Frenkel, Fermi, Joos, and Schottky, Sommerfeld Papers, DM.

75. W. Hume-Rothery to W. L. Bragg, 17 January 1929, Papers of Sir Lawrence Bragg, Royal Institution of Great Britain, file 77. We thank Stephen Keith for this reference.

76. Bethe interview (n. 48), 21.

77. H. A. Moe, secretary of the Guggenheim Foundation, to A. Sommerfeld, 19 May 1927, Sommerfeld Papers, DM.

78. W. V. Houston, interview with G. Phillips and W. J. King, 3 March 1964, AIP, 9–10.

79. While in Munich, W. V. Houston examined the problem of the temperature dependence of electrical resistivity: "Elektrische Leitfähigkeit auf Grund der Wellenmechanik," *Zs. Phys.* 48 (1928): 448–468; "The Temperature Dependence of the Electrical Conductivity," *Phys. Rev.* 34 (1929): 279–283. Realizing that the zero-point vibrations of the lattice scatter x rays, Houston attempted an analogous description of the scattering of electron "waves" in terms of the mean thermal displacements of individual atoms. The correct calculation, which Bloch would shortly carry out, required employing the actual phonon modes in the full Boltzmann equation. Houston recalls Sommerfeld describing his work as "the first decent treatment of the resistance law" (Houston interview [n. 78]). Sommerfeld wrote to Wentzel: "Houston has now really brought the theory of conductivity to a satisfactory conclusion; he has derived the temperature and pressure dependence quantitatively correctly, and has obtained the absolute magnitude reasonably correctly as well . . ." (AHQP, microfilm 66, sec. 5). See H. E. Rorschach, Jr., "The Contributions of Felix Bloch and W. V. Houston to the Electron Theory of Metals," *American Journal of Physics* 38 (1970): 897–904.

80. Moe (n. 77).

81. Kuhn et al. (n. 23), 144.

82. Nordheim interview (n. 52), 17–19.

83. E. Fermi to A. Sommerfeld, 18 February 1928; G. Joos to A. Sommerfeld, 15 May 1928; Sommerfeld to Joos, 9 June 1928; W. Schottky to A. Sommerfeld, 11 May 1928, Sommerfeld Papers, DM.

84. A. Sommerfeld and N. H. Frank, "The Statistical Theory of Thermoelectric, Galvano- and Thermomagnetic Phenomena in Metals," *Reviews of Modern Physics* 3 (1931): 1–42.

85. Sommerfeld and Bethe (n. 2).

86. Sommerfeld's lectures in Japan were at the Law School of Tokyo Imperial University in December 1928, on "Selected Problems on Wave Mechanics and Theory of Electrons." We thank Atsushi Katsuki and Shuntiki Hirokawa for a copy of notes from these lectures. During this year, Peierls went to Leipzig to work with Heisenberg while Bethe went to Frankfurt and then to Stuttgart to work as Ewald's *Assistent.*

87. Memorandum, J. A. Becker to N. H. Williams, University of Michigan, 15 May 1931, and memorandum, Becker to Bell Labs superiors, 15 October 1931, in W. Brattain's personal papers, Walla Walla, Wash.; see W. Brattain, interviews with A. Holden and C. Weiner, AIP; letters from Brattain to L. Hoddeson, January 1979, Urbana, Illinois. The historical link between Sommerfeld's course and the discovery of the first transistor is treated in L. Hoddeson, "The Entry of the Quantum Theory of Solids into the Bell Telephone Laboratories, 1925–40: A Case Study of the Industrial Application of Fundamental Science, *Minerva* 17, no. 3 (1980): 422–447, and "The Discovery of the Point-Contact Transistor," *Historical Studies in the Physical Sciences* 12, no. 1 (1981): 41–76.

88. J. Bardeen, interview with L. Hoddeson.

89. Bethe interview (n. 48), 21.

90. F. Bloch, interview with T. S. Kuhn, May 1964, AHQP, 21; J. Frenkel to A. Sommerfeld, 13 March 1928, AHQP, microfilm 14.

91. W. Schottky to A. Sommerfeld, 11 May 1928, microfilm 34, AHQP.

92. Peierls interview (n. 48).

93. A. Sommerfeld and F. Bopp, "Fifty Years of Quantum Theory," *Science* 113 (1951): 85–92, p. 91.

94. Sommerfeld, (n. 72), pt. I, 1.

95. Des Coudres (1862–1926) and Wiener (1862–1927) were typical classical physicists. For a picture of the period, see R. McCormmach, *Night Thoughts of a Classical Physicist* (Cambridge, Mass.: Harvard University Press, 1982).

96. On Jaffé's succession, Sommerfeld received a letter from the dean of the Leipzig Philosophical Faculty, Lichtenstein, on 19 January 1926, requesting recommendations. His reply suggested Heisenberg, Pauli, Herzfeld, and Wentzel. On 28 November 1926, Wiener asked Sommerfeld to suggest Des Coudres's succcessor, and Sommerfeld again suggested, among others, Pauli and Heisenberg. See the letters in Sommerfeld Papers, DM.

97. W. Pauli to G. Wentzel, 8 May 1926, *Scientific Correspondence I* (n. 34), 323. According to the minutes of the meeting of the Philosophical Faculty of Leipzig on 27 February 1926, the appointment list was ordered Heisenberg, Pauli, and Wentzel. Leipzig University archives (kindly supplied by D. Cassidy).

98. Minutes of the Council (Schulratsakten) of the ETH, Zurich, 17 December 1927. P. Debye to ETH president Rohn, 26 January 1927, 14 March 1927, 10 May 1927, 26 July 1927, archives of the ETH.

99. P. Debye to A. Sommerfeld, 10 June 1927, Sommerfeld Papers, DM. On 19 September 1927, Debye once again argued to the Saxon Ministry for Popular Culture for Heisenberg's appointment. Debye Papers, Max Planck Gesellschaft Archives, Berlin. We thank Dr. Rechenberg for providing access to these papers.

100. In 1927, Sommerfeld agreed to decline Planck's chair in Berlin, provided that his Munich position be improved by the addition of an *Extraordinariat* (associate professorship). He wrote to Heisenberg that he hoped he would remain "quietly in Copenhagen for a few years," then take this position when it became available, and eventually occupy Sommerfeld's own chair. However, he added, "I heard from Debye that you are apparently not prepared for a longer stay in Copenhagen. So my beautiful plan will probably fail. I persuade you, therefore, with a heavy heart, but at Debye's wish, to accept the Leipzig Ordinariat, which with Debye and Wentzel as colleagues is particularly attractive" (A. Sommerfeld to W. Heisenberg, 17 June 1927, Sommerfeld Papers, DM). Debye's acceptance of the Leipzig position is recorded in the Leipzig faculty minutes (n. 97) of 27 July 1927, where the appointment list for the theory professorship was ordered Heisenberg, Wentzel, and Pauli. Heisenberg's appointment followed on 28 October 1927. After Sommerfeld declined Planck's chair in Berlin, Schrödinger at the University of Zurich was called there, and Wentzel was appointed as Schrödinger's successor. Schrödinger's position in Leipzig was filled by Hund in the summer of 1929.

101. G. Wentzel to A. Sommerfeld, 26 May 1927, Sommerfeld Papers, DM.

102. W. Heisenberg, *Der Teil und das Ganze, Gespräche im Umkreis der Atomphysik* (Munich: Piper, 1969), 131.

103. W. Heisenberg to his parents, 9 November 1927 (supplied by D. Cassidy).

104. "Vorlesungsverzeichnis" (course catalogs), *Phys. Zs.* 28 (1927): 275, 743; 29 (1928): 215.

105. F. Bloch, interview by L. Hoddeson, 15 December 1981, AIP.

106. P. W. Bridgman to P. Debye, 1 March 1928, Debye Papers (n. 99).

107. W. Heisenberg to his parents, 7 May 1928 (supplied by D. Cassidy).

108. G. Wentzel to A. Sommerfeld, 12 May 1928, Sommerfeld Papers, DM.

109. R. E. Peierls, "Recollections of Early Solid State Physics," in *Beginnings,* 28–38, esp. 28. See also Peierls interview (n. 48).

110. Bloch knew Debye through the common physics colloquium of the university and ETH in Zurich. Schrödinger and Debye, professors he might have worked with, were leaving or planning to leave in the near future. Bloch interviews (nn. 90, 105); S. A. Blumberg and G. Owens, *Energy and Conflict: The Life and Times of Edward Teller* (New York: Putnam, 1976), 35–39; J. Slater, *Solid State and Molecular Theory: A Scientific Biography* (New York: Wiley, 1975), 6–7, 62–63.

111. Peierls interview (n. 48).

112. N. Mott and R. Peierls, "Werner Heisenberg, 1901–1976," *Biographical Memoirs of Fellows of the Royal Society of London* 23 (1977): 213–251, p. 225.

113. F. Bloch, "Reminiscences of Heisenberg," *Physics Today* 29, no. 12 (1976): 23–27, p. 25.

114. P. Debye to P. Ehrenfest, 12 May 1928, Debye Papers (n. 99).

115. This attitude was in keeping with the tradition of Leipzig, with an economic structure geared toward trade fairs and export.

116. P. Debye, ed., *Leipziger Vorträge,* 5 vols. (Leipzig: Hirzel, 1928–1933).

117. Bloch interview (n. 105).

118. W. Heisenberg, "Über die Spektra von Atomsystemenen mit zwei Elektronen," *Zs. Phys.* 39 (1926): 499–518.

119. Peierls interview (n. 48). W. Heisenberg to W. Pauli, 28 October 1926, *Scientific Correspondence I* (n. 34), 144, 349; Pauli to Heisenberg, 19 October 1926, *Scientific Correspondence I* (n. 34), 143, 340; Heisenberg to Pauli, 4 November 1926, *Scientific Correspondence I* (n. 34), 145, 352–353; W. Heitler and F. London, "Wechselwirkung neutralen Atome und Homöpolare Bindung nach der Quantenmechanik," *Zs. Phys.* 44 (1927): 455–472. Van Vleck, (n. 19), 337.

120. Bloch (n. 113); Bloch interviews (n. 90), 21, and (n. 105).

121. F. Hund, interview with M. Eckert, 2 April 1984, DM and AIP. Hund recalled the informal Copenhagen atmosphere and his almost daily discussions with Heisenberg during the winter of 1926/1927, most often during lunch.

122. F. Hund, "Zur Deutung der Molekelspektren. I," *Zs. Phys.* 40 (1927): 742–764 (submitted 19 November 1926).

123. F. Hund, scientific diary, DM. "Resonant wandering means tunneling from one potential well into the adjacent one. . . . This is how I had understood the simple molecule . . . we had the idea that maybe this is involved in conductivity . . . ," Hund recalled in his interview with Eckert (n. 121). W. Heisenberg to W. Pauli, 4 November 1926, *Scientific Correspondence I* (n. 34), 195, 352 ("Bei der Leitfähigkeit . . . kommt die Resonanzwanderschaft der Elektronen à la Hund in Frage").

124. Bloch interview (n. 105).

125. Bloch (n. 113), 23.

126. Heitler and London (n. 119).

127. Bloch interview (n. 105).

128. Bloch (n. 113), 23.

129. This form of the wave function was derived independently by E. Witmer and L. Rosenfeld, "Über die Beugung von de Broglieschen Wellen an Kristallgittern," *Zs. Phys.* 48 (1928): 530–540, in a description of the photoeffect, and submitted for publication five months before Bloch's paper. They realized that electrons in such states carried momentum, but they apparently did not appreciate the significance of the result for the problem of conduction.

130. W. Wien, in an interesting forshadowing of Bloch's work, "Zur Theorie der elektrischen Leitung in Metallen," *Sitz. Preuss. Akad.* (1913): 184–200, assumed that the electrons responsible for conduction traveled between the atoms in "guided" rather than random paths, determined by the structure of the solid. He specified that any imperfection of the lattice would break up the guided paths and thus increase the resistance. Wien introduced the idea that lattice vibrations scatter the electrons and that the scattering increases with the amplitude of the vibrations, a temperature-dependent quantity.

131. "Promotionsakten der Philosophischen Fakultät," Archive of the Karl-Marx-Universität Leipzig. We thank D. Cassidy for this reference.

132. Bloch (n. 4).

133. Bloch interview (n. 105).

134. F. Bloch, "Zum elektrischen Widerstandsgesetz bei tiefen Temperaturen," *Zs. Phys.* 59 (1930): 208–214.

135. Also significant was the work on conduction by Houston in Munich in the spring of 1928 (n. 79), and later by Nordheim. Despite their common interests, there seems to have been no significant interaction between Bloch and Houston in this period. Nordheim, in "Zur Elektronentheorie der Metalle. I, II," *Ann. Phys.* 9 (1931): 607–678, refined Bloch's work, including, for example, the more accurate "rigid ion" description of the interaction of electrons with ions, and extended it to describe further phenomena such as conduction in alloys and thermoelectric phenomena.

136. Bloch interview (n. 105).

137. J. Davisson and L. H. Germer, "Diffraction of Electrons by a Crystal of Nickel," *Phys. Rev.* 30 (1927): 705–740; Davisson and Germer, "Reflection of Electrons by a Crystal of Nickel," *Proceedings of the National Academy of Sciences* 14 (1928): 317–322. For a historical account of the Davisson and Germer experiments, see R. K. Gehrenbeck, "C. J. Davisson, L. H. Germer, and the

Discovery of Electron Diffraction" (Ph.D. diss., University of Minnesota, 1973), and "Electron Diffraction: Fifty Years Ago," *Physics Today* 31, no. 1 (1978): 34–41.

138. Bethe (n. 54). See also J. Bernstein, "Profiles—Master of the Trade," *New Yorker,* December 1979, 50–117.

139. *Promotionsakte,* 24 July 1928 (Munich University, Akte OC-N, Nr. 259), Munich University Archives; Bethe's dissertation was entitled, "The Theory of Diffraction of Electrons in Crystals."

140. H. Bethe, "Über die Streuung von Elektronen an Krystallen," *Die Naturwissenschaften* 15 (1927): 786–788; 16 (1928): 333–334. The final theory was published as "Theorie der Beugung von Elektronen an Kristallen," *Ann. Phys.* 87 (1928): 55–129.

141. P. P. Ewald, "Zur Begründung der Kristalloptik," *Ann. Phys.* 54 (1917): 517–597; Ewald, "Der Aufbau der festen Materie und seine Erforschung durch Röntgenstrahlen," in *Handbuch der Physik,* ed. H. Geiger and K. Scheel (Berlin: Springer, 1927), 24: 191–369. Later Bethe was to become Ewald's son-in-law.

142. Ewald (1917) (n. 141), 592ff.

143. P. M. Morse, "The Quantum Mechanics of Electrons in Crystals," *Phys. Rev.* 35 (1930): 1310–1324.

144. W. Heisenberg to W. Pauli, 3 May 1928, *Scientific Correspondence I* (n. 34) 192, 443; Bloch (n. 113).

145. Peierls interview (n. 48).

146. Sommerfeld and Bethe (n. 2), 366, 562.

147. Peierls (n. 109), 30.

148. R. E. Peierls, "Zur Theorie der galvanomagnetischen Effekte," *Zs. Phys.* 53 (1929): 255–266.

149. R. E. Peierls, "Zur Theorie des Hall-Effekts," *Phys. Zs.* 30 (1929): 273–274.

150. Bloch (n. 4).

151. Peierls (n. 148), 264; W. Pauli, "Über die Zusammenhang des Abschlusses der Elektronengruppen im Atom mit der Komplexstruktur der Spektren," *Zs. Phys.* 31 (1925): 765–783.

152. W. Heisenberg, "Zum Paulischen Ausschliessungsprinzip," *Ann. Phys.* 10 (1931): 888–904. Also see E. Spenke, *Electronic Semiconductors* (New York: McGraw-Hill, 1958), 58. Peierls recalls Heisenberg telling him that "this situation looks very similar to one I have encountered in atomic spectra, where the spectrum of an atom with one or two electrons in the last shell is very similar to that of an atom that has one or two electrons missing from the complete shell." Peierls is uncertain as to when this "significant conversation" took place, but reflects that were it at the time Heisenberg introduced him to the problem, "then he knew the answer from the beginning . . . and left me just to work out the mathematical details" ([n. 48], 38).

153. The physical picture of the Dirac hole first appeared in P.A.M. Dirac, "A Theory of Electrons and Protons," *Proc. Roy. Soc.* A133 (1930): 360–365; he discussed it further in "Quantized Singularities in the Electromagnetic Field," *Proc. Roy. Soc.* A133 (1931): 60–72.

154. Dirac (1931) (n. 153), 61.

155. Bloch interview (n. 105).

156. Heisenberg (n. 152).

157. Dirac (n. 153).

158. Peierls interview (n. 48); recording of Symposium on the History of Solid State Physics, Royal Academy of Sciences, London, 30 April–2 May 1979 (copy of tape in physics department, University of Illinois, Urbana; proceedings were published in *Beginnings*).

159. H. Bethe, private communication, April 1981.

160. Bethe interview (n. 48).

161. Heisenberg's visit to the University of Chicago on this trip resulted in the lecture volume, *The Physical Principles of the Quantum Theory* (Chicago: University of Chicago Press, 1930).

162. P. Scherrer to P. Debye, 5 November 1927, Debye Papers (n. 99).

163. Debye to Scherrer, 26 November 1927, Debye Papers (n. 99).

164. See the exchange of letters and telegrams between Pauli and Rohn (11 November 1927, 6 December 1927, 28 January 1928, 2 February 1928; 173a, 176a; 180a; 182a), *Scientific Correspondence II* (n. 34), 700–704. See also ETH Schulratsakten (n. 98).

165. W. Heisenberg to A. Sommerfeld, 6 February 1929, Sommerfeld Papers, DM.

166. Debye circular, 23 April 1928, Debye Papers (n. 99); Debye (n. 61).

167. *Scientific Correspondence I* (n. 34) is a helpful source for the relationship between these two. D. Cassidy furnishes a striking description of their opposite characters in his forthcoming biography of Heisenberg; we thank him for showing us parts of his manuscript prior to publication. See also D. Serwer, "Unmechanisher Zwang: Pauli, Heisenberg, and the Rejection of the Mechanical Atom, 1923–1925," *Historical Studies in the Physical Sciences* 8 (1977): 189–256. Also, K. Bleuler, interview with L. Hoddeson, January 1984; Bleuler, "Wolfgang Pauli and Werner Heisenberg" (1986, manuscript, copy at AIP).

168. Peierls (n. 109); Peierls interview (n. 48). Some weeks before his return to Leipzig, Heisenberg wrote to Pauli from America on the prospect of Peierls remaining in Zurich: "So you want to have Peierls as your *Assistent* next semester? To me that naturally seems completely correct in principle, but I think you should also send good physicists to me in L[eipzig] as compensation; I would especially like Bloch to come to L again for a while. Can this be done? I would find it very nice if we could arrange such an exchange of physicists between Zurich and L, but it must be mutual, for otherwise I would be left all alone" (1 August 1929, *Scientific Correspondence I* [n. 34], 234, 517). As Heisenberg requested, Bloch went to Leipzig as Heisenberg's *Assistent,* and Peierls stayed in Zurich to become Bloch's successor as Pauli's *Assistent.*

169. W. Pauli to N. Bohr, 6 May 1929, *Scientific Correspondence I,* (n. 34), 220, 496, 497 n. b.

170. W. Pauli to A. Sommerfeld, 16 May 1929, *Scientific Correspondence I* (n. 34), 225, 503.

171. Ibid.

172. See, for example, W. Pauli to H. Kramers, 19 December 1923, *Scientific Correspondence I* (n. 34), 52, 136. Also W. Pauli, "Über die Absorption der Reststrahlen in Kristallen," *Verhandlungen d. Deutscher Physikalische Gesellschaft,* 3d ser., vol. 6 (1925): 10–11.

173. R. Peierls, interview with L. Hoddeson and G. Baym, 20 May 1977, AIP.

174. R. E. Peierls, "Zur kinetischen Theorie der Wärmeleitung in Kristallen," *Ann. Phys.* 3 (1929): 1055–1101.

175. See L. Brillouin, "Revue d'une carrière scientifique preparée pour l'Institut Americain de Physique" (March 1962, manuscript [revised in English in 1964 and in French in 1966], iv–v, AIP).

176. The *Habilitationsschrift* put one on the road to becoming a professor, giving one the privilege of announcing formal lectures and inviting (paying) students to them.

177. Peierls interview (n. 48).

178. R. E. Peierls, "Zur Theorie der elektrischen und thermischen Leitfähigkeit von Metallen," *Ann. Phys.* 4 (1930): 121–148.

179. Peierls interview (n. 48).

180. F. Bloch, "Zur Suszeptibilität und Widerstandsänderung der Metalle im Magnetfeld," *Zs. Phys.* 53 (1929): 216–227.

181. Peierls interviews (nn. 48, 70).

182. Peierls (n. 178), 126, Fig. 1.

183. Peierls (n. 109) 31–32.

184. M.J.O. Strutt, "Zur Wellenmechanik des Atomgitters," *Ann. Phys.* 4th ser., vol. 86 (1928): 319–324. This paper follows two earlier ones, "Wirbelströme in elliptischen Zylinder," *Ann. Phys.* 84 (1927): 485–506, and "Eigenschwingungen einer Saite mit sinsförmige Massenverteilung," *Ann. Phys.* 85 (1928): 129–136, applying the Mathieu equation to various problems in modern physics.

185. Morse (n. 143).

186. P. M. Morse, *In at the Beginnings: A Physicist's Life* (Cambridge, Mass.: MIT Press, 1977), 92–100.

187. Ibid., 97–98.

188. L. Brillouin, *Les Statistiques quantiques et leurs applications* (Paris: Presses Universitaires de France, 1930).

189. L. Brillouin (n. 175). After leaving Germany in 1933, Bloch became acquainted with Brillouin during a short stay in Paris, during which Bloch lived with the Langevin family. Bloch recalls that Brillouin was a most interactive and lively person, with "a very sound grasp on reality" through

his engineering background, and a great expert on waves, especially radio waves. Bloch interview (n. 105).

190. Brillouin's complete citation on page 264, "Morse, *Phys. Rev.* t. (1929)" indicates that he had not at the time of writing of his book seen the final version of Morse's paper.

191. Peierls (n. 178).

192. Brillouin (n. 188), 294.

193. L. Néel, interview with L. Hoddeson and A. Guinier, 29 May 1981, AIP.

194. L. Brillouin, "Les Electrons dans les métaux et le rôle des conditions de réflexion sélective de Bragg, " *C.R.* 191 (1930): 198–100; Brillouin, "Les Electrons dans les métaux et le classement des ondes de de Broglie correspondantes," *C.R.* 191 (1930): 292–294.

195. Peierls (n. 178).

196. L. Brillouin, "Les Electrons libres dans les métaux et le rôle des réflexions de Bragg," *Journal de physique et le radium,* 7th ser., vol. 1 (1930): 377–400; Brillouin's refs. 1 on 283, 2 on 286, and 1 on 330 suggest a feeling of rivalry with Peierls.

197. The richness of the geometries of Fermi surfaces would first be studied by Bethe for the 1933 *Handbuch* review, where electron energies computed using Bloch's tight binding model were used to draw the ideal Fermi surfaces for a number of simple lattices—cubic, face-centered, and body-centered. Sommerfeld and Bethe (n. 2), 401. For the drawings, Sommerfeld and Bethe commissioned R. Rühle, who had done the figures for Jahnke and Emde's tables of functions. Bethe recalls that "it was clear to me . . . that it made a great difference whether [the Fermi surfaces] were nearly a sphere or were some interesting surface," and that for the problem of magnetoresistance, "it was very important how anisotropic the Fermi surface is" (Bethe interview [n. 48]).

198. Brillouin (n. 175), 17–18.

199. L. Brillouin, *Die Quantenstatistik und Ihre Anwendung auf die Elektronentheorie der Metalle* (Berlin: Springer, 1931).

200. Ibid, iv–v. However, by 1933 Brillouin would accept the validity of the *Umklapp* process. See L. Brillouin, "Le Champ self-consistent, pour des électrons liés; la supraconductibilité," *Journal de physique et le radium,* 7th ser., vol. 4 (1933): 333–361; Brillouin "Comment interpréter la supraconductibilité," *C.R.* 196 (1933): 1088–1090.

201. R. Kronig and W. G. Penny, "Quantum Mechanics of Electrons in Crystal Lattices," *Proc. Roy. Soc.* A130 (1931): 499–513. As they note, Van der Pol and Strutt, "On the Stability of the Solution of Mathieu's Equation," *Phil. Mag.* 5 (1928): 18–38, had previously considered the special case of the periodic square-well problem, but in a classical physics context. On Kronig's suggestion, Penney, whom he had met during a visit in Cambridge and London in 1929–1930, came in 1930 to Gröningen, where Kronig had obtained a permanent lecturership. R. Kronig, interview with M. Eckert, 11 February 1982, DM.

202. A. H. Wilson, "The Theory of Electronic Semi-conductors," *Proc. Roy. Soc.* A133 (1931): 458–91; Wilson, "The Theory of Electronic Semi-conductors—II," *Proc. Roy. Soc.,* A134 (1931): 277–287.

203. F. Bloch, "Zur Suszeptibilität und Widerstandsänderung der Metalle im Magnetfeld," *Zs. Phys.* 53 (1929): 216–227.

204. R. E. Peierls, "Das Verhalten metallischer Leiter in starken Magnetfeldern," in *Leipziger Vorträge 1930: Elektronen-Interferenzen,* ed. P. Debye (Leipzig: Hirzel, 1930), 78–85.

205. L. Landau, "Diamagnetismus der Metalle," *Zs. Phys.* 64 (1930): 629–637.

206. H. Bethe, "Change of Resistance in Magnetic Fields," *Nature* 127 (1931): 336–337.

207. A. H. Wilson, interview with C. Hempstead, undated transcript kindly supplied by G. Pearson, in L. Hoddeson files, University of Illinois.

208. Peierls (n. 204); Peierls, "Zur Theorie der magnetischen Widerstandsänderung," *Ann. Phys.* 10 (1931): 97–110.

209. Wilson interview (n. 207).

210. Ibid.

211. Ibid.

212. A. H. Wilson, "Opportunities Missed and Opportunities Seized," in *Beginnings,* 39–48, p. 45.

213. Wilson interview (n. 207).

214. Wilson (n. 212), 45–46.

215. F. Bloch, "Wellenmechanische Diskussionen der Leitungs- und Photoeffekte," *Phys. Zs.* 32 (1931): 881–886.

216. A. H. Wilson, interview with E. Braun and S. Keith, 1981.

217. "Über Halbleiter soll man nicht arbeiten, das ist eine Schweinerei; wer weiss, ob es überhaupt Halbleiter gibt" (W. Pauli to R. Peierls, 29 September 1931, *Scientific Correspondence II* [n. 34], 94).

218. Grüneisen (n. 17). Indeed, accepted experiments of H. J. Seeman, "Zür elektrischen Leitfähigkeit des Siliziums," *Phys. Zs.* 28 (1927): 765–766, and A. Schulze, "Umwandlungserscheinungen an sogenannten Halbleitern," *Zeitschrift für Metallkunde* 23 (1931): 261–264, indicated that pure silicon, in the absence of oxide films, was a good metal. That such metallic silicon, when covered with an oxide layer, could exhibit the observed increase of conductivity with temperature might be explainable, as Wilson further remarked (n. 202), by Frenkel's theory of transmission of thermally activated electrons across oxide films: J. Frenkel, "On the Electrical Resistance of Contacts between Solid Conductors," *Phys. Rev.* 36 (1930): 1604–1618. Frenkel also applied the theory to explain the temperature dependence of the conductivity of granular thin films with the charming analogy that "this relation can be illustrated by the fact that the gaps between adjacent rails in a railway line decrease in the summer and increase in the winter time, and not vice versa." The "canard" that silicon was a good metal would, as Wilson noted (n. 207), linger through the prewar period—for example, in A. H. Wilson, *Semi-Conductors and Metals* (Cambridge: Cambridge University Press, 1939), 44—and even into recent times. The *Handbook of Chemistry and Physics,* 51st ed. (Cleveland: Chemical Rubber Company, 1970), lists (F190) the electrical resistivity of Si at 0°C as only six times that of Cu at 20°C. The properties of silicon first began to be clarified by wartime research, particularly by F. Seitz and co-workers at the University of Pennsylvania, and J. Scaff, R. Ohl, and others at Bell Telephone Laboratories, see Hoddeson (1981) (n. 87); F. Seitz, interview with L. Hoddeson, January 1981, AIP; Seitz, as quoted in N. F. Mott, "Memories of Early Days in Solid State Physics," in *Beginnings* 56–66, p. 63–64.

219. Wilson (n. 202).

220. Heisenberg (n. 152).

221. Peierls (n. 148).

222. Wilson interview (n. 216).

223. Wilson (n. 202).

224. Peierls (n. 6).

225. W. Schottky and F. Waibel, "Die Elektronenleitung des Kupferoxyduls," *Phys. Zs.* 34 (1933): 858–864, pp. 862–863.

226. A. H. Wilson, "A Note on the Theory of Rectification," *Proc. Roy. Soc.* A136 (1932): 487–498.

227. Wilson (n. 218). This book was a textbook for experimental physicists; see Wilson interview (n. 216).

228. Hoddeson (1981) (n. 87).

229. Wilson interview (n. 216).

230. Pauli (n. 37).

231. See review in Van Vleck (n. 19), 100–102.

232. See n. 19.

233. For example, W. Sucksmith, "The Magnetic Susceptibility of Some Alkalis," *Phil. Mag.* 2 (1926): 21–29, notes the lack of "a satisfactory theory explaining the fact that a large number of elements exhibit a paramagnetic susceptibility independent of temperature," and the apparent smallness of the measured susceptibilities below 500°C compared with their values at the boiling points. For Na and K, see P. Pascal. "Propriétés magnétiques des métaux alcalins," *C.R.* 158 (1914): 37–39; for Cs and Rb, G. Joos, "Diamagnetismus und Ionengrösse," *Zs. Phys.* 32 (1925): 835–839. See also Van Vleck (n. 19), 222–223.

234. J. B. Sampson and F. Seitz, "Theoretical Magnetic Susceptibilities of Metallic Lithium and Sodium," *Phys. Rev.* 58 (1940): 633–639.

235. W. de Haas and P. M. van Alphen, "The Dependence of the Susceptibility of Diamagnetic Metals upon the Fields," *Koninklitje Akademie van Wetenschappen Amsterdam. Proceedings of the Section of Sciences* 33 (1930): 1106–1118 (*Communication* 212a, Physical Laboratory, Leiden). Pre–band theory attempts to explain the diamagnetism of bismuth were based upon bound elec-

trons having orbits that embraced several atoms. P. Ehrenfest, "Opmerkingen over het Diamagnetism nav vast Bismuth," *Physica* 5 (1925): 388–391; Ehrenfest, "Bemerkungen über den Diamagnetismus von festem Wismut," *Zs. Phys.* 58 (1929): 719–721.

236. Bloch interview (n. 105).

237. R. Peierls, taped replies to letter from L. Hoddeson, July 1981.

238. F. Bitter, "On the Diamagnetism of Electrons in Metals," *Proceedings of the National Academy of Sciences* 16 (1930): 95–98.

239. Landau (n. 205).

240. V. B. Berestetskii, "Lev Davydovitch Landau," *Uspekhi Fizica Nauk* 64 (1958): 615–623.

241. A. Livanova, *Landau, A Great Physicist and Teacher,* trans. J. B. Sykes (Oxford: Pergamon Press, 1980).

242. C. V. Raman, "Diamagnetism and Crystal Structure," *Nature* 123 (1929): 945; Raman, "Anomalous Diamagnetism," *Nature* 124 (1929): 412.

243. Ehrenfest (1929) (n. 235).

244. Peierls (n. 109).

245. De Haas and van Alphen (n. 235).

246. D. Shoenberg, "Chapter 1, Historical Introduction," from an unpublished manuscript. We thank Dr. Shoenberg for kindly showing Paul Hoch this manuscript, and thus making it available to the International Project on the History of Solid State Physics. See also D. Shoenberg, "The de Haas–van Alphen Effect," in *Ninth International Conference on Low-Temperature Physics,* ed. J. Daunt et al. (New York: Plenum, 1965), 665–676; Shoenberg, "Forty Odd Years in the Cold," *Physics Bulletin* 29, no. 1 (1978): 16–19; Shoenberg tells how around 1937 he visited Kapitza's laboratory in Moscow and observed the oscillatory variation in bismuth with Landau right on the spot for detailed interpretation (667). In this way, they made what Shoenberg believes was the first determination of the Fermi surface.

247. The single crystals of bismuth used in these experiments were grown by L. V. Schubnikov, who, like Landau, on graduating from Leningrad Polytechnical University, was sent by the Narkompros on an extended scientific visit to western Europe. Between 1926 and 1930, he worked in Leiden with de Haas. By improving on a method of Kapitza's, Schubnikov succeeded in producing single crystals of bismuth, which made possible the discovery of the Schubnikov–de Haas effect, the periodic change of electrical resistivity in bismuth as a function of magnetic field at low temperatures. This effect helped to motivate the de Haas–van Alphen experiments in 1930. Whether Landau made contact with Schubnikov during his visit to Europe in 1929 to 1930 is unclear. O. I. Balabekyan, "Lev Vasil'evich Schubnikov," *Soviet Physics: Uspekhi* 9, no. 3 (1966): 455–459.

248. Peierls (n. 109).

249. W. Pauli, "L'Electron magnétique," in *Le Magnétisme, rapports et discussions du Sixième Conseil de Physique tenu à Bruxelles du 20 au 25 Octobre 1930* (Paris: Gauthier-Villars, 1932), 183–190. Landau is referred to on pages 186 and 238. The active discussion of Pauli's paper gives a picture of the state of the application of quantum mechanics to magnetism at the turn of the 1930s.

250. Ibid., 243.

251. C. G. Darwin, in *Le Magnétism* (n. 249); Darwin, "The Diamagnetism of the Free Electron," *Proceedings of the Cambridge Philosophical Society* 27 (1931): 86–90. See also H. J. Seemann, "Magnetochimie der dia- und paramagnetischen Metalle und Legierungen," *Zeitschrift für technische Physik* 10 (1929): 309–408.

252. E. Teller, "Der Diamagnetismus von Freien Elektronen," *Zs. Phys.* 67 (1931): 311–319. Teller acknowledges discussions of the Landau work with Pauli, Van Vleck, and Peierls; a variant of Teller's argument, together with a pedagogical explanation of the connection with Landau's derivation was given the following spring by Van Vleck (n. 19), para. 81.

253. R. E. Peierls, "Zur Theorie des Diamagnetismus von Leitungselektronen," *Phys. Zs.* 33 (1932): 864.

254. R. E. Peierls, "Zur Theorie des Diamagnetismus von Leitungselektronen," *Zs. Phys.* 80 (1933): 763–791. Peierls derived the result here for the case of tight binding, but later work showed that the result is more general.

255. Peierls (n. 237).

256. For parabolic bands the diamagnetic susceptibility becomes essentially proportional to the

inverse of the electron effective mass m^* at the Fermi surface so that large diamagnetic suscepti-bilities arise when $E(\mathbf{k})$ has large curvature, as turned out to be the case in bismuth. H. Jones, "The Theory of Alloys in the Gamma-Phase," *Proc. Roy. Soc.* A144 (1934): 225–234; Jones, "Appli-cations of the Bloch Theory to the Study of Alloys and of the Properties of Bismuth," *Proc. Roy. Soc.* A147 (1934): 396–417. A similar form for the paramagnetic susceptibility, given by Bethe in Sommerfeld and Bethe (n. 2), 476, in terms of the electron density of states at the Fermi surface, would allow inclusion of lattice effects in the Pauli result; here the susceptibility turns out to be proportional to m^*, opposite to the behavior of the diamagnetic susceptibility.

257. Peierls interview (n. 70).

258. R. E. Peierls, "Zur Theorie des Diamagnetismus von Leitungselektronen. II," *Zs. Phys.* 81 (1933): 186–194.

259. Peierls (n. 237).

260. Peierls (n. 109), 36.

261. Peierls (n. 237).

262. Shoenberg (n. 246).

263. The de Haas–van Alphen effect was later exploited to obtain precise experimental infor-mation about electron energy bands. See Chapter 3.

264. *Scientific Correspondence I* (n. 34).

265. P. Robertson, *The Early Years: The Niels Bohr Institute, 1921–1930* (Copenhagen: Aka-demisk Forlag, 1979).

266. W. Heisenberg to W. Pauli, 5 May 1926, *Scientific Correspondence I* (n. 34), 132, 321.

267. Heisenberg interview (n. 25), 12–13.

268. Heisenberg ("Mehrkörperproblem") (n. 27).

269. W. Heisenberg, Über die Spektra von Atomsystemenen mit zwei Elektronen," *Zs. Phys.* 59 (1930): 208–214.

270. These papers are discussed in A. I. Miller, "Symmetry and Imagery in the Physics of Bohr, Einstein and Heisenberg," *Symmetries in Physics (1600–1980). Proceedings of the 1st Interna-tional Conference on the History of Scientific Ideas, 20–26 September 1983,* ed. M. G. Doncel, A. Hermann, L. Michel, and A. Pais (Singapore: World Scientific, 1988), 299–327.

271. "I was interested in two electrons and not in many electrons. Therefore I could forget about Bose and Fermi statistics" (Heisenberg interview [n. 25], 13ff).

272. W. Heisenberg to W. Pauli, 28 October 1926, *Scientific Correspendence I* (n. 34), 144, 349.

273. Pauli to Heisenberg, 19 October 1926, *Scientific Correspondence I* (n. 34), 143, 340.

274. Heisenberg to Pauli, 4 November 1926, *Scientific Correspondence I* (n. 34), 145, 352–353.

275. For the discovery of spin see, M. Jammer, *The Conceptual Development of Quantum Mechanics* (New York: McGraw-Hill, 1966), 146–153 and references therein.

276. W. Heisenberg to W. Pauli, 24 December 1925, *Scientific Correspondence I* (n. 34), 112, 271. Earlier, Heisenberg was not ready to accept Kronig's prior suggestion of the spin; see N. Bohr to R. de L. Kronig, 26 March 1926, AHQP, Bohr Scientific Manuscripts (hereafter referred to as AHQP-BSM).

277. P. Galison, "Theoretical Predispositions in Experimental Physics: Einstein and the Gyro-magnetic Experiments, 1915–1925," *Historical Studies in the Physical Sciences* 12, no. 2 (1982): 285–323. The topic was a major subject in the 1930 Solvay Conference, where P. Weiss reviewed "Les phénomènes gyromagnétiques," concluding that experiments, mainly by S. J. Barnett and L.J.H. Barnett ("Magnetisation of Ferro-Magnetic Substances by Rotation and the Nature of the Elementary Magnet," *Proceedings of the American Academy of Arts and Sciences* 60 [1925]: 127–216) provided "une présomption très forte en faveur de l'attribution du ferromagnétism à l'électron pivotant." See Instituts Solvay, *Rapports et discussions du 6ième Conseil de Physique* (Paris: Insti-tuts Solvay, 1932), 354.

278. Barnett and Barnett (n. 277).

279. W. Heisenberg, "Zur Theorie des Ferromagnetismus," *Zs. Phys.* 49 (1928): 619–636.

280. See also W. Heisenberg to W. Pauli, 10 May 1928, *Scientific Correspondence I* (n. 34) 195, 451; W. Heisenberg, "Zur Quantentheorie des Ferromagnetismus," in Debye (n. 61). He writes that "the orbital moments in the crystal are not freely orientable; on the whole they compensate and do not contribute to the crystal's magnetism" (115).

281. P. Weiss, "L'Hypothèse du champ moléculaire et la propriété ferromagnétique," *Journal*

de physique, 4th ser., vol. 6 (1907): 661–690. See also Weiss, Über die rationalen Verhältnisse der magnetischen Momente der Moleküle und das Magneton," *Phys. Zs.* 12 (1911): 935–952; Weiss, "La Constante du champ moléculaire. Equation d'état magnétique et calorimétrie," *Journal de physique,* 7th ser., vol. 1 (1930): 163–175.

282. P. Langevin, "Magnétisme et théorie des électrons," *Annales de chimie et physique,* 8th ser., 5 (1905): 70–127.

283. J. A. Ewing, "Contributions to the Molecular Theory of Induced Magnetism," *Proc. Roy. Soc.* 48 (1890): 342–358; Ewing, "Magnetic Qualities of Iron," *Philosophical Transactions of the Royal Society of London* A184 (1893): 985–1040; P. Weiss and G. Foëx, *Le Magnétisme* (Paris: Armand Colin, 1926).

284. W. Heisenberg to W. Pauli, 4 November 1926, *Scientific Correspondence I* (n. 34) 145, 352–353. Heisenberg refers here to the concept of tunneling, or "resonant wandering," which Hund, also in Copenhagen at this time, had just introduced to understand molecular spectra (n. 122). See also n. 123.

285. Heisenberg to Pauli, 4 November 1926, *Scientific Correspondence I* (n. 34), 145 n. a, 352–353.

286. Heisenberg to Pauli, 15 November 1926, *Scientific Correspondence I* (n. 34), 146, 354.

287. W. Lenz, "Beitrag zum Verständnis der magnetischen Erscheinungen in fest Körpen," *Phys. Zs.* 21 (1920): 613–615; E. Ising, "Beitrag zur Theorie des Ferromagnetismus," *Zs. Phys.* 31 (1925): 253–258; see also S. G. Brush, "History of the Lenz-Ising Model," *Reviews of Modern Physics* 39 (1967): 883–893. Ising correctly showed that the one-dimensional model would not exhibit ferromagnetism; however, because of his erroneous arguments that the three-dimensional "spatial model" also would not exhibit it, the Lenz–Ising model was not seriously worked on until the mid-1930s.

288. W. Pauli to N. Bohr, 26 February 1926, *Scientific Correspondence I* (n. 34), 122, 296; W. Pauli to H. Kramers, 8 March 1926, *Scientific Correspondence I* (n. 34), 125, 307.

289. W. Heisenberg to W. Pauli, 27 January 1926, *Scientific Correspondence I* (n. 34), 116, 281.

290. W. Pauli to A. Landé, 2 June 1926, *Scientific Correspondence I* (n. 34), 134, 327; E. Schrödinger to A. Sommerfeld, 11 May 1926; Mrs. Schrödinger to Mrs. Sommerfeld, 6 May 1951, Sommerfeld Papers, DM. We have not found published reports from the Zurich conference.

291. Heisenberg (n. 102).

292. *Vorlesungsverzeichnis der Universität München,* summer semester 1926.

293. They went through the classic papers by Wigner and von Neumann and used as a text A. Speiser, *Die Theorie der Gruppen von Endlicher Ordnung* (Berlin: Springer, 1923); Bethe and Peierls interviews (n. 48). The major papers (by Wigner, Hund, Heitler, and London) are listed by Heisenberg (n. 279).

294. Heisenberg interview (n. 25), 19. See also Bloch interview (n. 90), 22. Bloch recalls that the whole Leipzig group, including Heisenberg, Wentzel, and the students, was "quite frequently at colloquia in Berlin."

295. Sommerfeld, in recommending Becker for this chair in 1926, wrote, "Such a happy combination for theoretical physics and technology you could hardly find in another candidate" (Sommerfeld to von Rottenburg, 4 June 1926, Sommerfeld Papers, DM). Later on, Becker and his school would play an important role in the technological application of the theory of ferromagnetism.

296. Also at this institute at the time were Landau and J. Dorfman. See V. J. Frenkel, "Yakov (James) Il'ich Frenkel (1894–1952): Materials for his Scientific Biography," *Archive for the History of the Exact Sciences* 13, no. 1 (1974): 1–26.

297. J. Frenkel, "Elementare Theorie magnetischer und elektrischer Eigenschaften der Metalle beim absoluten Nullpunkt der Temperature," *Zs. Phys.* 49 (1928): 31–45.

298. J. Frenkel to A. Sommerfeld, 8 March 1928 and 13 March 1928, Sommerfeld Papers, DM.

299. Frenkel credits Dorfman for the idea of coupling between the spins of the free electrons and those bound in nonclosed shells ([n. 297] 35).

300. Heisenberg, (nn. 279, 280).

301. A. Sommerfeld to G. Joos, 9 June 1928, Sommerfeld Papers, DM.

302. W. Heisenberg to W. Pauli, 3 May 1928, *Scientific Correspondence I* (n. 34), 192, 443.

303. Weiss (n. 281).

304. Heisenberg to Pauli, 7 May 1928, *Scientific Correspondence I* (n. 34), 193, 447.

305. This branch of spontaneous magnetization, together with another (unmentioned by Heisenberg) present for all z, appears to arise from an extraneous power of the exchange interaction parameter that Heisenberg slipped into the last term of the fluctuation correction that he reported to Pauli; it was corrected in the published version.

306. Heisenberg to Pauli, 7 May 1928, *Scientific Correspondence I* (n. 34), 194, 451.

307. W. Pauli to N. Bohr, 16 June 1928, *Scientific Correspondence I* (n. 34), 201, 462.

308. Pauli to Bohr, 14 July 1928, *Scientific Correspondence I* (n. 34), 203, 464.

309. W. Heisenberg to W. Pauli, 10 May 1928, *Scientific Correspondence I* (n. 34), 195, 451.

310. Heisenberg to Pauli, 13 May 1928, *Scientific Correspondence I* (n. 34), 196, 455.

311. Heisenberg to Pauli, 14 May 1928, *Scientific Correspondence I* (n. 34), 196, 455.

312. Heisenberg to Pauli, 21 May 1928, *Scientific Correspondence I* (n. 34), 198, 457.

313. Heisenberg (n. 279).

314. The familiar "Heisenberg model" Hamiltonian of ferromagnetism did not appear in Heisenberg's paper, but was given later by Dirac (without reference to Heisenberg) in "Quantum Mechanics of Many-Electron Systems," *Proc. Roy. Soc.* A123 (1929): 714–733, p. 731.

315. Heisenberg's account of fluctuations, although inadequate, does give a noticeably better predicted critical temperature than a "mean field" calculation for $z \geq 8$. For example, for $z = 12$, mean field gives a critical temperature T_c divided by J_0 of 6, the modern "exact" result is 4.02, whereas Heisenberg's calculation yields 4.73; for $z = 8$, the results are 4, 2.53 and for Heisenberg, 2; for $z = 6$, the results are 6 and 1.68, but Heisenberg finds no ferromagnetism.

316. W. Heisenberg to W. Pauli, 13 June 1928, *Scientific Correspondence I* (n. 34), 200, 460; Heisenberg to Pauli, 31 July 1928, *Scientific Correspondence I* (n. 34), 204, 466.

317. Heisenberg (n. 280).

318. Bloch (n. 180).

319. F. Bloch, "Bemerkung zur Elektronentheorie des Ferromagnetismus und der elektrischen Leitfähigkeit," *Zs. Phys.* 57 (1929): 545–555.

320. Bloch interview (n. 105).

321. J. Dorfman and R. Jaanus, "Die Rolle der Leitungselektronen beim Ferromagnetismus. I, " *Zs. Phys.* 54 (1929): 277–288; J. Dorfman and I. Kikoin, "Die Rolle der Leitungselektronen beim Ferromagnetismus. II," *Zs. Phys.* 54 (1929): 289–296. Their result however is inconclusive. See Van Vleck (n. 19), 345 n. 43; E. Stoner, "Free Electrons and Ferromagnetism," *Proceedings of the Leeds Philosophical Society* 2 (1930): 50–55; Stoner, "The Interchange Theory of Ferromagnetism," *Proceedings of the Leeds Philosophical Society* 2 (1930): 56–60.

322. Bloch interview (n. 105).

323. Bloch (n. 319).

324. A result that Wigner and Seitz later cited in their calculation of the energy of the interacting electron gas (1934) (n. 7), 512 n. 5.

325. The problem of ferromagnetism of a uniform electron gas would be shortly revisited by Stoner (n. 321). As Stoner noted, a model of ferromagnetism due to free electrons would yield Curie temperatures much higher than those observed.

326. J. C. Slater, "Theory of Complex Spectra," *Phys. Rev.* 34 (1929): 1293–1322.

327. Slater says in his biography that he first met Bloch in Zurich at a conference, the July ETH "lecture week," that year on x rays and quantum theory. See report on meeting in *Phys. Zs.* 30 (1929): 513–515; W. Pauli and P. Scherrer to R. de L. Kronig, 6 May 1929, *Scientific Correspondence I* (n. 34), 221, 497. The datings of the Slater paper (received 8 June, and sent off, Slater says, before he went to Europe) and the Bloch paper (submitted 10 June), and the fact that both acknowledge the other, suggest that they in fact first interacted in early June. Slater (n. 110), 62, 123.

328. D. R. Hartree, "The Wave Mechanics of an Atom with a Noncoulomb Central Field. Part I. Theory and Methods. Part II. Some Results and Discussions. Part III. Term Values and Intensities in Series in Optical Spectra," *Proceedings of the Cambridge Philosophical Society* 24 (1928): 89–110, 111–132, 426–437; see also C. G. Darwin, "Douglass Rayner Hartree, 1897–1958," *Biographical Memoirs of Fellows of the Royal Society of London* 4 (1958): 103–116. Hartree was also present at the July meeting at the ETH (n. 327).

329. J. C. Slater, interview with T. S. Kuhn, AHQP; Slater, "The Current State of Solid-State and Molecular Theory," *International Journal of Quantum Chemistry* 1 (1967): 37–102, p. 52.

330. See n. 314.

331. The variational, now called Hartree–Fock, method of improving Hartree's approach was suggested independently in November 1929 by V. Fock and Slater. V. Fock, "Näherungsmethode zur des quantenmechanischen Mehrkörperproblems," *Zs. Phys.* 61 (1930): 126–48; J. C. Slater, "Note on Hartree's Method," *Phys. Rev.* 35 (1930): 210–211. See also V. Fock to J. C. Slater, 3 April 1930, Slater Papers, American Philosophical Society Library, Philadelphia.

332. Slater (n. 110), 62.

333. Bethe interview (n. 48).

334. Bloch interview (n. 105).

335. Slater (n. 110), 126. Hund had always favored the one-electron approximation, considering ferromagnetism a matter of multiplet level splitting and a competition between two energies of the same order of magnitude, that for excitation to states of the necessary symmetry, and the gain by level splitting. He later remarked, "Heinsenberg's understanding of ferromagnetism by means of the exchange integral was not to my taste" (Hund interview [n. 121]).

336. J. C. Slater, "Cohesion in Monovalent Metals," *Phys. Rev.* 35 (1930): 509–529.

337. See n. 314.

338. See n. 335.

339. J. C. Slater, "Atomic Shielding Constants," *Phys. Rev.* 36 (1930): 57–64.

340. Bloch interview (n. 105).

341. Ibid.

342. F. Bloch, "Zur Theorie des Ferromagnetismus," *Zs. Phys.* 61 (1930): 206–219. The discovery of spin waves was made simultaneously and independently by Slater, who in the concluding section of his paper on the cohesion of metals illustrated the effect of spin fluctuations in the normal state with a calculation of the excited states of a fully aligned chain of spins, computing the single flipped spin state exactly and the multiply flipped states approximately. Slater did not, however, draw the implications of spin waves for ferromagnetism.

343. F. Bloch, "Über die Wechselwirkung der Metallelektronen," in *Leipziger Vorträge 1930: Elektronen-Interferenzen,* ed. P. Debye (Leipzig: Hirzel, 1930), 67–74.

344. Peierls (n. 204).

345. Pauli (n. 249).

346. Bethe to Sommerfeld, from Capri, 30 May 1931, Sommerfeld Papers, DM.

347. H. Bethe, "Zur Theorie der Metalle. I. Eigenwerte und Eigenfunktionen der lineare Atomkette," *Zs. Phys.* 71 (1931): 205–226.

348. Sommerfeld and Bethe (n. 2) 607–618.

349. F. Bloch to R. Peierls, 6 November 1931. A copy was kindly given to us by Bloch.

350. F. Bloch, "Zur Theorie des Austauschproblems und der Remanenzerscheinung der Ferromagnetika," *Zs. Phys.* 74 (1932): 295–335.

351. Bloch interview (n. 90).

352. Bloch interview (n. 105). See also Peierls interview (n. 48).

353. J. Schwinger, "On Angular Momentum," AEC Report NYO-3071 (1952); see also E. Wigner, *Group Theory and Its Application to Quantum Mechanics of Atomic Spectra* (New York: Academic Press, 1959), ch. 14.

354. Bloch interview (n. 105).

355. Ibid.

356. See Chapter 8.

357. Ibid.

358. Sommerfeld and Bethe (n. 2), 555.

359. The experimental situation in superconductivity around 1928 was discussed by Grüneisen (n. 17); for the early 1930s, see W. Meissner, "Supraleitfähigkeit," *Ergebnisse der exacten Naturwissenschaften* 11 (1932): 219–263; H. Grayson Smith and J. O. Wilhelm, "Superconductivity," *Reviews of Modern Physics* 7 (1935): 237–271. A more complete discussion of the early period of superconductivity, as well as further references, is given in Chapter 8.

360. See, H. Kamerlingh Onnes, "The Imitation of an Ampere Molecular Current or of a Permanent Magnet by Means of a Supra-conductor," *Communications from the Physical Laboratory of the University of Leiden* 140b (1914): 9–18; W. Tuyn, "Quelques essais sur les courant persistants," *Communications from the Physical Laboratory of the University of Leiden* 198 (1929): 3–

18. W. J. de Haas and J. Voogd refined the characterization of superconductors in 1931: "We therefore regard the vanishing of the resistance within a few hundreths of a degree as the most characteristic phenomenon of supraconductivity in pure metals" ("On the Steepness of the Transition Curve of Supraconductors," *Communications from the Physical Laboratory of the University of Leiden* 214c [1931]: 17–33).

361. H. Kamerlingh Onnes, "Report on Research Made in the Leiden Cryogenic Laboratory Between the Second and Third International Congress of Refrigeration," *Communications from the Physical Laboratory of the University of Leiden,* suppl., 34b (1913): 35–70; W. Tuyn and H. Kamerlingh Onnes, "Further Experiments with Liquid Helium. AA. The Disturbance of Superconductivity by Magnetic Fields and Currents," *Communications from the Physical Laboratory of the University of Leiden* 174a (1925): 3–39.

362. For a description of McLennan's laboratory, see J. C. McLennan, "The Cryogenic Laboratory of the University of Toronto," *Nature* 112 (1923): 135–139. Work of the laboratory is reviewed in E. F. Burton, *The Phenomena of Superconductivity* (Toronto: University of Toronto Press, 1934).

363. W. Meissner, "Uber die Heliumverflüssigungsanlage der PTR und einige Messungen mit Hilfe von flüssigen Helium," *Phys. Zs.* 26 (1925): 689–694.

364. Ibid., 691.

365. W. Meissner, "Supraleitfähigkeit von Tantal, " *Phys. Zs.* 29 (1928): 897–904; "Meissner, Supraleitfähigkeit von Kupfersulfid," *Zs. Phys.* 58 (1929): 570–572.

366. G. I. Sizoo, "Untersuchungen über den supraleitenden Zustand von Metallen" (Ph.D. diss., Leiden University, 1926).

367. W. Meissner, "Der Widerstand von Metallen und Metallkristallen bei der Temperatur des flüssigen Heliums," *Phys. Zs.* 27 (1926): 725–730.

368. W. J. de Haas and H. Bremmer, "Thermal Conductivity of Lead and Tin at Low Temperatures," *Communications from the Physical Laboratory of the University of Leiden* 214d (1931): 35–52.

369. W. Meissner and R. Ochsenfeld, "Ein neuer Effekt bei Eintritt der Supraleitfähigkeit," *Die Naturwissenschaften* 21 (1933): 787–788; W. Meissner, "Bericht über neuere Arbeiten zur Superleitfähigkeit," *Zeitschrift für technische Physik* 15 (1934): 507–514.

370. Mrs. G. L. de Haas-Lorentz had in fact already discussed the penetration depth of superconductors, starting from the question of whether a magnetic field held completely outside a superconductor by screening currents can exert influence on the superconductor: "Iets over het Mechanisme van Inductieverschijnselen," *Physica* 5 (1925): 384–388. Having been published in Dutch, the work remained largely unnoticed prior to the discovery of the Meissner–Ochsenfeld effect.

371. For example, W. H. Keesom and J. A. Kok, "On the Change of the Specific Heat of Tin When Becoming Superconductive," *Communications from the Physical Laboratory of the University of Leiden* 221c (1932): 27–32; W. H. Keesom, "Das kalorische Verhalten van Metallen bei den tiefsten Temperaturen," *Zeitschrift für technische Physik* 15 (1934): 515–520.

372. P. Ehrenfest, "Phasenumwandlungen im üblichen und erweiterten Sinn, classifiziert nach den entsprechenden Singularitäten des thermodynamischen Potentiales," *Communications from the Physical Laboratory of the University of Leiden,* suppl., 75B (1933): 8–13; A. J. Rutgers, "Note on Superconductivity," *Physica,* 1st ser., vol. 2 (1934): 1055–1058; C. J. Gorter and H. G. B. Casimir, "Zur Thermodynamik des supraleitenden Zustandes," *Phys. Zs.* 35 (1934): 963–966; Gorter and Casimir, "Zur Thermodynamik des supraleitenden Zustandes," *Zeitschrift für technische Physik* 15 (1934): 539–542; Gorter and Casimir, "On Superconductivity," *Physica,* 1st ser., vol. 2 (1934): 306–320.

373. Assorted early theories were reviewed by E. Kretschmann, "Kritischer Bericht über neue Elektronentheorien der Elektrizitäts- und Wärmeleitung in Metallen," *Phys. Zs.* 28 (1927): 565–592. This article, written as Sommerfeld was developing the semiclassical free-electron theory of metals, did not benefit from the perspective of that theory.

374. F. A. Lindemann, "Note on the Theory of the Metallic State," *Phil. Mag.,* 6th ser., vol. 29 (1915): 127–141.

375. J. J. Thomson, "Further Studies on the Electron Theory of Solids. The Compressibilities of a Divalent Metal and of the Diamond. Electric and Thermal Conductivities of Metals," *Phil. Mag.,* 6th ser., vol. 44 (1922): 657–679.

376. A. Einstein, "Theoretische Bemerkungen zur Supraleitung der Metalle," in *Het Natu-urkundig Laboratorium der Rijksuniversiteit te Leiden in de Jaren 1904-1922* [*Gedenkboek H. Kamerlingh Onnes*] (Leiden: Eduard Ijdo, 1922), 429–435; see also the discussion in B. E. Yavelov, "On Einstein's Paper . . . ," in *Historical Problems of Science and Technology* (Moscow: Nauka, 1980), 3(67)–4(68): 46–55.

377. N. Bohr to W. Heisenberg, end of December 1928, from Hornbaek, to be published in *Niels Bohr Collected Works.* D. Cassidy kindly alerted us to this communication. According to Bloch, "Heisenberg tried it [superconductivity] at one point . . . [with] some idea of condensation in angular momentum space" (Bloch interview [n. 105]). Heisenberg did not publish on superconductivity in this period, but did return to the problem after the war. See W. Heisenberg, "Zur Theorie der Supraleitung," *Zeitschrift für Naturforschung* 2a (1947): 185–201; Heisenberg, "Das elektrodynamische Verhalten der Supraleiter," *Zeitschrift für Naturforschung* 3a (1948): 65–75; Heisenberg, *Electron Theory of Superconductivity: Two Lectures* (Cambridge: Cambridge University Press, 1949), 27–51.

378. In his 1932 paper on superconductivity sent to *Die Naturwissenschaften* (n. 402), which he withdrew in proof, Bohr remarks, immediately after discussing Bloch's conductivity work, "On the basis of the independent electron picture, no explanation can be given for Kamerlingh Onnes's discovery."

379. Bloch interview (n. 105).

380. Since Bloch's work was never published, our knowledge of it comes from references by others —for example, from a short description of the idea "Bloch and Landau suggested" by Bethe in the *Handbuch* article (Sommerfeld and Bethe [n. 2], sec. 44, 555–558); from L. Landau, "Zur Theorie der Supraleitfähigkeit. I, " *Physikalische Zeitschrift der Sowjetunion* 4 (1933): 43–49; from L. Brillouin, "Supraconductivity and the Difficulties of Its Interpretation," *Proc. Roy. Soc.* A152 (1935): 19–21; as well as from Bloch's interviews and retrospective articles (e.g., nn. 382, 390).

381. Landau (n. 380). Referring in his paper to his 1929 work, Landau wrote "similar thoughts were also simultaneously expressed by Bloch." Although Bloch was based in Leipzig at the time, he did turn up for visits in Zurich, where Landau worked briefly in late 1929, possibly giving the two opportunity to discuss their similar explanations of superconductivity. Bloch recalls that he conceived of his idea before he came to work with Pauli, but that he and Landau did not communicate about their related notions for some time. Bloch interview (n. 105).

382. F. Bloch, "Memories of electrons in crystals," in *Beginnings,* 24–27.

383. Bloch interview (n. 105).

384. W. Pauli to N. Bohr, 16 January 1929, *Scientific Correspondence I* (n. 34), 214, 485.

385. W. Pauli to O. Klein, 16 March 1929, *Scientific Correspondence I* (n. 34), 218, 494.

386. Pauli to Bohr, 25 April 1929, *Scientific Correspondence I* (n. 34), 219, 496.

387. W. Pauli to A. Sommerfeld, 16 May 1929, *Scientific Correspondence I* (n. 34), 225, 503.

388. W. Pauli to R. de L. Kronig, 2 June 1929, *Scientific Correspondence I* (n. 34), 226, 504.

389. Bloch interview (n. 105).

390. F. Bloch, "Some Remarks on the Theory of Superconductity," *Physics Today* 19, no. 5 (1966), 27–36, p. 27.

391. Brillouin (n. 380).

392. L. Brillouin, "Sur la stabilité du courant dans un supraconducteur," *Journal de physique et le radium,* 7th ser., vol. 4 (1933): 677–690.

393. For example, F. London, "Macroscopical Interpretation of Supraconductivity," *Proc. Roy. Soc.* A152 (1935): 24–34, p. 25; Bethe interview (n. 48).

394. Landau (n. 380).

395. F. London and H. London, "The Electromagnetic Equations of the Supraconductor," *Proc. Roy. Soc.* A149 (1935): 71–88.

396. W. J. de Haas and J. Voogd, "Measurements. . . of the Magnetic Disturbance of the Supraconductivity of Thallium," *Communications from the Physical Laboratory of the University of Leiden* 212d (1931): 37–44.

397. J. Frenkel, "On a Possible Explanation of Superconductivity," *Phys. Rev.* 43 (1933): 907–912.

398. H. Bethe and H. Fröhlich, "Magnetische Wechselwirkung der Metallelektronen. Zur Kritik der Theorie der Supraleitung von Frenkel," *Zs. Phys.* 85 (1933): 389–397.

399. Sommerfeld and Bethe (n. 2), 556.

400. N. Bohr to F. Bloch, 15 June 1932, AHQP-BSM 1932.

401. N. Bohr to M. Delbrück, dated 5 July 1932, AHQP-BSM.

402. The proofs of Bohr's article, with the heading "Naturwissenschaften, 11.7.32 (art. 492. Bohr)" are preserved in AHQP-BSM.

403. Note by Bloch, "Zur Supraleitung," AHQP-BSM 1932.

404. F. Bloch to N. Bohr, 12 July 1932, AHQP-BSM.

405. Bloch to Bohr, 26 July 1932, AHQP-BSM.

406. Bohr to Bloch, 7 September 1932, AHQP-BSM.

407. R. de L. Kronig, "Zur Theorie der Superleitfähigkeit. I, II," *Zs. Phys.* 78 (1932): 744–750; 80 (1932): 203–216.

408. N. Bohr to R. de L. Kronig, 17 October 1932, AHQP-BSM.

409. McLennan's experiments showed a decrease of the superconducting transition temperature in the presence of an alternating current, J. C. McLennan, A. C. Burton, A. Pitt, and J. O. Wilhelm, "The Phenomena of Superconductivity with Alternating Currents of High Frequency," *Proc. Roy. Soc.* A136 (1932): 52–76; McLennan et al., "Further Experiments on Superconductivity with Alternating Currents of High Frequency," *Proc. Roy. Soc.* A143 (1932): 245–258. The issue was whether these experiments indicated that "superconductivity is not, as assumed, a property in a metal at low temperatures independent of the presence of a current, but that it must be closely knitted to the current mechanism itself" (Bohr, "Om Supraledningsevnen" [n. 412]) and thus does not arise simply from a structure such as an electron lattice, described in terms of a modulation of the ground-state wave function. The second McLennan paper (received 12 September 1932) discussed the high-frequency currents as "confined to the outside of the . . . wire" by the skin effect, unlike the first paper, where a connection with the skin effect was dismissed; the fact that McLennan's effect was not a bulk phenomenon seemed to Bohr to have made the experiments a less crucial test of his conception. It was not realized before the Meissner effect that supercurrents are actually carried in the surface rather than by bulk motion of the electrons.

410. N. Bohr to R. de L. Kronig, 26 March 1926, AHQP-BSM.

411. Kronig to Bohr, 18 October 1932, AHQP-BSM.

412. Shortly after Kronig's visit, Bohr drafted several versions of an addendum to his article to explain the differences between his and Kronig's theory. N. Bohr, manuscripts, "Om Supraledningsevnen" (in Danish), and untitled (in German), 1932, AHQP-BSM.

413. Kronig to Bohr, 15 November 1932, AHQP-BSM.

414. F. Bloch to N. Bohr, 17 November 1932, AHQP-BSM.

415. Kronig to Bohr, 25 November 1932, AHQP-BSM.

416. Kronig to Bohr, 14 December 1932, AHQP-BSM; Kronig (n. 407).

417. Bohr to Kronig, 27 December 1932, AHQP-BSM. Here Bohr countered Kronig's argument against his wave function by explicitly constructing a small current state for a single electron in a one-dimensional periodic potential. This result, Kronig pointed out to Bohr, did not agree with exact solutions for his model with Penney; Kronig to Bohr, 8 January 1933, AHQP-BSM. This last argument of Kronig clearly lingered with Bohr, for as J. R. Schrieffer recalls, Bohr remarked to him in Copenhagen in 1958, during a discussion of the recently developed BCS theory of superconductivity, in essence, "We must go back to fundamentals, and first understand the Kronig-Penney model" (private communication).

418. F. Bloch to N. Bohr, 30 December 1932, AHQP-BSM.

419. Sommerfeld and Bethe (n. 2), 556–557.

420. J. Frenkel was one of the few to attempt to extend Kronig's electron chain theory, describing supercurrents as a collective electron motion—"like a chain gliding over a toothed track" ("The Explanation of Superconductivity," *Nature* 133 [1934]: 730).

421. Brillouin (n. 200).

422. In this paper, Brillouin finally recognized *Umklapp* processes, "transition anormales de Peierls," as a possible relaxation mechanism.

423. Brillouin (n. 380).

424. W. Elsasser, "Über Strom und Bewegungsgrösse in der Diracschen Theorie des Elektrons," *Zs. Phys.* 75 (1933): 129–133.

425. Sommerfeld and Bethe (n. 2), 556.

426. R. Schachenmeier, "Wellenmechanische Vorstudien zu einer Theorie der Supraleitung," *Zs. Phys.* 74 (1932): 503–546.

427. E. Kretschmann, "Beitrag zur Theorie des elektrischen Widerstandes und der Supraleitfähigkeit der Metalle," *Ann. Phys.* 13 (1932): 564–598. According to Peierls, Kretschmann seemed fair game for criticism; he was "generally known to be somebody who quibbled about the current theory without really understanding it. I don't know whether that's a fair assessment, but that was our impression at the time" (Peierls interview [n. 48]). Sommerfeld and Bethe (n. 2), 558.

428. H. Bethe to A. Sommerfeld, 25 February 1934, Sommerfeld Papers, DM.

429. W. Henneberg to A. Sommerfeld, 27 February 1934, Sommerfeld Papers, DM.

430. Bethe to Sommerfeld, 25 February 1934, Sommerfeld Papers, DM; H. Bethe, "Zur Kritik der Theorie der Supraleitung von R. Schachenmeier," *Zs. Phys.* 90 (1934): 674–679; see also Schachenmeier's reply; R. Schachenmeier, "Zur Theorie der Supraleitung. Entgegnung an Herrn Bethe," *Zs. Phys.* 90 (1934): 680–692.

431. Sommerfeld and Bethe (n. 2), 558.

432. "Vorlesungsverzeichnis für das Wintersemester 1930/1931," *Phys. Zs.* 31 (1930): 982. Heisenberg's handwritten lecture notes, "Quantentheorie der festen Körper," in the Heisenberg Papers, presently at the Max Planck Institute of Physics in Munich, to be deposited in the Archives of the Max-Planck Gesellschaft, Berlin, and in AHQP, microfilm 45, sec. 7.

433. "Vorlesungsverzeichnis," *Phys. Zs.* 33 (1932): 790.

434. We thank E. Wigner for sending us a copy of these notes, which are in his personal collection at Princeton.

435. E. Wigner, interview with L. Hoddeson and G. Baym, with F. Seitz present, January 1981, AIP.

436. For example, the 1927 Solvay Conference, "Electrons and Photons," included a talk by W. L. Bragg on x-ray diffraction and reflection in crystals. Instituts Solvay, *Rapports et discussions du Cinquième Conseil de Physique* (Paris: Gauthier-Villars, 1928).

437. The organization was undertaken by Pauli and Scherrer. See their correspondence, 6 May 1929, *Scientific Correspondence I* (n. 34), 497.

438. H. Bethe to A. Sommerfeld, 7 May 1934 and 9 June 1934, Sommerfeld Papers, DM. Further conferences on solid-state topics took place in 1935 at Stuttgart and in 1937 at Zurich. In planning for the 1937 Zurich meeting, Pauli again expressed his dislike of solid-state theory, writing to Sommerfeld on 6 December 1936: "I'm happy that you'll come to Zurich in January and I hope to meet you—independently of the congress on solid bodies, which doesn't interest me very much . . ." (*Scientific Correspondence II* [n. 34] 486).

439. See n. 6.

440. Slater recalls that in the spring of 1930, when the editor of *Reviews of Modern Physics,* John Tate, asked him to write on the electronic structure of metals, he was "trying to get a unified point of view regarding the whole subject of the electronic constitution of crystals, particularly of metals" ([n. 110], 192).

441. Bethe ("My Life") (n. 54).

442. Sommerfeld apparently never tried to understand the material Bethe covered, nor did he, as Bethe recalled, discuss the writing of these later sections much with him, although the two did go over the initial outline. The review was essentially unedited, and printed "pretty much in the way I had written it" (Bethe interview [n. 48]; A. Sommerfeld to W. L. Bragg, n.d. [ca. March 1933], Sommerfeld Papers, DM).

443. See n. 6.

444. First published in French (1930) and then substantially expanded in German (1931) and Russian.

445. A. H. Wilson, *The Theory of Metals* (Cambridge: Cambridge University Press, 1936).

446. N. F. Mott and H. Jones, *Theory of the Properties of Metals and Alloys* (Oxford: Oxford University Press, 1936), 320. This contains a more complete listing of texts and review articles.

447. See n. 48.

448. Peierls (n. 6).

449. Bloch (n. 6).

450. H. Bethe to A. Sommerfeld, from Rome, 9 April 1931, Sommerfeld Papers, DM.

451. Peierls interview (n. 173).

452. Bloch (n. 6), 238–239.

453. Brillouin (1931) (n. 6), v.

454. See, D. Pines, "Elementary Excitations in Quantum Liquids," *Physics Today* 34, no. 11 (1981): 106–131.

455. Wigner and Seitz (n. 7).

456. For their method, see Chapter 3. Bethe read the Wigner–Seitz work while in the last stages of preparing the *Handbuch* article and quickly added a section on it. Bethe interview (n. 48).

457. Peierls (n. 109), 34.

458. W. Pauli to R. Peierls, 22 May 1933, *Scientific Correspondence II* (n. 34), 163–65, p. 163. Pauli's attitude was certainly not unfamiliar to Peierls. For example, when in 1931 Peierls showed Pauli a now famous calculation dealing with the residual resistance, Pauli responded, "I consider it harmful when younger physicists become accustomed to order-of magnitude physics. The residual resistance is a dirt effect and one shouldn't wallow in dirt" ("Ich halte es für schädlich, wenn die jüngeren Physiker sich an die Grössenordungsphysik gewöhnen. Der Restwiderstand ist ein Dreckeffekt, und im Dreck soll man nicht wühlen") (W. Pauli to R. Peierls, in Ann Arbor, 1 July 1931, *Scientific Correspondence II* [n. 34] 85–87, p. 85).

459. L. Brown and L. Hoddeson, "The Birth of Elementary Particle Physics," in *The Birth of Particle Physics,* ed. L. Brown and L. Hoddeson (New York: Cambridge University Press, 1983), 3–36.

460. Nordheim interview (n. 52). D. Shoenberg, interview with P. Hoch, 10 February 1981.

461. Bethe interview (n. 48). H. Bethe to A. Sommerfeld, 1 August 1936, Sommerfeld Papers, DM.

462. Peierls interview (n. 48).

463. Bloch interview (n. 105).

464. Bethe to Sommerfeld, 1 August 1936, Sommerfeld Papers, DM. The experimental and team-orientation characteristic of American physics, compared with Continental physics in this period, was discussed further in Bethe's interview with C. Weiner, 1966–1972, AIP; Bethe interview (n. 48); and Brown and Hoddeson (n. 459).

465. S. T. Keith and P. K. Hoch, "Formation of a Research School: Theoretical Solid State Physics at Bristol, 1930–1954," *British Journal for the History of Science* 19 (1986): 19–44. We thank the authors for communicating their manuscript to us prior to publication.

466. See J. C. Slater, "History of the M.I.T. Physics Department 1930–48" (manuscript, Slater Papers, American Philosophical Society Library, Philadelphia). Direct ties between Slater's department and Sommerfeld's institute were built by earlier visits to Munich (e.g., by Frank, Morse, and William Allis in Slater's new department). Slater (n. 110).

467. Wigner interview (n. 435).

468. See Hoddeson (n. 87); G. Wise, *Willis R. Whitney, General Electric, and the Origins of U. S. Industrial Research* (New York: Columbia University Press, 1985).

469. Indeed, this emergence has the character of a "secondary scientific revolution," in the sense of T. S. Kuhn, *The Structure of Scientific Revolutions* (Chicago: University of Chicago Press, 1962).

470. A. H. Wilson, private communication, May 1979.

471. Casimir (n. 32).

472. See n. 458.

3 | *The Development of the Band Theory of Solids, 1933–1960*

PAUL HOCH
(WITH CONTRIBUTIONS FROM KRZYSZTOF SZYMBORSKI)

At the core of the modern theory of solids is the idea that the conduction electrons in solids can be represented as occupying a series of energy bands that correspond to the electronic shells of atoms. This picture, as discussed in Chapter 2, grew out of Felix Bloch's theory of conduction in metals (1928),[1] although Bloch had considered only the ground-state band, limiting himself to a single electron in a fixed periodic potential and ignoring interactions between electrons. A number of workers clarified this "one-electron" model during the next three years, most notably, Peierls,[2] Brillouin,[3] and A. H. Wilson.[4] They found the energy to be a multivalued function $E(\mathbf{k})$ of the conduction electron's wave vector $\mathbf{k}$, whose values are restricted to the inside of a polyhedron, the Brillouin zone in three dimensions. The set of succeeding branches of $E(\mathbf{k})$ for the inside of the polyhedron makes up the energy-band structure of the particular solid.

With Wilson's formulation of the band concept in the spring of 1931 (Chapter 2), it became clear that in terms of the energy bands one could answer many time-honored questions about the properties of solids: For example, why does a metal conduct, whereas an insulator does not? Conduction at absolute zero occurs if one or more of the outer bands are not fully occupied, or, in other words, if what was then called "the surface of the Fermi distribution"—the surface $E(\mathbf{k}) = E_{\text{Fermi}}$ that separates the occupied electron states of the Fermi distribution from the unoccupied ones at the temperature of absolute zero—falls within the outer bands, so that these outermost, or conduction, electrons have adjacent energy states available to move into. Insulator behavior occurs when the Fermi surface coincides with the top

182

of the outermost band, so no adjacent energy states are available for the conduction electrons. The unusual conduction properties of semiconductors could also be explained by assuming that their outer band is filled, as with insulators, but that one or more unfilled electronic levels are close enough to the next allowed band that excitation by heat from the filled band to the unfilled band can be promoted relatively easily.[4]

Solutions of such long-standing puzzles as what causes the intrinsic difference between metals and insulators were promising features of the new band concept, and they contributed to general confidence in the new quantum theory of solids. But until 1933, there was no manageable way to calculate the actual band structure of any particular substance, the function $E(\mathbf{k})$, or even the surface $E = E_{\text{Fermi}}$ for any metal. Thus one could not connect the theory with experimental observations made on real solids in more than a qualitative way.[5] Pictures of ideal Fermi surfaces corresponding to abstract models of crystal structures and unit cells had been published as early as 1933.[6] However, adequate representation of the Fermi surfaces of real materials would require experimental and theoretical techniques not available until after the Second World War (e.g., microwave techniques, computer hardware), and the ability to prepare sufficiently pure samples.[7] Thus by 1933, the theory of the electronic structure of metals had sound conceptual foundations, but was still almost entirely expressed in terms of ideal rather than real materials.[8]

However, a new intellectual development opened up in this year, just as the rise of a Nazi government in Berlin was effecting a dramatic institutional break in the development of modern physics, including the dispersal of major German centers to successors in Britain and America.[9] The impetus came from an apparently unambitious paper, written by a visiting Hungarian physicist, Eugene P. Wigner, working half-time at Princeton University, and Wigner's first graduate student, Frederick Seitz. This paper presented a simple model for computing the bands of a particular monovalent metal, sodium.[10] If the crystal were viewed as a network of identical cells surrounding single metallic ions, the conduction electron in each cell could be seen as acted on only by its "own" ion's field. This and other simple assumptions led to surprisingly good results for some of the properties of sodium, and the work soon suggested more complicated extensions of the method to other materials. Indeed, Wigner and Seitz's prototypal calculation was followed by detailed studies of real band structures, eventually permitting comparisons of theory and experiment for the properties of many different crystals, both metallic and nonmetallic. Thus began the modern era of solid-state theory of real materials.[11] As this chapter discusses, the first important lines of development out of the Wigner–Seitz breakthrough emerged at Princeton University, around Wigner; at the Massachusetts Institute of Technology (MIT), around John Slater; and at the University of Bristol in Britain, around Nevill Mott and Harry Jones.

The Second World War brought the development to a near halt for several years, but resulted in substantial increases in knowledge and new technology that nurtured the field of band structure in the postwar decades. The war also led to new funding of science and an entirely new scale of operations. Not long after the war, many new techniques introduced during military research were finding important application, and by the late 1950s studies of the electronic structure of solids climaxed with very accurate determinations, arising from both experiment and the-

Eugene Wigner and Gerhard Dieke. (AIP
Niels Bohr Library, Goudsmit Collection)

ory, of the band structures and Fermi surfaces of a wide range of solids. By this time,
the primary concern was no longer to create a new theory, but to develop an already
established one, to join the theory to experimental observations, and, in the process,
to unite solid-state physics with the study of materials in both industrial and academic institutions.

3.1   The 1930s: The First Realistic Band Structure Calculations

Princeton

In 1929, Princeton University decided to strengthen itself in mathematical physics
by hiring the Hungarian-born mathematician John von Neumann, then based in
Germany. This move was part of a long-term effort to synthesize the approaches of
related disciplines such as mathematics and physics, an effort supported and backed
financially by the Rockefeller Foundation.[12] This funding enabled Princeton to
bring in, from the autumn of 1928, such exceptional mathematical physicists as
Edward U. Condon, the relativity specialist Howard P. Robertson, as well as von
Neumann and—almost as an afterthought—his Hungarian colleague Wigner. In
his letter of invitation to von Neumann, Mathematics Department chairman
Oswald Veblen noted that "considerable interest has [also] been expressed in the
work of Dr. Wigner who has collaborated with you in some of your recent papers.
Would it not be a good idea to invite him to lecture here at the same time as your-
self?" Von Neumann agreed, noting in his reply that he had spoken to Wigner and
that "he too would very much like to come to America and especially to Princeton."
And so the arrangement was made.[13]

Wigner had received his scientific education in Germany, as did a number of
other Hungarian-refugee physicists who contributed importantly to solid-state

research.[14] He had been a student and then a staff member at the Technische Hochschule in Berlin, where he worked with such people as the Hungarian physical chemist Michael Polanyi. He spent 1927 and 1928 as a postdoctoral assistant to the German mathematician David Hilbert. In this period, he also wrote his important book on the applications of group theory to quantum mechanics.[15]

Princeton University's role as a center for research on the electron theory of metals in the 1930s was to grow out of the interests of Wigner, who between 1930 and 1936 helped to train several of the key theorists of the first solid-state generation. These included Wigner's graduate students Seitz, John Bardeen, and Conyers Herring, and such European postdoctoral fellows as Louis Bouckert, Roman Smoluchowski, and Gregory Wannier. Princeton had built its reputation in science earlier in the century, following the addition to its staff in 1905 and 1906 of the prominent British physicists James Jeans and Owen W. Richardson.[16] In the 1930s, it was still a rather small but elite school. A typical graduate science class at the time Seitz took his courses would not exceed 12 students.[17] In terms of its social habits, Princeton was, perhaps, a more conventional place than the majority of American universities—"you were expected to wear a tie all the time or be ostracized,"[18] as Seitz recalled. All the same, the small size of the science departments enabled easy and frequent personal contacts between students and faculty. The Physics Department, located in Palmer Lab, and the Mathematics Department, in Fine Hall, were closely tied, and at Fine Hall "every day at 4:30, everyone who could walk or go on crutches met [for tea] in what was called the social room and spent about 20 minutes to a half-hour talking."[19] There were also regular weekly seminars of both the Physics and Mathematics departments, occasionally attended by foreign visitors like Paul Dirac, Ralph Fowler, and Erwin Schrödinger.

Seitz came to Princeton at the end of 1931, after completing undergraduate work in physics at Stanford. He had met his initial Princeton research supervisor, Condon, while attending the latter's special summer course on modern physics at Stanford in 1930. At this point, as Seitz recalled recently, Condon was "one of the few individuals on the East coast with a working knowledge of quantum mechanics." When Condon became immersed in writing his book *The Theory of Atomic Spectra*,[20] Wigner assumed the responsibility for Seitz's thesis on the application of group theory to crystal lattices.[21] In his Princeton years, Seitz maintained his earlier acquaintance with John Slater's research student William Shockley, whom he had met in a course Seitz was attending at the California Institute of Technology given by the Sommerfeld-trained William V. Houston. Seitz recalls, among other interactions, driving across the country one summer with Shockley and stopping off to look in on the summer school in theoretical physics at the University of Michigan.[22]

Bardeen, who was to share in two Nobel Prizes, for his discoveries in semiconductor electronics (Chapter 7) and superconductivity (Chapter 8), obtained his introduction to quantum theory in 1928–1929, while a first-year graduate student in electrical engineering at the University of Wisconsin. His teacher was John H. Van Vleck, who had also, at the Universtiy of Minnesota, given undergraduate instruction in quantum theory to Walter Brattain. Brattain was to be Bardeen's collaborator in developing the first point-contact transistor. Van Vleck was to become the central theorist in the United States in the area of magnetism.

After working from 1930 to 1933 as a geophysicist at the Gulf Oil Company

John Bardeen (b. 1908). (AIP Niels Bohr Library, Physics Today Collection)

Frederick Seitz (b. 1911). (AIP Niels Bohr Library, Physics Today Collection)

Research Laboratories in Pittsburgh, Bardeen decided to return in 1933 to graduate school in mathematics at Princeton University, "not only because of [Princeton's] excellence in physics and mathematics, but also because the [independent] Institute for Advanced Study had recently been established there" in the same building as the Mathematics Department.[23] Bardeen's thesis problem under Wigner was to calculate the work function of a monovalent metal, the energy required to remove an electron from inside to outside, a matter that was to be of considerable importance at the end of the 1940s to the development of the transistor. Bardeen subsequently refined this approximate calculation, using a Hartree–Fock method to obtain the effect on the work function of the surface double layer, including the higher contributions due to electron correlation and exchange in a "jellium" model, one in which the positive charges of the ions are represented by a uniform distribution.[24]

Both Bardeen and Seitz belonged to a small informal group of theoretically oriented students at Princeton, which had formed originally around Condon and met from time to time in the afternoon to discuss some current topic in theoretical physics and then spent the rest of the evening over a glass of beer in the Nassau Tavern. This informal seminar was joined in 1934 by Herring, who after graduating in astronomy from the University of Kansas and spending a year at Caltech, decided to switch from astronomy to physics and work for his Ph.D. with Wigner.[25]

The seminal Wigner–Seitz calculation of the band structure of sodium grew, as Seitz recalls, out of Wigner's initial idea that the reason metals cohered was that their wave functions were probably smoother than atomic wave functions. (An atomic wave function is under the constraint that it has to go to zero at large distances, whereas in a metal it doesn't.) It was a qualitative notion and we started out to see if we could do something rough and ready that would indicate this.[26]

They chose the monovalent metal sodium for simplicity and because, as a literature search showed, W. Prokofjew, a Russian physicist from the Leningrad State Optical Institute, had worked out an expression for its atomic field.[27] Prokofjew had followed a method suggested by H. A. Kramers and employed in 1927 for calculations of sodium's atomic properties[28] by a Japanese physicist working with Kramers in Copenhagen, Yoshikazu Sugiura.

By defining an orderly procedure whereby a free-electron model could be applied to calculate particular chemical properties of real, not just ideal, metals, this first Wigner–Seitz paper marked the beginning of the application of the electron theory of metals to real materials. The authors noted that whereas previous attempts at a theory of metals involved either the hypothesis of free or "nearly free" conduction electrons to explain the conductivity properties, or approximations based on valence forces to deal with chemical properties, they would employ a free-electron picture to investigate such chemical properties as the equilibrium lattice spacing, heat of vaporization, and compressibility. One reason for preferring a free- or one-electron approximation even for the calculation of chemical properties was that "the application of the usual methods [expressing crystal states as linear combinations of atomic or molecular states] to calculate valence forces becomes more and more difficult as the number of atoms increases."[29]

The free-electron model assumed that the many-body problem, involving all the conduction electrons and the ionic cores of the crystal lattice, could be broken down into the motions of single, essentially non–interacting, conduction electrons in some localized crystal field. As discussed in Chapters 1 and 2, this approach was originally the appropriation by Paul Drude and H. A. Lorentz of the kinetic theory of the ideal gas of noninteracting particles to explain the properties of a metal, and was later carried almost in its entirety into the Pauli–Sommerfeld theories of the late 1920s. In the skillful hands of Wigner and Seitz and their successors, this approximation, with appropriate corrections, was to result in moderately satisfactory explanations of the cohesive forces, elastic constants and compressibilities, ordinary thermal and electrical conductivities, and most optical properties of many metals. But it was inherently incapable of explaining satisfactorily such collective phenomena as ferromagnetism, superconductivity, and second-order phase transitions, which depend strongly on the details of the interactions between electrons.

Wigner and Seitz noted that their approximation was basically the same as the one proposed in Leipzig by Friedrich Hund for molecules,[30] and the one applied at Hamburg by Sommerfeld's former student Wilhelm Lenz and his younger coworker Hans Jensen for ionic lattices.[31] It was also stimulated by a calculation at Bristol by J. E. Lennard-Jones and his research student H. J. Woods, using an ideal two-dimensional metallic lattice.[32] In theory, after using the approximation to calculate the motion of a particular conduction electron in its "own" ion's field, one could follow the prescriptions of Hartree and Fock,[33] plugging this result back into the charge distribution used to construct the approximate crystal potential, and in successive steps make the potential field more self-consistent, thereby gradually improving the accuracy of the approximation. A variation of this method was employed by Lenz, although not by Wigner and Seitz. Refining such approximations was to lead to a series of further papers on how better to adapt the method to sodium and other alkali metals.[34]

The potential that Wigner and Seitz used to calculate the motion of a single conduction electron was not derived from a self-consistent field calculation, but was derived phenomenologically by Prokofjew from a study of sodium's atomic spectra. In order to apply this potential to the situation considered by Sugiura, of conduction electrons in a metallic sodium crystal, it was necessary first to divide the crystal into small atomlike "cells," obtained by drawing the planes that bisect the lines between adjacent ions, enclosing only the volumes closest to a particular sodium ion. Assuming that the potential acting on the single sodium conduction electron within a particular "Wigner–Seitz cell" is spherically symmetric, it was then possible, with certain further assumptions, to set up and separate the one-electron Schrödinger equation for the motion of a particular conduction electron, the $k = 0$ ground state of the conduction band, in a spherical potential, just as Hartree had done for isolated atoms. Wigner and Seitz then solved the radial wave equation by numerical integration, employing the method of finite differences (also used by Prokofjew), to obtain the energy of this free conduction electron in its lowest bound state. From that, they derived an expression for the lowest energy level as a function of the lattice spacing. For infinite spacing, one got the wave function and lowest energy level of an isolated atom (as one must, since Prokofjew and Kramers had devised their potential to lead to the correct energy levels for an isolated atom). Since a conduction electron is, by definition, one that can move through a metallic crystal, its wave function, although similar to that of the corresponding atomic valence electron in the immediate vicinity of the atom, did not drop to zero after its maximum but was much smoother and, for the state of lowest energy, continued periodically through the whole crystal.

When this free-electron energy was plotted against radial distance from the center of the ion core, the radial coordinate of the energy minimum (i.e., at which binding energy is greatest), multiplied by an appropriate constant, gave the equilibrium lattice constant at the absolute zero of temperature. The depth of this energy minimum below that for the free atom, after subtracting a small correction of about 2.5 kcal for the energy gained by the inner ion shells in the course of being inserted in a crystal, then gave approximately the energy difference between the gaseous and solid states at absolute zero—that is, the heat of vaporization per atom. The curvature of the energy minimum could also be related to the experimentally observed compressibility. To be sure, as Slater later commented, "there were so many approximations that it is hard to accept the numerical results very seriously. However for the first time they had given a useable method for estimating energy bands in actual crystals."[35]

In following papers,[36] Wigner and Seitz worked to derive better approximations for the effects of correlation in motion between different conduction electrons. In their first formulation of the theory, they had assumed only correlations necessitated by the Pauli exclusion principle—that the wave function must change sign when the coordinates of any two electrons of parallel spin are interchanged. In this version, electrons of antiparallel spin were uncorrelated, their motion being affected only in an average way by their Coulomb repulsion. But Wigner and Seitz soon recognized that it was necessary to introduce additional correlations to obtain a high enough binding energy: "You had to take into account the actual electron repulsion to get anywhere near the right answer."[37] Wigner then made his 1934 calculation of the correlation energy of an electron gas in a jellium model.[38]

Seitz notes that "the concept of correlation energy then got absorbed into the whole of physics. It was recognized that it's a phenomenon that occurs in nuclei as well—that is, the proton repulsion is an important constituent in determining the energy of nuclei—so the thing permeated many-particle physics" (Chapter 8).[39] Because his appointment was not renewed, Wigner temporarily moved in 1936 from Princeton to the University of Wisconsin, where he had been offered a permanent position. This transition was associated with Wigner's earlier meeting that year at Princeton with the visiting Wisconsin theorist Gregory Breit and their work on nuclear physics, a field in which Wigner would continue to work for many years.

In the meantime, the effort to calculate band structures in the mid-1930s spread to the University of Rochester, where Seitz had taken a teaching post in 1935. With D. H. Ewing, a graduate student, Seitz spent the next two years extending the Wigner–Seitz method to self-consistent field calculations of the wave functions for two ionic crystals: lithium hydride and lithium fluoride.[40] Seitz also attempted, together with another graduate student, Albert G. Hill, to calculate the band structure of the divalent metal beryllium. However, Hill was forced temporarily to give up the calculation because he concluded that the Wigner–Seitz method was unable to deal with divalent metals. A year later, after moving to MIT, he would take up the problem again with Herring, who was then temporarily at MIT on a National Research Council fellowship.[41]

In the 1930s, the Rochester Physics Department was chaired by Lee DuBridge, who had gained a reputation for his work on the photoelectric effect and had a small group of students working on solid-state problems. It was at Rochester that Seitz started work on his seminal book, *The Modern Theory of Solids,* which was to be published in 1940,[42] and was partly based on a series of popular articles he co-authored for the *Journal of Applied Physics* during a summer at General Electric. His co-author was G.E.'s Ralph Johnson, a former graduate student at MIT.[43]

MIT and Harvard

The sodium calculation by the method of Wigner and Seitz was soon taken up at MIT by Slater,[44] who in 1934 published a significant review paper summarizing progress in the whole field of the band theory of metals. This paper was illustrated by plots of sodium one-electron wave functions as a function of distance and energy contours in momentum space.[45] Since 1930, Slater had been chairman of the MIT Physics Department, which was closely tied to Princeton through the personal contacts of Karl T. Compton. In 1929, Compton had left his post as chairman of Princeton's Physics Department to become president of MIT. He received strong financial backing from the Rockefeller Foundation, including permission to use a portion of the foundation's endowment for MIT for ongoing research.[46]

Slater had been awarded his doctorate in 1923 at Harvard for experimental work

John Clarke Slater (1900–1976), around 1937. (Massachusetts Institute of Technology. Courtesy of AIP Niels Bohr Library)

on a solid-state problem, the compressibility of alkali halide crystals. His supervisor had been Percy Bridgman, a pioneer in studies of high-pressure effects in solids; a fellow Harvard graduate student was John Van Vleck.[47] Slater then changed course for about seven years, turning to fundamental problems of quantum mechanics. During the academic year 1923–1924, he went on a traveling fellowship to Europe, spending time at the Cavendish Laboratory at Cambridge with Ralph H. Fowler, and then at Copenhagen with Bohr and Kramers, with whom he collaborated on a paper that would be basic to Heisenberg's development of matrix mechanics. When Slater again went abroad, he renewed his interests in the solid state, especially in magnetism (Chapter 2), during some months spent at Leipzig in 1929 on a Guggenheim fellowship.[48] Shortly after his return to Harvard, Slater accepted Compton's offer to become chairman of what was to be a rapidly expanding MIT Department of Physics.

Close contacts continued between the Physics Departments at MIT and Princeton during the 1930s, and there was a continual interchange of students, colloquium speakers, and visiting staff members between the two centers. Contacts between Wigner's student Seitz and Slater's student Shockley have already been mentioned. Bardeen, after two years at Princeton, accepted a three-year junior fellowship at Harvard, arranged with the help of Van Vleck, and also taught part-time in Slater's department at MIT. In 1937, Slater spent a visiting semester at Princeton at the Institute for Advanced Study, and brought back with him Herring, who, after completing a thesis under Wigner exploring a variety of symmetry and degeneracy properties of electronic energy bands, obtained a two-year National Research Council postdoctoral fellowship. Such interpenetration and collaboration contributed to the progress of both centers.

Some of the funds for the expansion of MIT's physics program came from industrial sources—for example, a substantial grant from George Eastman of the Kodak camera company to build the George Eastman Research Laboratory for the

Departments of Physics and Chemistry. "If there had not been such a fund available," Slater later noted, "it is not likely that the Institute would have been able to go ahead with the new building so soon after the 1929 stock market crash."[49] It would be misleading to suppose that this donation was in any simple manner motivated by hopes of financial return, but it is a fact that Kodak's photographic processes were to benefit substantially from ongoing research in solid-state physics. Compton also obtained sizable funding from the Rockefeller Foundation.

An important reason for the new MIT department's success, particularly in solid-state electronics, was the close relations it was able to achieve with outside corporate laboratories; with the related MIT Departments of Electrical Engineering, Metallurgy, and Mathematics; and with the post–Second World War Research Laboratory of Electronics at MIT. The Physics and Electrical Engineering departments developed particularly close relations with Bell Telephone Laboratories, where Slater's outstanding student William Shockley later contributed pivotally to the invention of the first transistors.[50]

Band structure calculations were a prominent part of the postgraduate theoretical physics program at MIT in the 1930s. Jacob Millman, a research student with Slater, in 1935 worked out an approximation for the energy bands of lithium,[51] and this work was followed by similar band calculations by other students or postdoctoral associates of Slater—for example, Harry M. Krutter, George E. Kimball, William Shockley, Marvin Chodorow, Millard F. Manning, and Jack B. Greene.[52] Krutter worked out the bands of copper, and Kimball dealt in 1935 with the band structure of diamond, duplicating in broad detail the cellular method calculation of Hund and his student B. Mrowka at Leipzig, who had also studied the bands in various compounds.[53] Along with Ewing and Seitz's work on lithium compounds in 1936 at Rochester, another of the first band calculations for compounds in the United States was made by Shockley, for sodium chloride, between 1935 and 1937 at MIT.[54] Shockley's work showed the large band gap characteristic of insulator behavior. He then went on to check the validity of such cellular method calculations by what has since been called the "empty lattice" test, which involved checking what the method would give for a free electron in a constant potential. The answer should have been a plane wave with a continuous relation between energy and wave vector. Instead, the calculated diagram of energy versus wave vector contained energy gaps almost as large as had been obtained for sodium by the cellular method. It was thus clear that to achieve more realistic results it would be necessary to improve the approximation. Unfortunately, such improvement was no easy matter with the comparatively rudimentary mechanical calculators available in the 1930s.

During his sabbatical semester in 1937 at the Princeton Institute for Advanced Study, Slater developed what became known as the augmented plane-wave (APW) method, employing what is popularly known as a "muffin tin" potential.[55] The basic idea is that in a crystal like sodium, the potential inside a Wigner–Seitz cell, although approximately spherically symmetric over the inscribed sphere, would probably be relatively unchanging in the outer region, remote from the ion cores, that falls inside the cell but outside this inscribed sphere; in this region, the potential can be taken as approximately constant. One finds the same form of wave function solution within the inscribed sphere, and the solution must then be matched across

Conyers Herring (b. 1914), around 1960. (AIP Niels Bohr Library)

the spherical boundary against the much simpler plane-wave solutions for the region of constant potential. In 1939, Chodorow made preliminary calculations with this method on copper, using a simple desk calculator; his office adjoined that of Herring, who remembers many involved conversations.[56] But it was not until the large digital computers arrived after the war that the method became fully practical.

Toward the end of the 1930s, Herring, then a postdoctoral fellow at MIT with Slater, elaborated what became known as the orthogonalized plane-wave (OPW) method for band structure calculations, in which one uses as a trial wave function a linear combination of a few low-lying atomic orbital wave functions belonging to inner electron shells, provided they do not appreciably overlap, plus a plane wave with the same wave vector, imposing the condition that this wave function be orthogonal to that of the inner electrons.[57] "To my surprise," Herring recalled, "even a single plane wave of this sort often gave a remarkably good quantitative approximation to the correct crystal wave function."[58] Using this method, Herring and Hill, in 1940, accomplished their calculation of the energy bands for beryllium,[59] the first energy-band calculation in the United States of cohesive energy for a noncubic crystal. They included exchange and electron correlation contributions as well as a discussion of the diamagnetism of beryllium.

Herring also had strong contacts with Bardeen in this period. Bardeen, as previously noted, had gone from Princeton to Harvard in 1935/1936, as a junior fellow. The arrangement was free from teaching duties, the idea, then as now, being simply "to give promising young students free time to do whatever they wanted in their research."[60] Working mostly on his own, Bardeen interacted closely with the MIT group under Slater and, especially during the first of his three years as junior fellow, with Shockley,[61] who was, like Bardeen, interested at that time in the electronic properties of metal surfaces. During this period, Bardeen also became interested in the high-pressure experimental work carried out by Bridgman.[62] "I thought," he would reminisce later, "that I ought to try to apply some of those methods that I learned at Princeton to some of the problems that they had at Harvard."[63]

Using the Wigner–Seitz cellular method, a more refined and accurate procedure for calculating the effective mass at the bottom of the band (i.e., the curvature of the plot $E(\mathbf{k})$ versus $\mathbf{k}$), and a simple desk calculator, Bardeen computed the binding energy at a given volume for sodium and lithium. He found "results for sodium [that] checked very well with experiment over the entire pressure range, but there were small but significant departures (ca. 15% in compressibility) for lithium."[64] Wigner's 1934 expression was used for the interelectron correlation energy, and Seitz's 1935 field for the lithium ion potential. "The calculations provided a good test of the Wigner–Seitz methods. They showed that one could calculate accurately the binding energy of alkali metals and the volume dependence with no other input than the atomic spectra of an atom."[65] In the autumn of 1938, Bardeen left Harvard to take up what was in effect Van Vleck's old post as resident theorist in the Physics Department at the University of Minnesota. Later, after the Second World War, he would move to Bell Labs.

At the brink of the Second World War, "the national situation of solid state physics in the United States was very limited," according to Seitz, the principal centers being at MIT, Harvard, and Columbia, "where Shirley Quimby was involved in the accurate determination of the macroscopic parameters of single crystals of minerals and metals."[66] In addition, Cornell had taken a giant step forward in both solid-state and nuclear physics by its appointment in 1935 of Hans Bethe, who almost immediately "played a key role in several [Cornell] study sessions devoted to solid state aspects of electronics,"[67] and who was soon to begin work on an improved cellular method calculation, aided by a bit of group theory, with his graduate student F. C. von der Lage. The department at that time also had a group under Sommerfeld's former postdoctoral student at Munich, Lloyd Smith. Smith, who had originally met Bethe during a postdoctoral period in Munich, was responsible for his invitation to Cornell.[68] Smith's group worked on various aspects of electron theory, particularly thermionic emission. Yet another group worked on x-ray crystallography. Bell Labs, although emphasizing the development of practical devices, was also becoming a strong solid-state research center even in the mid-1930s, especially after hiring Shockley, who did important work in this period on order–disorder transitions in alloys. Shockley would return to the band structure calculations of his postgraduate period only briefly in the late 1940s and early 1950s.

Bristol

The third main center for band theory of metals in the early 1930s, along with Princeton and MIT, was the relatively new University of Bristol in England, which had attained university status only in 1909. Although a certain amount of relevant research on the theory of metals had gone on there even in the early 1920s, particularly in the area of magnetism, it was only with the appointment of J. E. Lennard-Jones in 1925 as reader in theoretical physics and the completion of a physics building in 1927 that these efforts began to attain focus. Another key step was the appointment in 1930 of the young Harry Jones, educated at Leeds and Cambridge, to serve as Lennard-Jones's assistant, under a Department of Scientific and Indus-

Wills Laboratory, University of Bristol, 1934. *(first row)* Heitler, Jones, Powell, Coslett, Suck-smith, Potter, Teller; *(second row)* [?], Stoner, Simon, Prins, [?], Darwin, Gerlach, Mott, Tyndall, Marks, [?], Bernal, Peierls, Skinner; *(third row)* [?], F. London, Fröhlich, Smoluchowski, Bates, Jackson, Gurney, Appleyard, Piper; *(fourth row)* Shoesby, Wilson, Sykes, Mrs. Harper, Lovell, Crowther, Kurti; *(fifth row)* Hoave, Fincks, [?], Harper. (AIP Niels Bohr Library)

trial Research (DSIR) grant to apply the new electron theory of metals to problems that, it was hoped, would be of ultimate practical importance.

The DSIR grant was awarded on the basis of a proposal that Lennard-Jones was invited to submit, entitled "A Theoretical Investigation of the Physical Properties of the Solid State of Matter." The proposal, in its broadest terms, was initiated by Frederick Lindemann of the Advisory Council for Scientific and Industrial Research (ACSIR) of the Privy Council office, which directed the allocation of DSIR grants. Lindemann, director of Oxford's Clarendon Laboratory, suggested that theoretical research on the structure and properties of metals and alloys be undertaken as an alternative to the metallurgical methods employed at the National Physical Laboratory.[69] The Industrial Grants Committee of the ACSIR subsequently expressed the hope that "one or more of our leading physical laboratories could be induced to undertake work of this character."[70] Indeed, the chairman of this important committee, Sir William Larke, then director of the British Iron and Steel Federation and later chairman of the Industrial Research Committee of the Federation of British Industries, insisted that "every possible encouragement should be given to such research in the national interest, at least until the stage is reached where the possibility of determining, and, as suggested by Professor Lindemann, even forecasting the qualities of different metals can be ascertained with some degree of accuracy."[71]

The matter was then passed on to a meeting on 12 February 1930 of the Scientific Grants Committee of the ACSIR, which directed the allocation of funds for university-based research. It was agreed to encourage theoretical research of this kind on "the correlation of the physical properties of metals and alloys with their atomic constitution."[72] Lindemann was then delegated to contact Lennard-Jones, who had done some work in this area,[73] and communicated the "considerable anxiety" of the Advisory Council "at the lack of work in Britain on the theoretical aspects of the solid state," adding that any proposals would be very favorably considered.[74] Consequently, Lennard-Jones submitted an application for the then considerable sum of £1150 to employ a fairly senior man as research assistant for three years to undertake work in this area. Harry Jones was appointed, because he combined "in a remarkable way a familiarity with experimental technique and an intimate knowledge of mathematical theory."[75] Jones had by then taken an experimental physics Ph.D. in spectroscopy at Leeds, followed by another in mathematical physics with R. H. Fowler at Cambridge.

Bristol's new focus on the electron theory of metals threatened to evaporate when in 1932 Lennard-Jones left for a professorship in theoretical chemistry at Cambridge.[76] He was replaced the following year as professor of theoretical physics by 28-year-old Nevill Mott, then known almost entirely for his contributions to atomic and nuclear physics.[77] Mott had spent four student years, 1924 to 1928, in Rutherford and Fowler's Cavendish Laboratory at Cambridge, obtained further theoretical training in the winter of 1928 with Bohr's group at Copenhagen, briefly attended Max Born's institute at Göttingen, lectured in 1929 and 1930 in the crystallographer W. L. Bragg's department at Manchester, and then lectured for three years in mathematics at Cambridge.

At about the same time that Wigner and Seitz were writing their pivotal papers on sodium, Mott, under the stimulation of Harry Jones's important application of the electron theory to alloys, was turning his attention toward the theory of metals.[78] Mott recalls, "I was fascinated to learn that quantum mechanics could be applied to problems of such practical importance as metallic alloys and it was this as much as anything else that turned my interest to the problems of electrons in solids."[79]

The work of Jones in turn drew on the Oxford metallurgist William Hume-Rothery's rules for alloy formation and on the Manchester crystallographer A. J. Bradley's determination of the structure of gamma-brass using x-ray powder techniques developed by Westgren in Sweden (Chapter 1, "The Powder Method"). Jones recalls that he first learned of these matters in the course of attending a talk given by Lawrence Bragg in 1933.[80] Clearly, one important basis for studies of the solid state was crystal structure analysis. As the French physicist André Guinier has recently stressed:

> One of the fundamental bases of solid state physics is the model of the solid "on the atomic scale," and [the subject] could therefore not begin until after the determination of crystalline structures by x rays. This properly began around 1920, and by 1930 was completed for the simpler crystals. Knowledge of the structure of metals and alloys was particularly important: the work done in Sweden during the period 1920–30 [by Westgren and others] was of considerable importance.[81]

One consequence was the close working association that developed among x-ray crystallographers, physicists, chemists, and metallurgists speculating on the relationship between the structure and properties of metals and alloys.[82]

Consolidation of Bristol as a center for band theory was facilitated by the post-

1933 integration of refugee physicists from central Europe and close ties with groups at Cambridge and Göttingen. Strong financial support came from the Rockefeller Foundation and from the Wills tobacco family, which built the H. H. Wills Physical Laboratory, opened in 1927, and contributed a major endowment grant for its operation.[83] By the late 1930s, the department was also attracting funding from the British Electrical and Allied Industries Research Association, and soon developed close relations, based on the application of solid-state analysis to the photographic process, with Kodak.

On this unusually sound financial basis, Bristol was well placed to add more research staff, promote others, like Jones, to permanent posts, and secure the services of some key refugee theorists after the rise of Nazism in Germany. After January 1933, Bristol attracted at least eight major refugee physicists, including the theorists Herbert Fröhlich, Walter Heitler, Hans Bethe, Klaus Fuchs, and Heinz London. As Mott later put it, "It was this immigration, together with the exodus of physicists from Rutherford's school at Cambridge to other places, which really got British physics going nationwide from 1933 onwards."[84] Lennard-Jones, the spectroscopist H.W.B. Skinner, Jones, and Mott had all come to Bristol from Cambridge. The relationship with Göttingen was based on the close ties that existed between the director of the Bristol laboratory, Arthur M. Tyndall, and the Göttingen mathematical physics professor Born. Among the outstanding Göttingen physicists who came to Bristol even before 1933 were the future Nobel laureates Gerhard Herzberg and Max Delbrück and, after 1933, Born's former assistant Walter Heitler. The Rockefeller Foundation had also financed postdoctoral visits between Göttingen and Bristol, including a six-month visit to Göttingen in 1929 by Lennard-Jones.

Clarence Zener, a former member of the Princeton group in the early 1930s, moved in 1931/1932 to a research post at Bristol, where he worked closely with Harry Jones on the electron theory of metals, and later with Mott on the optical properties of metals, and on perfecting the theory of what would later be known as Zener breakdown in insulators. Zener had taken his Ph.D. at Harvard and then spent 1929 and 1930 as a postdoctoral fellow in Europe at Leipzig and Göttingen during the period that Lennard-Jones was there. Lennard-Jones was eventually to arrange for Zener's fellowship two years later at Bristol. Unlike Slater, who later claimed that Harvard was perhaps the best place in the world to learn quantum physics, Zener believes that he only got into the mainstream of the subject, and particularly its applications to metals, at Bristol.[85]

Mott's own accession to the Bristol chair of theoretical physics was somewhat problematic. When Lennard-Jones resigned, the Rockefeller Foundation had not definitely specified that the endowed chair be allocated to any particular subfield, and Tyndall took the view that they should try to get the best possible theoretical physicist regardless of his area.[86] Consulting Fowler at Cambridge, Tyndall was informed that the best available mathematician was Alan H. Wilson.[87] However, Fowler also recommended for his versatility the equally young but already extremely accomplished Nevill Mott. Although Mott had contributed little to the main fields in which his laboratory was active, Tyndall offered him the job. At first, Mott, anxious to remain in Cambridge, politely declined.[88] Lennard-Jones then pushed strongly for the German molecular chemist Erich Hückel as his successor,

Nevill Mott and F. Simon in Bristol, around 1937. (Courtesy Nevill Mott)

warning Tyndall privately that "Skinner and Co. [including Mott] will probably press for Bethe."[89] Further divisive debate was avoided when Mott changed his mind and accepted.[90]

For the Wills Laboratory, this was a period when work on the metallic state was advancing on both theoretical and experimental fronts. Zener and Jones had made important contributions to the theory while Skinner was off for a year on a Rockefeller fellowship to MIT, measuring in cooperation with Henry O'Bryan, the soft x-ray spectra of various light metals. Willie Sucksmith had completed impressive experimental work on the gyromagnetic ratios of certain paramagnetic solids and had studied the structure and magnetic properties of single crystals of iron and nickel. Other solid-state work in the laboratory included S. H. Piper's x-ray work on long-chain compounds and L. C. Jackson's measurements of the low-tempera-ture variation of the magnetic susceptibility of certain paramagnetic salts. Thus Mott, whom many thought of mainly as an expert in scattering problems, came into a laboratory with fields of interest quite different from his own:

> No-one could possibly have criticised him if he had continued with his special-ity. . . . Instead he decided to switch his own interests to metal theory, on which he said his ignorance was profound. . . . He arrived in Bristol in the Autumn of 1933, and within six months he was publishing work in this field.[91]

Even before this, while at Cambridge, Mott had become involved in the publi-cation difficulties of Zener and Jones in a way that forced him to read up on the new band theory of metals. In May 1933, Tyndall wrote to Mott[92] that Zener had requested the formal submission to the Royal Society of his paper with Harry Jones on metallic conduction. Tyndall refused on the ground that Jones was still technically Lennard-Jones's assistant and that Lennard-Jones "would be very hurt if [the paper] were communicated by anyone else."[93] Zener, however, felt aggrieved by a previous incident in which he felt that Lennard-Jones had held up one of his publications for an unnecessarily long time, and suggested handing the paper first to Mott for his approval, as a way of hurrying up Lennard-Jones. Mott accepted the task and, as he later admitted, knowing comparatively little about the subject, passed it along to A. H. Wilson, who was far more accomplishd in this field.[94] Wilson, in turn, made a number of serious criticisms that encouraged Mott to make "a real effort to learn something about the electron theory of metals, and really work through the paper."[95]

Mott's growing interest in the field coincided with the publication of the monumental review by Bethe and Sommerfeld in the *Handbuch der Physik* on "Elektronentheorie der Metalle,"[96] a comprehensive article (discussed in Chapter 2) that provided "a basis on which other workers [such as Mott, who often expressed his indebtedness] could build."[97] But Mott personally, as he recalled, needed the stimulus of Jones's work on alloys and "the visual evidence provided by Skinner's work to make me (at any rate) see that there was something here [in the theory of metals] that we really had to explain."[98]

Jones's application of band theory to alloys started from the problem of establishing the conditions under which two metals melted together will mix and, upon cooling, result in a homogeneous solid alloy.[99] Assuming that the brass and bronze alloys could perhaps be conceived in the same way as a metallic crystal with a conducting free-electron "gas" to which both alloy constituents contributed, Jones suggested that under particular conditions of temperature and pressure the equilibrium phase structure adopted would be the one for which the total energy is a minimum. He argued, in a review written much later, that this would occur at those energies for which the density of electron states function was a maximum.[100] In the approximation in which the Fermi surface enclosing the occupied electron states at absolute zero is regarded as spherical, Jones found that the first maximum in the density of states function for the body-centered cubic lattice occurred when the ratio of valence electrons to atoms is 1.48, which was very close to the empirical value of 1.5 for alpha-brass. Similarly, he found that for the gamma-brass structure, the ratio of valence electrons to atoms that emerged from this model was very close to the empirical 21/13 value put forward by the metallurgist Hume-Rothery in his famous phase rules. Jones was able to demonstrate that if the density of states functions for the face-centered and body-centered lattice structures were compared, the first maxima occurred at electron-to-atom concentrations of 1.36 and 1.48, respectively, implying that in the region between these electron concentrations the body-centered cubic structure would gradually become more stable than the face-centered one. This was the first convincing theoretical treatment of why particular concentrations of different metals crystallized in a particular alloy structure.

In his later review, Jones also noted that the reason for this "peculiarly nonchemical" behavior of metals and alloys could even then be best understood in terms of the relation between the bands of energy discontinuity in k-space and the surface of the Fermi distribution (discussed later). Forbidden energy bands correspond to Bragg reflections of conduction electrons off particular crystal planes, the effect being, seemingly, to depress the allowed energy values for particular values of the electron's wave vector: "Thus, for a given density the structure which has the lowest energy [highest binding energy] will be that which gives rise to planes of energy discontinuity wrapping round the Fermi surface as closely as possible." The lowering of total energy was found to be most pronounced when the plane of energy discontinuity (or Brillouin zone boundary) just touched the Fermi surface without overlap. This was found to be the structure most likely to occur, assuming that size factors of the ion cores, temperature and pressure, and so on permit it. In general, Jones found that as the concentration of divalent zinc atoms to monovalent copper ones increases in potential brass alloys, the situation where the zone boundary touches the Fermi surface becomes more likely, and he was able to correlate this state with especially stable phases that arise for particular compositions.

Peierls had previously derived an expression for the diamagnetic susceptibility of conduction electrons, which implied large values of the susceptibility where the curvature of the energy surfaces was large.[101] As Jones later recalled, this result suggested that near the zone boundaries of a particular alloy (i.e., the surfaces of energy discontinuity in k-space), one might expect large contributions to the diamagnetic susceptibility, a result in very good agreement with the Japanese physicist Y. Endo's experiments on gamma-brass.[102] Similarly, the American metallurgist Cyril Stanley Smith demonstrated at about the same time that several other alloys with the gamma-brass structure had large diamagnetic susceptibilities,[103] a rather striking (if inadvertent) confirmation of the conjectures of Peierls and Jones. Jones subsequently also examined two alloy phases with a close-packed hexagonal structure.[104] He also did important work on the galvanomagnetic properties of metallic bismuth,[105] whose understanding, as we shall see, later proved important for the successful determination of Fermi surfaces with the aid of magnetoresistance measurements.

The Skinner–O'Bryan experiments of 1933 at MIT had yielded soft x-ray emission bands corresponding more or less closely with the distribution of conduction electrons predicted by the single noninteracting electron theory expounded in the Bethe–Sommerfeld article, even though the actual interelectron interactions were large and expected by many people to have a sizable effect. Within six months of his arrival at Bristol, Mott, in collaboration with Jones and Skinner, had submitted an important paper explaining this surprising result. Confirming implicitly the physical reality of the Fermi distribution for conduction electrons, this work represented one of the first applications of many-body considerations to metals.[106] Throughout the 1930s and 1940s, important work was also done on the emission and absorption spectra of various metals in the Laboratory of Physical Chemistry at the University of Paris by a group whose most active member was Yvette Cauchois. A general review and theoretical discussion of the results of this work were subsequently given in 1949 by Cauchois and Mott, who also considered the results for the semimetals gallium and arsenic, which are now of considerable commercial interest.[107]

While Mott and Jones made their pioneering contributions to band theory in the early 1930s, they also worked on preparing their influential text, *The Theory of the Properties of Metals and Alloys* (1936).[108] The wide impact of this work was due particularly to its concise physical arguments, often presented by means of simple models chosen in preference to cumbersome mathematical calculations. By its close attention to experimental results, the book also offered considerable guidance to laboratory practice. As Seitz later recalled, "The book of Mott and Jones . . . opened a new area of speculative work, devoting substantially less effort to the establishment of fundamentals than to the use of approximate models to systematize important areas of empirical knowledge."[109] Some might argue that such an approach reflected a long tradition in British physics, yet it was undoubtedly influenced heavily by the close contact between theoretical and experimental work at Bristol and the continuing prominence in the early development of the field of both physical metallurgists, like Hume-Rothery, and crystallographers, like W. L. Bragg. As Mott later noted in reference to his own approach, "if by making approximations and neglecting even large terms . . . one could account for something that had been observed, the thing to do was to go ahead and not to worry. I remember many discussions with Hume-Rothery . . . who shared my point of view."[110] Nevertheless, as one of Mott's contemporaries has pointed out, the approach of the "English school" of Mott and Jones was not universally popular among the wider international community:

> Mott . . . and his colleagues horrified us by their "simple," visualizable, and seemingly uncomplicated models and mathematics. . . . In the 1940's and early 1950's, Mott's almost eclectic point of view was often criticised as being unsound and contrary to the need of a unified all-embracing solid state theory. I remember very well the heated arguments starting with the Metals Conference in Bristol in 1934 and continuing later at nearly all such gatherings including the famous Varenna School in 1952. On those occasions Mott stood up, or preferably walked in his stilt-like manner, and shaking his head slowly from side to side, explained patiently his model, comparing it with experiment. To many of us this was not *gründlich* enough. Slowly, slowly the influence of Mott and his students and the success of their approach became so evident that no one questioned seriously their value.[111]

An exception was Slater, who remained critical, maintaining as a principle the superior value of calculating all integrals that could be calculated and avoiding simplifying assumptions for which there was no warrant in first principles.[112]

Electrostatic calculation of elastic constants was another aspect of the electron theory of metals considered at Bristol. Like the compressibility calculated by Wigner and Seitz, these constants are second derivatives of the total energy as a function of strain for a particular distortion. The basic papers were written by Klaus Fuchs, Mott's research student in the mid-1930s.[113] Using a Wigner–Seitz cellular approximation, Fuchs first considered the band structure and cohesive forces of copper, obtaining reasonable agreement with experimental values for the lattice spacing and the heat of vaporization, but much too large a compressibility. Since the size of the individual positive copper ions (containing a $3d$ shell) is much larger than in sodium, Fuchs reasoned that a repulsive interaction between overlapping ion cores could have much more effect on the compressibility, and that this sort of interaction would increase quite rapidly as the internuclear distance decreased at higher pressures. Taking account of this interaction lowered the compressibility to a more reasonable figure.[114]

In 1938, Mott, and Walter Schottky at Siemens, independently took up the problem of using the new band theory to develop a better theory of rectification (Chapter 7) at a semiconductor-to-metal surface. They attributed the rectification to the experimentally observed potential barrier at the interface, acting in much the same way as does the region in a vacuum-tube diode between the filament and the plate. In metals, the electrons occupy energy levels up to the Fermi level, whereas in the semiconductor the Fermi level is, in equilibrium, approximately halfway between the impurity levels and the contact barrier, if the electrons acquire enough energy through thermal excitation. The rectification, according to the theory, was due to the fact that the applied voltage in the back direction effectively increased the barrier, whereas that in the forward direction lowered it.

By the end of the 1930s, the study of the electronic structure of solids had developed into a well-established subfield of physics. Important textbooks were beginning to appear on this subject, starting with Brillouin's *Quantenstatistik* (1931), which was to be followed by Mott and Jones's *Theory of the Properties of Metals and Alloys* (1936), and such books as Alan H. Wilson's *Theory of Metals* (1936) and Fröhlich's *Elektronentheorie der Metalle* (1936). Also appearing in 1936 was William Hume-Rothery's *Structure of Metals and Alloys,* a follow-up on his 1931 book *The Metallic State,* which had provided a review of pre–quantum-mechanical electron theories and experimental data for the practicing metallurgist, but whose appendix had also contained an edited English version of Bloch's 1928 quantum-mechanical treatment of an electron in a periodic field, prepared with the help of Bethe. The year 1939 brought Wilson's *Semiconductors and Metals,* and the following year Mott and Gurney's *Electronic Processes in Ionic Crystals.* In 1940, Seitz published his *Modern Theory of Solids,* bringing together the band theories of ionic crystals and of metals, and providing the overall theoretical coherence for the research that would develop into modern solid-state physics.

Impact of the Second World War

During the Second World War, theoretical band structure studies shrank, as the principal researchers were for the most part diverted into war work—for example, Harry Jones to problems of explosion and fragmentation;[115] Mott and Kimball to operations research;[116] Slater, Hill, and Bethe to radar work for the Radiation Laboratory at MIT,[117] and Bethe later to theoretical problems of the atomic bomb project;[118] Zener to improving armor at the Watertown Arsenal.[119] On the other hand, many scientific, technological, and institutional developments resulting from the war had a significant positive impact on work in the late 1940s and through the 1950s and 1960s.

In terms of postwar impact on theoretical band studies, a most significant wartime development was the progress in computers, used during the 1950s and after to calculate bands to a significantly higher degree of accuracy. The war led to improvements in experimental techniques using neutrons and microwaves, with the latter proving a particularly powerful tool for studying Fermi surfaces via microwave absorption at solid surfaces, particularly in relation to such phenomena as cyclotron resonance and the anomalous skin effect, as discussed later. The development of new means for producing purer samples of materials—particularly the vitally important semiconductor crystals silicon and germanium, considered for use in radar receivers—enabled experiments to achieve more accurate results capable of being compared with the more sophisticated theory available by then.

New postwar institutes for the study of solids grew out of the war: for example, the interdisciplinary Institute for the Study of Metals at the University of Chicago, the solid-state department at Bell Telephone Laboratories, Karl Lark-Horovitz's semiconductor research group at Purdue University, the MIT Research Laboratory of Electronics[120], and the greatly expanded Physics Department at the University of Illinois. The Bell solid-state group was a result of recognition by Bell's management that in the postwar period solid-state devices would be at the center of the "new market."[121] The Purdue University group grew directly out of wartime radar work focused on the properties of silicon and, especially, germanium.[122] The University of Illinois group was built up due to the efforts of Wheeler Loomis, the head of the Physics Department, and Louis Ridenour, who was recruited from the wartime MIT Radiation Laboratory to serve as dean of science. Ridenour subsequently helped to recruit for Illinois one of those who had worked on silicon during the war, Frederick Seitz.

3.2   The 1950s: More Realistic Bands

By the 1950s, a strong movement for further refinement of basic theory was coming not only from academic centers but also from industrial laboratories specializing in the new semiconductor electronics initiated by the invention of the transistor. In the early 1950s, Frank Herman, a young RCA electrical engineer, was thinking of doing a Ph.D. in the theory of solids at Columbia University. He had done experimental work on the semiconductor germanium, which had led to an RCA patent. In the process, he had read Seitz's text "from cover to cover." He had also searched the current literature on the electronic properties of germanium, but "it became clear that we really knew very little about its electronic [band] structure."[123] Given the decisive importance of germanium and silicon for the new electronics, as well as the prime importance of their band structure in determining their physical properties, Herman found it astonishing that understanding of these bands was still so rudimentary.

At RCA, as at most major electronics firms around 1950, the semiconductor electronics revolution was only beginning. The company had made its reputation in television cathode-ray tubes, vacuum-tube technology, electronic circuitry, thermionic emission, and photoconductivity. However, the practical significance of the quantum theory of solids began to be appreciated, and by 1950 it was clear that semiconductor physics was going to be important. Many senior scientists and engineers seriously began studying quantum mechanics and its applications to the solid state in informal seminar-type courses.

No one at Columbia was then working on the theory of the solid state. The Physics Department's solid-state specialist, Shirley Quimby, whose knowledge of the new quantum mechanics was limited, devoted most of his research to experimental determinations of the elastic properties of copper. After some preliminary discussions with the faculty, Herman eventually persuaded Henry M. Foley, an atomic physicist with a strong background in quantum mechanics, to become his dissertation adviser for a project aimed at calculating the band structure of diamond. This work was to be the first step toward dealing with the somewhat similar, if more complicated, bands of germanium.

Diamond had been considered in the 1930s by Hund at Leipzig, Kimball at MIT, and, more recently, A. Morita at Tohoku University, although in those precomputer days the results tended to be of more qualitative than quantitative significance. For example, although Kimball's calculation demonstrated why diamond was an insulator, the theoretical value for the band gap between the filled valence band and the empty conduction band came out around 50 eV, an order of magnitude off from the experimental value of 5.4 eV. By 1950, Kimball had long since left the field of band structure and was a professor of chemistry at Columbia. When Herman asked his advice, Kimball admitted that new calculational devices could yield more realistic results, but cautioned that the use of the same cellular method would involve a great deal of computational effort. Herman recalled recently, "So I proceeded to examine other methods, and I rejected one after another because they required excessive computation or were based on parameters that could not be estimated reliably. . . . Finally, I came across Conyers Herring's beautiful 1940 paper on the orthogonalized plane wave method (OPW)."[124]

The only computational devices Herman had at his disposal at RCA were a variety of slide rules and desk calculators, which were not an essential improvement over those available to Kimball at MIT in the mid-1930s. He was therefore forced to run off his detailed computations in the Thomas Watson IBM Computing Laboratory at Columbia, where he spent a few months in 1952 on electromechanical computers that were rudimentary by today's standards. By coincidence, there he met Léon Brillouin, then on the staff at the Watson Lab. In their conversation, Brillouin recalled the steps that had led him to the zone concept (Chapter 2), and expressed his delight "that a purely mathematical construct could play such an important role in determining the electronic structure of solids . . . [with applications] ranging all the way . . . to transistor action."[125] Some time after beginning his calculations, Herman also met Herring, now at Bell Telephone Laboratories in Murray Hill, New Jersey, a short drive from Columbia and the RCA Laboratories in Princeton, New Jersey. Herring was working on a range of topics, including thermionic emission theory, sintering, spin waves, and diffusion in alloys. Herring readily agreed to give advice as needed. Herman's mother helped by dealing with some of the higher-order secular equation calculations.[126]

By the fall of 1952, Herman had defended his dissertation before a committee including Brillouin, Kimball, and Charles Townes. The dissertation covered both the diamond and germanium band structures. His new calculations indicated clearly that the electronic energy bands of those materials differed in important respects from those calculated by Kimball for diamond.[127] The results led to a "gold rush" into silicon and germanium band calculations. Truman Woodruff, then at Hughes Aircraft Company, applied the OPW method to the computation of the electronic structure of silicon,[128] and many similar papers followed. Herring, too, began to take a renewed interest in such calculations, especially as he could see their relevance to Bell Labs' program in semiconductor electronics.[129] In November

1953, he delivered an invited paper, "Correlation of Electronic Band Structure with the Properties of Silicon and Germanium," at the American Physical Society meeting in Chicago.

The orthogonalized plane-wave method was not the only one in use for band structure calculations in the early postwar period. For example, the Bethe–von der Lage modification of the cellular method (which was "finished in 1941, but because of World War II . . . published only in 1947")[130] was taken up and applied by other workers, most notably Harry Jones and his Imperial College student D. J. Howarth, who made an improved calculation of the bands of sodium.[131] Subsequently, during a postdoctoral period with Slater at MIT, Howarth applied the APW method to copper, using the earlier potential developed in the mid-1930s by Slater's student Chodorow.[132] B. Shiff was another student with Jones in the early 1950s making cellular method calculations of the bands of lithium, before eventually moving on to the Weizmann Institute in Israel.[133] Similar calculations were made in this period on diamond by V. Zehler at Giessen,[134] and on silicon by D. K. Holmes at the Carnegie Institute of Technology[135] and E. Yamaka and T. Sugita in Japan.[136] G. G. Hall, a theoretical chemist at Cambridge University, calculated the electronic structure of diamond, using the equivalent orbitals method, and obtained results similar to Kimball's.[137] Important work was also being done in this period by a group headed by the former Italian refugee Leo Pincherle[138] at the Radar Research Establishment at Malvern in England; originally they worked on lead sulfide (a semiconductor-type substance used at Malvern in wartime infrared detectors), and later on silicon.

Another important advance in band structure analysis during this period was the development by J. Korringa at Leiden and by Walter Kohn and N. Rostoker at Carnegie Tech, independently and by different approaches, of a method for calculating band structures in which the interaction of the electron wave with one atomic center is separated from its multiple scattering by the array of atoms taken together. In this method, called "KKR," each ion core is surrounded by a sphere of radius R, and the Schrödinger equation is solved in the interstitial Swiss cheese–like region, with an appropriate boundary condition on the surface of the spheres. Each individual atom can be treated as a black box, and the electronic structure is completely determined once it is specified how strongly any incident wave from outside R is scattered into any given direction.[139] Further improvements in the KKR method were made in the 1950s and early 1960s by Slater's group. Also perfected by the MIT group, using the electronic computers of the early 1950s, was the APW method, developed by Slater in 1937.[140]

Just after the war, Slater devoted himself mainly to applying his wartime microwave techniques to the construction of particle accelerators. But he realized by the late 1940s that the transistor "was going to turn solid state physics into a major endeavor,"[141] and thus turned back to his earlier solid-state interests. In 1951 he set up at MIT a small, independent research group called the Solid State and Molecular Theory Group (SSMTG), with the long-term purpose of unifying the whole field of materials science, which was then being developed more or less separately by various MIT departments and laboratories. Initially, SSMTG focused on electronic structure calculations, especially of the kind needed to deal with ferromagnetic and antiferromagnetic materials, in which Slater had long been interested. In retrospect, he claimed he was "not willing to settle for any scheme (such as Herring's OPW method) that did not seem to be well adapted to such problems.[142]

Fortunately, in 1953 one of Slater's students, Alvin Meckler, found that Whirlwind I, the new research computer developed by the Department of Electrical Engineering at MIT, could be used for APW computations. Melvin Saffren, who in 1953 had just joined SSMTG as a graduate student, was encouraged to work through an APW calculation on the computer for sodium. The program was later adapted to the IBM 704 computer by another research student, John Wood. Slater would subsequently claim that Wood's Ph.D. work on the iron crystal "was really the beginning of the practical calculation of energy bands in crystals involving transition elements."[143]

In the early 1950s experimental efforts on germanium and silicon began to yield important results. In 1951, Bell Labs researchers Gerald L. Pearson and Harry Suhl published the results of an extensive study of the anisotropy of magnetoresistance in germanium[144] and concluded that "no electronic theory yet worked out is entirely consistent with experiment."[145] Subsequent measurements by Pearson on silicon caught the attention of Herring and resulted in a joint paper, presented by Pearson at the Amsterdam Conference on the Physics of Semiconductors in 1954, that derived the conduction band edge structure and the mass anisotrophy from the electrical measurements.[146] In the brief history of experimental and theoretical studies of the electronic structures, especially of semiconductors, this conference marked an important turning point. For the first time, the band structure of semiconductors, which would in Herman's words be "one of the major scientific themes for the next decade and half,"[147] was a principal focus of conference discussion. Herman delivered the main invited paper, although he at first had difficulty convincing the RCA management of the need for his attendance.[148] Also presented were papers on the use of the technique of cyclotron resonance to yield information on the detailed electronic structure of semiconductors.

The idea that one could observe in solids the resonant absorption of electromagnetic radiation by electrons whose energies were quantized by an external magnetic field has been considered independently by several physicists. In 1951 the Russian physicist Jakov Dorfman made a suggestion to this effect.[149] Simultaneously, R. B. Dingle, then a research student with David Shoenberg at Cambridge, considered in his Ph.D. dissertation the possibility of such "cyclotron" resonance for the electrons in metals.[150] Because of considerable experimental difficulties, such a phenomenon was generally considered to be unobservable.

Soon afterward, the question of measuring cyclotron resonances in semiconductors was taken up by Shockley[151] at Bell Labs. He had become interested in Herman's calculations and recognized the possible technological implications of a more thorough knowledge of the electronic band structure of semiconductors. Shockley's colleagues at Bell Labs, Suhl and Pearson, tried to observe cyclotron resonance in germanium early in 1953, but at the temperature they worked at, 77K, the mean free path of the electrons was too small to give resonance.[152] The feat was eventually accomplished later that year at Berkeley by G. Dresselhaus, A. F. Kip, and Charles Kittel, whose equipment enabled them to work at liquid-helium temperatures.[153] A few months later, more complete measurements of cyclotron resonance in germanium were independently reported by Benjamin Lax, H. J. Zeigler,

Kamerlingh Onnes Memorial Conference in Leiden, 22–26 June 1953, which included many of the physicists referred to in Chapters 2 and 3 (Dirac, Kronig, Mendelssohn, Heisenberg, Pippard, Bohr, Shoenberg, Casimir, Peierls, Bloch, F. London, and Pauli). The solid-state physicists were clearly now a community. (AIP Niels Bohr Library)

peratures.[153] A few months later, more complete measurements of cyclotron resonance in germanium were independently reported by Benjamin Lax, H. J. Zeigler, R. N. Dexter, and E. S. Rosenblum at Lincoln Laboratory, a military-research center at MIT.[154] Band structures, especially of semiconductors, were by now considered of military importance. Indeed, Lincoln Laboratory and the Office of Naval Research provided the major funding for Slater's SSMTG efforts.[155]

By the mid-1950s, solid-state theorists had at their disposal a variety of methods (among them the cellular, KKR, APW, and OPW methods) that could be used to solve the wave equation in a periodic crystalline potential. However, the numerical problem of solving a Schrödinger equation is only one of two major aspects of the band structure calculations. The other is to establish the proper expression for the potential used in the equation. Use of the OPW and APW methods greatly simplified the numerical difficulties, but progress was hampered by problems with choice of the potential, which involved approximations of interelectron effects. A significant advance was the development of the pseudopotential approach, which quickly began to win ground in the late 1950s and the 1960s. This method simplified the mathematical problems involved in the application of the OPW and the APW methods. Since the period of its full development and successful application to closing the gap between basic principles and quantitative data extends beyond the time span of this chapter, our discussion of this technique and its impact on the development of the band theory will focus only on its genesis.

The approach was founded on the assumption that the core electrons are inert:[156] they are the same in the solid as they are in an atom. Therefore, to a good approximation, a solid may be considered as consisting of a periodic array of cores composed of nuclei associated with their core electrons immersed in a sea of valence electrons. The pseudopotential describes the interaction of the valence electrons with cores. Its exact shape may be fitted to reproduce the empirical data, but it is not altogether arbitrary. According to the "cancellation theorem,"[157] the procedure is theoretically justified because the valence electrons experience a repulsive potential (due to the Pauli exclusion principle) near the core, and this repulsion cancels much of the attractive Coulomb forces. The result is a net pseudopotential that is relatively weak in the core region.

The pseudopotential concept had a relatively long and complicated history, for it was revived or rediscovered several times, and in some cases there was little communication among people developing its applications to the problems of solid-state physics. The first calculation using the pseudopotential approach was carried out in the mid-1930s by Enrico Fermi, who used it to explain the shift of alkali atomic spectral lines due to the scattering of electrons by foreign gas atoms.[158] Fermi approximated this scattering using an effective wave function and a scattering length. Later, he applied a similar approach to problems of nuclear physics.[159]

About the same time, independently of Fermi, the advantages of representing the core effects by a repulsive contribution to a pseudopotential were discussed in papers published in 1935 and 1936 by the Russian theorists from Karpow Institute for Physical Chemistry in Moscow, Hans Gustav Hellmann and W. Kassatoschkin.[160] Hellmann's approach, which he called the "combined approximation procedure," consisted essentially of treating simultaneously the atom cores by the "Thomas–Fermi" model, and the conduction electron's motion according to a Schrödinger equation. In the Wigner–Seitz method, one integrated the Schrödinger wave equation for the conduction electrons in the field of the ion cores and obtained wave functions that had, generally, several nodal surfaces near each nucleus; Hellmann and Kassatoschkin, however, used wave functions for the conduction electrons that had no such nodes near nuclei and could therefore be approximated fairly well by single plane waves.[161]

The idea that the condition underlying the OPW method—orthogonality of the wave functions of the valence electrons to those of the ion electrons—may be replaced by a repulsive potential determined statistically was also developed, in some respects parallel to the work of Hellmann, by the Hungarian school of Pál Gombás.[162] In 1939 he used this approximation to calculate the position and width of the energy bands in some alkali metals. Gombás's principal area of interest was atomic physics, but beginning in the mid-1930s he published papers dealing with the theory of metals.[163] Gombás and his school continued to be prolific after the end of the war, further refining their version of the pseudopotenial approach and applying it to other materials.[164] Ladányi carried out a detailed analysis of the difference between Hellmann's semiempirical and Gombás's statistical approach,[165] showing that it amounts essentially to the fact that Hellmann's potential is the same for both the lowest states s and p, while it follows from the general statistical derivation that different repulsive potentials correspond to states having the same n but different l (where s, p, n, and l are all quantum numbers). Another collaborator of Gombás, Peter Szepfalusy, in papers published in 1955 and 1956, discussed the quantum-mechanical justification for the use of a repulsive potential.[166]

Starting in the early 1950s, the effort of the Hungarian school in the area of application of the repulsive potential in the quantum theory of solids was paralleled by the work of the Prague theorist Emil Antoncik.[167] Antoncik used the repulsive potential approximation to calculate the energy values in some special points of the Brillouin zone of several alkali metals, as well as silicon, diamond, and lithium.

In the West, the work on the pseudopotential concept, although stimulated by the same initial impulse, developed independently of the research of the Hungarian and Czechoslovak schools. Herring came across Hellmann's work in the late 1940s while surveying theoretical literature for a review paper on thermionic emission. He soon realized that the pseudopotential idea promised solutions to some of the problems faced by solid-state theory.[168] In a paper presented at the 1954 photoconductivity conference[169] at Atlantic City, New Jersey, Herring suggested that the use of a "fictitious repulsive potential around each neighbor atom, so chosen as to represent the energetic effect of the nodes and loops necessitated by orthogonalization" might lead to a "reasonable estimate of the energy."[170] But it was not until 1957 that real progress in this direction was made in the United States.

Many of those who worked on pseudopotential theory were in one way or another connected with the Institute for the Study of Metals (ISM) at the University of Chicago, which also, as we will see, played an important role in the experimental study of the Fermi surface. This institute, renamed the James Franck Institute in 1967, traced its origin to the Manhattan Project's Metallurgical Laboratory (a code name for the laboratory in which the first nuclear reactor was developed). The founding director of ISM, Cyril Stanley Smith, formerly chief metallurgist of the Manhattan Project at Los Alamos Laboratory, sought to bring together in approximately equal proportions physicists, chemists, and metallurgists on a sufficiently broad front to deal with the fundamental "structure or properties of any atomic aggregates."[171] Indeed, some 40% of the papers published by scientists at ISM between 1952 and 1954 had little, if anything, to do with metals. The diversification of research subjects reflected the organization of the institute, which has been likened to "benevolent anarchy."[172] From its beginning, ISM attracted industrial support, and by 1954 it had 17 industrial sponsors contributing (in 1954–1955) approximately $260,000. Additional support came from the Office of Naval Research and the War Department, which cosponsored a research program on the deformation of metals ($184,000 per year).[173] The physics section of the institute consisted at the outset mainly of Clarence Zener—who studied internal friction, anelasticity, fracture mechanisms, and diffusion—and Andrew Lawson, who conducted studies of the specific heats of ionic crystals.

In 1951, Zener left to direct the Westinghouse Research Laboratory. The vacant position was offered to Morrel Cohen, then 25 years old, who had just written his Ph.D. dissertation under Kittel's supervision at Berkeley on the magnetic resonance spectrum of ferroelectrics. During his first years at ISM, Cohen worked on a variety of solid-state problems, including the interaction of electrons with grain boundaries. He became involved in band structure calculations during his work at Cambridge University in 1957 and 1958 as a Guggenheim fellow, where he collaborated closely with Volker Heine, with whom he was to publish important papers on band structures of metals.[174] Cohen had visited Cambridge in 1954 after attending a conference in Bristol. This visit resulted in his inviting the Cambridge experimentalist Brian Pippard to spend the 1955/1956 academic year at ISM. We shall see that this invitation had important consequences for band structure studies.

Just before Cohen left for his second Cambridge visit, one of his first graduate students, James C. Phillips,[175] defended his dissertation and took a job at Bell Labs. There Phillips came into contact with Herring, whose office was but a few doors away. During conversations with Herring, Phillips learned of Hellmann's and Kassatoschkin's work, and gradually developed his own applications of the pseudopotential approach. In a paper published in November 1958, Phillips constructed two-parameter pseudopotentials that gave good agreement with Herman's OPW calculations for diamond and silicon as well as with experiment.[176] Phillip's technique consisted of fitting the Fourier coefficients or form factors of the potential to experiment. Only a few terms in the Fourier expansion were necessary to obtain a satisfactory approximation, due to the cancellation of the core potential in the core regions. Although this technique greatly simplified calculation procedures, the physical argument on which the approach was based remained less than rigorous. He pointed out that "the best justification for an interpolation scheme lies in the results that can be obtained from it,"[177] but in the conclusion to this paper he cautioned that his results "should not be taken too seriously."[178]

Phillips took an important step toward clarifying his approach soon after his move from Bell Labs to Berkeley in the autumn of 1958. In a paper published the next year, he and Leonard Kleinman showed that for many cases the pseudopotential approximation could be justified in terms of an effective potential associated with the requirements of the Pauli principle.[179] Taking the OPW method as their point of departure and making several approximations (e.g., a local approximation and the retention of only a few couplings), they showed that the requirement of the orthogonality of the valence electron states to the core electron states could be replaced by a potential. This repulsive potential in effect keeps the valence electrons out of the core; added to the attractive Coulomb force from the core, it results in significant cancellation of the two potentials in the core region.

The work of Phillips and Kleinman demonstrated why the nearly free–electron model worked: it provided a scheme "in which a weak potential and a smooth wave function which can be expanded in plane waves yields the real eigenvalues for the valence electrons in a crystalline solid."[180] The empirical pseudopotential method, in which these few parameters were adjusted to fit experiment, grew out of this approach. It proved to be particularly useful in dealing with semiconductors and insulators.

Meanwhile, while Phillips was preparing his 1958 paper on the pseudopotential method, Cohen worked at Cambridge University with Heine on the electronic band structure of alkali and noble metals. On his return to Chicago, Cohen got in touch with Phillips to make comments on his, yet unpublished, paper with Kleinman. Afterward, Cohen continued to think about the pseudopotential problem, later discussing the issue with Heine. These thoughts and discussions resulted in their key paper "Cancellation of Kinetic and Potential Energy in Atoms, Molecules and Solids," providing more rigorous justification for Phillips's method.[181]

The pseudopotential theory of metals developed somewhat later and independently of the empirical pseudopotential method. It arose from the finding of the Cambridge researcher A. V. Gold[182] that the Fermi surface of lead resembles the free-electron sphere as if disrupted by a very weak potential. This argument was extended to all the polyvalent metals by Walter A. Harrison of General Electric

Laboratory,[183] who in the early 1960s went on to make corrections to the free-electron Fermi surface in terms of the weak coupling between orthogonalized plane waves.[184] Harrison then developed a general theory of simple metals using perturbation methods.[185]

Pseudopotentials proved useful for treating many solid-state physics problems, such as electron–lattice interactions, superconductivity, surfaces, electronic transport, lattice vibrations, and bonding. This new approach was even able to deal with the properties of liquid metals, which lacked the crystalline regularity that had been the very basis of earlier band theory. These applications were developed after 1960 and thus fall beyond the scope of our presentation.

3.3. Fermi Surface Studies[186]

The most straightforward way of differentiating metals from nonmetals and the properties of one metal from another is through their Fermi surface, an aspect so characteristic of a particular solid that it has often been called its "face." This concept was not one that emerged with definitive meaning or importance at one time; it was, rather, a notion whose changing significance and articulation reflected the progress and increased credibility of a particular theory.

By the late 1920s, as discussed in Chapter 2, it was clear that within the Pauli–Sommerfeld one-electron theory such important metallic properties as the magnetic susceptibility and the specific heat depend, at sufficiently low temperatures, only on the electrons in states within a small "distance" in quantum-mechanical electron wave-vector space—or more simply within a small energy difference—of what was then called "the surface of the Fermi distribution." However, it was by no means clear then what the effect of interelectron interactions would be in an actual metal and even less clear whether in view of them such an ideal surface would represent a meaningful and useful physical concept.[187]

Pictures of ideal Fermi surfaces had been published as early as 1933 by Bethe and Sommerfeld.[188] These drawings of the "Flächen konstanter Energie im Raum der Wellenzahlen" (surfaces of constant energy in wave-vector space) for simple cubic lattices were commissioned from R. Rühle of the Technische Hochscule in Stuttgart, who had also done the figures for Jahnke and Emde's classic tables of functions. Bethe recently recalled that "it was clear to me that [these surfaces] would be important," and especially for the problem of magnetoresistance "it made a great difference whether they were nearly a sphere or were some other interesting surface."[189]

This very question had been dealt with a bit earlier by both Bethe and Peierls, following the experimental work on magnetoresistance by Peter Kapitza at Cambridge in the late 1920s.[190] Indeed, Bethe's work on this problem arose directly out of his interaction with Kapitza during Bethe's first visit to Cambridge in the autumn of 1930.[191] Dmitri Blokhintsev and Lothar Nordheim, during the latter's postdoctoral year in Moscow (1932–1933), considered the magnetoresistance problem for the simple case in which the surface of the Fermi distribution was a quadratic function of the components of the wave vector. Subsequently, Jones and Zener at Bristol put forward an approximate solution, which they claimed was "valid when this is not the case."[192]

Also interesting about the Jones and Zener's paper is the facility with which the phrase "surface of the Fermi distribution" in the earlier part of the paper[193] was abbreviated to "Fermi surface" by the end,[194] which may suggest a subtle shift in meaning, with the first term more suggestive of a *mathematical* distribution function within a particular theory of noninteracting electrons, and the second closer to the present meaning of the term, more suggestive of the "real existence" of a particular *physical* entity, a surface whose contours correspond to definite physical properties. The evolution of such shades of meaning is in part due to the confidence with which such an apparently unrealistic model as the modified "gas" or "one-electron" model could be used to give an accurate description of the properties of actual metals or alloys.

Cambridge and Moscow

In 1932 a young research student, David Shoenberg, began work in Kapitza's Royal Society–supported Mond Laboratory, initially on magnetostriction, the change of length in an applied magnetic field, of a bismuth sample.[195] Shoenberg was Russian by birth, and Kapitza, as one of the very few other Russians in Cambridge, was a friend of the family. This interaction was to have significant consequences for Shoenberg, for his research with Kapitza introduced him to the physical peculiarities of bismuth. In particular, it introduced him to the rather puzzling oscillation of its magnetic susceptibility at low temperature, which had been discovered at the University of Leiden in 1930 by W. de Haas and his student P. M. van Alphen,[196] following similar work at Leiden in 1929 by the Soviet experimenter Lev Schubnikov[197] on the oscillation of bismuth's electrical resistivity in the presence of a magnetic field—the so-called Schubnikov–de Haas effect. Schubnikov was subsequently to organize the low-temperature laboratory at the Ukrainian Physico-Technical Institute (PTI) at Kharkov. He was imprisoned during the Stalinist purges and died in confinement. In any case, the "de Haas–van Alphen effect," under the joint probings of Cambridge experimentalists and Landau's school of theorists in Kharkov, was to become the most important of the techniques for determining the Fermi surface.

As early as 1932, while still in Continental Europe, Peierls had worked out a semiquantitative treatment of the bismuth oscillations. The possibility of such oscillations, not only in bismuth but in every other metal, had already been mentioned by Landau in his earlier classic paper on diamagnetism in metals, published in 1930, but had been discounted as unobservable because the inhomogeneity of the fields then used to measure magnetic susceptibility would have been sufficient to damp out the effect (Chapter 2).[198] Peierls's interest in diamagnetism arose to a large extent out of discussions with Landau during the latter's 1930 visit to Zurich, where Peierls was working as Pauli's assistant.[199]

In 1933, Peierls went on to Cambridge, by way of Fermi's laboratory in Rome, arriving in late spring. By then, the Nazi government had come to power in Berlin, and he knew he would not be returning to Germany. However, there was little time to discuss experiments on magnetism, for he left after a few months to take up an Academic Assistance Council research fellowship in W. L. Bragg's department at Manchester. Peierls did not return to the Mond Laboratory at Cambridge for another two years, by which time Kapitza had been forced to remain in the Soviet Union. Kapitza's Royal Society Messel Professorship funds were then used to bring into the Mond Laboratory a theoretical consultant, Peierls, and a Canadian low-temperature experimentalist from Toronto, Jack Allen, later professor at St. Andrew's in Scotland. Peierls's previous work on the bismuth oscillations then stimulated an improved calculation by a visiting research student to Cambridge, Moses Blackman, whom Born had brought with him to Cambridge after he had been dismissed from his chair at Göttingen.[200]

By this time, Shoenberg had obtained his Ph.D. and, in collaboration with an Indian graduate student, Zaki Uddin, was well on his way to confirming the main aspects of the Peierls–Landau theory of the de Haas–van Alphen effect. One problem was that the measurements, as they were then being done, required an inhomogeneous magnetic field across the bismuth sample to produce a force proportional to the magnetization, then measured on a special kind of Curie balance. But such an inhomogeneous field would produce oscillations in varying phases across the sample, oscillations that would rapidly damp out. At about this time, the Indian experimentalist K. S. Krishnan visited Cambridge and described in detail the methods he had developed at Dacca and Calcutta for examining magnetic anisotropy by measuring the torque, not the force, on a magnetized crystal. Shoenberg adapted this technique to his own work on bismuth, thus eliminating the need for the problem-causing inhomogeneous magnetic field.[201]

In 1936, Shoenberg went to the Soviet Union on a short visit, during which he consulted with Landau about a translation he was doing of Landau and E. M. Lifshitz's book *Statistical Physics.* Kapitza then tentatively invited him to spend the 1937/1938 academic year at the Institute for Physical Problems, which he was establishing. As Shoenberg recalls:

> [T]he torque technique [for measuring the bismuth oscillations] was an ideal experiment to take up, because it was really very simple. . . . It all got fixed up in a matter of weeks [after arrival in 1937 in Moscow] . . . and it was now possible to follow the oscillations to much lower fields, and study the effect in more detail. I was very lucky that the resident theoretician at the Institute was Landau (who had previously been at Kharkov), then not yet thirty, but already with an enormous reputation. . . . When I told him about my experiments he mentioned rather casually that he had just worked out how to calculate the oscillations explicitly, and there and then wrote down [on the back of an envelope] a rather complicated formula describing how the oscillations should depend on orientation, on field and on temperature . . . and said, "You might try to see if that will fit your results." In fact he'd worked out a more practical form of the theory, a formula you could actually check your results against.[202]

Within six months, Shoenberg was thereby able to do a pioneer experiment on the de Haas–van Alphen effect in bismuth, using Landau's model to check what was thus the first rough construction of a Fermi surface.

But when he came to write up his results, intricate political problems emerged:

Landau had been arrested—that was the height of the Stalinist purges, a particularly dangerous time for someone as incautious and sharp-tongued as Landau. . . . Sometime in the spring of that year, he just disappeared; he'd been taken away. Kapitza managed to do something eventually. He persuaded someone at a very high level that Landau's arrest would be a serious blow for Soviet physics, and he was eventually released about six months after I'd gone. In the meantime I'd written up the paper and had it translated—in those days the routine was that if you wanted to publish something outside Russia, it had to be published simultaneously in Russian. [The problem was that] in this paper I'd put in an acknowledgement of Landau's help, but the administrative director of the lab called me in and demanded how I could possibly express my thanks to "an enemy of the people."[203]

As a result, quite ironically, this important work by Landau is impossible to quote as a Russian paper. One needs to quote the English version of Shoenberg's paper, which included in an appendix a reconstruction by Peierls of the Landau work.[204]

In a letter to Peierls, dated 12 February 1939, Shoenberg reasserted the importance of publishing such an appendix, but added, "The only question is whether it may not damage anyone at the other end." He then asked Peierls's advice about "whether my acknowledgements in reference to Landau are in a suitable (i.e., not too warm) form," adding that "they were all deleted before I sent the manuscript from Moscow." In the same letter, he mentioned that in that year "Kapitza and Fock were the only physicists elected finally to the [Soviet] Academy."[205]

Yale and Cambridge

Landau's original suggestion of the possibility of de Haas–van Alphen oscillations in all metals was startlingly confirmed by experiments on zinc after the Second World War by Jules Marcus at Yale.[206] These experiments overthrew a good deal of tacit assumption that it was only the special electronic band structure of the "rogue metal" bismuth that had made the oscillations in its magnetic susceptibility observable. This result stimulated Shoenberg to restart his experiments, and from 1949 onward he found that "almost every metal crystal put into a copy of my torque apparatus produced a bewildering variety of oscillations." On the other hand, his results, as well as those of B. G. Lazarev and B. I. Verkin's group at Kharkov, were inconsistent with the assumption of ellipsoidal Fermi surfaces made in Landau's 1938 formulation in Moscow. Although an ellipsoidal shape might be a sufficiently good approximation for bismuth, the Fermi surfaces of other metals were proving to be much more complicated.

This was basically the situation until at least the 1950/1951 academic year, when the Norwegian-American quantum chemist Lars Onsager visited Shoenberg's Mond Laboratory on a year's sabbatical from Yale. As Shoenberg recalls:

For quite a while [even] before then he had made cryptic remarks . . . about a simple interpretation of the de Haas-van Alphen periodicities in terms of Fermi surface areas. But I could never understand just what he meant. It was only after repeated requests that he should write down his ideas that, practically on the eve of his return to Yale, he produced a three page paper for *Phil. Mag.,* which has since become a classic. He showed that the period [of de Haas–van Alphen oscillations] was inversely proportional to the extremal area of cross-section of the Fermi surface. It turned out that I. M. Lifshitz had had the same idea independently but had not published it; eventually it was I. M. Lifshitz and his students who developed the general theory in detail.[207]

Lifshitz's original contribution had come as early as 1950 "at a session of the Soviet Academy of Sciences [of the Ukrainian SSR in Kiev] when he read a paper that showed how the movement of an electron following a complex law of dispersion in a magnetic field can be quantized."[208] As Shoenberg's former research student Brian Pippard recalls, Shoenberg and Onsager had shared an office during the latter's visiting year at Cambridge. But even after Onsager had written his paper, at least for a year or two, Cambridge physicists tended to give it little importance. Pippard recalls:

> There wasn't a lovely lot of algebraic quantities and integrals and things which you could evaluate [in this paper] because Onsager was talking in *geometrical* terms— and I think David [Shoenberg] was disappointed to see "so little" coming out and failed to realize that Onsager had provided the complete clue. So nothing happened. The paper was published and nobody in Cambridge took any notice. They went on measuring the de Haas-van Alphen effect and fitting it with ellipsoidal shapes [in which the relation between energy and wave vector is assumed to be quadratic].[209]

That was the situation at the end of 1952.

Pippard had begun his research at Cambridge, first as a graduate student just after the Second World War, investigating certain low-temperature anomalies of the penetration of microwaves into the surface "skin" of metallic copper. Like Slater,

Alfred Brian Pippard (b. 1920) and Peter L. Kapitza (b. 1894) in Cambridge, early 1960s. (AIP Niels Bohr Library)

he had learned to work with microwaves in the course of the wartime radar program. Just before the war, Heinz London at Bristol had observed a somewhat higher electrical resistance than expected in an applied microwave field at temperatures sufficiently low that the mean free path between electron collisions is long compared with the penetration depth of the microwaves into the metal. London interpreted these anomalies of skin resistivity along the same lines that his former Bristol colleague Fuchs had used in discussing surface scattering of electrons in thin metallic films. The effect had first been observed at the turn of the century by J. J. Thomson.[210] Fuchs had recently moved on to work with Born at Edinburgh.

Pippard investigated the resistance under the same circumstances that London had and concluded that, although London was correct in thinking the anomalies arose from a mean free path effect, it was not as simple as that:

> I began to try to write down the mathematical formulation of the skin effect in long mean free path conditions—what I call the anomalous skin effect—but I was extremely inexperienced and got bogged down in the mathematics. That's when I asked Ernst Sondheimer [who was then a research student at Cambridge working with Alan H. Wilson] for help. He was able to formulate the problem, neatly and tidily, setting up the integral equation which had to be solved [which involved setting up a Boltzmann transport equation along the lines used by Fuchs for thin films in 1938]. He then sent for his friend, the professional mathematician [G.E.H.] Reuter (then Lecturer in Mathematics at Manchester) to deal with it—and that's how the Reuter-Sondheimer paper on the anomalous skin effect came about.[211]

Pippard recalls a visit to Cambridge, in the year after Onsager's visit, by the MIT theorist Paul Marcus, who had worked with Slater and Emanuel Maxwell on the anomalous skin effect in the late 1940s.[212] On this occasion, Pippard complained to Marcus about the difficulties of using Fermi–Dirac statistics for electrons in a metal compared with classical statistics, in which you could treat each electron separately:

> Marcus said I was making an awful lot of fuss about nothing—all you had to do was follow what deformations of the Fermi surface occurred, follow the electrons at the Fermi surface and see what happened to them, and the rest would have to follow suit. And that sentence, as far as I was concerned, was a very significant clue as to how to work out these problems easily. So, when about a year later, I tackled the general problem of the anomalous skin effect, I set out an arbitrary shape of a Fermi surface and followed the behaviour of the electrons at the surface and then simply integrated to get the current. That was the point at which, I remember, all the parameters cancelled out, just leaving the curvature lying there and I saw straight away how we could get at the Fermi surface.[213]

But he also recollected somewhat dimly, "Surely Onsager had said something which involved geometry, hadn't he?"[214] Reexamining Onsager's result, Pippard could at last see that Onsager had provided the fundamental insight. He then went to see Shoenberg,

> and we realized that what we had to do was to look at de Haas-van Alphen results not as things giving us parameters of ellipsoids, but to look at them with totally fresh lights and ask ourselves what [Fermi] surfaces are we dealing with that can be interpreted in the light of Onsager's results?[215]

Pippard concluded that the surface resistivity under anomalous skin effect conditions was determined by the electrons moving so nearly parallel to the surface that they had a good chance of completing a free path in the skin layer: one could simply remove from consideration those electrons that did not have their complete path in the surface.[216] By about 1954, he had realized that the results could help to find the detailed shape of the Fermi surface:

> The inspiration for this came from Sondheimer who had done some [further] work on the anomalous skin effect in metals with Fermi surfaces which were assumed to be ellipsoidal [rather than spherical], which he showed me. I took his results home, and wondered what my approximate techniques could do on this. To my astonishment, I got exactly the same answer [as Sondheimer], but using essentially geometrical considerations. Identical. And that made me think I could apply this method to the more general problem [of metals with an arbitrary shaped Fermi surface]. And I got out a most beautiful result: that the answer depended on the curvature of the Fermi surface and on no other parameter. That meant that if one systematically studied the anomalous skin effect in crystals cut in different orientations, one could see how the geometry of the Fermi surface was built up. But, that meant getting a lot of samples which were cut in different crystal orientations and also very well polished—they had to be strain-free because one was measuring surface phenomena in 500 Angstrom layers—so that meant electrolytic polishing of the surface which I didn't know how to do for copper [which he was then working on].[217]

In the mid-1950s, Pippard made use of this result to make an actual measurement of the Fermi surface of copper, this work being performed during the 1955/1956 academic year at the Institute for the Study of Metals at Chicago, about 10 years after his initial experiments. Pippard had accepted Morrel H. Cohen's invitation to visit ISM after he learned that "the Institute was very good at polishing copper surfaces."[218] This ability to produce highly polished single crystals came partly out of the Chicago Metallurgical Laboratory's wartime involvement in the Manhattan Project.[219] As Pippard recalls, "They were extraordinarily helpful; they'd bought the necessary microwave equipment, they bought the cryostat—everything was ready and they'd even begun to prepare the single crystals of copper and were ready to carve and polish them when I arrived."[220] Pippard's measurement of the Fermi surface of copper was the first of many precise subsequent investigations of the Fermi surfaces of metals, further stimulating Shoenberg's ongoing de Haas–van Alphen experiments, which were by then also profiting from the insights of Onsager and the Lifshitz school at Kharkov.

Shoenberg had in the meantime been working intermittently on the de Haas–van Alphen oscillations in simple monovalent metals such as the alkalis (sodium, potassium, and lithium) and the noble metals (copper, silver, and gold), where the Fermi surfaces were thought to be not too complicated. The initial difficulty, as he remembers, "was that the Fermi surfaces were expected to be large and the oscillations would therefore be too rapid to be resolved at the low fields I was using." This led him to consider using higher fields, "but in those days there were no superconducting magnets and our own water-cooled coils didn't go high enough and had rather unsteady fields."

He was eventually driven to apply

> Kaptiza's old technique of pulsed fields [but using a condenser instead of a dynamo as an energy source]. . . . Not only could I get much higher fields this way [5 or 10 times those of conventional magnets], but I could exploit the inevitable time variation of the field to detect oscillations in the E.M.F. induced in a pick-up coil.

This technique eventually enabled him to observe oscillations of much shorter period, corresponding to major pieces of the Fermi surfaces that had hitherto been undetectable, particularly in the noble metals. Robert Chambers, later a senior professor at Bristol, took a leading role in designing the electronics. Although this technique would eventually be used by Andrew Gold and Michael Priestley to determine the Fermi surfaces of lead and tin, Shoenberg recalls that

> we still found nothing in copper and sodium which I thought should be the simplest monovalent metals to try. In retrospect, this was bad judgement since copper seems to have the smallest amplitude [of oscillations] of the noble metals, and sodium is the most difficult of the alkalis (because of its martensitic transformation). It was rather like trying to catch a black cat in a dark room—completely negative results gave little guidance as to what to do next.[221]

Indeed, even after Pippard's successful anomalous skin effect determination of the Fermi surface for copper at Chicago, Shoenberg was still having great difficulty with the de Haas–van Alphen measurements needed to confirm the result. A critical difficulty, as Pippard recalls, was that "copper was a very difficult metal to get in crystalline form." Just at that time, however, Shoenberg learned that people had been making copper whiskers at the General Electric Company in Schenectady, New York, and so he got hold of some whiskers and set up his experiments. He got the results. Pippard recalls, "I was off at a conference and a telegram arrived from David saying the necks [in the Fermi surface for copper] I'd predicted [from anomalous skin effect measurements] were not there." It turned out eventually that "a crystallographic error had been made in the orientation of the whiskers. But when he did finally get hold of the right whisker, the necks showed up beautifully."[222]

Shoenberg himself recalls the fine morning of 18 December 1958, when he and his assistant first saw the fast oscillations in copper. They broke off the experiments to go to lunch, but he was so excited he came back early to continue the work:

> But alas I wasn't quite familiar with the circuit, and instead of seeing pretty oscillations when I closed the switch I heard a colossal bang. I had in fact made a direct short circuit of the condenser bank, completely by-passing the solenoid. Luckily the damage was not too serious—though I have been a little deaf ever since—and we soon got a good idea of the Fermi surfaces of copper and also of silver and gold. Incidentally it turned out that the whiskers were a bit irrelevant; it was perfectly possible to make good copper crystals by conventional means, but our methods had been too crude.[223]

The work of Pippard and Shoenberg on copper and the noble metals was the signal for a virtual explosion of Fermi surface studies, both experimental and theoretical, particularly in the United States and the Soviet Union, with perhaps the most important subsequent experimental work on these metals being done by Robert Morse and his collaborators at Brown University, Arthur Kip and his group at Berkeley, and N. S. Alekseevskii and his fellow experimentalists at Kapitza's Institute of Physical Problems in Moscow. The predominantly theoretical calculations, which had begun largely with Pauli and Sommerfeld and were developed further by Peierls and Landau, came in the 1950s to be perhaps even more centered in Landau's old school at Kharkov, particularly in relation to Fermi surface studies based on magnetoresistance and cyclotron resonance techniques.

The early experimental work on magnetoresistance, as previously noted, was done by Kapitza[224] in the late 1920s, in the days of Ernest Rutherford, in what was then called the Department of Magnetic Research at the Cavendish Laboratory at Cambridge. Kapitza noted the presence of magnetically induced resistance when the magnetic field was either perpendicular or parallel to the direction of electric current flow, and also the surprising result that in a polycrystalline specimen consisting of randomly aligned single crystals, this magnetoresistance increased, apparently without limit, in proportion to the strength of the magnetic field at high fields.

Subsequently, E. Justi and H. Scheffers at the Physikalisch-Technische Reichsanstalt in Berlin found that the magnetoresistance of a single crystal specimen depends strongly on the direction of the current relative to the applied field and reaches a maximal "saturated" value for particular directions of sufficiently high fields. This finding could apparently be understood by the familiar model in which a charged particle moving through an applied magnetic field is bent into a curved orbit whose radius is inversely proportional to the transverse component of the applied field. For sufficiently high fields, this radius approaches zero, and any further increase in the field strength cannot further retard the electron's motion or, in particular, contribute to the electrical resistance (which would thus "saturate"). However, this apparently straightforward expectation clashed with the rather puzzling Justi–Scheffers finding that for certain crystal orientations the magnetoresistance does not saturate, but increases without limit.

Kharkov

In 1939, B. I. Davydov and Isaac Pomeranchuk at Kharkov advanced a theory that explained the nonsaturation of the magnetoresistance in bismuth by assuming that conductivity is caused by equal numbers of holes and electrons.[225] This could occur when more than one band was partly filled—that is, when (although the theory was not phrased in these terms) the Fermi surface in a partly filled zone touches the zone boundary. The former Soviet theorist Mark Azbel recalls[226] that Landau considered this theory unsound, but no one knew quite how to deal with the problem. Indeed, when in the early 1950s Azbel suggested doing work on it, his teacher at Kharkov, Ilya Lifshitz, told him it was an impossible problem.

But Azbel persisted and managed to solve the problem so far as to demonstrate the saturation of the magnetoresistance for a closed Fermi surface: "Landau then joined in and had the idea that there may be open Fermi surfaces. Lifshitz then elaborated this and the idea of different kinds of Fermi surfaces."[227] The result was a famous paper by Lifshitz, Azbel, and Kaganov suggesting various possibilities of open electron orbits with certain directions of applied field. For these directions, the electrical resistivity does not saturate but grows without limits at high fields as the square of the applied field.[228] Although this method cannot in itself give the detailed curvature of the Fermi surface in this region, these particular field directions for nonsaturation indicate a region of contact between the Fermi surface and the zone boundary and thus give information about the geometry of the Fermi surface. Experimental observations of this effect on copper, silver, and gold were made at Kapitza's Institute for Physical Problems in Moscow by Alekseevskii and his student Gaidukov in 1959 and 1960.[229] Indeed, Evgeni S. Borovik had studied this effect experimentally in Lazarev's laboratory at Kharkov in the 1940s.[230] A complete topological theory for the case of open Fermi surfaces was subsequently developed by Lifshitz and his student Peshanskii,[231] who later showed that the Soviet magnetoresistance results for copper were consistent with the Fermi surface shape proposed by Pippard.[232] Also in 1959, Michael Priestley correlated the Soviet magnetoresistance data with the Fermi surface results from de Haas–van Alphen measurements.[233]

Each of the three types of experiments so far mentioned resulted in important information on the shape of the Fermi surface in a characteristic manner: anomalous skin effect measurements gave information on its curvature, de Haas–van Alphen oscillations gave the extremal cross-sectional areas, and magnetoresistance indicated possible contacts with Brillouin zone boundaries. It was later realized that the linear dimensions across the Fermi surface could often be inferred by resonances of sound absorption by the metallic specimen in the presence of an external magnetic field, a "magnetoacoustic effect" studied in considerable detail at Bell Labs and at Brown University.

The attenuation of ultrasonic waves by conduction electrons was first observed by H. E. Bommel in 1955 and further examined the same year by W. P. Mason and R. W. Morse.[234] In normal metals, a magnetic field H applied transverse to the direction of propagation causes oscillations of the attenuation of the sound waves that, like the de Haas–van Alphen effect, are periodic in $1/H$. The attenuation oscillates as the orbit diameter of the electrons passes through successive multiples of the sound wavelength—a sort of spatial resonance. By varying the directions of incoming sound waves and of the applied magnetic field, experimenters could eventually attempt a point-by-point measurement of linear dimensions of the Fermi surface, since one is at any point measuring the conduction electron momentum of an extremum on the Fermi surface perpendicular to both the external magnetic field and the incoming sound waves. The theory of the effect was first put forward by Pippard in 1957, more completely in 1960 by Morrel H. Cohen and his collaborators, and independently by the Soviet theorist V. L. Gurevich at the Ioffé Physico-Technical Institute in Leningrad.[235] The gradual disappearance of this attenuation below the transition temperature of a superconductor proved to be an important early check of the BCS theory of superconductivity (Chapter 8).

The velocity of an electron at the Fermi surface of a metal or semimetal was eventually inferred by a technique called Azbel–Kaner cyclotron resonance, after the Kharkov theorists who first suggested its possible utility.[236] Such an effect, analogous to the cyclotron resonance in semiconductors discussed earlier in this chapter, had been conceived as arising when the magnetic field was normal to the surface, so the electrons could execute their orbits within the skin layer. R. G. Chambers, however, had shown that under anomalous skin effect conditions nothing should be observed in this case, and the essential novelty of Azbel's proposal was that the magnetic field be oriented parallel to the surface of the metal.[237] Then the conduction electrons are bent into helical orbits passing into and out of the skin depth for microwave penetration, and can achieve resonant absorption of microwave energy from the applied alternating electromagnetic field when the field's frequency is an integral multiple of that of the electrons in their helical orbits, the "cyclotron" frequency.

The discovery of cyclotron resonance in metals opened up a new line of investigation of the Fermi surface of metals.[238] Azbel remembers[239] that when he first presented the idea at a seminar attended by Landau, the latter immediately raised a series of objections, and that only after two days of thrashing out the various possibilities was Landau convinced. Subsequently, Azbel and his younger Kharkov colleague E. A. Kaner showed how this technique could be used to determine Fermi surfaces.[240] If the shape of the Fermi surface was known from other experimental methods, a study of the cyclotron resonance frequency could give the velocity of conduction electrons at every point on the Fermi surface, a quantity of fundamental importance in all problems involving electron transport.

The first experiments to give a clear indiction of Azbel–Kaner cyclotron resonance were made in 1956 on tin and copper by Pippard's former research student Eric Fawcett. They were succeeded the following year by the more elaborate experiments on tin by Arthur Kip's group in Berkeley. Azbel recalls meeting Kip at a conference in Moscow and persuading him to look for the effect. Both groups subsequently studied the effect in aluminum, and the Berkeley group also examined copper.[241] Important work was also done by P. A. Bezuglyi and A. A. Galkin at the Institute of Radiophysics in the Ukraine, which had split off from Lazarev's laboratory at the Kharkov PTI, and by M. S. Khaikin at the Institute for Physical Problems in Moscow, mainly on tin, with the Ukrainian group also examining lead.[242]

The late 1950s and early 1960s were also the period when the first attempts were made to develop simple mathematical and modeling techniques to approximate the more complex surface shapes using parametric representations. Federico Garcia Moliner, then a research student at Cambridge with John Ziman, conceived the idea of expressing the Fermi surface in terms of a limited number of spherical harmonics and then adjusting the parameters in the expansion on the basis of results from de Haas–van Alphen measurements.[243] This was the approach used in 1965 by Douglas Roaf to specify the Fermi surface of copper, silver, and gold on the basis of Shoenberg's de Haas–van Alphen results. As previously noted, Andrew Gold, then a research student of Shoenberg doing measurements on lead, noticed that his de Haas–van Alphen data indicated a Fermi surface in some respects quite close to a sphere, which would result for noninteracting electrons, but completely carved up and rearranged.[244] This discovery was puzzling because it suggested that the conduction electrons in a crystal lattice, even of a polyvalent metal such as lead, could be considered as nearly free—as if the lattice potential were only a small perturbation—even though the actual potential must be quite substantial, especially close to the positive ion cores. This finding gave further encouragement to development of the theory of the pseudopotential.[245]

By this point, with the tremendous growth of theoretical and experimental techniques in the late 1950s, the investigation of the Fermi surface had become almost

International Conference on the Fermi Surface in Metals, Cooperstown, New York, August 1960. (AIP Niels Bohr Library)

a field unto itself—Fermi surface studies—with its own "invisible college" of fellow practitioners in widely scattered centers. The members talked about their research mainly to one another in their own particular language, inscrutable to others. The principal centers were Cambridge, Bristol (where Ziman and Chambers later moved), Kharkov, Moscow, Chicago, Berkeley, Bell Labs, and the Canadian National Research Council laboratory in Ottawa. Several other American laboratories were by then showing considerable interest. The founding convention of the field was a 1960 conference on the Fermi surface in Cooperstown, New York,[246] not far from the General Electric Laboratories in Schenectady, which along with the U.S. Air Force Office of Scientific Research sponsored the conference. The meeting drew a hundred or more of the main practitioners from the United States, Britain, and Canada, as well as a handful of others from France, Japan, Norway, and one or two other countries, but no Soviet scientists attended. Both the sponsorship and the degree of industrial and governmental participation reflected the direct physical relevance that the Fermi surface and the band theory were coming to have for the calculation of the properties of industrially important metals and alloys.

3.4 Conclusion

This chapter covers over three decades, during which the band theory of solids reached maturity. Its conceptual foundations were laid in the late 1920s when the solid state was considered by many physicists as a proving ground for quantum mechanics. After Wigner's and Seitz's work on sodium in 1933, this new theoretical tool was applied to real materials, and during the subsequent decades the theory came to be viewed as a useful tool for solving problems of the solid-state trade. The internal development of the theory of solids was now characterized less by conceptual breakthroughs than by steady improvement of approximate calculational methods, leading gradually to better fit of empirical data and theoretical interpretations.

In 1973, James Phillips observed:

> For thousands of years man has been curious about the composition and structure of the material world. The first quarter of the twentieth century saw the invention of two fields that have made it possible to turn curiosity into knowledge and knowledge into understanding. The first field is x-ray and electron diffraction, and the second field is the quantum theory of electrons. In principle the structures of atoms, molecules, and solids can be understood in great detail by combining these two approaches, one theoretical and one experimental. In principle there is little more to do in this area of human knowledge than to apply known methods in a routine manner. In principle the structure of matter is a solved problem. In practice, of course, the situation is quite different.[247]

Long before the Second World War, experimental studies of galvanomagnetic properties of materials, their x-ray emission and absorption spectra, diamagnetism, and compressibility—to name only a few areas of research—provided theorists with information relevant to the electronic structure of solids, supplementing data obtained from x-ray diffraction measurements. The application of quantum theory to the interpretation of experimental results was anything but routine and could only rarely be solved by exact calculations based on first principles. Many approximate procedures—among them the cellular, augmented plane-wave, and orthogonalized plane-wave methods—were developed and applied with varying degrees of success to band structure calculations. Even then, the computational difficulties were enormous, and theoretical results were at best in only qualitative agreement with experiment.

The war halted theoretical work on the electronic band structure of solids for almost a decade, but at the same time it provided physicists with new, powerful tools of the trade that had a strong impact on both experimental and theoretical studies of the solid state, including band structure studies in particular. Microwave techniques, greatly improved during the war work on radar, were used to measure cyclotron resonance in semiconductors and later in metals, enabling direct determination of the effective mass of electrons and holes and providing information about the shape of Fermi surfaces. Low-temperature techniques became more widely used due to the wartime development of improved helium cryostats. The technology of growing single crystals of high purity was considerably improved. Electronic digital computers, developed originally for military applications, became an invaluable tool for solid-state theorists. In the postwar period, progress in the study of the electron band structure of solids was achieved through close interaction between experimental and theoretical research.

Further improvements in theoretical methods continued to be based on a one-electron model approximation, which proved perfectly adequate for treating most phenomena occurring in crystals, down to the finest details, except for collective phenomena such as ferromagnetism, superconductivity, and second-order phase transitions. A major conceptual problem faced by solid-state theory was: Why did the nearly free–electron models used in all band structure calculation work? The degree of success of this model was surprising for two reasons. First, since the kinetic energy of the conduction electrons was about the same as their mutual Coulomb energy, one expected the electrons to be strongly correlated by their electrostatic interactions. Second, the electron–ion interaction was expected to be approximately 10 eV, much too great for the electrons to behave as a gas of nearly free particles.

The absence of strong correlations was explained in the late 1950s by Landau's theory of Fermi liquids (Chapter 8), and the relative weakness of the electron–ion interactions found its explanation in the pseudopotential concept. Landau showed that in a system of strongly interacting fermions the excitations of the system can be considered as "quasi-particles," resembling free fermions, an idea subsequently used to develop an elementary excitation model of the solid that formally solved the puzzle of how supposedly highly correlated electrons can act as if they were free.

The progress of the band theory in the period covered in this chapter involved to a considerable extent the development of increasingly sophisticated theoretical models and increasingly powerful computational hardware, without both of which it would have been very difficult to obtain theoretical correlations with important properties of metals and semiconductor crystals. The field has come today almost to the point where one may expect the theory—say, in its pseudopotential development—to facilitate the prediction of the technologically significant properties of whole classes of materials.

Notes

This work was supported by grants from the Leverhulme Trust and the British Academy, with travel support from the British Council and the Soviet Academy of Sciences. Acknowledgment is gratefully given for the many important editorial contributions of Lillian Hoddeson, who also took a leading part in the overall conception of the chapter. The authors and editors thank many solid-state physicists, including Marvin L. Cohen, Walter A. Harrison, Frank Herman, Conyers Herring, Harry Jones, Walter Kohn, Sir Nevill Mott, James C. Phillips, Sir Brian Pippard, David Shoenberg, and John Ziman, for their interest and collaboration at various stages of the work, from sharing their recollections to reading the drafts of this chapter. They are, of course, not responsible for its imperfections.

1. F. Bloch, "Über die Quantenmechanik der Elektronen in Kristallgittern," *Zs. Phys.* 52 (1928): 555–600.

2. R. E. Peierls, "Zur Theorie des Hall-Effekts," *Phys. Zs.* 30 (1929): 273–274; Peierls, "Zur Theorie der galvanomagnetischen Effekte," *Zs. Phys.* 53 (1929): 255–266; Peierls, "Zur Theorie der elektrischen und thermischen Leitfähigkeit von Metallen," *Ann. Phys.* 4 (1930): 121–148.

3. L. Brillouin, "Les Electrons dans les métaux et le rôle des conditions de réflexion sélective de Bragg," *C.R.* 191 (1930): 198–201; Brillouin, "Les Electrons dans les métaux et le classement des ondes de de Broglie correspondantes," *C.R.* 191 (1930): 292–294; Brillouin "Les Electrons libres dans les métaux et le rôle des réflexions de Bragg," *Journal de physique et le radium* 1 (1930): 377–400; Brillouin, *Statistiques quantiques* (Paris: Presses Universitaires, 1930); Brillouin, *Die Quan-*

tenstatistik und ihre Anwendung auf die Elektronentheorie der Metalle, vol. 12 of *Struktur der Materie in Einzeldarstellungen,* ed. M. Born and J. Franck (Berlin: Springer, 1931).

4. A. H. Wilson, "The Theory of Electronic Semi-conductors," *Proc. Roy. Soc.* A133 (1931):458–491; Wilson, "The Theory of Electronic Semi-conductors. II," *Proc. Roy. Soc.* A134 (1932): 277–287; Wilson, "A Note on the Theory of Rectification," *Proc. Roy. Soc.* A136 (1932): 487–498; F. Bloch, "Wellenmechanische Diskussionen der Leitungs- and Photoeffekte," *Phys. Zs.* 32 (1931): 881–886.

5. H. Bethe, interview with L. Hoddeson, May 1981, AIP.

6. A. Sommerfeld and H. Bethe, "Elektronentheorie der Metalle," in *Handbuch der Physik* (Berlin: Springer, 1933), 24, pt. 2: 333–622, p. 401.

7. C. Herring to L. Hoddeson, 7 June 1984.

8. See, for example, F. Bloch, "Elektronentheorie der Metalle," in *Handbuch der Radiologie* (Leipzig: Akademische Verlagsgesellschaft, 1933), 6, pt. 1: 226; R. Peierls, "Elektronentheorie der Metalle," *Ergebnisse der Exakten Naturwissenschaften* 11 (1932): 264–322.

9. See, for example, P. Hoch, "The Reception of Central European Refugee Physicists of the 1930's: USSR, UK, USA," *Annals of Science* 40 (1983): 217–246.

10. E. Wigner and F. Seitz, "On the Constitution of Metallic Sodium. I, II," *Phys. Rev.* 43 (1933): 804–810; 46 (1934): 509–524.

11. "It was the first time we had energy bands accurate enough to verify the general conclusion that metals should come from partly filled bands, insulators from filled bands with a large gap above, and then empty bands; and semiconductors from a similar situation with a much narrower gap" (J. Slater, *Solid-State and Moleculer Theory: A Scientific Biography* [New York: Wiley, 1975], 185).

12. As early as 1925, the Foundation's General Education Board had awarded $1 million to five Princeton science departments, including Physics and Mathematics, for the development of graduate education and research, stipulating that another $2 million must be raised from other sources. Half of the total of $3 million was originally devoted to endowing six research professorships, but except for the Göttingen mathematician Hermann Weyl, who stayed for only one year, the Mathematics and Physics departments were initially unable to secure the services of suitable candidates. "Minutes of the Department of Physics (Permanent Staff), March 19, 1930," Department of Physics, Princeton University, Princeton, New Jersey; see also the papers of physics chairman Karl T. Compton at the Department of Physics, Palmer Physical Laboratory, Princeton University, and those of mathematics chairman Oswald Veblen in the Library of Congress, Washington D.C., where the negotiations with the foundation may be found in box 29. General Education Board, Annual Report 1925–1926, p. 6, Rockefeller Foundation Archives Center, Pocantico Hills, North Tarrytown, New York; Veblen Papers, boxes 5 and 29, especially in box 29 two important memoranda from Karl T. Compton et al.: "Memorandum for Dr. Wickliffe Rose, President of the General Education Board, in Support of the Application to the General Education Board for its Support of the Fundamental Sciences at Princeton University" (undated, probably early 1925) and "Memorandum of Conversation with Dr. Wickliffe Rose on the subject: 'Support for Research in the Fundamental Sciences at Princeton University,'" (undated, probably late May 1925).

13. O. Veblen to J. von Neumann, 15 October 1929; von Neumann to Veblen, 13 November 1929, Veblen Papers (n. 12).

14. Hungarian refugees include Dennis Gabor, Nicholas Kurti, Egon Orowan, Michael Polanyi, and Laszlo Tisza, and such nuclear physicists as Leo Szilard and Edward Teller. Biographical details on Wigner are in E. P. Wigner, interview with T. S. Kuhn, 3 December 1963, pp. 21–26, AHQP.

15. E. P. Wigner, *Die Gruppentheorie und ihre Anwendungen auf die Quantenmechanik der Atomspektren* (Braunschweig: Vieweg, 1931).

16. D. Kevles, *The Physicists* (New York: Vintage Books, 1979), 77–78.

17. F. Seitz, interview with L. Hoddeson, 26 January 1981, AIP.

18. Ibid., 14.

19. Ibid., 18.

20. Edward U. Condon, with George M. Shortley, *The Theory of Atomic Spectra* (Cambridge: Cambridge University Press, 1935).

21. Seitz interview (n. 17); F. Seitz, "Biographical Notes," in *Beginnings,* 84–89, p. 87.

22. Seitz ("Biographical Notes") (n. 21), 85, 87.

23. J. Bardeen, "'Reminiscences of Early Days in Solid State Physics," in *Beginnings,* 77–83, pp. 77–79.

24. E. P. Wigner and J. Bardeen, "Theory of the Work Functions of Monovalent Metals," *Phys. Rev.* 48 (1935): 84–87; J. Bardeen, "Theory of the Work Function. II: The Surface Double Layer," *Phys. Rev.* 49 (1936): 653ff.

25. C. Herring, interview with Lillian Hoddeson, 23 July 1974, p. 11, AIP; C. Herring, "Recollections," in *Beginnings,* 67–76, pp. 67–69.

26. Seitz interview (n. 17), where Seitz adds that while the calculation they did might sound simple now, "at the time it was quite a struggle. . . . Today you would put the thing on a computer and press a button . . . but then [the numerical integration] had to be done point by point with an old Monroe calculator that rattled and banged."

27. Ibid.; Seitz ("Biographical Notes") (n. 21), 87; W. Prokofjew, "Berechnung der Zahlen der Dispersionszentren des Natriums," *Zs. Phys.* 58 (1929): 255–267; H. A. Kramers, "Wellenmechanik und halbzahlige Quantisierung," *Zs. Phys.* 39 (1926): 828–840.

28. Seitz interview (n. 17), 35; Y. Sugiura, "Applications of Schrödinger's Wave Functions to the Calculation of Transition Probabilities for the Principal Series of Sodium," *Phil. Mag.* 4 (1927): 495–504.

29. Wigner and Seitz (n. 10), 509.

30. F. Hund, "Zur Deutung der Molekelspektren I," *Zs. Phys.* 40 (1927): 742–764; as applied to crystals, Hund, "Zur Theorie der schwerflüchtigen nichtleitenden Atomgitter," *Zs. Phys.* 74 (1932): 1–17. Hund, who had taken his Ph.D. in 1922 with Max Born and then served as his assistant, spent 1926–1927 in Copenhagen with Niels Bohr and 1927–1929 as *Extraordinariat* at Rostock before taking a permanent post alongside Heisenberg at Leipzig (where one of the visiting Americans that first year was Slater).

31. W. Lenz, "Über die Anwendbarkeit der statistischen Methode auf Ionengitter," *Zs. Phys.* 77 (1932): 713–721; H. Jensen, "Die Ladungsverteilung in Ionen und die Gitterkonstante des Rubidiumbromids nach der statistischen Methode," *Zs. Phys.* 77 (1932): 722–745. During the period in which this work was done, Lenz, who had taken his Ph.D. in 1911 with Sommerfeld, was professor and director of the Institute of Theoretical Physics at Hamburg, where Jensen also worked. Pauli was also a *Privatdozent* in this institute from 1923 to 1928 before going on to the ETH in Zurich.

32. J. Lennard-Jones and H. J. Woods, "The Distribution of Electrons in a Metal," *Proc. Roy. Soc.* A120 (1928): 727–735.

33. See, for instance, D. R. Hartree, "Distribution of Charge and Current in an Atom Consisting of Many Electrons Obeying Dirac's Equation," *Proceedings of the Cambridge Philosophical Society* 25 (1929): 225–236.

34. Wigner and Seitz (n. 10); E. Wigner, "On the Interaction of Electrons in Metals, *Phys. Rev.* 46 (1934): 1002–1011; F. Seitz, "The Theoretical Constitution of Metallic Lithium," *Phys. Rev.* 47 (1935): 400–412; H. A. Bethe, "Quantitative Berechnung der Eigenfunktion von Metallelektronen," *Helvetica Physica Acta* 7, suppl. 2 (1934): 18–23. Wigner's paper was less a refinement of the Wigner–Seitz method than a way of "going beyond it." See Herring to Hoddeson (n. 7).

35. Slater (n. 11), 176.

36. Wigner (n. 34); Seitz (n. 34).

37. Seitz interview (n. 17).

38. Wigner (n. 34).

39. Seitz interview (n. 17).

40. Seitz ("Biographical Notes") (n. 21), 89; D. H. Ewing and F. Seitz, "On the Electronic Constitution of Crystals; LiF and LiH," *Phys. Rev.* 50 (1936): 760–777.

41. A. G. Hill, interview with P. Hoch, 4 August 1982, to be deposited at AIP.

42. F. Seitz, *The Modern Theory of Solids* (New York: McGraw-Hill, 1940).

43. F. Seitz and R. P. Johnson, "Modern Theory of Solids. I, II, III," *Journal of Applied Physics* 8 (1937): 84–97, 186–199, 246–260.

44. J. Slater, "Electronic Energy Bands in Metals," *Phys. Rev.* 45 (1934): 794–801.

45. J. Slater, "Electronic Structure of Metals," *Reviews of Modern Physics* 6 (1934): 209–280.

46. Karl Compton to Dr. Trevor Arnett, 6 December 1935, and other documents in General

Education Board records (n. 12), file #2508, box 691, folder 7125, Rockefeller Foundation Archives Center.

47. Van Vleck in the late 1940s was also to do some band structure work with his student Thomas S. Kuhn. T. S. Kuhn and J. H. Van Vleck, "A Simplified Method of Computing the Cohesive Energies of Monovalent Metals," *Phys. Rev.* 79 (1950): 382–388. This was a condensation of Kuhn's Ph.D. dissertation with Van Vleck.

48. Slater (n. 11), 95. Unless otherwise noted, biographical details on Slater can be assumed to come from this work.

49. Ibid., 167. Like the Wills tobacco family at Bristol, which built that university its Royal Fort laboratory building only a couple of years earlier, the Eastman family had for many years been associated with financing new building.

50. J. Stratton, untaped interview with P. Hoch, 4 August 1982; Hill interview (n. 41). See also L. Hoddeson, "The Entry of the Quantum Theory of Solids into the Bell Telephone Laboratories, 1925–40: A Case-Study of the Industrial Application of Fundamental Science," *Minerva* 18 (1980): 422–444, esp. p. 438, on Bell Labs' participation in the MIT "electrical engineering cooperative course"; Hoddeson, "The Discovery of the Point-Contact Transistor," *Historical Studies in the Physical Sciences* 12, no. 1 (1981): 41–76.

51. J. Millman, "Electronic Energy Bands in Metallic Lithium," *Phys. Rev.* 47 (1935): 286–290.

52. The significant papers are H. M. Krutter, "Energy Bands in Copper," *Phys. Rev.* 48 (1935): 664–671; G. E. Kimball, "The Electronic Structure of Diamond," *Journal of Chemical Physics* 3 (1935): 560–564; W. Shockley, "Electronic Energy Bands in Sodium Chloride," *Phys. Rev.* 50 (1936): 754–759; Shockley "Energy Bands for the Face-Centered Lattice," *Phys. Rev.* 51 (1937): 129–135; M. F. Manning and M. I. Chodorow, "Electronic Energy Bands in Metallic Tungsten," *Phys. Rev.* 56 (1939): 787–798; M. I. Chodorow and M. F. Manning, "Energy Bands in the Body-Centered Lattice," *Phys. Rev.* 52 (1937): 731–736; M. F. Manning, "Electronic Energy Bands in Body-Centered Iron," *Phys. Rev.* 63 (1943): 190–202; J. B. Greene and M. F. Manning, "Electronic Energy Bands in Face-Centered Iron," *Phys. Rev.* 63 (1943): 203–210.

53. F. Hund, "Zustände der Elektronen in einem Kristallgitter," *Zeitschrift für Technische Physik* 16 (1935): 494–497; F. Hund and B. Mrowka, "Über die Zustände der Elektronen in einem Kristallgitter, insbesondere beim Diamant," *Mathematische-physikalische Berichte Sächsische Akademie der Wissenschaft* (Leipzig) 87 (1935): 185–206; Hund and Mrowka, "Über die Zustände der Elektronen in einem Kristallgitter," *Mathematische-physikalische Berichte Sächsische Akademie der Wissenschaft* 87 (1935): 325–350.

54. Ewing and Seitz (n. 40); Shockley (1936) (n. 52).

55. J. Slater, "Wave Functions in a Periodic Potential," *Phys. Rev.* 51 (1937): 846–851.

56. Herring ("Recollections") (n. 25), 69.

57. C. Herring, "A New Method for Calculating Wave Functions in Crystals," *Phys. Rev.* 57 (1940): 1169–1177.

58. Herring ("Recollections") (n. 25), 69.

59. C. Herring and A. G. Hill, "The Theoretical Constitution of Metallic Beryllium," *Phys. Rev.* 58 (1940): 132–162.

60. J. Bardeen, interview with L. Hoddeson, 12 May 1977, AIP.

61. Ibid.

62. Bardeen, (n. 23), 81.

63. Bardeen interview (n. 60).

64. J. Bardeen, "An Improved Calculation of the Energies of Metallic Li and Na," *Journal of Chemical Physics* 6 (1938): 367–371; Bardeen (n. 23), 81.

65. Subsequently, as Bardeen explains, Herring found that this Seitz field did not yield a good fit to the experimentally observed atomic spectra, and in answer to a query about this, Seitz found that he had inadvertently published not the final version of the empirical field but one he used at an earlier stage in his calculation (n. 23). The discrepancies for lithium were not removed until 1953 when Harvey Brooks obtained a close empirical fit using the Kuhn–Van Vleck quantum defect method (n. 47). See H. Brooks, "Cohesive Energy of Alkali Metals," *Phys. Rev.* 91 (1953), 1027–1028.

66. Seitz interview (n. 17), 91. He adds that one of the major required readings for Quimby's

group was W. Voigt's classic monograph *Lehrbuch der Kristallphysik* (see Chapter 1). Quimby had entered the field as a result of his work on piezoelectric crystals during the first World War.

67. Seitz interview (n. 17), 90.

68. H. A. Bethe, "Recollections of Solid State Theory, 1926–33," in *Beginnings,* 49–51, p. 50. F. C. von der Lage and H. A. Bethe, "A Method for Obtaining Electronic Eigenfunctions and Eigenvalues in Solids with One Application to Sodium," *Phys. Rev.* 71 (1947): 612–622; H. A. Bethe, interview with C. Weiner, 17 November 1967, AIP.

69. W. J. Larke, "Memorandum: Research in Relation to Alloys," 23 January 1930, DSIR file/ 2/382, Public Record Office, London. Apart from Larke and Lindemann, the committee was made up of former Prime Minister Arthur Balfour, Sir W. H. Bragg, and Sir David Milne-Watson. See Stephen Keith and Paul Hoch, "Formation of a Research School: Theoretical Solid State Physics at Bristol, 1930–54" *British Journal for the History of Science* 19 (1986): 19–44. Much of this section on the Bristol School follows this article.

70. Keith and Hoch, (n. 69).

71. Ibid.

72. Advisory Council, Scientific Grants Committee, "Minutes of the Meeting Held on Wednesday, 12th February, 1930," DSIR file/1/16 (n. 69). This committee consisited of F. G. Donnan (chair), V. H. Blackman, Sir Alfred Ewing, F. Lindemann, Lord Rayleigh, and Sir James Walker.

73. Lennard-Jones and Woods (n. 32). This paper concludes by giving support to the by then thoroughly discredited theory of electrical conductivity in crystals put forward some years earlier by F. A. Lindemann, "Note on the Theory of the Metallic State," *Phil. Mag.* 29 (1915): pp. 127– 140.

74. F. A. Lindemann to J. E. Lennard-Jones, 25 March 1930, Cherwell Papers, Nuffield College, Oxford.

75. Scientific Grants Committee Meeting of 18 June 1930, "Memorandum: Application from Prof. J. E. Lennard-Jones," DSIR file/2/386 (n. 69).

76. A. M. Tyndall, "A History of the Department of Physics in Bristol, 1876–1948, with Personal Reminiscences," manuscript, August 1956, p. 26, A. M. Tyndall Papers, University of Bristol Physics Archives (kindly made available by J. M. Ziman and Norman Thompson).

77. For example, N. F. Mott and H.S.W. Massey, *The Theory of Atomic Collisions* (Oxford: Clarendon Press, 1933).

78. N. F. Mott, "Electrons in Crystalline and Non-Crystalline Metals: From Hume-Rothery to the Present Day," *Metal Science* 14 (1980): 557–561, p. 557.

79. Ibid.

80. H. Jones, "Notes on Work at the University of Bristol, 1930–37," in *Beginnings,* 52–55, p. 52.

81. A. Guinier to E. Braun, March 1981, History of Solid State Physics Project Files, Technology Policy Unit, University of Aston in Birmingham, United Kingdom.

82. See, for example, W. Hume-Rothery, "Applications of X-ray Diffraction to Metallurgical Science," in *Fifty Years of X-ray Diffraction,* ed. P. P. Ewald (Utrecht: International Union of Crystallography, 1962), 190–211.

83. The £50,000 research endowment secured from the Rockefeller Foundation in 1930 was contingent on Bristol finding another £25,000 from local sources to permanently endow what was then the only chair of theoretical physics in the country, then occupied by Lennard-Jones and soon to be held by Mott. The additional money was then contributed by Melville Wills, after whom the chair was subsequently named. Tyndall notes that Melville Wills left him with the impression he made his contribution "with reluctance, feeling he was almost blackmailed by the Foundation" ([n. 76], 26).

84. Nevill Mott, interview with P. Hoch and E. Braun, 15 January 1981, AIP.

85. Slater (n. 11), 5; C. Zener, interview with L. Hoddeson, 1 April 1981, AIP; H. Jones and C. Zener, "A General Proof of Certain Fundamental Equations in the Theory of Metallic Conduction," *Proc. Roy. Soc.* A144 (1934): 101–117; Jones and Zener, "The Theory of the Change of Resistance in a Magnetic Field," *Proc. Roy. Soc.* A145 (1934): 268–277; N. F. Mott and C. Zener, "The Optical Properties of Metals," *Proceedings of the Cambridge Philosophical Society* 30 (1934): 249–270; Zener, "A Theory of the Electrical Breakdown of Solid Dielectrics," *Proc. Roy. Soc.* A145 (1934): 523–529.

86. Tyndall (n. 76), 26.

87. N. F. Mott, "Notes on My Scientific and Professional Career," unpublished memoir, undated, in Mott's personal file at the Royal Society, London (kindly made available by him).

88. Ibid.; N. F. Mott to A. M. Tyndall, 17 August 1932, Bristol University Physics Archives.

89. J. E. Lennard-Jones to A. M. Tyndall, 22 August 1932, Bristol University Physics Archives.

90. A decisive factor was R. H. Fowler's warning that if he stayed in Cambridge, the heavy demands of college teaching might cause his research to suffer. In contrast, one of the purposes of the endowed Wills chair was to free the occupant from teaching and administration. See Mott (n. 87), 10.

91. Details on the work in the laboratory and quote from Tyndall (n. 76), 27.

92. A. M. Tyndall to N. F. Mott, 12 May 1933, Bristol University Physics Archives.

93. Ibid.

94. Mott to Tyndall, 25 May 1933, Bristol University Physics Archives.

95. Ibid.

96. Sommerfeld and Bethe (n. 6).

97. N. F. Mott, "Memories of Early Days in Solid State Physics," in *Beginnings,* 56–66, p. 57.

98. Ibid., 58. H. M. O'Bryan and H.W.B. Skinner, "Characteristic X-rays from Metals in the Extreme Ultraviolet," *Phys. Rev.* 45 (1934): 370–378.

99. H. Jones, "The Theory of Alloys in the γ Phase," *Proc. Roy. Soc.* A144 (1934): 225–234.

100. H. Jones. "Electrons in Metals and Alloys," *Physics Bulletin* 19 (1968): 176–181. The key paper in this earlier work is H. Jones, "The Phase Boundaries in Binary Alloys. I: The Equilibrium Between Liquid and Solid Phases," *Proceedings of the Physical Society (London)* 49 (1937): 243–249. Jones also recalls contacts with Slater in the 1930s. He and Mott also had a great deal of productive contact with the Oxford metallurgist Hume-Rothery, whom he remembers as "very keen on using theoretical physics to interpret the properties of metals." During a year on staff at the Mond Laboratory in 1937 to 1938, Jones recalls a visit from Wigner. H. Jones, interview with P. Hoch, 20 January 1981, AIP.

101. R. Peierls, "Zur Theorie des Diamagnetismus von Leitungselektronen. I, II," *Zs. Phys.* 80 (1933): 763–691; 81 (1933): 186–194.

102. H. Jones, "Notes on Work at the University of Bristol, 1930–37," in *Beginnings,* 52–55, p. 52.

103. C. S. Smith, "Magnetic Susceptibility of Some Copper Alloys of Gamma Brass Structure," *Physics* 6, no. 1 (1935): 47–52.

104. H. Jones, "Applications of the Bloch Theory to the Study of Alloys and of the Properties of Bismuth," *Proc. Roy. Soc.* A147 (1934): 396–417.

105. H. Jones, "The Theory of Galvanomagnetic Effects in Bismuth," *Proc. Roy. Soc.* A155 (1936): 653–663. Jones's paper included a discussion of the sensitivity of such properties to small impurities.

106. H. Jones, N. F. Mott, and H.W.B. Skinner, "A Theory of the Form of the X-ray Emission Bands and Metals," *Phys. Rev.* 45 (1934): 379–384.

107. Y. Cauchois, "Nouvelle mesures et observations relatives au spectre L d'emission du platine," *C.R.* 201, pt. 2 (1935): 598–602; Y. Cauchois and Iona Manescu, "Spectres d'absorption et niveaux caractéristiques de l'uranium, du platine et du tungstene," *C.R.* 210, pt. 1 (1940): 172–174; Y. Cauchois and N. F. Mott, "The Intepretation of X-ray Absorption Spectra of Solids," *Phil. Mag.* 40 (1949): 1260–69.

108. N. F. Mott and H. Jones, *The Theory of the Properties of Metals and Alloys* (Oxford: Clarendon Press, 1936).

109. Seitz ("Biographical Notes") (n. 21), 89.

110. Mott (n. 97), 57.

111. R. Smoluchowski, "Random Comments on the Early Days of Solid State Physics," in *Beginnings,* 100–101.

112. Slater (n. 11), 191; N. F. Mott, interview with P. Hoch, 28 December 1981, AIP.

113. K. Fuchs, "A Quantum Mechanical Investigation of the Cohesive Forces of Metallic Copper," *Proc. Roy. Soc.* A151 (1935): 585–602; Fuchs, "A Quantum Mechanical Calculation of the Elastic Constants of the Monovalent Metals," *Proc. Roy. Soc.* A153 (1936): 622–639; Fuchs, "The Elastic Constants and Specific Heats of the Alkali Metals," *Proc. Roy. Soc.* A157 (1936): 444–450.

A few years later, at the start of the Second World War, Fuchs, as an "enemy alien," was interned and imprisoned in Canada, being subsequently released to take part in the American atomic bomb project.

114. In the early 1950s, Fuchs's elastic constant calculations were considerably refined and extended by Harry Jones to magnesium and beta-brass, and by his Imperial College colleague E. P. Wohlfarth to diamond. H. Jones, "The Effect of Electron Concentration on the Lattice Spacings in Magnesium Solid Solutions," *Phil. Mag.* 41 (1950): 663–670; Jones, "A Calculation of the Elastic Shear Constants of Beta-brass," *Phil. Mag.* 43 (1950): 105–112; Jones, "Structural and Elastic Properties of Metals," *Physica* 15 (1949): 13–22; E. P. Wohlfarth, "Electrostatic Contribution to the Elastic Constants of Solids with a Diamond Structure," *Phil. Mag.* 43 (1952): 474–476.

115. Jones interview (n. 100).

116. Mott interview (n. 112); J. Slater, "A History of the MIT Physics Department 1930–48," 24, John Slater Papers, American Philosophical Society, Philadelphia.

117. Slater (n. 11), 209–216; Hill interview (n. 41); Bethe interview (n. 68), 153–156.

118. Bethe interview (n. 68), 157–159.

119. Zener interview (n. 85).

120. Stratton interview (n. 50).

121. M. Kelly, "A First Record of Thoughts Concerning an Important Post War Problem of the Bell Telephone Laboratories and Western Electric Company" (1 May 1943), 1–2; memorandum, M. Kelly to O. Buckley, 29 July 1949, "Five Year Program for Control of Laboratories," 5–6, Records of the Bell Telephone Laboratories, cited as ref. 43 of Hoddeson (1981) (n. 50).

122. P. W. Henriksen, "Solid State Physics Research at Purdue," *OSIRIS,* 2nd ser., vol. 3 (1987): 237–260.

123. F. Herman, interview with K. Szymborski, 17 June 1982, AIP, where Herman also noted that he became interested in the band theory principally through the influence of three texts: Seitz's *Modern Theory of Solids,* Brillouin's *Wave Propagation in Periodic Structures,* and Wilson's *Theory of Metals.* See also F. Herman, "Elephants and Mahouts—Early Days in Semiconductor Physics" (based on interview by Szymborski), *Physics Today* 37, no. 6 (1984): 56–63.

124. Herman interview (n. 123); Herring (n. 57).

125. Ibid.

126. Ibid.

127. Herman had collaborated on the germanium stucture with a Princeton graduate student, Joseph Callaway; their paper on germanium appeared in the same period as Herman's paper on diamond. F. Herman and J. Callaway, "Electronic Structure of the Germanium Crystal," *Phys. Rev.* 89 (1953): 518–519; F. Herman, "Electronic Structure of the Diamond Crystals," *Phys. Rev.* 88 (1952): 1210–1211.

128. T. Woodruff, "Solution of the Hartree-Fock-Slater Equations for Silicon Crystal by the Method of Orthogonalized Plane Waves," *Phys. Rev.* 98 (1955): 1741–1742; Woodruff, "Application of the Orthogonalized Plane-Wave Method to Silicon Crystal," *Phys. Rev.* 103 (1956): 1159–1166.

129. Herring ("Recollections") (n. 25), 70.

130. Bethe ("Recollections") (n. 68), 50.

131. D. J. Howarth, and H. Jones, "The Cellular Method of Determining Electronic Wave Functions and Eigenvalues in Crystals, with Applications to Sodium," *Proceedings of the Physical Society (London)* A65 (1952): 355–368.

132. D. J. Howarth, "Electronic Eigenvalues of Copper," *Proc. Roy. Soc.* A220 (1953): pp. 513–529.

133. B. Shiff, "A Calculation of the Eigenvalues of Electronic States in Metallic Lithium by the Cellular Method," *Proceedings of the Physical Society (London)* A67 (1954): 2–8; H. Jones to P. Hoch, 14 January 1982.

134. V. Zehler, "Die Berechnung der Energiebander im Diamantkristall," *Ann. Phys.* 13 (1953): 229–252. This calculation gave a quite realistic value of 5.89 eV for the forbidden energy gap.

135. D. K. Holmes, "An Application of the Cellular Method to Silicon," *Phys. Rev.* 87 (1952): 782–84.

136. E. Yamaka and T. Sugita, "Energy Band Structure in Silicon Crystal," *Phys. Rev.* 90 (1953): 992.

137. G. G. Hall, "The Electronic Structure of Diamond," *Phil. Mag.* 43 (1952): 338–343.

138. D. G. Bell, D. M. Hum, L. Pincherle, D. W. Sciama, and P. M. Woodward, "The Electronic Band Structure of PbS," *Proc. Roy. Soc.* A217 (1953): 71–91; D. G. Bell, R. Hensman, D. P. Jenkins, and L. Pincherle, "A Note on the Band Structure of Silicon," *Proceedings of the Physical Society (London)* A67 (1954): 562–563; see also D. P. Jenkins and L. Pincherle, "A Variational Principle for Electronic Wave Functions in Crystals," *Phil. Mag.* 45 (1954): 93–99.

139. W. Kohn, "Variational Methods for Periodic Lattices," *Phys. Rev.* 87 (1952): 472–481; W. Kohn and N. Rostoker, "Solution of the Schrödinger Equation in Periodic Lattices with an Application to Metallic Lithium," *Phys. Rev.* 94 (1954): 1111–1120. J. Korringa, "On the Calculation of the Energy of a Bloch Wave in a Metal," *Physica* 13 (1947): 392–400.

140. J. Slater, "An Augmented Plane Wave Method for the Periodic Potential Problem," *Phys. Rev.* 92 (1953): 603–608; J. C. Slater and M. M. Saffren, "An Augmented Plane Wave for Periodic Potential Problem. II," *Phys. Rev.* 92 (1953): 1126–1128.

141. Slater (n. 11), 227.

142. Ibid., 256. The OPW method is not suitable in its original form because the *d* bands of transition metals, unlike the valence bands of nontransition metals, do not anywhere resemble single plane waves or combinations of two or three such.

143. Ibid., 261–262; J. H. Wood, "Energy Bands in Fe via Augmented Plane Wave Method," *Phys. Rev.* 126 (1962): 517ff.

144. G. L. Pearson and H. Suhl, "The Magneto-Resistance Effect in Oriented Single Crystals of Germanium," *Phys. Rev.* 83 (1951): 768–776.

145. Ibid., 768.

146. Herring ("Recollections") (n. 25), 70; G. L. Pearson and C. Herring, "Magneto-resistance Effect and the Band Structure of Single Crystal Silicon," *Physica* 20, no. 11 (1954): 1–4.

147. Herman interview (n. 123).

148. "The management's initial response was that young scientists like myself did not get sent to international conferences. Such honors were reserved for senior scientists and managers. It wasn't a question of money. It was a question of tradition" (ibid). The paper was reprinted as F. Herman, "Some Recent Development in the Calculation of Crystal Energy Bands—New Results for the Germanium Crystal," *Physica* 20 (1954): 801–812.

149. J. Dorfman, "Paramagnetic and Diamagnetic Resonance of Conduction Electrons," *Doklady Akademiia Nauk* 81 (1951): 765–766.

150. R. B. Dingle, "Magnetic Properties of Metals" (Ph.D. diss., Cambridge University, 1951); Dingle, "Some Magnetic Properties of Metals. III. Diamagnetic Resonance," *Proc. Roy. Soc.* A212 (1952): 38–47.

151. W. Shockley, "Cyclotron Resonances, Magnetoresistance, and Brillouin Zones in Semiconductors," *Phys. Rev.* 90 (1953): 491.

152. H. Suhl and G. L. Pearson, "Faraday Rotation in Germanium," *Phys. Rev.* 92 (1953): 858ff.

153. G. Dresselhaus, A. F. Kip, and C. Kittel, "Observation of Cyclotron Resonance in Germanium Crystals," *Phys. Rev.* 92 (1953): 827; Dresselhaus et al., "Spin-Orbit Interaction and the Effective Masses of Holes in Germanium," *Phys. Rev.* 95 (1954): 568–69; Dresselhaus et al., "Cyclotron Resonance of Electrons and Holes in Silicon and Germanium Crystals," *Phys. Rev.* 98 (1955): 368–384.

154. B. Lax, H. J. Zeigler, R. N. Dexter, and E. S. Rosenblum, "Directional Properties of the Cyclotron Resonance in Germanium," *Phys. Rev.* 93 (1954): 1418–1420.

155. John C. Slater Papers, American Philosophical Society. An interesting discussion of this period is contained in C. Kittel to K. K. Darrow, 28 June 1954, Darrow Papers, AIP.

156. M. L. Cohen, "Fifty Years of Pseudopotentials," *International Journal of Quantum Chemistry* 17 (1983): 583–595. My discussion here draws on an unpublished draft history kindly provided by Professor W. Harrison.

157. See, for instance, V. Heine, "The Pseudopotential Concept," *Solid State Physics* 24 (1970): 1–36, esp. pp. 20–26.

158. E. Fermi, "Displacement by Pressure of the High Lines of the Spectral Series," *Nuovo Cimento* 11 (1934): 157–166; M. L. Cohen, "The Fermi Atomic Pseudopotential," *American Journal of Physics* 52 (1984): 695–703.

159. E. Amaldi, O. D'Agostino, E. Fermi, B. Pontecorvo, F. Rasetti, and E. Segré, "Artificial Radioactivity Produced by Neutron Bombardment. II," *Proc. Roy. Soc.* A149 (1935): 522–558.

160. H. Hellmann, "A New Approximation Method in the Problem of Many Electrons," *Journal of Chemical Physics* 3 (1935): 61; H. Hellmann and W. Kassatoschkin, "Metallic Binding According to the Combined Approximation Procedure," *Journal of Chemical Physics* 4 (1936): 324–325.

161. Hellmann and Kassatoshckin (n. 160).

162. See, for instance, P. Gombás, *Statistische Theorie des Atoms* (Vienna: Springer-Verlag, 1949), 19, 35.

163. P. Gombás, "Über die metallische Bindung," *Zs. Phys.* 94 (1935): 473–488; Gombás, "Zur Bestimmung der Verteilung der Metallelektronen in Alkalimetallen," *Zs. Phys.* 108 (1938): 509–522; Gombás, "Über eine Methode zur Berechnung der Lage und Breite des Energiebandes der Valenzelektronen in Alkalimetallen," *Zs. Phys.* 111 (1938): 195–207; Gombás, "Bestimmung der Lage und Breite des Energiebandes der Valenzelektronen der Metalle Na, K, Rb und Cs," *Zs. Phys.* 113 (1939): 150–160.

164. The binding energy of aluminum was, for instance, calculated by R. Gáspár, "Über die Bindung des Metallischen Aluminum," *Hungarica Acta Physica* 2 (1952): 31–46.

165. K. Ladányi, "Zur Theorie der Edelmetalle," *Hungarica Acta Physica* 5 (1956): 361–380.

166. P. Szepfalusy, "Über die Orthogonalität der Wellenfunktionen von Atomelektronen," *Hungarica Acta Physica* 5 (1955): 325–338; Szepfalusy, "Die Hartree-Focksche Methode im Falle eines nichtorthogonalen Einelektronenwellenfunktionen-Systems," *Hungarica Acta Physica* 6 (1956): 273–292.

167. E. Antoncik, "A New Formulation of the Method of Nearly Free Electrons," *Czechsolovak Journal of Physics* 4 (1954): 439–452; Antoncik, "The Repulsion Potential of Unoccupied States," *Czechoslovak Journal of Physics* 7 (1957): 188–9; Antoncik, "The Use of the Repulsive Potential in the Quantum Theory of Solids," *Czechoslovak Journal of Physics* 9 (1959): 291–305; Antoncik, "Approximate Formulation of the Orthogonalized Plane-Wave Method," *Journal of the Physics and Chemistry of Solids* 10 (1959): 314–320. Antoncik, "On the Approximate Formulation of the Orthoganalized Plan-Wave Method," *Czechoslovak Journal of Physics* 10 (1960): 22–27.

168. C. Herring and M. M. Nichols, "Thermionic Emission," *Reviews of Modern Physics* 21 (1949): 185–270; C. Herring to J. C. Phillips, 24 November 1980 (kindly made available by Herring).

169. C. Herring, "Theoretical Ideas Pertaining to Traps or Centers," in *Proceedings of the Atlantic City Conference on Photoconductivity* (New York: Wiley, 1956), 81–110.

170. Ibid., 86.

171. C. S. Smith, "Institute for the Study of Metals, University of Chicago, Report of the Director to the Committee on Physics, Chemistry, and the Institutes," manuscript, 17 December 1954, AIP (courtesy of C. S. Smith).

172. Editor's introduction, "Twelfth Quarterly and Third Annual Report to Sponsors of the Institute for the Study of Metals, February 1, 1949–April 30, 1949," 1, AIP, Urbana.

173. Smith (n. 171), 4.

174. M. H. Cohen and V. Heine, "Electronic Band Structures of the Alkali Metals and of the Noble Metals and Their Alpha-Phase Alloys," *Advances in Physics* 7 (1958): 395–434.

175. He had decided to enter solid-state physics after asking Fermi's advice. According to Cohen, Fermi told Phillips, "Particle physics is quiet and very difficult now, but solid state physics seems to be booming" (M. M. Cohen, interview with L. Hoddeson, 5 June 1981, AIP).

176. J. C. Phillips. "Energy-Band Interpolation Scheme Based on a Pseudo-potential," *Phys. Rev.* 112 (1958): 685–695.

177. Ibid., 686–687.

178. Ibid., 695.

179. J. C. Phillips and L. Kleinman, "New Method for Calculating Wave Functions in Crystals and Molecules," *Phys. Rev.* 116 (1959): 287–294.

180. Cohen (n. 158).

181. Cohen interview (n. 175); M. Cohen and V. Heine, "Cancellation of Kinetic and Potential Energy in Atoms, Molecules, and Solids," *Phys. Rev.* 122 (1961): 1821–1826.

182. A. V. Gold, "An Experimental Determination of the Fermi Surface in Lead," *Philosophical Transactions of the Royal Society of London* A251 (1958): 85–112.

183. W. A. Harrison, "Fermi Surface in Aluminum," *Phys. Rev.* 116 (1959): 555–561; Harrison, "Electronic Structure of Polyvalent Metals," *Phys. Rev.* 118 (1960): 1190–1208.

184. W. A. Harrison, "Band Structure and Fermi Surface of Zinc," *Phys. Rev.* 126 (1962): 497–505.

185. W. A. Harrison, *Pseudopotentials in the Theory of Metals* (New York: Benjamin, 1966).

186. A substantial portion of this section appears in P. Hoch, "A Key Concept from the Electron Theory of Metals: History of the Fermi Surface 1933–60," *Contemporary Physics* 24 (1983): 3–23. However, the present version has been edited and revised substantially in a number of spots.

187. Jones interview (n. 100).

188. Sommerfeld and Bethe (n. 6).

189. H. A. Bethe, interview with L. Hoddeson, 29 April 1981, AIP.

190. P. L. Kapitza, "The Change of Electrical Conductivity in Strong Magnetic Fields, Part I—Experimental Results," *Proc. Roy. Soc.* A123 (1929): 292–341. Both Bethe and Peierls then pointed out that variations of magnetoresistance with crystal orientation were due to departures from the isotropic free-electron model—that is, to departures from a spherical Fermi surface. H. A. Bethe, "Change of Resistance in Magnetic Fields," *Nature* 127 (1931): 336–337; R. Peierls, *Leipziger Vorträger* (Leipzig: Hirzel, 1930), 75; Peierls, "Zur Theorie der Magnetischen Widerstandsänderung," *Ann. Phys.* 10 (1931): 97–110.

191. Bethe ("Recollections") (n. 68), 50–51; P. L. Kapitza, untaped interview with P. Hoch, November 1982.

192. D. Blokhintsev and L. Nordheim, "Zur Theorie der anomalen magnetischen und thermoelektrischen Effekte in Metallen," *Zs. Phys.* 84 (1933): 168–194; Jones and Zener ("Theory of the Change") (n. 85). Very important subsequent work was done by S. Titeica, "Über die Widerstandsänderung von Metallen in Magnetfeld," *Ann. Phys.* 22 (1939): 129–161, who discussed the peculiar oscillations of the magnetoresistance that are somewhat similar to the de Haas–van Alphen oscillations in the magnetic susceptibility; and M. Kohler, "Transversale und longitudinale Widerstandsänderung von zweiwertige Metallen Kubischraumzentrierter Kristallstruktur," *Phys. Zs.* 39 (1938): 9–23; Kohler, "Magnetische Widerstandsänderung und Leitfähigkeitstypen. II," *Ann. Phys.* 5 (1949): 99–107. Thus, in a sense, the magnetoresistance effect could be said to have been potentially the oldest technique of Fermi surface study, although it was not actually applied until the detailed theoretical work of Ilya Lifshitz and his school in the late 1950s and the experimental results of Alekseevski and Gaidukov. See R. G. Chambers, "Magnetoresistance," in *The Fermi Surface: Proceedings of an International Conference held at Cooperstown, New York on Aug. 22–24, 1960,* ed. W. A. Harrison and M. B. Webb (New York: Wiley, 1960), 100–124.

193. Jones and Zener ("Theory of the Change") (n. 85), 268, 269.

194. Ibid., 275, 277.

195. Biographical details on Shoenberg, unless otherwise noted, come from his interview with P. Hoch, 10 February 1981, to be deposited at AIP, and from his article, "Forty Odd Years in the Cold: Reminiscences of Work in Low Temperature Physics," *Physics Bulletin* (January 1978): 16–19.

196. W. de Haas and P. M. van Alphen, "The Dependence of the Susceptibility of Bismuth Single-crystals upon the Field," *Communications from the Kammerlingh Onnes Laboratory of the University of Leiden* 220d (1932): 17–22; de Haas and van Alphen, "The Dependence of the Susceptibility of Diamagnetic Metals on the Field," *Communications from the Kammerlingh Onnes Laboratory of the University of Leiden* 208d (1930): 31–33.

197. H.B.G. Casimir, *Haphazard Reality, Half A Century in Science* (New York: Harper & Row, 1983), 335–336.

198. R. Peierls ("Theorie II") (n. 101); L. D. Landau, "Diamagnetismus der Metalle," *Zs. Phys.* 64 (1930): 629–637, where Landau notes that "it would hardly be possible to observe this effect experimentally, since on account of the inhomogeneity of available [externally applied magnetic] fields there would always be an averaging." Peierls's theory required a very sharp curvature of the Fermi surface in bismuth near the zone boundary, a crucial detail explained by Jones (n. 105).

199. R. Peierls, interview with J. L. Heilbron, 18 June 1963, p. 11, AHQP; Peierls, interview

with L. Hoddeson, 20 May 1977, p. 14, AIP; R. Peierls, "Recollections of Solid State Physics," in *Beginnings,* 28–37, pp. 35–36.

200. M. Blackman, "On the Diamagnetic Susceptibility of Bismuth." *Proc. Roy. Soc.* A166 (1938): 1–15.

201. Shoenberg (1978) (n. 195), 16.

202. Shoenberg interview (n. 195).

203. Ibid.

204. D. Shoenberg (with appendix by R. Peierls), "The Magnetic Properties of Bismuth, III. Further Measurements on the de Haas–van Alphen Effect," *Proc. Roy. Soc.* A170 (1939): 341–64.

205. D. Shoenberg to R. Peierls, 12 February 1939, Peierls Papers, Bodleian Library, Oxford.

206. J. A. Marcus, "The de Haas–van Alphen Effect in a Single Crystal of Zinc," *Phys. Rev.* 71 (1947): 559.

207. Shoenberg (1978) (n. 195), 18–19; L. Onsager, "Interpretation of the de Haas–van Alphen Effect," *Phil. Mag.* 43 (1952): 1006–1008. For an overview of work done in this period, see I. M. Lifshitz and A. M. Kosevich, "Theory of Magnetic Susceptibility in Metals at Low Temperatures," *Soviet Physics JETP* 2 (1956): 636–645.

208. M. I. Kaganov, "On the History of the Electron Theory of Metals," in *Razvite Kriogeniki na Ukraine,* ed. B. G. Lazarev et al. (Kharkov: Ukrainian Physico-Technical Institute Low Temperature Laboratory, Ukrainian Academy of Sciences, 1978), 125–138, kindly sent to me by Professor V. Ya' Frenkel. Lifshitz's 1950 report is also cited by I. M. Lifshitz and M. I. Kaganov, "Geometric Concepts in the Electron Theory of Metals," in *Electrons at the Fermi Surface,* ed. M. Springford (Cambridge: Cambridge University Press, 1980), 3–45, p. 9. The first widely published Soviet development of the theory was by Lifshitz and his student at Kharkov, A. M. Kosevich, "Towards the Theory of the de Haas–van Alphen Effect for Particles with an Arbitrary Dispersion Law," *Doklady Akademiia Nauk* 91 (1954): 795ff. Subsequently, Lifshitz and the Kharkov topologist A. V. Pogorelov in 1954 showed that if the Fermi surface is convex and has a center of symmetry, its shape may be calculated from the dependence of the de Haas–van Alphen periods on the direction of the applied magnetic field. I. M. Lifshitz and A. V. Pogorelov, "Determination of the Fermi Surface and Velocities in a Metal from the Oscillations of the Magnetic Susceptibility," *Doklady Adademiia Nauk* 96 (1954): 1143ff.

209. A. B. Pippard, interview with P. Hoch and E. Braun, 22 January 1981, AIP. All biographical details on Pippard come from this interview.

210. H. London, "The High-Frequency Resistance of Superconducting Tin," *Proc. Roy. Soc.* A176 (1940): 522–533. J. J. Thomson, "On the Theory of Electric Conduction," *Proceedings of the Cambridge Philosophical Society* 11 (1901): 120–122; K. Fuchs, "The Conductivity of Thin Metallic Films According to the Electron Theory of Metals," *Proceedings of the Cambridge Philosophical Society* 34 (1938): 100–108.

211. Pippard interview (n. 209); G.E.H. Reuter and E. H. Sondheimer, "The Theory of the Anomalous Skin Effect in Metals," *Proc. Roy. Soc.* A195 (1948): 336–364. Fuchs and London were refugees. Both Reuter and Sondheimer were born in Germany, sent over to England after 1933 by their parents as children, and later attended Trinity College, Cambridge.

212. Pippard interview (n. 209); E. Maxwell, P. M. Marcus, and J. C. Slater, "Surface Impedance of Normal and Superconductors at 24.000 Megacycles per Second," *Phys. Rev.* 76 (1949): 1332–1347.

213. Pippard interview (n. 209).

214. Ibid.

215. Ibid.

216. This development was discussed later by R. G. Chambers, "The Fermi Surface," *Science Progress* (Oxford) 54 (1966): 163–191, p. 180.

217. Pippard interview (n. 209), E. H. Sondheimer, "Theory of Anomalous Skin Effect in Anisotropic Metals," *Proc. Roy. Soc.* A224 (1954): 260–272; A. B. Pippard, "Theory of Anomalous Skin Effect in Anisotropic Metals," *Proc. Roy. Soc.* A224 (1954): 273–282; Pippard, "Experimental Methods for Determining the Fermi Surface," *Proceedings of the Tenth Solvay Conference,* 123–157, originally published as Instituts International Solvay, *Les Electrons dans les métaux, rapports et discussions du Dixième Conseil de Physique* (Brussels: Stoops, 1955).

218. Pippard interview (n. 209); Cohen interview (n. 175). The key paper is A. B. Pippard, "An Experimental Determination of the Fermi Surface in Copper," *Philosophical Transactions of the Royal Society of London* A250 (1957): 325–357.

219. Pippard interview (n. 209).

220. Ibid.

221. Shoenberg (1978) (n. 195), 19; A. V. Gold, "An Experimental Determination of the Fermi Surface in Lead," *Philosophical Transactions of the Royal Society of London* A251 (1958): 85–112; A. V. Gold and M. G. Priestley, "The Fermi Surface in White Tin," *Phil. Mag.* 5 (1960): 1089–1104. For a careful review of this work, see Gold, "The de Haas–van Alphen Effect," in *Solid State Physics: The Simon Fraser Lectures: 1. Electrons in Metals,* ed. J. F. Cochran and R. R. Haering, (New York: Gordon and Breach, 1968), 39–126.

222. Pippard interview (n. 209).

223. Shoenberg (1978) (n. 195), 19.

224. Kapitza (n. 190).

225. B. I. Davydov, I. Pomeranchuk, and O. Vlianii, "On the Influence of Magnetic Fields on the Conductivity of Bismuth at Low Temperatures," *Zhurnal Eksperimentalnoi i Teoreticheskoi Fiziki* 9 (1939): 1294ff. A very similar theory for the nonsaturation of the magnetoresistance of bismuth had been advanced by Jones (n. 105). The complexity of some of the electron orbits needed to explain nonsaturation was considered in 1950 by Shockley, who was perhaps the first to explicitly relate such complexities to the geometry of the Fermi surface. W. Shockley, "Effect of Magnetic Fields on Conduction—Tube Integrals," *Phys. Rev.* 79 (1950): 191–192.

226. Mark Ya. Azbel, interview with P. Hoch, 30 March 1982, to be deposited at AIP.

227. Ibid. Kaganov, untaped interview with P. Hoch, 15 November 1982, did not recall this idea as coming from Landau, but suggested that the latter may have mentioned it to Lifshitz privately. I am extremely grateful to Academician Kapitza for making possible my interview with Kaganov by inviting me to the Institute of Physical Problems as an official guest for November 1982. Professor Viktor Ya' Frenkel has also supplied me with much information on Soviet work in this period.

228. I. M. Lifshitz, M. Azbel, and M. I. Kaganov, "The Theory of Galvanomagnetic Effects in Metals," *Soviet Physics JETP* 4 (1957): 41–54.

229. N. E. Alekseevskii and Iu. P. Gaidukov, "Anisotropy of the Electrical Resistance of a Gold Monocrystal in a Magnetic Field at 4.2°K," *Soviet Physics JETP* 8 (1959): 383–386; Alekseevskii and Gaidukov, "Measurement of the Electrical Resistance of Metals in a Magnetic Field as a Method of Investigating the Fermi Surface," *Soviety Physics JETP* 9 (1959): 311–313; Alekseevskii and Gaidukov, "The Anisotropy of Magnetoresistance and the Topology of Fermi Surfaces of Metals," *Soviet Physics JETP* 10 (1960): 481–484; Iu. P. Gaidukov, "Topology of the Fermi Surface for Gold," *Soviet Physics JETP* 10 (1960): 913–920.

230. Kaganov (n. 208). Indeed, it was Borovik's experiments that had stimulated the interest of the Kharkov PTI's theorists.

231. I. M. Lifshitz and V. G. Peshanskii, "Galvanomagnetic Characteristics of Metals with Open Fermi Surfaces. I," *Soviet Physics JETP* 10 (1959): 875–883.

232. I. M. Lifshitz and V. G. Peshanskii, "Galvanomagnetic Characteristics of Metals with Open Fermi Surfaces. II," *Soviet Physics JETP* 11 (1960): 137–141.

233. M. G. Priestley, "Magnetoresistance of Copper, Silver and Gold," *Phil. Mag.* 5 (1960): 111–114.

234. H. E. Bommel, "Ultrasonic Attenuation in Superconducting and Normal-Conducting Tin at Low Temperatures," *Phys. Rev.* 100 (1955): 758–759; W. P. Mason, "Ultrasonic Attenuation Due to Lattice-Electron Interaction in Normal Conducting Metals," *Phys. Rev.* 97 (1955): 557–58; R. W. Morse, "Ultrasonic Attenuation in Metals by Electron Relaxation," *Phys. Rev.,* 97 (1955): 1716–1717. For a review, see Morse, "The Fermi Surfaces of the Noble Metals by Ultrasonics," in Harrison and Webb (n. 192), 214–223. In the Soviet Union, important work was done at the Institute of Radiophysics in the Ukraine by A. A. Galkin and A. P. Korolyuk, "Anisotropy of the Absorpotion of Ultrasound in Metals in a Magnetic Field,"*Soviet Physics JETP* 9 (1958): 925–927; Galkin and Korolyuk, "Oscillations in the Absorption Co-efficient of Sound in Tin at Low Temperatures," *Soviet Physics JETP* 10 (1960): 219–20; Galkin and Korolyuk, "Absorption of Ultrasound in Zinc at Low Temperatures," *Soviet Physics JETP* 11 (1960): 1218–1222. For a

review of the Soviet contributions, see the valuable survey by D. ter Haar, J. M. Baker, R. Berman, and C. T. Walker, *Solid State and Low Temperature Physics in the USSR: A Survey* (Paris: OECD, 1963), esp. the chapter "Fermi Surfaces."

235. A. B. Pippard, "A Proposal for Determining the Fermi Surface by Magneto-Acoustic Resonance," *Phil. Mag.* 2 (1957): 1147–1148; M. H. Cohen, M. J. Harrison, and W. A. Harrison, "Magnetic-Field Dependence of the Ultrasonic Attenuation in Metals," *Phys. Rev.* 117 (1960): 937–952; V. L. Gurevich, "Ultrasonic Absorption in Metals in a Magnetic Field. I, II," *Soviet Physics JETP* 10 (1960): 51–58, 1190–1197.

236. M. Azbel and E. A. Kaner, "Theory of Cyclotron Resonance," *Soviet Physics JETP* 5 (1957): 730–744; Azbel and Kaner, "Cyclotron Resonance in Metals," *Journal of the Physics and Chemistry of Solids* 6 (1958): 113–35.

237. R. G. Chambers, "Cyclotron Resonance under Anomalous Skin Effect Conditions," *Phil. Mag.* 1 (1956): 459–465; M. Ya. Azbel, "The Skin Effect in a Magnetic Field," *Doklady Akademiia Nauk* 100 (1955): 437–440; one should consult also the treatment in his earlier paper, "Films in a Longitudinal Magnetic Field," *Doklady Adademiia Nauk* 99 (1954): 573–588, which lays the basis for the 1955 treatment.

238. Kaganov (n. 208).

239. Azbel interview (n. 226).

240. Azbel and Kaner (1957) (n. 236).

241. E. Fawcett, "Cyclotron Resonance in Tin and Copper," *Phys. Rev.* 103 (1956): 1582–1583; A. F. Kip, D. N. Langenberg, B. Rosenblum, and G. Wagoner, "Cyclotron Resonance in Tin," *Phys. Rev.* 108 (1957): 494–495; D. N. Langenberg and T. W. Moore, "Cyclotron Resonance in Aluminum," *Physical Review Letters* 3 (1959): 137–138; E. Fawcett, "Cyclotron Resonance in Aluminum," *Physical Review Letters* 3 (1959): 139–141; J. K. Galt et al. subsequently examined the effect in zinc and bismuth at Bell Labs: "Cyclotron Resonance Effects in Zinc," *Physical Review Letters* 2 (1959): 292–294, and "Cyclotron Absorption in Metallic Bismuth and Its Alloys," *Phys. Rev.* 114 (1959): 1396–1411; M. Ya. Azbel and E. A. Kaner, "A New Resonance Effect in Metals at High Frequencies," *Soviet Physics JETP* 12 (1961): 283–291. Important theoretical discussions were subsequently given by, among others, V. Heine, "Theory of Cyclotron Resonance in Metals," *Phys. Rev.* 107 (1957): 431–437; R. G. Chambers, "Line-Shapes in Azbel-Kaner Cyclotron Resonance," *Proceedings of the Physical Society of London* 86 (1965): 305–308. Other important theoretical discussions were given by D. C. Mattis and G. F. Dresselhaus, "Anomalous Skin Effect in a Magnetic Field," *Phys. Rev.* 111 (1958): 1616–1620; J. C. Phillips, "Cyclotron Resonance: Theory," in Harrison and Webb (n. 192), 154–158.

242. For a review of Soviet work, see Azbel and Kaner (n. 241).

243. F. G. Moliner, "On the Fermi Surface of Copper," *Phil. Mag.* 3 (1958): 207–208.

244. Gold (n. 221).

245. Pippard interview (n. 209).

246. The proceedings were published in Harrison and Webb (n. 192).

247. J. Phillips, "Quantum Structure of Solids," *Surface Science* 37 (1973): 2–23, p. 2.

4 | *Point Defects and Ionic Crystals: Color Centers as the Key to Imperfections*

JÜRGEN TEICHMANN AND KRZYSZTOF SZYMBORSKI

Point defects are localized positions of atomic dimensions in a crystal in which the
perfection of the crystal lattice is interrupted—say, because an atom is out of place.
One especially interesting class of these imperfections came to be called "color centers." There are several reasons why ionic crystal point defects in general, and alkali
halide color centers in particular, occupy an important place in the history of solid-
state physics. To some degree their significance is a reflection of a larger issue—the
role played by crystal imperfections in determining the properties of real materials.
Several of these properties, some of fundamental technological importance (such
as the electronic conductivity of semiconductors and the mechanical properties of

metals), are indeed controlled by minute additives of foreign atoms or by irregularities in crystalline structure, such as grain boundaries and dislocations. Of all crystal imperfections, point defects are the simplest.

A vacancy where an atom would normally be, a foreign atom replacing the usual type in a lattice position, and an interstitial atom are all examples of point defects. Despite small concentrations, in a largely undisturbed lattice these local defects can have a strong effect on many macroscopic crystal phenomena, such as optical properties, electrical conduction, and diffusion. Like free gas atoms or like ions in diluted solutions, to a first approximation the point defects do not influence one another. They proved to be relatively easy to examine experimentally and to handle theoretically. Thus the systematic study of real crystals began with point defects.

No exact starting date can be set for this work, since many nineteenth-century observations proved relevant to subsequent studies of imperfect crystals. Before the physics of imperfect crystals emerged as an established research specialty toward the end of the 1930s, it had developed along three main lines. For simplicity, we will call these the thermodynamic-chemical, the empirical-physical, and—the last line of development, building on the second—the quantum physics approaches. The first approach grew out of an interest in ionic conductivity and diffusion in ionic crystals. Stimulated by the hypothesis of interstitial ions suggested by Ioffé in 1916, this line of work led to the development about a decade later by Frenkel, Jost, Wagner, and Schottky of a general thermodynamic treatment of point defects. The underlying concept was the realization that, even in a very pure real crystal, a certain number of intrinsic structural defects are always present for thermodynamic reasons. Further, four types of simple defects were identified as a priori possible in ionic crystals (these were combinations of vacancies and/or interstitial ions). Similar processes played an important role in research on the technologically important semiconductor Cu_2O.

Meanwhile, optical and electrical studies of insulating crystals were begun around 1920 by several research centers, most notably by Robert Wichard Pohl's school in Göttingen. Although the Göttingen school did not restrict itself to alkali halides, these proved the most suitable model materials for the study of many properties: dispersion, absorption, luminescence, ionic and electronic conductivity, diffusion, and thermoelectric effects. Unlike metals and semiconductors, pure, perfect alkali halide crystals do not conduct electricity and are transparent within a wide spectral range. However, their electrical and optical properties are extremely sensitive to imperfections; therefore, well-developed, precise measurement techniques can be used for the study of their point defects. From a purely pragmatic point of view, alkali halide crystals were easily available and inexpensive. Artificial crystals could be grown with relative ease and with a controlled impurity content. Alkali halide color centers, a class of imperfections essentially consisting of current carriers (e.g., electrons) trapped at simple structural defects, were the first lattice imperfections to be studied in great detail.

The third major line of development, which merged with the thermodynamic and empirical approaches to make a specialty of color center physics at the end of the 1930s, was a series of attempts to apply quantum mechanics to imperfect crystals and especially to the results of the Pohl school. The contribution of Nevill Mott's Bristol school was particularly significant in this area.

A historical process is rarely a sequence of logically interrelated events. The actual development of point defect and color center physics was a more complex process than our sketchy model of three-pronged advance suggests. An emerging research specialty usually does not have clear-cut boundaries as it first takes shape, and, of course, before it begins to take shape, the boundaries are even less definite. A specialty's development has many ramifications that can be distinguished from the mainstream only by hindsight. Schools other than Pohl's were active before the Second World War, and Pohl and his co-workers studied a wide range of materials, including diamond and silver halides. In our study, we begin by singling out the Göttingen research on alkali halides and their intrinsic defects, especially color centers, because this work would prove to have crucial consequences. The history of other intrinsic defects, such as vacancies, interstitial atoms, and aggregate centers, especially the structural explanation of these defects, is considered later, in connection with the subsequent development of color center research.

This chapter has two parts, each by one of the authors, dealing with a different period and focusing on different geographical regions. The first section is concerned with pre–Second World War developments and concentrates on the contributions made by the Göttingen school; the second describes postwar events and is written from a predominantly American perspective. Although some bias in both sections is probably unavoidable, the authors believe that this approach is justified by history and that these were indeed the respective loci of the most significant progress during the two periods.

4.1   To 1940: The Predominance of Germany and Experiment

Prehistory: Isolated Research on Insulators to 1923 (Rock Salt Coloration, Luminescence, Electrical Conduction)

Long before the discovery of the first point defects, investigation of insulator crystals followed several lines of development—three can readily be distinguished in the nineteenth century—that are important for our study, although at first they seemed unrelated. Each line, with a surprising internal coherence, gathered its own material independently of the others and sought its own interpretations.[1]

The *first* important path from classical crystal physics to the subsequent physics of point defects began with the examination of discoloration in certain crystals. Mineralogists were especially familiar with these phenomena. As early as 1830, it was known that clear fluorspar could be discolored by electric discharges close to the surface (i.e., by ultraviolet light).[2] Also, sodium chloride and potassium chloride, in the form of rock salt or sylvite, respectively, were sometimes blue instead of colorless and transparent. A comparable coloration could be produced artificially—for example, electrolytically or chemically—by melting transparent crystals together with a particular metal. This was investigated around 1860. The chemical explanation appeared obvious: the coloration was caused by an excess of metal whereby the original compounds were transformed into "lower chlorine comounds."[3] Around 1900, research on these phenomena intensified.

It is characteristic of this period of rapid economic development in Germany that a combination of industrial interests and a new invention quickly led to new scientific results. For example, colloid research was of practical interest in connection with the coloration of porcelain and glass.[4] This research was furthered when an ultramicroscope was invented in 1903 by the Zeiss scientist Henry F. W. Siedentopf (who by 1907 was the head of the firm's microscope department) and the chemist Richard A. Zsigmondy. Otherwise imperceptible small particles, down to about 1 nm in size, could now be discerned by their diffraction patterns in dark-field illumination. Siedentopf examined the discoloration of rock salt with the ultramicroscope and compared it with the coloration of ruby glass *(Goldrubinglas)* and other materials.[5]

First he divided all previous rock salt research into two classes: additive coloration by alkali metals (here he described an experimental arrangement for discoloration with alkali vapor) and subtractive coloration "by ionization."[6] Then he demonstrated that the blue tint of rock salt was not produced by lower chlorine compounds, but was due to the precipitation of colloidal sodium particles. Rock salt contained pale yellow hues as well. Since the ultramicroscope revealed no diffraction patterns in this case, Siedentopf assumed that particles of atomic dimensions were the cause. Almost 20 years later, the physics of real crystals would receive a crucial stimulus from this yellow coloration.

The temperature dependence of coloration was already well known. This applied to the bleaching out of color as well.[7] A first connection with photoelectric properties was also known: negatively charged colored salts lost their charge faster in light than in darkness.[8] Siedentopf also saw that his colored colloidal particles were concentrated along "crystal cracks" and, from the many particles in his samples, concluded that "an immense number of vacant edges and corners must exist in the interior of the rock salt crystal," down to the "smallest observable elements of space."[9] (After 1945, this process was called decoration along dislocations.) Siedentopf's crystal cracks would soon play an important role in the development of the first theories of crystal defects.

The physics of luminescence represented a *second* and independent area of research. (*Luminescence* is a modern word, including both phosphorescence, which involves an afterglow, and fluorescence, which does not.) The mysterious glowing of illuminated materials drove an increasing number of researchers to extensive examination of the phenomena.[10] The significance of small amounts of impurities for these physical properties was recognized in 1886: solid luminescent materials glowed because of the small amounts of heavy metals they contained.[11] Moreover, in 1891 a connection was found between the external photoelectric effect and luminescence—that the brighter a phosphor glows, the stronger is its photoelectric effect.[12]

The broad interest in these phenomena got nuclear physics off to an early start, although due to a false conclusion. The discovery of x rays in 1895 was initially seen as a continuation of the luminescent spectrum of the x-ray tube's glass; therefore, Antoine Henri Becquerel, whose father, Edmond Becquerel, was one of the most famous luminescence researchers of the nineteenth century, searched for x rays in luminescent materials, including a potassium uranyl sulfate crystal, and thereby discovered the radioactivity of uranium in 1896. The emergence of nuclear physics was thus directly indebted to crystal physics. The great sensation caused by the newly discovered radiation, especially from radium, led to the founding of the

Vienna Institute for Radium Research. Under the direction of Karl Przibram, the institute, in 1916, produced crystal coloration and luminescence by radiation (using radioactivity and electrons) and began intensive investigations of these phenomena.[13]

The various classical physical and physical-chemical theories of the luminescence mechanism around 1890 were rendered obsolete in 1900 by the discovery of the connection with the photoelectric effect and by the simultaneous development of the first atomic models based on cathode-ray research.[14] In 1901 the Dutch chemist Louis E. O. Visser was the first to suggest that electrons were released when stimulated by light. According to his theory, the foreign ions in the crystal are split up into electrons. Their subsequent vibrations produce phosphorescence.[15] Visser's theory was overshadowed by comprehensive investigations and hypotheses of Philipp E. A. Lenard and his co-workers; in particular, the later color center concepts in Göttingen were directly influenced by this work. Through his research on cathode rays, Lenard became an originator of the new atomic model of negative electrons embedded in a positive space charge,[16] and from 1904 on he applied this concept to the connections between luminescence and the photoelectric effect.

Among other results, Lenard and his co-workers showed that stimulating luminescent materials with cathode rays caused electron emission as secondary radiation.[17] This was closely coupled with the simultaneous production of luminescence.[18] In 1904 they had formulated a photoelectric theory of luminescence, in which the decisive impurities together with the surrounding atoms of the base material would form microcrystalline regions, called emission centers. Light stimulation would cause electrons to emerge from foreign atoms, and their return would cause "phosphorescence."[19] This process, which Lenard certainly understood as taking place inside the "center," is comparable with the later concept of electron stimulation. These theoretical concepts predated Bohr's atomic model and remained very different from it in regard to the mechanism of energy transfer.

Along with the important and detailed empirical material that Lenard and his co-workers gathered, these concepts ensured that luminescence acquired the character of a model for many solid-state theories after Bohr's ideas were accepted. For example, this was the case for Karl Przibram in Vienna and for Robert Wichard Pohl, first in Berlin and later in Göttingen, as described here. Many no longer doubted that the stimulation of luminescence and the photoelectric effect were causally connected, even though the experiments had shown only parallels between the two.[20]

A *third* important root for the subsequent development of point defect physics was the discovery of ionic and electronic conductivity in solid insulators. (This sounds paradoxical, but there was still a great quantitative difference between these substances and ordinary electrical conductors.) Michael Faraday had found that ionic crystals (e.g., silver sulfide) conduct electricity. The conduction exhibited properties that are today ascribed to semiconductors, such as negative temperature coefficients.[21] A number of nineteenth-century researchers arrived at similar results. In 1908, Karl Bädeker diffused iodine vapor through copper iodide and measured the "minimal concentrations" in the crystal, concentrations that showed a remarkable lowering of resistance as a function of the iodine content. With the help of the Hall effect, he was even able to show that the currents were carried by electrons.[22]

This would later be recognized as the discovery of the influence of "doping" on electronic conductivity. Bädeker's research on copper iodide was subsequently seen as a direct forerunner of Pohl's experiments on the additive coloration of alkali halides, but Bädeker's results were scarcely noticed at first. It was ionic conduction that attracted the interest of researchers.

Before 1914, the chemist Carl Tubandt had begun to measure conductivities and transport numbers that demonstrated pure ionic conduction—for example, with very pure silver and copper halides. These measurements became well known and determined the course of all further research in ionic conduction. In 1920 he confirmed the great influence of iodine-vapor impurities on copper iodide, but only as a disturbance in the pure ionic conduction. In contrast to Bädeker's 1908 approach, Tubandt had no specific interest in this impurity effect.[23] From this point on, the significance of the smallest amounts of impurities was increasingly recognized, both by those involved in luminescence research and by those outside it. Beginning in 1920, Georg von Hevesy at the Vienna Institute for Radium Research examined diffusion and ionic conduction in ionic crystals (with the help of radioactive isotopes he measured diffusion constants down to 10^{-17} cm^2 sec^{-1}).

How could such transport processes be possible in a fixed crystal lattice? As an explanation, in 1922 Hevesy published a hypothesis of "loosened [*aufgelockerten*] positions" in the crystal lattice. The loosening of the lattice due to thermal agitation or, alternatively, to absorption of light and other "disturbances" should make possible an exchange of ion positions, giving, for example, electrical conduction.[24]

Piezoelectricity, discovered in 1880 in quartz,[25] became important for subsequent developments through the work of Abram Fedorovich Ioffé, working under Röntgen in Munich. The great reputation that Röntgen had in Russia as well as in western Europe had drawn the Russian engineer Ioffé, in 1902, to study physics in Munich. With the aid of piezoelectricity, Ioffé examined the "elastic after-effect" in quartz, especially the fact that the quartz plates continued to deform linearly despite constant mechanical stress. In 1904 he found that single quartz crystals lost their piezoelectric charge after irradiation by radium, x rays, or ultraviolet radiation; that is, the irradiation increased the conductivity. Diamond and rock salt experienced a similar increase in conductivity under the influence of radiation. After some hesitation, Röntgen accepted Ioffé's discovery as true and buried himself in increasingly intensive investigations; he finally published the results in 1921 in an excessively careful and detailed article, almost 200 pages long, on rock salt. This was the immediate inspiration for the Pohl school's intensive attention to alkali halides. Ioffé later said that for 10 years he had discussed with Röntgen how careful this publication would have to be, and then the First World War came between them and delayed everything yet again.[26]

Independently of Röntgen, Ioffé carried out further experiments in Leningrad, the results of which he presented at international conferences and during visits throughout the 1920s—for example, to America in 1927. Beginning in 1916, before Hevesy's work, Ioffé developed the first theory of ionic conduction in solid bodies based on his own experiments; he assumed an interstitial "evaporation" of ions to explain these strange transport phenomena. Ioffé's theory gained great influence through the work of Jacov Ilyich Frenkel in 1926. Following this line, the problem of ionic conduction in solid bodies would lead directly to the first concept of point

Robert Wichard Pohl (1884–1976) at his seventieth birthday. (DM. Courtesy of AIP Niels Bohr Library)

defects that was correct at the atomic level: interstitial atoms. Within this concept "vacancies" also had to be introduced, as we will see. The physical research in Leningrad on "dielectrics" was strongly supported by the interest of the young Soviet Union in electrical engineering—for example, in insulator problems.[27]

The Göttingen School of Experimental Solid-State Physics: From the External Photoelectric Effect to F′-centers, 1909–1940

At the end of the First World War, a research program devoted almost exclusively to experimental crystal physics was pursued at the First Physical Institute at the University of Göttingen under the direction of Robert Wichard Pohl.[28] This program, the first of its kind, became more and more comprehensive as the years passed, while being virtually ignored by the rest of the world. The Pohl institute can be described as the first great school of solid-state physics.

From X Rays to the External Photoelectric Effect. As a student, Pohl had been deeply impressed by the "world sensation" caused by x rays in 1895. He kept this enthusiasm for important and striking phenomena throughout his life. By his own admission, one of Pohl's guiding principles came from his first university teacher, Georg H. Quincke in Heidelberg. Quincke, famous for his optical and acoustical research, among other things, is supposed to have said, "Theories come and go; facts remain." On the other hand, Pohl had listened with enthusiasm in Berlin to Max Planck's lectures on theoretical mechanics,[29] but he never abandoned his reserve toward theoretical physics. Only in 1933, when the Nazis attacked theoretical physics in particular, did Pohl stand on the side of theory, in an article commemorating the centenary of Carl Friedrich Gauss and Wilhelm Weber's telegraph:

Experimental and theoretical physics cannot be separated from each other and evaluated individually. One cannot esteem music while rejecting the notes because they are not understandable to everyone. All physical knowledge, the most important basis of technological advance, can be furthered only through the application of two methods: through experiment and through its mathematical evaluation.[30]

In the end, it was not Heidelberg but Berlin that left the deepest impression on Pohl, along with his friend and subsequent Göttingen colleague James Franck. In Berlin, Pohl became acquainted with Paul Drude, Heinrich Rubens, Emil Warburg, and Walther Nernst and their scientific work. Here Pohl had to referee Lenard's luminescence work in the scientific colloquium and had to master Lenard's difficult style, which consisted mainly of footnotes. During the holidays Pohl was home in Hamburg, conducting experiments in the search for x-ray diffraction; he began publishing in this field in 1908. Thus he had not forgotten the theme that sparked his youthful enthusiasm in 1895. However, Pohl's first three publications in 1905, before his dissertation, dealt with the emission of light in gas discharges. This corresponded to the "atomic fever" of the time. He worked intensely through Joseph John Thomson's book *Conduction of Electricity Through Gases,* and also knew the work of Johannes Stark and Lenard on emission of light in gases.[31] The physics of gas discharges was in fact the midwife for Pohl's solid-state physics as well as for Lenard's luminescence physics.

Why did Pohl move from ionization of gases and x rays to a new theme, the external photoelectric effect, around 1909? Gudden reported later that Pohl wanted to examine the polarization of x rays by means of secondary electrons (this polarization was demonstrated by Barkla in 1908). There were other possibilities. Analogies could be assumed between the ionization of a gas and the liberation of electrons in the surface layers of solids.[32] Moreover, the photoelectric effect was of great current technological interest for photoelectric cells.

Together with his colleague Peter Pringsheim, Pohl began research, at first on optically perfect mirrors of platinum and copper, and later on mirrors of solid alkali metals. To manufacture these mirrors, Pohl and Pringsheim mastered the excellent new method of cathode sputtering.[33] During their investigations, they closely examined the various influences of the state of polarization and the wavelength of light on the electron emission of these mirrors. Their attention was drawn to certain wavelength-dependent sharp maxima in the electron production, the geometric shape of which looked like resonance curves. As subsequently demonstrated in Göttingen, they were in fact making a photoelectric measurement of absorption spectra for thin layers.[34] From experiments with various enriched alloys, Pohl and Pringsheim concluded that these maxima would have to be "connected with the molecular binding of the alkali atoms" and would not be characteristic properties of the atoms themselves. The widths of the resonance bands were explained by the perturbation effects of neighboring atoms.[35]

Thus they were adopting and extending Lenard's concept of microcrystalline "centers"—the source of the electrons in his theory connecting the external photoelectric effect with luminescence. For Lenard as well as for Pohl and Pringsheim, the photoelectric effect was "the primary cause of luminescence." But the latter pair extended the molecular interpretation to cases independent of luminescence.[36]

Their interest in further pursuing the phenomena in solids, with analogies to the physics of gas discharges, became more and more evident.

Pohl and Pringsheim also developed new experimental techniques during their investigations—for example, the evaporation of reflecting metal surfaces in a vacuum with a pressure of 10^{-3} to 10^{-4} mm.[37] Their measurements of quantum yield (electrons emitted per light quantum absorbed) produced a result that, in 1920, would be decisive for the shifting of Pohl's research to the interior of solids. Rather than the expected 100% "efficiency," which would have definitively demonstrated the phenomenon as a quantum effect, they obtained only 2 to 3%. They were also disturbed by "a series of little known or completely unknown peculiarities of the metal surface." Even the 2 to 3% efficiency was obtained only by using electrical discharges to change the metal surface (obtained, for example, by evaporation) into a colloidal variant. In their interpretation, many electrons, which would otherwise be captured in the vast contiguous metal area, could now jump out of very small metal spheres.[38]

The Internal Photoelectric Effect and the Problem of Single Crystals. In 1918, Pohl came to Göttingen. The war had ended his work in Berlin. Like many others, he had been employed in war research, and was assigned to help develop the new radio technology. Afterward, he caricatured the military's lack of basic scientific knowledge with the example of a military contract given to an aeronautical physicist, demanding that a gas be invented that was lighter than hydrogen and not inflammable.[39] In Göttingen, Pohl began investigating the internal photoelectric effect, with the first publication coming in March 1920.

The reason that he gave much later for his change to the interior of solids was that a high vacuum could not be obtained in impoverished Göttingen. However, the written sources show that the unsolved physics problem of the small electron surface emission must have been at least partly responsible. It was impossible to demonstrate the "Einstein quantum equivalence law"—that is, 100% quantum yield—with the external photoelectric effect. This quantum yield had been found by Lenard in 1914 for luminescence. But finely divided metal could be examined inside a crystal—that is, in solid solution![40] It was certainly also important that Pohl had to take on a graduate student who worked on mineralogical topics. Pohl was very fond of this student, who soon became a close collaborator, working with Pohl on an equal footing and supplying many ideas. Their individual contributions cannot be distinguished, as Pohl himself admitted. This student was Bernhard Gudden, who spread the gospel of the Göttingen school by taking up in 1925 the position of full professor in Erlangen. His reviews of electrical conduction in nonmetalic solids became standard literature in the field.[41]

In their first publication, Gudden and Pohl obtained a considerable increase in the dielectric constant by using powdered zinc sulfide containing 0.01% copper.[42] The increase depended on the wavelength of the incident light. They still did not completely believe that free electrons were being produced by light; instead, they suggested the "possibility" that the effect was caused "by the finely distributed Cu." However, almost immediately they also observed, by means of sufficiently high external fields, free electrons even in substances without metal additives.[43] After examining various larger sulfide crystals instead of powdered ones, they chose diamond as a new test material. According to their published account, they decided to use diamond because of its simple structure and to demonstrate that sulfur was not responsible for the photoelectric conduction of the sulfide crystals. They introduced Lenard's term *centers* during these investigations: the slow changes in the conductivity of diamond under illumination could be caused only by "transpositions in finely distributed volume elements . . . which we want to call centers for short."[44] Here, as with Lenard, "centers" were not atomic structures but microcrystalline regions.

The role of the electron was now of great interest from the standpoint of the atomic physics of the time. Pohl and Gudden constantly referred to Lenard's model of the connection between the photoelectric effect and luminescence, and they also referred to Einstein's light quanta. They used Lenard's model in a more general sense for every photoelectric process, even for processes not connected with luminescence. Although less frequently than Lenard, they also referred to the increasingly influential leader of the new atomic physics of this period, Niels Bohr; they hoped "to gradually find out something about the position of Bohr's electron orbits in the crystalline structure." In retrospect, one could say that they wanted to apply Bohr's atomic physics to solids.[45]

The investigation of electrons in crystals with the aid of their optical and electrical effects became Pohl and Gudden's main theme of research. They published nine articles in 1920, and that was not an easy time! As Pohl later remarked, a great discovery, such as finding electrical current in the classic insulator diamond, had to be celebrated with rolls, margarine, and marmalade.[46] Germany had been hit very hard by the First World War, and experimental work got under way slowly; it would scarcely have been possible without help from private foundations.

It was no wonder that in these years applications of science became more interesting than ever. The *Zeitschrift für Technische Physik* was founded in the same year, 1920, while Pohl and other physics professors drew up a proposal for a course of study in *technische Physik* (applied physics).[47] Pohl strongly supported this development, as well as the funding of corresponding institutes, although he personally did not want to do applied physics and, as he often later remarked, refrained from all patent applications. But he understood the importance of physics and physicists for technology and knew that industry could be helpful to science. He sent many of his students to industry, and industry co-sponsored the research of his institute. Of course, there were difficulties of communication between the financially dependent basic researcher and the industrialist who thought along other lines, as indicated by Pohl himself in a nice anecdote. A physicist wanted to apply for support by enthusiastically demonstrating how he could embrace an extensive field of phenomena in the exceedingly simple formula $f(x) + f(y) = 0$; he was asked why, if everything was zero, did it still interest him.[48]

The practical significance of the photoelectric effect in selenium may have given Pohl and Gudden some external stimulus for their research; at any rate, Pohl made frequent reference in his subsequent publications to selenium, which at this time was scientifically very complicated but technologically of great interest. Even before the First World War, and still more afterward, as the first mass-produced electrical photometers were marketed by Gossen in Erlangen, the internal photoelectric effect had a strong appeal for research in industry. All told, Pohl's switch from electronic surface phenomena to the interior of a solid can be seen as an omen of the technological revolution from vacuum tubes to transistors that would take place much later. Pohl, with his feel for technical developments, may even have foreseen this. He often emphasized the close relation between gas discharges and solid-state effects. The complicated photoelectric effect in selenium was comparable to the "discharges in gas, before cathode rays had been recognized as the simple primary process of these exceedingly intricate and colorful phenomena."[49]

By 1922, Pohl and Gudden had already offered a "primary" process for the internal photoelectric effect. However, their explanation was restricted to insulator crystals, which for the first time were explicitly termed a model for the more complicated semiconductor. An internal photoelectric effect, for which the number of electrons released would increase in proportion to the absorbed light intensity, was primary for Pohl and Gudden. This proportionality was measurable as long as the exposure time and radiation density were sufficiently small.

In Pohl's mind, the so-called primary current in insulators corresponded to the induced discharge in rarefied gases. Currents measured under other conditions were called "secondary currents." From a modern point of view, the primary current can be compared with the initial value of the photoelectric current, whereas the secondary current represents a complicated behavior over time. An experimental result—that if only a thin strip of the crystal was illuminated, electrical currents were measured in unexposed parts of the crystals as well—was especially important for the analogy between cathode rays and primary current. Strictly local processes thus seemed to be ruled out.[50] By this time, it was scarcely doubted, both inside and outside Pohl's institute, that these currents were caused by the movement of electrons, although only the "great migration velocity" and the strong influence of "disturbances of all kinds" supported this.[51] The work to distinguish primary and secondary current with respect to the photoelectric effect was certainly notable and influential as "first attempts to obtain a consistent theory of photoconduction in crystals," even if from a modern viewpoint this distinction was only temporarily useful in mastering the complex relationships.[52] Pohl probably realized this. He is supposed to have jestingly defined the secondary current in conversation as everything that was not understood.[53]

The search for the quantum equivalent, proof that the release of electrons through light absorption was a quantum process in solids as well as in gases, became a special problem from 1922 on. To measure the yield, one had to measure the energy absorbed in the interior of the crystal. In the crystals examined at the time, diamond and zinc blende, photoelectric conduction existed only in the long-wave region of the spectrum. That is, it could be measured only in a region of weak absorption, near the edge of the absorption band. Thus thick layers of the natural specimens used would have been needed. But the Göttingen workers had only very small pieces, a few cubic millimeters. This also caused technical problems for the experiment, such as with mounting "water" electrodes. These transparent electrodes were necessary in order to beam the light rays in the direction of the electric field. Because of all these difficulties, the first results, published after long delays, did not appear certain enough for Pohl and Gudden. And their attempts "to grow the required crystal material synthetically . . . for better measurements . . . did not make any headway."[54] An attempt to calculate the considerable influence of crystal impurities had already been made. The "interspersed foreign atoms" were supposed to "act as electron traps or to interrupt a conduction chain."[55] Purer, perhaps synthetic, crystals were urgently needed.

From April 1923 on, the search for purer crystals was mentioned in Pohl and Gudden's publications. In 1922 they had received a large sum of money for this purpose from the Helmholtz Society. However, since research with alkali halide crystals did not begin until 1924, they could not have been concerned with growing these crystals at this time. The first synthetic rock salt specimens, although not produced in their own laboratory, could finally be examined and compared with natural ones by the end of 1924.[56] Rudolf Hilsch, a co-worker of Pohl since 1925 and soon afterward the leading experimentalist of the Göttingen school, was also occupied with this problem. Pohl finally gained the services of the physical chemist Spiro Kyropoulos from the neighboring institute of the chemist Gustav Tammann. Kyropoulos, building on Tammann's work, published his process for growing crystals at the beginning of 1926. The principle was that when an alkali halide crystal suspended in a melt of the same substance is cooled, a single crystal is formed around this seed crystal.[57] The success of this growth process contributed considerably to the subsequent turn in Pohl's physics toward alkali halides.

Despite all problems, by 1924 Pohl was convinced that "photoelectric phenomena can be handled uniformly in all crystals, as far as the mechanism of conduction is concerned."[58] An insulator was an insulator because, without additional treatment of the crystal, there were no electrons to be released within it, not because there were no electron paths.

Alkali Halides as the "Model Substance" of a "Physics of the Solid State."[59]

The most important starting point for the Göttingen and Vienna alkali halide research was Röntgen's excessively long and detailed 1921 publication on rock salt.[60] Certain of Röntgen's results were essential for subsequent developments (e.g., his findings that rock salt discolored by x rays showed photoelectric conductivity 40,000 times larger than before irradiation, and that this conductivity depended strongly on the wavelength of light). There were even sharply "pronounced maxima of the photoelectric effect" in various crystals (for yellow rock salt around 470 nm). The photoelectric "activation" was produced not only by x rays, but also by artificial coloration caused by metal vapor. However, Röntgen mainly used naturally colored crystals and investigated their blue tint. For the spectral position and bandwidth of the maximum photoelectric current, he determined essential differences between naturally colored rock salt (i.e., blue) and rock salt colored by x rays (i.e., yellow). Röntgen attributed this difference to the "differences in the state or behavior of the Na-particles effective in both cases." In the case of yellow rock salt, particles were not recognizable through diffraction in an ultramicroscope. Röntgen clearly considered that the electrical conduction caused by the coloration was "movement of electrons." He believed that only absorbed light produced this effect, but he still found the optical determination of absorption maxima, which might be identical with the photoelectric effect maxima, too difficult.

At the time, few recognized the significance of this careful and complete survey. The first reference by Gudden and Pohl to Röntgen's work came immediately after its publication, but they referred to only the coloration caused by ultramicroscopic (i.e., colloidal) particles. This "strange coloration" was supposed to be the cause of the photoelectric effect.[61] In order to classify these phenomena, which were new to them, they had to introduce two different groups of crystals. The first group included those handled up until that time, for which the light absorption took place in the base material (or in base material with additives—for example, zinc sulfide with "phosphorescence centers"). In the second group, as in rock salt, color was absorbed only by colloidal admixtures, which were produced "additively" (i.e., by metal vapor) or "subtractively" (i.e., by radiation).[62] The first thorough examination of rock salt, finally published in 1924 and clearly drawing on Röntgen's work, spoke of "ultramicroscopic or submicroscopic lattice disturbances . . . these disturbances need not be optically demonstrable."[63] Thus even smaller noncolloidal centers were first taken into consideration in Göttingen three years after Röntgen's publication. Intensive research on alkali halides must have begun during 1924, for the publications date from the beginning of 1925.[64] Perhaps the availability of synthetic crystals was the determining factor? But in any case, alkali halides could be obtained as fairly large natural crystals.[65]

Within two years, Röntgen's methods and results from almost 20 years of research were extended in many directions and improved or distilled down to their essentials. Research was quickly limited to noncolloidal absorption, for which Röntgen had already found much sharper maxima than for the colloidal case. Measured photoelectric data became more reproducible, partly because the method of "primary current" measurement prevented interference from space charges. These data were compared with the results from synthetic crystals. Exact optical measurements were also taken of the absorption curves, and the area of investigation was extended to ultraviolet (with double spectrometers, called double monochromators [Fig. 4.1], to reduce the dispersed light) and infrared: from 186 to 2000 nm, using quartz, fluorspar, and rock salt optics. Further, the quantum yield was determined photoelectrically. This was now done for natural alkali halides as well, which

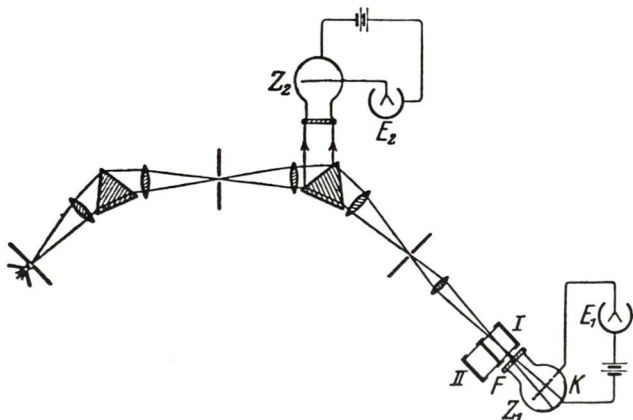

Fig. 4.1   The principle of the double monochromator used for exact absorption spectroscopy in Pohl's institute. Z_1 and Z_2 are vacuum photocells (Z_2 only for controlling the constancy of the light intensity). (From R. Pohl, "Zum optischen Nachweis eines Vitamines," *Die Naturwissenshaften* 15 [1927]: 433)

was much easier than for diamond or zinc blende, since the photoelectric current now arose exactly in the area of the discoloration absorption bands. (This may also have been an important reason for Pohl to shift his research to the alkali halides, and the result—a quantum yield proportional to wavelength—clearly suggested quantum absorption.) The experimental efforts involved can scarcely be seen in the published statements. For example, in order to take infrared measurements, one had to sit as motionless as possible behind the galvanometer for hours at night so that small disturbances would not influence the sensitive instrument.[66]

The idea of "centers" gained more and more significance through new theories and experimental results. Thus light was now absorbed by "excited centers" as well as by the centers considered up until that time, now termed "unexcited." (The latter were what would subsequently be called *F*-centers, after the German *Farbe* (color); the "excited" ones would be called *F'*-centers.) However, the absorption spectra of both still appeared "identical," except that the spectrum of the excited centers was extended to longer wavelengths.[67] The crystal was excited by irradiating it with a wavelength within the normal unexcited absorption spectrum.

The temperature dependence of the absorption was investigated down to 20 K— the temperature of liquid hydrogen. In order to reach these temperatures, the Berlin Physikalisch-Technische Reichsanstalt was brought into the picture, where the low-temperature physicist Walther Meissner helped out.[68] It was also impossible to produce x rays in Pohl's institute during the first years. Crystals were sent out of Göttingen for irradiation or exposed to x rays in the university women's clinic alongside the patients. Because the discoloration did not remain constant over time, for low-temperature experiments one had to rush with steaming dewar flasks onto the bus to the women's clinic, wait next to the irradiation room until the procedure was finished for the patient and crystal, and return at once to the physics institute for measurements. This was viewed with extreme suspicion by the uncomprehending citizens on the bus.[69] By 1925 this research yielded several results. These included the narrowing of the absorption curve, as measured photoelectrically, with decreasing temperature, from 300 K through 81 K down to 21 K, and the independence of the positions of the maxima from the purity and origin of the crystal, although these maxima shifted with temperature.

The starting point of the Göttingen interpretation of these results was the hypothesis that

> the carriers of the coloration, termed centers for short, are non-microscopic structures with a simple characteristic absorption spectrum. The curves of this spectrum are strongly reminiscent of . . . resonance curves. A connection between these electrically or optically determined absorption spectra and any eigenfrequencies of the alkali metal or halide atoms has not been found so far.[70]

The researchers wanted very much to find this connection. A significant narrowing and shifting of the bands with lowered temperature was sought as a sign of a lessening of the influence of the surrounding crystal. Thus they undertook to extend the rather complicated measurements at the temperature of liquid hydrogen. However, the very pronounced narrowing and shifting obtained at 21 K was an error of measurement, as later experiments showed.[71]

Nevertheless, the following comment remained valid: "After discoloration by x-

Fig. 4.2   An artificial KCl crystal, grown in Pohl's institute, 1936. (From *Angewandte Chemie* 49 [1936]: 69)

rays, all natural and synthetic NaCl crystals of arbitrary degrees of purity always exhibit the same discoloration absorption band positions. This suggests that neutralized Na atoms are centers of this absorption."[72] The hypothesis of atomic centers had already been advanced by others, notably by Karl Przibram in 1923,[73] but the young Göttingen school was the first to provide thorough experimental evidence. They immediately conceived an extensive "alkali halide" research program built around their first results with rock salt.[74]

In 1925 they introduced additive crystal coloration, which was more constant than x-ray coloration, and soon explained the phenomenon as the inward diffusion of metal atoms and their absorption on inner surfaces. However, until 1931 the additive method had a great disadvantage, for the light absorption by larger colloidal particles in the region of longer wavelengths had to be accepted as well. By this time, all alkali halides were available as synthetic single crystals in usable sizes and purities (Fig. 4.2). Desired additives, such as thallium, silver, lead, and copper could be included, especially for investigations of luminescence.

Of course, this increasingly focused line of alkali halide research took place within a broader context. Pohl had interests in technology, as discussed already, but he also searched for contacts in other directions outside his research program. In principle, he was against an "encapsulation of the physics business." In this principle he included theoretical physics, although that was not as visible; for example, theoretical questions were discussed with James Franck and, occasionally, Friedrich Hund.[75] Pohl was further interested in making physics instruction more attractive. As an excellent demonstrator and teacher, he began not only to make his classroom and lecture experiments into a model of physics education (it was called "Pohl's circus" in a mixture of respect and ridicule) but to write highly significant textbooks as well.[76] For example, the first edition of his electricity textbook appeared in 1927, and it already contained references to the problems of crystal physics. Until his death in 1976, 51 editions of his textbooks had appeared, including a mechanics-acoustics-heat textbook, the electricity textbook, and an optics–atomic physics textbook, with each edition thoroughly revised. These textbooks

The experimental lecture room in Pohl's institute in Göttingen, without fixed table and with his famous shadow projection of experimental apparatus. (DM. Courtesy of AIP Niels Bohr Library)

decisively influenced university education in Germany and were translated into several foreign languages.

Through Pohl's efforts to make contact with other institutes related to physics, his institute was able to act as a midwife for a biochemical discovery, vitamin D, for which he was awarded the 1928 Nobel Prize. While skiing in the Black Forest around the end of 1925, Pohl was introduced to the problem of a chemist colleague, Adolf Windaus. Windaus had searched in vain for chemical changes in cholesterol, which when irradiated by ultraviolet was thereby made effective against rickets. Pohl considered it possible that purely physical processes, such as the displacement of electrons, occurred in the liquid, similar to the excited centers of his crystals (the subsequent F'-centers), and he put the spectroscopic experience of his institute at Windaus's disposal. This brought absorption spectroscopy with a detection limit of 1 particle in 10^8 into biochemistry and thereby helped discover an impurity that was 1 part in 6000 of the nonirradiated cholesterol. The absorption bands caused by this impurity disappeared under ultraviolet irradiation. The chemists finally showed this foreign substance to be the provitamin ergosterin.[77] Thus developments in crystal physics became fruitful for a completely different branch of research. Even if the analogies had only a short-term effect, the transfer of methods of measurement was permanent. By this time, double monochromators—copies of Pohl's apparatus—were being manufactured by the Göttingen firm Spindler & Hoyer.

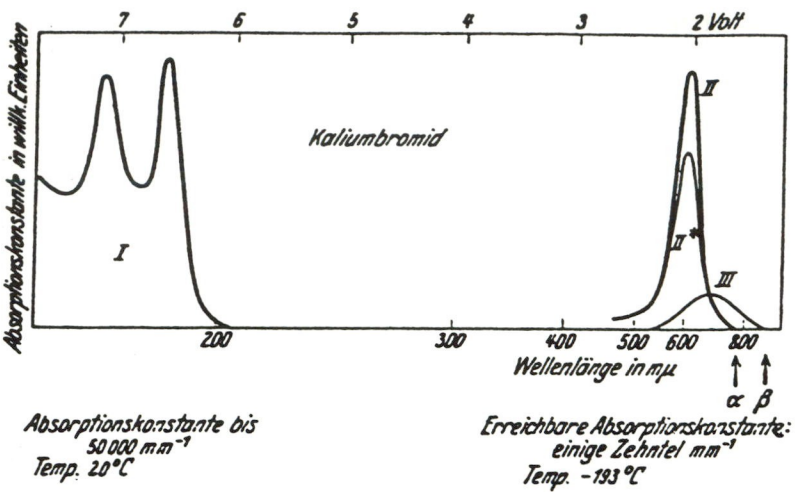

Absorptionskonstante in willk. Einheiten

7   6   5   4   3   2 Volt

Kaliumbromid

I

II

II*

III

200   300   400   500 600 800

Wellenlänge in mμ

α β

Absorptionskonstante bis
50 000 mm⁻¹
Temp. 20°C

Erreichbare Absorptionskonstante:
einige Zehntel mm⁻¹
Temp. –193°C

Fig. 4.3   The absorption spectrum of KBr in UV-region (I) and in visible light (II, III) as measured by Hilsch and Pohl in 1931. II is the absorption of *Farbzentren;* III, the absorption of *Erregungszentren,* the latter called F'-centers. (From R. Hilsch and R. W. Pohl, "Über die Lichtabsorption in einfachen Ionengittern und den elektrischen Nachweis des latenten Bildes," *Zs. Phys.* 68 [1931]: 721)

Pohl's institute had broad interests indeed; there was even a member who did research on spider eyes! Reputable publications were produced there too.

Pohl had many other contacts within the physics community beyond Göttingen. He fostered an especially intensive relationship with Ioffé's institute in Leningrad, the only institute in the world that worked in a similar area and at a comparable level to Göttingen, although the Leningrad workers concentrated strongly on electrical measurements and were not as interested in optics. Pohl was in Leningrad at least twice, letters were exchanged, and several of Ioffé's colleagues were research guests at Pohl's institute.[78] Pohl also took part frequently in the annual German physics convention and at conferences of the regional groups *(Gauvereine),* reporting on his research. However, with the exception of Walter Schottky, almost no one took notice of his work in Germany. On the other hand, Pohl took an interest in developments in atomic physics. Thus he discussed, among other things, possible experimental tests of de Broglie's wave–particle idea before general interest in this was aroused in 1927 by electron diffraction experiments.[79] Pohl grew more and more into the role of *spiritus rector* and organizer of his institute, disseminating the results and caring for the further placement of his people. Rudolf Hilsch took over the leading role in the intensive experimental work after Gudden departed in 1925.

Absorption experiments in the ultraviolet under 250 nm date from 1926. Initially, thin crystal plates only a few tenths of a millimeter thick were still used. It is very likely that, following the example of atomic physics, all wavelength regions were searched for laws. But the reason for the work that stands out in publications was the significance of sylvite and rock salt as optical materials for spectral apparatus as well as for luminescent crystal base material. (The various sulfide phosphors were available only in powder form.)[80] To search for "ultraviolet eigenfrequencies" of the crystal—analogous to H. Rubens's "energy levels" in the infrared—did not appear reasonable at first, for it was believed that superposition led to an effectively continuous absorption band. Thus the discovery in 1927 of alkali halide phosphors (with thallium and lead additives) with surprisingly narrow bands between 190 and 300 nm was all the more astonishing. This also certainly spurred on the ultraviolet investigation of crystals without additives. For Hilsch and Pohl, the phosphor with additives, as a model substance for solid aggregate states that showed sharp "energy levels in solids," served as an experimental tool, similar to the tool provided by low concentrations of gases and vapors in optical investigations of gaseous aggregate states.[81] However, "compared to gases and vapors, we still know exceptionally little about the energy levels of solids, i.e., of crystals." This comment was the introduction to a report of the discovery of the first sharp absorption bands in the ultraviolet range for pure alkali halides, discovered in 1929 in potassium iodide crystals at 219 nm.

Here an extremely thin crystal layer (around 10–100 nm) was needed, because the characteristic absorption was very strong. These layers were soon manufactured by evaporation on a quartz window in a high vacuum. The measurements were also extended to other alkali halide crystals, advancing down to 160 nm—that is, into vacuum spectroscopy.[82] That was also something new for the times, and using a double monochromator in a vacuum was extraordinary. An important result in 1930 was that ultraviolet absorption spectra from different alkali halides, but with the same anion, appeared very similar. Thus various bromides exhibited two sharp bands whose maxima lay approximately at the same wavelength, a tantalizing clue to the underlying mechanism (Fig. 4.3).

There may have been another reason for performing these ultraviolet experiments. By 1921, an attempt had been made outside the Pohl school to explain photochemical processes in silver halides from a purely electronic standpoint, as the transfer of an electron from halide to metal ions. The process in the silver halide layer was a matter of major practical importance for photography, and it was still a great mystery.[83] This was a great incentive for Pohl, perhaps also stimulated by his industrial contacts, to use his alkali halide model to explain the "mythical" photochemical processes, which could conserve an invisible and unmeasurable latent image for an astonishingly long time. In his many publications on this problem, we also find a popular reference to the Andrée polar expedition, which at the close of the nineteenth century had ended disastrously in the ice of the polar sea. The explorers' exposed photographic plates had just been found undamaged, and even at that late date could be developed perfectly to document the tragedy.[84]

In 1929 it was reported from Göttingen that light absorption in the first absorption band of the characteristic alkali halide absorption corresponded to the "elementary photographic process" in silver halides. Ultraviolet light absorption releases an electron to wander from anion to cation, neutralizing both.[85] The analogy between this crystal light absorption and absorption by an individual atom—for example, the role of the lattice energy—was discussed with Franck and Born.

The crystal absorption bands in the visible wavelength region (now also examined in silver halides), or rather the responsible "color centers," were identified in 1930 with the neutral alkali or silver atoms brought about by the wandering electron in the "elementary photographic process." These atoms had to be "somehow bonded with the lattice." The name "color centers" was used here for the first time.[86] These centers were now supposed to cause the latent photographic image.

However, the color center concentration in the very thin (20 μm) photographic layers used commercially was not sufficient to make the latent image directly visible; a single-crystal silver halide plate, 700 times thicker, had to be used. Thus a correspondingly longer exposure was needed. Furthermore, the separation of the energy levels in the ultraviolet range was much better in alkali halides, and there were other differences as well. Yet the analogy between alkali halides and silver salts was very fertile for research on photochemical processes in the 1930s, even if the conclusions drawn from it in the area of atomic physics subsequently turned out to be incorrect. Pohl disseminated this analogy in lectures and publications, especially because of his enthusiasm for such a visible demonstration of "mysterious" processes. His preference also yielded experimental results that subsequently became fruitful in other areas. For example, in contrast to silver salts, no electrical current in the range of the "elementary photographic process" was found for alkali halides. This was later seen as important support for the suggestion that the corresponding absorption stimulated excitons, as will be described at the end of this chapter.[87]

By 1927, the number of "centers" involved in discoloration had been calculated from measurements of anomalous dispersion and of different absorption variables. Most important here were the half-width value and the maximum absorption constant. Alkali halide luminescent materials and rock salt irradiated by x rays yielded values on the order of 10^{15} color centers per cubic centimeter. For this "discoloration of the first kind," as it was called in Göttingen, photochemical interest now pushed the quantitative side once again into the foreground, especially for the determination of the quantum yield. In 1930, Alexander Smakula published a formula, subsequently named after him, for calculating the number of "discoloration centers" (as he still called them). This formula was calculated from classical dispersion theory and experimental values, much as in 1927. As Smakula himself admitted, his work was provoked by the work of Hilsch and Pohl on measurements of the absorption spectrum of the silver halide latent picture in unexcited and excited states.[88] Working from optical measurements, he could now demonstrate with some certainty the quantum equivalence as a limiting case. The Smakula formula was soon simplified for practical application. A calibration came in 1936 with the help of chemical analysis, and a correction of about 20% was shown to be required.[89] With this formula, a basis for a quantitative physics of crystal defects was created. The different absorption measurements could now be compared precisely with one another on an atomic basis.

An important shortcoming of the results of research on color center spectral bands was that no regular relation could be found to "any other kind of variable characterizing the lattice or its component parts." This was in contrast to the first ultraviolet absorption band, whose energy level could apparently be calculated exactly in terms of the transfer of an electron from anion to cation. Given the simplicity of the color center bands, a search for comparable simple regularities seemed the obvious next step.

In 1931, Erich Mollwo, since 1930 a doctoral candidate at the institute and Pohl's most important collaborator aside from Hilsch, succeeded in finding a surprisingly simple relation between the frequency ν of the absorption maximum and the lattice constant d of the respective crystal at room temperature: vd^2 = constant. This relation still bears Mollwo's name. In his search for this relation, he repeated all the previous absorption measurements of the Pohl school on color centers. He used new crystals, however, which he had to grow himself (as was typical for Pohl's institute).

During discussions with colleagues, to explain this relation a dipole effect was suggested, leading to the first clue: the quantum energy of the electron making a transition between two energy bands had to be inversely proportional to a length. How this length was related to the lattice constant and how the atomic model should appear remained unclear. This was eagerly discussed in the institute, without Pohl's participation, and in theoretical publications outside Pohl's group. The discord that followed was yet another reason why Pohl made no exception to his aversion to theoretical speculation. None of the models considered at his institute were published outside.[90] "The institute was his kingdom; he did not govern—he dominated everything."[91]

At about the same time, Carl Wagner and Schottky began research stimulated by semiconductor problems and by the earlier work of Friedrich Wilhelm Jost and Frenkel. Wagner and Schottky discussed the hypothesis of crystal "vacancies" on a thermodynamic-chemical basis. Mollwo studied several theories at this time, including the new band theory, but he soon opposed the latter, assuming instead that localized atomic impurities had to be responsible. In this regard, he had already used the hypothesis of an electron in an anion "vacancy" for the wave-mechanical explanation of the vd^2 relation. Given the censorship Pohl exercised over every sentence, however, the corresponding published remarks had to be formulated very cautiously and were no longer recognizable to outsiders, although these and other hypotheses were discussed within the institute.[92]

A second discovery by Mollwo at this time was of more direct importance for experimental research. Since the frequency of the absorption maximum decreased with rising temperature while the lattice constant had to increase, exact measurements of the temperature dependence of this frequency were of interest. However, the color of the irradiated crystal was unstable at high temperatures. Stability could be achieved by the additive coloration method, heating the crystals in alkali vapor. But in certain temperature regions, a colloidal coagulation occurred. Mollwo was able to prevent this by quenching the crystals—that is, taking them suddenly from high to low temperatures.

Thus for the first time additive coloration became a general tool of color center research.[93] The method also led to Mollwo's third discovery of this year—although it was neither explicitly stated nor clearly enough investigated—a proportionality between the vapor pressure and the number of color centers in the crystal exposed to the vapor.[94] This relationship, finally published in 1937, would then lend more weight to the "vacancies" interpretation of color centers, as described later. The exclusive dependence of color on an exactly determinable variable, here vapor pressure (for crystals largely free of defects), could be achieved only by additive coloration. For x-ray coloration, the researchers had to deal with what was at the time readily ridiculed as "semiphysics," with its usual problems—the strong effect of even minute amounts of defects.

Thus it was Mollwo who was in a position to state conclusively in 1933 that color centers had to be intrinsic atomic defects in the crystal. He said nothing about where the corresponding electrons should be situated.[95] In an experiment on fluorspar, he demonstrated that the crystal density was measurably reduced at high color center concentrations (here more than 10^{22} per cubic centimeter could be reached, in contrast to the maximum concentration in alkali halides of 10^{18} per cubic centimeter). Mollwo commented that "the difference is created by the outward migration of fluoride-ions, which leave as incoming electrons enter."[96] Yet the possibility that color centers might be identical to electrons in vacancies was not explicitly stated, even in subsequent publications. At this time, density measurements were used as an aid for crystal defect determination in crystallography and physical chemistry. Later, for a short time after the war and before the introduction of the electron spin resonance method, these measurements became famous for proving the existence of vacancies in alkali halides containing color centers.

At the same time, there were other developments that made color center physics interesting for the broader aspects of solid-state physics and technology, greatly influencing how the *F*-centers were interpreted in Göttingen. In 1932, Ostap Stasiw found that in a KCl crystal discolored by the additive method at 1000 V and 400°C temperature, the blue color centers wandered to the anode and left the crystal colorless toward the cathode. The rear boundary of the color center distribution was sharply outlined, while the front boundary remained unclear. When Stasiw reversed the polarity, a blue residual band wandered back along the same path to the former cathode.[97] This experiment was varied in subsequent publications and demonstrations. In particular, the "blue cloud" was produced directly by applying an electric field to a pointed cathode in a previously clear crystal and was made to wander to the flat anode. Gudden had already described similar phenomena in 1924 as "growth" from "sodium threads" in rock salt, but at that time he had considered the phenomena only as a problem interfering with the measurement of ionic conduction in the crystal.[98]

This visible electron migration (in the institute electrons were jokingly described as blue) was a triumph for Pohl and his procedure of combining electrical and optical experimental methods as much as possible. Color centers were now interpreted as loose combinations of cations and electrons, which could be separated by heat. By the application of a field, the displacement velocity *(Platzwechselgeschwindigkeit)* of the electrons was observed. Only electrons should diffuse into the lattice, even for additive discoloration. The metal ions of the vapor reacted only on the outer surface, or "possibly also in cracks" (i.e., on inner surfaces);[99] this conclusion was confirmed by subsequent developments. But the decisive fact that halide ions diffused out and left vacancies behind was not mentioned. In fact, at first it remained an open question where the reaction partners for the metal ions on the outer surface came from. The model of the visible electron conduction was, after all, a question of interpretation; the motion of electrons was linked inseparably with that of ions, as subsequent work would show.[100]

The "Stasiw-cloud" certainly made an impression on those who paid attention to science—it was as impressive as solid-state physics could be at that time. The effect was reproduced in many institutes, sometimes only as a demonstration, such as Pohl masterfully conducted everywhere he went. Hypotheses about the phenomena were exchanged. In particular, semiconductor scientists like Schottky and Wagner suddenly became interested in alkali halides. They had almost simultaneously produced increased conductivity with an "additive" excess of a component in semiconductor crystals such as cuprous oxide (a material that unfortunately was not optically accessible). Moreover, with its single, pointed electrode the Stasiw crystal possessed a different forward and backward resistance, analogous to semiconductor rectifiers. Pohl immediately called attention to possible future applications.[101]

In March 1933 a further discovery was made in Pohl's institute—*U-centers*. This discovery would cause the alkali halide research to wait in suspense for three years, until *U*-centers were surprisingly explained as impurity defects. Experimentally, these centers appeared to be a new absorption band in the ultraviolet.[102] The band was accidently discovered during the manufacture of crystal ultraviolet filters for demonstration purposes.[103] This band also appeared to be caused by intrinsic atomic defects, similarly to the *F*-band. In particular, no influence by foreign atoms could be demonstrated. *U*-centers could also be manufactured in the crystals by electrolysis. Further, under certain conditions *U*- and *F*-centers could be converted into each other. Irradiation of the *U*-centers caused them to decompose and produce *F*-centers. Crystals could thus be photochemically "sensitized" by *U*-centers!

A study of the influence of this misinterpretation—the *U*-centers as intrinsic defects—on subsequent developments is a philosophically interesting case of how science may proceed. The interpretation was immediately accepted everywhere, showing how well the results of Pohl's school were trusted and how little comparable experimental crystal research was carried out in other places. As usual, the Pohl school devoted itself more to description than to interpretation, setting out to develop a "band spectrum" *(Termschema)* for the "giant molecule" of the crystal. The *U*-bands appeared to be a new piece of the puzzle that had to be examined in connection with other phenomena. Efforts at this time to publish "images" and "concepts" of the atomic nature of *U*-centers did not appear as important. However, these efforts existed, and they throw light on the research methods and interactions of the institute members.

In 1933, for the first and only time, Hilsch and Pohl published comprehensive atomic hypotheses on the nature of *F*- and *U*-centers.[104] Here not the *F*-centers but the *U*-centers were supposed to be electrons in crystal lattice vacancies! The stimulation had come from outside; the paper cited Wagner (with respect to vacancies), Przibram, and Smekal. This bold stroke *(Husarenstreich),* as Mollwo later called it,[105] sharply contradicting Pohl's aversion to theory, was never discussed in the institute. It appears to have been inspired by Hilsch alone, who may not have been merely an experimentalist in some of his collaboration with Pohl.[106] However, this interpretation of the centers (which occasionally sounded different in subsequent publications) never completely excluded the possibility that the vacancies were occupied by atomic or molecular impurities. In the end, the motto "Theories come and go; facts remain," so characteristic of the success and the boundaries of the Pohl school, remained stronger than this theoretical episode. Although the hypothesis of *U*-centers as intrinsic defects seemed to receive further confirmation, and competent theorists attempted to follow up this idea, the experimental mistrust of theory remained, so other possibilities were held open.

In 1936, Hilsch could telegraph to Pohl (who was in Cairo) that hydrogen had been found to be the cause of the *U*-bands. The further possibility that this hydrogen "replaces" halide ions was soon published. This was first completely accepted at the discussion in Bristol in 1937: hydrogen ions sat in the lattice in place of halide ions. The splitting-off of electrons in U-centers because of ultraviolet light absorption would cause the neutral hydrogen to diffuse away and leave an F-center—that is, a vacancy with an electron.

A further important result at this time was Mollwo's 1935 discovery of bands in the ultraviolet region, caused by heating alkali salts in halide vapor instead of alkali vapor. Frederick Seitz later called these V-bands. Mollwo explained the movement of the corresponding color cloud by the hypothesis of "electron replacement conduction"—that is, defect electrons corresponding to the electronic conduction of the Stasiw cloud. Schottky and Wagner's classification of semiconductors, at that time understood as including the alkali halides, into "excess" and "defect" conductors had made this even more interesting for Mollwo.[107]

However, the V-bands were less significant than research on what happened, when F-centers were transformed into F'-centers, to the product of the quantum yield and the mean range. This mean range *(Schubweg)* was defined as the component of an electron's path in the direction of the field during the life of the electron. Here the dependence on temperature and on color center concentration were of interest. G. Glaser and W. Lehfeldt repeated Gudden and Pohl's 11-year-old "primary current" measurements on additively discolored NaCl (and afterward on further crystals and in different temperature regions) and, among other things, found an astonishing reduction of the aforementioned product for the lowest temperatures; for NaCl it was reduced to about $\frac{1}{1000}$ below $-150°C$. Moreover, the product, at constant temperature, was inversely proportional to the color center concentration.[108] The remarkable drop with temperature now demanded an examination of the influence of temperature on the quantum yield and the *Schubweg* separately.

In 1936, Heinz Pick received the assignment of studying the quantum yield. After some early unsatisfactory results in which it remained constant, Pohl could announce happily on 30 June 1937, "Pick has now clearly shown that the quantum yield drops with falling temperature and that the first explanation suggested in Glaser and Lehfeldt's work [the suggestion of a temperature influence only on the quantum yield] is basically correct."[109] Pick now found—and it was "surprising"— a maximum value of 2 for the quantum yield of the transformation from F-centers to F'-centers. (At the time, the value 2 could only be calculated from the experimental data for KCl, and held down to $-100°C$.)[110] The one thing that was determined in this way was the number of color centers that vanished per light quantum absorbed. Even before the full research on the inverse reaction of F'-centers into F-centers, first published in 1940, all the experimental pieces of the puzzle could be put together to form a clear picture.

Naturally, only one color center per light quantum could be destroyed by light absorption in the color centers. The liberated electron, according to Pohl,

> leaves in its place a positive alkali ion. The electron then diffuses in the electrical field, and at the end of a finite range ω is captured by a positive ion in the neighborhood of another Farbzentrum [color center]. Thus it forms a new alkali atom. Both of the neighboring alkali atoms (which are in a region of negative charge) have

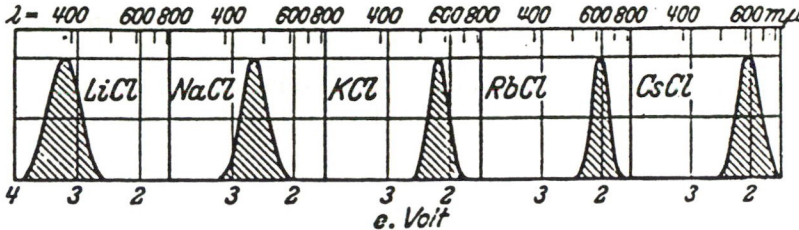

Absorptionsbande der Farbzentren bei 20° C.

Absorption peaks for different halide crystals, as determined by a double monochromator, at Pohl's institute. (From R. W. Pohl, "Zusammenfassender Bericht über Elektronenleitung und photochemische Vorgänge in Alkalihalogenid kristallen, " *Phys. Zs.* 39 [1938]: 36–54)

an absorption spectrum somewhat different from that of two single Farbzentren [color centers], and appear optically in the form of two F'-centers.[111]

This would mean that one light quantum destroys two color centers. The model of a color center as an electron in a halide ion vacancy, coming from the discussions in Bristol in 1937, made the matter still clearer. An electron coming from a halide ion vacancy would be captured by a second vacancy already containing an electron. This vacancy, now with two electrons, was the F'-center. The number of color centers, now effective as electron traps, naturally limited the mean free path of the electrons. This explained the inverse proportionality between the color center concentration and the product of the quantum yield times the *Schubweg*.

A famous review that Pohl gave of the results from his institute at the 1937 Bristol conference, a meeting dealing with electrical conduction in insulators, semiconductors, and metals, from which the previous quotation comes, was the high point (and the end point) of the almost exclusive influence his school held over the beginnings of color center physics. Along with the work described in the foregoing, there were many other results, notably on diffusion of color centers, on "secondary current," and on crystal luminescent materials. There was other important research in the institute as well—for example, leading to a model of a crystal amplifier in 1938 and to the detection of the "Z-bands" in 1939, in connection with previous analogous experiments on electrolytic conduction by Wagner and co-workers.[112] Many of these activities were slowed down by the Second World War. It was other scientific centers, in England and the United States, that continued and extended Pohl's research.

Experiments and Reflections on Intrinsic Defects, 1916–1940

Early Research: Frenkel's Defects and Smekal's Pores. Why did so much time pass before alkali halides became generally recognized outside the Göttingen school as the model substance of "a physics of the solid state"? They had also been used for special problems by other researchers, such as Erwin Madelung, and even before Pohl, Przibram in Vienna during the 1920s had done intensive research on coloration. Were there other comparably important investigations and models of intrinsic defects in ionic crystals? If so, were they used in other connections outside the Göttingen school? In order to answer these questions, we must return to the early history.

The 1916 experiments by Ioffé and his co-workers on the "electric conductivity of pure crystals" were carried out to determine whether impurity ions or ions of the crystal lattice itself were responsible for this conduction.[113] For example, extreme purification of the crystal yielded a constant, called the self-conductivity, independent of the sample used, of 2×10^{-15}/cm Ω (for ammonium alum), where Ω is the resistance. Thus the ions of the crystal itself had to be responsible for conduction, and they were assumed to be "completely removed from equilibrium." By 1923 at the latest, this assumption had developed into a hypothesis of a partial electrolytic dissociation, analogous to pure water, with interstitial movement of ions through the crystal as through a "sieve."[114] For quartz, an analogy to the electrical conduction of gases was discussed. Ioffé calculated the degree of crystal lattice dissociation from measurements of the electrical conductivity of quartz and showed, in connection with previous experiments, that impurity ions could wander through a crystal lattice without destroying it.

In 1925, Frenkel, from Ioffé's Leningrad institute, developed a theory of the defects later named after him, assuming that only a small percentage of all ions were mobile—in opposition to Hevesy's hypothesis that all crystal ions were mobile. Considering especially the mobility of impurity ions, he was driven to conclude that surrounding the lattice atoms there was sufficient free "interstitial space," corresponding to the "sieve" model. Frenkel calculated the probability, dependent on the temperature, of "interstitial evaporation" of the crystal ions—that is, their movement from regular crystal locations to "irregular" equilibrium positions. The places thus vacated could now be occupied by dissociated neighboring ions, a process that he wanted "at best to conceive of as an elementary displacement of the vacant place."[115] By 1923 the existence of interstitial atoms (thought of as impurities) was in any case made evident through crystallographic work on alloys (Chapter 1).

Frenkel's ideas were first noticed by chemists, who had long been occupied with the problem of diffusion and ionic conduction in solid materials. This was especially true for Tubandt's school in Halle, where Wilhelm Jost, in his 1926 dissertation, examined conductivity and diffusion in ionic crystals and tested their theoretical relationship. He also wanted to confirm Tubandt's results that only one type of ion moves during electrolytic conduction. In the theoretical part of his thesis he cited Frenkel, but only with respect to a "similar" application of the concept of degree of dissociation; this concept he still understood in terms of the "fraction of ions of one type . . . capable of changing position." These would be ions with a given minimal energy. Otherwise, Jost thought that "at the moment little can be said definitively about the mechanism of ionic position change."[116]

In this, Jost was vigorously opposed to another theory that the atomic theoretician Adolf Smekal had developed from Hevesy's concepts and experimental results in 1925, and that for almost 10 years was to have a great influence on the ideas of many physicists, especially in the Pohl school. Smekal's theory never became systematically quantitative or well developed. From the beginning, it remained more of a qualitative, suggestive first attempt. Smekal's idea was to treat submicroscopic disturbances in "real crystals" as fundamental for many "structure sensitive" properties of solid bodies; "structure insensitive" properties would be well described by the theory of the "ideal crystal."[117]

In 1912, Smekal had enrolled at the Technical University in Vienna, where he studied mineralogy, physics, chemistry, and astronomy. Studying in Berlin from 1917 to 1919, he was also in contact with the Kaiser Wilhelm Institute for fiber chemistry. In 1923 he became a lecturer *(Honorardozent)* at the newly founded section for applied physics at the Vienna Technical University. By 1928 he was a full professor for theoretical physics in Halle, remaining there until 1940, and despite his title he was able to carry out a great deal of experimental work at his institute. Smekal was interested in basic physical laws, especially for quantum physics, an interest that led to his prediction in 1923 of what was subsequently named the Smekal–Raman effect in atomic spectroscopy. He also had mineralogical and technological interests from the very beginning. His first encounter with solids dated from 1922, when he discussed the theory of fracture, following A. A. Griffith and investigations of Michael Polanyi.

This interest may have led Smekal to his important 1925 hypothesis of "crystal lattice pores," which were supposed to explain self-diffusion, electrolysis, and the strength of solids. The problem of strength was described by Smekal as "perhaps the most interesting and pressing application of the lattice pore concept."[118] At the beginning of the work, he also mentioned Hevesy, with his atomic-theoretical ideas about ionic conduction and self-diffusion in ionic lattices. For Smekal, these considerations were still tied only too closely to the concept of ideal crystals. Along with Ioffé and Jost, he considered such ideas unusable, especially because x-ray experiments indicated to him that the lattice constant would not vary greatly even at high temperatures. Smekal concluded from this that there would not be space for paired interchange reactions of equivalent lattice ions. However, in contrast to Ioffé and Jost, he noted, "It therefore seems extraordinarily obvious to completely exclude the idea of a volume process here and to understand in principle the phenomena mentioned (self-diffusion, etc.) as processes on the interior and outer surfaces."[119]

In this context he assumed "pores" and "pore passages" of variable size (also "impurity enclosures"), which separated ideal crystal regions from one another. He considered the Siedentopf experiment among others to be experimental proof of this; as mentioned, Siedentopf had observed additively introduced sodium and potassium colloids on "crystal cracks" within NaCl and KCl crystals. Smekal also referred to Siedentopf in subsequent publications. Working from these observations, Smekal, by 1925, had estimated the upper limit for the diameter of his pores at around 4×10^{-5} cm. He believed this pore hypothesis would be usable for "almost all properties of the crystalline state which have been left unexplained by the Born lattice theory up to now," and thus would also explain the thermal and electrical conductivity of metals. In 1927 he improved this "pore" concept by replacing it with a picture of "spongy," or "porous," regions *(Lockerstellen)*.

The breadth of the claims of Smekal's hypotheses, even if he did not pursue the last two examples, was the main reason for a great immediate interest among physicists as well as for an immediate stout resistance. In the light of subsequent knowledge, it is easy to see the limitations of his hypothesis. With a single concept, he wanted to explain many real crystal processes, but different processes were caused by very different types of defects. He would probably have been more successful (although possibly less) had he decided to limit himself to mechanical problems (e.g., the theory of fracture), as he did after 1940.

Wagner and Schottky, having their roots in chemistry and thermodynamics, soon joined Jost as opponents of Smekal. They stressed that atomic point defects could offer hypotheses that were quantitatively more useful for dealing with the processes of diffusion and electric conduction. Indeed, in 1926 Jost had expressly rejected Smekal's hypothesis with respect to the explanation of the motion of ions. There is actually a great deal of interaction between different types of defects—for example, between dislocations and point defects—but these interactions were slowly clarified only after 1945.[120]

At first, Smekal's influence on certain scientists was great. This holds, for example, for Przibram's concepts in the Vienna Radium Institute and especially for the work of the young scientist J. H. de Boer at the Philips company (until his contact with Schottky in 1936). Pohl, too, was influenced by Smekal, although he handled these hypotheses, like all theoretical concepts, very cautiously.[121] Pohl was indeed the only one who still used "pore" ideas after the 1937 Bristol conference. Even before Pohl and his school, Przibram had quickly sided with his young colleague Smekal; it was, after all, in Przibram's institute that Hevesy's investigations and theories, which partly inspired Smekal, had begun. Przibram successfully carried out thorough experiments on the characteristic coloration of alkali halide crystals, such as rock salt, by irradiating them with radium. As noted, by 1923 he had mentioned atomic centers, and in 1926 he was the first to use the term *Farbzentren,* combining Smekal's ideas with the Gudden–Pohl results to form the first precise hypothesis on the atomic structure of a color center:

> In agreement with the other observations of the speaker and his collaborators and with the views of A. Smekal on porous regions [*Lockerstellen*] in the crystal structure, the yellow coloration can be ascribed to Na atoms which are subject to the influence, first considered by Gudden and Pohl, of the lattice at disturbance points [*Störungsstellen*], whereas the blue coloration is caused by free Na-particles, which dominate when exposed to light, thanks to their greater stability.[122]

In contrast to the Pohl school, the experiments from Przibram's circle, which was more interested in mineralogy, did not form a systematic program of research into the optical and electrical properties of alkali halides; these experiments were often far less detailed and usually were inexact as well. In general, only broad parts of the spectrum were examined and compared in studies of coloration, and likewise in extensive studies of luminescence, without differentiating narrow lines of wavelength.

The Smekal school also lacked a broad research program. In particular, Pohl found very suspect Smekal's and Przibram's quick and often short-lived interpretations for every experiment. An example of the limited experimental program and the short life of the corresponding explanations is given by the 1936 attempts by Smekal's school to explain the F'-centers in terms of band theory and crystal strain alone.[123]

Theoretical Attempts in Germany and Russia. Other theorists showed interest in alkali halide research at this time, even though there was no experimental work on it within their own institutes. Thus in 1933, H. Fröhlich attempted to explain the Mollwo relation on the basis of wave mechanics by assuming weakly bound electrons and without using the hypothesis of localized atomic defects. In 1936, A. von Hippel published an extensive article in which he tried to explain F-centers and F'-centers by a rearrangement of electrons that produced neutral atoms; the surrounding ideal lattice was supposed to be correspondingly disturbed. He also incidentally mentioned vacancies, in order to explain U-centers and what were later called V-centers.[124] Soviet physicists occupied themselves further with insulator crystals and alkali halides. In Leningrad, research was concentrated mainly on ionic conduction and the photoelectric effect, but also on diffusion, elasticity, and the strength of solids.[125]

A few theoretical concepts of electrical and optical processes in insulator crystals became important for the international research community. These concepts were stimulated primarily by the Pohl school's results, which were often used to test the concepts, or by the Ioffé–Frenkel publications. In a short communication around 1933, Lev Landau outlined his idea of an electron that was trapped by "an extremely distorted part" of the lattice, and hinted that this could explain the NaCl coloration. Landau's concept was the first one close to what was later called the polaron, even though he did not mention any polarization of the lattice by the electron itself. Frenkel soon discussed this as if it had been part of Landau's idea. The ideas of A. von Hippel, mentioned earlier, were also related to Landau's ideas.[126]

Another important concept, the exciton, was developed by Frenkel from 1931 to 1935. This was a freely movable or trapped atom in a state of excitation, in contrast to ionization; it thus corresponded to a bound electron–hole pair. He was in touch with R. Peierls, who in 1932 suggested a localized stimulation at a given atom, which was somewhat different from Frenkel's first 1931 ideas.[127] Frenkel also cited D. Blokhintsev's 1935 attempt to explain the fundamental ultraviolet absorption of alakli halide crystals and the well-known model of the transfer of an electron from a halide ion to a metal ion (here Pohl was referred to as well); Blokhintsev's concept was comparable to the exciton, for it also involved a pair consisting of an electron and a hole.[128] However, in contrast to Peierls, Blokhintsev had suggested that the electrons and holes moved independently, which he hoped could explain the conductivity. Thus he apparently did not fully know Pohl's results, which had shown no photoelectric conduction for the fundamental ultraviolet absorption.

Now Frenkel attempted to explain the consequences of "electrically active" light absorption. He referred to Pohl's "trapping" of charges as part of an explanation of the internal photoeffect. Extending his exciton concept, Frenkel suggested two possibilities for the electron and hole produced by light absorption: either both were free mobile particles, or they formed "a 'trapped' or practically fixed state, the light particle carrying as it were the burden of the displacements of the heavy nuclei." The latter shifting of the neighboring atoms was supposed to fix the free electron or hole at an atom—trapping it, so to speak. With this, Frenkel again opposed Smekal's attempted explanation of crystal defects in terms of "physical inhomogeneities (small cracks and the like), or chemical impurities," but he also opposed the 1933 concept of Landau, who had "confined himself to a rather vague scheme involving a physical inhomogeneity (potential valley) produced by the electron itself."[129] Despite Frenkel's objections, Landau's hypothesis of a fixed polaron would have a strong influence on the early color center concepts that the Mott school would develop, but Frenkel's exciton too would have enduring value, in explaining the fundamental ultraviolet absorption and the "energy migration" during luminescence.[130]

A 1934 publication by Frenkel dealing directly with the problem of color centers took up his 1926 model of interstitial atoms, without any reference to the vacancy discussion that Wagner and Schottky had begun in the meantime. Other Soviet physicists supported Pohl's vacancy hypothesis, in the form that Hilsch and Pohl had used in their erroneous first attempt to explain the U-centers.[131] All this work did have the value of pushing forward again the idea of intrinsic atomic imperfections, at a time when most theoretical physicists were reluctant to abandon the concept of an ideal lattice.

Schottky's Discussion of Cu_2O and Alkali Halides. From 1932 on, Pohl was well informed on the concept of vacancies through an ongoing discussion with Schottky and occasionally with Wagner. In 1927, Schottky, never an enthusiastic university teacher, had left his Rostock professorship in theoretical physics to return to the Siemens firm in Berlin, where he was captivated by the puzzle of the Cu_2O rectifier effect. He soon became the most important mediator between technology and physics in the area of point defects, enthusiastically combining the standpoints of an "electrical engineer" and an "atomic theoretician." Since he was an exceptionally careful manager of his thoughts and statements, preserving every small note, we have found much more historical material on Schottky's interaction with the Pohl school underneath the surface of the publications and interviews than we have found for other researchers (of course, this can also distort the historical picture). Schottky's meticulous care and reserve were his strength and his problem; his immediate reactions in conversation were not as impressive as his subsequent thinking, deep but also often complex and hard to understand.[132]

Schottky met Wagner for the first time at the end of the 1920s during a discussion on thermodynamic problems at one of Fritz Haber's famous colloquia, where many Berlin physicists and physical chemists met. Schottky pressed Wagner at once to become co-author of his book on thermodynamic problems, and they became close friends, soon cooperating on Cu_2O problems.[133] Around 1930, Jost was just settling a bitter feud with Smekal, who opposed thermodynamic treatment in favor of "interior surfaces,"[134] but the further development of Jost's ideas was no longer a point of conflict for Wagner and Schottky. Their large general work of 1930 on the "theory of ordered compound phases" now systematically examined different types of lattice defects: interstitial atoms, vacancies, and substitution of one atomic component of the compound crystal by the other.[135] In particular, they noted that these defects could be caused by deviations in the stoichiometric composition of a crystal. Yet even with an ideal stoichiometric composition, there would always be some defects present, depending on the temperature. The dissociation of pure water was referred to as an analogy. The significance of these concepts for "solid salts" was only mentioned as an aside.

This 1930 publication followed important experimental work on the conductivity of Cu_2O, undertaken to clarify its barrier layer effect. In developing his theses about the great physical significance of defects in this crystal, Schottky was evidently much influenced by Gudden's ideas. In 1932, Wagner and his co-worker Heinz Dünwald found that raising the external oxygen pressure around Cu_2O crystals caused the conductivity to rise. Transport measurements showed that copper ions were in motion, although these were only a small fraction of the charge carriers in comparison with the hole current. It was already clear to them that increased conductivity involved an increased number of defects. They therefore concluded that instead of excess oxygen ions, the lattice in an oxygen-rich environment had to have excess copper ion vacancies. These arose from the migration from the interior of four copper ions per external oxygen molecule; the copper recombined with oxygen on the outer surface in an extended regular lattice.[136]

Schottky and Wagner were acquainted with the contemporary crystallographic work from 1930 on. Thus Wagner referred to the substitution of FeO by O in wustite (an iron oxide mineral); the substitution idea came from experiments combining x-ray lattice constant determinations with density measurements. For him, this was a case of the existence of vacancies.[137] Conversely, the crystallographers took notice of the work of the physical chemists.[138] In 1936, Wagner himself undertook density measurements of halides (see Mollwo's density measurements on fluorspar already noted), using silver halides because he knew that the normal degree of disorder in alkali halides was too small for such measurements to be fruitful. He did not find the expected 16% density variation for AgBr and took that as proof for the existence of Frenkel's defects.[139] The continuation of these experiments yielded an influential method of producing a cation vacancy increase by the addition of bivalent ions, which among other things allowed in 1937 a determination of the mobility from measurements of conductivity.[140] Meanwhile in 1933, apparently simultaneously with Wagner, the Pohl school adopted surface cultivation of certain outwardly diffusing crystal components (anions recombining with metal vapor atoms), although without using the vacancy hypothesis to explain the Stasiw-cloud.

In 1933, Schottky applied the vacancy hypothesis to the alkali halides.[141] Wagner felt that this case, with simultaneous anion and cation conduction, was still too complicated for simple interpretation. But Schottky probably became aware, through the Stasiw-cloud, that alkali halides could be the simplest model of ionic crystal processes.[142] In 1934 he was the first to clearly state the connection between an electron and an anion vacancy in order to explain color centers, but this was in a letter to Pohl, not in a publication. Schottky supposed the electron was bound to a neighboring cation, opposing the hypothesis of an electron that "made its own bed" (he had already discussed and rejected this before Landau's concept); he also rejected the hypothesis of trapping by "random, chemically undefined lattice defects," and Pohl's hypothesis of excess alkali atoms:

> Because of your latest work, and even partly in complete agreement with your interpretation, I now believe that *every color center,* whether it is visible or in the ultraviolet, whether in a pure alkali halide or in one crystallized with impurities and then discolored, *will always be found in the immediate neighborhood of a halide lattice vacancy.* Put more exactly, I believe that it arises from the neutralizing valence electron which is deposited at one of the alkali or impurity metal ions in the immediate vicinity of the lattice vacancy.[143]

It is unfortunate that Mott was not Schottky's partner in this discussion, for Pohl was apparently not influenced in a lasting way by the interaction with Schottky. The problem of *U*-centers, which Schottky also had to deal with, added complications

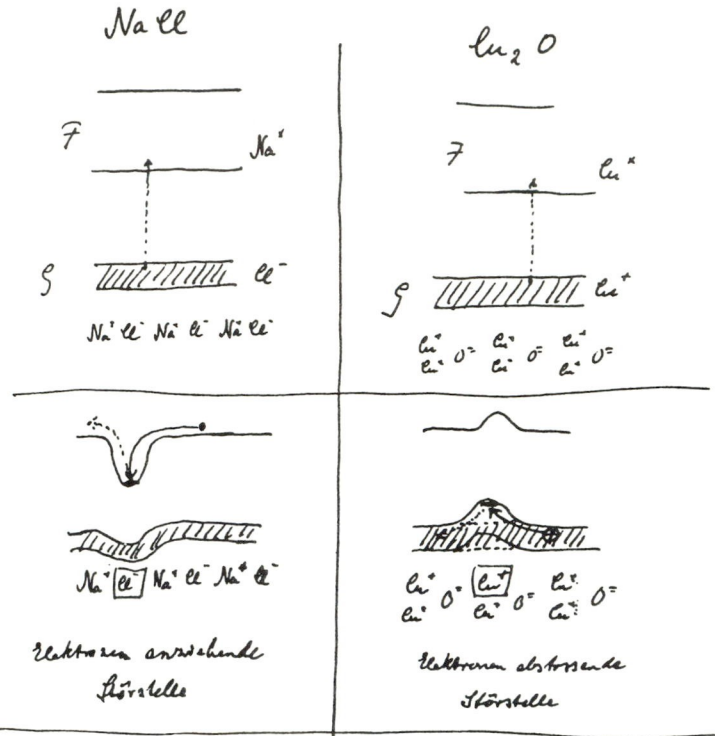

Fig. 4.4　Schottky's explanation of color centers in NaCl and equivalent intrinsic defects in Cu_2O by means of vacancies and of the band model. (Manuscript, 1936. Courtesy of Siemens Museum, Munich)

to the whole matter. Schottky had little taste for the optical side of Pohl's research; for example, he never discussed the Mollwo relation, which from a quantum-mechanical standpoint certainly pointed clearly toward the potential well model (as discussed by Mollwo inside Pohl's institute) rather than toward a structure similar to hydrogen.

An important 1935 publication by Schottky introduced "Schottky disorder," as Jost called it—what later became famous as the two-vacancy "Schottky defect." This work owed its existence to alkali halide research, or, more exactly, to the problem of ionic conduction in alkali halides. Schottky took his reflections on vacancies in these crystals as an opportunity to discuss possible types of disorder once again, this time specifically in ionic crystals of the NaCl type. In this regard, he cited Frenkel and, especially, Jost. What was new here, in comparison with his 1930 work, was the assumption that equal numbers of positive and negative (anion and cation) vacancies actually exist in crystals such as NaCl. Schottky proved this through energy calculations. As an aside, he put forth a hypothesis of anion vacancies in the proximity of cation electrons in order to explain color centers, and also noted a second possibility: the hypothesis of "electron traps" near associated anion and cation vacancies.[144] He also sent this work to Pohl.

Schottky went still further in unpublished discussions with his friend Wagner

(Fig. 4.4). He stimulated Wagner during 1935 to 1937 to examine thermodynamically the idea of electrons in anion vacancies of the alkali halide lattice. But Wagner did not especially favor this hypothesis. Schottky connected thermodynamical and quantum-mechanical considerations in this regard, and tried some graphic representations of a potential well model.[145]

During Pohl's talk at the local meeting of the German Physical Society in the autumn of 1936 at Hanover, where he explained *U*-centers as hydrogen impurities, Schottky entered the discussion, especially concerning the proportionality between alkali vapor pressure and color center concentration. He claimed this was compatible with his concept of halide ion vacancies coupled to alkali atoms, but thought that there would also be the same proportionality in case of what he called "excess" lattice electrons. However, about six months later (around three weeks before the Bristol conference began) he saw the proportionality results as proof that color centers could not be interpreted as "excess" lattice electrons: for such an explanation, the square root of the color center concentration would have to be proportional to the vapor pressure.[146]

The Bristol Conference and Its Consequences. None of Schottky's and Wagner's thoughts were published, although publication was occasionally considered. Schottky was not present in Bristol. Along with Wagner, he was more interested in semiconductors, an area where optical investigations played no role, and in thermodynamically established hypotheses on ion conduction. Pohl, who could have played the middleman between Schottky and Mott, remained silent. He relied on his own method of cautious images *(Bilder)* and experimental intuition, developed from accepted basic assumptions of classical crystal physics and early quantum considerations. That his caution was justified we may see, with historical hindsight, from the model of *F'*-centers held by Schottky and others, a model subsequently refuted by Pick's measurements in 1937.

How much of this division in the development of thought on color centers was due to Pohl and how much to the isolation of Nazi Germany must remain uncertain. Schottky would certainly have taken part had an international conference like the Bristol one been held in Göttingen, which, given Pohl's importance, would probably have happened in a democratic Germany. Schottky might even have helped plan the conference, as he had planned with Gudden and Pohl the section on "electronic and ionic conduction in non-metallic solid bodies" at the 1935 German Physicist's Conference in Stuttgart. This was the first joint public appearance of semiconductor and insulator research, influenced by the significance of the band theory and the defect hypotheses. But there was no international middleman for the Jost, Wagner, and Schottky hypotheses—with one exception.

This exception was the Dutchman J. Hendrik de Boer, employed in the Philips research laboratory in Eindhoven, and known for his thorough work on electronic phenomena in dielectric materials. His book on this topic appeared in English in 1935 and in German in 1937. The German edition was radically altered because of Schottky's review of the English edition and Pohl's publications.[147] In particular, Schottky criticized the basic hypotheses of porous regions and interior surfaces (taken from Smekal), which seemed improbable for crystals in thermodynamic equilibrium. In his 1937 publication, de Boer came closer to Schottky's views, in particular with respect to the temperature dependence of the concentration of color centers, the invariant uniformity of absorption bands with respect to different specimens, and other results of the Pohl school. He rewrote his alkali halide chapter completely, changed the semiconductor chapters a great deal, and reduced the two 1935 Schottky *F*-center hypotheses to a model of a neutral alkali atom with a coupled anion vacancy. This German translation, with a foreword by Schottky, had no international response before the July 1937 Bristol conference (it appeared in June 1937 at the earliest), but de Boer had an immediate influence at the conference itself. Jost also participated at the conference, but could contribute little to the question of electrons, in contrast to ionic conduction and diffusion.

This conference saw the first important international cross-fertilization of all these ideas. R. W. Gurney had already published a work on color centers in the beginning of 1933, obviously stimulated by results from the Pohl school (possibly the Stasiw experiment in particular) and by Wilson's theory of bands.[148] Gurney accepted Smekal's hypothesis of internal disturbances as "submicroscopic cracks and glideplanes," which would produce new "localized levels" for electrons. He never referred to the thermodynamical defect theory, but examined the idea of missing ions qualitatively as an example of a "submicroscopic crack." In 1935, Gurney came to Bristol, where Mott had worked since 1933.[149]

Until 1936, Mott's main interests were metal and alloy problems, but Gurney had worked especially on ions in solutions, and the color center work of the Pohl school had also made Mott aware of insulators and semiconductors. For the Bristol conference (which can be traced to Mott's initiative)[150] Gurney and Mott presented a paper on the atomic structure of color centers, again based on an ideal crystal lattice. Referring to Landau, Frenkel, and von Hippel, they developed the model of an "electron trapped by digging its own potential hole." They did not take into consideration the corresponding positive charge.[151]

The comprehensive talk by Pohl, described earlier, seemed especially important in Bristol. During the discussion, de Boer called the Göttigen research very important for other solid phenomena. For exactly this reason he held that a discussion of color center hypotheses was opportune, in opposition to Pohl's advice to wait for experimental decisions in this area. On the basis of Schottky defects, the discussion group accepted the model of an electron "trapped in the neighborhood of a point where a negative ion is missing." One reason was the proportionality of the number of color centers to the alkali vapor pressure, which spoke against the Landau model. Furthermore, various results described by Pohl, such as the transmutation of *U*-centers into *F*-centers, could be easily explained by this model. Smekal's "cracks" were definitively excluded because the *F*-center spectrum was invariable. The vacancy center as the point of lowest energy and the extension of the wave function of the electron to the six neighboring ions were finally accepted. Moreover, using additional oral information from Pohl, the discussion group developed a model for the *F'*-center as two electrons in the potential well of a vacancy.[152]

All this was given depth by subsequent publications out of Bristol. These provided an improvement of the Jost–Schottky energy calculations on a quantum-mechanical basis. An end product was Mott and Gurney's classic 1940 ionic crystal book, which definitively put the "theoretical clothes" on Pohl's work and showed "that the phenomena observed in alkali halides shed a great deal of light on the more complex behavior of substances of greater technical importance, such as semiconductors, photographic emulsions and luminescent materials."[153]

In fact, technological interests became increasingly important for ionic crystal research. For example, in 1938 Mott contacted the Kodak company, which was interested in work on the photographic latent image. Mott and Gurney's models for the latent image had great influence as the first theoretical impulse for intensive research on this problem.[154] In 1938, Pohl's collaborator Stasiw went to the Zeiss-Ikon firm in Dresden for similar reasons. Also, a Hilsch and Pohl crystal amplifier experiment of 1938 (current amplification by a three-electrode alkali halide crystal) had been stimulated by inquiries from the Allgemeine Elektrieitätsgesellschaft (AEG).[155] And Bauer's earlier discovery of reflection-reducing surface layers was soon applied by Smakula at the Zeiss firm for an antireflective lens coating. Independent of the main stream of Göttingen research during the 1930s was applied development of luminescent lamps, which was also a crucial stimulus for important physical research—for example, on the trapping mechanism.[156] The Second World War got further developments under way, especially applications of semiconductors in radar work.

Yet the book by Mott and Gurney, along with Seitz's book, which appeared the same year, serve as witnesses to the fact that a stage had been completed even before the war began.[157] The development of the most important basis for insulator physics, especially for atomic intrinsic defects, was completed by 1940.

4.2 After 1940: The United States and the Merging of Theory with Experiment

Between Princeton and Urbana

By the beginning of the 1930s, the new generation of American physicists had virtually filled the gap in quality between European and American research. As John C. Slater proudly recalled more than three decades later, even European scholars of the highest reputation were coming over "to learn as much as to instruct."[158] And with the Nazis' rise to power in Germany, an increasing number of them were coming to stay. One of the places that attracted some of the best was Princeton, New Jersey, with its university and its Institute for Advanced Study, created in 1932. Two Hungarian theorists, John von Neumann and Eugene Wigner, joined the faculty in 1930.

In January 1932, Frederick Seitz arrived at Princeton with the intention of working with Edward U. Condon, but eventually, at Condon's suggestion, he ended up as a graduate student of Wigner. Their collaboration resulted in the series of papers on the electronic structure of sodium, discussed in Chapter 3.

Scitz had already been familiar with the physics of solids before he started his work with Wigner. What he recalled as his "first serious encounter with solid state science"[159] were his readings of P. P. Ewald's *Kristalle und Röntgenstrahlen* and Léon Brillouin's *Statistiques quantiques,* which he had selected for his elementary language courses as an undergraduate student. His first paper published with Condon was inspired by Frenkel's book dealing with electromagnetic aspects of crystal optics.[160] Especially interesting, however, for the purpose of this story was that in

1932 Seitz, still a graduate student, at a weekly colloquium of the Physics Department at Princeton gave a talk summarizing the work carried out on F-centers in Pohl's Göttingen laboratory.

By the early 1930s, very little research related to color centers had been done in the United States. The work of J. O. Perrine and of P. L. Bayley[161] at Cornell had not been continued, and a study on colored rock salt carried out at the Chemical Laboratory of the University of Illinois by T. E. Phipps and his collaborators[162] did not attract much attention from physicists. Although interest was growing in some aspects of what was later to be called solid-state physics, the scope of research was limited. Even a few years later, in 1936, when A. L. Hughes wrote his review article "Photoconductivity in Crystals,"[163] the only American contribution he referred to was Foster Nix's work on red mercuric oxide. The bulk of the reviewed material was provided by the Göttingen school.

American scientists were generally more mobile than their European colleagues. People came and went, discussed, exchanged ideas, engaged in collaborations, published joint papers, and traveled across the continent, stopping off wherever something interesting in physics was going on.[164] This open and easygoing atmosphere may have been in part a reflection of the job market. In Europe, young physicists often had to spend years teaching in high school or assisting a professor before they won a full-fledged university post. In the United States, as Sommerfeld once told his students, "every young man who could smoke a pipe became an assistant professor."[165] Besides job openings at academic institutions, a growing number of positions were offered by industry, and interaction between the two was, much more than in Europe, a two-way street.

During the 1930s, Seitz moved from Princeton through the University of Rochester and the General Electric Laboratories to the University of Pennsylvania, and then in 1942 to the Carnegie Institute of Technology, along the way engaging in collaborations with researchers of RCA, DuPont, and Westinghouse. He was getting to know almost everybody active in solid-state research.

In these early years of solid-state research, it was still possible, at least for someone of Seitz's capabilities, to avoid narrow specialization and keep track of the development of the whole field. He started his work in solid-state theory by attacking fundamental problems: his and Wigner's treatment of the constitution of metallic sodium soon became a classic. He then worked extensively on the symmetry properties of crystals, applying group theory; on the interpretation of infrared absorption spectra of crystals; on the quantum theory of valence; and on the quantum theoretical treatment of ionic crystals (Chapter 3). By the late 1930s, he had turned to such problems as luminescence, lattice defects in silver halides, electron diamagnetism, dielectric breakdown, diffusion, dislocations, neutron diffraction, radiation damage, and color centers in alkali halides.

It was during his work at the General Electric Research Laboratory in Schenectady, New York, that Seitz became seriously involved with lattice defects. G.E. was about to start manufacturing luminescent lamps, and a research group at RCA headed by Humboldt Leverenz, with whom Seitz engaged in close collaboration, was studying materials for possible use in cathode-ray tubes. Looking through the literature, Seitz renewed his acquaintance with the work of Pohl's school, this time on luminescent alkali halide crystals activated with thallium. In 1938 and 1939, he published a series of articles providing a detailed and comprehensive theoretical interpretation of their results.[166] His discussions with Mott in Pittsburgh in 1938, along with the fact that the luminescence of these materials depended on the presence of impurities, made him aware of the critically important role of crystalline lattice defects.[167] About the same time, he made an attempt to find out what the prevalent type of lattice defect in silver bromide crystals might be, by interpreting results of measurements by Wagner and Lehfeldt in Germany.[168] He concluded, however, that available density and x-ray lattice constant measurements were not precise enough to decide whether the defects were of Frenkel or Schottky type.

With the coming of the Second World War, basic research in such areas as ionic conductivity and color centers suffered a setback. The research effort was thoroughly reorganized, tending to concentrate on a few problems having direct military application. This included some aspects of the physics of metals and semiconductors, and the problem of radiation damage in the graphite used as a moderator in nuclear reactors. Knowledge about the defects in ionic crystals did not seem to be very relevant to the war effort. There was, however, one field in which it proved useful: the development of "dark-trace" cathode-ray tubes for radar indicators.

The crash program of wartime development of radar had multiple effects on solid-state research. Perhaps the single most spectacular technological breakthrough in microwave radar was the development of the cavity magnetron, a powerful source of microwave energy, by Mark Oliphant's group in Birmingham, England. The ensuing substantial improvement of microwave technology would lay the foundations for rapid postwar progress in the field of magnetic resonance, an experimental technique later used with remarkable success in studying a number of problems related to the solid state. Moreover, vacuum tubes proved useless in the high-frequency range, and had to be replaced by silicon point-contact rectifiers; therefore, much effort was concentrated on the improvement of semiconductor diodes, and important progress was made in the technology of silicon and germanium.

Another area in which the perfection of radar performance depended on progress in a domain of solid-state physics was cathode-ray tubes. Most display tubes used in radar sets had luminescent screens on which the signal appeared as a bright spot on a dark background. However, for some applications, particularly in naval operations, luminescent screens had shortcomings. Because of poor contrast, observations had to be made in a darkened room; moreover, excess light excited luminescence over the whole screen, thus further diminishing the contrast. This was a highly undesirable feature because, as Robert B. Windsor observed in an unpublished dissertation defended at MIT in 1944, "for tactical use the information must be presented on a plotting table in a well-lighted operations room where several officers can view it and make other observations."[169]

A way to eliminate this inconvenience was suggested by a British researcher, A. H. Rosenthal, of the Scophony Laboratories, working in the budding field of commercial television. In 1940 he proposed that the phenomenon of color centers be used to design a cathode-ray tube with a screen covered by a thin layer of polycrystalline alkali halide.[170] When the idea of the "skiatron," as Rosenthal named his new kinescope, came to the attention of the British military, further research on the invention was promptly classified and taken over by the military-research establishments.[171] In 1941 the first British skiatrons, called dark-trace tubes on the other side of the Atlantic, arrived at the MIT Radiation Laboratory, the center of American radar work.

In the United States, production of dark-trace tubes was started at the RCA plant at Harrison, New Jersey, by the General Electric Research Laboratories, and later by the Allen B. DuMont Laboratories. At the same time, further development of the design and technology was assigned to a team of Rad Lab researchers belonging to Group G2 (indicators) of Division 6 (receiver components), supervised by Wayne Nottingham. Nottingham, taking advantage of his rare privilege of holding a joint appointment at the Rad Lab and at the MIT Physics Department, put four of his graduate students to work on dissertations related to dark-trace tubes.[172] Understandably, their research was strongly focused on applied aspects of the problem, and, as one of them would admit somewhat apologetically in his dissertation, "Since the research has been dictated by military needs, a number of problems of physical interest which have no immediate military application have had to be postponed."[173]

As it happened, some of these physical problems proved quite relevant to the operational quality of dark-trace tubes. The signal displayed on the screens was more persistent than in luminescent tubes. Initially, this feature was considered a major merit of dark-trace tubes, but it turned out to be their most serious shortcoming. It was soon found that although the traces left by fast-moving objects could be easily erased by heating the screen and illuminating it with bright light, signals left by fixed objects, displayed repeatedly at the same spot at the screen, were fading away very slowly. As spectroscopic measurements made by James Buck showed, the absorption of the transient traces was essentially identical with the F-band observed by Pohl and his collaborators, but as the darkening of the screen became more intense, the shape of the absorption curve changed and could not be accounted for by either F- or F'-bands.

Further research carried out at the Rad Lab concentrated on practical methods of accelerating the decay of these persistent absorption bands of unknown origin. Meanwhile, Seitz, who had moved to Carnegie Tech in Pittsburgh late in 1942 to chair the Department of Physics, set up a solid-state group to study more fundamental physical aspects of dark-trace tubes under a contract with the Rad Lab. In addition to Robert Maurer and James Koehler, who had worked with Seitz at the University of Pennsylvania and followed him to Pittsburgh, the Carnegie Tech group included two outstanding experimentalists who were refugees from Nazi Germany: Otto Stern and Immanuel Estermann; both had been at Carnegie Tech before Seitz's arrival. Koehler and Maurer were already seasoned solid-state researchers—Koehler had been mainly interested in dislocations, and Maurer had worked on electronic phenomena and defects in crystals in von Hippel's laboratory at MIT from 1940 to 1941[174]—but Stern and Estermann had gained their reputations working on molecular beams.[175] Now, all but Koehler joined Seitz in the dark-trace tube project. With the persistent coloration of dark-trace identified as a crucial problem, Seitz's group concentrated its efforts on attempts to explain the physical mechanism of the phenomenon.[176] Although these efforts did not bring immediate success, their long-term impact on color center physics would be considerable.

In fact, color center absorption bands other than the F- and F'-bands had been observed years earlier by Pohl's school at Göttingen. In 1928, for example, R. Ott-mer registered an additional absorption band lying on the long wavelength side of the *F*-band,[177] and in 1935 Mollwo reported another band on the ultraviolet side of the *F*-band upon introducing a stoichiometric excess of iodine into iodine halide crystals.[178] However, Pohl and collaborators, quite rightly, chose to concentrate their attention on the most basic phenomena and had been remarkably successful in establishing experimental conditions in which *F*-centers appeared in almost pure form and could be studied in depth. They were aware that some spurious absorp-tion bands might be due to impurities and that, in any case, until the nature of the *F*- and *F'*-centers was definitively elucidated, not much could be said about what seemed to be secondary optical features of colored crystals.

As a result of Pohl's low-keyed attitude toward Ottmer's discovery, after a series of additional absorption bands had been reported in 1940 by J. P. Molnar, Seitz took it for granted that Molnar was the first to observe them.[179] Accordingly, Seitz proposed to name the new bands as follows: *M*-band (for Molnar), *V*-bands (for violet), and *R*-bands (for red). This nomenclature was accepted by Rad Lab researchers working on dark-trace tubes, and after the war most researchers adopted it.

While the Rad Lab workers were concentrating on technical aspects of dark-trace tube production while trying to get rid of the persistent coloration of the screens—for instance, by doping calcium chloride with aluminum—Seitz focused on unrav-eling the microscopic structure of the bothersome centers and the mechanism of their formation. He summarized his conclusions in a paper published in 1946 in the *Reviews of Modern Physics*.[180] Meanwhile, Stern and Estermann became inter-ested in an even more basic problem: the nature of the *F*-center itself. The vacancy model worked out at the Bristol conference in 1937 was still considered tentative and lacked unequivocal corroboration. Stern, somewhat skeptical about the model's validity,[181] conceived an ingenious experimental technique that apparently might be used to test it. The underlying idea was that if the vacancy hypothesis was correct, then coloration of a crystal would result in a slight but measurable decrease in its density. This idea was not new. Density measurements had been carried out since the 1930s, as described earlier. But now the change of density could be mea-sured with extreme precision using the flotation method, in which the crystal was immersed in a solution of density exactly matching its own. Working with Ester-mann and graduate student William J. Leivo, Stern was indeed able to show that when crystals had been darkened by x rays to the point of containing about 2×10^{18} *F*-centers per cubic centimeter, their density decreased by about 1 part in 10^4. This was in fairly good agreement with the vacancy theory.

Stern, Estermann, and Leivo began their experiments in September 1943, but the exigencies of war, and other unexpected circumstances, delayed publication of their results. In 1943 Stern received the Nobel Prize, although the public announcement of the award was postponed until the end of the war. Soon thereafter, he decided to retire and left for California. Estermann and Leivo carried on with the measure-ments, but did not send a paper to the *Physical Review* until 1949,[182] long after their results had already been publicized by Seitz in his *Reviews of Modern Physics* paper. Seitz also discussed the question of whether these results were necessarily a con-vincing proof of the validity of the vacancy model.

Seitz's review paper, in which the current state of the art in the field of color centers was thoroughly updated, covering both prewar Göttingen contributions and wartime progress at home, appeared in July 1946.[183] The issue of color centers in ionic crystals was seemingly becoming more complex, while at the same time its important place in solid-state physics was firmly acknowledged. "Almost every field of physics," wrote Seitz in the introduction to his paper,

> possesses a few problems which merit particular attention, both because they occupy a central position and because one has reasonable hope that, as a result of their inherent simplicity, they may eventually be understood in a complete fashion. In the field of solids, the properties of the alkali halides have an enduring interest, since these crystals have continuously yielded to persistent investigation and have gradually provided us with a better and better understanding of some of the most interesting properties of all solids.[184]

Seitz could assume inherent simplicity behind the apparently complicated phenomena of crystal coloration, since the paradigm of their explanation was at hand. In the best-known and simplest case, the F-center, it had for some time been "generally agreed that de Boer's interpretation in terms of electrons bound to vacant halogen-ion sites had overwhelming support."[185] (This was one part of the Bristol discussion already described.) There then followed the Bristol explanation of F'-centers as entities formed by two electrons occupying the same negative ion vacancy. As for the rest of the observed bands, interpretations imposed themselves: an electron or electrons trapped by aggregates of two or more vacancies, and a hole or holes trapped by a positive ion vacancy or aggregate of vacancies.

In this way, color center physics became an inherent and key part of the physics of crystalline lattice imperfections. The subject generated studies of problems such as the formation and diffusion of defects, their energy of formation, crystalline lattice relaxation processes, and the structure of the color centers that give rise to each particular absorption band. Accordingly, a good part of Seitz's paper was devoted to discussion of the lattice imperfections and their treatment in the 1930s by Schottky, Wagner, Koch, Mott, and Littleton. He also presented his proposals concerning models of V-, R-, and M-centers.

Assuming strict symmetry between negative and positive ion vacancies, Seitz suggested that V-bands were due to holes trapped by positive ion vacancies; R-bands might be elementary aggregates of F-centers, consisting of a pair of them, possibly a pair that had lost one electron; an M-center could be an L-shaped triple vacancy, in fact a combination of an F-center with a supposedly very mobile pair of vacancies (a Schottky defect).

Most of Seitz's tentative proposals would not be borne out by careful experimental scrutiny. As new experimental evidence turned up, perhaps the most persistent controversy continued to focus on the issue of the M-center model. It is therefore worthwhile to look more closely at the combination of experimental observations and theoretical premises that prompted Seitz to set forth his particular hypothesis.

Molnar's work and the research carried out by the dark-trace tube project strongly suggested that the M-band was due to some type of aggregate center; it was the first pronounced new absorption band to appear when a crystal containing a high concentration of F-centers was bleached. The process of F-center coagulation occurred, as the Carnegie Tech group found, even at room temperature. Since *F*-centers—negative ion vacancies with trapped electrons—were presumably immobile at such low temperatures, Seitz started looking for a possible mobile intermediary agent that could be involved in the mechanism of aggregation. No such agent was discovered by optical measurements, and its nature could be deduced only from theoretical calculations of the energies of formation and diffusion activation.

At this point, Seitz was helped by one of his former colleagues from the University of Pennsylvania, Hillard Huntington, who had joined the Rad Lab crystal rectifier group. In 1942, Huntington had written his Ph.D. dissertation at Princeton University on the subject of self-diffusion in copper, using theoretical techniques that now seemed useful for the solution of the *M*-center riddle. Seitz and Huntington "put their heads together"[186] and, checking each other's calculations, concluded that the mobile agent was a neutral pair of negative and positive ion vacancies, and that an *M*-center was formed when such a pair associated with an *F*-center. Huntington wrote to Seitz on 1 September 1943, including a sketch of a triple L-shaped aggregate of two negative and one positive ion vacancies, with the comment

> This configuration is important in the motion of a dipole. It indicates that this is really the stable configuration for two vacancies of opposite sign. I *think* this is right, though I was greatly surprised at first. It has taken me about two days to convince myself that it is possible.[187]

It would take some 15 years to prove that this was not the *M*-center.

Notwithstanding the eventual failure of Seitz's models, his 1946 paper had the great merit of pulling the subject of color centers together into a coherent whole. It was widely read by physicists the world over, with the result that many got interested and started to do experiments and develop theories related to color centers. In Robert Maurer's words,

> the influence went far beyond what success he had. What he really did . . . was bring to the surface all of the basic phenomena which were scattered throughout lots of papers and inadequately appreciated. It probably would not have been as influential in this country if it had not been done just after the war when a lot of people were going back to their laboratories asking themselves, "now, what do we do?"[188]

Thus the postwar growth of interest in color centers on the part of American physicists was, in considerable measure, an aftermath of a relatively marginal wartime project dealing with the development and improvement of dark-trace tubes used as radar indicators by the navy.[189] However, the issue of radiation-induced defects in solids also emerged as a major problem in the context of another big wartime project—the production of plutonium for atomic bombs at the Chicago Metallurgical Laboratory.

In the middle of 1942, months before the first self-sustained chain reaction was observed by Fermi and his collaborators, warnings had been voiced that extremely high levels of radiation inside a reactor might lead to changes in the mechanical properties of the reactor materials.[190] The problem gained notoriety (within the restricted circle of people cleared for work on the highly secret Manhattan Project) when it became a serious preoccupation of Wigner and Leo Szilard. In December 1942, Wigner suggested that fast neutrons would cause a potentially dangerous accumulation of lattice defects in the graphite used as a moderator in atomic piles; a few months later, Szilard followed with a warning that the energy stored in interstitial atoms knocked out of their regular sites might be released in a violent and uncontrollable way. These dangers, which came to be called the "Wigner disease" and the "Szilard complication" by their Met Lab colleagues,[191] were given serious consideration. Several physicists tried to estimate theoretically the amount of energy stored in irradiated graphite;[192] however, a special group of physical chemists headed by Milton Burton was organized into a radiation chemistry section to study the problem in depth.

When, in the second quarter of 1943, a large fraction of the theoretical physicists working at the Met Lab were transferred to Los Alamos, Seitz was called from Carnegie Tech to Chicago to continue the theoretical study of radiation damage. Subsequently, the whole Carnegie Tech group switched from dark-trace tubes to radiation-damage research. Meanwhile, separate groups working on radiation damage in graphite were established at Oak Ridge, Tennessee, and in mid-1944 at Hanford, Washington.

This concentrated effort was rewarded in the sense that, "with an ample seasoning of intuition and hope,"[193] the risk associated with the "Wigner disease" was determined to be tolerable. Measurements carried out at Hanford, and verified by Maurer at Chicago, indicated that energy accumulated in irradiated graphite at the rate of 2 cal/g per day. After the rate of release of this energy was established by John Archibald Wheeler, strict recommendations were issued concerning temperature control in the Hanford reactors.[194]

After the war, studies of radiation damage and color centers became interconnected to a certain extent. The initial studies made on graphite were later extended to cover a wide variety of materials. For example, the radiation damage studies on graphite, oxides, and metals initially carried out in the Met Lab were extended to alkali halides, which appeared to offer the best chance for understanding some of the elementary processes of interaction of radiation with solids.[195]

As for theory, the F-center began to be viewed as an elementary system, a "hydrogen atom of solid state,"[196] and attracted the attention of a number of theorists.[197] Seitz, himself a theorist, was naturally interested in the basic character of the F-center, but his review was also addressed to experimentalists, suggesting possibilities for new discoveries. Another factor favoring the growth of interest in the field was that single crystals of alkali halides were cheap and relatively easy to obtain. Finally, for some researchers studies of color centers were just a convenient point of entry into the field of solid-state physics.

After the war, the Carnegie Institute of Technology became one of the leading American centers of ionic crystal research. Among the graduate students working with Maurer were Clifford Klick, who carried out studies of optical properties of color centers in alkali halides[198] and the Hall effect in diamond; Howard Etzel, who completed a dissertation on the concentration and mobility of vacancies in sodium chloride;[199] and Dillon Mapother and Nelson Crooks, who measured self-diffusion of sodium in sodium chloride and sodium bromide, using radioactive sodium tracers supplied by nuclear reactors at the new Oak Ridge and Argonne National Laboratories.[200]

Seitz spent the 1946/1947 academic year at Oak Ridge, where Wigner headed a school on reactor physics, and gave lectures on radiation effects in solids. Meanwhile, he continued his interest in ionic crystals and supervised Ph.D. dissertation work by G. J. Dienes, who tried to develop a more rigorous theory of the diffusion of coupled pairs of vacancies—the hypothetical mobile factor involved in the process of coagulation of F-centers.[201] At the same time, two papers on the self-trapped electron recently published by S. I. Pekar, a Russian physicist from Kiev, came to Seitz's attention.[202] Pekar had a hypothesis of an electron trapped by interaction with an ideal ionic lattice, what he called a polaron; this was built on earlier ideas, especially those of Landau, but Pekar was able to obtain more quantitative results by using an approximation in which the crystal lattice was treated as a continuous medium.[203] In his review, Seitz had discussed Landau's hypothesis of electron self-trapping. Now he took up the problem once again and, jointly with Jordan Markham (a participant in the Oak Ridge Training Program at the time), wrote a paper[204] that concluded that the self-trapping, if possible at all, would be observed only at very low temperatures.[205]

In 1949, Wheeler Loomis, who after the end of the war had returned to the University of Illinois at Urbana to head its Department of Physics, offered Seitz a position at Urbana with a chance to establish a relatively large solid-state group. Loomis and Seitz had gotten to know each other during the war when Seitz worked on contracts with the Rad Lab, where Loomis had played an important role as Lee DuBridge's right-hand man for administrative and personnel problems. Although solid-state physics at Carnegie Tech was flourishing, around 1948 it became clear that, as Maurer put it, "the industrial people around the Pittsburgh area didn't have the concept of developing Carnegie into a competitor of MIT. The future looked limited," and there was little chance of expanding the solid-state physics program at the institute.[206] Loomis's offer looked promising enough to make Seitz accept it with little hesitation. Another Rad Lab veteran and an old friend of Seitz's, Louis J. Ridenour, had been appointed the dean of the Graduate College at the University of Illinois, and from this strategic position he was able to support Seitz's efforts to develop a strong solid-state physics center in Urbana.[207] It was to become one of the strongest of its kind in any university.

The University of Illinois had not benefited from the wartime boom of military research. On the contrary, it had suffered a drain of personnel and equipment. Now, thanks to Loomis and Ridenour, it was quickly catching up. Seven "founders" of solid-state physics and low-temperature physics at Illinois came to the department between 1949 and 1951.[208] Seitz brought Maurer and Mapother with him from Carnegie Tech, and in the same year David Lazarus and Charles P. Slichter, "with the ink scarcely dry on their Ph.D. diplomas," joined the department as instructors.[209] For Slichter, who had earned his Ph.D. at Harvard while working on electron spin resonance with Edward Purcell, a chance to do research with the solid-state group at Urbana provided a strong incentive for coming to Urbana. Seitz was already a symbolic figure for his generation—"Mister Solid State Physics."[210] In 1950, James Koehler came, and in 1951 John Bardeen was persuaded to move from the Bell Telephone Laboratories to Illinois.

Openness and diversity were the characteristic features of solid-state research at

Urbana. Many lines of research were pursued, and a variety of techniques and approaches were applied. Each faculty member worked more or less independently with his group of students, but overlapping interests led to strong and fruitful interactions. Mapother established a low-temperature physics laboratory; Bardeen, together with research associate Leon Cooper and Ph.D. candidate Robert Schrieffer, worked on the theory of superconductivity; Lazarus mastered the technique of using radioactive tracers for measurements of diffusion in metals; and Koehler continued his research on dislocations. There were many others, some visiting for only a year or two, some staying longer.

Color centers were chiefly Maurer's domain. He and his co-workers studied optical properties of KCl, using thermoluminescent techniques to observe "glow curves" produced by the release of trapped carriers when a crystal irradiated at low temperature was heated up.[211] In 1953 to 1955 a Swiss physicist, Werner Känzig, joined Maurer's group as a visiting research assistant professor. Collaborating with Slichter and one of his graduate students, Theodore G. Castner, Känzig measured electron spin resonance spectra of KCl irradiated with x rays at liquid-helium temperature, discovering a new type of color center that had not been observed by optical methods. The key to his success was a simple, sensitive electron paramagnetic resonance (EPR) spectrometer he built; the EPR technique, of which we will say more later, was becoming a more and more powerful tool for the study of color centers since Clyde A. Hutchison first measured the paramagnetic resonance absorption of F-centers in 1949.[212] Känzig initially associated the well-resolved multiline hyperfine spectrum he discovered with the V_1-center, which, according to Seitz's suggestion,[213] was supposed to be an antimorph of the F-center: a hole trapped by a positive ion vacancy.[214] However, more careful analysis of experimental data subsequently carried out by Castner and Känzig showed that the model of the new center involved no vacancy whatsoever, but could most appropriately be described as a self-trapped hole.[215] This new center came to be known as the V_k-center.

The search for centers formed by trapped holes was one of the main preoccupations of color center physicists in the 1950s, and the center discovered by Känzig through paramagnetic resonance methods was also studied by Charles Delbecq, Philip Yuster, and Bernard Smaller at Argonne. Essentially corroborating Castner and Känzig's conclusions, they identified the new center as a $(Cl_2)^-$ molecule ion and were able to measure its optical absorption spectrum.[216]

As for Seitz, there was hardly a field of solid-state research he was not involved in. His papers on color centers, two review papers in particular, became a standard reference along with the classic Göttingen work. He was also involved in an extremely extended network of personal, informal scientific contacts. His name is to be found in acknowledgments for "helpful discussions" in a good part of all papers dealing with color centers published in and outside of Urbana. In return, in his second review paper on color centers, published in 1954, he referred to more than 20 unpublished and to-be-published papers by people from all over the United States, England, Japan, and Germany.[217]

During the late 1950s, Seitz and Maurer became gradually less active in pursuing their personal scientific works. "Seitz began to spend more and more time in Washington, around the country," Maurer recalled. "The size of the solid state physics group got larger and larger, we had more and more contracts. I found myself spending more and more time just doing unofficial administrative work."[218] The University of Illinois continued to make outstanding contributions to the physics of crystalline lattice defects and color centers, but a new generation took over. From 1949 to 1960, 24 researchers at the University of Illinois (many of them foreign visitors from Italy, Germany, France, and Japan) published more than 30 papers dealing with color centers. On the American scene, Urbana remained one of the major centers of point defect research. No single research group, however, had a truly dominant position comparable to that of Pohl's Göttingen school in prewar German solid-state research. A look at some of the other American groups will show how widespread and diverse the institutional base for solid-state physics became in the decade after the war.

The Wider American Scene

In the late 1940s, American physics was undergoing a process of transition. Peace brought deep changes in the direction and organization of research as thousands of scientists, demobilized from military projects, moved to new places and new subjects. Normal academic life was restored, and the universities expanded to teach masses of students who until recently had been soldiers. Color center physics was a new field of research that aroused growing interest, partly related to the increasing importance of radiation-damage studies for the rapidly developing nuclear science and industry. Having accomplished its mission, the Chicago Metallurgical Laboratory, where these studies were pioneered, ceased to exist, giving birth to new scientific institutions, among them the Argonne National Laboratory, created in 1946. In ensuing years, the laboratory occupied new permanent quarters at Argonne, Illinois, on the outskirts of Chicago. Oliver C. Simpson, a Carnegie Tech physicist who had been on leave to the Met Lab since 1944, was invited to stay on at the new laboratory and contribute to its work in solid-state physics.[219] He appreciated the need to supplement Met Lab and early Argonne studies of radiation effects on solids by basic work on alkali halides, a class of materials whose physical and chemical properties could be reasonably well specified.

Further encouragement to go ahead with the extension of radiation-effects studies came through the unexpected availability of Peter Pringsheim, then on a temporary appointment at the University of Chicago. Pringsheim was an outstanding German experimentalist known principally for his work on photoelectric effect and luminescence. He had done research on the photoelectric effect in collaboration with Pohl in Berlin before the First World War; spent the war period in Australia, where he was interned as an enemy alien by the British; and later worked at Berlin. When Hitler came to power, Pringsheim, having enough Jewish blood to be subject to the Nuremberg Laws, left for Belgium, his wife's native country. Then, only a step ahead of the Germans, he passed across France to Spain and from there to the United States.[220] His friend James Franck brought him to Chicago.

When Pringsheim came to Argonne, he gradually changed his research program from luminescence to color center studies. From 1949 on, with Pringsheim setting the pace, a color center group focused its efforts on alkali halides. Before he came to Argonne, Pringsheim had been at least somewhat familiar with Pohl's work, so his transition to the new field was not difficult. Further, Yuster, one of his first collaborators at Argonne, recalled that

> he had a supply of reprints that exceeded anything we had seen before, especially of the German work. He had those reprints organized, and not in the way we keep our reprints, thrown in a pile here and there. He had boxes with labels, because he wrote books and he knew how to take care of documents.[221]

In addition to Pringsheim, the core of the Argonne color center group consisted of Yuster and Delbecq, whom we mentioned earlier in connection with their work on the V_k-center. They had both been educated as chemists, which would prove advantageous for their studies of alkali halides. Yuster obtained a Ph.D. in 1949 with S. I. Weissman at Washington University in St. Louis. His thesis dealt with luminescence, and he initially intended to continue research in the field when he joined Argonne Laboratory after an interview by Pringsheim. Delbecq, who was hired by Simpson about the same time as Yuster, had graduated in physical chemistry from the University of Illinois.

In spite of the difference of age and status, straightforward cooperative relationships were soon established among the three scientists.[222] Pringsheim did some experiments himself, but mostly he supervised Delbecq and Yuster's work. For example, he was the first to notice an absorption that indicated the presence of unknown bands in the ultraviolet, α- and β-bands, but he handed the problem over to Yuster and Delbecq, whose detailed studies led to the Argonne group's first major discovery. Pringsheim had also designed a Dewar flash that made it possible to irradiate a crystal and measure its absorption spectrum in low temperature without ever warming up the crystal. He intended to use this apparatus in the search for the trapped-holes centers (*V*-centers), which since the publication of Seitz's review paper were a hot issue in color center physics.

In 1955, Pringsheim retired and returned to Europe, leaving at Argonne a rather small but remarkably active color center group. Yuster and Delbecq occasionally collaborated with other physicists, but their own scientific careers became inseparable and continued in this way for 25 years until Yuster's retirement in 1980. In 1956, Argonne hosted the first international color center symposium, an event to which we will return shortly.

In contrast to Argonne, the Naval Research Laboratory (NRL) in Washington, D.C., was the site of by far the largest single group of physicists whose interests included color centers as a major subject. For some, time there had been two independent groups at the NRL, headed respectively by Elias Burstein and James Schulman, doing research in this area. Both groups developed from the Crystal Branch, part of the Sound Division of NRL, formed during the war with the purpose of growing crystals for use in sonar underwater detecting devices. After the war, the branch was transformed into a research laboratory consisting of three sections headed by three young physicists: Burstein, Sam Zerfuss, and Schulman.[223]

Burstein joined the Crystal Branch in 1945 before the end of the war, and worked on sonar during his first year. Like Yuster and Delbecq, his educational background was in chemistry, but he gradually turned into a solid-state physicist. After graduating from the University of Kansas in 1941, he shifted to physics while doing defense-related research with Hans Mueller at MIT. At NRL, Burstein became interested in photoelastic and infrared properties of dielectric crystals, diamond and magnesium oxide among others. At that time, around 1948, he was, as he would later say, "looking for a place to dig in deeply";[224] having read Seitz's review paper, he decided that color centers might be the right place to dig. Since his previous experience was with infrared techniques, Burstein in a sense approached the problem from the long-wavelength side of the absorption spectrum. In 1949 he published with J. J. Oberly a paper on the infrared color center bands in the alkali halides, in which they reported the discovery in KCl of a new absorption band, the *N*-band, lying in the far red, and tentatively identified it as due to a higher aggregate of *F*-centers.[225] Burstein published a few more papers on color centers, but about 1953 the NRL Physics Section, which he headed, shifted still farther into the infrared region and turned to the study of the photoconductivity of doped semiconductors, particularly narrow-gap indium antimonide.

Schulman entered the field a little later but stayed in it much longer. A former collaborator of von Hippel, under whom he carried out his Ph.D. dissertation in 1941, Schulman came to NRL in 1946 to head the luminescence group. The beginnings of the group were modest. When Clifford Klick joined it in December 1949, their optical equipment consisted of a student glass spectroscope "which they put a motor on for a drive and made a sort of homemade dial with marks on it, so that one could begin to get a little bit of an automatic reading of emission spectra. . . ."[226] The luminescence section slowly grew into a branch of its own, cooperating at times with Burstein's group but also competing with it "for prominence, strength, and money."[227]

One of the great advantages of doing research at NRL was the fact that one of the first Collins helium liquefiers was installed in the laboratory. Schulman's group was quick to move into low-temperature studies of emission processes. With luminescence as the major subject, Schulman investigated the properties of various heavy-metal centers in alkali halides, developing an experimental setup for both emission and absorption measurements at low temperatures.

Klick came to NRL in 1948 with a fresh Ph.D. obtained under Maurer at Carnegie Tech. In 1950 he started to work on the luminescence of color centers,[228] and also searched for *F*-center emission.[229] Among those he worked with was George Russell, a graduate student of Maurer at the University of Illinois, who was doing research for his dissertation while on assignment with the navy.[230] In addition to Klick, Howard Etzel and Dale Compton, who joined Schulman's group in 1950 and 1955, respectively, had also carried out dissertation work under Maurer at Carnegie Tech and Urbana.

The microscopic structure of color centers was another fundamental issue that the NRL researchers worked on. John Lambe and Dale Compton studied the symmetry of the *F*-aggregate centers,[231] and Herbert Rabin contributed to a number of research projects aimed at solving some basic problems dealing with the structure and the mechanism of formation of various color centers.[232]

Along with basic research, Schulman and his group worked on the application of color centers to the practical purpose of radiation dose measurements. Two different classes of dosimeters were designed. One was based on the change of absorption of metal-activated glasses under the impact of ionizing radiation.[233] The other was based on the photoluminescence of crystals using centers produced by irradiation. After the (already obsolete) dark-trace tubes, these were perhaps the only successful practical applications of color centers developed during the 1950s.

This short account of the color center research carried out at NRL only skims the surface, since by 1960 about 60 papers related to color center research were published by more than 25 authors working at the laboratory.

The color center groups at the University of Illinois, Argonne National Laboratory, and Naval Research Laboratory were among the most prolific. There were, however, many others. By the end of the 1950s almost 30 American laboratories were researching color centers in alkali halides. Large single crystals of alkali halides were easily available, and the experimental techniques used were relatively inexpensive. Solid-state physics in general had reached a stage in which the understanding of the role and properties of crystal imperfections had become more important than the studies of ideal, perfect crystals. Color centers in alkali halides were the model type of lattice defects that could be successfully studied with a wide variety of methods—mechanical, electrical, optical (in both absorption and emission), and electron spin resonance techniques. It was seldom that immediate technological demands inspired so many people to explore so many byways of color centers, but a hope that the results would eventually prove useful somehow, combined with the plain desire to solve puzzles, to come up with notable discoveries, and eventually to understand the entire architecture of imperfect materials.

After the war, the United States found itself, at least temporarily, the world's leading scientific power. While Europe and Japan were slowly recovering from wartime destruction and political upheavals, American science was reaping the fruits of its military triumphs. The specialty of color center physics benefited from the generally favorable situation, but also from the fact that the two review papers published by Seitz presented a state-of-the-art survey of the field and outlined a consistent research program for ongoing studies. The problems he identified included both dynamical aspects of defect formation and evolution and their static structural characteristics. While progress in understanding the dynamics of point defects was slow and largely based on speculation, during the 1950s color center physics focused on unraveling the structure of a variety of types of centers, whose number approached 30 by 1960.[234]

A count of the number of papers related to color centers in alkali halides published between 1946 and 1960 shows that before 1958 approximately half the papers originated in the United States (Table 4.1). The domination by American color center physics began to decline toward the end of the 1950s as research in Japan, western Europe, and the Soviet Union gained momentum. However, the contribution of non-American researchers had already been significant.

The International Color Center Community

During the war, the intensity of color center research at Göttingen University had decreased considerably. Heinz Pick, for example, became involved in applied research, working on infrared detection techniques at Kiel.[235] A Bulgarian physicist, S. Petroff, visiting Göttingen in the early 1940s, studied the thermal and optical bleaching of *F*-centers and observed all the long-wavelength absorption spectra described earlier by Molnar, but the results of his measurements remained unpub-

Table 4.1 Approximate number of published papers on color centers in alkali halide crystals, 1946–1960

Year	Worldwide	United States
1946	1	1
1947	5	1
1948	12	2
1949	22	8
1950	39	15
1951	27	16
1952	32	21
1953	80	42
1954	59	24
1955	42	20
1956	36	18
1957	79	39
1958	66	28
1959	84	23
1960	106	36

lished until 1950.[236] At the end of the war, all research was suspended. When Seitz visited Göttingen in mid-1945 as a civilian member of the Field Intelligence Agency Technical (FIAT)—an organization set up to study wartime research activities in Germany—he found Pohl's institute closed and its equipment deteriorating for lack of maintenance. The building was temporarily off-limits to Germans.[237] In 1948, Pick returned to Göttingen and together with Pohl and Mollwo rebuilt the laboratory and resumed the study of color centers. By 1949, Pohl's institute was working full blast, producing a steady flow of publications.[238]

Two distinct methodological tendencies became apparent in Göttingen work in the early 1950s. Pohl, approaching retirement age and less directly involved in experimental research, was still very much in control of the scientific program of the institute.[239] His scientific style had not changed, and he remained true to his extremely empirical approach, avoiding any reference to band theory and to speculation about structural models of color centers. He favored the extension of experimental studies of color centers to crystals other than the usual ones (i.e., alkali halides, alkali halide mixed crystals, or alkali halides doped with divalent metal ions or alkali hydrides). Others, Pick and Lüty in particular, were more interested in some still unresolved questions concerning color centers: excited states of the *F*-center, fluorescence of the *F*-center, and the structure of *F*-aggregate centers. This tendency was closer to the research program outlined by Seitz.

Göttingen practically ceased to be an active center of color center research when in 1954, after more than 30 years as head of the institute and leader of a scientific school, Pohl retired and was succeeded by Rudolf Hilsch, who was called to Göttingen from Erlangen. Pick accepted a position at Stuttgart, where, with some of his former students who followed him from Göttingen, he established his own very productive school, which made important contributions to color center physics.

In the late 1940s and early 1950s, several groups of Japanese physicists also became interested in the study of color centers. Unlike that of Germany, where the tradition of experimental research in the field had been very strong, the contribution from Japan was mainly theoretical, at least initially. Toshinesuke Muto, considered one of the "father founders of Japanese solid state physics,"[240] became interested in the theory of *F*-centers between 1946 and 1954, and in the early 1950s theoretical studies of color centers were carried out in Tokyo by three independent groups.[241] Experimental studies were started at Kyoto University by Yoichi Uchida, who during the war worked under a military contract on developing methods of growing large single crystals of alkali halides to be used in infrared spectroscopy.[242]

Exchanges and collaborations among groups in different countries were gradually becoming closer, and direct personal contacts played an important role in the emergence of an international color center community. After meeting Pohl while visiting Germany in 1945 with the FIAT mission, Seitz met him again in 1949 when he and Maurer attended a conference in ionic crystals organized by Pohl at Göttingen.

Two years later, Pohl paid a short visit to Urbana. In November 1951 a meeting of the American Physical Society was convened at Chicago during the two days following Thanksgiving, and after a day's break a smaller, more private meeting was organized some 150 miles to the south at Urbana. The University of Illinois Conference on Physics of Ionic Crystals, as it was called, was attended by about 70 physicists and physical chemists, and Pohl, whose presence in America had prompted the meeting, was treated as a guest of honor.[243] The two-day conference was chaired by Seitz and Maurer, but it was Maurer who actually "ran the show."[244]

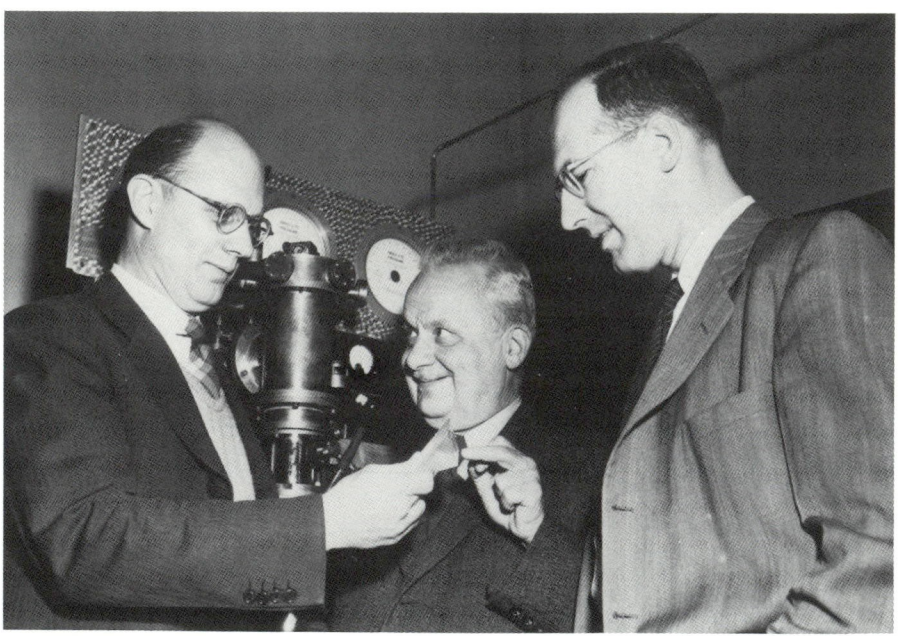

Frederick Seitz, Robert Wichard Pohl, and Robert Joseph Maurer at the Conference on Ionic Crystals, University of Illinois at Urbana, October 1951. (University of Illinois Newsphoto. Courtesy of AIP Niels Bohr Library)

Scientifically it was an important gathering, for several discoveries had just been made and were now reported in public for the first time. The 19 speakers included Jordan Markham, announcing his and Duerig's discovery of the *H*-center; Allen Scott from Oregon State College, discussing experiments on the optical and thermal coagulation of *F*-centers; E. Burstein, J. Krumhansl, and Robert Sproull from Cornell University; and L. Apker from General Electric Laboratories. But perhaps more important than the formal discussions was that six years after the end of a devastating war that had stalled most research in the field and had severed all international contacts, a representative group of investigators involved with ionic crystals could meet face to face and get to know one another.

As David Dexter wrote in his report from the meeting: "With this small group it was possible for most of those present to engage in active discussion between formal papers. Also, the several social gatherings and intermissions encouraged informal conversations that helped make the conference of particular value to the participants."[245] We have no record of these informal conversations, of course, but a quick glance at the list of participants shows that they must have been interesting. On the one hand, American solid-state physics was represented by some of its best workers, with John Slater, John Van Vleck, and John Bardeen present. On the other hand, the conference was an occasion for some former European collaborators to meet after separations of many years. For example, Pohl, who, as usual, gave a well-prepared and beautiful lecture, had a chance to talk to Pringsheim, with whom he had worked in Berlin on the photoelectric effect some 40 years and two wars earlier. There was also a symbolic Japanese presence, for although no one came from Japan, Seitz was notified by correspondence of an ingenious experiment (as yet unpublished) carried out by Masayasu Ueta, a collaborator of Uchida at the University of Tokyo, concerning the symmetry of the *M*-center.[246]

Physicists working in the relatively narrow field of color centers had not been complete strangers to one another even before meeting at Urbana, but this gathering accelerated the process of forming an international community encompassing all those active in the field, perhaps with the exception of Russian researchers, who did not participate in such an international conference until the second half of the 1960s. For the rest, as Klick reminisced,

> you got to know people personally, on a first name basis. We were reading their papers, they were reading our papers; you sent preprints to many of them and you got that sort of thing in return; you tried to make trips abroad whenever there was a good enough excuse to do it, or they came to visit you. Whenever we did a paper we knocked off 25 copies, and once it was cleared for publication we'd send those to people we thought were interested.[247]

The importance of these informal, personal contacts for the development of the field is difficult to assess. However, if one looks at the acknowledgments included in papers appearing after 1951, one can find the names of visitors from other laboratories, sometimes from abroad, who took part in "helpful discussions," suggested a problem, or communicated some work done elsewhere. The practice of exchanging samples also became quite common, and some crystals traveled all the way from Germany to the United States.[248] Further, as will become obvious when we proceed to the discussion of technical problems, direct personal contacts helped to concentrate efforts on certain specific issues, to unify the terminology used by different groups, and, in short, to make color center physics an international venture.

It took some years for these contacts to evolve to the point of becoming institutionalized. The first Urbana meeting was not a color center conference in a strict sense, and there was no formal provision for the contacts to be continued on a regular basis. Five years elapsed before an appropriate occasion arose to convene a new international color center symposium. Pringsheim was retiring and would return to Europe, so Yuster and Delbecq decided to call a meeting at Argonne in his honor.[249] With advice from Seitz, they drew up a list of invited speakers and finally admitted 37 papers. The three-day meeting started on 31 October 1956.[250]

This time it was a truly international conference. S. Amelinckx from Ghent gave a paper on the dislocations in alkali halides; Fausto Fumi and M. P. Tosi, theorists from the University of Palermo, Italy, presented a paper on point imperfections; J. H. Simpson, a former student of Mott at Bristol who had since returned to Canada, spoke on his refined calculations of the wave function and thermal dissociation energy of a color center in NaCl. For some foreign visitors the trip was easier to make, since they happened to be in the United States on longer visits. But by a special arrangement, it was also possible to bring some people from farther away; the Military Air Transport Command provided free transportation for one person from Japan (the organizers decided to invite Yoichi Uchida) and one from Germany. The latter opportunity was given to Fritz Lüty, Pick's assistant from Stuttgart. Pick came too, and the two German physicists made a tour of the laboratories in the eastern United States before arriving at Argonne. Among other places, they visited the Naval Research Laboratory, RCA, General Electric Laboratories, and Cornell University, talking mostly to people involved in color center research. For Lüty these encounters had additional significance: by giving talks, he literally earned his return ticket, since the arrangement with MATC worked only one way. This first visit to the United States had an important long-run consequence for Lüty: he liked the country so much that some years later he would move permanently to America to work at the University of Utah.[251]

Following the Argonne symposium, as a natural development, a routine was established whereby color center conferences were held every three years, each time at a different place. These were always rather informal, relatively small gatherings of people who knew one another's work but always found it invigorating to get together and discuss the most recent developments.

The next symposium, held in September 1959, was hosted by Allen B. Scott at the Oregon State College at Corvallis; once again Pohl, now 75 years of age, was the guest of honor. Werner Martienssen, Pohl's last assistant at Göttingen, paid him homage in a short introductory address. Fifty-four years had passed since Pohl published his first scientific paper, and almost 40 since he began his work on color center physics, "which had without interruption bound them together as though by contract."[252]

Color center physics had grown to such an extent by the time of the Corvallis symposium that 60 papers were scheduled to be read by speakers representing nine countries, including England and Israel. Not all the authors could actually be present. Russians did not come, although A. A. Shatalov, Th. B. Lushchik, and N. P. Kalabuchov had been expected to talk; such last-minute cancellations by Soviet authorities were common at postwar scientific meetings. A paper by A. Bohun and his collaborators from Prague was read by Maurer.

This was the last color center conference in which Pohl actively participated. But the next conference, at Stuttgart in 1962, would be attended by another pioneer of color center physics, Karl Przibram from Vienna. Over many decades, Przibram, with remarkable persistence, had studied naturally colored rock salt, and he now reminded his younger colleagues that some of the questions asked decades earlier had not yet been satisfactorily answered. "A few questions are indicated," reads the summary of his talk,

> which seem to deserve further study. Is there a continuous transition from F-centers by way of higher centers and the vacancies containing colloidal particles to microscopic and macroscopic cavities? In natural rock salt, all of these lattice disturbances, except the F-centers, tend to be located in certain layers of the crystal.[253]

This sounded like a voice from a different epoch, for nobody seemed to be interested in natural alkali halides any more.

Experimental Techniques

The measurements of the optical absorption, the luminescence, and the electrical conductivity and photoconductivity of irradiated or additively colored crystals— all three techniques mastered by Pohl's school—remained the most widely used methods in color center research throughout the 1950s. It was the transparency of most pure ionic crystals in the visible range of the optical spectrum, in contrast to the opacity of metals and semiconductors, that made these crystals such attractive materials for the study of lattice defects. Using newly improved optical instruments as well as ingenuity and perseverance, physicists continued to learn new things about the properties of the F-center and to discover new absorption bands.

Besides absorption measurements, optical methods included thermoluminescence as well as the use of polarized light. Those color centers that lacked cubic symmetry, like Seitz's hypothetical M- and R_1-centers, were expected to show dichroism in polarized light: out of a population of randomly oriented centers, those with their dipoles oriented along the plane of polarization of the incident light would absorb it and consequently be bleached. The crystal's absorption would therefore become different for different directions of polarization. This method, pioneered by Ueta,[254] helped Delbecq, Smaller, and Yuster[255] to identify the V_k-center as a chlorine molecular ion. The method was extensively used in studies of the M-center, which we will discuss later.

The studies of the relationship between mechanical properties of crystals and their coloration, initiated by Przibram, Smekal, and Hartly,[256] were continued and yielded some important results. Alongside the methods used by Estermann, Leivo, and Stern to measure changes in density upon irradiation, discussed earlier, other methods were developed during the 1950s. These consisted, for example, of measurement of the change in capacitance when one plate of a condenser was moved by the expansion of the irradiated crystal;[257] observation of anomalous birefringence induced by the strains caused by partial x-raying of a crystal surface;[258] and changes of the lattice constant as seen with x-ray diffraction measurements.[259]

At the Carnegie Institute of Technology, a group headed by Roman Smoluchowski carried out extensive studies of the changes of volume and mechanical properties of alkali halide crystals irradiated with x rays or with high-energy protons produced in a synchrocyclotron. Smoluchowski had been interested chiefly in the physics of metals when, in 1946, he left General Electric to take a position as a professor of physics and metallurgical engineering at Carnegie Tech. There he became involved in radiation-damage studies and soon realized that alkali halides might be the most suitable materials for the study of the mechanism of formation of point defects.[260] His group (including W. Leivo, D. A. Wiegand, and M. F. Merriam) greatly improved the sensitivity of the measurement of density changes in irradiated alkali halides.[261]

Of all the new techniques used in color center research after the Second World War, the most powerful tool turned out to be electron spin resonance. It had long been realized that if the Bristol model of the F-centers was valid, unpaired electrons should manifest themselves by turning the colored crystal paramagnetic. As early as 1939, P. Jensen, working at the Göttingen Second Physical Institute, showed that the magnetic susceptibility of an alkali halide crystal changed following the crystal's coloration.[262] His measurements could not, however, provide any information concerning the microscopic structure of the paramagnetic centers. The situation changed significantly in the mid-1940s, when the first electron spin resonance (ESR) experiments were successfully performed by E. Zavoisky in the Soviet Union, R. L. Cummerow and D. Holliday at Pittsburgh, and B. Bleaney and R. P. Penrose in England.[263]

This extremely useful technique, which developed alongside nuclear magnetic resonance (NMR), found applications in various fields of physics, chemistry, and biology. It owed much to the wartime radar research on microwaves. Those physicists, such as Purcell, who had been involved with this research during the war had a clear advantage in developing magnetic resonance methods for peaceful methods. By the end of the 1940s, when the first wave of postwar graduate students had completed their Ph.D.s, the application of new techniques was spreading like wildfire.

The feeling that magnetic resonance would provide important information about color centers was very much in the air. Early in 1947, Herbert Fröhlich had suggested at a colloquium at Oxford University that the F-centers should have an effect on nuclear relaxation—a process by which a magnetic system reaches equilibrium following a change in the external magnetic field.[246] In fact, Rollin and Hatton, working at the Clarendon Laboratory, found when irradiating calcium fluoride crystals that the coloration led to a significant reduction of the relaxation time.[265] But ESR was more promising. Slichter, for one, came to Urbana in the autumn of 1949 with the clear idea that the first thing he would do would be to look for the F-center's electron spin resonance.[266]

This newly developed experimental technique was particularly suitable for the study of the unpaired electrons that the researchers expected to find in at least some of the color centers. These electrons, because of their uncompensated magnetic moments, would exhibit paramagnetic properties. In the presence of an external magnetic field, their energy levels corresponding to the two different spin states would be split, and the resonance absorption due to the transition between these states could be measured. The ESR spectrum was potentially a source of fundamental and very precise information concerning the electronic structure of a paramagnetic center. The spectroscopic splitting factor, called the *g*-factor, which is the measure of the separation of the spin levels, equals 2 for a free electron. An electron in a crystal is not entirely free, and its *g*-factor, in general, differs from the free-electron value; the magnitude and the sign of this difference Δg provides information on the interaction of the electron with the lattice, including the tightness of its binding, and could reveal whether the center consisted of an electron or a hole.

Shortly after his arrival at Urbana, Slichter was already in the process of assembling the apparatus when he learned that Beringer and Castle at Yale had built a very sensitive microwave rig using a bolometer as a detector and had already looked unsuccessfully for an *F*-center resonance. That discouraged him, and he temporarily turned away from color centers.[267]

About the same time, however, a successful measurement was made by Clyde A. Hutchison, a physical chemist at the University of Chicago.[268] Hutchison collaborated with the color center group at Argonne, and Simpson had suggested the problem to him. Making his measurements on LiF crystals irradiated with neutrons, Hutchison found that the paramagnetic resonance of the *F*-center consists of a single line. After publishing his first short note, he went on with more detailed studies, setting one of his graduate students at the University of Chicago, Gordon Noble, to work on a dissertation dealing with the spin resonance of color centers. Meanwhile, papers on the same subject by Michael Tinkham and Arthur Kip of MIT[269] and E. E. Schneider and T. S. England of King's College (University of Durham), Newcastle-upon-Tyne,[270] appeared.

Soon, the attention of the researchers turned to the interpretation of the exact shape of the resonance lines, which, in addition to the *g*-factor itself, proved to be quite revealing. The electron bound to the point defect interacts not only with the applied external magnetic field, but also with crystalline fields and the spins of surrounding nuclei. This last interaction, called hyperfine interaction, leads to an additional splitting of electron energy levels into $2I + 1$ levels if the nucleus has a spin quantum number I. In practice, the hyperfine lines quite often overlap and cannot, for example in the case of KCl, be resolved; their weighting factors produce a single broad resonance with an approximately Gaussian envelope. When Hutchison and Noble discussed their results, the role of the hyperfine interaction was by no means clear, and they misinterpreted the cause of the line broadening, attributing it to the dipole–dipole interaction. That conclusion had, in a sense, a beneficial effect, because it made Charles Kittel of Berkeley get involved in color center research.

Kittel's group played an important part in American solid-state physics and deserves special attention. The beginning of the 1950s was a difficult time for Berkeley physics as a result of controversy over the loyalty oath demanded by the state of California. The faculty lost its four theoretical physicists, who decided to leave California in protest.[271] The protesters were mostly nuclear physicists, whereas some of those who joined the faculty around that time were solid-state physicists. Perhaps more important in changing the whole profile of Berkeley research was the fact that some graduate students working or intending to work on theoretical theses were left without advisers, so when Charles Kittel decided to move from Bell Labs to Berkeley they were quick to join him to form a strong theoretical solid-state group. (One of them, Albert Overhauser, had written his dissertation under Kittel even before the summer of 1951 when the latter finally came to California.)[272] Thus Kittel became the senior theorist of the department and had six or so graduate students working with him.

Experimental research did not lag behind, since along with Kittel came Arthur Kip, who already had engaged in close collaboration with him while working as an assistant professor at MIT. Kittel had worked on ferromagnetic resonance at Bell Labs, had followed Purcell's and Bloembergen's work, and had a clear sense that electron spin resonance was a promising and powerful research tool. Following his suggestion, Kip had carried out ESR measurements at MIT and brought experience as well as some hardware with him to Berkeley. The collaboration between the experimental and theoretical groups was very close. As Portis describes it:

> Kip and Kittel were closely associated during that period and Kittel would come by and follow what we were doing. Of course, once we had the spectrometer going, he became quite actively interested in what we put into it. He wanted to follow the results on a day-to-day basis. Everything that we looked at—I think, without exception—was because of the ideas he had. . . . More than anything he followed the literature. He looked at every journal that came in. He scanned it for ideas. He looked at what experimental people were doing, he looked at the analysis, and he decided what was promising and what wasn't, what kind of interpretation would hold water and what wouldn't.[273]

When Hutchison and Noble's pioneering paper on F-center resonance was published in 1952, Kittel naturally read it and immediately realized that they were wrong in attributing the linewidth to the dipole–dipole interaction. Kittel thought the line broadening could have arisen from the hyperfine interaction, and it occurred to him that a simple experimental test was possible if one used separate isotopes of potassium with different values of nuclear spin. Such isotopes had been available for some years from the nuclear laboratories at Oak Ridge. The appropriate specimens were therefore obtained, irradiated with x rays to produce F-centers, and subjected to ESR measurements by the team of Kittel, Kip, Portis, and Robert Levy. The results were conclusive, fully supporting Kittel's theoretical predictions. According to the theory, if one takes into account only the interaction of an F-center electron with the first surrounding shell of six potassium nuclei, each having a spin of $\frac{3}{2}$, the result is 19 unresolved lines.[274] Each of these is, in turn, broadened by interactions with nuclei lying in more remote shells surrounding the center. The result is a single broad line observed in ESR spectra.

Further improvement of magnetic resonance techniques soon made it possible to resolve even the line structure due to the interaction with these distant nuclei. One of the Berkeley graduate students, George Feher, had carried out his undergraduate studies in electrical engineering, and when he started to work with Kip and Kittel he proved to have exceptional skills in instrument design. He did his dissertation at Berkeley working on electron spin resonance in metals; later, after he went to Bell Labs, he invented a new, ingenious, and extremely sensitive technique called electron nuclear double resonance (ENDOR). It was first tested on color centers in

1957.[275] This technique, involving stimultaneous application of microwave and radio frequencies to the sample, made it possible to detect the interaction of the *F*-center electron with nuclei lying in shells as much as seven concentric layers distant, where the *F*-center electron density is decreased by a factor of 10^6 compared with its density at the middle of the center. With this technique, the resolution was improved by about four orders of magnitude. Not only could the spatial charge-density distribution of the *F*-center electron be precisely measured, but the asymmetry of the electron's wave function could be detected by changing the orientation of the sample in relation to the applied magnetic field.

Electron spin resonance studies showed that the Bristol model of the *F*-center was valid beyond any doubt. It turned out that in KCl, the trapped electron wave function is about 60% *s* type with an admixture of *p* type centered at the halogen ion vacancy and extending over a range of several atomic spacings.[276]

It was beyond the reach of the theory to match this degree of precision. The calculations of transition energies for a relatively simple system such as the *F*-center gave results within 10 to 20% of the observed energies, but to obtain better agreement between theory and experiment, theorists often resorted to fitting of scaling parameters. Some physicists working in the field (Jordan Markham, for instance) considered the theory of the *F*-center to be "one of the biggest challenges faced by the solid state theorist," and made considerable efforts to match experimental and theoretical results.[277]

Theoretical Progress

Color center physics had been founded by Przibram and Pohl in the 1920s as an empirical science, and for many years it remained so. Once the theory started working, it developed along four lines. One had been initiated in 1933 by Lev Landau. As described previously, Landau thought that the phenomenon of crystal coloration might be explained in terms of a free electron interacting with a perfect lattice with no point defects. Mott initially adopted Landau's concept of a "self-trapped electron," but in 1937 he converted to the model consisting of an electron occupying a negative ion vacancy, and instigated a second line of development, the so-called semicontinuum model. The third line, represented mostly by the Japanese school, stemmed from viewing the *F*-center as a kind of "hydrogen atom of solid state."[278] This approach treated the trapped electron together with surrounding ions as a kind of huge molecule, and used a linear combination of atomic orbitals (LCAO) method to calculate its energy levels. A fourth approach, called the localized-mode approximation, was in some ways more phenomenological than the first three, but provided a useful picture of absorption and emission processes occurring at the *F*-center. We will not discuss all four approaches in detail, but we will outline some of their underlying assumptions, since they seem characteristic of the state of the theory. Although each method aimed at giving an account of the properties of exactly the same physical system, they differed vastly in describing its basic features.[279]

The approximation first investigated by Mott and his Bristol students S. R. Tibbs[280] and J. H. Simpson[281] was based on a conceptually simple (indeed, obviously oversimplified) picture. Largely ignoring the periodic nature of the crystal, they viewed the *F*-center as an electron trapped in a spherically symmetric cavity (halogen ion vacancy) surrounded by a continuous medium with an appropriate dielectric constant. Mathematically, the problem was reduced to solving the Schrödinger equation for an electron in a potential well. The success of the method depended, essentially, on the choice of the well's depth and shape.

The method was refined and used by various authors to calculate the transition energies and activation energies for ionization in different alkali halide crystals.[282] An important modification was proposed in 1957 by Gourary, Luke and Adrian.[283] Their method, called the point ion lattice approximation, was partly inspired by the empirical observation that the absorption energy of the *F*-band varied monotonically with the lattice parameter.[284] Gourary and Luke suggested that some physical properties that did not change in this way with the lattice parameter, such as electron polarization or exchange and overlap, could be neglected in the first-order calculation of *F*-center transition energy. Later Gourary and Adrian used this principle to develop a method of determining the potential that traps the electron: they replaced the ions by point charges and evaluated their spherically symmetric contribution to the potential at the vacancy site. The calculated transition energies agreed with the experimental values only to within 15%, but they did succeed in reproducing the general dependence of energy on interionic distance.

The so-called continuum approximation, developed by the Ukrainian physicist Pekar,[285] mentioned earlier, and independently by Kun Huang and Avril Rhys[286] at the University of Liverpool, was based on a totally different physical picture. As in the previous approach, the crystal was treated as a dielectric continuum, but the short-range interactions of the *F*-center electron with the neighboring ions were neglected and the lattice was replaced by a system of linear oscillators, all of equal frequency. The *F*-center was viewed as a static charge distribution whose field polarized the lattice and influenced its vibrational states, an elaboration of Landau's earlier idea, now named a polaron.

Landau's concept seems to have had strong appeal to many theorists, Mott and Seitz among others, because it considered the phenomenon of color center formation as an intrinsic property of an ideal crystal and, therefore, a problem of fundamental and general significance for solid-state theory. After the vacancy model had been accepted, the idea of a "self-trapped electron" continued to live its own life; Pekar's work was essentially an attempt at a theoretical solution of this problem. This formal approach, however, turned out to be general enough to treat the *F*-center as a particular system involving the electron–lattice interaction.

Despite its limitations, Pekar's theory was remarkably successful in predicting some observable properties of the *F*-center. However, the effective mass of the electron, which Pekar treated as an adjustable parameter, yielded an unrealistically high value. Models similar to that of Pekar and Huang and Rhys were nevertheless adopted later by several other theorists.[287]

Meanwhile, the so-called localized mode approximation attracted attention. This was, in a way, a complement to Pekar's continuum approximation approach, for it assumed that the interaction of the electron with six nearest-neighbor alkali ions was of paramount importance. This approach was not used to calculate the *F*-center properties from first principles, but to describe relatively realistically what happens when the *F*-center absorbs or emits a photon.[288] A physical model was developed (based on some earlier ideas of von Hippel[289] and Seitz[290]) that used the assumption that the electron, which is strongly localized in the vacancy, interacts with the collective vibration of six surrounding ions. An important feature of this model, in contrast to the continuum approximation, was that it assumed that the frequencies of the lattice vibrations with which the electron interacts are different in the ground and excited states. This approach was developed mainly by Clifford Klick and George Russell at the Naval Research Laboratory.[291] The model was rather phenomenological, being used to describe the relations between absorption and emission energies, the bandwidths, and their dependence on temperature. It was, as Klick put it, treated by most theorists as "an oversimplified view of life."[292]

The last of the theoretical methods that we mentioned earlier, the molecular approximation, was based on the assumption that the *F*-center electron is well localized and interacts significantly with only its closest neighbors, so that the system may be considered to resemble a molecule. The LCAO technique, developed for the study of molecules, could therefore be applied to the *F*-center. Several Japanese theorists devoted considerable effort to applying this method to various color centers, starting with the work of Muto[293] in 1949 and Inui and Uemura[294] in 1950. The method was also used by Kip, Kittel, Levy, and Portis[295] in their interpretation of the ESR measurements already described, which was perhaps the most successful collaboration of experimenters and theorists in the history of color center research.

New Bands, New Centers

We have focused so far on the emerging color center community, its institutional background, and the development of research techniques. We will now take a closer look at some particular substantial developments in color center research that occurred in the period under consideration.

It is usually true that when a new field or specialty of physics emerges, two opposite but not conflicting processes are under way: differentiation and integration. On the one hand, the new problem area generates specific issues calling for specific research techniques. As the subject matter and methodology become distinct, so does the subcommunity formed by researchers involved in the new field. On the other hand, to become a legitimate part of the universal scientific endeavor, the new field has to establish clear relations with the mainstream of the science of which it is a part and remain integrated with this broader discipline.

This general pattern shows up in the development of color center physics. During the 1920s and 1930s, color centers were treated by Pohl's school as just one aspect of ionic crystals, not as a clearly distinct self-contained research subject. At the same time, there was relatively little interaction between Pohl's school and other groups working on the physics of crystals. The situation changed in the late 1930s due to people like de Boer and Mott, who appreciated the universal importance of color center research for the entire field of physics of solids and tried to integrate color centers into a general theory of the crystalline state.

The growing appreciation of the critical importance of crystalline lattice defects for many significant properties of crystals encouraged the postwar rise of interest in color centers and brought the study of such centers into contact with areas of great practical significance, such as radiation damage, diffusion in metals, and the properties of semiconductors. The concept of the exciton and polaron that originated from the study of ionic crystals (in the latter case, from the study of color centers) provided another link between color center physics and the electronic properties of crystals in general. The physics of imperfect crystals (or "nearly perfect crystals," as they were sometimes called) eventually became subdivided into many branches, with the field of color centers in alkali halides becoming simply one of its constituents. Thus the field ultimately acquired a distinct identity, but it continued to occupy a special place in solid-state research, near the forefront of the physics of point defects.

The shift of focus from perfect to not-so-perfect crystals may be exemplified by the evolution of Seitz's interests, discussed earlier in this chapter. His involvement with color centers had a lasting impact on both the external circumstances and the internal development of the field. By the time he published his first review paper on the subject, his scientific reputation was well established, and the mere fact that he became interested in this particular field made the study of color centers respectable. In addition, his two review papers contained new ideas that consolidated the field conceptually.

It had long been known that the F-band is just one example of the whole class of optical effects due to imperfections in alkali halide crystals. Some of the other bands, like the F'-band and U-band, had already been the subject of systematic study by the Göttingen school in the late 1930s, but Seitz was the first to try to systematize the acquired empirical knowledge about all possible types of color centers and to propose tentative models of defects responsible for their occurrence. His proposals are shown in Table 4.2.

Table 4.2 Summary of imperfections in alkali halides, with illustrative data for wavelength of maximum absorption in KCl

Imperfection	Proposed Description by Seitz, 1954	Angstroms
F-center	Anion vacancy with associated electron	5400
F'-center	Anion vacancy with two associated electrons	7000 (broad)
R_1-center	2 neighboring anion vacancies with an associated electron	6500
R_2-center	2 neighboring anion vacancies with 2 associated electrons	7300
M-center	2 neighboring anion vacancies with a neighboring cation vacancy and an associated electron	7750
N-center	Unidentified unit containing 3 neighboring anion vacancies	9900
V_1-center	Cation vacancy with an associated hole	3560
V_2-center	2 neighboring cation vacancies with 2 associated holes	2300
V_3-center	2 neighboring cation vacancies with an associated hole	2120
V_4-center	2 neighboring cation vacancies with a neighboring anion vacancy and an associated hole	2540
H-center	Pair of neighboring anion and cation vacancies with an associated hole	3540
U-center	H ion substitutionally replacing a halide ion	2100

As it turned out, virtually all of Seitz's guesses missed the target, although sometimes narrowly. New phenomena he did not envisage were discovered, and many of his ideas had to be modified. Nevertheless, it is to his credit that, starting in the late 1940s, both experimental and theoretical efforts became more convergent and directed toward the solution of clearly defined problems.

We will now discuss some of the developments in color center physics in the late 1940s and through the 1950s from a more "internal," physics-oriented viewpoint.

Trapped Hole Centers. It is not unusual in experimental research that physicists searching for a specific effect come across another, equally important, phenomenon. This happened more than once in the history of color center research, and was the case, for instance, for the discovery of the *H*-band by Jordan Markham and William Duerig.[296] Markham and Duerig had been introduced to the field of color centers by Seitz. Like many others, Markham, as we mentioned earlier, met Seitz at Oak Ridge, where they wrote a paper on the theory of the polaron. Theoretical treatment of the problem led them to the conclusion that low-temperature optical measurements of ionic crystals might possibly provide experimental evidence for the existence of polarons.[297]

A few years later, Markham joined the Applied Physics Laboratory of Johns Hopkins University, where he met Duerig, then a part-time graduate student at the nearby University of Maryland. Duerig set up low-temperature equipment extending the accessible range to liquid-helium temperatures, with the aim of pursuing the search for hypothetical polarons. But when Duerig and Markham irradiated with x rays a number of alkali halide crystals at liquid-helium temperature, what they discovered was not a polaron absorption band. Contrary to the conviction shared at the time by many physicists, they found that the *F*-band could be formed easily by x-ray irradiation even at this low temperature. Moreover, they found a new large absorption band on the ultraviolet side of the *F*-band, in the region where at higher temperatures the system of *V*-bands had been observed; the new band bleached upon warming to 78 K. In his doctoral dissertation, Duerig referred to it as the *H*-band—*H* for helium and for Hopkins.[298]

The *H*-band was later observed by other workers and was generally believed to be due to one of the trapped hole centers.[299] "What are the *H*-centers?" asked Seitz rhetorically in his 1954 review paper. "The writer would like to propose that they arise from holes trapped at neutral vacancy pairs, the hole in each center of this type being associated with the halogen ions adjacent to the positive-ion vacancy in the pair." The evidence, however, was scanty, and Seitz admitted that "it is also formally possible that the H-band arises from a positively charged center which is the antimorph of the *F'*-center."[300]

A different model was proposed the same year by J.H.O. Varley, who considered the *H*-center to be a hole trapped at an interstitial halogen atom.[301] The final verification of the *H*-center model (as well as some of the *V*-center models) was an outgrowth of spin resonance studies coupled with polarized light measurements. In 1958, Compton and Klick showed that none of the models proposed to date had been adequate.[302] They observed the dichroism produced in the *H*-band after bleaching with polarized *H*-band light and found that the optical dipole moment of the center is parallel to the crystal plane designated by the triad of integers 110.

Seitz's model predicted that the axis of symmentry would lie along the 100 or 111 plane, whereas, according to Varley, the *H*-center should have tetrahedral symmetry.

Compton and Klick also showed that the *H*-centers in KBr, and to a much lesser extent in KCl, can be reoriented by absorption of light at liquid-helium temperature. Stimulated by these results, in 1959 Känzig and Woodruff, then at General Electric, carried out paramagnetic resonance experiments that confirmed the symmetry deduced by Compton and Klick from their optical studies, and went on to suggest a detailed model of the *H*-center configuration.[303] Känziz and Woodruff concluded that this center was made of an X_2^- halogen molecule-ion situated at a single halide ion site, along with two halide ions adjacent to this site, all lined up in the 110 plane.

The *H*-center configuration, called crowdion for obvious reasons, while bearing no resemblance to the model Seitz had proposed for it, is closely related to the Varley model and is sometimes referred to in the literature as a halogen interstitial center.[304] Compared with the perfect lattice, the defect contains an extra atom, produced along with a halogen vacancy during irradiation. The interstitial position for a halogen ion or atom seems, however, to be highly unstable in an alkali halide lattice, so an X_2^- molecule-ion located at a single halide ion site is created. The *H*-center is complementary to the *F*-center: the two annihilate each other if they are brought together, reconstituting a perfect lattice structure. The creation of the *F*–*H* pair by irradiation at very low temperatures is, therefore, an elementary process of fundamental importance for understanding the dynamics of point defect formation. This process became the focus of attention for many researchers in the 1960s and 1970s.

The unraveling of the V_k-center (named by Seitz in honor of Känzig's discovery) was also a triumph both of the ESR technique and of Känzig, who for years had applied this technique to the study of trapped hole color centers with remarkable success. As mentioned earlier, this center was proved to consist, rather surprisingly, of two normal lattice halogen ions that had trapped a hole, or, in other words, had lost an electron. Thus physicists had finally found a center (in this case a self-trapped hole) resembling the self-trapped electron envisaged by Landau in 1933. Actually, the kind of mechanism involved in the center's formation can be looked on as more chemical than physical. What formally could be viewed as a trapped hole can be looked on as a halogen molecular ion, a rather stable chemical entity, occupying two halogen lattice sites, but slightly contracted compared with the normal separation between the negative ions in the lattice.

Exciton Bands. In the course of searching for the absorption bands due to trapped holes or *V*-centers, Delbecq, Pringsheim, and Yuster discovered what turned out to be the absorption bands arising from excitons trapped at point defects.[305] Studying potassium iodide, they found that an additively colored crystal, containing primarily *F*-centers, exhibits an absorption peak just on the edge of the fundamental absorption band in the ultraviolet part of the spectrum. This peak, which they named the β-band, was found to be closely associated with the *F*-band. Its intensity decreased in proportion to the decrease of the *F*-band on bleaching, and at the same time a new peak, which they named the α-band, appeared adjacent to the β-band on its long-wavelength side.

When the Argonne physicists made their discovery, the exciton was still a hypothetical entity. Its theory had been formulated some 20 years earlier by Frenkel, Peierls, and Wannier, but its existence had not been proved beyond doubt.[306] According to the theory, an exciton was a bound pair consisting of a conduction-band electron and a valence-band hole traveling through the crystal together. Delbecq, Pringsheim, and Yuster analyzed the results of their measurements and promptly concluded that the new bands could not be ascribed to *V*-centers. Although they did not use the term *exciton* in the discussion of their results, they came close to it by proposing that the new absorption bands were closely related to the first fundamental absorption band, "which is supposedly due to perturbations caused by the various lattice defects."[307] The Argonne group's first experimental results were confirmed by their subsequent research on potassium bromide[308] as well as by Werner Martienssen, Pohl's assistant at Göttingen.[309]

Before the paper of Delbecq, Pringsheim, and Yuster was published, D. L. Dexter, then at the University of Illinois at Urbana, learned about their work from their mutual colleague W. R. Heller. Dexter promptly got in touch with Pringsheim, gained access to the experimental data, and elaborated the models proposed by the Argonne physicists. He was interested in the theoretical interpretation of the absorption spectra of ionic crystals, having just published a paper dealing with pure and colored alkali halide crystals in which he ascribed their fundamental absorption band to excitons.[310] He now made an attempt to calculate oscillator strengths of α- and β-bands, obtaining results in qualitative agreement with the experimental data.[311]

Apker and Taft, in turn, were informed about the discovery of α- and β-bands by Seitz, who kept in touch with most people active in color center research. They applied to the β-band their techniques of measuring the photoelectric effect on thin films, and observed emission that could be associated with the band in a fairly unambiguous way.[312]

Higher Excited States of the F-*center.* It had been suggested by Mott and Gurney[313] that the absorption band found in 1936 by Kleinschrod[314] as a shoulder on the short-wavelength side of the *F*-band in KCl (and accordingly named the *K*-band by Seitz)[315] might be associated with transitions of the *F*-center to higher excited levels, perhaps to the crystal's conduction band. This conjecture was further supported by an investigation of the ratio of the peak intensities of the *K*- and *F*-bands in KCl, carried out by Duerig and Markham in conjunction with the low-temperature studies described earlier. They found that no significant deviation appeared at different levels of crystal darkening, which suggested a close association between *K*- and *F*-centers.[316]

However, studies by Etzel and Geiger at the Naval Research Laboratory showed that the *K/F* band ratio is somewhat variable, and for several years the problem of the nature of the *K*-band remained unsolved.[317] Then at the end of the 1950s, Lüty, Pick's collaborator in Stuttgart, undertook a systematic study of the absorption spectra of the face-centered cubic alkali halides on the high-energy side of the *F*-

bands. The results were surprising. In this thoroughly studied region of the spectrum, he discovered three new absorption bands: the L_1-, L_2- and L_3-bands.[318] They were of very low intensity, with oscillator strengths two orders of magnitude smaller than that of the F-band; to measure them, Lüty used thick, densely colored samples. They also had to be free of V-centers, which appear in the same spectral region. When Lüty measured the K-band as well, he found that the L_1/F and K/F band ratios were constant over a wide range of the density of coloration. He concluded that all these bands are due to transitions of the F-center to higher excited states.[319]

F-aggregate Centers. The problem of unraveling the nature of centers formed by the coagulation of two or more F-centers was a long story, so for brevity we will focus on some of the most important features.

The absorption bands associated with F-aggregate centers had already been observed in the 1920s by Göttingen researchers,[320] and although it was too early to ascribe the bands to any definite defects (the nature of the F-center itself was still a mystery) many researchers understood that these bands were the result of coagulation of defects.[321] More empirical knowledge about the new absorption bands was accumulated after the outbreak of the Second World War by researchers in the Soviet Union, Germany, and the United States,[322] who did not know of one another's work and carried out their investigations in complete isolation. When Seitz published his 1946 review article,[323] in which he put forward the first tentative models of the centers responsible for the R_1-, R_2-, and M-bands, he was not aware of the relevant work that had been done in Germany and the Soviet Union.

Confidence in Seitz's insights and his personal popularity might have been responsible for some misinterpretations of experimental data. Here are a few examples. In 1953, W. A. Smith and A. B. Scott of Oregon State College at Corvallis found a linear relationship between the concentrations of F- and M-centers in thermal equilibrium;[324] this seemed to give support to Seitz's model, showing that for every M-center formed, one F-center is destroyed. However, the relationship would later be found to be not linear but quadratic.[325] Again, according to Seitz, the M-center was a one-electron defect and should therefore be paramagnetic; measurement of the ESR signal of the M-center, although not very straightforward, was indeed reported by some researchers.[326] It would, however, turn out to be spurious.

As for Seitz's models of the R_1- and R_2-centers, R. Herman, M. C. Wallis, and R. F. Wallis of the Johns Hopkins Applied Physics Laboratory suggested in 1956 that, contrary to Seitz's proposal, the two bands originate from a single center.[327] Nevertheless, Lambe and Compton[328] defended Seitz's models of the R_1- and R_2-centers in 1957, although in a paper written with Klick a year and a half later, Compton decided that this view had to be rejected.[329]

It was perhaps not accidental that the first outspoken objection to Seitz's M-center model was voiced in neither the United States nor Japan, but in Europe. In 1956, C. Z. van Doorn and Y. Haven, working at the Philips Research Laboratories in the good tradition of de Boer, published a paper on the absorption and luminescence of color centers in KCl and NaCl. They wrote, "We are inclined to believe that an M-center is formed by the coagulation of two F-centers according to the relation

$$F + F \rightarrow F_2(M)_1$$

leading to a model of the M-center consisting of two neighboring Cl vacancies with two electrons."[330]

More empirical evidence was needed before the van Doorn and Haven model would replace Seitz's triple vacancy M-center model. The crucial experiment that unequivocally refuted Seitz's model was conceived by Albert Overhauser and carried out by him and Hugo Rüchardt at Cornell University in 1958.[331] Earlier experiments, like that performed in 1952 by Ueta, [332] who induced dichroism in the M-band by bleaching it with polarized light, had shown that the center lacks spherical symmetry. However, both F_2 and triple vacancy models were anisotropic, so the technique Ueta used could not distinguish between them. Overhauser had concluded that Stark effect measurements could yield more precise information on the symmetry of the aggregate centers, and proved in his experiment with Rüchardt that the M-center had axial symmetry and no permanent electric dipole moment.

Before the issue of the M- and R-center models was finally settled, Knox made an attempt to salvage Seitz's M-center model by modifying it so as to make it axially symmetric.[333] By that time, however, there was mounting evidence that the center consisted of two negative-ion vacancies with two electrons (F_2-model). In 1960, van Doorn and Faraday, and the next year Rabin and Compton, showed that in thermodynamic equilibrium the concentration of the M-centers is proportional to the square of the F-center concentration.[334] At the same time, M. Gross and M. C. Wolf[335] of Stuttgart carefully investigated the electron spin resonance of the M-center and concluded that it is diamagnetic.[336] The final models of the M-, R-, and N-centers were proposed in 1960 by Heinz Pick.[337] According to him, they were, respectively, double, triple, and quadrupole aggregates of F-centers.

Thus by the beginning of the 1960s, the structure of all the essential color centers was known. That meant the end of a certain stage of development of the physics of color centers—what we might call the static or structural stage. In the following years, the emphasis would gradually shift to investigations of the dynamic processes of color center formation and transformation.

4.3   Conclusion: Experiment and Theory

As a case study in the history of physics, the story of point defects, and of color centers in particular, touches on general issues related to the process of growth of scientific knowledge. For instance, an important aspect of this process is the role of hypotheses. Especially illuminating are hypotheses that turn out to be of limited validity or that are completely wrong. Such hypotheses make clear the complexity of each historical situation and help call into question the seemingly logical stringency of scientific progress. They show how important it is at crucial times to have the boldness to radically simplify problems (occasionally simply closing one's eyes when facts do not fit) and, on the other hand, how easy it is with hindsight to determine what the limits of the simplification should have been.

Many such hypotheses, some sweeping and others less so, are put forward in the course of any scientific development. Pohl's hypothesis of perceptible electron movement in the "Stasiw-cloud" was one example; another was Pohl's analogy between color centers and the photographic latent image in silver salts, which despite its limited validity stimulated some interest in color center studies. Smekal's hypothesis that porous regions *(Lockerstellen)* or microcracks were responsible for all "structure-sensitive" properties of crystals gained some popularity, but it had limited heuristic value and its impact was correspondingly limited. The hypothesis was vague, failing to suggest experiments that could unequivocally confirm or refute Smekal, and it attempted to explain too much—mechanical, thermodynamical, optical, and electrical phenomena—with too little. For example, it soon became evident that Smekal's hypothesis did not account for the crystal conduction and diffusion processes in thermodynamical equilibrium.

In general, the first two decades of color center studies, dominated by the personality of Robert Wichard Pohl, did not include strong interaction between experiment and theory. By the late 1940s, however, the relations between experimental and theoretical color center studies had significantly changed. The field that had grown largely by the accretion of experimental results now had a consistent theoretical framework, and a few clearly formulated general issues controlled a more or less orderly progress. Although this progress was still, as in the prewar period, dominated more by experimental discoveries than by incisive theoretical insights, theorists now played an important role in interpreting experimental results, integrating color center studies into the rest of physics, and posing new problems.

If there is a moral to this story, it is that the usefulness of theories cannot be judged solely from their immediate validity. We have seen how, in the prewar period, theorists advanced from one unreliable hypothesis to another like travelers in a swamp who step from one tussock to another, abandoning each just as it sinks, yet not without making some forward progress. And in the postwar period, many theoretical models set forth by Seitz, who more than anyone else influenced the development of color center physics in those years, missed their target; eventually refuted, these models nevertheless were dynamic ideas that challenged experimenters and theorists alike, providing guidance for work in the field.

Why does one "false" theoretical proposition prove useful and make an impact while another does not? We will not attempt a general answer to this important question. However, we hope that the history of color center studies has given the reader some insight into the matter.

Notes

Jürgen Teichmann is the author of Section 4.1; Krzysztof Szymborski is the author of Section 4.2. Section 4.1 was translated by Mark Walker. A detailed monograph by J. Teichmann, *Zur Geschichte der Festkörperphysik-Farbzentrenforschung bis 1940,* was published in 1988 (Stuttgart: Steiner). Thanks to those who helped with advice and sources for Section 4.1: Dr. V. Ya. Frenkel, Mrs. R. Gerlach, Prof. Dr. G. Glaser, Dr. H. Goetzeler (Siemens Museum), Prof. Dr. K. Hecht, Mrs. A. Hilsch, Prof. Dr. W. Jost, Prof. Dr. B. Karlik, Prof. Dr. G. v. Minnigerode, Prof. Dr. H. Pick, Mrs. A. Pohl, Prof. Dr. N. Riehl, Mr. M. Schottky, Prof. Dr. O. Stasiw, Prof. Dr. F. Stöckmann, Dr. J. Teltow, and special thanks to Prof. Dr. E. Mollwo for intensive critical advice.

1. The prehistory of insulator crystals in the nineteenth century must be examined more closely; at any rate the subjects reported in this article are often cited in the literature. See, for example, K. Przibram, *Verfärbung und Lumineszenz* (Vienna: Springer, 1953).
2. T. J. Pearsall, "On the Effects of Electricity upon Minerals which Are Phosphorescent by

Heat," *Journal of the Royal Institution* 1 (1830–1831): 77, 267. Pearsall's English article was immediately translated into French and German and often cited. See n. 1.

3. H. Rose, "Über eine neue Reihe von Metalloxyden," *Ann. Phys.* 120 (1863): 15. Rose cites here earlier results of R. Bunsen.

4. E. Hückel, *Ein Gelehrtenleben* (Weinheim: Chemie, 1975), 104.

5. H. Siedentopf, "Ultramikroskopische Untersuchungen über Steinsalzfärbungen," *Phys. Zs.* 6 (1905): 855–866.

6. Ibid., 856–857. He cites Rose (n. 3), F. Kreutz, and F. Giesel for the additive coloration, and E. Goldstein (x rays, ultraviolet), E. Becquerel (radiation), G. Holzknecht (x rays), and T. J. Pearsall ("electric sparks") for subtractive coloration.

7. "Die Thätigkeit der Physikalisch-Technischen Reichsanstalt in der Zeit vom 1. April 1895 bis 1. Februar 1896," *Zeitschrift für Instrumentenkunde* 16 (1896): 211; E. Goldstein's observations are written under "Arbeiten der wissenschaftlichen Gäste."

8. J. Elster and H. Geitel, "Über eine lichtelectrische Nachwirkung der Kathodenstrahlen," *Ann. Phys.* 59 (1986): 487–496, pp. 487–490. Also see the remark by A. Becker on increased conductivity through cathode rays (his own experiments before 1906) in B. Gudden and R. Pohl, "Über lichtelektrische Leitfähigkeit," *Phys. Zs.* 22 (1921): 529–535, p. 535. For the significance of Elster and Geitel for the history of the photoelectric effect, see K. H. Wiederkehr, "Zur Geschichte der lichtelektrischen Wirkungen," *Technikgeschichte* 40 (1973): 93–103.

9. Siedentopf (n. 5), 858.

10. *Luminescence* as a generic term for fluorescence and phosphorescence came from E. Wiedemann in 1888. For the history, see E. N. Harvey, *A History of Luminescence. From the Earliest Times Until 1900* (Philadelphia: American Philosophical Society, 1957). H.G.J. Kayser, *Handbuch der Spectroscopie,* vol. 4 (Leipzig, 1908), has an extensive description of the literature up until 1908. P. Pringsheim, *Fluoreszenz und Phosphoreszenz,* 3d ed. (Berlin: Springer, 1928), has an extensive description of the literature from 1908 to 1928, which is continued in the English edition (New York: Wiley Interscience, 1949).

11. Harvey (n. 10), 347–348. According to Harvey, the first was probably A.V.L. Verneuil.

12. J. Elster and H. Geitel, "Über die durch Sonnenlicht bewirkte electrische Zerstreuung von mineralischen Oberflächen," *Ann. Phys.* 44 (1891): 722–736.

13. S. Meyer, "Die Vorgeschichte der Gründung und das erste Jahrzehnt des Institutes für Radiumforschung," *Sitzungsberichte der Österreichischen Akademie der Wissenschaften, Math. nat. wiss. Klasse* 159 (1950):1–26; K. Przibram, "1920 bis 1938," *Sitzungsberichte der Österreichischen Akademie der Wissenschaften* 159 (1950): 27–34. B. Karlik, "1938 bis 1950," *Sitzungsberichte der Österreichischen Akademie der Wissenschaften* 159 (1950): 35–41.

14. See also E. J. Adirowitsch, *Einiqe Fragen zur Theorie der Lumineszenz der Kristalle* (Berlin: Akademie, 1953), 41. Adriowitsch does not mention these reasons, but only causes from the internal history of luminescence. However, he obviously did not use primary sources.

15. L.E.O. Visser, "Essai d'une théorie sur la phosphorescence de longue durée, spécialment sur celle des sulfures alcalinoterreux," *Recueil des travaux chimiques des Pays-Bas et de la Belgique* 20 (1901): 435–456, pp. 450–453. This could have been stimulated by the Zeeman effect thesis.

16. J. Teichmann, "Kathodenstrahlen und Elektron," in *Das Experiment in der Physik,* ed. F. Fraunberger and J. Teichmann (Braunschweig: Vieweg, 1984), 203–220.

17. P. Lenard, "Über sekundäre Kathodenstrahlung in gasförmigen und festen Körpern," *Ann. Phys.* 15 (1904): 485–508, pp. 496–498.

18. P. Lenard and S. Saeland, "Über die lichtelektrische und aktinodielektrische Wirkung bei den Erdalkaliphosphoren," *Ann. Phys.* 28 (1909): 476–502, for example, pp. 447, 489.

19. P. Lenard and V. Klatt, "Über die Erdalkaliphosphore (Schlussteil)," *Ann. Phys.* 15 (1904): 633–672, pp. 671–672. See also, without reference to luminescence, B. R. Wheaton, "Philipp Lenard and the Photoelectric Effect," *Historical Studies in the Physical Sciences* 9 (1978): 299–322.

20. B. Gudden, "Phosphoreszenz," in *Müller–Pouillets Lehrbuch der Physik,* 11th ed., ed. A. Eucken et al. (Braunschweig: Vieweg, 1929), 2, pt. 3: 2326–2350, esp. pp. 2335–2336.

21. M. Faraday, "Experimental Researches in Electricity—Fourth Series," *Philosophical Transactions of the Royal Society* (1833): 507–522, pp. 518–520 (article no. 429–438).

22. K. Bädeker, "Über eine eigentümliche Form elecktrischen Leitvermögens," *Phys. Zs.* 9

(1908): 431–433; Bädeker, "Über eine eigentümliche Form electrischen Leitvermögens bei festen Körpern," *Ann. Phys.* 29 (1909): 566–584. See also W. Kaiser, "Karl Bädeker's Beitrag zur Halbleiterforschung," *Centaurus* 22 (1978): 187–200.

23. See W. Jost, "Some Remarks on the Historic Development and the Understanding of Kinetic Phenomena in Crystals and Gases," manuscript of a lecture given in Moscow, 1981, copy in DM. Until 1926, Jost was a graduate student under C. Tubandt. Citations for the important work of Tubandt can be found in B. Gudden, "Elektrizitätsleitung in kristallisierten Stoffen unter Ausschluss der Matalle," *Ergebnisse der exakten Naturwissenschaften* 3 (1924): 116–159. See also C. Tubandt, "Leitfähigkeit and Überführungszahlen in festen Elektrolyten," in *Handbuch der Experimentalphysik*, ed. W. Wien and F. Harms (Leipzig: Akademische Verlagsgesellschaft, 1932), 12, pt. 1: 383–469.

24. On G. von Hevesy, see Jost (n. 23). For citations of the work of Hevesy as of 1922, see Gudden (n. 23). See also Gudden (n. 20). For the influence of Hevesy (and A. Smekal) on early developments, see, for example, R. W. Pohl, *Einführung in die Elektrizitätslehre* (Berlin: Springer, 1927), 192–193. For radioactive indicator methods, see, for example, W. Jost. *Diffusion und chemische Reaktion in festen Stoffen* (Leipzig: Steinkopf, 1937), 94–95.

25. Despite a few preliminary observations in the nineteenth century, the Curie brothers were the first to see piezoelectricity unambiguously, in 1880. See W. Voigt, *Lehrbuch der Kristallphysik* (Berlin: Teubner, 1910), 801. All investigations of piezoelectricity were closely connected to the pyroelectricity of crystals, which seemed especially important in the nineteenth century.

26. W. C. Röntgen, "Über die Elektrizitätsleitung in einigen Kristallen und über den Einfluss einer Bestrahlung darauf," *Ann. Phys.* 64 (1921): 1–195. Röntgen designated this publication as the second part of a work, with the same title, that he had already begun: *Ann. Phys.* 41 (1913): 449. The explanation that this was done "in part together with A. Joffe [Ioffé]" was made more precise in 1923: "by far the greater part of the experiments on rock salt" had been carried out by Röntgen. See A. Ioffé, "Elektrizitätsdurchgang durch Kristalle," *Ann. Phys.* 72 (1923): 461–500, p. 461 n. 1. This footnote obviously comes from Röntgen himself. See letter from Röntgen to Ioffé, 12 November 1922, in a new Russian edition (Leningrad: Nauka, 1983), with 114 letters to Ioffé, of A. F. Ioffé, *Begegnungen mit Physikern* (Basel: Pfalz, 1967). Further, see V. Ya. Frenkel, "Abram Fedorovich Joffe [Ioffé] (Biographical Sketch)," *Soviet Physics Uspekhi* 23 (1980): 531–550. Electronic "conductivity" of luminescent materials was discovered by Lenard. See Lenard and Saeland (n. 18), 496.

27. N. Semenoff and A. Walther, *Die physikalischen Grundlagen der elektrischen Festigkeitslehre* (Berlin: Springer, 1928), iii. See also H. Kant, "Aus der Biographie Joffes [Ioffé] und Betrachtungen zu seinem wissenschaftsorganisatorischen und -strategischen Wirken," *Kolloquien des Instituts für Theorie der Wissenschaften an der Akademie der Wissenschaften der DDR* 23 (1981): 123–138.

28. Beginning in 1928, R. Pohl continually referred to himself as R. W. Pohl.

29. For this and the following early biographical statements, see R. W. Pohl, interview with T. S. Kuhn and F. Hund, 25 June 1963, pp. 1–2, AHQP. Also, "Statement from R. Pohl to C. A. Hempstead, 25 July 1944," in *Beginnings,* 112–115. For a positive evaluation of "new things" in theoretical physics by Pohl, see also Pohl to W. Wien, 7 December 1921, DM.

30. R. W. Pohl, "Zur Jahrhundertfeier des elektromagnetischen Telegraphen von Gauss und Weber" (talk, 18 November 1933), *Mitteilungen des Universitätsbundes Göttingen* 15, no. 2, (1934): 1–8, p. 8. See also E. Mollwo, interview with J. Teichmann, 26 November 1982, tape 2, p. 2, DM.

31. The first published work was R. Pohl, "Über das Leuchten bei Ionisation von Gasen. Zur Deutung der Versuche des Hrn. B. Walter," *Ann. Phys.* 17 (1905): 375–377. Here he cited J. J. Thomson, *Conduction of Electricity through Gases* (Cambridge: Cambridge University Press, 1903), along with J. Stark and P. Lenard. Among Pohl's personal papers in Göttingen there is a used copy of this book, which has been worked through intensively with many handwritten notes.

32. B. Gudden, "R. W. Pohl zum 60. Geburtstag," *Die Naturwissenschaften* 32 (1944): 166–169, p. 166.

33. The first publication was R. Pohl, "Über den lichtelektrischen Effekt an Platin und Kupfer im polarisierten ultravioletten Licht," *Verh. Ges.* 11 (1909): 339–359. Regarding cathode sputtering work of G. Leithäuser from 1908, see ibid., 340.

34. R. Pohl, "Zur quantenhaften Lichtabsorption in festen Körpern," *Die Naturwissenschaften* 14 (1926): 214–219, p. 217 (this summarizes Pohl's work as of 1910). For optical measurements, see R. Fleischmann, "Eine selektive Lichtabsorption in dünnen Alkalimetallschichten," *Nachr. Gött.* (1931): 252–256. R. W. Pohl, "Lichtelektrische Erscheinungen," in *Handwörterbuch der Naturwissenschaften,* 2nd ed. (Jena: Fischer, 1931), 6: 239–255, p. 254.

35. R. Pohl and P. Pringsheim, "Die lichtelektrische Empfindlichkeit der Alkalimetalle als Funktion der Wellenlänge" (second report), *Verh. Ges.* 12 (1910): 349–360, p. 360. See also Pohl and Pringsheim, "Der selektive lichtelektrische Effekt an K-Hg-Legierungen," *Verh. Ges.* 12 (1910): 697–710, p. 697. For resonance width see, for example, R. Pohl, "Über den selektiven und den normalen Photoeffekt," *Die Naturwissenschaften* 1 (1913), 618–621, p. 620.

36. R. Pohl, "Über eine Beziehung zwischen dem selektiven Photoeffekt und der Phosphoreszenz," *Verh. Ges.* 13 (1911): 961–966, p. 962. Pohl and Pringsheim (n. 35), see 697. The name "center" was used only once before 1920 and was applied very unspecifically.

37. R. Pohl and P. Pringsheim, "Über den selektiven Photoeffekt des Lithiums und Natriums," *Verh. Ges.* 14 (1912): 46–59, pp. 46–49, 51; Pohl and Pringsheim, "Über die Herstellung von Metallspiegeln durch Destillation im Vakuum," *Verh. Ges* 14 (1912): 506–507.

38. R. Pohl and P. Pringsheim, "Der selektive Photoeffekt bezogen auf absorbierte Lichtenergie," *Verh. Ges.* 15 (1913): 173–185, pp. 179–183.

39. *R. Pohl, Gedächtniskolloquium am 29. November 1976* (Göttingen: Musterschmidt, 1978), especially E. Mollwo, "Gedächtnisrede auf Robert Wichard Pohl," 13–19, p. 15.

40. For Pohl's later comments, see Pohl interview (n. 29); R. W. Pohl, interview with E. Braun, 16 December 1975, DM. For the published sources, see, among others, B. Gudden and R. Pohl, "Das Quantenäquivalent bei der lichtelektrischen Leitung," *Zs. Phys.* 17 (1923): 331–346: "The wish to free ourselves from these (unclear conditions on the surface) caused us to investigate the light-electric effect three years ago, no longer on the surface of solid and liquid bodies but rather in the interior of crystals" (332).

41. Gudden (n. 23); B. Gudden, *Die lichtelektrischen Erscheinungen* (Berlin: Springer, 1928); Gudden, "Elektrische Leitfähigkeit elektronischer Halbleiter," *Ergebnisse der exakten Naturwissenschaften* 13 (1934): 223–256.

42. B. Gudden and R. Pohl, "Zur Kenntnis des Sidotblendephosphors," *Zs. Phys.* 1 (1920): 365–375, p. 375. See also R. W. Pohl, interview with H. Pick, 25–26 June 1974, in possession of A. Pick, Stuttgart, transcript in DM, tape 1, p. 4.

43. B. Gudden and R. Pohl, "Litchtelektrische Beobachtungen an Zinksulfiden," *Zs. Phys.* 2 (1920): 181–196. Gudden and Pohl, "Lichtelektrische Beobachtungen an isolierenden Metallsulfiden," *Zs. Phys.* 2 (1920): 361–372.

44. B. Gudden and R. Pohl, "Über lichtelektrische Leitfähigkeit von Diamanten," *Zs. Phys.* 3 (1920): 123–129, p. 124. For the reasons for choosing diamond as an object of research see ibid., 123; Gudden and Pohl (n. 43), 372. The well-known story about a diamond destroyed during investigation of the Hall effect (see, for example, Pohl interview [n. 42], p. 6, and R. Hilsch, manuscript of a talk in United States, 1965, p. 2, in possession of A. Hilsch, Göttingen, copy in DM), which abruptly ended this experiment, cannot be supported by documentary evidence. On the contrary, a letter from Pohl to A. Ahearn, Bell Telephone Laboratories, 10 October 1947, reported that a pure ZnS crystal was destroyed during Hall-effect investigations by a magnet's loose pole tips. Pohl Papers, I. Physical Institute, Göttingen University. The Hall effect was examined in diamonds. See B. Gudden and R. Pohl, "Über elektrische Leitfähigkeit bei Anregung und Lichtemission von Phosphoren," *Zs. Phys.* 21 (1929): 1–8, p. 1 n. 4.

45. R. Pohl, "Über lichtelektrische Leitfähigkeit in Kristallen," *Phys. Zs.* 21 (1920): 628–630, p. 630. B. Gudden and R. Pohl, "Neuere Beobachtungen über den Zusammenhang elektrischer und optischer Erscheinungen," *Die Naturwissenschaften* 11 (1923): 348–354, p. 348 (lecture on 10 November 1922 in the administration building of the Siemens-Schuckert Works, Berlin). See also Pohl interview (n. 42).

46. See Pohl interview (n. 42), tape 1, p. 6.

47. J. Stark to W. Wien, 2 May 19 [21?]; Pohl to Wien (n. 29).

48. R. W. Pohl, "Die Physik und ihre Anwendungen (Festrede zur Reichsgründungsfeier der Georg-August-Universität zu Göttingen am 18. Januar 1928)," in *Kulturgeschichtliche Studien*

und Skizzen aus Vergangenheit und Gegenwart, Festschrift zur 400Jahrfeier der Gelehrtenschule des Johanneums in Hamburg (Hamburg: Broschek, 1929), 387–399, p. 393.

49. Gudden and Pohl (n. 45), 349; B. Gudden and R. Pohl, "Über lichtelektrische Leitfähigkeit," *Phys. Zs.* 22 (1921): 529–535, p. 532, para. 7.

50. The first remarks in this regard are from B. Gudden and R. Pohl, "Über lichtelektrische Leitfähigkeit von Zinkblende," *Zs. Phys* 5 (1921): 176–181, p. 180. A more extensive treatment by the same authors was "Über den zeitlichen Anstieg der lichtelektrischen Leitfähigkeit," *Zs. Phys.* 6 (1921), 248–256. Also Gudden and Pohl, "Über den Mechanismus der lichtelektrischen Leitfähigkeit," *Zs. Phys.* 7 (1921): 65–72, p. 65–66.

51. Gudden (n. 23), 149–150.

52. G.F.J. Garlick, "Photoconductivity," in *Handbuch der Physik*, ed. S. Flügge (Berlin: Springer, 1956), 19: 316–395, p. 317; F. Stöckmann, "Photoconductivity—A Centennial," *Physica status solidi* al5 (1973): 381–390; Stöckmann, "Theorie der lichtelektrischen Leitung in Mischleitern," *Zs. Phys* 128 (1950): 185–211.

53. O. Stasiw, interview with J. Teichmann, 5 April 1983, tape 1, DM. But see the later work on quantitative explanation of the secondary current, especially R. Hilsch and R. W. Pohl, "Eine quantitative Behandlung der stationären lichtelektrischen Primär- und Sekundärströme in Kristallen, erläutert am KH-KBr-Mischkristall als Halbleitermodell," *Zs. Phys.* 108 (1937): 55–84, esp. see pp. 56, 66.

54. Gudden and Pohl (n. 45), 352; B. Gudden and R. Pohl, "Das Quantenäquivalent bei der lichtelektrischen Leitung," *Zs. Phys.* 17 (1923), 331–346, pp. 332, 345.

55. B. Gudden and R. Pohl, "Über lichtelektrische Wirkung und Leitung in Kristallen," *Zs. Phys.* 16 (1923): 170–182, p. 180. For "lattice disturbances," see also Gudden and Pohl (n. 54), 340. Because of these experiments, Gudden in particular appears to have reflected more and more on the significance of very small imperfections in solid bodies. See Gudden (n. 23), 150. Pohl interview (n. 42), tape 3, p. 6. Gudden wrote to Pohl from Erlangen, thus after 1925, about the decisive significance of small amounts of additives, both among the alkali halides and in other substances.

56. The experiments were mentioned in B. Gudden and R. Pohl "Lichtelektrische Leitung und chemische Bindung," *Zs. Phys.* 16 (1923): 42–45, p. 43; also Gudden and Pohl (n. 54), 332; Gudden (n. 23), 159. See also the document of the Helmholtz Society, 1922, Wien Papers, DM; R. W. Pohl, "Zusammenfassender Bericht über Elektronenleitung und photochemische Vorgänge in Alkalihalogenidkristallen," *Phys. Zs.* 39 (1938): 36–54, 36 (in English, "Electron Conductivity and Photochemical Processes in Alkali Halide Crystals," *Proceedings of the Physical Society* 49, suppl. [1937]: 3–31). The first synthetic alkali halide crystals used in Pohl's institute were manufactured by G. Veszi in Tamman's institute from molten "purissimum" NaCl. Z. Gyulai, "Zur lichtelektrischen Leitung in NaCl-Kristallen," *Zs. Phys.* 31 (1925): 296–304, pp. 300–301. They were not very usable. At around the same time (December 1924) Pohl and Gudden were still waiting for synthetic crystals. See B. Gudden and R. Pohl, "Über den lichtelektrischen Primärstrom in NaCl-Kristallen," *Zs. Phys.* 31 (1925): 651–665, p. 657. But only a half year later, "natural and synthetic crystals of arbitrary grades of purity" were taken for granted. See B. Gudden and R. Pohl, "Zur lichtelektrischen Leitung bei tiefen Temperaturen," *Zs. Phys.* 34 (1925): 249–254, pp. 253–254.

57. S. Kyropoulos, "Ein Verfahren zur Herstellung grosser Kristalle," *Zeitschrift für anorganische und allgemeine Chemie* 154 (1926): 308–313. This process was improved by R. Hilsch and G. Bauer so that single crystal cubes with edge length 10 cm had been obtained. See R. Hilsch and R. W. Pohl, "Über die ersten ultravioletten Eigenfrequenzen einiger einfacher Kristalle," *Nachr. Gött.* (1929): 73–78, p. 76. A discussion of the improved process was given by K. Korth, "Dispersionsmessungen an Kaliumbromid und Kaliumjodid im Ultraroten," *Zs. Phys.* 84 (1933): 677–685, pp. 678–679.

58. B. Gudden and R. Pohl, "Zum Mechanismus des lichtelektrischen Primärstromes in Kristallen," *Zs. Phys.* 30 (1924): 14–23, p. 23.

59. The earliest written evidence of this concept being used by the Pohl school came in 1936: "In general it is always very valuable for me to have gained a view of the 'physics of solids' [*Physik der Festkörper*] and their photochemistry in your institute" (in K. Hecht to R. W. Pohl, 30 December 1936, in possession of K. Hecht, Kiel; copy in DM).

60. Röntgen (n. 26). Among the Pohl papers in Göttingen is a bound reprint of Röntgen's work, studied intensively and supplied with comments.

61. Gudden and Pohl ("Über den Mechanismus") (n. 50), 65 n. 2; Gudden and Pohl (1921) (n. 49), 533, para. 10.

62. Gudden and Pohl (n. 55), 170–171.

63. J. Bingel, "Über lichtelektrische Wirkung in Steinsalzkristallen," *Zs. Phys.* 21 (1924): 229–241. The article was sent off during October 1923. This first publication must have stood isolated at first. Pohl no longer mentioned it in his reminiscences. Pohl (1938) (n. 56), where the works on alkali halides were all counted as starting with 1925. Pohl interview (n. 42) tape 3, p. 1.

64. For the first works, see Gyulai (n. 56); Gudden and Pohl ("Über den lichtelektrischen Primärstrom") (n. 56).

65. Gyulai (n. 56), 300–301. Also Z. Gyulai, "Zum Absorptionsvorgang in lichtelektrisch leitenden NaCl-Kristallen," *Zs. Phys.* 33 (1925): 251–260, p. 252.

66. Hilsch (n. 44), 5.

67. Gyulai (n. 65), 251, 260. See also Z. Gyulai, "Zum Quantenäquivalent bei der lichtelektrischen Leitung in NaCl-Kristallen," *Zs. Phys.* 32 (1925): 103–110, p. 106.

68. Gudden and Pohl ("Zur lichtelektrischen Leitung") (n. 56), 252.

69. A. Arsenjewa, "Über den Einfluss des Röntgenlichtes auf die Absorptionsspektra der Alkalihalogenidphosphore," *Zs. Phys.* 57 (1929): 163–172, p. 164. See especially Mollwo interview (n. 30), tape 1, p. 2.

70. Gudden and Pohl ("Zur lichtelektrischen Leitung") (n. 56), 249.

71. R. Hilsch and R. W. Pohl, "Zur Photochemie der Alkali- und Silberhalogenidkristalle," *Zs. Phys.* 64 (1930): 606–622, p. 609.

72. Gudden and Pohl ("Zur lichtelektrischen Leitung") (n. 56), 254.

73. Przibram (n. 1), 55. Przibram cited A. Deauvillier's 1920 work as even earlier evidence. For Przibram's 1923 work, see Z. Gyulai, "Lichtelektrische und optische Messungen an blauen und gelben Steinsalzkristallen," *Zs. Phys.* 35 (1926): 411–420, p. 419.

74. Gudden and Pohl ("Zur lichtelektrischen Leitung") (n. 56), 249, 254.

75. Mollwo interview (n. 30), tape 2, p. 2. F. Hund, interview with M. Eckert, H. Rechenberg, H. Schubert, J. Teichmann, and G. Torkar, 18 May 1982, DM.

76. Hund interview (n. 75). F. Hund to J. Teichmann, 10 February 1982, DM. See also W. Gerlach, "Robert Wichard Pohl, 10.08.1884–05.06.1976," *Jahrbuch der Bayerischen Akademie der Wissenschaften* (Munich: B. Akademie der Wissenschaften, 1978), 214–219, pp. 216–217. The Gerlach school represented another influential part of experimental physics in Germany, a rival of Pohl, and completely different in research content, methodology, and style of leadership.

77. R. Pohl, "Zum optischen Nachweis eines Vitamines," *Die Naturwissenschaften* 15 (1927): 433–438; Pohl interview (n. 42), tape 3, p. 6.

78. Ioffé (1967) (n. 26), 75; Pohl (n. 48), 392. For work by Ioffé's collaborators at Pohl's institute, see also R. W. Pohl to A. F. Ioffé, 11 March 1928, in Ioffé (1983) (n. 26), 177; a copy of the original letter is in DM.

79. Hund interview (n. 75). For the electron diffraction experiments in Pohl's institute after 1927, see also R. W. Pohl to A. F. Ioffé, 11 March 1928, 24 December 1928, and 16 January 1929, in Ioffé (1983) (n. 26), 177, 186, 187.

80. R. Hilsch and R. Ottmer, "Zur lichtelektrischen Wirkung in natürlichem blauen Steinsalz," *Zs. Phys.* 39 (1926): 644–647; R. Hilsch, "Über die ultraviolette Absorption einfach gebauter Kristalle," *Zs. Phys.* 44 (1927): 421–430.

81. A. Smakula, "Einige Absorptionsspektren von Alkalihalogenidphosphoren mit Silber und Kupfer als wirksamen Metallen," *Zs. Phys.* 45 (1927): 1–12; R. Hilsch and R. W. Pohl, "Über die ersten ultravioletten Eigenfrequenzen einiger einfacher Kristalle," *Zs. Phys.* 48 (1928): 384–396.

82. Hilsch and Pohl (n. 57), 73. Hilsch and Pohl knew the work from A. H. Pfund, "Metallic Reflection from Rock-Salt and Sylvite in the Ultra-violet," *Phys. Rev.* 32 (1928): 39–43, who had already found ultraviolet absorption bands for NaCl and KCl, with the help of the Rubens residual radiation *(Reststrahlen)*. This work probably stimulated them to take better measurements. See R. Hilsch and R. W. Pohl, "Die in Luft messbaren ultravioletten Dispersionsfrequenzen der Alkalihalogenide," *Zs. Phys.* 57 (1929): 145–153, p. 145; Hilsch and Pohl, "Einige Dispersionsfrequenzen der Alkalihalogenidkristalle im Schumanngebiet," *Zs. Phys.* 59 (1930): 812–819, p. 812.

83. In 1938, Pohl referred to S. E. Sheppard and A. P. Trivelli, "Relation between Sensitiveness and Size of Grain in Photographic Emulsions," *Photographical Journal* 61 (1921): 400–403. See

Pohl (1938) (n. 56), 50. However, the direct stimulation for Hilsch and Pohl appears to have come from work by K. Przibram in 1923 or K. Fajans in 1926. See Hilsch and Pohl (n. 71), 622; R. Hilsch and R. W. Pohl, "Vergleich des photographischen Elementarprozesses in Alkali- und Silbersalzen," *Zs. Phys.* 77 (1932): 421-436, p. 423, n. 4. There were others working on this problem in 1926. See P. Lukirsky, N. Gudris, and L. Kulikowa, "Photoeffekt an Kristallen," *Zs. Phys.* 37 (1926): 308-318, esp. pp. 315-317; H. Pick, "Der photographische Elementarprozess," *Die Naturwissenschaften* 38 (1951): 323-330.

84. R. Hilsch and R. W. Pohl, "Über die Ausnutzung des latenten Bildes bei der photographischen Entwicklung," *Nachr. Gött.* (1930): 334-337, p. 334.

85. Hilsch and Pohl (1929) (n. 82). For attempts by Hilsch and Pohl to calculate this process by lattice theory (with useful results in practice, although theoretically incorrect), see W. Beall Fowler, *Physics of Color Centers* (New York: Academic Press, 1968), 31-32.

86. Hilsch and Pohl (n. 71), esp. 618.

87. The lack of electron current was definitely clear in 1931. See R. Hilsch and R. W. Pohl, "Über die Lichtabsorption in einfachen Ionengittern und den elektrischen Nachweis des latenten Bildes," *Zs. Phys.* 68 (1931), 721-734, p. 723. See also F. Stöckmann, "Das atomistische Bild und das Bändermodell der Kristalle," *Zeitschrift für Physikalische Chemie* 198 (1951): 215-231, p. 228; F. Seitz, "Color Centers in Alkali Halide Crystals. II," *Reviews of Modern Physics* 26 (1954): 7-94, p. 27.

88. M. A. Bredig, "Über anomale Dispersion in Alkalihalogenidphosphoren," *Zs. Phys.* 46, (1927); 73-79, pp. 77-79; A. Smakula, "Über Erregung und Entfärbung lichtelektrisch leitender Alkalihalogenide," *Zs. Phys.* 59 (1930): 603-614, esp. p. 604.

89. Hilsch and Pohl (n. 87), 733; E. Mollwo, "Über die Farbzentren der Alkalihalogenidkristalle," *Zs. Phys.* 85 (1933): 56-67, p. 59; F. G. Kleinschrod, "Zur Messung der Zahl der Farbzentren in KCl-Kristallen," *Ann. Phys.* 27 (1936): 97-107. See also Pohl (1938) (n. 56), 40.

90. E. Mollwo, "Über die Absorptionsspektra photochemisch verfärbter Alkalihalogenid-Kristalle," *Nachr. Gött.* (1931): 97-99. The Mollwo relation was examined more closely by H. F. Ivey and others after 1945. See Fowler (n. 85); E. Mollwo, "Zur Vor- und Frühgeschichte der Farbzentrenforschung," manuscript, 1981, in Mollwo's possession, Erlangen, copy in DM. Mollwo interview (n. 30).

91. Gerlach (n. 76), 6.

92. E. Mollwo, "Zur additiven Färbung der Alkalihalogenidkristalle," *Nachr. Gött.* (1932): 254-260, pp. 259-260. Mollwo (n. 89), 65-67; Mollwo, "Über Elektronenleitung und Farbzentren in Flussspat," *Nachr. Gött.* (1934): 79-89, pp. 79-80, 86. It can be proved that the formulations here come from Pohl. See R. W. Pohl to R. Hilsch, 26 March 1934, in possession of A. Hilsch, Göttingen, copy in DM. See also Mollwo (1981) (n. 90); Mollwo interview (n. 30). See also Hilsch (n. 44). Most of the original work on color centers after 1930, particularly by Mollwo and Stasiw, was published in the not widely distributed *Nachr. Gött.* For unpublished notes of Mollwo's attempts to give a wave-mechanical solution of the problem of color centers (potential wells with infinitely high walls), see Mollwo (1981) (n. 90). Laboratory notes also exist for Mollwo's time in Göttingen, see Mollwo Papers, Erlangen, partial copy in DM.

93. E. Mollwo, "Das Absorptionsspektrum photochemisch verfärbter Alkalihalogenidkristalle bei verschiedenen Temperaturen," *Nachr. Gött.* (1931): 236-239, p. 238. For a thorough account, see Mollwo (1932) (n. 92).

94. Mollwo (n. 89), according to Pohl (1938) (n. 56), 38-39. However, the proportional connection was not explicitly expressed by Mollwo, and from the diagrams it is evident only for NaCl with Na vapor. Vapor pressure and crystal temperature were not varied independently because the vaporized alkali metal came from colloidal particles inside the crystal. As late as 1936, Pohl spoke of the "parallelism of color center concentration and external vapor pressure of the metal" as no more than an aside. See R. W. Pohl, "Über die Absorptionsspektren der Alkalihalogenidkristalle," *Acta Physica Polonica* 5 (1936): 349-360, p. 360. The proportional connection was explicitly examined by H. Rögener, "Über Entstehung und Beweglichkeit von Farbzentren in Alkalihalogenidkristallen," *Ann. Phys.* 29 (1937): 386-393.

95. Mollwo (n. 89), 56. In 1930, Hilsch and Pohl still considered extrinsic defects necessary for color centers ("foreign color of the first type"). Hilsch and Pohl (n. 71), 618.

96. Mollwo (1934) (n. 92), 86.

97. O. Stasiw, "Die Farbzentren des latenten Bildes im elektrischen Felde," *Nachr. Gött.* (1932): 261–267. See also Stasiw interview (n. 53).

98. Gudden (n. 23), 119.

99. Stasiw (n. 97), 267; R. W. Pohl, "Das latente photographische Bild," *Die Naturwissenschaften* 21 (1933): 261–264, p. 263.

100. For the problem of the interpretation as electrons, see W. Schottky to R. W. Pohl, 30 June 1934, p. 4, Siemens Museum, Munich. See also E. Mollwo, "Über Elektronen- und Ionenleitung in Alkalihalogenidkristallen," *Nachr. Gött.* (1943): 89–102.

101. Stasiw interview (n. 51); Pohl (n. 30), 7.

102. The discovery probably took place between 3 and 12 March 1933. See R. Hilsch, laboratory notebooks (1925–1939), here notebooks 1925–1933, p. 128, and 1933–1934, p. 1. The centers were called "UV-centers" at first. See R. Hilsch to R. W. Pohl, 19 March 1933, p. 2, all in possession of A. Hilsch, Göttingen, copy in DM. R. Hilsch and R. W. Pohl, "Eine neue Lichtabsorption in Alkalihalogenidkristallen," *Nachr. Gött.* (1933): 322–328.

103. E. Mollwo, interview with J. Teichmann, 25 June 1982, tape 1, p. 2, DM.

104. R. Hilsch and R. W. Pohl, "Zur Photochemie der Alkalihalogenidkristalle," *Nachr. Gött.* (1933): 406–419.

105. Mollwo interview (n. 103), tape 1, p. 2.

106. R. Hilsch, laboratory notebooks (1925–1933), pp. 120–121, 21 February 1933; R. Hilsch to R. W. Pohl, 19 March 1933; Pohl to Hilsch, 9 March 1933, p. 1 (n. 102).

107. E. Mollwo, "Sichtbare Elektronen-Ersatzleitung in Alkalijodidkristellen," *Nachr. Gött.* (1935): 215–220.

108. G. Glaser and W. Lehfeldt, "Der lichtelektrische Primärstrom in Alkalihalogenidkristallen in Abhängigkeit von der Temperatur und von der Konzentration der Farbzentren," *Nachr. Gött.* (1936): 91–108; G. Glaser, "Weitere Versuche über den Einfluss der Temperatur auf den lichtelektrischen Primärstrom in Alkalihalogenidkristallen," *Nachr. Gött.* (1937): 31–44.

109. R. W. Pohl to G. Glaser, 30 June 1937, in possession of G. Glaser, Stuttgart, copy in DM. For preliminary developments, see also R. Hilsch to R. W. Pohl, 6 October 1936, p. 3, in possession of A. Hilsch, Göttingen, copy in DM. See also Pohl to Glaser, 3 December 1936, DM.

110. H. Pick, "Über den Einfluss der Temperatur auf die Erregung von Farbzentren," *Ann. Phys.* 31 (1938): 356–376. See also H. Pick, interview with J. Teichmann and G. Torkar, 2 October 1981, tape 1, p. 2, DM.

111. Pohl (1937) (n. 56), 13; Pohl (1938) (n. 56), 45.

112. R. Hilsch and R. W. Pohl, "Steuerung von Elektronenströmen mit einem Dreielektrodenkristall und ein Modell einer Sperrschicht," *Zs. Phys.* 111 (1938): 399–408; H. Pick, "Über die Fabrzentren in KCl-Kristallen mit kleinen Zusätzen von Erdalkalichloriden." *Ann. Phys.* 35 (1939): 73–83. See also F. Seitz, "Color Centers in Additively Colored Alkali Halide Crystals Containing Alkaline Earth Ions," *Phys. Rev.* 83 (1951): 134–140.

113. A. Ioffé and M. Kirpitchewa, "Electric Conductivity of Pure Crystals" (in Russian), *Journal of the Russian Physical-Chemical Society* 48 (1916): 261–262, reference from A. F. Ioffé, *The Physics of Crystals* (New York: McGraw-Hill, 1928), 80; Frenkel (n. 26), 539; Ioffé (1923) (n. 26), 461–500.

114. Ioffé (1923) (n. 26), 499.

115. J. Frenkel, "Über die Wärmebewegung in festen und flüssigen Körpern," *Zs. Phys.* 35 (1926): 652–669.

116. W. Jost, *Über Diffusion in festen Verbindungen* (Halle: Gebauer-Schwetschke, 1926), 34ff, copy in DM.

117. A. Smekal, "Über den Einfluss von Kristallgitterporen auf Molekülbeweglichkeit und Festigkeit," *Phys. Zs.* 26 (1925): 707–712. The first work to distinguish between structure sensitive and structure insensitive properties was A. Smekal, "Elektrizitätsleitung und Diffusion in kristallisierten Verbindungen," *Zeitschrift für Elektrochemie* (1928): 472–480. See also A. Smekal, "Strukturempfindliche Eigenschaften der Kristalle," in *Handbuch der Physik,* ed. H. Geiger and K. Scheel (Berlin: Springer, 1933), 24, pt. 2: 795–922. L. Flamm, "Adolf Gustav Smekal" (obituary), *Österreichische Akademie der Wissenschaften, Almanach für das Jahr 1959* 109 (1960): 421–427. See also P. Forman, "A. Smekal," in *Dictionary of Scientific Biography* (New York: Scribner, 1975), 12: 463–465—although containing little on solid state matters. K. Przibram, "Smekal's

Leistungen auf dem Gebiet der Festkörperphysik," manuscript of lecture, 26 May 1959, in possession of B. Karlik, Institute for Radium Research, Vienna, copy in DM. H. Rumpf, "Zur Entwicklungsgeschichte der Physik der Brucherscheinungen—A. Smekal zum Gedächtnis," *Chemie-
Ingenieur-Technik* 31 (1959): 697–705.

118. Rumpf (n. 117), 698; Smekal (1925) (n. 117), 710.

119. Smekal (1925) (n. 117), 708.

120. Seitz (n. 87), 25–26, 80–81.

121. For the opposition of Ioffé to Smekal, and vice versa, see Ioffé (1967) (n. 26), 83; V. Frenkel
to J. Teichmann, 16 July 1984, DM.

122. K. Przibram, "Die Verfärbung des gepressten Steinsalzes," *Verh. Ges.* 8 (1927): 11–12, p.
11. The name "color centers" appears in K. Przibram, "Zur Theorie der Verfärbung des Steinsalzes
durch Becquerelstrahlen," *Sitzungsberichte der Akademie der Wissenschaften in Wien, Math.-
Nat. Klasse, Abt. IIa* 135 (1926): 197–211, for example, p. 197.

123. A. Smekal, "Absorptionsbanden und Energiebänder von Alkalihalogenid-Kristallen," *Zs.
Phys.* 101 (1936): 661–679, pp. 669–672. For the different group characteristics of the Pohl, Smekal, and Przibram institutes, see K. Symborski, "The Physics of Imperfect Crystals—A Social History," *Historical Studies in the Physical Sciences* 14, pt. 2 (1984): 317–355.

124. H. Fröhlich, "Über die Lage der Absorptionsspektra photochemisch verfärbter Alkalihalogenid-Kristalle," *Zs. Phys.* 80 (1933): 819–821. See also Mollwo (1981) (n. 90). A. von Hippel,
"Einige prinzipielle Gesichtspunkte zur Spektroskopie der Ionenkristalle und ihre Anwendung auf
die Alkalihalogenide," *Zs. Phys.* 101 (1936): 680–720.

125. For example, N. A. Brilliantow and J. W. Obreimow, "On Plastic Deformation of Rock
Salt. III," *Physikalische Zeitschrift der Sowjetunion* 6 (1934): 587–602 (pts. I and II were published
by I. W. Obreimow and L. W. Schubnikoff, as "Über eine optische Methode der Untersuchung
von plastischen Deformationen in Steinsalz," *Zs. Phys.* 41 [1927]: 907–930).

126. L. Landau, "Über die Bewegung der Elektronen im Kristallgitter," *Physikalische Zeitschrift der Sowjetunion* 3 (1933): 664–665. See also N. F. Mott and R. W. Gurney, *Electronic Processes in Ionic Crystals,* 2nd ed. (Oxford: Oxford University Press, 1948; reprint, New York: Dover,
1964), 87; N. F. Mott, "Memories of Early Days in Solid State Physics," in *Beginnings,* 56–66, p.
62; C. T. Walker and G. A. Slack, "Who Named the -On's?" *American Journal of Physics* 38
(1970): 1380–1389, p. 1385.

127. J. Frenkel, "On the Absorption of Light and the Trapping of Electrons and Positive Holes
in Crystalline Dielectrics," *Physikalische Zeitschrift der Sowjetunion* 9 (1936): 158–186. Here
Peierls's publication "Zur Theorie der Absorptionspektren fester Körper," *Ann. Phys.* 13 (1932):
905–952, is also discussed. See also Walker and Slack (n. 126), 1384–1385; R. Peierls, "Recollections of Early Solid State Physics," in *Beginnings,* 28–38, p. 35. In contrast to Peierls's decription
here, he already knew of and had criticized Frenkel's concepts by 1931. See K. von Meyenn, ed.,
Wolfgang Pauli. Wissenschaftlicher Briefwechsel mit Bohr, Einstein, Heisenberg u.a., vol. 2: *1930–
1939* (Berlin: Springer, 1984), 85, 89, 94 (letters to Peierls).

128. D. Blokhintsev, "Zur Theorie der Lichtabsorption in heteropolaren Kristallen," *Physikalische Zeitschrift der Sowjetunion* 7 (1935): 639–651.

129. Frenkel (n. 127), 171–172; Mott (n. 126), 62.

130. For this and later developments, see F. Seitz, *The Modern Theory of Solids* (New York:
McGraw-Hill, 1940), 414ff; W. Shockley, J. H. Hollomon, R. Maurer, and F. Seitz, eds., *Imperfections in Nearly Perfect Crystals* (New York: Wiley; London: Chapman and Hall, 1952), 22; Seitz
(n. 87) 27ff. The "wandering of energy" *(Energiewanderung)* during luminescence was discovered
and named by N. Riehl, "Aufbau und Wirkungsweise leuchtfähiger Zinksulfide und anderer Lumi-
nophore," *Ann. Phys.* 29 (1937): 636–664. For the explanation, see N. Riehl and M. Schön, "Der
Leuchtmechanismus von Kristallphosphoren," *Zs. Phys.* 114 (1939): 682–704. See also N. Riehl,
interview with J. Teichmann, 19 May 1982, 30 July 1982, two tapes plus notes by Riehl, DM.

131. J. Frenkel, "Über die Wanderungsgeschwindigkeit der Elektronenfarbzentren in Kristallen," *Physikalische Zeitschrift der Sowjetunion* 5 (1934): 911–918; P. Tartakowsky and W. Pod-
dubny, "Über die Natur der U-Zentren in Alkalihalogenidkristallen," *Zs. Phys.* 97 (1935): 765–
773.

132. For "electrical engineer" and "atomic theoretician," see W. Schottky, "Zur Theorie der
lichtelektrischen Sekundärstrome," manuscript, 1938, pp. 7, 31, DM. For the man and his work,

see A. Seeger, interview with E. Braun and G. Torkar, 3 June 1982, two tapes, DM. See also F. Seitz, interview with L. Hoddeson, 26 January 1981, p. 123, AIP. H. Rothe, E. Spenke, and C. Wagner, "Zum 65. Geburtstag von Walther Schottky," *Archiv der elektrischen Übertragung* 5 (1951): 306–313.

133. See the preface to W. Schottky, H. Ulich, and C. Wagner, *Thermodynamik* (Berlin: Springer, 1929), viii–ix; W. Schottky to J. Jaenicke, 24 November 1954, DM. According to W. Jost, verbal communication to J. Teichmann, 1 July 1982, the colloquium mentioned was held in 1928. The speaker was E. Kordes on mixed phases.

134. For example, W. Jost, "Zum Mechanismus der Ionenleitung in 'gutleitenden' festen Verbindungen," *Zeitschrift für physikalische Chemie B* 7 (1930): 234–242; Jost, "Bemerkung der Redaktion zur Diskussion Smekal-Jost," *Zeitschrift für physikalische Chemie B* 7 (1930): 243.

135. C. Wagner and W. Schottky, "Theorie der geordneten Mischphasen," *Zeitschrift für physikalische Chemie B* 11 (1930): 163–210. For further publications ("Theorie der geordneten Mischphasen. II, III"), from Wagner alone, see *Zeitschrift für physikalische Chemie, Bodenstein Festband* (1931): 177–186; *Zeitschrift für physikalische Chemie B* 22 (1933): 181–194.

136. H. Dünwald and C. Wagner, "Untersuchungen über Fehlordnungserscheinungen in Kupferoxydul und deren Einfluss auf die elektrischen Eigenschaften," *Zeitschrift für physikalische Chemie B* 22 (1933): 212–225. See also W. Schottky, "Bericht über die Kupferoxydularbeiten von Dünwald and Wagner" (internal Siemens report), 2 December 1932. See also Schottky manuscripts of 5, 6, and 7 December 1932; W. Schottky to C. Wagner, 7 December 1932; C. Wagner to W. Schottky, 12 December 1932, all in Schottky Papers, Siemens Museum, Munich.

137. Wagner (1933) (n. 135), 185.

138. For example, this conclusion could be drawn from G. Hägg and J. Sucksdorff, "Die Kristallstruktur von Troilit und Magnetkies," *Zeitschrift für physikalische Chemie B* 22 (1933): 444–452, p. 451, at beginning of the last section. On the other hand, see also references to crystallographic work on mixed systems in Wagner and Schottky (n. 135).

139. C. Wagner and J. Beyer, "Über die Natur der Fehlordnungserscheinungen in Silberbromid," *Zeitschrift für physikalische Chemie B* 32 (1936): 113–116.

140. E. Koch and C. Wagner, "Der Mechanismus der Ionenleitung in festen Salzen auf Grund von Fehlordnungsvorstellungen. I," *Zeitschrift für physikalische Chemie B* 38 (1938): 275–324. For an overview of these and other experiments, especially of intrinsic defects, see C. Wagner, "Platzwechselvorgänge in festen Stoffen und ihre modellmässige Deutung," *Berichte der Deutschen Keramischen Gesellschaft* 19 (1938): 207–227; Pick (n. 112). Pick examined optically what Koch and Wagner had done electrochemically.

141. W. Schottky and F. Waibel, "Die Elektronenleitung des Kupferoxyduls," *Phys. Zs.* 34 (1933): 858–864, p. 864. The authors assumed Cu^+ vacancies in an oxygen-enriched Cu_2O lattice. In addition, they suggested transferring "a part" of the discussion to "the processes in alkali halides." Others understood such phrases in this article as a discussion of the vacancy hypothesis (now of anion vacancies) for alkali halides. See Pick (n. 112), 80. However, this hypothesis was discussed explicitly only in unpublished notes by W. Schottky in 1933, for example in manuscript page, 10 June 1933, Schottky Papers, Siemens Museum, Munich.

142. W. Schottky, "Zu Stasiw, 14. Oktober 1932," manuscript, Schottky Papers, Siemens Museum, Munich. Here, and already by 24 August 1932, Schottky described a mechanism of electrons wandering from Cl ions to Na ions, still without the assumption of lattic defects. In earlier manuscripts there is no reference to the alkali halides.

143. Schottky's emphasis. W. Schottky to R. W. Pohl, 30 June 1934, Schottky Papers, Siemens Museum, Munich. The entire letter may refer directly to the work of Hilsch and Pohl (n. 104), also cited.

144. W. Schottky, "Über den Mechanismus der Ionenbewegung in festen Elektrolyten," *Zeitschrift für physikalische Chemie B* 29 (1935): 335–355. For the stimulation given Schottky by the alkali halide problems, see p. 342.

145. W. Schottky, manuscript, 19 May 1936, drawing A, NaCl—Cu_2O, Schottky Papers, Siemens Museum, Munich.

146. W. Schottky to C. Wagner, February 1937, p. 1 at bottom: manuscript addition by Schottky on 22 June 1937, Schottky Papers, Siemens Museum, Munich.

147. J. H. de Boer, *Electron Emission and Adsorption Phenomena* (Cambridge: Cambridge Uni-

versity Press, 1935); de Boer, *Elektronenemission und Adsorptionserscheinungen* (Leipzig: Barth, 1937). Critique by W. Schottky, *Die Naturwissenschaften* 23 (1935): 657–659. See also J. H. de Boer, "Über die Natur der Farbzentren in Alkalihalogenid-Kristallen," *Recueil des travaux chimiques des Pays Bas* 56 (1937): 301–309, p. 302.

148. R. W. Gurney, "Internal Photoelectric Absorption in Halide Crystals," *Proc. Roy. Soc.* A141 (1933): 209–215.

149. Mott (n. 126), 61; N. Mott to J. Teichmann 29 May 1984, DM.

150. "Report of Conference on the Conduction of Electricity in Solids held at Bristol July 1937," *Proceedings of the Physical Society* 49, suppl. (1937).

151. Mott (n. 126), 62; R. W. Gurney and N. F. Mott, "Trapped Electrons in Polar Crystals," *Proceedings of the Physical Society* 49, suppl. (1937): 32–35.

152. For Pohl's lecture, see Pohl (1937) (n. 56); "Discussion of the Papers by Pohl and by Gurney and Mott Reported by N. F. Mott," *Proceedings of the Physical Society* 49, suppl. (1937): 36–39. The definitive *F*-center model with the electron in the center of the vacancy as the place of lowest energy appears to have been suggested first by Mott in a "subsequent discussion" with Verwey, ibid., 37. See also R. W. Gurney and N. F. Mott, "Conduction in Polar Crystals. III. On the Colour Centers in Alkali-Halide Crystals," *Transactions of the Faraday Society* 34 (1938): 506–511, p. 506. This names J. H. de Boer with his work ("Über die Natur der Farbzentren") (n. 147), 301–309, as the originator without reference to Schottky, although de Boer had only taken up Schottky's hypothesis; this historically incorrect statement has been frequently repeated. F. Seitz, "Color Centers in Alkali Halide Crystals," *Reviews of Modern Physics* 18 (1946): 384–408, p. 401 and Fig. 19.

153. Mott and Gurney (n. 126), preface.

154. J. W. Mitchell, "Dislocations in Crystals of Silver Halides," in *Beginnings,* 149–159, p. 149.

155. Hilsch and Pohl (n. 112). E. Mollwo, private communication, 3 December 1983: AEG was interested in the semiconductor amplifier patent of O. Heil, 1934. For this, see H. Goetzeler, "Zur Geschichte der Halbleiter-Bausteine der Elektronik," *Technikgeschichte* 39 (1972): 31–50, pp. 39–40.

156. N. Riehl, *Physik und technische Anwendungen der Lumineszenz* (Berlin: Springer, 1941).

157. Mott and Gurney (n. 126); Seitz (n. 130).

158. J. C. Slater, "Quantum Physics in America Between the Wars," *Physics Today* 21, no. 1 (1968): 43–51.

159. F. Seitz, "Biographical Notes," in *Beginnings,* 84–99, p. 84.

160. Ibid. Seitz probably referred to I. I. Frenkel, *Wave Mechanics: Elementary Theory* (Oxford: Clarendon Press, 1932).

161. J. O. Perrine, "A Spectrographic Study of Ultraviolet Fluorescence Excited by X-rays," *Phys. Rev.* 22 (1923): 48–57; P. L. Bayley, "Coloration of the Alkali Halides by X-rays," *Phys. Rev.* 24 (1924): 495–501.

162. T. E. Phipps and W. R. Brode, "A Comparative Study of Two Kinds of Colored Rock Salt," *Journal of Physical Chemistry* 30 (1926): 507–512.

163. A. L. Hughes, "Photoconductivity in Crystals," *Reviews of Modern Physics* (1936): 294–315.

164. Seitz (n. 159), 84.

165. D. Kevles, *The Physicists* (New York: Vintage Books, 1979), 220.

166. F. Seitz, "Interpretation of the Properties of Alkali Halide-Thallium Phosphers," *Journal of Chemical Physics* 6 (1938): 150–162; Seitz, "An Interpretation of Crystal Luminescence," *Transactions of the Faraday Society* 35 (1939): 74–85.

167. R. Maurer, interview with K. Szymborski, 22 December 1981, AIP.

168. F. Seitz, "The Nature of Lattice Defects in Silver Bromide Crystals," *Phys. Rev.* 54 (1938): 1111–1112.

169. R. B. Windsor, "Production and Properties of Skiatron, Dark-Trace Cathode Ray Tubes with Microcrystalline Alkali Halide Screens" (Ph.D. diss., Massachusetts Institute of Technology, 1944), 2, MIT Archives.

170. A. H. Rosenthal, "A System of Large-Screen Television Reception Based on Certain Electron Phenomena in Crystals," *Proceedings of the Institute of Radio Engineers* 28 (1940): 203–209.

171. British research on the skiatron was carried out by a group led by R.A.R. Ricker at the Admiralty Signal Extention Establishment at Bristol and by the Telecommunication Research Establishment at Malvern. P.G.R. King and J. F. Gittins, "The Skiatron or Dark-Trace Tube," *Journal of the Institution of Electrical Engineers* 93:3A (1946): 822–831. In Germany, research along similar lines initiated by H. Kurzke and J. Rottgardt, "Über die Entfärbung von Alkalihalogenidkristalliten," *Ann. Phys.* 39 (1941) 619–632, led to the development of a device called Blauschrift-Röhre.

172. In addition to Windsor (n. 169), they included J. G. Buck, "Absorption Phenomena in Polycrystalline Potassium Chloride Cathode Ray Tube Screens and the Operating Characteristics of Skiatrons" (Ph.D. diss., Massachusetts Institute of Technology, 1942); R. W. Hull, "A Study of Optical and Electronic Properties in Various Skiatron Screens" (Ph.D. diss., Massachusetts Institute of Technology, 1943); H. F. Ivey, "Optical and Electronic Properties of Polycrystalline Potassium Chloride Skiatron Screens" (Ph.D. diss., Massachusetts Institute of Technology, 1944), MIT Archives.

173. Windsor (n. 169), 5.

174. Maurer interview (n. 167).

175. In 1943, Stern was awarded a Nobel Prize for Physics for his contributions to the development of the molecular ray method and his discovery of the magnetic moment of the proton.

176. Out of 12 reports written by the group for Division 14 of the National Defense Research Committee, 4 dealt with theoretical aspects: F. Seitz, O. Stern, I. Estermann, R. Maurer, S. Lasof, and G.I. Kirkland, "The Theory of Dark-Trace Tubes," NDRC Division 14 Report No. 131, 27 April 1943; Seitz et al., "The Theory of Dark-Trace Tubes. III," Report No. 257, 6 April 1944; F. Seitz, "Theory of Dark-Trace Tubes IV," Report No. 265, 8 May 1944, as cited in Windsor (n. 169).

177. R. Ottmer, "Zur Kenntnis der Absorptionsspektra lichtelektrisch leitender Alkalihalogenide," *Zs. Phys.* 46 (1928): 798–813.

178. E. Mollwo, "Sichtbare Elektronen-Ersatzleitung in Alkalijodidristallen," *Nachr. Gött.* (1935): 215–220.

179. Molnar was an MIT graduate student working in von Hippel's laboratory. J. P. Molnar, "The Absorption Spectra of Trapped Electrons in Alkali-Halide Crystals" (Ph.D. diss., Massachusetts Institute of Technology, 1940), MIT Archives.

180. Seitz (n. 152).

181. J. Koehler, private communication, 24 February 1982.

182. I. Estermann, W. J. Leivo, and O. Stern, "Change in Density of Potassium Chloride Crystals upon Irradiation with X-rays," *Phys. Rev.* 75 (1949): 627–633.

183. Seitz (n. 152).

184. Ibid., 384.

185. Ibid., 401.

186. F. Seitz to K. Szymborski, 13 June 1983.

187. H. Huntington to F. Seitz, September 1, 1943, courtesy of Seitz (in his possession).

188. Maurer interview (n. 167).

189. Dark-trace tubes soon became obsolete, partly because the problem of the persistent coloration of the screens was never satisfactorily resolved and partly due to progress in electronic systems with built-in memories, which enabled construction of luminescent tubes of greatly superior operational performance.

190. According to F. Seitz, "Radiation Effects in Solids," *Physics Today* 5, no. 6 (1952): 6–9. The first to point out this danger were Frank Spedding and Edward Teller.

191. W. P. Eatherly, "Nuclear Graphite—the First Years," *Journal of Nuclear Materials* 100 (1981): 55–63, p. 56; James Koehler, interview with L. Hoddeson, 6 March 1982, AIP.

192. Those included, in addition to Wigner and Szilard, James Franck and Edward Teller.

193. Seitz (n. 190), 7.

194. Eatherly (n. 191), 58. This was not the end of the radiation damage story. Extensive research, mostly classified, was continued after the war.

195. O. C. Simpson, interview with K. Szymborski, 13 August 1982; P. Yuster and C. Delbecq, interview with K. Szymborski, 24 March 1982, AIP.

196. Y. Toyozawa, interview with K. Szymborski, 3 June 1982, AIP.

197. This was particularly true of the Soviet Union—where S. J. Pekar developed Landau's earlier idea into a theory of polarons and *F*-centers: S. J. Pekar, "Theory of *F*-Centers" (in Russian), *Zhurnal Eksperimentalnoi i Teoreticheskoi Fiziki* 20 (1950): 510–523—and of Japan: T. Muto, "Theory of the F-Centers of Colored Alkali Halide Crystals," *Progress of Theoretical Physics* 4 (1949): 181–192, 243–250.

198. C. C. Klick and R. J. Maurer, "Optical Absorption Bands in Alkali Halide Crystals," *Phys. Rev.* 72 (1947): 165.

199. H. W. Etzel and R. J. Maurer, "The Concentration and Mobility of Vacancies in Sodium Chloride," *Journal of Chemical Physics* 18 (1950): 1003–1007.

200. D. Mapother, "Self-Diffusion and Ionic Conductivity of the Sodium Ion in Sodium Chloride," Carnegie Institute of Technology, Technical Report, no. 8 (1949); H. N. Crooks, "Self-Diffusion and Ionic Conductivity of the Sodium Ion in Sodium Bromide," Carnegie Institute of Technology, Technical Report, no. 12 (1949), as cited in *Chemical Abstracts*.

201. G. J. Dienes, "Activation Energy for the Diffusion of Coupled Pairs of Vacancies in Alkali Halide Crystals," *Journal of Chemical Physics* 16 (1948): 620–632.

202. S. I. Pekar, "Local Quantum States of an Electron in an Ideal Ionic Crystal," *Journal of Physics of the USSR* 10 (1946): 341–346; Pekar, "Autolocalization of an Electron in a Dielectric Inertially Polarizing Medium," *Journal of Physics of the USSR* 10 (1946): 347–350.

203. He later extended his approach including also the treatment of the *F*-center. S. I. Pekar, *Untersuchungen über die Elektronentheorie der Kristalle* (Berlin: Akademie-Verlag, 1954). See also n. 197.

204. J. J. Markham and F. Seitz, "Binding Energy of a Self-Trapped Electron in NaCl," *Phys. Rev.* 74 (1948): 1014–1024.

205. Following this conclusion, Markham would later study optical properties of alkali halides in liquid helium temperatures and discover new types of color centers, called *H*-centers. See n. 296.

206. Maurer interview (n. 167).

207. G. M. Almy, "A Century of Physics at the University of Illinois" (Paper presented at the History of Science Society, December 1967; manuscript in University of Illinois Archives).

208. As Seitz reminisced in a recent interview, Loomis "had held off building up the department in the immediate postwar period, because he felt he couldn't get the kinds of people he wanted. Practically all education had stopped at the Ph.D. level in 1940, with a few exceptions, but not many. So there was a big scramble for the people who had degrees. And he felt that if he tried to build up the department at that time, he'd get too many second raters. So he held off, temporized, felt that we wanted more than high energy physics, something to complement it, and saw a good chance to build a group" (F. Seitz, interview with L. Hoddeson, 27 January 1981, AIP).

209. Almy (n. 207), 17.

210. When Slichter was offered a job by Loomis, who at the beginning of 1949 was visiting laboratories around the country and looking for people to hire, he asked only two questions: "Is it true that Seitz is coming to Illinois?" and "May I wait and give my answer when we know?" (C. P. Slichter, interview with K. Szymborski, 29 January 1982, AIP).

211. D. Dutton and R. Maurer, "Color Centers and Trapped Charges in KCl and KBr," *Phys. Rev.* 90 (1953): 126–130; K. Teegarden and R. Maurer, "V_1 and H-Centers in KCl," *Zs. Phys.* 138 (1954): 284–289.

212. C. A. Hutchison, Jr., "Paramagnetic Resonance Absorption in Crystals Colored by Irradiation," *Phys. Rev.* 75 (1949): 1769–1770.

213. Seitz (n. 87).

214. W. Känzig, "Electron Spin Resonance of V_1-Centers," *Phys. Rev.* 99 (1955): 1890(L).

215. T. G. Castner and W. Känzig, "The Electronic Structure of V-Centers," *Journal of Physics and Chemistry of Solids* 3 (1957): 178–195.

216. C. J. Delbecq, B. Smaller, and P. H. Yuster, "Optical Absorption of Cl_2-Molecule-Ions in Irradiated KCl," *Phys. Rev.* 111 (1958): 1235–1240.

217. Seitz (n. 87).

218. Maurer interview (n. 167).

219. Simpson interview (n. 195).

220. Yuster and Delbecq interview (n. 195).

221. Ibid.

222. Yuster reminisced: "We worried at first, because we argued with him. We were young and he was a distinguished scientist. And we later found out that he thought it was wonderful!" (ibid.).

223. E. Burstein, 1 April 1982, C. Klick, 2 April 1982, and D. Compton, 20 April 1982, interviews with K. Szymborski, AIP.

224. Burstein interview (n. 223).

225. E. Burstein and J. J. Oberly, "Infrared Color Center Bands in the Alkali Halides," *Phys. Rev.* 76 (1949): 1254.

226. Klick interview (n. 223).

227. Ibid.

228. C. C. Klick, "Luminescence of Color Centers in Alkali Halides," *Phys. Rev.* 79 (1950): 894–889.

229. C. C. Klick, "Search for F-Center Luminescence," *Phys. Rev.* 94 (1954): 1541–1545.

230. D. L. Dexter, C. C. Klick, and G. A. Russell, "Criterion for the Occurrence of Luminescence," *Phys. Rev.* 100 (1955): 603–605; G. A. Russell and C. C. Klick, "Configuration Coordinate Curves for F-Centers in Alkali Halide Crystals," *Phys. Rev.* 101 (1956): 1473–1479.

231. J. Lambe and W. D. Compton, "Luminescence and Symmetry Properties of Color Centers," *Phys. Rev.* 106 (1957): 684–693.

232. Among others, H. Rabin, "X-ray Expansion and Coloration of Undoped and Impurity-Doped NaCl Crystals," *Phys. Rev.* 116 (1959): 1381–1389; H. Rabin and J. H. Schulman, "F-Band Structure in Cesium Bromide," *Phys. Rev. Letters* 4 (1960): 280–282; H. Rabin and C. C. Klick, "Formation of F-Centers at Low and Room Temperature," *Phys. Rev.* 117 (1960): 1005–1010.

233. W. D. Compton and J. H. Schulman, *Color Centers in Solids* (Oxford: Pergamon Press, 1962).

234. Ibid.

235. H. Pick, "Meine Göttinger Zeit, eine Rückerinnerung," recollections for the International Project in the History of Solid State Physics, January 1983, DM.

236. S. Petroff, "Photochemische Beobachtungen an KCl-Kristallen," *Zs, Phys.* 127 (1950): 443–454.

237. Seitz (n. 159), 97.

238. By 1954, Pohl, Pick, and their students and collaborators, H. V. Harten, H. Dorendorf, H. Hummel, W. Martienssen, E. Miessner, H. Rüchardt, and F. Lüty, had published about 30 papers dealing mainly with color centers in alkali halides.

239. F. Lüty, interview with K. Szymborski, 4 June 1982, AIP.

240. Toyozawa interview (n. 196).

241. Muto (n. 197). In addition to Muto, a group was led by Taturo Imu and another by Ryogo Kubo. Toyozawa interview (n. 196).

242. Y. Uchida, M. Ueta, and Y. Nakai, "A New Absorption Band in Colored Rocksalt Crystals," *Journal of the Physical Society of Japan* 4 (1949): 57–58.

243. D. L. Dexter, "University of Illinois Conference on Ionic Crystals," *Science* 115 (1952): 199–200.

244. Klick interview (n. 223).

245. Dexter (n. 243).

246. Ibid., 200.

247. Klick interview (n. 223).

248. See, for example, J. J. Markham, R. T. Platt, and J. L. Mador, "Bleaching Properties of F-Centers in KBr at 5°K," *Phys. Rev.* 92 (1953): 597–603.

249. Yuster and Delbecq interview (n. 195).

250. Argonne Color Center Symposium, Program, 1956, Argonne Archives.

251. Lüty interview (n. 239).

252. W. Martienssen, Corvallis Color Center Symposium, Program, 1959. See also Pohl's lecture at Corvallis, DM (copy courtesy of R. Maurer).

253. Stuttgart Color Center Symposium, Program, 1962, p. 24, in R. Maurer's possession.

254. M. Ueta, "On the Bleaching of Color Centers (M-Center) in KCl Crystal with Polarized Light," *Journal of the Physical Society of Japan* 7 (1952): 107–109.

255. Delbecq et al. (n. 216).

256. Z. Gyulai and D. Hartly, "Elektrische Leitfähigkeit verformter Steinsalzkristalle," *Zs. Phys.* 51 (1928): 378–387.

257. L.-Y. Lin, "Expansion of Potassium and Sodium Chloride Crystals Due to X-ray Irradiation of Weak Intensities," *Phys. Rev.* 102 (1956): 968–974; H. Rabin, "X-ray Expansion and Coloration of Undoped and Impurity-Doped NaCl Crystals," *Phys. Rev.* 116 (1959): 1381–1389.

258. W. Primak, C. J. Delbecq, and P. H. Yuster, "Photoelastic Observations of the Expansion of Alkali Halides on Irradiation," *Phys. Rev.* 98 (1955): 1708–1715.

259. D. Binder and W. J. Sturm, "Equivalence of X-ray Lattice Parameter and Density Changes in Neutron Irradiated LiF," *Phys. Rev.* 96 (1954): 1519–1522; C. Berry, "Change in KCl Lattice by Soft X-rays," *Phys. Rev.* 98 (1955): 934–936; A. Smakula, J. Kalnajs, and V. Sils, "Densities and Imperfections of Single Crystals," *Phys. Rev.* 99 (1955): 1747–1750; M. N. Podaschewsky, "Über die Photoelektrische Methode zur Bestimmung der Elastizitätsgrenze eines Röntgenisierten Steinsalzkristalles," *Physikalische Zeitschrift der Sowjetunion* 7 (1935): 399–409; Podaschewsky, "Über die Wirkung der Photochemischen Verfärbung auf die Streck- und Festigkeitsgrenze der Steinsalzmonokristalle," *Physikalische Zeitschrift der Sowjetunion* 8 (1935): 81–92.

260. R. Smoluchowski, interview with K. Szymborski, 16 August 1982, AIP.

261. W. Leivo and R. Smoluchowski, "X-ray Coloring of 400-MeV Proton-Irradiated KCl," *Phys. Rev.* 93 (1954): 1415–1416; D. A. Wiegand and R. Smoluchowski, "Evidence for Interstitials in Low-Temperature X-ray Irradiated LiF," *Phys. Rev.* 110 (1958): 991–992; M. F. Merriam, R. Smoluchowski, and D. A Wiegand, "Enhanced Thermal Expansion in X-rayed Rocksalt," *Phys. Rev.* 125 (1962): 52–64; Merriam et al., "High-Temperature Thermal Expansion of Rocksalt," *Phys. Rev.* 125 (1962): 65–67.

262. P. Jensen, "Die magnetische Suszeptibilität von Kaliumbromidkristallen mit Farbzentren," *Ann. Phys.* 34 (1939): 161–177.

263. E. Zavoisky, "Paramagnetic Relaxation of Liquid Solutions for Perpendicular Fields," *Journal of Physics of the USSR* 9 (1945): 211–216.

264. B. V. Rollin and J. Hatton, "Nuclear Paramagnetism at Low Temperatures," *Phys. Rev.* 74 (1948): 346.

265. Ibid.

266. Slichter interview (n. 210).

267. Ibid.

268. Hutchison (n. 212); C. A. Hutchison, Jr., and G. A. Noble, "Paramagnetic Resonance Absorption in Additively Colored Crystals of Alkali Halides," *Phys. Rev.* 87 (1952): 1125–1126.

269. M. Tinkham and A. F. Kip, "Paramagnetic Resonance Absorption in Crystals Containing Color Centers," *Phys. Rev.* 83 (1951): 657–658.

270. E. E. Schneider and T. S. England, "Paramagnetic Resonance at Large Magnetic Dilutions," *Physica* 17 (1951): 221–233.

271. A. Overhauser, 22 February 1982, A. Portis, 15 June 1982, and A. Kip, 16 June 1982, interviews with K. Szymborski, AIP.

272. Overhauser interview (n. 271).

273. Portis interview (n. 271).

274. A. Kip, C. Kittel, R. A. Levy, and A. M. Portis, "Electronic Structure of F-Centers. Hyperfine Interaction in Electron Spin Resonance," *Phys. Rev.* 91 (1953): 1066–1071.

275. G. Feher, "Observation of Nuclear Magnetic Resonance in the Electron Spin Resonance Line," *Phys. Rev.* 103 (1956): 834–835; Feher, "Electronic Structure of F-Centers in KCl by the Electron Spin Double Resonance Technique," *Phys. Rev.* 105 (1957): 1122–1123.

276. Kip et al. (n. 274).

277. J. Markham, *F-Centers in Alkali Halides* (New York: Academic Press, 1966), 293.

278. Toyozawa interview (n. 196).

279. For a review of the theory of the *F*-center, see, for instance, Compton and Schulman (n. 233); B. S. Gourary and F. J. Adrian, "Wave Functions for Electron-Excess Color Centers in Alkali Halide Crystals," in *Solid State Physics*, vol. 10, ed. F. Seitz and D. Turnbull (New York: Academic Press, 1960), 127–247.

280. S. R. Tibbs, "Electron Energy Levels in NaCl," *Transactions of the Faraday Society* 25 (1939): 1471–1484.

281. J. H. Simpson, "Charge Distribution and Energy Levels of Trapped Electrons in Ionic Solids," *Proc. Roy. Soc.* A197 (1949): 269–281.

282. W. A. Smith and A. B. Scott, "A Calculation of Electron Energy Levels in F-Centers," *Bulletin of the American Physical Society* 28, no. 2 (1953): 21–29. J. A. Krumhansl and N. Schwartz, "The Calculation of F-Center Energy Levels," *Phys. Rev.* 89 (1953): 1154–1155.

283. B. S. Gourary and P. J. Luke, "Approximate Wave Functions for the M-Center by the Point Ion Lattice Method," *Phys. Rev.* 107 (1957): 960–965; B. S. Gourary and F. J. Adrian, "Approximate Wave Functions for the F-Center, and Their Application to the Electron Spin Resonance Problem," *Phys. Rev.* 105 (1957): 1180–1185.

284. Gourary and Adrian (n. 279).

285. Pekar (n. 197, 202).

286. K. Huang and A. Rhys, "Theory of Light Absorption and Non-Radiative Transitions in F-Centres," *Proc. Roy. Soc.* A204 (1950): 406–423.

287. M. Lax, "The Franck-Condon Principle and Its Application to Crystals," *Journal of Chemical Physics* 20 (1952): 1752–1757; R. C. O'Rourke, "Absorption of Light by Trapped Electrons," *Phys. Rev.* 91 (1953): 265–270; H.J.G. Meyer, "Theory of Radiationless Transitions in F-Centers," *Physica* 20 (1954): 181–182; Meyer, "On the Theory of Transitions of F-Center Electrons," *Physica* 20 (1954): 1016–1020.

288. Dexter et al. (n. 230).

289. A. von Hippel, "Einige prinzipelle Gesichtspunkte zur Spektroskopie der Ionenkristalle und ihre Anwendung auf die Alkalihalogenide," *Zs. Phys.* 101 (1936): 680–720.

290. F. Seitz, "An Interpretation of Crystal Luminescence," *Transactions of the Faraday Society* 35 (1939): 74–85.

291. Russell and Klick (n. 230).

292. Klick interview (n. 223).

293. Muto (n. 197).

294. T. Inui and Y. Uemura, "Theory of Color Centers in Ionic Crystals," *Progress of Theoretical Physics* 5 (1950): 252–265, 395–404.

295. Kip et al. (n. 274).

296. W. H. Duerig and J. J. Markham, "Color Centers in Alkali Halides at 5°K," *Phys. Rev.* 88 (1952): 1043–1049.

297. Markham and Seitz (n. 204).

298. Actually, Duerig and Markham observed two separate bands, which they named H_1 and H_2, but later in the literature the letter H was retained for the first one (lying at longer wavelength), the latter being somehow forgotten, and possibly spurious. Markham (n. 277), 151.

299. For example, K. Teegarden and R. Maurer, "V_1 and H-Centers in KCl," *Zs. Phys.* 138 (1954): 284–289.

300. Seitz (n. 87).

301. J.H.O. Varley, "A Mechanism for the Displacement of Ions in an Ionic Lattice," *Nature* 174 (1954): 886–887.

302. W. C. Compton and C. C. Klick, "Some Low-Temperature Properties of the R, M and N-Centers in KCl and NaCl," *Phys. Rev.* 112 (1958): 1620–1623.

303. W. Känzig and T. O. Woodruff, "Electron Spin Resonance of H-Centers," *Phys. Rev.* 109 (1958): 220–221; Känzig and Woodruff, "The Electronic Structure of an H-Center," Report No. 58-RL-2080, General Electric, 1958.

304. See, for example, P. D. Townsend and J. C. Kelly, *Color Centers and Imperfections in Insulators and Semiconductors* (New York: Crane, Russak, 1973), 55.

305. C. J. Delbecq, P. Pringsheim, and P. Yuster, "A New Type of Short Wavelength Absorption Band in Alkali Halides Containing Color Centers," *Journal of Chemical Physics* 19 (1951): 574–577.

306. For a review of the literature on excitons see, for instance, R. S. Knox, *Theory of Excitons*, Solid State Physics Series, suppl. 5, ed. F. Seitz and D. Turnbull (New York: Academic Press, 1963).

307. Delbecq et al. (n. 305), 576.

308. C. J. Delbecq, P. Pringsheim, and P. Yuster, "α- and β-bands in Potassium Bromide Crystals," *Journal of Chemical Physics* 20 (1952): 746–747.

309. W. Martienssen, "Photochemische Vorgänge in Alkalihalogeniden," *Zs. Phys.* 131 (1952):

488–504; W. Martienssen and R. W. Pohl, "Zur photochemischen Sensibilisierung," *Zs. Phys.* 133 (1952): 153–157. The Argonne group discovery came as a surprise to Pohl, and when he met Yuster and Delbecq at the 1951 Urbana meeting he admitted that he had at first been quite skeptical about their results (n. 195). It turned out, however, that he had not been far from discovering α- and β-bands some 10 years earlier. In 1940 one of his collaborators had carried out an experiment in which a potassium bromide crystal containing V-centers was bleached at liquid nitrogen temperature. H. Thomas, "Zur Photochemie des KH-KBr-Mischkristalles," *Ann. Phys.* 38 (1940): 601–608. The bleaching gave rise to a new band, which was assumed to be somehow associated with the U-band and was given the name U'. It later proved to be identical with the α-band.

310. D. L. Dexter, "Note on the Absorption Spectra of Pure and Colored Alkali Halide Crystals," *Phys. Rev.* 83 (1951): 435–438.

311. D. L. Dexter, "Oscillator Strengths for the α- and β-bands in Alkali Halide Crystals," *Phys. Rev.* 83 (1953): 1044–1046.

312. A. Taft and L. Apker, "Photoelectric Emission Associated with β-band Absorption in Alkali Iodides," *Journal of Chemical Physics* 20 (1952): 1648–1653.

313. Mott and Gurney (n. 126).

314. Kleinschrod (n. 89).

315. Seitz (n. 87), 7.

316. Duerig and Markham (n. 296).

317. H. Etzel and F. E. Geiger, Jr., "K-Band in Additively Colored KCl," *Phys. Rev.* 96 (1954): 225–226.

318. F. Lüty, "Höhere Anregungszustände von Farbzentren," *Zs. Phys.* 160 (1960): 1–15.

319. Lüty attributed the variation of the K/F band ratio observed earlier by Etzel and Geiger to changes in the absorption bands that peak at longer wavelength than the F-band.

320. Ottmer (n. 177).

321. Ottmer wrote that "the first thought was, naturally, that the side maxima are only of secondary nature originating from the coagulation of the alkali metals to form bigger particles" (ibid., 805).

322. Soviet Union: for instance, N. Kalabuchov, "Some Observations on the Inner Photo-Effect in Crystals of Alkali-Halide Salts," *Journal of Physics of the USSR* 9 (1945): 41–44; Germany: Petroff (n. 179) (n. 236); United States: Molnar.

323. Seitz (n. 152), 384.

324. W. A. Smith and A. B. Scott, "Thermal Equilibrium between F-centers and M-centers in Alkali Halides," *Journal of Chemical Physics* 21 (1953): 2096.

325. C. Z. van Doorn, "Thermal Equilibrium Between F and M Centers in Potassium Chloride," *Physical Review Letters* 4 (1960): 236–237; B. J. Faraday, H. Rabin, and W. D. Compton, "Evidence for the Double F Model of the M Center," *Physical Review Letters* 7 (1961): 57–59.

326. N. W. Lord, "M-Center Spin Resonance and Oscillator Strength in LiF," *Phys. Rev.* 106 (1957): 1100–1101.

327. R. Herman, M. C. Wallis, and R. F. Wallis, "Intensities of the R_1 and R_2 Bands in KCl Crystals," *Phys. Rev.* 103 (1956): 87–93.

328. J. Lambe and W. D. Compton, "Luminescence and Symmetry Properties of Color Centers," *Phys. Rev.* 106 (1957): 684–689.

329. Compton and Klick (n. 302).

330. C. Z. van Doorn and Y. Haven, "Absorption and Luminescence of Color Centers in KCl and NaCl," *Philips Research Reports* 11 (1956): 479–488, p. 487.

331. A. W. Overhauser and H. Rüchardt, "Inversion Symmetry of M and R-Centers," *Phys. Rev.* 112 (1958): 722–724.

332. Ueta (n. 254).

333. R. S. Knox, "Inversion Symmetry of the M Center," *Physical Review Letters* 2 (1959): 87–88.

334. Van Doorn (n. 325); Faraday et al. (n. 325).

335. H. Gross and H. C. Wolf, "Elektronenspin-Resonanz-Untersuchungen an M-Zentren in KCl," *Naturwissenschaften* 48 (1961): 299–300.

336. H. Gross, "Hat das M-Zentrum in KCl ein magnetisches Moment," *Zs. Phys.* 164 (1961): 341–358.

337. H. Pick, "Farbzentren-Assoziate in Alkalihalogeniden," *Zs. Phys.* 159 (1960): 69–76.

5 | *Mechanical Properties of Solids*

ERNEST BRAUN

5.1   Plastic Deformation of Metals Before 1934

The utility of metals is based on many of their properties, but pride of place is taken by the combination of strength and ductility that most metals possess. For unless metals could be shaped by pressing, rolling, or forging in a hot or cold state, their utility as general-purpose materials of construction for tools, weapons, machines, bridges, and buildings would be greatly impaired. The term *ductility* implies that the material can be permanently deformed—or shaped—by applying a force to it. This plastic deformation is distinct from elastic deformation, where strain follows stress but the material returns to its original shape when the external force is removed. Thus plastic deformation is one of the most desirable and technically important properties of metals, yet for a very long time posed extremely difficult puzzles. Indeed, so difficult were the puzzles that science had practically no answer, and engineering developments proceeded without the aid of pure science.

From the earliest days of civilizations based on metals, the preparation of metals and metallic implements was in the hands of craftsmen who learned the traditions of their trade and gradually improved on them. It was only with the onset of the industrial revolution, when demand for iron and steel and nonferrous metals began to soar and not only greater quantities but also better grades and higher reliability were demanded, that the old craft methods were rapidly and systematically improved. The developments included the use of coke instead of charcoal for the production of pig iron in blast furnaces and the introduction of separate converters where crude and brittle pig iron, which previously could be improved only by mechanical working, was converted to ductile steel through heat treatment. The first such converter was introduced in 1856 by Henry Bessemer and was rapidly followed by improved methods. It was nevertheless still characteristic of the period that improvements, both quantitative and qualitative, were made by empirical methods, although the scale of things had increased tremendously and large entrepreneurial capitalist enterprises, instead of small cottage industries, had emerged. This fact meant also that attempts at large steps in technical developments could now substitute for the gradual step-by-step improvements in traditional methods handed down from master to apprentice.

It has been said that the art of metalworking grew into the science of metallurgy, and one of the founders of this science was Henry Sorby, who in 1861 began to apply the optical microscope to the systematic investigation of metals.[1] As ever-increasing demands were made on the structural qualities of cast and wrought iron, they gradually were replaced by steel, with its superior ductility and elasticity. One of the historical landmarks in this process was the Tay Bridge disaster of 1879, when the wrought-iron railway bridge collapsed in high wind with the loss of 78 lives.[2] It is thought that this disaster contributed to the establishment of the engineering subject of metallurgy as a proper academic subject of applied research and teaching.

The introduction of the optical microscope into metallurgy is a good example of what has been called the instrument bond between science and technology.[3] The optical microscope was initially a product of pure research into optics; it was developed by engineering design and then applied both in pure research, such as botany, and in engineering research, development, and quality control, such as in metallography. When we say that metallurgy developed from an art into a science in the second half of the nineteenth century, we must clarify the different meanings of the word *science.* Metallurgy certainly moved along with the rest of engineering to become more science-based and scientific; it moved from the art and craft of making things, to organized and abstracted knowledge—from technique to technology. But when we speak of the science of engineering or technology, we do not mean the same thing as when we speak of a pure science, such as physics. Although both engineering—at least scientific engineering as practiced from the second half of the nineteenth century—and pure science use the scientific method, they ask fundamentally different questions and have fundamentally different aims. Engineering asks how to make things, how to solve practical problems; while science asks why: Why do solids behave as they do? Why do we obtain atomic spectra? Of course, the distinctions are a little blurred, and there is much controversy about the nature of the scientific method, but in essence the distinction holds and there is a rough consensus that science uses experiment and observation to support or falsify theoretical edifices. In science, these edifices are fairly abstract and aim to answer the question why; in engineering, the theoretical edifices are more descriptive and prescriptive, telling the practitioner how to do things. Both science and engineering use knowledge accumulated in either branch and are thus linked by the "common pool of knowledge bond."[4]

It is remarkable that science in the pure sense did not enter metallurgy until a very late stage, the past few decades, and even today it remains somewhat peripheral. Indeed, the very success of the engineering science of metallurgy in developing better alloys for every purpose, better methods of extraction and preparation, and better quality controls enabled it to dispense with pure science. Metallurgy did, of course, use systematic experimentation and documentation and was supported extensively by the use of scientific instruments. The bond between engineering and science is particularly strong in the common use of instruments, based on scientific knowledge and developed by engineering expertise. The success of metallurgy came when it used scientific methods of experiment, measurement, and documentation, yet without leaving the realm of application in practical tasks. The very success of metallurgy as an engineering subject made the intrusion of pure physics into it more difficult. A profession that can supply the world's industry with all the materials it needs at a price it can afford does not look to pure science for help.

Pure science works without a definite time horizon: it postulates hypotheses and theories, and these remain on permanent probation, as it were, accepted as correct until shown otherwise. Engineering, on the other hand, must provide solutions today; it cannot wait for better methods or materials, but solves today's problems with today's methods. In the abstract world of science, theories are seemingly permanent, albeit probationary, edifices; in the real world of engineering, products and solutions tend to be obsolescent even at birth. The scientific theory exists indefinitely, until shown to be wrong or useless, which can be a long time, whereas the engineering product is under constant pressure from its successor. This difference is apparent in the teaching of the subjects and in the research underpinning them. While dislocation theory took decades to develop but was accepted by scientists as "essentially correct" when it seemed merely plausible, metallurgists began to believe it only when tangible proofs were obtained, and even then regarded it as peripheral to their interests because it provided no practical prescriptions, no improved recipes for the preparation of practical materials.

The very success of metallurgy in producing reliable engineering materials barred the entry of physics and kept metallurgy firmly within the realm of engineering. This is inevitable, since the task of engineering metallurgy is to produce suitable materials at an acceptable price. Physics is called on only when necessary, such as in the development of very specialized materials for semiconductor electronics or for nuclear reactors. In the normal course of events, physics merely serves the purposes of underpinning practical knowledge with theoretical understanding and providing suitable instrumentation.

The success of metallurgy delayed not only the entry of physics into the art of materials manufacture, but perhaps even any interest physicists might have had in problems of the structure, preparation, and mechanical properties of metals. Even the complex systems of precipitation-hardened aluminum alloys, the family of alloys whose ancestor was duralumin, were discovered and developed empirically. This proved no obstacle to their widespread use in airship and aircraft construction.[5] These alloys obtain their strength by a distribution of small precipitates, a mechanism known as dispersion hardening. The alloys could be controlled and used, but the mechanism of hardening was not understood. Eventually, this manmade mystery proved as much of a puzzling challenge to science as the mysteries of nature. Thus perhaps yet another mechanism of cooperation and interdependence between physics and engineering is operative, a bond in which engineering systems pose puzzles to the scientist and scientific knowledge may or may not help to develop better engineering systems.

This is rather similar to the mechanism previously called the innovation bond.[6] In this mode, physics and engineering work together on the invention and development of a product or process intended to be marketed as a technological innovation. In this cooperation, engineering solutions may pose puzzles to pure science, which are taken up with the intention of improving the engineering solutions by a deeper understanding of the underlying mechanisms. The innovation bond became distinctly and consciously operative when materials had to be developed that were able to withstand the ravages of nuclear radiation in nuclear reactors or when materials for semiconductor electronics had to be obtained.[7] Until that time, physics and metallurgy went largely their own separate ways, and it must be admitted that the path of metallurgy was the more successful. The situation was summed up by Mott:

> But if three decades ago physicists had been asked to say why the addition of a little of one metal improves the properties of another, they would have had to say that they were as ignorant of it as they were of why salt improves the taste of meat. . . . The history of metallurgy has been the opposite, then, of the history of the newer skills, such as electronics and nuclear energy, in which an understanding of what an electron or nucleus is and why it behaves as it does preceded the technology of how to use it.[8]

It must be added that metallurgy was by no means exceptional in this respect and that many branches of engineering developed to a very high level without the aid of pure science. In some cases, technology gets far ahead of science—metallurgy is a good example—and then poses scientific problems rather like nature. In other cases, science offers opportunities for engineering developments; microelectronics is a good example of this. In all cases, much collaboration between the two kinds of activity occurs.

The beginning of modern physical understanding of the mechanical behavior of metals can be dated to the period following the First World War. Progress was made mainly through the application of new scientific experimental techniques, particularly the use of x rays for the study of crystal structure and the growing of single crystals of metals. These two techniques, together with the extensive use of optical metallography and stress-strain measuring equipment, formed the basic armory for an attack of physics on mechanical properties of metals. Thus two older metallurgical techniques combined with two new scientific developments to begin an era that we would be justified in calling the era of metal physics.

Two major monographs summarize the state of knowledge of plastic deformation of metals acquired in that era and before the advent of dislocation theory, roughly from 1918 to 1934. C. F. Elam completed her book *Distortion of Metal Crystals* in the summer of 1934 and added a note on Taylor's paper introducing dislocations in proof prior to publication of the book in 1935.[9] The other major book came from the German tradition and was written by E. Schmid and W. Boas.[10]

Perhaps the most important first step from engineering metallurgy to metal physics was the use of single metal crystals instead of polycrystalline materials for experimentation. The three main methods of crystal growth were (1) solidification from the liquid, (2) deposition from the vapor, and (3) recrystallization in the solid. Many names are associated with these early successful attempts at growing single crystals and thus being able to determine definite relationships between directions of stress and of strain, something that cannot be done in a polycrystalline solid in which the crystallites are randomly oriented. The method of growth from the melt, developed between 1918 and 1924, is associated with, among others, Czochralski (1918), Bridgman (1923), and Obreimow and Schubnikow (1924).[11] Recrystallization in the solid is achieved by a suitable combination of stress and heat treatment, which causes some grains of a polycrystalline specimen to grow preferentially at the expense of others. The method was described by Carpenter and Elam in 1921.[12] It seems likely that the method was based on previous work by Daniel Hanson, who probably grew aluminum crystals as early as 1920 at the National Physical Laboratory (NPL). A cache of such crystals was found at the NPL bearing the date 1920.[13] Growth from the vapor phase was described by Elam for single crystals of brass and by van Arkel, de Boer, and Fast for the growth of single-crystal tungsten.[14]

In the years prior to 1934, a tremendous amount of experimental work on the plastic deformation of such single crystals of a variety of substances was carried out. Gradually, a description of the behavior of crystals under stress was built up, but no plausible theoretical explanation was apparent. A wealth of experimental results and a variety of theoretical explanations existed, but the first truly satisfactory hypothesis, which was to provide a common basic explanation for all the known phenomena, was not put forward until 1934, when dislocation theory was proposed.

When a material is subjected to a small load, the strain is proportional to the stress; when the load is removed, the material returns to its original dimensions. This phenomena is known as Hooke's law, and is obeyed to a good approximation. When the material reaches its elastic limit, it either deforms plastically or suffers brittle fracture. Plastic deformation is characterized by the fact that a change in dimensions is retained even when the load is removed from the specimen. The majority of metals can be plastically deformed, and there are essentially two mechanisms by which this happens: slip on specific glide planes and twinning, although the latter is also brought about by a combination of slip on many glide planes simultaneously.

A great deal of experimental effort was devoted to detailed observations of slip bands and to the geometric relationships between the crystallographic orientation of the glide planes and the direction of stress. The conclusion was that there were certain preferred glide planes and that slip occurred on those of the family of preferred glide planes on which the shear stress owing to the applied load was greatest. The slip planes are normally planes of high atomic density. It was established that the crystal structure remains largely unchanged during deformation, although the orientation of the crystal may change gradually relative to the direction of stress. A sometimes observed elongation of the Laue spots indicates a slight bending of the crystal planes, a phenomenon known as asterism, to which we shall return later.

The alternative mechanism of plastic deformation, by twinning, can also be explained as slip. In this case, however, slip must occur on many planes simultaneously, rather than on just a few selected planes. Combinations of twinning and slip are also sometimes observed, with the former facilitating the latter.

Three features of plastic deformation of metal crystals appeared especially puzzling and thus provided avenues for further investigation. The first is that crystals deform and fracture under stresses that are many orders of magnitude smaller than those predicted for the strength of ideal crystals from the theory of cohesive forces. The second is that a plastically deformed crystal shows increased resistance to further deformation, a phenomenon known as work hardening. Finally, it was hard to understand why slip should occur on many planes, forming systems of slip bands, rather than leading to fracture in the plane that first started to slip.

Several scientists attempted to calculate the theoretical strength of materials, using the known cohesive energy. As early as 1903, Traube, for example, calculated the theoretical strength of iron in the then common unit of kgf mm^{-2} and found a value of 3330, whereas the experimental breaking strength of the best steels was of the order of only 300 kgf mm^{-2}. Zwicky performed similar calculations for sodium chloride, based on earlier estimates by Polanyi, on the basis of the Born theory of cohesive forces in ionic crystals.[15]

A very influential, often quoted, calculation of the strength of ideal solids was performed by Frenkel. Unlike Zwicky, who calculated the force required to separate two crystal planes, Frenkel calculated the strength by assuming two planes, each with periodic potential, to slide across each other.[16] Despite the different approach, the result remained the familiar one; real crystals are several orders of magnitude weaker than they should be according to theory.

This fact posed a true challenge to science, a "puzzle" in the sense in which Kuhn uses the term, and we shall see how eventually, after many predecessor theories, the theory of dislocations resolved this question.

A major preoccuption during the period after the First World War was the calculation of stress components in different crystal planes and hence a prediction of systems of glide planes. Much theoretical effort was thus devoted to purely geometric considerations. So, for example, charts were constructed by von Goler and Sachs for predicting the strength and extension of aluminum crystals as a function of orientation relative to the axis of the test piece.[17]

The most puzzling feature of all in the process of plastic deformation is work hardening. In this context, too, the mechanism of glide of one crystal plane over another became the center of attention. It was clear that resistance to plastic deformation could be increased by previous deformation (work hardening) or, more commonly and technically effectively, by alloying. Neither mechanism was properly understood, although both seemed somehow to make the gliding of planes over each other harder: "It is quite clear that in the present state of our knowledge no simple explanation is likely to be found to account for all the results of plastic deformation."[18] Two main explanations for all or some of the phenomena observed were put forward and variously developed. The first theory, proposed mainly by Beilby, postulated localized melting of the crystal in the glide plane. According to this theory, glide between crystal planes occurs in a locally liquid region that allows limited movement and then solidifies into a glassy phase. This hard amorphous layer locks adjacent planes together and gives rise to work hardening.[19] One of the counterarguments against this theory was that the molten region could be expected to lead to slip in an arbitrary direction, preferably closer to the direction of stress, rather than permit slip in the glide planes only as observed in experiments. Nevertheless, Beilby's theory was extremely influential, and it was generally believed that plastic flow occurred in temporarily liquid regions of the crystal. It was equally widely held that the actual strength of work-hardened, alloyed, or polycrystalline materials was determined by the amorphous layers lying between regions of crystalline perfection. According to N. P. Allen,

> By 1920 the average metallurgical student was as fully convinced of the existence of amorphous metal as his modern equivalent is of the existence of dislocations,

and one of the main objects of metallurgical research was to determine separately the properties of the crystals, and the properties of the boundaries.[20]

Beilby's original terminology is more colorful and more oriented toward his special interest, which he pursued over 20 years of experimental observations, on "the micro-structure and physical properties of solids in various states of aggregation." The full flavor is best obtained from his own words:

> According to this theory, hardening results from the formation at all the internal surfaces of slip or shear of mobile layers similar to those produced on the outer surface by polishing. These layers only retain their mobility for a very brief period and then solidify in a vitreous state, thus forming a cementing material at all surfaces of slip or shear throughout the mass.[21]

The most fruitful early theories of plastic deformation all assumed some sort of distorted or otherwise defective lattice structure. Thus as early as 1920, the principle of lattice defects (i.e., imperfections accidentally present in real crystals) was vaguely enunciated. While Beilby suggested temporary liquefaction and amorphous solidification, Griffith suggested that crystals contained very fine cracks and calculated the strength of materials by equating the loss in elastic energy caused by the extension of a crack to the gain in its surface energy.[22] These calculations show that realistically small forces can lead to the propagation of a crack and thus to plastic deformation. The postulate was a result of Griffith's preoccupation with calculations of stress concentrations at the end of cracks and generally in association with sharp edges. He was led to such considerations while working in the Royal Aircraft Factory in Farnborough during the First World War. In practice, sharp corners in aircraft and engine structures led to mechanical failures. Rounding the corners, which theoretically still left large stress concentrations, seemed in practice sufficient to restore the strength of the structure. Considerations of this kind led Griffith not only to think in terms of cracks, but also to realize that there was more to cracks than just stress concentration. Griffith's theory of crack propagation as a mode of plastic deformation was one of the direct predecessors of dislocation theory. A personal factor, which perhaps eased the transition from crack theory to dislocation theory, was that Taylor, one of the fathers of dislocation theory, worked with Griffith at Farnborough during the early part of the First World War and was therefore thoroughly acquainted with the theory of cracks.[23]

Although Griffith and, in a sense, Beilby postulated the existence of a defect that was responsible for the experimentally observed weakness of real crystals, it was Smekal who introduced the more general concept of defects. He postulated that crystals exhibited some properties that were insensitive to the existence of defects— such as absorption spectra, specific heat, and thermal or electrical conduction in metals—and other properties that were structure-sensitive in the sense of being influenced by deviations from the ideal crystal structure. Thus a range of properties sensitive to defects was recognized by Smekal, and among these is the mechanical strength of crystals.[24] Indeed, Smekal recognized that all real crystals contain defects and are therefore weak. Other structure-sensitive properties are luminescence, electrical conductivity of semiconductors and insulators, and color centers.[25]

As far as plasticity is concerned, Smekal preferred "pores" to cracks and suggested that all "real crystals have pores and other defects, such as inclusions of impurities."[26]

"The increased strength during glide can be explained by the decreasing size of those pores which contribute most to the weakening of the crystal."[27] Elsewhere, Smekal suggested that the ideal regions of real crystals each contain about 10^4 molecules, which is much less than the mosaic blocks postulated by Darwin and Ewald, who concluded from x-ray investigations that structural faults occurred between blocks containing about 10^{10} molecules.[28]

Zwicky suggested that mosaic blocks were formed in crystals by contraction during solidification, rather on the pattern of cracks forming in drying mud.[29] The misfit between mosaic blocks eases their movement when the crystal is stressed and thus explains the ease of plastic deformation. The sliding of such mosaic blocks during deformation gives rise to bent lattice planes, a phenomenon called by Mark, Polanyi, and Schmid *Biegegleitung*—that is, a combination of glide and bending.[30] The authors refer to earlier observations on bending glide by O. Mügge in 1895.

Mosaic structure was a common component of theories of plastic deformation and of work hardening; indeed, the mosaic block survived and later became part of dislocation theory. It was plausible to think that building blocks that did not fit together perfectly would yield to applied stresses more readily than a perfect lattice. The mosaic block theory could also explain work hardening if the blocks piled up against one another and locked together, thus making further movement more difficult. The influential theoretical physicist Richard Becker suggested a thermally activated mechanism of plastic flow, which was not widely accepted, but summed up the widespread view of work hardening in terms of piled up mosaic blocks:

> The cause for work hardening is now widely seen in the blocking of glide by a defective lattice. The various concepts, such as blocking of glide planes, crumpling [*Verknüllung*], internal stresses, displacement [*Verlagerung*], etc., all have the view in common, that in the work hardened crystal the ideal lattice, with its perfectly smooth slip planes, no longer exists.[31]

Even Griffith himself, in an internal Royal Aircraft Establishment report with B. Lockspeiser, argued for a variant of mosaic structure as a mechanism of stress concentration in crystals.[32] Griffith and Lockspeiser considered two- and three-phase systems and argued, from equilibrium conditions, that

> A. Crystals containing only single-phase and two-phase regions. . . . Under any given type of shearing load the crystal possesses an elastic range. If the elastic limit be exceeded the material yields under stress concentrations existing at the edges of the two-phase surfaces. . . .
>
> B. Crystals containing single-phase, two-phase and three-phase regions. . . . permanent deformation . . . occurs under any finite shearing load, however small.[33]

Yet in the same paper, the authors speak of sheared cracks in elastic solids. It would appear that Griffith regarded cracks as important in elastic deformation and brittle fracture, as in his experiments on glass rods, whereas for plastic deformation he regarded a "polyphasic" structure as necessary for stress concentration. Griffith's followers seem to have been less worried about the difference between elastic and plastic crystals in this respect.

Becker's theory of thermally induced glide has some historic significance in that it was Becker in Berlin who introduced the young student Orowan to problems of plastic deformation by asking him to prove that all crystals become brittle at very low temperatures.[34] This work eventually led Orowan to the postulate of dislocations. The test itself became unnecessary, because Meissner, Polanyi, and Schmid showed that some metals retained their plasticity at very low temperatures even before Orowan had completed his preparations for an experimental test of Becker's theory.[35]

Although many attempts were made to explain work hardening and plastic deformation, none of them was entirely successful. Indeed, the theoretical explanations of the phenomena associated with plastic deformation of crystals fell well short of being satisfactory, and hence the field remained open to new hypotheses. Taylor showed that both the crack theory and the mosaic block theory suffered from inconsistencies.[36] He showed that it was necessary to assume that deformation caused an extension and sealing up of cracks, thus leading to an increase in their number. This, in turn, led to an increase in the ratio of the mean stress to the maximum stress at the end points of cracks. It was clear that the major obstacle to the plausibility of any of these theories was the fact that, particularly in cubic crystals with an abundance of possible slip planes, the amount of hardening obtainable by any of the postulated mechanisms was insufficient to account for observations. Taylor used the argument particularly against the mosaic block theory and preferred the crack theory, although, as is clear with hindsight, only as a stepping stone toward dislocations.[37]

The two people who perhaps came closest to the concept of dislocations, yet without formulating it in a generally useful and acceptable way, were Prandtl and Dehlinger. They assumed, as did many others, that a crystal consisted of a mosaic of small blocks separated by some kind of subboundaries. Prandtl, in a paper describing a model rather than a real crystal, explained work hardening and slip by assuming that sheets of atoms would jump irreversibly from one position of equilibrium to another.[38] Such jumps would leave behind defective sturctures. Dehlinger came even closer to the concept of the dislocation in postulating an imperfection that he called *Verhakung*.[39] In this concept, an imperfection is confined to a small region of a surface of misfit, while farther away from this region the atoms are in near-perfect alignment. Thus he combined the postulate of mosaic structure (i.e., a surface of misfit) with a localized defect consisting of atoms displaced from their normal lattice positions. The effect of such nonuniform displacement of atoms is the setting up of internal strains, and such strains were assumed to be the cause of work hardening.

Prandtl suggested his model mainly to explain mechanical hysteresis and "aftereffects," meaning presumably work hardening and possibly creep, and to calculate the speed of plastic flow. His model consists of two rulers, one with mass points and the other with a crystal field of forces.

He did not really introduce the concept of a dislocation, but the paper and model became influential on two counts. Prandtl calculated the linear movement of mass points in a periodic field of force and showed that an irregular distribution of mass points along a line was plausible and could be used for model calculations of real phenomena of plastic deformation. The paper also helped in the process of making

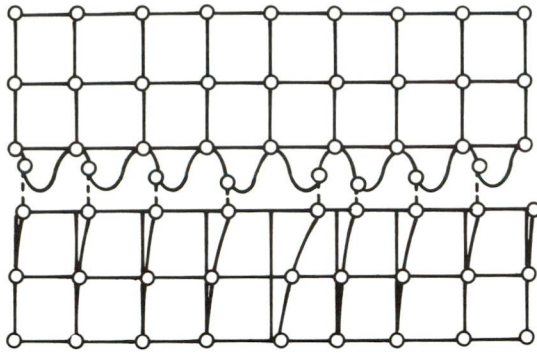

Fig. 5.1 Diagram of a *Verhak-ung* (Redrawn from U. Dehlin-ger, "Zur Theorie der Rekrastal-lisation reiner Metalle," *Ann. Phys.* 2 [1929]: 768)

the strength of materials "respectable" among physicists. "In the meantime, because of advances in knowledge of crystal structures and atomic physics, the study of the physics of the strength of materials has become modern again (for pure physicists it has become modern for the first time)."[40] The highly respected Prandtl certainly contributed to this new popularity.

Ulrich Dehlinger, in his *Habilitationsschrift*[41] in Stuttgart, refers to Prandtl as the originator of his own concept of *Verhakung,* although there really is only a loose similarity of the *Verhakung* with Prandtl's model (Dehlinger read Prandtl's paper at a late stage of his own work) and a close similarity with the later concept of the dislocation (Figure 5.1). The common and independent assumption of the two models is a periodic field of forces acting on a row of atoms in a crystal lattice.

Dehlinger's attention was focused mainly on recrystallization. He saw the *Verhakung* as the link between regions of the crystal with different orientations—that is, within a mosaic substructure resulting from plastic deformation. The word *Verhakung,* derived from *Haken* (hook), was introduced at a time when German physicists did not yet look over their shoulders toward their Anglo-Saxon colleagues and has no English translation. Although the *Verhakung* looks very similar to a dislocation, its function is merely to take up lattice distortions and not to ease glide. Dehlinger argued against Czochralski, who concluded from x-ray studies of plastically deformed crystals that much of the lattice is destroyed and that only islets of perfect crystal remain.[42] Dehlinger concluded from his x-ray studies that most of the lattice remains crystalline, but is broken up into small mosaic blocks bounded by regions of imperfections. The *Verhakung* is a "deformation which emanates from the boundary between two regions and penetrates into both, but does not extend throughout the interior."[43] Recrystallization consists of the elimination of these defects so that regions of different orientation become unified into larger grains.

Dehlinger assumed that plastic deformation starts at points of easy glide *(Gleit-lockerstellen)*—some kind of defect, consisting of a loose region that predisposes the crystal toward slip. During deformation, those defects are eliminated and the new defect, the *Verhakung,* is formed. The mechanism of formation is described as follows: "When two rows of atoms are pulled past each other, some atoms may, because of irregularities in the glide plane, be left behind the others and eventually be caught by the attractive force of an atom in the opposite plane other than in correspondence with its normal position."[44] The *Verhakung,* although similar in appearance to a dislocation, thus played the role of accommodating deformation, not that of allowing it in the first place. Dehlinger assumed that slip ceased because of internal stress fields.

The word *dislocation* was introduced into the theory of mechanical properties of solids by A.E.H. Love in an appendix to Chapter VIII and IX, apparently added in the third edition (published in 1920), of his famous *Treatise on the Mathematical Theory of Elasticity.*[45] Love translates as *dislocation* the Italian word *distorsione,* used by Volterra and others to describe a many-valued displacement in an elastic medium. The term was thus originally introduced for a displacement in a continuum, not in a crystal. A relatively straightforward way of introducing linear discontinuities in an elastic medium, which have the geometric properties of dislocations, is described by Friedel.[46]

5.2. First Introduction of the Modern Idea of Dislocations, 1934–1946

The year 1934 was the *annum mirabilis* of dislocation theory. Three papers published in that year can all be seen as originators of the modern idea of a linear crystal defect—the dislocation—that is the prime cause of all phenomena related to the plastic deformation of crystalline solids.[47] The most extensive and most influential of the three papers was that by G. I. Taylor, a Royal Society Professor at Cambridge and a most eminent figure in British science. Taylor's paper begins with a critical review of experimental observations and possible theories of strain hardening. The first mechanism mentioned is that of the formation of blocks during deformation and the locking of blocks, holding up slip. He argued against this mechanism for all those crystals that possess more than one direction of easy glide. In the second mechanism, blocks are not formed by deformation, but the mosaic structure is inherent in the crystal. The argument against the effective operation of a locking mechanism is much the same as in the first case.

In these two theories, the bulk of the crystal is inherently weak and mosaic imperfections increasingly strengthen it. Taylor did not mention the various speculations why a crystal should be weak, such as the theory of weak points *(Lockerstellen),* discussed previously. The third theory, which is in better accord with calculations of inherent strength, assumes the perfect crystal to be able to withstand very large stresses, and the observed weakness of crystals is attributed to stress concentrations by internal cracks. Taylor referred to his own paper of 1928 to argue that increasing deformation leads to an increasing number of cracks.[48] He then considered the mechanism of slip in detail and came up against the difficulty of accounting for unique directions of slip without requiring whole faces to slip simultaneously. This would need very large forces and would make it difficult to account for work hardening. Taylor therefore suggested—and this is the crux of his argument from which everything else follows—that slip might not occur "simultaneously over all atoms in the slip plane but over a limited region, which is propagated from side to side of the crystal in a finite time." He then suggested that the stresses near the edge of the region which has already slipped make "the propagation of slip . . . readily understandable and . . . analogous to the propagation of a crack."[49]

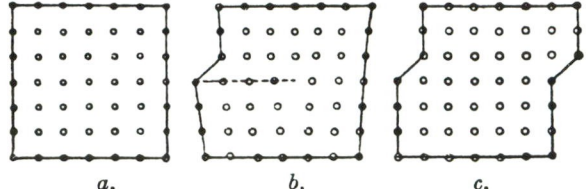

Fig. 5.2   (a) The atoms in the lattice of a crystal block; (b) the condition of this block when a slip of one atomic spacing has been propagated from left to right into the middle; (c) the block after the unit slip, or dislocation, has passed through from left to right. (From G. I. Taylor, "The Mechanism of Plastic Deformation of Crystals," *Proc. Roy. Soc.* A145 [1934]: 369)

The actual definition of a dislocation is given in the text describing a figure in Taylor's paper (Figure 5.2).

For a detailed discussion of the atomic mechanism involved, Taylor reached the following conclusions as first approximations:

1. Above a certain critical temperature T_D any stress, however small, will cause dislocations to travel,
2. "At temperatures below T_D the centres will not move till the shear stress attains some finite value."[50]

Thus the weakness of real crystals can be explained if slip occurs by a dislocation mechanism.

Taylor assumed that crystals do not a priori contain dislocations, but that these are formed either at surfaces or by thermal fluctuations. Although there is some similarity with Becker's work, there is no reference to it and no reason to assume any influence. Taylor certainly thought that dislocations would form quite readily as a result of thermal activation and could then move easily under the influence of even small stresses. As platic deformation progressed, more and more dislocations—or centers of dislocations, as he sometimes called them—would form and the stress field associated with them would make further movement more difficult. Work hardening was thus obtained by the stress fields set up by an increasing density of dislocations. Taylor supported his qualitative arguments by approximate calculations and obtains the correct orders of magnitude.

> If the crystal is initially perfect, the first few dislocations migrate through it under the action of a shear stress which may be regarded as infinitesimally small. As the distortion proceeds, however, the number of dislocations will increase and the average value of d will decrease so that the resistance to shear will increase with the amount of distortion.[51]

(He assumed a temperature above a certain critical temperature T_D; d is the spacing of dislocations in a regular row.)

In the second part of his paper, Taylor tried to explain many observations with the aid of dislocation theory. He ran up against the difficulty of ill-defined specimens, met by all early solid-state physicists, and recognized the cause of the problem.

Another difficulty is that the large changes which very small amounts of impurity can cause in the stress-strain curves make it difficult to be certain that the observed curves are really those which would be found if the test were made with a perfect crystal of a metal containing no impurities.[52]

In this part of his paper, Taylor came to grips with the problem of mosaic structures and explained that the distance a dislocation can travel is given not by the size of the crystal specimen, but by internal boundaries.

The length L must be a length connected with the structure of the crystal. It represents the distance through which a dislocation can travel freely along a slip plane under the influence of a small shear stress before being held up by some fault in the perfection of the regular crystal structure, or surface of misfit. . . . If the faults or surfaces of misfit are everywhere opaque to dislocations, L would be the linear dimension of a cell of a superstructure or a mosaic.[53]

He went on to argue that the opaqueness of subboundaries in "Darwin's type of mosaic" (no reference given) should vary with temperature and interprets the results of Schmid and Boas (1931) on the deformation of aluminum crystals over the range of -185 to $600°C$. He dismissed his own attempts to explain plasticity at very low temperatures as "speculative until some method can be found for calculating the internal stresses which might be expected."[54]

The second extensive paper introducing the concept of dislocations was based on work done by Egon Orowan under the direction of R. Becker at the Technical University of Berlin.[55] Unhappily, the work was interrupted by the political events of 1933, when Orowan had to return to his native Hungary as a first stage of his flight from Nazi persecution. The first two parts of Orowan's paper deal with temperature dependence and dynamics of plastic deformation, and only the third part deals with the mechanism of glide (pp. 634–659). Orowan realized that the dislocations *(Versetzungen)* he suggested were identical with those suggested by Polanyi. Interestingly, Orowan assumed some form of cracks or notches *(Kerben)* as the cause of weakness of crystals and saw dislocations only as a mechanism of glide.

We may say that the weakening effect of notches present in the crystal spreads by dislocations through the glide plane. . . . The transition between the part of the plane which has slipped and that which has not slipped yet, which has here been characterized by added stress is, from the point of view of the lattice geometry, identical with the dislocations described by Polanyi.[56]

Polanyi's paper is the shortest of the three, but in some ways the one that proved most acceptable in its arguments.[57] He found cracks unconvincing and argued that whatever defect causes the weakness of crystals, it ought to be of a "crystallographic" nature. Polanyi referred to Prandtl and Dehlinger as forerunners of his own dislocation *(Gitterversetzung),* in which "10 atoms on one side are opposed by 11 atoms on the other side of the line." Polanyi admitted that he had no explanation for work hardening and realized that continuous creation of dislocations is necessary for extended deformation. He had no clear mechanism of formation, but thought that "dislocations ought to form preferentially at the crystal surface."

Although all three papers were published in the same "miraculous" year, and all three contained virtually identical descriptions of what we now call an edge dislocation, the papers had very different origins and their authors imagined different mechanisms of plastic deformation.

G. I. Taylor, at that time the most established and most eminent of the three, had close acquaintance with Griffith's theory, which was the theory of plasticity with the best mathematical foundation current at the time. Although Taylor's main interest lay in other parts of physics, particularly hydrodynamics, he took an occasional interest in mechanical properties of solids: first in 1914, when he worked with Griffith, and then again in 1922, when he was asked for advice by Carpenter and Elam.[58] Taylor himself recalls a meeting at the Royal Society in 1922, when Carpenter and Elam presented some of their results on the deformation of aluminum crystals, which led to collaboration between him and C. Elam.[59] This work led to further elucidation of stress-strain relationships and the geometry of glide. In fact, the British school around Carpenter and Elam, with which Taylor and others were associated, and the German school around Polanyi, Schmid, Mark, Boas, and others elucidated in very careful experimentation the geometric, crystallographic, and quantitative properties of plastic deformation in single crystals of a variety of metals and other materials. The results were summed up in the two major monographs of 1935 mentioned earlier.[60] What finally led Taylor to his 1934 paper seems to have been two things: his dissatisfaction with the nonmathematical nature of existing theories, and the realization "that if the material on the two sides of a Griffith shear crack joined together again after the passage of its ends the stress round each end would become independent of that at the other as they separated and could therefore exist in isolation."[61] In this way, the problem became one of solving the stress distribution around a singularity, and the solution was available in Lowe's textbook on elasticity. Thus the mathematical theory of cracks led Taylor directly to the concept of a dislocation.

Michael Polanyi's career in solid-state physics started in 1920, when he joined the Institute of Fiber Chemistry in Berlin, and lasted for three years, when he "plunged back into reaction kinetics."[62] He was soon joined by other gifted young men, such as H. Mark, E. Schmid, and K. Weissenberg. The atmosphere must have been stimulating and active. According to Polanyi, "we had a glorious time," and his then colleague Erich Schmid confirmed 60 years later that "Polanyi was perhaps the most inspiring man I ever met."[63]

It was a time when a substantial problem—a white patch on the map—became explorable by the new availability of a suitable tool, x-ray diffraction. It was also a time of new hope for science after a terrible war—an emergence from darkness into light. The light turned out to be but a short-lived flicker, and German science was soon to be smashed to smithereens. Yet the brief period of enthusiastic creativity caused a great leap forward in knowledge of solid-state physics, and happy were those bright young men who were there at the right time and place. "The strength of solids had now become my principal interest. The technological purpose of the Institute had thrown this great problem into my lap. I saw that the characteristic feature of the solid state, namely its solidity, was yet unexplained, and indeed, hardly explored by physicists."[64] This is very much an echo of the words of Becker. Indeed, the subject was becoming respectable only because the newly developed x-ray technique, together with the availability of good single crystals, made proper physical measurements and interpretations of deformation processes possible.

Polanyi and his colleagues used cylindrical single crystals to eludicate the crystallographic laws of plastic flow in zinc and tin, while Taylor and Elam did the same for aluminum. The process of work hardening was also discovered during this period. "I was deeply struck by the fact that every process that destroyed the ideal structure of crystals . . . increased the strength of crystalline materials."[65] This seemed to Polanyi to refute theories based on Griffith's cracks.

> However, the idea of dislocations causing a high degree of plasticity did gradually take shape in my thoughts. I gave a full account of this theory in April 1932 in a lecture addressed to the members of Joffé's [Ioffé's] institute in Leningrad, who received it well. On returning to Berlin I talked about my theory to Orowan who told me that he had developed a similar idea in his thesis about to be submitted for a degree. He urged me to publish my paper without considering his rival claim, but I preferred to delay this until he too was free to publish.[66]

Alas, Polanyi too had left Germany by the time his paper was published, and he had joined the stream of German scientists who found refuge in England. Polanyi's own account and his 1934 paper show that in his view dislocations could account for the weakness of crystals without recourse to cracks. Thus his paper differs in important respects from that of Orowan and, in other respects, from Taylor's paper, with its emphasis on work hardening and on mathematical formulation.

Orowan was the youngest of the three inventors of the edge dislocation and thought of the dislocation to solve yet another aspect of the problem of plastic deformation, particularly so-called jerky creep, also known as the yield phenomenon. Orowan's introduction to plasticity came through Becker, who asked the young undergraduate to check his theory of plasticity experimentally. To Orowan, a student of electrical engineering, "plasticity was a prosaic and even humiliating proposition in the age of De Broglie, Heisenberg and Schrödinger, but it was better than computing my sixtieth transformer, and I accepted with pleasure."[67] The phenomenon of jerky creep was discovered by accident, when Orowan had dropped his last useful crystal and used the bent and straightened specimen for stress-strain measurements. It was the ready propagation of slip once started, which led Orowan to the concept of the dislocation.

Orowan's account of why he and Polanyi published separate but simultaneous papers is slightly, but not substantially, at variance with that of Polanyi.

> I wrote to Polanyi, with whom I discussed them several times, suggesting a joint paper. He replied that it was my bird and I should publish it; finally we agreed that we should send separate papers to Professor Scheel, editor of the *Zeitschrift für Physik,* and ask him to print them side by side.[68]

The year 1934 was distinguished by yet another event of significance for solid-state physics. An international conference on physics held in London contained a substantial section entitled "The Solid State of Matter." At this conference, two further schools of some significance in the field of plastic deformation were represented: the Netherlands school, with an important paper by W. G. Burgers, and the Soviet school, represented by A. Ioffé.[69]

Burger's paper deals mainly with recrystallization, but one or two of his remarks may shed light on the immediate reception of dislocation theory. He referred to the different recent theories and suggested that they all share the view that the macroscopic glide process consists of a combination

> of "local glide steps," the occurrence of which is closely related to the presence, even in the undeformed crystal, of some kind of deviations from the "ideal" lattic structure. Although the various theories differ apparently regarding the exact nature of these deviations [*Lockerstellen* (Smekal), "dislocations" (Taylor), *Versetzungen* or *Verhakungen* (Polanyi, Orowan, Dehlinger)] and so to the way in which they initiate slip (as a consequence of local stress concentration or of special conditions of potential energy), their conceptions agree in so far that local glide steps occur under the influence of shear forces, which, while being originally small compared to that expected for an "ideal" lattice, increase with the number of deviations.[70]

In another section of his paper, Burgers dealt with the vexed question of bent lattice planes in deformed crystals and extended the work by Taylor and by Yamaguchi.[71] He argued that limited broadening of Debye–Scherrer lines proves that "only a relatively small proportion of all the atoms has undergone irregular displacements from ideal lattice-positions. It seems probable, therefore, that a local curvature corresponds more or less to a broken-up crystal region . . . which are glued together by severely distorted regions."[72] Thus the mosaic theory continued to serve, and indeed still serves, although with an improved experimental and theoretical base.

At the same conference, Schmid, at this time also an absentee from Germany, albeit only temporarily and for reasons of shaken finances in the German research system, presented a paper that put work hardening on an even firmer footing. He also reaffirmed the crystallographic nature of slip, as opposed to creep: "Slip is in the first place characterized by the *strictly crystallographic selection of the slip elements,* slip planes and direction."[73]

> Just as the yield point, which depends on the position of the slip elements, can be expressed by a single constant, the critical shear stress, so the course of the stress-strain curve, which depends on the orientation, can be expressed in a single curve, the "hardening curve" *(Verfestigungskurve).* In this the shear strength of the operative slip-system is plotted as a function of the shear (relative displacement of two slip planes at unit distance apart); from it the hardening (increase in strength) of the crystal can be read directly.[74]

The word *dislocation* is not mentioned in the paper; reference is, however, made to the three dislocation papers of 1934 with the comment that the nonsimultaneous slip on a plane is hardly likely to be the correct model for twinning. "In this case, the attainment of stable equilibrium necessitates that all the atoms in the neighborhood of a particular atom shall change their positions simultaneously."[75] This view does not seem to have stood the test of time, as Cottrell argued that mechanical twinning as a deformation mechanism can also be explained by a movement of dislocations.[76]

From the early days of research on the strength and plasticity of solids, much speculation centered on the surface. After all, the surface is an unavoidable defect in the crystalline structure, and its role was speculatively that of a source of dislocation or, experimentally, a determinant of strength. Some of the experiments were reported by A. Ioffé and discussed by A. Smekal at the 1934 London conference. "In the case of rock salt, two experiments were performed by Levitskaya and myself to demonstrate the importance of surface conditions. Dissolving the surface by hot water during the loading experiment, we found strengths exceeding the usual value by a factor of twenty or even more."[77] Ioffé was well aware of the many attacks on his conclusions, but had arguments against all his detractors. He made no mention of dislocations and regarded "surface crevices" as the cause of weakness of crystals. "The practical weakness is sufficiently explained by the Griffith crack theory."[78]

In the discussion, E. N. da C. Andrade pointed out that surface effects had also been found by Roscoe in cadmium crystals.

> With a superficial layer of oxide one or two molecules thick, the critical shear stress may be as high as twice the value obtained with a clean surface. . . . No doubt with a clean surface the initial slip which takes place at the critical shear stress is due to minute cracks in the surface, much as has been described by Dr. Orowan for rock salt, and the effect of the oxide layer is to round off and strengthen the sharp edges at which the stress is high.[79]

To modern ears, in these days of ultra-high vacuum, the claim of a clean metallic surface seems rather lightly made. Historically, these matters are of interest for two reasons. First, surfaces played a large speculative role and exercised many minds 50 years ago, yet the full elucidation of the role of surfaces has probably yet to be achieved. Second, it is fascinating to observe that rock salt, that classic ionic crystal which served as a model substance in matters of cohesion of solids and of point defects, also served as a proxy for the study of plastic properties of metals.

Alkali halide crystals are taken up by Smekal at the very same 1934 conference in London. In fact, Smekal argued that changes in the coloration properties of sodium chloride crystals are the most sensitive of all measures of plastic deformation. He defined a "photochemical limit of elasticity." Because the plastic properties of rock salt are strongly temperature-dependent, Smekal argued that "*heat motion participates energetically in the local overcoming of lattice cohesions, and in the propagation of the stress maxima. This results from the influence of the occurrence of high stress maxima on self-diffusion processes within the flaws.*"[80] Smekal obviously believed in local flaws *(Lockerstellen),* and in thermally assisted propagation of slip. He applauded Taylor's 1934 paper only insofar as it accepted some parts of older theory and ignored Polanyi and Orowan altogether.

It should not be imagined that dislocation theory, as soon as it had been enunciated, swept all before it. Metallurgists largely ignored dislocations and indeed hardly believed in them until a much later period, when they had been experimentally observed and their existence proved with as much certainty as anything can be proved in science.[81] Physicists accepted the mechanism as a plausible and possible one, but it was no more than yet another hypothetical explanation in a field that had been experimentally reasonably well documented but in which plausible explanations abounded and yet none fully convinced. Dislocation theory gradually gained acceptance as it proved the most plausible, most fruitful, and, eventually, experimentally best-proven theory. This was not a scientific revolution, no heroic change of paradigm, but a gradual emergence from confusion and multiplicity of explanations of complex and varied phemomena. Neither was it a rejection of older theories by falsification, but an eventual preponderance of one theory and the atrophy of others.

Johannes Martinus Burgers (1895–1981), with his father (*left*) and Paul Ehrenfest (1880–1933), around 1917. (AIP Niels Bohr Library)

A number of physicists made further contributions to dislocation theory in the years between 1934 and the end of the Second World War. The brothers W. G. and J. M. Burgers published review articles on viscosity and plasticity in which they pointed out that dislocations could pile up against obstacles and thus exert a back-stress that could cancel the applied stress, thus explaining work hardening.[82]

W. G. Burgers had been trained as a chemist and was introduced to x-ray diffraction and crystal physics during a period of postgraduate research at the Royal Institution in London from 1924 to 1927.[83] On his return to the Netherlands, he was offered a post at the Philips Research Laboratory, where he worked on problems of recrystallization. This was a characteristic case of fundamental research in a field of interest to a manufacturer of incandescent filaments. Burgers came to read the paper by Dehlinger on recrystallization, and this started him reading more on plastic deformation. When he had problems in reconciling the 1934 papers by Taylor and by Orowan, he turned for help to his brother J. M. Burgers, a professor of fluid mechanics. In their 1935 paper, they attempted to reconcile the theories of Taylor and of Orowan and this led them to their own version of work hardening.

It was the fluid mechanics expert J. M. Burgers who went back to the mathematical theories of dislocations in elastic media and described the dislocation in terms of a vector, now generally known as the Burgers vector. This mathematical description also led to the screw dislocation, first presented at the 1939 Bristol Conference on Internal Strains in Solids, published in the Netherlands in the same year and in England in 1940.[84] "Volterra's work referred to continuous media in which the Burgers vector was not related to a lattice vector but could be arbitrary and Burger's work of 1939–1940 was based upon an adaptation of Volterra's ideas to lattices of discrete points. . . ."[85] Alas, the work by J. M. Burgers came too shortly before the war to make much of an impact until many years later. During the war, British physicists either did not work on mechanical problems or, if they did work on strength of materials, were concerned with the cold drawing of shell cases or with the resistance of armor plate to projectiles.[86] Similarly, German physicists had to worry about much more mundane things than dislocations. Apart from munitions and armor, they had to produce substitute alloys for materials unavailable in besieged Germany.[87]

Although it is always emphasized that war enhances the progress of science, this is true only in selected fields. As the more abstruse parts of the theory of strength of materials, such as dislocation theory, certainly showed no immediate promise of helping to win the war, they were cast aside. Only those parts of science deemed of direct benefit to the war effort were pursued; the others lay dormant until peace returned. Dislocation theory did not receive the dubious benefit of enhancement by war, nor can one think of any indirect benefits that the war effort might have bestowed on it.

Two young British researchers somehow seem to have bridged the gap between the prewar and postwar period in dislocation research. One was F.R.N. Nabarro, whom N. F. Mott invited to Bristol in 1938, when he had completed his degree in Cambridge and had done another year's work in mathematics. Nabarro was given a theoretical problem that arose directly out of Mott's interest in G. I. Taylor's 1934 paper. Mott had a model in which particles of precipitate would produce shear stresses of various signs in different regions of the crystal. The dislocation could move only if the sum of the stresses favored its movement, and thus precipitation hardening would arise. Nabarro was to work out a proper quantitative theory based on Mott's idea. His results for spherical particles went counter to experimental observations on age-hardening, but the results for lamellar particles showed great promise and were communicated by Mott to the Royal Society.[88] Nabarro then had to abandon dislocations in order to "work out the consequences of Solly Zuckerman's experiments on the penetration of shell fragments into telephone directories and legs of beef."[89] Nevertheless, he provided some continuity of dislocation research in Bristol, for Mott arranged for him to come back at the end of the war to continue the work that had been so rudely interrupted.

Young British students of subjects deemed useful to the war effort—and physics was certainly among these—were usually allowed to spend up to two years at university and often did accelerated degree courses. John Nye was one such student; he took his final examinations in Cambridge in December 1943. He then expected to be called up, but instead was employed on war work at the Cavendish Laboratory, then under the direction of W. L. Bragg, and was assigned to assist Orowan. The latter left Hungary in 1937, as the threat of war and fascism became apparent, and joined Peierls in Birmingham. Two years later, he was invited by Bragg to come to Cambridge, where he stayed until 1950.[90] Orowan was working on the theory of brittle fracture, a problem that caused great anxiety in welded ships, for the Armaments Research Department of the Ministry of Supply, a group headed by N. F. Mott. Nye did some very practical work on this problem, and a comprehensive report was prepared in 1945, too late to influence the course of the war. Nye was in close contact with Orowan and learned a great deal about plastic deformation, fracture, and dislocations. He also knew about a bubble model of dislocations that Bragg had built.[91]

On Bragg's own evidence, the idea came to him when he was filling a two-stroke

Nevill Mott (AIP Niels Bohr Library. Photograph by Lotte Meitner-Graf, courtesy of Sir Nevill Mott)

lawn mower with a mixture of petrol and oil. Bragg hated the preoccupation with war work and administration and, when the war was over, was extremely keen to get back to real research. This pent-up desire to get back to research was a most common experience and shows the fallacy of thinking that war was a godsend to physicists and offered wonderful opportunities for research. Most of them were frustrated with mundane practical tasks that bore little relation to their real interests. When at the end of the war Bragg needed somebody to help him to develop the bubble raft, he "borrowed" Nye to do the practical work. Apart from "lending" his student, Orowan made one other contribution to the bubble raft model. When he saw a technician blowing bubbles in a soapy solution in the raft and they all came out different shapes, he suggested turning the nozzle upside down and all was well.[92] "The work on bubble rafts was not that easy. Defects could not be made to order. I made large rafts, took photographs with a quarter-plate camera and enlarged the bits of interest."[93] Nye, who received a first-hand introduction to dislocations while still a young student, was very conscious of the fact that they were a mere hypothesis. A common question among metallurgists and the new breed of metal physicists was, Do you believe in dislocations? An affirmative answer was the exception.[94]

The work of Bragg and Nye was quickly completed and published.[95] More quantitative aspects were later considered by Bragg's research student Lomer.[96] The advocacy of dislocations by a man as famous as Bragg, a frequent guest speaker at many a gathering, armed with the intuitively convincing slides and cine-film of the bubble model, did a great deal to gain adherents to the idea of the dislocation. The film made a great impression at the 1947 Bristol conference and helped to convince many physicists. It took a great deal longer to convince practical metallurgists. It was to be a few more years before faith in dislocations could be experimentally supported to a sufficient degree to convince the doubters.

But just as metallurgists lacked faith in dislocations and doubted the blessings bestowed on them by physicists, so many members of the physics community doubted the legitimacy of the newly emerging metal physics. "I suppose there was a question whether it was really physics. . . . Orowan's work had a metallurgical engineering slant and many people at the Cavendish would regard it as slightly suspect on these grounds."[97] The doubt was not shared by Bragg and Mott, who saw good physics emerging and welcomed the fact that the new knowledge might prove useful to engineering. This, however, proved an uphill struggle. "It is a bit like trying to improve the flavour of a dish in cookery by using knowledge from chemistry."[98]

A small flicker of continuity in dislocation research in Britain during the war was provided by occasional meetings of a few people interested in the topic. They came from the three main centers for this work: Birmingham, Bristol, and Cambridge. Orowan recalls one such colloquium where he spoke on the possibility of locking of dislocations by foreign atoms as a cause of the yield phenomenon.[99]

In the United States, where the Second World War hit a little later, three significant papers on dislocations were published during the early part of the war. They were all associated with Frederick Seitz, then at the University of Pennsylvania. Seitz, somewhat like Mott, took an interest in everything connected with the physics of solids. He was most influential in making solid-state physics not only respectable, but even recognizable as a field of research in the United States. His book *Modern Theory of Solids* was the first comprehensive theoretical treatment of this topic and, in a very real sense, helped to define and create the subject.[100]

Seitz joined the Physics Department of the University of Pennsylvania in the spring of 1939 and was allowed to build up a small solid-state group. "I had become convinced by that time that if the field of solid state physics was to come into its own in the United States, it was essential to develop a number of centers in university physics departments where a combination of experimental and theoretical work could be undertaken."[101] Seitz had worked at the General Electric Research Laboratory before coming to the University of Pennsylvania and had earlier been E. U. Condon's research student at Princeton. It was therefore not surprising when Condon, who was by then head of the Westinghouse Research Laboratories, invited Seitz to visit the laboratory in the summer of 1939 to discuss the work of some postdoctoral fellows. One of them, Thomas A. Read, was studying internal friction in metals. Seitz thought that the work "gave convincing proof that some of the dissipation was associated with the motion of dislocations, a concept which was still relatively new and unexplored."[102]

As a result of this facet of his many interests in solid-state physics, Seitz and members of his research group published several papers before war work interrupted these activities in the United States also.[103] Huntingdon computed the electrostatic energy between atoms in a dislocation in rock salt to obtain the free energy of a dislocation. Because the dislocation represents a discontinuity, the energy can be calculated only by assuming a small hollow core of radius $r_0 = 10^{-7}$ cm, with most of the energy residing outside this core.[104]

When Koehler first arrived at the University of Pennsylvania, Seitz suggested that he work on the effect of dislocations on electrical resistivity. This proved rather difficult, and Koehler took up Seitz's second suggestion of examining work hardening. This proved more successful.[105]

5.3   The Period 1946 to 1953

Although the foundations to dislocation theory had been laid and some details worked out by 1941 or so, they remained an obscure hypothetical part of physics and were mostly disregarded during the war. In the balmy days of postwar enthusiasm, when all the pent-up desire for pure research was given free rein in a world that believed in progress through science, dislocations rapidly established themselves as a fruitful and valid subfield of solid-state physics.

One of the means of establishing an international community in a field is to organize conferences. They are useful for reporting progress, establishing a consensus, and, above all, making contacts and exchanging views in informal discussion. Friedel lists the major conferences at which dislocations formed a topic, starting with the 1946 London Symposium on Internal Stresses and ending with the 1961 Berkeley Conference on Transmission Electron Microscopy of Metals—24 conferences in all.[106]

Another important feature of the international physics community is summer schools. At these, young scientists have a chance to meet the "grand old men" of the field and to learn the latest state-of-the-art. Of possibly even greater importance is the fact that the up-and-coming young scientists meet one another and establish links of friendship and collaboration. The best of all possible worlds was had at Bristol in 1947, when a conference on the strength of solids was held at the conclusion of a summer school on the same topic.

The first session of the conference dealt with creep and plastic flow, and the first paper was given by Mott and Nabarro. Dislocations had come a long way by then. "In most modern theories of slip, it is assumed that slip begins at one end of the crystal and travels across it, thus avoiding the concept of simultaneous slip."[107] Mott and Nabarro may have been the first to describe a dislocation as a "line discontinuity," thus opening the way for treating it as a geometric entity with certain properties. This further step into abstraction, treating the dislocation as a line rather than going back in each consideration to the atomic arrangement of the crystal, proved of major importance.

Much effort was devoted to calculating the forces required to move a dislocation[108] and to determining the mechanisms of creating dislocations. Orowan, in his 1934 paper, had written that one glide plane could "infect" neighboring planes, and thus vaguely foreshadowed some multiplying mechanism for dislocations. Yet Mott and Nabarro regarded the problem of formation of dislocations as unsolved. "The problem of their formation is at present the furthest from solution in the whole theory. . . . It seems highly probable, as will be seen below, that one dislocation, once it starts moving, generates others."[109] It would seem that the problem of generation of dislocation is obscure to this day. Although several mechanisms have been suggested and shown to operate, the question is still not completely clarified.[110]

Mott and Nabarro thought that work hardening was caused by dislocations "getting stuck," and devoted most of their paper to calculating stresses around precipitates and the flexibility of dislocation lines. On the one hand, we see a direct continuation of their prewar work; on the other, a higher level of abstraction in viewing the dislocation as a line with elastic properties. They concluded that "a dislocation will approximate to a smooth curve rather than to a square zig-zag."[111]

At this point perhaps a small digression is necessary, for the 1947 Bristol conference showed not only the state-of-the-art, but also the state of research centers at that time. Clearly, Bristol had reestablished itself as a leading school in solid-state physics and, more especially, in the physics of plastic properties of crystals. At the end of the war, Mott managed to attract several people to Bristol to work in this area and arranged, with the active support of the vice chancellor, Philip Morris, to supply them with positions and funds. Some had worked with him before, like Nabarro, and some he had known during the war, like F. C. Frank and J. W. Mitchell. Other researchers later joined the group, which proved to be of decisive international significance and a true nucleation center for research on solids. The other center was at Cambridge, with people like Bragg and Orowan in senior positions and Nye, Lomer, and Cahn as the up-and-coming young men. Soon another generation was to take over, with Peter Hirsch as its leading member. The third center, and the one we have hardly mentioned, was at Birmingham.

The Department of Metallurgy at Birmingham was headed by D. Hanson, who had worked with W. Rosenhain at the National Physical Laboratory and was continuing the tradition of trying to introduce an atomic viewpoint into metallurgy. In other words, Rosenhain and Hanson were among the first metal physicists who were introducing a new slant to the old subject of metals. When the war was over and people could begin to think about free research again, Hanson set up two research groups, funded with Department of Scientific and Industrial Research (DSIR) money. One, headed by Geoffrey Raynor from Oxford, was to look into the constitution of alloys; the other, headed by Hanson's former student Alan Cottrell, was to look into strength and plasticity. Cottrell had been introduced to dislocations as an undergraduate in metallurgy, when Taylor's 1934 paper was required reading for all of Hanson's final-year students. "While I hoped, somehow eventually to prove the existence of dislocations, the only obvious way forward at that time was to start growing and straining single crystals of pure metals, in the manner of Andrade and Orowan."[112]

In the United States at that time, the interest in dislocations continued on a modest scale. The group around Seitz and Koehler was now at the Carnegie Institute of Technology and its interests had, through wartime involvement with nuclear reactor materials, shifted somewhat toward radiation damage and hence to the properties of point defects. On the other hand, the new solid-state group at Bell Laboratories was pursuing the subject of dislocations, although, of course, semiconductors took pride of place. Shockley himself took some interest in dislocations and was supported by the electron microscopist R. D. Heidenreich and the theoretician Thornton Read.

The other groups at that time were at Philips in the Netherlands, with W. G. Burgers working directly on dislocations and J. L. Snoek working on problems of internal friction. In France, P. Lacombe at the Centre Nationale de Recherche Scientifique laboratory in Vitry sur Seine was working on crystal boundaries and subboundaries in the tradition of A. Guinier and C. Crussard. German work on dislocations did not restart until later; the ravages of war made scientific work impossible until the early 1950s. In the Soviet Union, dislocations were not spoken of until much later, in fact not until the late 1950s.[113]

We must return to the 1947 Bristol conference. The first session included, apart from the paper by Mott and Nabarro, papers by Andrade on creep, and Sir Lawrence Bragg on the yield point of a metal, two more papers by Nabarro, and papers by Cottrell and Frank that we must look into in some detail.

Cottrell's paper on the effect of solute atoms on the behavior of dislocations was the first of a line of work on hardening by solute atoms and therefore on yield in real materials, such as steel. "The big prize was to understand yield in steel."[114] The idea developed in an interesting to-and-fro between Cottrell and Nabarro. Nabarro had written a paper on pinning of dislocations by solute atoms,[115] and Cottrell had developed the idea by considering that solute atoms were mobile and could be precipitated on dislocations. Nabarro picked up Cottrell's idea during a colloquium in Bristol and developed it further.[116] The essence of Cottrell's paper presented at the Bristol conference is best expressed in his own abstract.

> It is shown that solute atoms differing in size from those of the solvent can relieve hydrostatic stresses in a crystal and will thus migrate to the regions where they can relieve the most stress. As a result they will cluster round dislocations forming "atmospheres" similar to the ionic atmospheres of the Debye-Hückel theory of electrolytes. The conditions of formation and properties of these atmospheres are examined and the theory is applied to problems of precipitation, creep and the yield point.[117]

Cottrell insisted that he could not have developed this theory but for the power cuts of the crisis winter of 1946/1947. Only when all experimental apparatus had to be shut down and he wondered what to do, did he find time to teach himself "some elasticity theory and then to use this to try to calculate the elastic forces between solute atoms and dislocations."[118] This is splendid confirmation of Orowan's dictum that "the genesis of dislocation theory provides no exception from the fairly general rule that radical steps in science as in art required in the past some breakdown, malfunctioning, or inefficiency of social life which gave to a few persons opportunities to ramble instead of marching in a prescribed direction."[119] No wonder it was Orowan whom Cottrell consulted before embarking on his postwar research program!

Charles Frank's contribution to the 1947 Bristol conference perhaps best illustrates the speculative stage of dislocation theory at that time. His paper is imaginative, plausible, and fascinating, but consists of pure fanciful speculation. Dislocations and everything connected with them were still pure hypothesis. Frank assumed that dislocations reach velocities of the order of the velocity of sound. If so, they can be reflected at free boundaries and traverse the crystal many times, thus increasing the slip. Glide will cease in a plane when it becomes too heavily laced with dislocations.

> We consider an edge dislocation moving back and forth between boundaries parallel to itself. Now we suppose that for some reason a small length of the dislocation at one end fails to reflect. That end of the remainder of the dislocation which lies in the interior of the crystal now makes a trail of screw dislocations at rest behind it; one portion of the crystal glides while an adjacent portion does not. On return,

the edge dislocation has opposite sign and opposite direction. If the whole of it were reflected, it would double the pitch of the screw. It is unlikely to do this; instead we may expect the stresses from the screw to prevent reflection of a further length of the edge dislocation, the result being that a second similar screw is drawn parallel to the first. In the continuance of this process, the whole of the moving edge dislocation will be used up and we shall be left with a parallel grid of screw dislocations, all of the same sense and distance apart.[120]

Not only is this a piece of imaginative writing, it is also one of the early uses of the screw dislocation that had lain dormant throughout the war.

The paper by Heidenreich and Shockley is remarkable in that it represents one of the early applications of electron microscopy to the study of dislocations and is thus a forerunner of the final vindication of the theory by observation. The paper also presented to an international forum the half-dislocation, which is often called a Shockley dislocation, and the extended dislocation. In face-centered cubic lattices, a unit dislocation is unstable and dissociates into two half-dislocations. Slip proceeds by the two half-dislocations moving together as an extended dislocation.[121] Frank later introduced another partial dislocation, known as the sessile dislocation. Sessile dislocations cannot glide freely and by their interaction with normal dislocations form an obstacle to glide.[122]

It is perhaps noteworthy that the idea of the half-dislocation came quite independently to Bragg when he thought of slip in face-centered cubic lattices. He even wrote a paper and sent it to *Nature*, but withdrew it when he learned from Lomer that Shockley had presented the same idea at a meeting in the United States.[123]

The second session of the 1947 Bristol conference dealt with grain boundaries and recrystallization. We shall briefly describe only two contributions, those of P. Lacombe and W. G. Burgers. Lacombe's paper dealt with subboundary and boundary structures in high-purity aluminum. He had prepared single crystals by the method of Carpenter and Elam and found in them a micromosaic structure of blocks with dimension 0.1 μm. The micromosaic structure resides within a macromosaic structure with much larger blocks and misorientations between them of 1° to 3°. Lacombe rejected the old 1912 hypothesis of W. Rosenhain and A. Ewing, which assumed that grain boundaries were amorphous, and suggested instead that they consist of a transition lattice. The structure of the boundary depends on the relative orientation of adjacent crystalline blocks.[124]

It is only a small step from this statement to the introduction of dislocations as an element of grain boundaries. It remained for Burgers to take this step in his paper on recovery and recrystallization as processes of dissolution and movement of dislocations. "Single-crystal, polycrystal and cold-worked states differ only in degree; they are all built up of lattice blocks, eventually in a stressed condition, connected and separated by transition layers of dislocated atoms."[125] It thus became possible to construct every kind of boundary layer by a suitable array of dislocations.

It may be justified to pause at this point and draw up an intermediate balance sheet, to the end of 1947, with a few brief statements:

1. A small band of a new breed of solid-state or metal physicists had emerged, to whom it was obvious that dislocations played a role in the plastic deformation of crystals.

2. There was little preoccupation with the old question of cracks and weak regions; it was simply assumed that dislocations were there and could easily be generated to provide both the cause of easy glide and its mechanism.
3. Dislocations were now viewed as lines with elastic properties and with forces acting on them and between them.
4. Dispersion hardening as well as solute hardening were ascribed to dislocations, and work hardening was explained by locking, piling up, and interactions of dislocations.
5. Grain boundaries and subboundaries were beginning to be viewed as arrays of dislocations.
6. Partial dislocations, screw dislocations, and stacking faults had been considered.
7. Precipitation on dislocations had been predicted but not observed.
8. Finally, it must be reemphasized that dislocations remained a hypothesis without direct proof.

5.4   The Direct Observation of Dislocations

As Frank pointed out, the postulate of dislocations, up to the stage at which they were actually observed, is related to four physical phenomena.[126] The first, described here in some detail and the prime cause for the postulate of dislocations, is the weakness of real crystals, allied with the phenomenon of work hardening. The second, mentioned briefly, is the mosaic structure of real solids. This has been postulated in various forms to explain either the mechanical behavior of solids or the intensities of x-ray diffraction patterns from real solids. The postulated mosaic patterns can best be explained in terms of dislocations. The third body of evidence relates to the rate of growth of crystals. The real crystal grows from a slightly supersaturated vapor many orders of magnitude faster than an ideal crystal theoretically should, and this discrepancy was, as we shall see, explained with the aid of dislocations. Finally, Frank mentioned the rate of creation of lattice vacancies and *F*-centers in irradiated crystals as evidence for a process of internal formation that was eventually traced to the presence of dislocations.

We have dealt with the first of these four aspects in some detail. Two others form part of the process of making dislocations visible, and this process we shall now pursue. Regretfully, we shall disregard the aspect connected with radiation damage.

The story of crystal growth and dislocations is inseparable from the name of Charles Frank. Frank had met Mott during the war, and when the war came to a close, Mott asked him to come to Bristol. After a period of part-time attachment before his final release from war work, Frank finally arrived at Bristol in January 1947. There he shared a room with Nicolas Cabrera and J. Burton and became closely acquainted with Frank Nabarro and Jack Mitchell. At first, he had not settled on any definite research program, but Burton and Cabrera talked to him about crystal growth, Nabarro about dislocations, and Mitchell about point defects. It was Nabarro who drew his attention to the paper by J. M. Burgers on screw dislocations. Mott asked Frank to lecture on crystal growth, and Frank used as his source Max Volmer's book *Kinetik der Phasenbildung,* which he obtained during his postdoctoral period with Debye in Berlin. Volmer described the growth of iodine crystals from the vapor at 1% supersaturation. Burton calculated that in a perfect crystal the supersaturation required for the observed rate of growth would have to be 50%. At about the same time, Mitchell asked Frank why sodium chloride crystals, under certain circumstances, grew in only one direction. As is Frank's wont, he came up with an immediate hypothesis. Although the hypothesis did not fit the case quoted by Mitchell, it fitted the case of iodine crystals and was the general answer to the justified doubts about the previous theories of rates of crystal growth raised by Burton and Cabrera.

The difficulty about crystal growth from the vapor is that if an atom attaches itself to a perfect surface, it is much more likely to leave the surface again than to be joined by a second atom on the same site. If, however, one atom attaches itself to a surface step, it is held there by cohesive forces in two dimensions and forms a stable nucleation center for the next atom to attach itself. Thus a surface step is a precondition for crystal growth at reasonable rates at low supersaturation. A screw dislocation emerging at a surface can provide just such a step, and the crystal can grow in a spiral, which perpetuates the required step while it grows.

When a conference on crystal growth was held in Bristol under the auspices of the Faraday Society in April 1949, Frank presented a paper that made no mention of the screw dislocation idea. He had thought of it between writing the paper and presenting it at the conference.[127] According to Mitchell, Frank actually mentioned his new theory at the conference and was told that nobody had observed spiral growth.[128] In the autumn of the same year, another summer school-cum-conference was held in Bristol, organized by Frank. "Bristol had set itself the missionary task to teach solid state physics."[129] From the start, a young researcher from S. Tolansky's laboratory, J. Griffin, tried to tell something to Frank, who was busy organizing the meeting and put him off repeatedly. When Griffin finally managed to get Frank's attention, the effect was shattering. Griffin had examined surfaces of beryl crystals in phase contrast microscopy and observed spirals that looked precisely like those Frank had conjectured. In the first flush of enthusiasm, nobody noticed that the observed features were much larger than atomic dimensions. The pictures were shown to all and aroused great excitement. It was the first visible experimental triumph for the theory of dislocations and a fine example in which two theories, those of nucleation and of dislocations, came together to produce an experimentally verifiable prediction.

Later, it was somewhat sobering to discover that the steps seen in the pictures were too large to be single-growth steps but were actually the result of impurity adsorption and etching. When John Forty, then Frank's research student, found similar spirals on magnesium and silver crystals, it was clear that monatomic steps were indeed essential for rapid growth of crystals from the vapor, but the steps were made visible by chemical attack or deposition on them.[130] Sometimes the necessary corrosive agent was provided by the vapor from the plasticine used to mount the crystal for observation. This differential etching or, more particularly, differential deposition became known as decoration.[131] Many more observations were made by many researchers on all kinds of crystals, using either optical methods or electron microscope replica techniques. The field has been well summarized by S. Amelinckx and by D. W. Pashley.[132]

As Frank pointed out at the 1949 conference, in reply to a question, the reverse of the growth process should also occur, so that dislocations emerging at surfaces should be the sites of preferential etching. "Dissolution . . . should proceed in a manner closely equivalent to growth . . . by unbuilding at molecular terrace lines ending on screw dislocations."[133]

Horn, in 1952, was the first to show that preferential etching indeed occurred at the site of the emergence of a screw dislocation.[134] In 1953, Vogel and his co-workers at Bell Labs initiated an important application of the etching technique by producing conical etch pits on (100) and (111) surfaces of germanium.[135] These etch pits formed rows that corresponded to the emergence of a tilt boundary. The boundary is formed by a suitable array of edge dislocations. In this way, it was established that indeed tilt boundaries consisted of parallel edge dislocations and that these could be revealed by etching.[136] This work was highly significant for three reasons:

1. It firmly established a practical method of revealing dislocations that became widely used in, for example, quality control of semiconductor crystals used for the manufacture of transistors and, later, integrated circuits.
2. It provided important evidence and became a useful tool in research on a whole range and hierarchy of crystalline subboundaries.
3. Last, but not least, etch pits provided fairly tangible proof for the existence of dislocations. They were an important step on the road to general acceptance of this theory.

A great deal of work was done by many researchers on the study of plastic deformation, using etch pits. Mitchell and Amelinckx summarize some of this work in their review papers.[137] Certainly, by the mid-1950s, so many unmistakable traces left by dislocations had been observed that practically nobody doubted their existence and the understanding of processes of plastic deformation had improved greatly. Only the very first puzzle remained somewhat puzzling, despite a great deal of effort: the work-hardening curve was still not fully understood. As Frank said in 1958, "The full truth of this complex matter has yet to be said."[138]

Growth steps and etch pits are manifestations of dislocations emerging at crystal surfaces, and, perhaps unsurprisingly, they provided the first visual evidence for the reality of dislocations. The decoration of dislocations in the interior of crystals was to follow very soon.

Mitchell, another postwar arrival in Bristol, picked up Mott's old interest in the theory of the photographic process. When he found hundreds of optically flat glass disks in the laboratory, left from some wartime optical work, he used them to grow thin single-crystal sheets of silver bromide to conduct experiments on photographic sensitivity.[139] In all this work, he was greatly helped by the scientifically trained, brilliant glass blower John Burrows and by a long line of research students.

The photographic process is one of the several areas of technological–scientific endeavor in which technology has long been ahead of science. In an age when photographic emulsions for every purpose could be made with a good deal of sophistication, knowledge of the fundamental physical process of latent image formation was still fairly rudimentary. Mitchell set about to change all that and has kept up his interest and his contributions for many years.[140]

Fascinating as the story of the photographic latent image is, it is peripheral to our

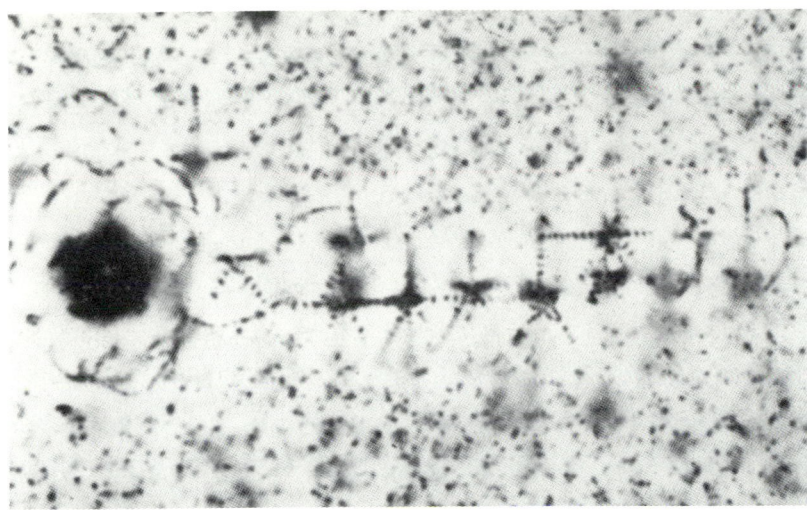

Fig. 5.3 Prismatic dislocations produced by Hysil glass sphere. Systems of similar disloca-tions on a much smaller scale, with axes along ⟨110⟩ directions, surround the larger deco-rating silver particles. ×2400. (From A. S. Parasnis and J. W. Mitchell, "Some Properties of Crystals of Silver Chloride Containing Traces of Copper Chlorides," *Phil. Mag.,* 8th ser., vol. 4 [1959]: Fig. 7, pl. 34)

main interest. It deserved mention, though, because the occupation with the latent image and his interaction with Frank and others at Bristol pushed Mitchell toward the discovery of decorated dislocations in the interior of crystals and into the whole field of research into the plastic deformation of crystals.

Mitchell and Hirsch and many others have given expression to one of the basic truths about scientific research: pursue some line of research and keep your mind open; then you will discover something, although it may not be what you set out to discover.[141] Discoveries do not happen by accident, but those who seek sometimes accidentally find what they had not sought, provided that their knowledge and existing theory make them receptive to what they see. This is the true nature of much-praised serendipity.

Mitchell set out to discover the mechanism of latent image formation and dis-covered the visible world of dislocations in the interior of crystals; "they were sim-ply there."[142] The first systems of subboundaries, consisting of rows of dislocations, were made visible under the optical microscope by photolytic silver depositions along these boundaries in crystals of silver bromide and silver chloride exposed to light.[143] These observations were continued for several years by a succession of research students, and much work was done on crystals of silver and alkali halides that were heat treated, deformed, sensitized, and decorated in different ways. Mitchell gave a brief review of the state-of-the-art to about 1960 and told his own story of dislocations in silver halides in the book *Beginnings of Solid State Phys-ics.*[144]

Figure 5.3 shows a beautiful example of a system of dislocations caused by the inclusion of a tiny glass sphere in a crystal of silver chloride;[145] Figure 5.4 shows slip planes in a crystal of potassium bromide.[146]

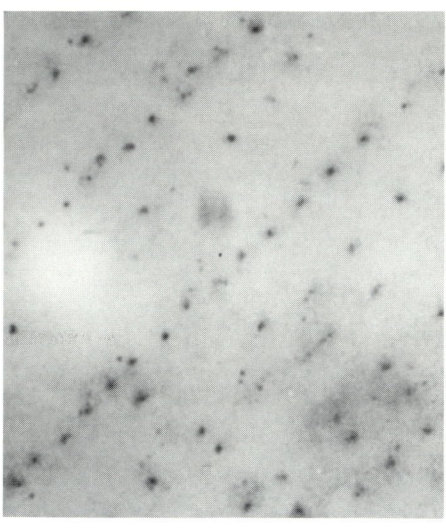

Fig. 5.4 Edge dislocations on slip planes in a crystal of potassium bromide. (From D. J. Barber, K. B. Harvey, and J. W. Mitchell, "A New Method for Decorating Dislocations in Crystals of Alkali Halides," *Phil. Mag.,* 8th ser., vol. 2 [1957]: Fig. 1, pl. 24)

Another aspect of the internal decoration of dislocations was discovered by Dash.[147] He worked on silicon, which cannot be plastically deformed at room temperature, and studied high-temperature plastic deformation processes by the decoration of dislocations with copper diffused into the crystal. The copper can be seen in an infrared transmission microscope, and the interior dislocations can be matched with their points of emergence, the latter being shown by etch pits.

Etch pits, as we have seen, played a prominent part in early work on dislocations, and they form a bridge to a different set of experimental evidence for the existence of dislocations—polygonization. According to Robert Cahn, who kindly furnished the material for the following section, it was the elucidation of the phenomenon of polygonization that supplied some of the most firm early evidence for the existence of dislocations.[148]

The rearrangement of deformed crystals into a mosaic structure had been postulated by many authors from the early days of deformation research. Some of the early evidence for simultaneous gliding and bending was supplied by the phenomenon known as asterism, when the spots in a Laue diffraction pattern become elongated owing to lattice curvature. The elongated spots often break up into separate spots on annealing, indicating that the curved lattice had broken up into straight subgrains. This process of accommodating the damage from bend-gliding during annealing is called polygonization.

When Cahn joined Orowan as a research student in Cambridge in 1945, he was given the task of elucidating these phenomena. Dislocations were not mentioned to him in this context, but, of course, nobody in Orowan's group could be unaware of them.[149]

Cahn found a 1932 paper by Konobeevski and Mirer, who had observed subdivided asterism in bent and annealed crystals of rock salt, and decided to use bending instead of stretching for his own experiments.[150] This seemed a more straightforward method of producing a bent lattice, although asterism had previously been observed in a variety of stretched crystals.[151]

Cahn discussed the problem with his father-in-law, Daniel Hanson, professor of metallurgy in Birmingham, and with A. J. Cottrell of the same department. In a letter dated 18 July 1946, Cottrell proposed that an excess of dislocations of one sign, which must be present in a bent lattice, might rearrange themselves by thermal activation into walls normal to the slip plane and thus form subboundaries. Soon Cahn found a way of revealing these subboundaries (tilt boundaries) as etch pits.[152] Further work by various experimenters and the theory of small angle boundaries by Shockley and Read[153] firmly established the role of dislocations in small-angle subboundaries and made it "possible to interpret the observed tendency of many slender crystallites in a bent and annealed crystal to coalesce into a few fatter ones."[154] Orowan gave the phenomenon the name *polygonization* and, when he presented the work in Paris in 1947, found that Crussard had obtained similar results on his x-ray diffraction patterns from stretched aluminum crystals.[155] Lacombe and Beaujard established a substructure in aluminum grains and showed the subboundaries by etching, while Guinier and Tennevin measured the misorientation of the subgrains.[156] Dislocations were not mentioned in this French work.

While this work only failed to mention dislocations, W. A. Wood "stoutly resisted any suggestion that the cells he observed, especially in crept aluminum, were in any way connected with dislocations."[157] The discussions raged for some time, until Cahn, Bear, and Bell showed with the aid of microbeam x-ray diffraction that the intercell subboundaries in crept zinc were indeed tilt boundaries, and other evidence had shown that they were composed of an array of dislocations.[158]

The microbeam x-ray diffraction technique had been developed by Noel Kellar and Peter Hirsch, two research students in Cambridge working with W. H. Taylor, between 1946 and 1948. The idea for this equipment came from Bragg, and the purpose was to see whether Wood was right in thinking that cold-worked crystals broke up into a mosaic structure or whether H. Lipson was right in thinking that the broadening of diffraction lines from the cold-worked metals was caused by variations in the lattice parameter. Hirsch and his collaborators used the method to show that cold-worked crystals do break into smaller particles, but that most of the x-ray broadening is caused by internal strains. Of infinitely more importance was the fact that this work served as a stepping stone toward the final proof for the reality of dislocations—their direct observation by the electron microscope.[159]

Hirsch, too, went to the famous 1947 Bristol conference and heard the paper by Heidenreich and Shockley on partial dislocations. His interest at first was not aroused so much by this theory as by the fact that Heidenreich had used an electron diffraction facility in an electron microscope to look at misorientations in cold-worked metals. Later, when Heidenreich reported his electron microscope observations of small subgrains in cold-worked aluminum, Hirsch was struck by the fact that Heidenreich could obtain micrographs in seconds, while his x-ray microbeam diffraction patterns took many hours. Hirsch went to see another research student, J. W. Menter, who worked with a Metropolitan Vickers EM3 electron microscope in Bowden's Research Laboratory for the Physics and Chemistry of Rubbing Solids, as it then was called. Using the microscope, spotty electron diffraction rings were quickly obtained from cold-worked (beaten) thin foils of gold.[160] Electron micrographs were also taken, but their complex structure was to be described and analyzed in a further paper. The paper was never published because Hirsch thought that beaten foils were too complex to analyze and wanted to introduce the technique of electrolytic thinning, previously used by Heidenreich, instead. Heidenreich reported the observation of small subgrains and used the dynamical theory of diffraction to explain "extinction contours." Although individual dislocations were probably visible on Heidenrich's pictures, nobody saw them. This is one of the many episodes in science that illustrate how easy it is to miss things one is not looking for. The relationship between "looking" and "finding" is anything but straightforward.

In 1954, M. J. Whelan joined Hirsch as a research student and was to work on the direct observation of dislocations. By that time, Hirsch had set himself the task of doing just that. He began by developing an ion etching technique for thinning specimens, previously used by Castaing in work on Guinier–Preston zones. He eventually gave up on this technique and settled on an etching method.[161]

At that stage, Hirsch used his thorough knowledge of diffraction theory to deduce two things:

1. He though that the streaky images that he observed on beaten gold foil were caused by phase changes across stacking faults.
2. According to Heidenreich and Shockley, dislocations could be dissociated, with stacking faults between them. Hence, he thought, dislocations should be visible in the electron microscope. He also thought that the strain field around a dislocation might reduce "primary extinction," but had no way of estimating the effect on contrast.

In the meantime, a new electron microscope, the Siemens Elmiskop I, had been delivered to Coslett's electron microscopy group, and this instrument gave far superior resolution. In October 1955, Whelan and R. W. Horne started taking micrographs and diffraction patterns of cold-worked aluminum and gold. The subboundaries in the aluminum foils

> were found to consist of individual dots or short lines, which we thought were probably dislocations, since their spacing agreed well with the prediction of Frank's formula, from the measured values of misorientation across the boundaries. However, we could not be sure, since it was possible that the dots or short lines could be due to Moiré effects from two overlapping crystals. On 3 May 1956 Horne used the microscope in the "high resolution mode" with the double condenser system. He also pulled out the condenser aperture to increase the beam intensity. The lines were observed to move parallel to traces of (111) planes. This left us in no doubt that the lines were indeed images of individual dislocations, and that their glide was observed in the experiments.[162]

These experiments were widely reported, and the cine-film of moving dislocations, which demonstrated almost everything that theory had predicted, presented irresistible proof of the true existence of dislocations and the essential correctness of the theory describing their behavior.[163] Nevill Mott and G. I. Taylor were among the first to be shown the experiments and were, of course, delighted. Even metallurgists were forced to believe in dislocations now.

It is interesting to note that Hirsch's original idea that dislocations should be visible because of the stacking fault ribbon proved not to be appropriate to dislocations in aluminum for which the width of ribbon is too small to produce any observable effects on the image; in this case, the image contrast is effectively due entirely to the strains around the dislocation line. That this was actually the mechanism responsible for seeing dislocations in thin aluminum foils was already suggested in the original paper by Hirsch, Howie, and Whelan on the observation of dislocations.[164] The theory of image contrast from crystal defects was worked out only after the event, first for stacking faults by Whelan and Hirsch, and then for dislocations by Hirsch, Howie, and Whelan.[165] Nevertheless, Hirsch's "hunch," based on theoretical ideas as to why dislocations might be visible, sufficed to give him confidence in the search for an experimental result. Without such theoretical guidance, Hirsch might have neither have looked for dislocations nor found them. The new instrument proved vital: "I am sure that if Bob Heidenreich had taken his pictures in 1949 on a microscope with the resolution of the Siemens Elmiskop I and with double condenser illumination, he would have seen the individual dislocations."[166]

On the other hand, it is undisputed that Hirsch, Whelan, and Horne were not the only discoverers of the visible dislocation. Walter Bollman joined the Battelle Laboratory in Geneva in 1953 and worked first on a research contract that involved the deposition of germanium on germanium. He used an electron microscope and in 1956 took charge of the instrument. At that time, several things came together: Bollman's previous experience with high-voltage physics, acquaintance with the work of Castaing on transmission electron microscopy of metals, and a research contract on improved stainless steel alloys. An additional factor was experience in the laboratory with electropolishing of steel, which had been undertaken for the watch industry. Bollman put all these things together and examined thinned electropolished samples of stainless steel in the electron microscope. One day, in June 1956, he observed what to him were obviously dislocations.[167]

Although Bollman had read some papers on dislocations and, in particular, was familiar with the work of Heindenreich, he found visible dislocations without the same degree of theoretical anticipation as Hirsch. Here was a highly competent scientist working on a problem of allows and keeping his eyes open to allow serendipity to bestow its blessings. Bollman's findings were undoubtedly independent and only days later than Hirsch's, although it is possible that Crussard, who visited both groups, acted as a bee carrying some pollen of thought.[168]

Bollman attended the conference in Reading on 24–26 July 1956 at which Whelan first disclosed the Cambridge findings. He contacted Hirsch, Whelan, Pashley, and Coslett and showed them his own pictures of dislocations. They all agreed that Bollman and Hirsch and colleagues had made the same discovery independently and nearly simultaneously. Hirsch also agreed that the electrolytic polishing technique used by Bollman was superior and that stainless steel was a very interesting material for the study of dislocations. One result of the discussion among Hirsch, Whelan, and Bollman was a joint study that established that dissociation of dislocations and the formation of stacking faults by the movement of partial dislocations could be observed directly in stainless steel. The paper also refers to some of the background to the original discoveries by Hirsch, by Horne and Whelan, and by Bollman.[169]

The Origin of Dislocations

It is, perhaps, a travesty that we have said so much about the verification of the existence of dislocations and nothing at all about their creation. Certainly, the question of the mechanism of formation of dislocations was one that exercised many minds.

To put the problem in a nutshell: if slip occurs by the movement of dislocations and the contribution of each dislocation to slip ceases when it has either emerged at the surface of the crystal or piled up on a subboundary or some other obstacle, then, for continuous slip to occur, new dislocations must be created. The question was, How?

When Frank was thinking about this problem in his early days at Bristol, his main idea was that dislocations could move very fast, carrying sufficient kinetic energy to cause multiplication. On further reflection, he thought that their limiting velocity had to be the velocity of sound and was disappointed to discover that Frenkel and Kontorova had thought of that in 1938.[170] In the end, Frank thought that if dislocations traveled at 0.866 times the speed of sound, they could still multiply. With this and several other points in mind, he set out on a trip to the United States in the spring of 1950. He visited various laboratories and attended a conference on plastic deformation in Pittsburgh. The rest of the story of the birth of the Frank–Read source of dislocations is best told in Frank's own vivid words.

> When I got to Pittsburgh I received a letter from Jock Eshelby. . . . "Have you read Leibfried's paper which shows that the speed of dislocations under typical test conditions will not be greater than 0.07 of the speed of sound?" I had not seen Leibfried's paper. I went to the Carnegie Tech. library and said "Has the *Zeits. f. Phys.* (or whatever it was) come in?" They said no, but there is a parcel in from Germany today, it might be in that, so I waited around and they opened up the parcel and the requisite journal was in it, so I read Leibfried's paper and just had time to go and catch a train. I arrived at Ithaca at a time they did not expect; I arrived there before lunch and they had not expected me until tea time, so they dumped me in my hotel, said "Have lunch, we have a meeting this afternoon; we are very sorry, can you look after yourself until 5 o'clock." And I had read this paper, it might be wrong, it was not conclusive that he was right, but what do I do if he is right? While I was walking on the campus I had the idea, of course dislocations in the interior of a crystal are not awfully different, geometrically they are the same thing, as the growth step on the surface of a crystal and I already knew that that thing turned into spirals. So I said "Yes, dislocations can wind themselves up into spirals on the slip plane." So then I gave a lecture at Cornell about crystal growth theory and I went to a party and woke up next morning and went to Schenectady, and sitting on the floor in Hollomon's house with John Fisher and a few other people, drinking beer, I said, "Do you see anything wrong with this idea?" And John looked up and said, "No, I can't see anything wrong with that." The following day I rode with the General Electric people in a car to Pittsburgh and we all assembled in a hotel lobby and were introduced to each other. John Fisher brought Thornton Read. Thornton, as soon as he was introduced to me, said "Frank, there is something I want to tell you" and John Fisher replied, "Frank has got something to tell you." So we started talking and we found that we were telling each other what was in all basic principles the same. So I said, "When did you think

of that?" and he said, "When I was drinking my tea last Wednesday afternoon about 4 o'clock." I said, "I was walking on the Cornell campus from 3 till 5."

It was a remarkable coincidence. The preconditions were there, I was preparing for the same conference as Thornton Read and we both had good reasons to focus our thoughts on the same problem and we got them into focus precisely simultaneously.[171]

The Frank–Read source was described in a joint paper[172] and became widely known and accepted. Its description can be found in any textbook on dislocations. On the other hand, experimental verification for its operation has been found in only a few instances. One of these was the observation by Dash in crystals of silicon.[173] It is still plausible that in most cases dislocations form on surfaces, as postulated from the earliest days of dislocation theory.[174] "The first stage in the plastic deformation of a dislocation-free crystal involves the formation of dislocation loops at the surface."[175]

The Acceptance of Dislocation Theory

After the initiation of dislocation theory in Berlin and Cambridge, the center of gravity quickly shifted to Cambridge. Polanyi moved out of the field, and Taylor, Orowan, and Bragg were all in Cambridge. Further early centers were associated with Seitz and Mott in academe and the Philips concern in industry. After the war, dislocation work in Britain was concentrated in three centers: Cambridge, Bristol, and Birmingham. Bristol acted as a dissemination center by hosting summer schools, conferences, and many visiting scholars. Two of these scholars became very influential in spreading dislocation theory and experiment into Germany and France, respectively.

Alfred Seeger belongs to the immediate postwar generation of German metal physicists. Although he studied in Stuttgart, the home of Dehlinger, research under chaotic conditions of the immediate postwar period was impossible, and he longed to get out of the poverty-stricken claustrophobic atmosphere. His chance came when he obtained a scholarship to Bristol (1951–1952), where he not only learned the latest state-of-the-art on dislocations, but also became a member of the world scientific community. He took back to Stuttgart his knowledge and his friendships and soon formed the nucleus of the successful German research school on dislocations.[176] In fact, Seeger became the first incumbent of the first chair in solid-state physics in the Federal Republic, founded in Stuttgart in 1959, and wrote one of the important reviews on the subject of dislocations.[177]

Similarly, Jacques Friedel, a descendant of a family of distinguished French scientists, spent some time in Bristol and became fascinated by dislocations. Although there was a long tradition of research on crystalline subboundaries and mosaic structure, dislocations had not been studied in France. Friedel summed up his Bristol experiences, and greatly enlarged on them by his own theoretical elaborations, in his successful text.[178] He, too, became firmly established in the community of scholars working on dislocations, and his work was highly influential in spreading knowledge of dislocations in France.

Metallurgists resisted dislocations to the last, but could resist no longer when they had been so spectacularly substantiated in studies on surface spirals and decoration, and, as a final triumph, by direct observation with the electron microscope.

Although how important dislocation theory is to the practical metallurgist is hard to tell. Frank, one of the leaders of research on dislocations, has few illusions in this matter, at least as far as "normal" engineering is concerned.

> Dislocations may be important to people who make turbine blade alloys. I hope that dislocation theory has been a useful guide to people doing that sort of job.
>
> A man shapes a piece of malleable metal by hitting it with a hammer and the manner in which he hits it is scarcely modified by any consciousness that he is causing dislocations to move in the crystals.[179]

Undoubtedly, the understanding of the behavior of metals was greatly enhanced, which, in turn, probably helped in the solution of some advanced practical metallurgical problems, as in special alloys for aircraft, materials for nuclear reactors, and other extreme applications. Thus the practical utility of the knowledge of dislocations may not be overwhelming in the field that gave birth to it, but it is useful in what is now called high technology. Dislocations proved very important in materials for the semiconductor industry, an industry that makes notoriously high demands on everything it uses. It turns out that dislocations affect the electrical properties of semiconductor materials, such as silicon, and methods for growing virtually dislocation-free material had to be developed.

The following treatment of the topic of dislocations in semiconductors is based on a recent review by Hans J. Queisser.[180]

When the electrical properties of dislocations were first considered by Shockley in 1953, it was thought that they constituted dangling bonds. Under these circumstances, dislocations would have very serious consequences for the electronic behavior of semiconductors, and much research effort went into attempts at elucidation, on the one hand, and elimination, on the other. If dangling bonds from dislocations could capture electrons, the band structure would suffer local distortions and the doping level would become uncontrollable.[181] It was also considered likely that arrays of dislocations could form conduction channels and might thus short-circuit $p-n$ junctions.

> Initial observations seemed to corroborate this pessimistic prediction. The only clear correlation at that time between device failure and other material properties relied on the etch pit count, a measure for dislocation density. . . . This general trend of observations stimulated extensive research to improve the starting material, which culminated in the success of growing perfectly dislocations-free silicon single crystals.[182]
>
> The tetrahedrally bonded semiconductors Ge and Si can fortunately easily be grown dislocation-free for two reasons: the thermal conductivity is comparatively large, which reduces steep thermal gradients with their detrimental stresses; secondly the dislocation velocity in these materials is relatively slow.[183]

Unfortunately, the initial absence of dislocations is no guarantee for their continued absence, as the procedures of device fabrication undoubtedly reintroduce dislocations. Despite this, transistors could be made to work. The simple image of the role of dislocations as dangling bonds was therefore abandoned, and the decorated dislocation was revealed as both the culprit and the savior. Impurities migrate and segregate on dislocations, thus neutralizing and rendering harmless the dangling bonds. On the other hand, if the dislocation density becomes too great, inhomogeneities in the doping levels and local breakdown effects can occur.[184]

With the extremely small feature sizes of modern integrated circuits, any distortions caused by dislocations are intolerable, and silicon crystals with extremely low densities of dislocations are used for modern device fabrication. In noncritical regions of the material, dislocations can be used deliberately to "getter" unwanted impurities and so-called backside damage is often employed. Modern devices require such incredible accuracy that the motion of a few dislocations can suffice for significant mechanical distortion. The presence or absence of dislocations needs to be most carefully manipulated, and thus, 50 years after their discovery, dislocations play their greatest role in materials and in ways that in the 1930s had not been thought of. The strongly science-based semiconductor industry requires all available scientific insights for the manufacture of its materials, and thus the science of dislocations has made its practical contribution, albeit in an unexpected field.

Notes

1. W. Alexander and A. Street, *Metals in the Service of Man,* 2nd ed. (Harmondsworth: Penguin Books, 1954), 12.

2. Ibid. 116.

3. E. Braun, "Science and Technology as Partners in Technological Innovation," *Physics in Technology* 15 (1984): 80–85.

4. Ibid.

5. H. Y. Hunsicker and H. C. Stumpf, "History of Precipitation Hardening," *The Sorby Centennial Symposium on the History of Metallurgy,* ed. C. S. Smith (New York: Gordon and Breach, 1965), 271–311.

6. Braun (n. 3).

7. N. F. Mott, "The Rutherford Memorial Lecture 1962," *Proc. Roy. Soc.* A275 (1963): 149–160. E. Braun and S. Macdonald, *Revolution in Miniature,* 2nd ed. (Cambridge: Cambridge University Press, 1982), chaps. 1, 2.

8. Mott (n. 7), 151.

9. C. F. Elam, *Distortion of Metal Crystals* (Oxford: Clarendon Press, 1935).

10. E. Schmid and W. Boas, *Kristallplastizität* (Berlin, Springer, 1935).

11. For a summary of methods of crystal growth, see ibid., 24–37.

12. H.C.H. Carpenter and C. F. Elam, "The Production of Single Crystals of Aluminum and Their Tensile Properties," *Proc. Roy. Soc.* A100 (1921): 329–353.

13. R. Cahn, interview with S. Keith, 24 July 1981.

14. C. F. Elam, "The Diffusion of Zinc in Copper Crystals," *Journal of the Institute of Metals* 43 (1930): 217–231; A. E. van Arkel, J. H. de Boer, and J. D. Fast, "Darstellung von reinem Tatanium, Zirkonium, Hafnium und thorium Mettall," *Zeitschrift für Anorganische und Allgemeine Chemie* 148 (1925): 345–350.

15. F. Zwicky, "Die Reissfestigkeit von Steinsalz," *Phys. Zs.* 24 (1923): 131–137. The reference to Traube is given by Elam (n. 9).

16. J. Frenkel, "Theory of the Elasticity Limits and Rigidity of Crystalline Bodies," *Zs. Phys.* 37 (1926): 572–609. This work has achieved wide popularity and is often quoted; see A. J. Dekker, *Solid State Physics* (London: Macmillan, 1962), 83–85.

17. G. von Göler and G. Sachs, "Das Verhalten von Aluminiumkristallen bei Zugversuchen," *Zs. Phys.* 41 (1927): 103–115.

18. Elam (n. 9), 165.

19. G. Beilby, *Aggregation and Flow of Solids* (London: Macmillan, 1921).

20. N. P. Allen, "Fifty Years' Progress in the Understanding of Metals," *Journal of the Birmingham Metallurgical Society* (1953): 89–109, p. 102.

21. Beilby (n. 19), 123.

22. A. A. Griffith, "The Phenomena of Rupture and Flow in Solids," *Philosophical Transactions of the Royal Society,* A221 (1920): 163–198, G. I. Taylor, "Note on the Early Stages of Dislocation Theory," in Smith (n. 5), 355–358.

23. Taylor (n. 22).

24. A. Smekal, "Structure Sensitive Properties of Crystals," in *Handbuch der Physik* (Berlin: Springer, 1933), 24, pt. 2: chap. 5.

25. These matters are discussed in more detail in the appropriate chapters of this volume.

26. A. Smekal, "Über den Einfluss von Kristallgitterporen auf der Molekülbeweglichkeit und Festigkeit,"*Phys. Zs.* 26 (1925): 707–712. For a discussion of Smekal's contributions to the theory of plasticity and fracture, see also H. Rumpf, "Zur Entwicklungsgeschichte der Physik der Bruch-erscheinungen," *Chemie-Ingenieur-Technik* 31 (1959): 697–705.

27. Smekal (n. 26), 711.

28. A. Smekal, "Über die Grössenordnung der ideal gebauten Gitterbereiche in Realkristallen," *Ann. Phys.* 83 (1927): 1202–1206.

29. F. Zwicky, "Imperfections of Crystals," *Proceedings of the National Academy of Sciences* 15 (1929): 253–259.

30. H. Mark, M. Polanyi, and E. Schmid, "Vorgänge bei der Dehnung von Zinkkristallen," *Zs. Phys.* 12 (1923): 58–116.

31. R. Becker, "Über die Plastizität amorpher und kristalliner fester Körper," *Phys. Zs.* 26 (1925): 919–925, pp. 923–924.

32. A. A. Griffith and B. Lockspeiser, "The Mechanism of Stress Concentration in Plastic Crystals," Report No. H. 1200, Royal Aircraft Establishment, December 1927.

33. Ibid.

34. E. Orowan, "Dislocations in Plasticity," in Smith (n. 5), 359–376.

35. K. Meissner, M. Polanyi, and E. Schmid, "Messungen mit Hilfe von flüssigem Helium. XII Plastizität von Metallkristallen bei tiefsten Temperaturen," *Zs. Phys.* 66 (1930): 477–489.

36. Taylor (n. 22).

37. G. I. Taylor, "Resistance to Shear in Metal Crystals," *Transactions of the Faraday Society* 24 (1928): 121–125.

38. L. Prandtl, "Ein Gedankenmodell zur kinetischen Theorie der festen Körper," *Zeitschrift für angewandte Mathematik und Mechanik* 8 (1928): 85–106.

39. U. Dehlinger, "Zur Theorie der Rekristallisation reiner Metalle," *Ann. Phys.* 2 (1929): 749–793.

40. Prandtl (n. 38), 87.

41. *Habilitationsschrift* is an extended postdoctoral dissertation requirement for becoming a fully fledged teacher in German universities. Note 39 is the published version of the *Habilitationsschrift*.

42. I. Czochralski, "Verlagerungshypothese und Röntgenforschung," *Zeitschrift für Metallkunde* 15 (1923): 60.

43. Dehlinger (n. 39), 758.

44. Ibid, 763–764.

45. A.E.H. Love, *A Treatise on the Mathematical Theory of Elasticity,* 4th ed. (New York: Dover, 1944).

46. J. Friedel, *Dislocations* (Oxford: Pergamon Press, 1964), 4.

47. G. I. Taylor, "The Mechanism of Plastic Deformation of Crystals," *Proc. Roy. Soc.* A145 (1934): 362–415 (in three parts). M. Polanyi, "Über eine Art Gitterstörung, die einen Kristall plastisch machen könnte," *Zs. Phys.* 89 (1934): 660–664. E. Orowan, "Zur Kristallplastizität," *Zs. Phys.* 89 (1934): 605–659 (in three parts).

48. Taylor (n. 37).

49. Taylor (n. 47), 368.

50. Ibid., 375.

51. Ibid., 383.

52. Ibid., 388.

53. Ibid., 389.

54. Ibid., 399.

55. Orowan (n. 47).

56. Ibid., 640.

57. Polanyi (n. 47).

58. E. Orowan, interview with S. Keith, 4 October 1981.

59. Taylor

60. Elam (n. 9); Schmid and Boas (n. 10).

61. Taylor (n. 22), 358.

62. M. Polanyi, "My Time with X-rays and Crystals," in *50 Years of X-ray Diffraction,* ed. P. P. Ewald (Utrecht: International Union of Crystallography, 1962), 629–636.

63. E. Schmid, interview with E. Braun, summer 1982.

64. Polanyi (n. 62), 632.

65. Ibid., 633.

66. Ibid., 636.

67. Orowan (n. 34).

68. Ibid., 365.

69. W. G. Burgers, "Shear Hardening and Recrystallization of Aluminum Single Crystals," in *International Conference on Physics,* vol. 2: *The Solid State of Matter* (London: Physical Society, 1935), 139–160. A. Ioffé, "On the Cause of the Low Value of Mechanical Strength," in *International Conference on Physics,* 2: 72–76; Ioffé "On the Mechanism of Brittle Rupture," in *International Conference on Physics,* 2: 77–80.

70. Burgers (n. 69), 140.

71. Taylor (n. 37); K. Yamaguchi, "The Internal Strain of Uniformly Distorted Aluminium Crystals," *Scientific Papers of the Institute of Physical and Chemical Research, Tokyo* 11 (1929): 151–169; Yamaguchi, "Slip-Bands of Compressed Aluminium Crystals," *Scientific Papers of the Institute of Physical and Chemical Research, Tokyo* 11 (1929): 223–241.

72. Burgers (n. 69), 143.

73. E. Schmid, "On Plasticity, Crystallographic and Non-Crystallographic," in *International Conference on Physics* (n. 69), 2: 161–170, p. 163.

74. Ibid., 166.

75. Ibid., 169.

76. A. H. Cottrell, *Dislocations and Plastic Flow in Crystals* (Oxford: Clarendon Press, 1953), 86–89.

77. Ioffé ("On the Cause") (n. 69), 75.

78. Ibid., 80.

79. E. N. da C. Andrade, in *International Conference on Physics* (n. 69), 2: 173.

80. A. Smekal, "The Structure-Sensitive Properties of Salt Crystals," in *International Conference on Physics* (n. 69), 2: 93–109, p. 108.

81. J. Koehler, interview with E. Braun, 24 October 1983.

82. J. M. Burgers, "Second Report on Viscosity and Plasticity," *Verhandelingen der Koninklijke Nederlandische Akademie van Wetenschappen, Afdeeling Nataurkunde, Reeks* 1 (1938): 1–28; W. G. Burgers and J. M. Burgers, "Plasticity of Rocksalt and the Taylor and Becker-Orowan Theories of Crystalline Plasticity," *Nature* 135 (1935): 960.

83. W. G. Burgers, "How My Brother and I Became Interested in Dislocations," in *Beginnings,* 125–130.

84. J. M. Burgers, "Some Considerations on Fields of Stress Connected with Dislocations in a Regular Crystal Lattice," *Proceedings of the Koninklijke Nederlandsche Akademie van Wetenschappen* 42 (1939): 293–325; Burgers, "Geometrical Considerations Concerning the Structural Irregularities to Be Assumed in a Crystal," *Proceedings of the Physical Society* 52 (1940): 23–33.

85. Burgers (n. 83), 130 (in an addendum).

86. For example, J. Mitchell in private conversation with E. Braun in October 1983; F.R.N. Nabarro, "Recollections of the Early Days of Dislocation Physics," in *Beginnings,* 131–135.

87. For example, Schmid interview (n. 63).

88. F.R.N. Nabarro, "The Strains Produced by Precipitation in Alloys," *Proc. Roy. Soc.* A175 (1940): 519–538; Nabarro (n. 86).

89. Nabarro (n. 86), 132.

90. Orowan interview (n. 58).

91. W. L. Bragg, "Discussion. Part I," *Proceedings of the Physical Society* 52 (1940): 54–55.

92. Orowan interview (n. 58).

93. J. Nye, interview with S. Keith, 28 June 1982.

94. Ibid.

95. W. L. Bragg and J. F. Nye, "A Dynamical Model of a Crystal Structure," *Proc. Roy. Soc.* A190 (1947); 474–481.

96. W. L. Bragg and W. M. Lomer, "A Dynamical Model of a Crystal Structure. II," *Proc. Roy. Soc.* A196 (1949): 171–181.

97. Nye interview (n. 93).

98. Ibid.

99. Orowan interview (n. 58).

100. F. Seitz, *The Modern Theory of Solids* (New York: McGraw-Hill, 1940).

101. F. Seitz, "Biographical Notes," in *Beginnings,* 84–99, p. 91.

102. Ibid., 92.

103. See, for example, F. Seitz and T. A. Read, "Theory of the Plastic Properties of Solids. I, II," *Journal of Applied Physics* 12 (1941): 100–118, 170–186; J. S. Koehler, "On the Dislocation Theory of Plastic Deformation," *Phys. Rev.* 60 (1941): 397–410; H. B. Huntington, "Dislocations in NaCl," *Phys. Rev.* 59 (1941): 942–943.

104. See also Cottrell (n. 76), 39.

105. Koehler interview (n. 81).

106. J. Friedel, *Dislocations* (Oxford: Pergamon Press, 1964) (based on the first French edition published in 1956).

107. N. F. Mott and F.R.N. Nabarro, "Dislocation Theory and Transient Creep" (Report of the Conference on Strength of Solids held at the H. H. Wells Physical Laboratory, University of Bristol, 7–9 July 1947), 1.

108. See, for example, F.R.N. Nabarro, "Dislocations in a Simple Cube Lattice," *Proceedings of the Physical Society* 59 (1947): 256–272.

109. Mott and Nabarro (n. 107), 3–4.

110. Koehler interview (n. 81); Mitchell interview (n. 86).

111. Mott and Nabarro (n. 107), 8.

112. A. H. Cottrell, "Dislocations in Metals: The Birmingham School, 1945–55," in *Beginnings,* 144–148, p. 144.

113. C. Frank, interview with E. Braun and P. Hoch, 31 July 1981.

114. A. Cottrell, interview with S. Keith, 15 December 1983.

115. F.R.N. Nabarro, "The Mechanical Properties of Metallic Solid Solutions," *Proceedings of the Physical Society* 58 (1946): 669–676.

116. Cottrell interview (n. 114); F.R.N. Nabarro, interview with E. Braun and S. Keith, 30 July 1981; Nabarro, "Deformation of Crystals by the Motion of Single Ions," in *1947 Bristol Conference* (London: Physical Society, 1948), 75.

117. A. H. Cottrell, "Effect of Solute Atoms on the Behaviour of Dislocations," in *1947 Bristol Conference* (n. 116), 30.

118. Cottrell (n. 112), 144.

119. Orowan interview (n. 58).

120. F. C. Frank, "On Slip Bounds as a Consequence of the Dynamic Behaviour of Dislocations," in *1947 Bristol Conference* (n. 116), 49.

121. R.D. Heidenreich and W. Shockley, "Study of Slip in Aluminum Crystals by Electron Microscope and Electron Diffraction Methods," in *1947 Bristol Conference* (n. 116), 71–74.

122. F. C. Frank, "Sessile Dislocations," *Proceedings of the Physical Society* 62 (1949): 202–203.

123. Nye interview (n. 93).

124. P. Lacombe, "Sub-boundary and Boundary Structures in High Purity Aluminum," in *1947 Bristol Conference* (n. 116), 91–94.

125. W. G. Burgers, "Recovery and Recrystallization as Processes of Dissolution and Movement of Dislocations," in *1947 Bristol Conference* (n. 116), 134. One wonders whether the word *eventually* is not a mistranslation and should read *possibly.*

126. F. C. Frank, "Dislocation Theory," *Nuovo Cimento* 7, suppl. (1958): 386–413 (proceedings of a summer school on solid-state physics held in 1957 in Varenna).

127. Frank interview (n. 113).

128. J. W. Mitchell, "Direct Observations of Dislocations in Crystals by Optical and Electron Microscopy," in *Direct Observation of Imperfections in Crystals,* ed. J. B. Newkirk and J. H. Wernick (New York: Wiley Interscience, 1962), 3–27.

129. Frank interview (n. 113).

130. A. J. Forty, "Growth Spirals on Magnesium Crystals," *Phil. Mag.* 43 (1952): 481–483; A.

J. Forty and F. C. Frank, "Growth and Slip Patterns on the Surfaces of Crystals of Silver," *Proc. Roy. Soc.* 217 (1953): 262–270.

131. Frank interview (n. 113).

132. S. Amelinckx, *The Direct Observation of Dislocations,* Solid State Physics Series, suppl. 6, (New York: Academic Press, 1964), 117–169. D. W. Pashley, "The Direct Observation of Imperfections in Crystals," *Reports on Progress in Physics* 28 (1965): 291–330.

133. F. C. Frank, "Crystal Growth," *Discussions Faraday Society* 5 (1949): 72.

134. F. H. Horn, "Screw Dislocations, Etch Figures, and Holes," *Phil. Mag.* 43 (1952): 1210–1213; Mitchell (n. 128), 7.

135. F. L. Vogel, W. G. Pfann, H. E. Corey, and E. E. Thomas, "Observations of Dislocations in Lineage Boundaries in Germanium," *Phys. Rev.* 90 (1953): 489–490.

136. Mitchell (n. 128), 8.

137. Ibid., Amelinckx (n. 132).

138. Frank (n. 126).

139. Mitchell (n. 128).

140. See, for example, J. W. Mitchell, "The Concentration Process in the Formation of Development Centres in Silver Halide Microcrystals," *Journal of Photographic Science* 31 (1983): 148–157.

141. J. Mitchell, in private conversation with E. Braun, 1954; P. Hirsch, interview with S. Keith, 10 January 1981.

142. Mitchell (n. 140).

143. J. M. Hedges and J. W. Mitchell, "The Observation of Polyhedral Sub-structures in Crystals of Silver Bromide," *Phil. Mag.* 44 (1953): 223–224; Hedges and Mitchell, "Some Experiments on Photographic Sensitivity," *Phil. Mag.* 44 (1953): 357–388.

144. Mitchell (n. 128); J. W. Mitchell, "Dislocations in Crystals of Silver Halides," in *Beginnings,* 140–159.

145. A. S. Parasnis and J. W. Mitchell, "Some Properties of Crystals of Silver Chloride Containing Traces of Copper Chlorides," *Phil. Mag.* 4 (1959): 171–179.

146. D. J. Barber, K. B. Harvey, and J. W. Mitchell, "A New Method for Decorating Dislocations in Crystals of Alkali Halides," *Phil. Mag.* 2 (1957): 704–707.

147. W. J. Dash, "Copper Precipitation on Dislocations in Silicon,"*Journal of Applied Physics* 27 (1956): 1193–1195.

148. R. W. Cahn, contribution to a conference held in London to mark the fiftieth anniversary of the discovery of dislocations (1984).

149. R. Cahn interview with S. Keith, 24 July 1981.

150. S. Konobejewski and I. Mirer, "X-Ray Determination of Elastic Tensions in Bent Crystals," *Zeitschrift für Kristallographie* 81 (1932): 69–91.

151. E. N. da C. Andrade and L. C. Tsien, "The Glide of Single Crystals of Sodium and Potassium," *Proc. Roy. Soc.* 163 (1937): 1–15.

152. R. W. Cahn, "Recrystallization of Single Crystals after Plastic Bending," *Journal of the Institute of Metals* 76 (1949): 121–143.

153. W. Shockley and W. T. Read, "Quantitative Predictions from Dislocation Models of Crystal Grain Boundaries," *Phys. Rev.* 75 (1949): 692; Shockley and Read, "Dislocation Models of Crystal Grain Boundaries," *Phys. Rev.* 78 (1950): 275–289.

154. Cahn (n. 148), 3.

155. C. Crussard, "Study of the Physical Aspects of Metals Conducted at the Centre de Recherches Metallurgiques of the Ecole Nationale Supérieur des Mines in Paris," *Revue de Métallurgie* 41 (1944): 45–48.

156. P. Lacombe and L. Beaujard, "The Application of Etch-Figures on Pure Aluminium (99·99%) to the Study of Some Micrographic Problems," *Journal of the Institute of Metals* 74 (1948): 1–16; A. Guinier and J. Tennevin, "X-Ray Study of the Fine Texture of Al Crystals," *C.R.* 226 (1948): 1530–1532.

157. Cahn (n. 148), 4.

158. R. W. Cahn, I. J. Bear, and R. L. Bell, "Some Observations on the Deformation of Zinc at High Temperatures," *Journal of the Institute of Metals* 82 (1954); 481–489; for example, Barber et al. (n. 146).

159. J. N. Kellar, P. B. Hirsch, and J. S. Thorpe, *Nature* 165 (1950): 554; P. B. Hirsch, "Direct

Observation of Dislocations by Transmission Electron Microscopy: Recollections of the Period 1946–56," in *Beginnings* 160–164; Hirsch interview (n. 141).

160. P. B. Hirsch, A. Kelly, and J. W. Menter, "The Structure of Cold Worked Gold," *Proceedings of the Physical Society* 68B (1955): 1132–1145. This paper was published with considerable delay.

161. Hirsch interview (n. 141).

162. Hirsch (1980) (n. 159), 162.

163. P. B. Hirsch, R. W. Horne, and M. J. Whelan, "Direct Observations of the Arrangement and Motion of Dislocations in Aluminium," *Phil. Mag.* 1 (1956): 677–684. This first publication reported results on aluminum only. It was preceded by a brief report given by Whelan at a conference in Reading in July 1956.

164. Ibid.

165. M. J. Whelan and P. B. Hirsch, "Electron Diffraction from Crystals Containing Stacking Faults: I.," *Phil. Mag.* 2 (1957): 1121–1142; Whelan and Hirsch, "Electron Diffraction from Crystals Containing Stacking Faults: I.," *Phil. Mag.* 2 (1957): 1303–1324; P. B. Hirsch, A. Howie, and M. J. Whelan, "A Kinematical Theory of Diffraction Contrast of Electron Transmission Microscope Images of Dislocations and Other Defects," *Philosophical Transactions of the Royal Society* A252 (1960): 499–529; A. Howie and M. J. Whelan, "Diffraction Contrast of Electron Microscope Images of Crystal Lattice Defects. III. Results and Experimental Confirmation of the Dynamical Theory of Dislocation Image Contrast," *Proc. Roy. Soc.* 267 (1962): 206–230; Hirsch interview (n. 141).

166. Hirsch (1980) (n. 159), 163.

167. W. Bollman, interview with E. Braun, summer 1981; Bollman, *Phys. Rev.* 103 (1956): 1588–1589.

168. Hirsch interview (n. 141).

169. M. J. Whelan, P. B. Hirsch, R. W. Horne, and W. Bollman, "Dislocations and Stacking Faults in Stainless Steel," *Proc. Roy. Soc.* 240 (1957): 524–538.

170. Frank interview (n. 113); J. Frenkel and T. Kontorova, "On the Theory of Plastic Deformation and Twinning," *Physikalische Zeitschrift der Sowjetunion* 13 (1938): 1–10.

171. F. C. Frank, "The Frank-Read Source," in *Beginnings,* 137.

172. F. C. Frank and W. T. Read, "Multiplication Processes for Slow Moving Dislocations," *Phys. Rev.* 79 (1950), 722–723.

173. W. Dash, "Generation of Prismatic Dislocation Loops in Silicon Crystals," *Physical Review Letters* 1 (1958): 400–402; Dash, "Growth of Silicon Crystals Free from Dislocations," *Journal of Applied Physics* 30 (1959): 459–474.

174. Mitchell (n. 140).

175. Mitchell (n. 128), 13.

176. A. Seeger, interview with E. Braun and G. Torkar, 3 June 1982.

177. A. Seeger, "Theorie der Gitterfehlstellen," in *Handbuch der Physik* (Berlin: Springer, 1955); 7: pt. 1, 383–665.

178. Friedel (n. 106).

179. Frank interview (n. 113).

180. H. J. Queisser, "Electrical Properties of Dislocations and Boundaries in Semiconductors," in *Defects in Semiconductors II,* ed. S. Mahajan and J. W. Corbett (New York: North-Holland, 1983), 323–341.

181. W. T. Read, "Theory of Dislocations in Germanium," *Phil. Mag.* 45 (1954): 775–796.

182. Dash (1959) (n. 173); E. Spenke and W. Heywang, "Twenty-five Years of Semiconductor-Grade Silicon," *Physica Status Solidi* 64 (1981): 11.

183. Queisser (n. 180), 326.

184. For example, H. J. Queisser, K. Hubner, and W. Schockley, "Diffusion along Small-Angle Grain Boundaries in Silicon," *Phys. Rev.* 123 (1961): 1245–1254.

6 | *Magnetism and Magnetic Materials*

STEPHEN T. KEITH AND PIERRE QUÉDEC

By the 1930s, the exploitation of the magnetic characteristics of materials formed a key element in some of the most significant modern technologies, including power generation and transmission and the growing industry of telecommunication. Nothing was more representative of the modern world than the new marvels of radio and television, which incorporated magnetic components as important features of their design. Yet, paradoxically, scientific understanding of magnetic materials was still rudimentary. Even such an apparently simple question as, Why are some solids ferromagnetic and others not? begged for a satisfactory answer. Regarding the commercial production of such a familiar item as a permanent magnet, it was, as one contemporary confessed, hard to find an industry more entangled in "occult molecular phenomena."[1] If gradual improvements in the technical characteristics of magnetic materials were forthcoming, they would be in the context of empirical methods practiced for centuries.

To be sure, there were some simple intuitive models of the kind produced by Alfred Ewing, and there were phenomenological rules that held over a fairly wide range of circumstances (see "Magnetism" section of Chapter 1). Yet for those concerned with materials fabrication and application, the nature of magnetic behavior, beyond the realms of bulk properties and measurements, remained a mystery. Sydney Evershed, an English electrical engineer and inventor-cum-entrepreneur who worked on magnetic materials, as recently as 1925 gave the opinion that "what magnetism is, no-one knows. We can only think of it as a peculiar condition created in space by the motion of electricity."[2] Such a view was over a century old, yet essentially correct in its assessment of the current state of knowledge.

In 1925 the subject matter usually dealt with under magnetism included matters ranging from the magnetic properties of individual atoms, as seen for example in gases, to terrestrial magnetism, with appeals to the more specifically magnetic parts of electromagnetic theory—essentially magnetostatics, following the rather formal treatment that had been developed at least 50 years earlier, much of which dated back more than 100 years.[3] Study of the magnetic properties of solids implied, almost exclusively, the magnetic properties of iron. Apart from the work of Ewing and investigations of a more technological kind on ferromagnetism, most of the major scientific contributions on magnetism from the end of the nineteenth century into the mid-1920s had originated in France and were not widely known or appreciated elsewhere.

The coming of quantum theory did not transform research in magnetism as suddenly and thoroughly as it did in many other areas of physics.[4] Classical theory and the earlier quantum theory remained adequate for most researchers in magnetism. There were certain problems, however, in which no progress was possible before the advent of quantum mechanics. Eventually, quantum mechanics would take much of the mystery out of the subject at the atomic level, providing an explanation for the previously obscure molecular field effects and a means for determining the magnitudes of the atomic moments of ferromagnetics.

In general, physicists could take two different viewpoints on magnetism (as noted in Chapter 1). The first was fixed on the atom and attempted to grasp all its properties before considering the whole edifice. The other, of broader scope, made up for ignorance of the details by describing overall effects. Because the magnetism of solids is generally a matter of the whole more than its parts—that is, it is a collective phenomenon—it can be most usefully approached from the viewpoint that encompasses a material as a whole. The other viewpoint would eventually prove its worth in leading atomic physicists into the nucleus. But first, in combination with quantum mechanics it did bring something to magnetism.

Here two lines of work were particularly important. The first was opened up by

Friedrich Hund, who in 1924 used quantum ideas to calculate the correct values of the magnetic moments of certain rare earths. Subsequent work by Wolfgang Pauli, Lev Landau, and others significantly advanced understanding, in terms of the quantum, of both diamagnetism and paramagnetism (see Section 2.6). But since the viewpoint here was that of the individual atom or ion, these results did not initiate any striking advance through the entire vast domain of the magnetism of solids.

The second important line, also developed in the late 1920s, came when Werner Heisenberg showed that the cause of ferromagnetism lies in the quantum-mechanical exchange interaction between electrons imposed by the Pauli exclusion principle (see Section 2.6). This opened a new way for physicists striving to solve the puzzle of ferromagnetism. Yet Heisenberg's 1928 article bore the stamp of continuity as well as change, since it followed up a step already taken by Pierre Weiss in 1907. Weiss had hypothesized that the interactions within a ferromagnetic substance combine to give the same effects as a fictional mean field, which he called the "molecular field." Heisenberg had identified the nature of these interreactions, making clear that they were short-range quantum effects; this brought new attention to the earlier French work and moved the magnetic properties of solids back into the mainstream of interest in physics. At the theoretical level at least, the subject was rejuvenated.

If it had been possible to develop rigorously all the consequences of Heisenberg's idea, Weiss's step forward would have been superseded. However, although quantum mechanics had removed one mystery, it had also brought its own. The mathematical treatment was more complicated, or at least less familiar, and the transition from symbolic representation to physical observables appeared much less direct. Classical theory gave a more immediate feeling of understanding, with a closer relationship to directly observable phenomena. Valid analogies could be drawn with physical processes that could be seen or readily visualized, and a sense of understanding was provided which, although qualitative and even illusory, was enormously helpful. With quantum mechanics, this feeling of understanding had disappeared to a degree. Ewing, a master of the classical approach, suddenly saw himself as "a hopelessly old-fashioned magnetician," mildly embarrassed at being asked to open a Physical Society meeting on magnetism in 1930. Nevertheless, he gave his listeners a clear lesson in the value of studying bulk properties:

> These reminders of the past will serve your purpose if they induce some of you, who are naturally engrossed by newer aspects of magnetic theory, to revert to a study of the ferromagnetic group. After all, it is there that we have the magnetism of the atom revealing itself on a relatively grand scale . . . there I think that we may most profitably search for clues to what is now obscure.[5]

One of his contemporaries, Evershed, was equally lost, if still more skeptical of the ways of a younger generation. The elderly engineer chided the "ultra-modern physicists" who were, in his view, tempted to believe "that Nature in all her infinite variety needs nothing but mathematical clothing" and who were "strangely reluctant to contemplate Nature unclad. Clothing she must have. At the least she must wear a matrix, with here and there a tensor to hold the queer garment together."[6]

Yet the "queer garment" had solved the long-standing mystery of the origins of ferromagnetism and provided a convincing explanation of paramagnetism for several groups of elements. In general, the theoretical breakthroughs of the late 1920s cleared up some outstanding difficulties, focused attention on more neglected aspects of magnetism, and helped to systematize a wealth of uncoordinated empirical data. This did not prevent one commentator from reflecting that magnetism in 1930 remained at a "pre-Newtonian" stage, dominated by:

> . . . cycle on epicycle, orb on orb
> with centric and eccentric scribbled o'er.[7]

For in truth, when it came to explaining the complex properties of real materials, quantum mechanics had merely pointed an educated finger. As Francis Bitter, then a young American industrial scientist, explained, despite the progress made, on the one hand, in learning empirically how to make new materials with important magnetic properties for commerical use and, on the other, in understanding the fundamental mechanisms responsible for magnetic behavior, "it remains one of the important tasks of the future to link together these two avenues of progress so that we may not only produce, but also understand how we produce the materials which we use."[8] The metallurgist and historian Cyril Stanley Smith has more recently noted that it is exactly such a joining of these two "avenues of progress" that in general marked the genesis of solid-state physics as a distinct professional activity.[9]

Following these two viewpoints will lead us at first to emphasize the work of Weiss and his school between the wars. We will show that from the moment that he threw himself into work on the elementary magnetic moment, Weiss focused almost exclusively on the atom, until he could no longer see the forest for the trees. The next step came from another direction through Louis Néel's work, in which Weiss's classical molecular field was examined in the new light of Heisenberg's short-range interactions; now it became a field created locally by the atoms that are nearest neighbors to the point in question.

Throughout the 1930s, the understanding that was growing on both the theoretical and experimental levels was still very much in qualitative terms. Theoretical models, both classical and quantum mechanical, were crude, by necessity incorporating simplifying approximations, which in the nature of such an underdeveloped field provided diversity and dissent. Arbitration through reference to the measured properties of real materials was beset with the general difficulty of determining quantitively the effects of relevant physical factors such as impurities, lattice imperfections and internal strains, and theoretical factors, including the lack of any detailed understanding of the structure of valence electrons in transition metals. As Edmund Stoner, one of the leading theorists in the field, put it in a review of magnetism in 1930, an explanation of the properties of a complex ferromagnetic material could hardly be given when there were still difficulties in interpreting the properties of a relatively simple iron crystal.[10] A later generation would discover that the magnetic properties of numerous alloys were surprisingly more amenable to a reasonably simple interpretation than were those of the pure metal. Even 50 years later, the properties of iron remained less than fully understood.[11]

The period of development before the Second World War, despite these difficulties, did provide an important impetus for what many have interpreted as a revolution in the understanding of magnetic phenomena and materials. Fairly rapidly in the postwar years came the confirmation of new classes of magnetic structures that, in turn, provided for significant innovations in the fabrication and application of magnetic materials. Scientific and technological development became related in a way unknown in the prewar era—a process we shall illustrate through the discovery of the "garnet" magnetic substances, which had great value for high-frequency electrotechnology; the development of ferrites, a new class of materials that revolutionized permanent magnet applications; and the development of the theory of single-domain effects in fine particles and thin films, which predicted the unusual and, again, technologically valuable properties of such materials. The postwar appearance of neutron scattering and magnetic resonance techniques in particular had impacts far beyond the investigation of atomic and nuclear magnetism, respectively. Both techniques paved the way for a more sophisticated understanding of magnetic behavior, through the detailed elucidation of complex magnetic structures. Along with further contributions from the wider arena of solid-state physics, the new techniques brought some long-awaited progress to a fundamental understanding of ferromagnetism at the atomic level.

The development of an understanding of the magnetic properties of solids is a complex and detailed story impossible to document in depth within the limitations of this study. Chapter 2 has already described the quantum-theoretical breakthroughs of the period, which, however, pointed more toward the future theory of collective phenomena in general than toward immediate understanding of the majority of real magnetic substances. We offer here two further approaches to the history. The study of magnetism in solids was complicated by the fact that a detailed theoretical scheme for magnetic properties had already been established before the introduction of quantum mechanics, a classical scheme that provided a considerable intellectual heritage, little altered or threatened by developments in atomic physics. In the first part of this chapter, Quédec describes the development of this early, predominantly French, magnetism tradition, which introduced concepts that still occupy a central place in modern magnetism. In a complementary approach, Keith, in the second part of the chapter, discusses how in the social structure of the study of magnetism and magnetic materials several strands of development can be identified, interacting and coalescing into part of a modern solid-state physics.

6.1   The Classical Traditions of Modern Magnetism
Curie's Laws

We cannot get into the history of magnetism without choosing when to begin, a choice necessarily unsatisfactory because it is in part arbitrary. In Section 1.2 we sketched a few early developments, simply to show magnetism as one of the independent strands that constituted the tangled physics of solids by the early twentieth century. To tell more fully the modern history of magnetism, it seems reasonable to begin at a moment when the main observations and theoretical ideas had been gathered by a single physicist, who classified and complemented them and then, in

Pierre Curie (1859–1906). (AIP Niels Bohr
Library, Weber Collection)

his explication of them, enabled his readers to make their own judgment and move
ahead.

July 1895, when Pierre Curie published a paper on the magnetic properties of
bodies at various temperatures, is one of those special moments. In this paper Curie
clarified the knowledge of his time and indicated directions that, followed by others,
would prove to be well chosen. Curie's important paper, published in the *Annales
de chimie et de physique,* was more than 100 pages long.[12] He set down the experimental facts reported by his predecessors and added the results of experiments that
he had made between 1892 and 1895 at the laboratory of the Paris Municipal
School of Physics and Chemistry. In classifying all this data, he was attempting to
disengage certain guiding principles, arising from a scrupulous analysis of the facts
together with a few intuitive ideas.

Curie's Options. It had long been recognized that substances with magnetic properties can be divided into three categories—paramagnetic, diamagnetic, or ferromagnetic—according to their behavior in magnetic fields.

In amorphous or otherwise isotropic substances, such as polycrystalline materials or powders, magnetization is parallel to the applied field. Thus the magnetic
susceptibility X of a substance is defined by the relation $M = \chi H$, where the magnetization M is the magnetic moment per unit volume of the substance and H is
the applied field. It is often preferable to use the specific magnetization I, which is
the magnetic moment per unit of mass; then the coefficient of specific magnetization K is defined as $I = KH$. A diamagnetic substance is characterized by a negative
coefficient K; a paramagnetic substance, on the other hand, has positive K. Each
type of substance loses its magnetization when the applied field is removed. Ferromagnetic bodies take on a far more intense magnetization than the other two and,
under certain conditions, can retain it in the absence of an external magnetic field.

Almost anything could be assumed about these types. Was there a single cause
that put a particular body in a particular category? Or could a body contain several
characteristics and reveal only the most striking of them to the observer? Was it
conceivable that a body could be classified differently depending on the conditions

under which it was observed? Did the three observed effects have three causes or fewer than three? Decades earlier, Michael Faraday, using the experimental methods he had refined, had shown that red-hot iron does not completely lose its magnetic properties, as coarser measurements had previously suggested. He thus established the possibility that the same body might appear strongly or weakly magnetic, demonstrating that there can be a transition region between two of the three recognized categories. This result spurred scientists to make more measurements, and more precise ones.

In another direction, the many questions that could be asked about the origin of the three phenomena could give rise to many hypotheses. Pierre Curie did not venture along this course of inquiry, but instead chose, like Faraday, to perform more observations, explaining that "independently of any theory, one feels that a phenomenon is known when our knowledge of the facts forms a continuous whole".[13] Recognizing that magnetic phenomena were far from satisfying this criterion, he devoted himself to building up such a "continuous whole" by studying the magnetic properties of various substances subjected to different temperatures and to magnetic fields of varying strengths. The experimental methods that he used, and described with scrupulous care, did not signal any break with methods employed in the preceding half-century; as he remarked, it was "not different in principle from the methods used by Becquerel and Faraday."[14]

The Three Categories of Magnetism.   Dealing with supposedly diamagnetic bodies, Curie found that out of 14 substances he studied, all but one (bismuth) confirmed that the coefficient of specific magnetization K was constant with variations in temperature and magnetic field strength. This property, together with the very low absolute value of the magnetization coefficient and its negative sign, allowed him to give a precise definition of the class of diamagnetic substances.

The "weakly magnetic" (i.e., paramagnetic) substances that he studied were few, but when examining them he learned his greatest lesson. For oxygen, whose magnetization coefficient Curie measured over a termperature interval of nearly 500° C, gave him an idea that would lead him to the first experimental law in the history of magnetism. At this stage, he wrote only that "the specific magnetization coefficient of oxygen varies inversely with absolute temperature," while observing that observed deviations from this law could well depend on the pressure of the gas.[15]

Dissolved magnetic salts had been the subject of experiments, and a comparison of these experiments had shown physicists not only that the coefficient of specific magnetization K fell with increasing temperature T, but also that the coefficient's rate of variation $(1/K)(dK/dT)$ had two curious features: it was the same for all salts; and, still more surprising, it had the same absolute value as the gas expansion coefficient α. Was this coincidence mere chance, or did it bear the stamp of a law of nature? The wisest course was to subject other salts to the same experiment. Several physicists did this, some of them with the secret hope of making further discoveries. The matter was of some importance, for if it was true that at any temperature the expression $(1/K)(dK/dT)$ was equal to $-1/T$, then the coefficient K should vary with absolute temperature according to a hyperbolic law, $K = A/T$.

Curie examined the results of his predecessors, supplementing them with his own findings over a wider range of temperatures. He became convinced that the product of K and T was not uniformly constant and that the expression $(1/K)(dK/dT)$ was not equal to the gas expansion coefficient α. However, around a dozen substances had a K that did indeed vary as the inverse of absolute temperature; thus the law formulated for oxygen could be extended to this group of substances and would serve to characterize it. These results can be rephrased in an elegantly simple form, what became known as Curie's law: the magnetization of a paramagnetic body is proportional to the intensity of the magnetic field divided by the absolute temperature. The constant of proportionality C was later named the Curie constant.

The third class, ferromagnetic materials, comprised only a few substances, but because of the many important practical questions raised by iron metallurgy, this class would long preoccupy physicists and chemists.

Curie, performing experiments that he again extended over a wider temperature range, classified the different forms of iron and assigned to each a specific magnetic character, which he made particularly clear by using logarithmic coordinates to display the variations of magnetization coefficient with temperature. This graphical representation, repeated for other ferromagnetic substances, helped him show that above a certain temperature each of these substances behaved similarly to those with a low magnetization coefficient that, like oxygen, obeyed the law he had formulated.

Curie's Lessons. "Independently of any theory" (following his own expression, already quoted), Curie thus affirmed the existence of three idealized categories of substances. He had convinced himself that diamagnetism has a unique character, but he showed that there was a relationship between the two other classes. Having developed the definition of the weakly magnetic class by formulating the first experimental law, he was able to recognize the same characteristics in the behavior of ferromagnetic materials above a certain temperature. From this point on, physicists would seek a trait common to these two phenomena but not shared by diamagnetism.

In addition, Curie had pointed out the way for other scientists to make measurements on additional substances, using the methods he had described in extensive detail. Further, whether or not a body obeyed his law could help to classify substances.

Curie noticed and compared the roles that pressure and magnetic field, and density and magnetization, play in a fluid and a magnetic body, respectively, although this analogy with fluid physics had not helped him. Did the liquefaction of a gas resemble the magnetization of a ferromagnetic material? And if this idea was taken over to molecular models, what type of mutual action, similar to the forces between constituents of a fluid, might exist between hypothetical "magnetic particles" in a ferromagnetic substance? We will see that the analogy was taken up without aiding in any way the invention of new concepts. For physicists who used the analogy, it was simply a language to assist in explaining other concepts or a pretext for getting them accepted.

Although he made use of them, Curie advised people "not to let themselves be blinded by these analogies to the extent of failing to see any importance in characteristic facts that disagree with them."[16] In demanding respect for these "characteristic facts," he may also have dissuaded physicists from searching too stubbornly for an absolute temperature scale in dissolved magnetic salts. Cases were found where an intensely wished-for coincidence was in the end no longer attributed to chance—where unconditional faith in a hypothesis led to measurements being adjusted in a Procrustean fashion.

In a memorandum written in 1902, Curie gave no reason for his research on magnetic bodies other than his intention to "elucidate the links and transitions that can exist between the properties of bodies when they are in the diamagnetic, weakly magnetic or ferromagnetic state."[17] These were the same reasons he gave in the introduction to his paper in 1895. However, in the memorandum he added a remark where the authoritative voice of a master showed through: "The laws that I have found . . . constitute the quite specific conditions that theories put forward to explain these phenomena must satisfy." Curie's lessons went beyond his initial ambition: he put a halt to decades of theoretical meanderings inspired by uncertain experimental results. The appearance of his definitive paper, which no "characteristic fact" has subsequently tarnished, plainly marks the beginning of the history of modern magnetism.

Langevin and the Theoretical Approach

Curie limited his study to isotropic substances, gases or liquids, or solids that, if not always strictly speaking isotropic in detail, at least appeared to be so. There remained a major gap in this area of knowledge: the magnetism of crystals. Pierre Weiss, whose work we will deal with at length, had already become interested in this question and was looking for the keys to magnetism in the structure of crystals; it would be natural to join him at this point in our story. However, we will leave Weiss to his experiments for a few years and will now turn to a specifically theoretical view, ignored by Curie, that Paul Langevin expounded in masterly fashion in 1905.

Langevin was a former pupil of Curie and retained a friendship with his mentor as well as with other French physical scientists, as part of an extensive network of personal relationsips that was sustained by frequent informal meetings and extensive correspondence. This network, reinforced by a common interest in the exciting discoveries that were transforming atomic physics, was a main factor in making France for a time a chief locale of research into magnetism and other properties of solids. There was no single laboratory or institute we can point to as the focus, but in a sense this Paris-centered network was an institution of its own, permitting a free interchange of materials and ideas. It seems likely that Langevin's interest in magnetism was spurred by Curie and other friends through private discussions as much as through their publications.

When Langevin took up the interpretation of magnetic phenomena, he had at his disposal a theoretical framework, constructed by H. A. Lorentz, based on a new conception of the constitution of matter in terms of electrons, and, as we have seen, he had a credible set of observations reported by Curie. These observations guided Langevin's work. Thus we find him working in turn on diamagnetic and paramagnetic phenomena, ending up, as did Curie, with some questions and some intuitive thoughts about ferromagnetism.

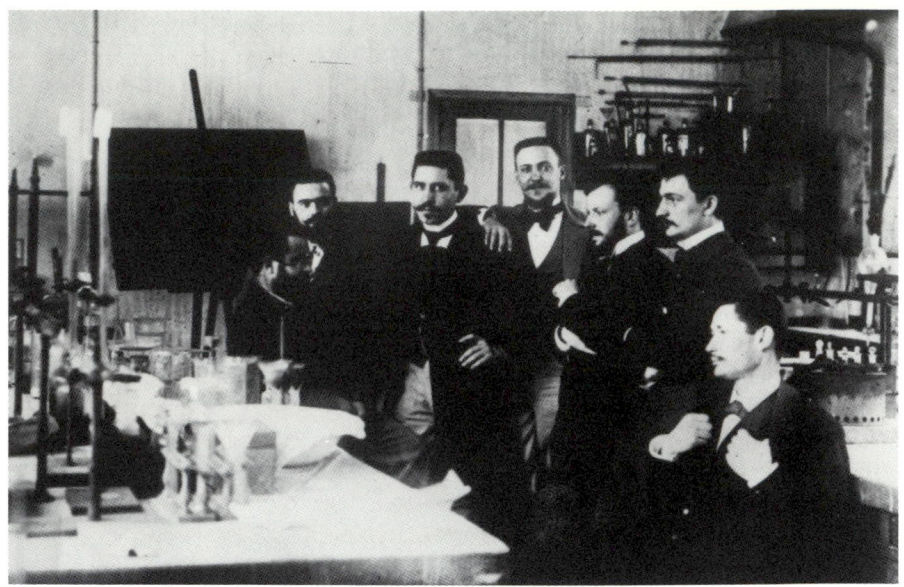

Paul Langevin (1872–1946) *(third from left)* as a student at the Ecole Normale Supérieure. (AIP Niels Bohr Library)

Diamagnetism. Langevin founded his theory on the existence of a "magnetic moment," whose existence in an isolated molecule he claimed he could justify.[18]

Lorentz's equations having made possible calculations of the magnetic field created by a moving electron, Langevin postulated a closed trajectory for such an electron, and further restricted his calculation to the effect produced at large distances. These two restrictions led to a result in which he recognized a dipole field such as a magnet could have created. With the effects of all the electrons in a molecule superimposed, a field at a large distance would show, according to whether it existed, if the molecule had a magnetic moment.

The absence of an effect at a distance did not exclude the possibility of a field existing closer in, and Langevin remarked, "I think it important to note that this field could play a considerable role in the cohesive action and in the mutual orientation of molecules that are very close together, which must be involved in determining the crystal structure."[19] The physical reality, in Langevin's eyes, was the "particle current" due to an electron moving in equilibrium around a stable orbit. The magnetic moment was no more than a convenient substitute for the set of electrons in the molecule when considered in terms of magnetic effects at large distances. This simplification allowed him to study the effects of a simple uniform field on a molecule. On the other hand, it made it impossible for him to deal in the same terms with the short-range interactions between molecules.

Remaining within the framework he had built for himself, Langevin showed that imposing a uniform field on the space occupied by a molecule would change the motion of the electrons so as to modify the net magnetic moment. The result of this modification would be diamagnetism.

Since the movement of electrons within the molecule did not seem to be much affected by variations in temperature—as was shown by the invariability of spectral lines—he concluded that diamagnetism "must vary hardly at all with temperature, in agreement with Curie's experimental results."[20] Curie had also noted more generally that diamagnetism is independent of the substance's physical or chemical state. Although Langevin's theory did not explain this fact, at least it did not contradict it, since it attributed diamagnetism to a property of the molecule.

Up to this point, there was no need for the molecule to have an intrinsic moment in the absence of an external magnetic field; diamagnetism was simply a result of the way electrons on fixed orbits would respond when a field was imposed. The "diamagnetic modification" was conveniently expressed in terms of a moment, whose sign and magnitude agreed with experimental results.

It remained to be shown that diamagnetism, brought into being along with the imposed field, would not disappear by some sort of relaxation process once a constant field had been established. With this in mind, Langevin for the first time used some of Boltzmann's statistical ideas. He explained that for molecules initially lacking a magnetic moment, application or a field would not modify their movements; once the field was established, collisions would continue as before and would not affect the mean distribution of electrons. The net properties of the movement of electrons would be conserved, and the diamagnetic moment would persist. The class of diamagnetic substances whose properties Curie had observed was thus explained in molecular terms in a way that contradicted no known fact and that both confirmed and clarified Curie's ideas. Indeed "all matter," Langevin concluded, "must possess the property of diamagnetism, whether or not the initial net moment of its molecules is zero".[21]

Paramagnetism. A similar analysis, carried out for molecules that all had the same magnetic moment μ before they were subjected to an imposed field, would explain paramagnetism. Langevin noted that in this case the movement of each molecule would be changed, but not in the same way for all of them. As a result, the energy of a given molecule would no longer be compatible with the temperature it had shared with all the other molecules at the outset, and now collisions would result in a "rearrangement." The result of this rearrangement would be paramagnetism.

The chief innovation in Langevin's approach to this problem was his use of the theory that Boltzmann had devised for a gas at uniform temperature. The result of Langevin's calculation was an expression for the specific magnetization I in the direction of the imposed field H as a function of the ratio of the intensity of this field to temperature, $a = \mu H/kT$. The expression was

$$I = I_0 \left(\coth a - \frac{1}{a} \right)$$

where $I_0 = N\mu$ is the specific magnetization that the substance would have if the N molecular moments per unit mass within it were aligned in the same direction. This formula would long remain the only key available to experimenters for getting access to magnetic quantities on a molecular scale (Figure 6.1).

Langevin learned several things from this result. In particular, the specific moment would vary linearly with the ratio of field to temperature for low values of

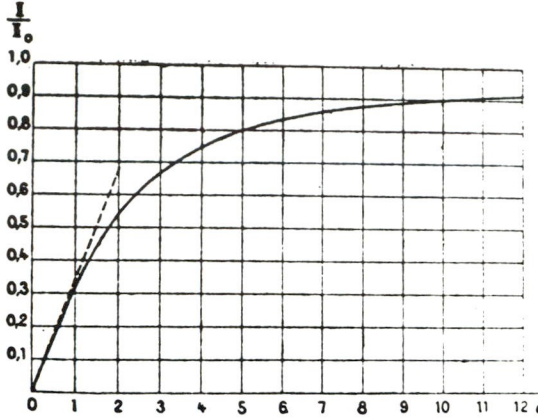

Fig. 6.1 Langevin's function, representing the law of magnetization of a paramagnetic substance as a function of $a = \mu H/kT$. (From P. Langevin, "Magnétisme et theorie des électrons," *Annales de chimie et de physique*, 8th ser., vol. 5 [1905]: 118)

this ratio. Thus his theory not only accounted for Curie's experimental law $I = C(H/T)$ but could make it more precise, for where Curie had found only a proportionality coefficient C, Langevin introduced the molecule's magnetic moment μ, related to Curie's constant by the formula

$$\mu = \sqrt{\frac{3kC}{N}}$$

That magnetic moment was the cornerstone of his theory. He also established that for high values of the ratio a, the specific magnetization would reach a limiting value $I_0 = N\mu$, whose measurement would allow calculation of the molecule's magnetic moment.

Finally, the appearance of the curve representing the variation in the specific magnetization was reminiscent of the curve for the magnetization of iron. Did paramagnetism and ferromagnetism share the same behavior? In support of this idea, Langevin used Curie's measurements to calculate the magnetic field strength needed to observe the beginning of saturation in oxygen, and found it was beyond the reach of experiment. Another estimate suggested that the limiting value of the specific magnetization of oxygen would be comparable with that of saturated iron. These two possibilities, both of them reasonable but unfortunately not verifiable, led him to explain paramagnetism and ferromagnetism by a single process: molecules possess a magnetic moment whose orientation by a field is disrupted by thermal agitation.

Langevin's 1905 theory fitted what was known about paramagnetism at the time. It also opened up a way toward explaining ferromagnetism, by offering an invitation to examine what had been left aside. Might the mutual interactions of molecules orient moments in iron in a way that no external field managed to do in oxygen? Langevin affirmed that "only these [interactions] make magnetic saturation possible."[22] The idea of a molecular interaction whose effects would be comparable with those of a field had been launched.

Langevin, like Curie, also took up the analogy with fluid thermodynamics. Regarding the unified explanation of paramagnetism and ferromagnetism, he wrote, "Curie's comparison of the transition between weak magnetism and ferromagnetism to the transition between the gaseous and liquid states, wherein mutual actions play an essential role, is fully justified."[23] The use of this common ground, more familiar to physicists, would gradually create an illusion that the understanding of magnetism had arisen from knowledge of changes of state in fluids.

In referring to the strong interactions that exist in ferromagnetic materials, Curie kept the vague term "magnetic particle." Langevin used the more precise concept of a molecule, with which he associated a moment in order to describe the effects of electron currents at large distances. But the origin of these moments remained uncertain. To refine the concept, Langevin had calculated, using Curie's experimental values, that a single electron would suffice to give the oxygen molecule its observed magnetic moment. A similar result could be supposed for iron, where saturation was of the same order of magnitude.

This did not at all clear up the problem, but Langevin was happy with his result, finding that it strengthened the contrast between paramagnetism and diamagnetism. He drew the conclusion that "only one or at most a few electrons contribute to producing [para]magnetic properties, whereas all the electrons cooperate in diamagnetism."[24]

Failure of the Classical Theory. For some years following the publication of Langevin's paper, no doubts were raised about the modification of molecular moments by a field. Yet knowledge was available at the time to contest this. Physicists failed to notice that, according to Langevin's approach, diamagnetic susceptibility should be strictly equivalent to paramagnetic susceptibility, and thus no magnetic effect could result. Only six years later, in 1911, Bohr demonstrated in his doctoral dissertation a statistical thermodynamic theorem that established in very general terms that in classical theory, no magnetization could appear due to the imposition of a field on a population of electrons in thermal equilibrium.[25] It now became evident that Langevin's theory had succeeded only because he imposed fixed electron trajectories, in effect smuggling a quantum condition into what had seemed to be a purely classical theory (see Section 1.3).

Had it come earlier, what would this demonstration have meant for the influence of Langevin's theory? Would the refutation have been considered final? Bohr's argument was severe, stripping the theory of all its consequences. Yet it could have been judged excessively destructive, since "independently of any theory," to return to Curie's phrase,[26] did not substances belong to the three classes stated by Curie? Although the molecular moment itself retained its secret, at least the theory could embrace in one structure the realities of diamagnetism—which it connected with the Zeeman effect—and those of paramagnetism while apparently remaining open to adaptation to the realities of ferromagnetism as well.

In the last analysis, the reprieve granted to classical theory, for lack of prompt criticism, was beneficial for the understanding of the mechanism of ferromagnetism. Langevin had remarked that in this regard he would study the mutual actions between molecules in crystalline bodies, "where the distribution of molecular magnets permits a simpler approach to the problem."[27] Pierre Weiss had already been working on just that for several years. In 1926, when Marcel Brillouin put Weiss forward for membership in the Academy of Sciences, he discerned two pieces of luck in Weiss's scientific career: the first was that he had found himself confronted with a crystal of the enigmatic mineral pyrrhotite, and the second was "that Langevin had published his excellent theory of diamagnetism and paramagnetism in 1905."[28] We may add another piece of luck: instead of pausing to consider the merits of Langevin's magnetic moment, Weiss could accept the theory as such and annex it to his own developing views.

Weiss's Molecular Field

Magnetism in Crystals. The research carried out by Curie and Langevin gave the development of the theory of magnetism a unity in accord with the simplicity of the chronological order of events. Nevertheless, we should recall that Curie's approach was essentially experimental and that the law he formulated was only the concise expression of observations, whereas Langevin put forward a theory. The concordance of their results appeared to unite their work in a single outcome, whose success posterity recognized while still making a distinction between "Curie's law" and "Langevin's theory of paramagnetism."

Had the passage of time been our only guide and a chronicle of events our only purpose, we would link the next major step, Weiss's hypothesis of the molecular field, to the theory of paramagnetism. This shortcut would seem to have been justified by Weiss himself, who, according to Edmond Bauer, confided that the idea of the molecular field "came to his mind by a subconscious association, and that he suddenly understood that it could provide the key to the properties of iron, nickel and cobalt."[29] The idea did indeed arise (in 1907) subsequent to Langevin's work, but only after long and laborious studies dating to 1896. Facing the problems raised by ferromagnetism, Weiss had originally undertaken, like Curie, a solid experimental research program. Yet his first publications on magnetism showed a speculative mind seeking to escape the austere areas Curie had explored. We will find him following, like Langevin, a theoretical inclination, throwing out various hypotheses, some of them bold indeed. A look at Weiss's earlier studies will show how this came about.

In his dissertation, presented in 1896, Weiss broke away from an idea on magnetism in crystals that was commonly held by physicists of the time. This idea, that magnetization in crystals was proportional to the applied field, until then had been entirely in agreement with the known facts. The assertion had been stated in 1851 by Lord Kelvin and had always been verified subsequently, so all cubic crystals were thought to be magnetically isotropic.[30] In studies of magnetite (iron oxide, the venerable lodestone), Weiss found the first example to contradict this rule. He showed that although magnetite crystallizes in a cubic system, it has magnetism whose magnitude varies with the direction of the field and whose direction is generally not that of the field. Magnetite also has the special feature of being ferromagnetic, whereas all the crystals that had supported Kelvin's statement had been paramagnetic. The isotropy and the paramagnetism of these, as opposed to the anisotropy and ferromagnetism of magnetite, indicated the value of combining the study of crystalline structure with that of magnetic properties. Although magnetite was ferromagnetic, its behavior was different from that of nickel, cobalt, iron, and their alloys; however, the observation made on these metals had always used heterogeneous industrial substances in which the small crystals had a variety of orientations, giving an illusion of isotropy.

Weiss's observations on magnetite were of descriptive value, contributing new facts about ordered matter to the catalog drawn up by Curie, but they offered no help in understanding the mechanism of ferromagnetism. Their main merit was that they convinced Weiss of the importance of magnetism in crystals, a conviction that certainly helped him when he confronted the disconcerting properties of crystals of pyrrhotite (iron sulfide, also called magnetic pyrite).

Following Brillouin, we may call the study of pyrrhotite Weiss's first lucky break, but one that "would have brought about utter defeat for a less perspicacious mind."[31] Much later Louis Néel showed that Weiss had not seen the true face of pyrrhotite, which makes his luck even greater. The circumstances of this work merit a closer look.

In 1898, Weiss published his first impressions, gathered from work on crystals from Brazil, whose quality exceeded those of crystals from other locales. He had obtained them from Joaquim Cándido Da Costa Sena, a mining engineer in Ouro Preto. A Brazilian, author of "Voyage d'études métallurgiques au centre de la province de Minas" (1881), Cándido belonged to more than 25 European scientific societies, including the Paris Mineralogical Society.[32] It was natural for Charles Friedel, a Paris mineralogy professor who was on familiar terms with Curie, Langevin, and Weiss, to contact this scientist from a province rich in minerals, in order to obtain fine pyrrhotite crystals. The crystals Friedel was given and handed to Weiss were, according to the latter, "perfectly homogeneous over a distance of several centimeters." This circumstance was perhaps not without significance for the history of ferromagnetism.

Armed with these rare and excellent crystals, Weiss reported his first observation.[33] It was simple: by placing a magnet in front of a piece of pyrrhotite, he found that if the plane of cleavage was perpendicular to the lines of magnetic field there was no attraction. On the other hand, the attraction was very great if the plane of cleavage contained the field lines. Weiss concluded, "There is thus a direction in which magnetization is impossible, and one is immediately led to generalize and assume that the material can only become magnetized in the plane perpendicular to this direction, what I shall call the magnetic plane."[34] In this plane, Weiss would later postulate the molecular field.

The Molecular Field. Having noted that the magnetic plane of pyrrhotite was itself anisotropic, Weiss explained the results of his observations by the action of a "demagnetizing field" in the direction in which magnetization was difficult. "Everything happens," he wrote, "as if a demagnetizing field, due to the substances' structure, cancelled a component of the field proportional to the component of magnetization in the direction difficult to magnetize, with the remaining component then parallel to the magnetization."[35] At this stage, Weiss was more interested in the crystal's anisotropy than in the mechanism of ferromagnetism. He set aside irreversible phenomena such as hysteresis, which he considered "a light piece of embroidery, superimposed on the irreversible part of the phenomenon and scarcely changing it."[36] Yet it was by moving his attention from the fundamental phenomenon to subsidiary ones that Weiss would take a second decisive step, which would

Pierre Weiss's laboratory in the Institute of Technology, Zurich, 1913: *(front)* K. F. Herzfeld, Otto Stern, Albert Einstein, E. Picard, Pierre Weiss, Miss Girgorjeff (translator); *(rear)* Paul Ehrenfest between two Dutch high-school teachers, Foex and Wolfers (two of Weiss's lab assistants). (AIP Niels Bohr Library, gift of K. F. Herzfeld)

lead him to picture a field, of internal origin like the "demagnetizing field," but operating in the direction where magnetization was easy.

Weiss tried to gather these two fields into a molecular scheme. Since 1902 he had been installed as director of the physics laboratory at the Zurich Eidgenössische Technische Hochschule, so he was not fully a part of the Paris scientific scene, although he remained in touch by letter. His article, published in December 1905, did not refer to the one Langevin had published in July; it seemed rather to be inspired by Ewing's model, in which a magnetic substance was imagined as a collection of tiny magnetized needles (Chapter 1). Although unrealistic, the model that Weiss proposed would help him to discover a new feature that illuminated his ideas on ferromagnetism.

Weiss imagined that the "magnetic plane" of pyrrhotite was formed of small magnetized needles, arranged in parallel rows in the direction of easy magnetization. These magnets acted on one another only if they belonged to the same row. Being free to turn in the magnetic plane, they would all orient themselves in the direction of their row. In the presence of a uniform applied magnetic field, they would find a new equilibrium under the conflicting actions of the applied field, which tended to turn them out of line, and their neighbors, which tended to keep them lined up. A magnetic element would thus be subject to an "internal field" from all the other elements in its row, whose component in the direction of easy magnetization would add to the component of the external field.

For the first time, an internal anisotropic field, whose components were proportional to the components of magnetization in the two distinguishable directions, was associated with a substance's ferromagnetism.

Weiss's excursion into the microscopic realm did not lead any farther. His hypothetical molecular scheme played only one role: it allowed him to complete the laborious construction of the "internal field." The unifying simplification that resulted helped him to understand the significance of a comment of Langevin, who had pointed out "the importance of the mutual actions between molecules, which alone make magnetic saturation possible [in ferromagnetic substances], a saturation that remains extremely distant, for the same external field, in the case of paramagnetic substances."[37]

At this point Brillouin pointed to Weiss's second stroke of luck, the publication of Langevin's theory. From this account of how his ideas grew, we can understand that Weiss was better placed than anyone else to continue Langevin's work. In any event it is certain, as Brillouin noted, that between 1905 and 1907 nobody but Weiss thought hard about internal magnetic fields.

The result was that Weiss published an article in the *Annales de physique*, two years after Langevin, entitled "The Hypothesis of the Molecular Field and Ferromagnetism."[38] The success of this hypothesis partly explains the fact that the work leading up to it was forgotten. Weiss may also have contributed to discouraging any exegesis when he confessed to Bauer that the idea came to him as an extraordinary intuition.

In his 1907 article, Weiss gathered into a single new concept, which he called the "molecular field," the characteristics of the magnetizing interactions between molecules, suggested by Langevin, and the characteristics of the internal anisotropic field that he himself had hypothesized. As we will see, this required considerable ingenuity. He assumed that the "crystalline edifice possesses three orthogonal planes of symmetry and that each component of the molecular field is proportional to the corresponding component of the magnetizing field, with a different positive coefficient $[N_1, N_2, N_3]$ for each of the three axes."[39] This simple hypothesis allowed him to explain the properties of the pyrrhotite crystal, to explain results concerning certain paramagnetic bodies that in 1895 had fallen outside Curie's experimental law, and finally to account for ferromagnetism.

Regarding pyrrhotite, Weiss only needed to say that the magnetization vector **I** aligns itself in the direction given by the superposition of the molecular field and the applied field **H**. If the latter is confined to the plane of the magnetization, the condition is expressed by an equation in terms of the coefficients N_1, N_2 of the molecular field in this plane: $HI \sin(\alpha - \phi) - (N_1 - N_2)I^2 \sin\phi \cos\phi = 0$, where ϕ and α are the angles between the axis of easy magnetization and the directions, respectively, of the magnetization and the applied field. This equation was virtually identical to the experimental law with which Weiss had summarized the properties of pyrrhotite: $HI \sin(\alpha - \phi) - NI^2 \sin\phi \cos\phi = 0$. In the reasoning that led to the first equation, $N_1 I \cos\phi$ and $N_2 I \sin\phi$ are the components of the molecular field projected, respectively, on the axis of easy magnetization and the axis of difficult magnetization. On the other hand, Weiss had established the second, experimental equation by imposing a demagnetizing field proportional to the component of magnetization in the direction of difficult magnetization: $N \sin\phi$. These two interpretations seemed mutually exclusive.

Weiss managed to reconcile them by formally identifying the coefficient N as the difference between N_1 and N_2. A loss of magnetization under the first hypothesis thus became a failure to increase under the second. The demagnetizing field was merely the deficit of the molecular field in the direction of difficult magnetization as compared with the direction of easy magnetization. The agreement that Weiss maintained between the two viewpoints was satisfactory, in particular because the first helped in the birth of the second. However, his view of ferromagnetism was now fundamentally changed. A body was not weakly magnetic because, as he had formerly believed, there was something in it that opposed the action of the external field, but because what it contained was not strong enough to be ferromagnetic.

Fortified by these clear ideas obtained using homogenous crystals, in his 1907 article Weiss returned, as he had intended from the beginning, to the study of properties of complex materials as laid out by Curie. We recall that experiments had been carried out in particular on samples of industrial iron where the confusion of interlocked crystals created an illusion of isotropy. Before tackling this problem, Weiss formulated his new hypothesis in a way suited to isotropic media, simply by removing the anisotropy from the molecular field: "I assume that each molecule experience, from the collection of molecules surrounding it, a force equal to that of a uniform field proportional to the intensity of magnetization and in the same direction."[40]

Weiss was only introducing an extra field $h = nI$, proportional to the magnetization I into the theory of paramagnetism, leaving the theory otherwise unchanged. Langevin's concept survived, except that now the effective field H_e was the sum of the applied field H and the molecular field h. The specific magnetization continued to obey Langevin's equation, $I = I_0 (\coth a - 1/a)$ (Figure 6.1), where now a was equal to $\mu(H + h)/kT$.

In the absence of an external field, Weiss showed that the bare existence of the molecular field implied a "spontaneous magnetization" I_s, defined by the intersection of the curve giving the variation of magnetization as a function of $a = \mu h/kT$ and the straight line of the equation $I = (kT/n\mu)a$ (Fig. 6.2). Above a certain temperature, this straight line, whose slope will increase with temperature, would not intersect Langevin's curve: the substance would lose its spontaneous magnetization. Thus Weiss explained the existence of the temperature marking the transition to the paramagnetic state, a temperature that he later decided to call, in agreement with Kamerlingh Onnes, the "Curie point."[41]

In his 1907 article, Weiss proceeded in reverse of the way we have discussed it:

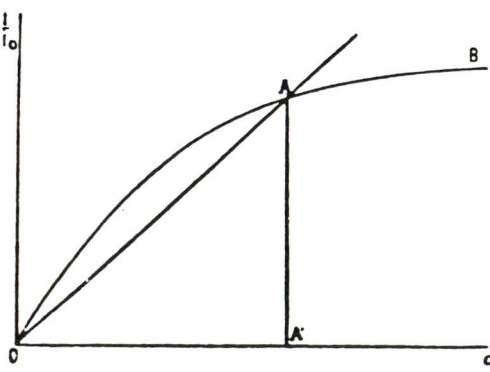

Fig. 6.2 The intersection of the straight line of slope $kT/\mu n$ with the Langevin function curve OB defines spontaneous magnetization. (From P. Weiss, "L'Hypothèse du champ moléculaire et la propriété ferromagnétique," *Journal de physique,* 4th ser., vol. 6 [1907]: 663)

he introduced the mathematically simple isotropic molecular field before the aniso-tropic field, with its three coefficients. We may suspect that concern for clear expla-nation guided his presentation, although he did stress that apparent isotropy is physically more complex than anisotropy in a single crystal. Or his approach may have been guided by tactical considerations. His hypothesis was certainly bold, and it would take him some time to get it accepted. By first introducing the isotropic molecular field, it was easier for him to claim that the idea had been "suggested by van der Waals's 'internal pressure,'" as he wrote in 1908.[42] The connection is doubtful, but it was respectable and reassuring in its invocation of the analogy with fluids. Weiss presumably thought that the connection would give more weight to his idea, even while reducing its apparent originality. Nearly two decades later, when the theory was well founded, its author was freer to be himself, and he stated that the idea of the molecular field had come from his own studies on pyrrhotite.[43] Certain opinions that put him forward as the leading theorist of collective phenom-ena would eventually induce Weiss to be a little more than himself.

Magnetism in 1907. The return to Curie's work is significant. Weiss had struck a rich vein in the molecular field. Inserting it into Langevin's theory, he used the best of both and founded the theory of ferromagnetism. Experimental verification of his initial conclusions was already present in the findings of Curie, who could not have reached these conclusions for lack of the guidance of the Langevin–Weiss theory. Three dates, 1895, 1905, and 1907, and three names thus remain associated in marking the appearance of clear ideas in an area where there had been nothing but confusion and inconsistency.

Both Curie and Langevin had firmly marched straight ahead along their chosen directions. Weiss had been more hesitant and had not succeeded at first. He had some help from luck, but he knew how to make the most of it, and all on his own. As we have seen, this investigator of pyrrhotite had no competitors.

We should not conclude that physicists now held the keys to magnetism and that research would easily forge ahead. Far from it. The molecular field, although it might be thought to contain a substantial share of truth, remained obscure in its origins. The molecular field's magnitude was disturbing, for no known interaction could equal it. Weiss, confident in this theory, foresaw that new hypotheses con-cerning the structure of the atom would shed light on these shadowy matters. He would make his own attempts in this direction. He also thought that quantitative study of magnetic phenomena as a function of temperature in certain alloys would better define the nature of the molecular field, for the addition of another element could modify the region of stability of an allotropic phase of iron. That is what he asked Néel to undertake 21 years later when he welcomed him to his laboratory.

For the time being, Weiss argued in favor of the molecular field. Although aware that this seemed an unlikely hypothesis, he assumed it to be uniform for the sake of simplicity. He also assumed that the actions between a molecule and its neigh-bors did not extend beyond a certain "sphere of activity" (it was this same assump-tion that made him prefer the term "molecular field" to "internal field," which he found too ambiguous). Should he then assume that the number of molecules con-tained in this sphere was large enough so that only an average effect was detectable? Or should he attribute the averaging to thermal agitation? Weiss chose to stick with an explanation that used nothing but the size of the "magnetic element"; if this were infinitely small, the field would appear uniform over its entire extent. This hypothesis contradicted that of Ewing, who had tried to explain residual phenomena—those "light embroideries"—by mutual actions where it was the finite size of magnetic elements with respect to the distance between molecules that played the leading role.[44] In France the success of Weiss's theory eclipsed Ewing's model, since no attempt was made to reconcile them. Néel later suggested that this rejection was an error that hindered progress on theories of hysteresis and applied magnetism.[45]

Consequences of Molecular Field Theory

The Specific Heat Anomaly and the Magnetocaloric Effect. From 1907 on, Weiss devoted himself to consolidating the hypothesis of the molecular field. It had already proved satisfactory for the study of the magnetic properties of bodies at various temperatures, of ferromagnetic crystals, and of metals that seemed isotropic. In the same year, it would score another success by accounting for a phenomenon that had long been known but had remained unexplained. It was known that specific heats of ferromagnetic materials exhibited an anomaly: near a particular temperature, the specific heat increases with temperature in a regular fashion, and then falls sharply to a normal value. Such a phenomenon could be connected with an exchange of latent heat. Weiss and Paul Beck showed that the true cause of this discontinuity is the existence of an energy associated with the molecular field, an energy that disappears at the Curie temperature.[46] Returning to heat measurements made decades earlier by Joseph Pionchon[47] and to Curie's magnetic measurements for iron, and supplementing them with their own measurements on magnetite and nickel, Weiss and Beck verified that for these three substances the theory was valid, yielding estimates of the magnitude of the discontinuity and of the temperature at which it appeared.

Considerably later, in 1918, a new phenomenon would show the usefulness of the molecular field hypothesis. Taking up the study of the isotherms of nickel, Weiss and Auguste Piccard observed that the creation of a magnetic field was accompanied by a sudden rise in the substance's temperature and, conversely, that the elimination of the field was accompanied by cooling.[48] The reversible nature of this variation and the order of magnitude of the quantity of heat involved excluded any link with hysteresis.

This new phenomenon, which Weiss proposed to call the magnetocaloric effect, is easily detected in a ferromagnetic body. It is also present in paramagnetic substances; in 1905, Langevin had predicted its existence and had even estimated its magnitude for oxygen gas (using the relation $c\Delta T = \chi(H^2/2)$, where c is the heat capacity).[49] In such a substance, the temperature variation is extremely small, only $\frac{1}{1000}$ of a degree in a field of 10,000 gauss at ordinary temperatures. However, Langevin had noted that this rise in temperature would increase, along with the magnetic susceptibility, in inverse ratio to the absolute temperature. Thus he specified one of two reasons why the effect is detectable in paramagnetic substances; he did not know the other reason—that the specific heat of solids, falling with decreasing temperature, significantly reinforces the phenomenon at very low temperatures.

Not until 1927 did W. F. Giauque explain the possibility of using adiabatic demagnetization to cool a substance.[50] The technology of refrigeration had to advance before it could take advantage of this effect, which from this standpoint seems ahead of its time. Indeed, Langevin had spoken only of a rise in temperature; the reversibility of the operation did not catch his attention.

Although it was not yet possible to reach the low temperatures required to make the effect significant, it was at least possible to imagine its importance in very strong fields. In this connection, Langevin remarked that in a field that was four times stronger than could then be reached, the temperature variation would be in the neighborhood of $\frac{1}{100}$ of a degree. If the field were 800 times stronger, then the temperature variation would certainly be noticeable, and such a field strength was just what the molecular field offered. Since 1907, both Langevin's thermodynamic argument and Weiss's hypothesis were known, so the stage was set to predict the magnetocaloric effect and estimate its magnitude. Yet this did not occur to Weiss. It was left to experiment to show him the heating of nickel, and he still avoided the hypothesis by assuming there was an experimental error ("we have made sure that the field has no effect on the electromotive force of the thermocouple used to measure the temperature," he wrote in the first note to the *Comptes rendus*), before he found in his own theory what it could have revealed 11 years earlier.[51] Even leaving aside the four years of the First World War, when Weiss and his colleagues were engaged in very different fields of research, the magnetocaloric effect was long in coming. We may again observe that the inventor of the molecular field indeed had no competitors.

Yet the magnetocaloric effect was discovered at an opportune moment. When it appeared, it gave valuable support to Weiss's hypothesis. It enabled values of the specific heat to be calculated on the basis of strictly magnetic measurements, values that could be compared with those measured by calorimetric procedures. The match between these results contributed to belief in the molecular field. Weiss concluded the 1918 memorandum by writing that "this new phenomenon is a striking confirmation of the theory of the molecular field, which could have predicted it and which accounts for all its details."[52]

Since the time he introduced it, Weiss's hypothesis has never been refuted, but neither has there been any decisive demonstration of the "spontaneous magnetization" that was required by the existence of the molecular field. This time it would become clear that the spontaneous magnetization could indeed exist by itself, since now calorimetric measurements had been carried out in the absence of an applied magnetic field. Thus the magnetocaloric effect came to the aid of Weiss, who since 1907 had persistently maintained that spontaneous magnetization exists even though it had escaped observation.

Remanent Magnetization, Spontaneous Magnetization, and "Domains." In 1920, Albert Perrier and G. Balachowsky supplied a new argument in favor of spontaneous magnetization.[53] They had studied ferromagnetic samples that lacked a demagnetizing field and had demonstrated that in this case the remanent magnetization (i.e., the magnetization remaining after an external field was removed) varied with temperature in a reversible way. This helped pin down the interpretation of remanent magnetization. If it was only an attenuated part of the magnetization caused by an external field, any loss of that magnetization would necessarily have been final unless one put the sample back into an external field. The experiment showed, on the contrary, that the remanent magnetization lost during an increase of temperature would return when the sample was cooled. It was therefore plausible to interpret remanent magnetization, as Weiss did, as the resultant of a disordered spontaneous magnetization. In 1926 this interpretation was confirmed by Weiss and Robert Forrer, who noted a remarkable proportionality between the variations with temperature of remanent magnetization and of spontaneous magnetization.[54]

Meanwhile, the concept of what would later be called domains was making solid progress. In his efforts to convince people that spontaneous magnetization was reasonable, Weiss had at first invoked the usual analogy with liquids "which can exist under zero external pressure."[55] But an analogy was far from a proof. Weiss also had to explain the fact that spontaneous magnetization could dwindle away, so he went on to propose that, since there was nothing to determine a particular direction, all directions were possible and indeed would be found in one or another small region of the material; these could cancel one another, so the material would be nonmagnetic overall.

The source of this idea lay in observations of the magnetization cycle of pyrrhotite. Only two values of magnetization should be found along the direction of easy magnetization—those corresponding to saturation in that direction or its opposite—but intermediate values were seen. Weiss concluded that a "certain fraction of the material is magnetized in the positive direction and the remainder in the negative direction and that it is the difference that is observed."[56] Actually, the magnetism of the "fraction" Weiss referred to represents only the persistence of an equilibrium, an unstable one, due to the reversal of an external field along the major axis of pyrrhotite.

Subsequently, examining substances such as iron that give the appearance of isotropy, he assigned a spontaneous magnetization to each microscopic "elementary crystal," and described the whole set as a tangle of crystals oriented in all directions. More specifically, "when the material is in the neutral state, the magnetization vectors of the different elementary crystals are distributed with constant density over a sphere."[57] In 1918, Weiss still held that the direction of bulk spontaneous magnetization "depended on the chance arrangement of the crystalline microstructure."[58] He made no use of energy considerations in his proposal, which might be regarded as a plain list of circumstantial evidence, although posterity would recognize it as prefiguring the concept of domains. He extended the concept of areas of uniform spontaneous magnetization to crystal grains, the sort that could be seen when an alloy was examined through a microscope, and this enabled Weiss to take another large step toward the domain concept: he explained that the role of the applied field is not to create magnetization, but to make it susceptible to observation. There is "no essential difference between a magnetized body and a body that can be strongly magnetized. Both are magnetized to saturation; the only difference is that in the one case the magnetization is coordinated and in the other it is disordered."[59]

In the following year, H. Barkhausen performed a striking experiment that still bears his name.[60] He demonstrated that the magnetization of a ferromagnetic substance increases in a series of tiny jumps when the substance is subjected to a continuously increasing magnetic field. These abrupt variations can be detected because they will induce a current in a coil surrounding the test piece, a current that when sufficienty amplified produces a crackling sound in an earphone.

Following in Barkhausen's footsteps, B. van der Pol studied various substances and verified that the phenomenon could be found in each of them, finding it particularly marked in the field interval where the magnetization varies rapidly.[61] He attributed these discontinuities to the simultaneous reversals of a number of atomic moments. From the number of discontinuities counted up to the saturation point, he estimated the number of atoms that were involved in each reversal process; for this he deduced that the volume in which each process takes place is small, about a fraction of a cubic millimeter.

Van der Pol reported a number of other interesting observations, and in 1922 Weiss set out to interpret them.[62] For instance, if two coils positioned on a bar several centimeters apart sent pulses to a galvanometer, these pulses were usually simultaneous. Therefore, the regions sustaining the same sudden variation in magnetization seemed to be quite broad despite their small volume, which led Weiss to say that they probably corresponded to "crystals drawn out by rolling."[63] In another observation, an iron wire with remanent magnetization produced the characteristic crackling in the earphone when it was flexed. This manipulation eventually removed the remanent magnetization, but the crackling continued, although attenuated.

A tempered iron wire that had not been magnetized would also emit sounds when it was flexed. Weiss saw in this a demonstration of the existence of spontaneous magnetization as foreseen by his theory of the molecular field. Convinced that the region of uniform magnetization was identical with a crystal grain, Weiss supposed that Barkhausen's discontinuities would be even more abrupt in a uniform crystal. To represent the phenomenon, he called again on the knowledge he had gained through the study of pyrrhotite. His observations on a plate of apparently homogeneous magnetite did not provide him with the expected results. However, based on the number of discontinuities, he concluded that "the particle size of the structure is probably fairly coarse."[64] The interpretation of Barkhausen's phenomenon was widely accepted, although still uncertain: the induced current pulses could be attributed to the reversal of moments in limited regions, but the extent of these regions remained uncertain. Walther Gerlach kept the entire question open by suggesting that the observed effect was linked with magnetostriction rather than being a direct consequence of stepwise magnetization.[65]

In 1924, E. P. T. Tyndall laid down in precise terms for the first time the lesson to be drawn from the Barkhausen effect.[66] He showed that his own results, obtained with silicon steel, supported the generally accepted interpretation but excluded the possibility that the region of uniform magnetization was identical to the extent of a single crystal grain. He concluded that

> it seems best to adopt the notion that for some reason, limited portions of the material magnetize by jumps, either partially or to saturation, and to leave open the question of what determines the size and shape of these portions in various materials and in various specimens of the same material which have undergone different treatments.

This introduced the domain as it is known today. Weiss gave it its name in 1926: "we shall call an 'elementary domain' a region that is small enough for magnetization to have the same direction throughout its extent. Elementary domains are much smaller than the elements of the crystalline structure.".[67]

A new direction was opened up for research, attracting attention in many nations. "The identification of these individuals," as Tyndall had written "and a better understanding of their behavior would give microscopic detail to pictures of the mechanism of magnetization."[68] In particular, in 1931 Francis Bitter demonstrated domains by observation of the microscopic patterns traced out by fine magnetic particles spread over the surface of a magnetized sample.[69] As for theory, J. Frenkel and J. Dorfman attempted to rationalize the existence of domains, but they failed to allow for the crystal anisotropy that is necessary for the formation of domains.[70]

At the General Electric laboratories, Irving Langmuir pointed out the importance of the role that must be played by the surface between a pair of domains, and this led K. J. Sixtus and L. Tonks to carry out experiments on drawn Permalloy wires, studying the displacement of the walls that separate domains.[71] Around the same time, Felix Bloch undertook the first quantum-mechanical calculations of the properties of a one-dimensional transition layer between two magnetized domains.[72] In 1935, Lev Landau and Israel Lifshitz, arguing for the existance of domains, calculated the properties of the layer between domains.[73]

The Analogy with Fluids. In the foregoing we have pointed out on occasion the comparison frequently made between magnetization phenomena and the physics of fluids. Curie was the first to introduce it, using it to help clarify the description of magnetic phenomena. Langevin then took it up, followed by Weiss, who searched for an analogue to the unprovable molecular field in the thermodynamics of fluids. Why could the ferromagnetic state not exist in the absence of an external magnetic field, when a liquid can exist under zero pressure? Yet his gradual construction of the molecular field concept owed nothing to the internal pressure of fluids, the analogy used for didactic purposes. Allied with the authority of van der Waals's classic works, the analogy gave credence to the idea that Weiss's theory was a simple transformation of van der Waals's hypothesis, and Langevin's theory also found a place in the wake of the ideas of D. Bernoulli.[74] This contrivance also resulted in bringing to Weiss's hypothesis a certain clarity that the original formulation lacked. In 1911, for instance, Langevin found it "quite natural to assume that the orienting action exerted by molecules on any particular one of them be determined by the extent of their parallelism."[75] Weiss, fired by the need to prove his hypothesis, was as eager to improve its formulation as he was to discover how the molecular field originated.

In 1918, Weiss defined the molecular field in terms of the partial derivative of the internal energy with respect to magnetization.[76] This molecular field, called the energy field, sidestepped any questions about the nature of the interactions that produce the field. The new definition incorporated the old formulation of the molecular field, seen as a "corrective" field, to which it reduced in the particular case where it did not depend explicitly on temperature. By this means, Weiss would later develop an elegant thermodynamic theory of ferromagnetism.[77]

Following his 1918 work, Weiss transposed the approach he had used for magnetic phenomena to problems of the compressibility of fluids.[78] He assigned an internal pressure, which up to then had been introduced only as a correction factor for external pressure, with a definition modeled on that of the molecular field, setting it equal to the partial derivative of internal energy with respect to volume. Weiss thus shed new light on some old questions that "long efforts by eminent physicists seemed to have exhausted."[79] As may be seen elsewhere in this book, the study of collective phenomena in solids, in this case ferromagnetism—probably the first such case—produced an idea that would be transposed to other fields of physics.

As for the origin of the molecular field, Weiss thought that it might be approached if one could study a given substance at different densities and thus find how the interactions produced by the molecular field vary as a function of distance. In 1914, basing his arguments on the way the Curie point of solid solutions of nickel and cobalt varied with their concentration, Weiss showed that the molecular interactions decreased with the reciprocal of the sixth power of distance, which excluded any purely magnetic or electrostatic origin for the molecular field.[80] He compared this result with experiments conducted by Charles Maurain on electrolytic deposits of iron in the presence of a magnetic field.[81] These demonstrated the existence of a "magnetizing action" exercised by the initially deposited layers on those that followed. This effect varied with the reciprocal of the sixth power of distance, indicating that Maurain's magnetizing action was identical to Weiss's molecular field.

In the same year, Weiss reported a discussion that he had had with Lorentz and Einstein about the difficulties that the concept of the molecular field would encounter if one assumed a magnetic origin. An electrostatic field, on the other hand, might be able to produce the effects required by the theory of ferromagnetism. The objection arising from the order of magnitude could be discarded, since, as Lorentz had pointed out to Weiss, for two electrons moving in parallel the ratio of their magnetic interaction to their electrostatic interreaction varies as the square of their velocity compared with that of light. The hypothesis of an electrostatic interaction required that molecules possess not only a magnetic moment, but also an electrical polarity. Weiss observed that the theory of dielectric phenomena, recently developed by Debye,[82] had a number of features in common with the theory of the molecular field, and found it reasonable to assume that magnetic bodies had dielectric moments, comparable to those of certain dielectrics. However, despite these advantages the hypothesis was finally rejected, for it did not conform to the law of variation as the reciprocal of the sixth power of distance. Weiss concluded that "in the last analysis, we are thus led to attribute the phenomena studied to natural forces as yet unknown."[83]

However, he reversed this conclusion in 1924, returning to the idea that atoms of ferromagnetic metals carried both a magnetic moment and an electric moment.[84] He had observed that the dipolar moments μ, calculated through the relationship

$$\mu = \sqrt{\frac{3kC}{N}}$$

were of the same order of magnitude as the moments measured in various other ways. This seemed to him an excellent argument in favor of an electrical origin for the molecular field.

Many other physicists kept an open mind about the fascinating possibility that

Pierre Weiss (1865–1940). (AIP Niels Bohr Library, Weber Collection)

ferromagnetism pointed to hitherto unknown natural forces. As seen in Chapter 2, it would not be until the 1930s that they would begin to suspect what these new forces were: exchange forces, a type of interaction that could not even be imagined until the theory of quantum mechanics had matured. Well into the 1920s, no physicist could hope to guess what direction of research would lead to an explanation of the molecular field. The direction that Weiss chose, guided by his interest in the moments of molecular elements, seemed eminently reasonable, but, as we will see, it led him increasingly astray.

The Magneton

Leiden Cryomagnetic Measurements. The theory of the molecular field, although seemingly confirmed beyond doubt by various facts, refused to agree with other observations that had been increasing since 1907. The most important contradiction, and the first to be found, was the variation in the relative saturation magnetizations of iron, nickel, and cobalt near a given temperature; unlike magnetite, these did not follow the curve predicted by Weiss's theory. New measurements on magnetic substances had expanded the range of observed behaviors. It had been hoped that the Curie-Weiss law—a refinement of Curie's law, stating that the inverse of magnetic susceptibility varies in proprotion to the difference between the absolute temperature and the Curie temperature—defined with or without a molecular field, in its simplicity could embrace all magnetic phenomena. But this law no longer seemed to reflect more than a part of reality. For instance, Honda had reported that certain substances were characterized by increasing paramagnetism with temperature, whereas others had constant paramagnetism.[85]

According to Weiss, his theory had been confirmed by too many triumphs for it to be abandoned. Setbacks only indicated a need for improvements, not a need to call into question anything fundamental. This excluded taking a new look at the (nonquantized) Boltzmann energy partition law as used by Langevin or at the existence of the molecular field.

For the latter, one might hypothesize a more complex relationship between field and magnetization; for instance, instead of the constant proportionality coefficient, one might try a function involving temperature and magnetization. Working in this direction, Weiss found he had to invent different functional forms for each substance, and the effort ended inconclusively.

The constant characterizing the interactions of molecules having thus been set aside, there remained two constants that could be used, both of them properties of the individual molecule: mass and magnetic moment. Weiss opted for the variability of mass, betting on the invariability of magnetic moments. He hoped to be able to describe the magnetic aspects of the atom in terms close to those used for its electrical aspects, looking in effect for a magnetic equivalent of the electron. That would be something that could give an intrinsic character to the atom's magnetic moment, independent—contrary to Langevin's theory—of the electrical charges and their movements.

Weiss might not have gone in this direction if he had not known and appreciated the work of one of his Swiss friends, the young physicist Walter Ritz. Ritz had sought to explain the phenomenological laws of atomic spectra by the supposition that each atom contained a chain of magnetized or neutral rods, which were responsible for the atom's magnetic characteristics. He felt that the length of the rods should be a constant, on the same fundamental level as the charge of the electron.[86]

It was in this frame of mind that Weiss went to Leiden in 1909 at the invitation of Kamerlingh Onnes, in order to carry out measurements of magnetization at the temperature of liquid hydrogen. When thermal agitation was frozen out, the elementary magnets would be aligned, and at saturation whatever magnetization remained would be simply the sum of whatever elementary invariant magnetic moments might exist.

The question could be stated simply. Given two different magnetic metals, such as iron and nickel, were the magnetic moments of a gram-atom of each in simple integral ratio to each other? If so, that would point to a submultiple, such as an elementary magnet within the atom would generate.

A short paragraph from Kamerlingh Onnes and Weiss's paper of 1910 hinted at a difference of opinions regarding their results. Considering whether simple ratios held between the magnetic moments of iron and of nickel, they answered that their table of values "shows that this is not the case."[87] But in the next sentence they added, "Of course, if proof of the primordial significance of this quantity escapes us in this case, nothing would authorize the converse conclusion either." We may surmise that the authors had arrived at opposite opinions, and rather than stating these fully in an article they signed jointly, they apparently chose to remain on neutral territory.

The Experimental Magneton. Once Weiss was disengaged from the joint enterprise and back in Zurich, he grasped again at the quantity that had evaded him. The following year, he published arguments proclaiming the success of his new hypothesis.[88] Had some new, decisive fact arrived in the meantime to sweep away doubts

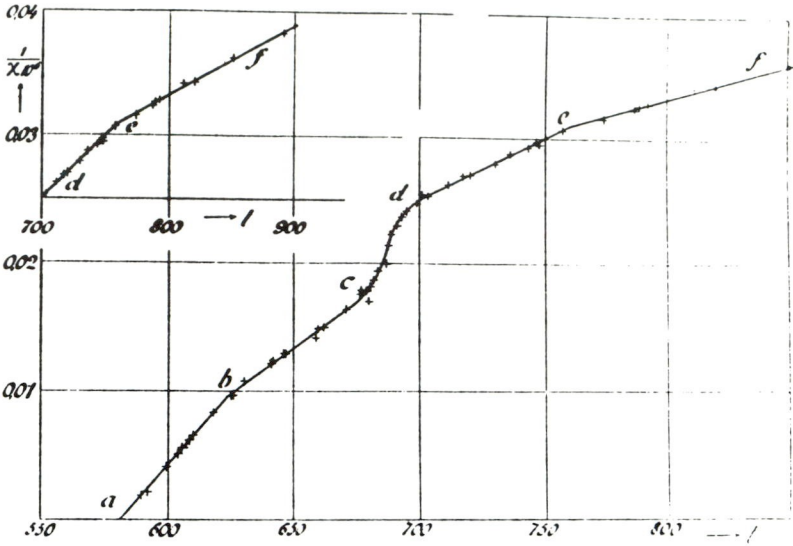

Fig. 6.3   Investigation of the curve of inverse magnetic susceptibility $1/\chi$ versus temperature for magnetite seemed to give five straight lines, pointing to magnetons. (From P. Weiss, "Sur une propriété nouvelle de la molécule magnétique," *C.R.* 152 [1911]: 80)

and establish the fundamental discovery Weiss had been seeking? No—or at any rate not in regard to measurements at low temperatures. Weiss nevertheless referred to the results that he had obtained with Kamerlingh Onnes a few months previously, but now he gave free rein to his comments. He made his conviction plain: in Leiden, the magnetic moments of a gram-atom of nickel and iron were indeed in a strict integral ratio, 3:11. From this ratio, he estimated the size of the quantity he had sought, what he later called the experimental magneton.

In the same article, Weiss also reported other experimental findings, all presented as strong evidence for the magneton. We will mention only one, which illustrates particularly well the method employed and which would also play a decisive role, frequently referred to by Weiss over the years.

Gabriel Foëx had measured the magnetic susceptiblity χ, finding that the curve of $1/\chi$ against temperature could be split into several connected straight-line segments, although a continuous curve rather than a set of segments might just as easily have been seen in the data points (Fig. 6.3). The slope of each segment gave a value for the Curie constant. And according to Langevin's theory, this constant was proportional to the square of the magnetic moment. Weiss and Foëx found that, give or take a little, the magnetic moments thus computed from the various straight-line segments bore simple ratios to one another. They concluded that the magnetic moment of the magnetite molecule varied abruptly by one or more times a characteristic quantity. Was this characteristic quantity the experimental magneton? The question was not asked, but at another point the authors remarked that the state of the data on magnetite tainted "the absolute values with various uncertainties."

Checking the Magneton Theory. Other measurements were undertaken. With some occasional help from supplemental hypotheses, these measurements kept producing the same submultiple of atomic moments. Finally, Weiss felt justified in giving this quantity an existence and a name, the magneton.

In particular, Weiss submitted numerous solid and dissolved paramagnetic salts to the test of his new unit. The success was patchy, for some magnetic moments stubbornly refused to obey the integer rule. Yet faced with results different from an integer by less than 20%, Weiss judged them to be "favorable to the existence of the magneton within the limits of experimental precision." Only three results were declared to be "in formal disagreement" with the rule.[89] On balance, these did not outweigh the favorable cases, as we may see from reading some of the statements in Weiss's 1911 paper:

> Considering the magnetic history of a sample, from now on I have no qualms in substituting for the phrase, "sample that does not follow the theory," the phrase, "sample that does not keep the same number of magnetons throughout the interval under study. . . ."

> One can state that as the electron has symbolized the new ideas on the discontinuous nature of electricity, so the magneton marks an analogous revolution in our representation of magnetic phenomena.[90]

From these quotes, one might imagine that Weiss was close to letting his pride run away with him. Possibly Aimé Cotton, a Paris scientist with whom Weiss kept up a correspondence, tried to moderate his friend's enthusiasm; in replying to a letter, Weiss confessed that "a touch of intoxication has perhaps led me to stray from what would have been strictly prudent."[91]

The magneton hypothesis was presented by Langevin at the first Solvay Conference, in October 1911.[92] He expressed no reservations and structured his argument around the remarkable relationship that he found between the magneton and the "quantum of action" recently suggested by Sommerfeld. Langevin's demonstration was simple. He calculated an electron's action for a complete revolution of its orbit and equated it, in accordance with Sommerfeld's idea, to a fraction of Planck's constant. He thus deduced a value for the magnetic moment that was very close to Weiss's experimental moment. Thus Sommerfeld's and Weiss's theories supported each other; theory and experiment seemed to agree on the existence of a magnetic moment quantum. Several others promptly derived a slightly different version from Niels Bohr's model of the atom; Bohr himself had found the news of the Weiss magneton of interest and spent some time on the question.[93] Unfortunately, Weiss's empirical magneton was about one-fifth of the value of the Bohr magneton. We shall see later how theorists approached the problem; Weiss himself had no taste for theory, and at the time (when quantum ideas were all uncertain) he felt no need to address it.

The magneton was announced with unusual force. Launched by the inventor of the theory of ferromagnetism, taken up by the author of the theory of paramagnetism, and presented by him at the summit of European physics, it was certainly very impressive.

One physicist well aware of this was Blas Cabrera, recently named director of the physics laboratory at the University of Madrid. Seeing a chance to give a new impulse to his research, he immediately undertook the study of paramagnetic salts, and in the following year traveled to Zurich to visit Weiss's laboratory and begin a collaboration that would last for more than 20 years. Much later, when the magneton theory was put in doubt, Cabrera would reaffirm, "From the point of view of the theory of magnetism, there is no more problem more worthwhile than that of definitively resolving whether Weiss's magneton is a reality or not."[94]

The magneton theory had a fundamental weakness: once quantum theory was a little better understood, no theoretical argument could be found to justify Weiss's value. But until quantum theory was developed much more fully, it was also unable to prove that Weiss's magneton did not exist. Its defenders, unable to interpret the idea, worked constantly to extend its experimental basis. The zeal with which they worked to sort out the magneton from innumerable measurements of magnetic moments on some 20 ions in 65 different compounds gave the hypothesis a long life. As a result, the magneton inspired a number of solid and valuable measurements, which had to wait for a correct analysis. We shall look at the most striking of them.

Empirical Laws of Magnetism. We first must take note of revisions in the empirical laws and their interpretation, and shall then examine the hypotheses put forward to justify them. The Curie–Weiss law, $\chi = C/(T - T_c)$, stating that χ varies inversely with the difference between the absolute temperature and the Curie temperature T_c (where C is Curie's constant), was found to err in many cases. The curves were still linear, but they intersected the temperature axis in the negative region. Accordingly, after a study of gaseous oxygen at high pressure, Kamerlingh Onnes proposed a new formula in which the inverse of the susceptibility was proportional to the sum of the absolute temperature and a new constant Δ, which for a given case could be either positive or negative: $\chi = C/(T + \Delta)$.[95] The Curie-Weiss law corresponded to a particular negative value for this constant. This generalized formula seemed to fit a substantial number of experiments, including those done on very dilute solutions. It could be concluded that the new constant's existence was not caused by the interaction of paramagnetic atoms alone, but was more generally related to the action of atoms, of whatever nature, surrounding the paramagnetic atom. This evidently modified the role played by the molecular field as conceived by Weiss.

Other results continued to fall outside the generalized law. To allow for measurements made on metallic palladium and platinum, W. Kopp proposed adding another constant $-\chi'$ to the expression for the susceptibility: $\chi = -\chi' + C/(T + \Delta)$. This new modification, minor in form, was associated with an important physical fact: it pointed out a term in the susceptibility that obeyed the Curie–Weiss law and another term $-\chi'$ that was independent of temperature. If this constant is positive, it may be attributed simply to diamagnetism, which is independent of temperature and is superimposed on paramagnetism, as Pierre Curie had already noted. Measurements by Kopp[96] and by Foëx[97] justified this interpretation.

A negative constant, on the other hand, is the manifestation of a supplementary paramagnetism. This feature was rightly linked with examples of susceptibilities that were strictly invariable with temperature, drawn from wide-ranging series of measurements on various elements and compounds. Further, by adding such a paramagnetism constant, Cabrera was able to linearize the results obtained from a number of rare-earth salts.[98]

With these two constants, Δ and χ', the generalized Curie–Weiss law would for a time cover every measurement, which later led John Van Vleck to say that it had certainly been the most "overworked formula in the history of paramagnetism."[99] It seemed to unify phenomena relating magnetism to temperature, although modification had to be made to account for a rich variety of behaviors. There developed a large field of descriptive magnetism, with which the name of Foëx became closely associated.

Weiss and his collaborators were now in Strasbourg, for in 1919 he was named director of the physics institute at the university, restored to France after the First World War. Weiss thus returned to his native land, Alsace, which he had left with his family in 1870 after it was seized by Prussia. In Strasbourg he established a laboratory modeled on the one he had organized in Zurich, contributing to the renovation and reestablishment of French dominance in the univeristy and the province. That was an important aim of the French government; therefore, support was not lacking for the establishment of an important research center, which Weiss dedicated to the study of magnetism.

Foëx, aiming to study magnetons further, used the concept of a "magnetic state" to extract magnetons from a variety of susceptibility curves. For each straight-line segment that he drew through his data, there was a corresponding Curie constant, which pointed to a particular magnetic state. Bends between two consecutive straight lines were not always clean, requiring a so-called transition zone between the two presumed states. Another complication was that measurements of magnetic moments on a given ion depended on the particular compound studied. Indeed, samples of a same substance having the same composition and even sometimes the same crystalline structure sometimes had different moments. After Foëx reported this observation, Weiss's group spoke of "magnetic varieties." In principle, they hoped not only to reinforce their proofs of existence of the magneton, but to use it to investigate the structure of solids.[100]

The Gadolinium Ion and Classical and Quantum Theories. Meanwhile, experiments were confirming most of the established theories more clearly than ever. As mentioned, Langevin had shown that in order to observe the beginning of saturation at ordinary temperatures, the field needed would be stronger than could be attained in the laboratory. Measurements by Kamerlingh Onnes, performed at very low temperatures on octohydrated gadolinium sulfate, for the first time verified the magnetization process Langevin had described.[101]

The agreement between experiment and theory was further improved when Langevin's function was replaced with a function obtained on the basis of quantum theory. The modifications due to the new theory essentially concerned the orientation of the elementary magnetic moment. In 1920, Wolfgang Pauli had shown that the initial susceptibility of a paramagnetic substance can be explained if one assumes there can be only a limited number of possible orientations for the elementary moment.[102] But at the time he made this proposal, the interpretation of the abnormal Zeeman effect was not accurate enough to give a correct choice for the number of possible orientations. But within a few years, the new quantum theory achieved spectacular success in explaining nearly every property of the isolated atom, including diamagnetism and paramagnetism; all that remained in doubt was what happened in substances where atoms were combined, as in a solid.

The rare earths offered a particularly complex example and became a test case for the quantum theory. In 1927, W. F. Giauque,[103] using a quantized empirical susceptibility formula developed two years earlier by Friedrich Hund,[104] plotted a theoretical magnetization curve for gadolinium in particular that fitted the experiments perfectly. Those like Weiss who held to basically classical ideas of magnetism could no longer find common ground with the creators of quantum theory, even though the theories came into agreement at high values of the quantum numbers.

Despite its accuracy, the study of gadolinium did not allow a final decision betwen the two viewpoints. The curves could be explained just as well in terms of classical elementary magnetons, which indeed seemed to give a more precise value of the moment. The defenders of the classical theory felt free to criticize the quantum theory, arguing that it wrongly assumed properties of extreme dilute substances—those studied by spectral analysis—in the dense substances that were the subject of magnetization measurements. True, the quantum theory was proving highly successful in the study of the paramagnetic salts of the rare earths. But to Weiss's group, the field of application of the quantum theory seemed to cover the rare-earth ions by chance because the magnetic layer was deep and thus, like the molecules of a gas, little affected by interactions with neighboring ions.

Further Experiments Between the Wars. Other work should be mentioned briefly at this point. Foëx started working again on paramagnetic substances, using the approach that Weiss had found so successful. He decided to study single crystals and chose an iron carbonate, siderite; with this, he became the first to obtain the thermal variations of the magnetization coefficients along each of a crystal's principal axes. The interpretation he gave these results, using nothing beyond the kinetic theory and the anisotropic molecular field, led him to conclude bluntly: "The result is that quanta did not appear to play any role in the magnetic phenomena as such".[105]

But other measurements promptly appeared to verify the results of quantum theory, even where the experiments had been motivated by considerations that were far from that theory. In particular, in the measurements made by Cabrera[106] and Stefan Meyer[107] on the dilute salts of rare-earth metals, Hund found verification of his quantum formula for susceptibility. There remained two ions, samarium and europium, that did not agree with Hund's theory. But in 1929, Van Vleck[108] predicted their thermal behavior by calculating second-order terms, and Cabrera[109] immediately confirmed the correctness of these views. Most physicists no longer doubted that the explanation of every magnetism phenomenon was to be found somewhere in quantum theory—an idea made still more plausible by Heisenberg's 1928 demonstration that even ferromagnetism and the molecular field itself could be explained as some sort of manifestation of quantum exchange forces (Chapter 2).

In Weiss's Strasbourg laboratory, the situation was vacillating, and opposing states of mind began to become evident. We may see this particularly by comparing work done there by Alice Serres and by Charles Sadron. We recall that supposed linear segments in the curve of $1/\chi$ with temperature for magnetite had been one of the arguments Weiss used in 1911 to justify the existence of the magneton. Subsequently, more precise measurements by Kopp on natural and artificial magnetites cast doubt on the idea that these straight-line segments existed.[110] What was clear was that the curve exhibited a marked concave shape toward the temperature axis, a curvature that remained unexplained. Serres hoped to cast light on the question by discriminating between the roles of ferrous iron and ferric iron.[111] She therefore started work on ferric oxide and a few ferrites—magnetites (normally Fe_2O_3–FeO) in which the normal ferrous iron is replaced by a bivalent ion of another metal M that may or may not belong to the magnetic group (Fe_2O_3–MO).

The thermomagnetic study of ferric oxide was nothing new. Following K. Honda and others, H. Forestier had shown that the stable variety of ferric oxide possesses a paramagnetism that is independent of temperature.[112] The constant value for susceptibility, verified by Serres over a temperature range going some 50° above the Curie point, enabled her to calculate quite precisely the susceptiblity attributed to the ferric iron ion. She used this result in an attempt to interpret the behavior of ferrites.

The curve of the thermal variation of $1/\chi$ for ferrites exhibits, like magnetite, a marked concavity toward the temperature axis.Serres postulated that a constant paramagnetism was superimposed on the ordinary paramagnetism and, moreover, that the paramagnetism of the ferric ion was the same in ordinary magnetite and in ferrites; she sought the underlying phenomenon by subtracting the constant she had measured from the curve measured by Kopp. The result seemed definitive: in her eyes, it was not a continuous curve but several connected straight-line segments that should be drawn through the data points (Fig. 6.4). Thus ferric oxide appeared to retain a fundamental paramagnetism even in ferrites, where it was added onto the ordinary paramagnetism.

Having thus restored the linear relationship, Serres could return to the classical theory to calculate the Curie constant and deduce the atomic moments. However, this time the moments did not turn out to be multiples of the magneton. And however attractive it might be, Serres's explanation could not be extended to the other ferrites. In the ferrite of zinc, for example, the variations were linear with no need to bring in corrections. Was it necessary to suppose that ferric oxide lost its paramagnetic character in this compound? The interpretation no longer seemed easy.

This remarkable experimental work was resolutely turned toward the past and Weiss's moribund magneton. Its interpretation was based on the "most overworked" formula for inverse susceptibility and otherwise on purely descriptive magnetism. Serre's subsequent labored attempts to explain her results in classical terms were an anachronism in 1932.

Nevertheless, the end of her paper showed the effects of new ideas from Néel, which we will address shortly, on the negative interactions between neighboring magnetic moments. She tried to link these ideas to her interpretation of ferrite behavior. Néel had felt it necessary to conclude that two neighboring moments, of unequal strength, could occupy antiparallel positions. Serres denied this, arguing that "if each [of the magnetic moments] is equal to a whole number of magnetons, that must likewise be the case for the difference between them. This is not what

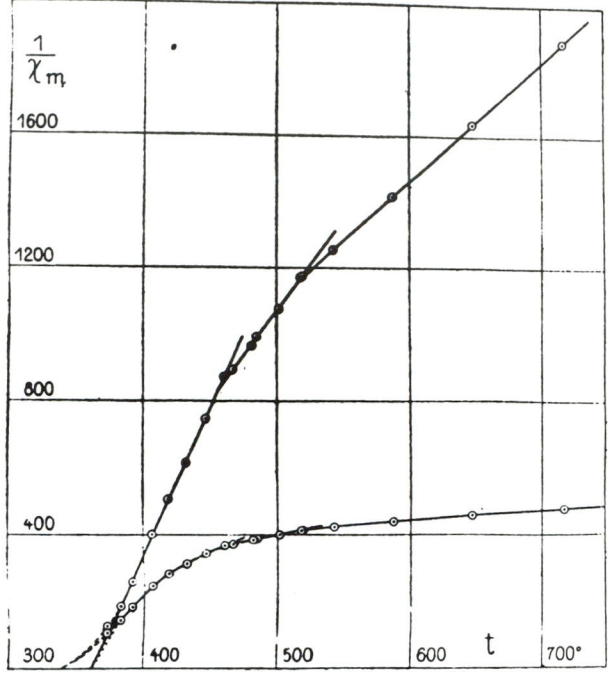

Fe²O³,MgO (2e échantillon).

Fig. 6.4 Variations of $1/\chi$ as a function of temperature in magnesium ferrite. *(below)* experimental curve; *(above)* result of correction attempted by Serres to recover Weiss's straight line segments. (From A. Serres, "Recherches sur les moments atomiques," *Annales de physique* 17 [1932]: 60)

experiment shows," as noted earlier.[113] She thus rejected with a stroke of the pen, in the name of the experimental magneton, an idea that would later be used, *mutatis mutandis,* to explain ferrimagnetism.

This school of magnetism died out with Serres. No further ideas emerged within the school, and there was no receptivity to ideas coming from outside. However, the researchers left a vast inheritance of accurate measurements that would prove highly useful to others.

We can observe a constrasting state of mind at Strasbourg in the work of Sadron.[114] His dissertation, published, like Serres's paper, in 1932, dealt with the magnetization of binary alloys of the iron group. The curve of the variation of the moment as a function of the concentration of dissolved metal at first is linear and then deviates from the straight line. To interpret this result, Sadron assumed that each atom of the same metal has a single moment, rather than a moment made up of a number of submultiples, as magneton theory would have it. The linear part of the curve corresponded simply to the ratio of the two different atoms in the mixture. He qualitatively explained the curvature for greater amounts of additive by invoking a "local field" created around an atom by its nearest neighbors. If the additional atoms took up random positions, at an early stage some of them could completely surround an atom of the solvent. Thus the solvent would be removed from influencing some of the atoms of its own type. If, in addition, the interaction between atoms of different types were negative, a deficit would be produced that would explain the observed direction of curvature. Using this model, Sadron not only calculated the fundamental magnetic moment of atoms, but also specified the sign of their magnetic interactions.

The origin of this work lay in earlier research on magnetic moments, motivated as always by the desire to improve the verifications of the existence of Weiss's magneton. This belonged to the past, but Sadron's approach was justified by new concepts that were being developed in the same laboratory by Néel. In his conclusion, Sadron gave the magneton little to stand on: "One cannot state with certainty whether a given new moment described in this work is or is not expressed by a whole number of experimental magnetons."[115] We shall leave Sadron at this point where he himself broke off. A Rockefeller Foundation grant took him temporarily away to the California Institute of Technology and, finally away from magnetism, into research on polymers.

The two papers we have just examined show the same reality, with its interpretation refused by one and accepted by the other. Although opposed in this, taken together they offered the first experimental demonstration of a negative coupling between two magnetic moments, which forced them to align parallel to each other but in opposite directions—what is now called ferrimagnetic coupling.

By introducing the local field, Néel had opened new vistas. Nobody joined him; he would inspect these vistas alone, with a precise and confident eye. Before discussing that work, we conclude the story of the Weiss magneton.

Epilogue to the History of the Magneton. We have organized the description of the main work from 1911 into the 1930s around the magneton theory because it was a pivot for much experimental research between the wars. It may seem surprising that this theory remained for so long in the balance; the reason was that its failures did not discourage its defenders. As early as 1913, Weiss had pointed out before anyone else that the moment of nitric oxide did not obey the rule of integral ratios.[116] and in 1918, following more accurate measurements, he had confirmed the exception and added the case of oxygen.[117] He flatly stated that the measurements he was reporting were "in contradiction with the magneton theory," but he added, "The theory otherwise rests on such an extensive set of consistent values that it does not seem justified to abandon it."[118]

This set of values was subjected to statistical analysis. Twice between 1927 and 1930, Cabrera surveyed the available findings, allocating each a weight of 1 to 3 according to the confidence that, in his opinion, it deserved.[119] Weiss also listed all the findings and then threw out those that he felt were not well established.[120] These Procrustean methods allowed both scientists to conclude that, as Cabrera put it, "the persistence with which the Weiss magneton appears in these cases as a natural unit of atomic moments cannot be attributed to pure chance."[121]

The Solvay Conference in 1930 was the first major occasion in which the whole edifice built up of straight-line segments, Cabrera's formula, and the experimental magneton was openly and severely criticized. Jakov Dorfman observed that the determination of the slopes of the line segments was sometimes inaccurate, that other curves could just as well have been drawn through the data points, and that, in the last analysis, the choice had not been made without hindsight.[122] Three years later, C. Gorter asserted that the straight lines owed their character to nothing more than an excessively restricted range of thermal variations.[123] It was from Weiss's own laboratory that the most serious blow against the magneton came. In 1934, André Lallemand measured the magnetic moments of ferric and manganous ions for which quantum theory predicted values that were not whole multiples of the Weiss magneton. His experimental results were incontrovertibly fractions, and the accuracy of the measurements precluded the sort of post hoc adjustments that had sometimes been done in Weiss's laboratory.[124] Néel, a witness to the event, recalled that on that day Weiss seemed to lose confidence that he would win his bet.[125]

The experimental magneton was the banner of a school that remained skeptical of quantum theory, a theory that to it seemed unable to move forward on the problem of magnetism in solids. Most other physicists were increasingly confident that ferromagnetism would soon yield all its secrets to the quantum theorists, and by the 1930s nearly everyone outside Weiss's and Cabrera's laboratories had lost interest in the magneton. After Weiss's death in 1940 it was, with minor exceptions, abandoned by all except Foëx. In 1951, in a revised edition of a book that he had coauthored with Weiss, Foëx let stand the following remark: "It is not certain that a more powerful theory . . . will not reveal, quite naturally, the role of the experimental magneton."[126] However, in 1962 when he proposed to the Paris Academy of Sciences the publication of Weiss's collective works to commemorate his centenary, Foëx wrote: "One might perhaps leave out those [works] that concern the magneton."[127] Whether in a complete or selected format, Weiss's works were never published. In this chapter, on the other hand, we have not shied away from this curious and instructive episode; with the magneton surgically removed, the history of magnetism would not have been properly told.

Magnetism as Seen by Néel

Interest in the Weiss magneton would have died away more quickly if its utility had been questioned further. Research in this field was limited to the questions of magnetic moments and of new saturation values, to obtain some experimental evidence for the magneton's existence, but results were obtained whose diversity took them beyond that particular question. It became impossible to believe that Weiss's hypothesis, even it were true, could clarify the whole array of evidence.

In examining the work of Alice Serres and of Charles Sadron, we have shown that their conclusions depended on the new ideas of Louis Néel. Between the rather stifling research on the magneton and the period when everything came under the sway of quantum mechanics, Néel opened a way in which the French school, after Weiss, could retain its personality.

Néel understood straight away that research at Strasbourg was deadlocked, and decided that the magneton would not be his problem. This was in 1928. Weiss, who was looking for a new assistant, had asked his colleagues at the Ecole Normale Supérieure in Paris to recommend a candidate. They recommended Néel, who was just finishing his studies. Néel accepted, at the same time distancing himself from Paris and specializing in magnetism. The consequences were lasting. Néel's entire career developed outside Paris, first in Strasbourg and then, after the 1940 armistice, in Grenoble. Aside from a visit to Clermont-Ferrand, where he tried his hand

Louis Eugène Felix Néel (b. 1904). (AIP
Niels Bohr Library, Weber Collection)

at geophysics, only to return after a few weeks to Strasbourg, his career would be
entirely devoted to magnetism.

One might, although at risk of making a derogatory simplification, bracket all
the theoretical research in Weiss's laboratory between 1908, when the molecular
field was discovered, and 1928, when Néel arrived. We have already mentioned the
immense amount of experimental work carried out during this period. Now we
wish to go back to the subject as it appeared in 1907, because at that point the history might have taken a different path.

On reading the admirable discussion that Weiss left as the conclusion of his article on the molecular field, one finds it rich with possiblities.[128] Puzzling over the
difficulty in understanding the incredible intensity of the molecular field, he wrote
that this must not be taken as an objection but as "pointing research toward new
hypotheses about the nature of the atom."[129] Of course, it was such a hypothesis
that Weiss made a little later in proposing the magneton. We emphasized in our
introduction that in the physics of magnetism it is dangerous to look only at the
atom, and more fruitful to consider the material as a whole. Weiss could still do
this. We may note his statement that "the hypothesis of the uniform nature of this
[molecular] field . . . is somewhat unrealistic . . . one could say that the uniformity
is only apparent and that it results as an *effect of averaging*" [italics added]. This
remark shows the path Néel would take, but Weiss turned away from it at once,
writing, "One can get along entirely without this argument by supposing that the
magnetic element is an infinitely small magnet since the couple exerted on it is the
same as if the field was uniform." Thus he concentrated again on the "magnetic
element." At the same time he abandoned the phenomena of hysteresis, writing
that "on the other hand, in Ewing's theory of magnetic cycles . . . the non-unifor-
mity of the magnetic field plays a central role."

This retreat shows Weiss's fatal choice, the decision to limit himself to a constant
field and give preference in his work to hypotheses on the nature of the atom.

From the outset, Néel was more fortunate in his choice. He was of course helped
by Weiss's mistakes, but he was also helped by his advice. It is interesting that, by
building on ideas that had not completely escaped Weiss and by carring out without
delay experiments that Weiss had only thought about, Néel acquired an overview
that would quickly become authoritative.

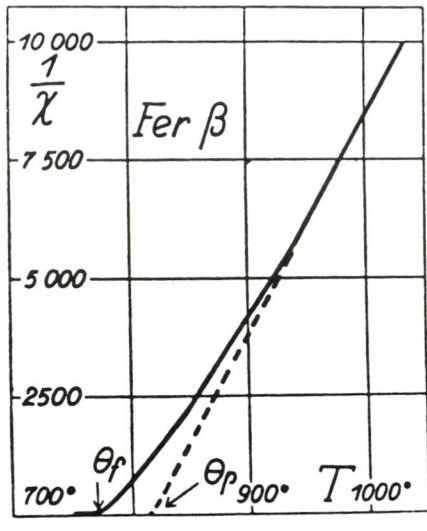

Fig. 6.5 Experimental curve giving the variation of $1/\chi$ for iron as a function of temperature. (From L. Néel, "Influence du champ moléculaire sur les propriétés magnétiques des corps," *Annales de physique* 18 [1932]: 98)

To see this, let us look again at the article that Weiss wrote in 1907. Here he noted that the presence of 2% of silicon in iron suppresses the gamma phase; from this fact he hoped to gain "some clarification of the nature of the molecular field."[130] This project lay dormant for two decades. Taking up now a paper by Néel, we find the following:

> The addition to iron of certain metals . . . such as silicon raises the transition temperature for beta to gamma and lowers that for gamma to delta, such that one needs only 2 percent of silicon to make the gamma phase completely disappear. . . . Professor Weiss has suggested using this fact to extend the study of paramagnetic properties of iron beyond 900°.[131]

What was only a sketch in 1907 was fully worked out in 1928. Now that we have brought these two dates together, we shall follow with Néel the theoretical development that might have come earlier.

Fluctuations in the Molecular Field.   Néel started with the curve representing the inverse of the susceptibility, $1/\chi$, as a function of temperature for various iron alloys. Since the work of R. Forrer, it was known that this curve is linear at temperatures sufficiently above the Curie point, but that it bends inward near this point. It is therefore convenient to define the paramagnetic Curie point θ_p given by the linear law and a ferromagnetic Curie point, noticeably lower, at which ferromagnetism appears (Fig. 6.5).[132] Weiss's theory does not make it possible to distinguish these two points. Since the latter described a ferromagnetic material as a paramagnetic material endowed with a uniform molecular field, Néel decided to explore the difference between theory and experiment by assuming a nonuniform field. He based his work on the fact that the interactions between the entities carrying the magnetic moments were short range; he was aware of Heisenberg's theory of short-range exchange interaction between nearest neighbors.[133]

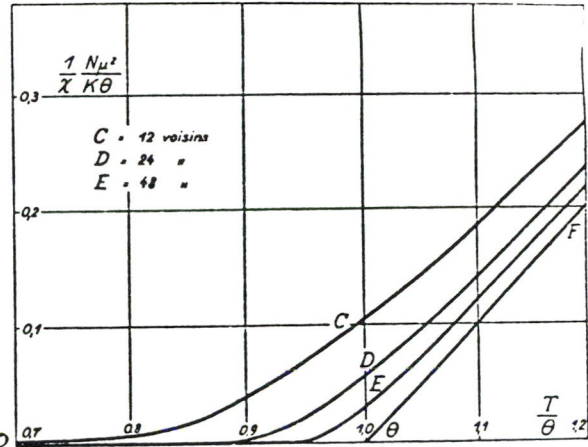

Fig. 6.6 Variation of the influence of the fluctuations of neighboring moments when one varies the number of neighbors used in the calculation of $1/\chi$. The curvature decreases when the number is increased; the broken line $00F$ represents the limit reached with Weiss's uniform molecular field. (From L. Néel, "Influence du champ moléculaire sur les propriétés magnétiques des corps," *Annales de physique* 18 [1932]: 35)

If he had not known this result, would he have begun in the same way? Asked this question, Néel replied, "Certainly."[134] One can easily agree by recalling that Weiss, confronted by various experimental results, concluded that the interaction between the elementary magnets fell off very quickly with distance d.[135] "If Weiss," Néel pointed out, "had thought out properly the consequences of an interaction of the type $1/d^6$, he would have realized that the molecular field could not be constant."[136] But Weiss at this time concentrated on another aspect of the problem— the origin of the molecular field. Néel saw things differently: "I am not making any hypothesis about the origin of the potential energy between two carriers [of magnetism]," he wrote in 1932; "I simply assume that it exists."[137] The nonuniformity of the molecular field was introduced by the separation of the environment of a carrier into two zones. The first zone is occupied by near neighbors. They exert a strong influence, and because of their small number their fluctuations have a large effect. The second zone contains more distant carriers of magnetism. Each one has only a weak influence, but because of their large number the total effect is not negligible. Here the fluctuations are unimportant, so Néel could describe their action by using the Weiss molecular field.

Néel then calculated, within the framework of statistical mechanics, the effect of the near neighbors. He showed that a modification of the classical theory due to the fluctuations could account for the two Curie temperatures (Fig. 6.6). He also predicted that the excess specific heat of a ferromagnetic material should persist above the ferromagnetic Curie point.[138] This latter prediction came before the corresponding observation; Weiss's co-workers, influenced by the theory of the molecular field, continued to find the discontinuity in the specific heat exactly at the Curie point.[139] Eventually, Néel was able to give the experimental proof of his prediction.[140] Also, through the fluctuations, Néel explained better than Weiss's theory the way in which spontaneous magnetization dropped as temperature rose toward the transition temperature. In the end, Néel could explain the behavior of ferromagnets over a range of 100° around the transition.[141]

This outcome convinced him that the right approach lay in defining a local molecular field, not necessarily the same in every point of the substance, since it should depend on the immediate environment of each point. This approach could be extended to substances containing several kinds of atoms. Néel thus introduced a binary alloy, consisting of atoms A and B, with three molecular field coefficients representing interactions between A and A, B and B, and A and B.[142] The action of the atoms that are neighbors to A is described by a local field h_A equal to the sum of terms proportional to the contributions to the magnetization J_A from atoms A and J_B from atoms B. Thus[143]

$$h_A = aJ_A + bJ_B$$
$$h_B = bJ_A + cJ_B$$

This model gave him results closer to experimental values for such alloys. It gave in particular a variation of $1/\chi$ that was hyperbolic with the temperature, in place of the sequence of straight-line segments.[144]

Moreover, this coupling between neighbors can be imagined so that the moments, instead of being aligned in the same sense, are aligned in the opposite sense. Néel proposed an example: in a body-centered cubic lattice, such a negative interaction would align moments located at the corners of cubes in one direction, and that at the center in the other. This lattice would comprise the superimposition of two simple cubic sublattices, with saturation magnetism in two directions, but as a whole magnetically neutral. Calculations showed that such a substance would have a susceptibility independent of the temperature. Néel attributed to such a mechanism the temperature-independent paramagnetism seen in manganese, whose order of magnitude was inexplicable in Pauli's theory.[145]

These ideas were not to the taste of the theorists. Néel remembers a 1929 visit to Strasbourg by C. J. Gorter, "a physicist much stronger than I was in quantum mechanics." Gorter said that Néel's ideas did not make much sense and were contrary to quantum mechanics. But Néel said to himself, "Thank God I do not know this quantum mechanics."[146]

Néel did not profess complete ignorance of quantum mechanics, for he used it where necessary to make corrections to his calculations.[147] But he would not be deflected from the personal approach that he had chosen since the beginning of his own work on magnetism. He continued to explore the subject, remaining faithful to his own insights, and in 1936 published a more complete study. In this, he showed that a material where the magnetic moments lay on two identical sublattices, giving overall neutrality, would lose this ordering at a higher temperature. The phenomenon would be analogous to the Curie point transition in a ferromagnet and would be accompanied by a discontinuity in the specific heat. Below the transition temperature, the susceptibility would be independent of the temperature; above, the moments would be disordered, and the susceptibility would follow the Curie–Weiss law.[148]

When Néel published his article, no known substance followed this behavior.

The phenomenon was observed two years later by H. Bizette in manganese monoxide (MnO).[149] In a theoretical study published in 1938, Francis Bitter proposed the name antiferromagnetism for this new phenomenon.[150] The hypothesis of two sublattices was confirmed in 1949, with the aid of neutron diffraction, by C. G. Shull and J. S. Smart in the same compound.[151]

Also in that year Néel suggested that the direction in which the magnetism of the two sublattices was aligned was determined by the crystal lattice. Taking into account a coupling energy between the magnetic moments and the lattice enabled him to account for a behavior very characteristic of an antiferromagnet: when the magnetic field is applied in the antiferromagnetic direction—the direction of the ordered moments—the susceptibility should remain zero until the field reaches a certain value. Then there should be a sudden decoupling between the moments and the lattice, and the susceptibility should take a finite value that does not change with the field.[152]

This discontinuity in the susceptibility was observed in 1952 by N. J. Poulis and G. E. G. Hardeman, two co-workers of Gorter at Leiden. The observation was made in $CuCl_2, 2H_2O$ using magnetic resonance. This material was the first insulator to show characteristic antiferromagnetic behavior. It was during this work that the authors first suggested what is now common practice, giving Néel's name to the temperature of the transition between the antiferromagnetic and the disordered states.[153]

Ferrimagnetism. The method used by Néel predicted the existence of the antiferromagnetic state before such materials had been observed. On the other hand, other magnetic substances, although thoroughly studied experimentally, continued to intrigue physicists. These were the ferrites, with the general formula Fe_2O_3MO, where M is a divalent metal. We have already taken note of these materials, which seem to have a ferromagnetic Curie point but differ from true ferromagnets in two ways. Below the Curie temperature, the magnetization is much weaker than we would expect from the magnitude of the elementary magnets, and at higher temperatures the susceptibility does not obey the Curie–Weiss law.[154]

The ferrites had an interest beyond fundamental research; in the 1930s they were a strong concern of the Philips Research Laboratories with, in particular, E. J. W. Verwey studying their chemistry and J. L. Snoek their physics. In 1936 the latter stated that he had produced a material that was an insulator for both electricity and ferromagnetism,[155] and this type of material was investigated for industrial electronic components. In this kind of research, the aim was not so much to obtain compounds that would turn out to have interesting properties as to prepare materials that would have desired properties specified in advance. Obviously, the key would be to relate the magnetic properties to the crystal structure. Consequently, these two workers tried to understand the chemical bonds and the distribution of ionic sites.

In 1947, Verwey published a paper that would be decisive for ferrite development. Until then, it had been thought that ferrites had the normal spinel structure, in which the sites occupied by the ions were divided into two classes. One comprised the tetrahedral sites A, each surrounded by four oxygen atoms; the other was the octahedral sites B, each surrounded by six oxygen atoms. A molecule of Fe^2O^3MO

includes one site A and two sites B. In normal spinels, the ions M^{2+} are on the sites A and the ions Fe^{3+} on the sites B. But there exist inverse spinels in which the sites A are all occupied by Fe^{3+} and the sites B occupied equally by M^{2+} and Fe^{3+}.

Through x-ray studies, Verwey showed that the ferrites are inverse spinels.[156] In 1948, Néel applied his model to this new structure and constructed a theory for the ferrites.[157] He used the "local field" in the same way as in 1932 for binary alloys, assuming that the interaction between ions A and B was dominant and antiferromagnetic.[158] In other words, he supposed that in a molecule the two ions Fe^{3+} neutralized each other and that the molecule gained its moment only from the ion M^{2+}. Observations of saturation of the magnetic moment fit the hypothesis perfectly, and three years later it was definitively confirmed by neutron diffraction.[159]

Néel then more generally proposed the term *ferrimagnet* for a material made up of two magnetized sublattices with spontaneous magnetizations of opposite direction and unequal intensity. An antiferromagnetic material is thus a ferrimagnet in which the opposed spontaneous moments cancel each other exactly.

As soon as it was understood, the theory of ferrimagnetism allowed rapid progress in the technology of ferrites. With the sought-for-link between crystallographic structure and magnetic properties understood, compounds with the desired properties could be manufactured (see the section "The War and Its Aftermath").

The clarifying power of Néel's concept of a local field can be appreciated further in connection with a phenomenon that Jean Becquerel in Paris, followed by W. J. de Haas in Leiden, had called metamagnetism. Becquerel had studied the magnetooptical properties of materials such as siderites; de Haas, the magnetism of the iron salts. The descriptions of their results were similar—an accord easy for them to establish, for Becquerel went regularly to Leiden to use the low-temperature facilities. Various explanations were attempted from 1939 on.[160] Becquerel wrote in his personal papers in 1944;

> I have been trying for four years to establish a theory of metamagnetism which I believe is caused by the formation, under the influence of an external field, of metastable states that remain in existence temporarily after the external field has been turned off. This is entirely different from ferromagnetism which is due to a magnetization on which the external field has no effect except to orient it.[161]

In fact, as Néel had shown at the Tenth Solvay Conference, these materials are antiferromagnetics that become ferromagnetic under the influence of a field, because the coupling between the two sublattices is weak.[162]

The Discovery of "Garnets." The theory of ferrimagnetism illuminated many unexplained observations on ferrites. In contrast, the theory of antiferromagnetism predicted a state of matter that had not yet been observed and had to wait two years for confirmation. We next describe the history of the "garnets," in which theory and experiment went together, theory interpreting or predicting observations by turns. The result was the discovery of substances whose industrial importance was greater even than that of the ferrites.

The genesis of these substances is rather murky. They first appeared as a source of strange properties in substances in which they lay concealed. Before they had been identified, Néel proposed an explanation for the observed phenomena; then

F. Bertaut discovered their existence and confirmed Néel's theory. After they had been obtained in the pure state, their thermomagnetic properties verified the consequences of the theory, and the whole was finally confirmed by neutron diffraction.

The history of the ferromagnetic garnets could be considered as beginning in 1950 when they arrived, undetected, in some new compounds. But their origin was earlier; it is part of a tradition. As we wish to show once again the permanence of many aspects in magnetism despite the upheavals in solid-state physics, we will return to 1928. In that year H. Forestier at Strasbourg developed a method for preparing ferrites.[163] The work of Alice Serres, already mentioned in another connection, was not entirely isolated, for Forestier was meanwhile studying the materials he produced. He referred to the work of this period when, 22 years later, he started work with G. Guiot-Guillain on new ferromagnetic compounds. The same method of preparation, applied to equimolecular mixtures of iron oxides with rare earths, provided some ferrites of a new kind, which they and they alone would study for three years.[164]

At the moment of their appearance in 1950, the five ferrites of rare earths (to which an yttrium ferrite was added) were announced as a new class of ferromagnets. However, some of their curious properties were not new; thus the ferrites of lanthanum and praseodymium were thermoremanent, a property that Forestier had described decades earlier after he observed it in the sesquioxide of iron.[165] There seemed to be no fundamental problems here. The thrust of Forestier and Guiot-Guillain's work was to extend the range of such products and to analyze them with x rays. The first results, published in 1951, showed that the ferrites of lanthanium and praseodymium have cubic structures of the perovskite type.[166]

The situation became more thoroughly complicated in 1952 when, from the group that includes, besides the ferrite of yttrium, nine other rare-earth ferrites, the two scientists isolated five materials that had the unprecedented property of *two* Curie point temperatures: "we are confronted by the new experimental fact that there are compounds that show, as the temperature is varied, two ferromagnetic transitions."[167]

One year later, Guiot-Guillain made a further remark: the transition temperatures associated with thermoremanence lie on a smooth curve when plotted against the ionic radius of the rare earth.[168] This observation reminded him of certain correlations between ferromagnetic properties and the spacing of the magnetic layers of atoms that Slater[169] and Néel[170] had noted previously. But at the end of 1953, no theoretical explanation had yet been attempted. In the face of the disconcerting observations, people referred to known theories in a conversational tone, without getting to the heart of the matter. One spoke of the two sublattices introduced by Néel[171] and mentioned casually the noncollinear structures recently proposed by Y. Yafet and Charles Kittel.[172]

It was at this point that work on the problem was no longer confined to Strasbourg. The fact that not only the two Curie points, but experimental features generally, were without theoretical support led the Grenoble laboratory to take up the rare-earth ferrites.

R. Pauthenet began by carrying out a detailed study of the magnetization σ of a gadolinium ferrite prepared in Grenoble.[173] Helped no doubt by earlier investigations on a single crystal of hematite, where the same combination of ferro- and paramagnetism appeared,[174] he quickly obtained an experimental law that clarified the problem: the isothermal specific magnetization varies linearly with the applied field ($\sigma = \sigma_0 + \chi H$). This showed that a paramagnetic behavior was superimposed on a ferromagnetic one. But the problem was not much simplified, for a group of remarkable new effects was revealed in studying the variations of the susceptibility χ and the spontaneous magnetization σ_0 with temperature. The most unexpected result was that the spontaneous magnetization first disappeared at a temperature that had not been suspected in Strasbourg, to reappear and disappear twice more at the temperatures they had noted, so thus there existed *three* Curie points.

Pauthenet's note was submitted to the Paris Academy of Sciences by Néel in July 1954. In another note sent in the same day, Néel interpreted his collaborator's results. Néel's idea was not far from thoughts he had derived from the study of hematite. He had been unclear about the origin of the ferromagnetism of this substance, but he could not exclude the possibility that there was "an imperfect cancellation of the two sub-lattices, that is, a kind of ferrimagnetism, related for example to a stoichiometric defect."[175] He took up again his model of equivalent ferric ions arranged on two sublattices to form a ferrimagnetic structure. To these two sublattices he added a third with paramagnetic behavior, formed by gadolinium ions, and proposed that a negative interaction coupled this sublattice to the ferrimagnetic structure.

This model did not explain all the observations, but neither did it contradict any of them. It accounted above all for the three Curie points. The two highest are those that the ferrimagnetic lattice shows under certain conditions.[176] Since the weakest of these is a temperature of inversion rather than a transition, it is clear that no thermoremanence should show itself; on the other hand, when this effect appears, it is at the highest temperature, which according to this explanation should correspond to a transition. The fact that this temperature varies uniformly with the radium of the rare-earth ions is not explained by the model but does not contradict it, because the effect should depend on a property of the lattice rather than on the nature of the ions.

As for the third Curie point discovered by Pauthenet, it would be a consequence of a weak coupling, which Néel thought to be negative, between the ion sublattice and the gadolinium. The paramagnetic sublattice of gadolinium would take a direction opposite to that of the ion that produced it. As the temperature dropped, their magnetizations would become equal, so the third point would be that at which the two opposed moments exactly cancelled each other.

Although he did not claim that this could prove at one stroke his three bold hypotheses, Néel noted that the meaning of the third point could be tested by experiment. He suggested the means: "a remanent magnetization previously given to the ferrite should change its direction, in the absence of a field, when the temperature crosses θ_3." At the moment when Néel was preparing this note, Guiot-Guillain and Forestier stated that they had been able to verify his insight for the ferrites of dysprosium and erbium.[177] They communicated their findings to the academy the next week. Everyone might have been satisfied, but Néel pressed on by further questioning the theory. Why do the ferric ions form a weakly ferrimagnetic sublattice, when it would seem natural for reasons of symmetry for them to take up a perfect antiferromagnetic arrangement?

A reply came 18 months later from Bertaut.[178] He looked at the problem as a crystallographer and as a chemist, although in both capacities he was guided by Néel's ideas. The only structural studies available at that time were those of Forestier, who had found that the ferrites of lanthanum and praseodymium have the perovskite structure. It was also known that some deformed perovskites crystallize in a group in which the iron occupies centers of symmetry. If this was the case for the ferrites of the rare earths, a perfect antiferromagnetic structure would seem inevitable and Néel's theory would be challenged.

Bertaut chose to cast doubt on the structure rather than on the theory. The fact that the ferrites of gadolinium studied at Strasbourg and those studied at Grenoble did not have the same spontaneous magnetization led him to suspect the presence of an impurity. He knew also, from Debye–Scherrer diagrams, that an increase in the proportion of iron would decrease the amount of perovskite in favor of another phase richer in iron than in gadolinium. The substance that since 1950 had been thought to be equimolecular could then have the properties of some unknown compound. Bertaut isolated this compound, gave it the formula ($5Fe_2O_3$, $3Gd_2O_3$) and stated in addition that it was an isotype of the garnets. The lack of symmetry of the two classes of site for the ferric ions was proved. The ferromagnetic garnets could now begin their remarkable career. Bertaut remembers, "This was the Christmas present for 1955 that we offered to Monsieur Néel."[179]

Pauthenet then studied separately the ferrites of gadolinium of the perovskite and garnet types.[180] The remarkable three Curie points found in the earlier work were explained: the two highest corresponded to the Curie points of garnet and perovskite, and the third to the inversion point belonging to garnet, in accord with Néel's theory. Each specimen owed its behavior to the proportions of the various phases it contained. A study of the structure completed the task. This study was carried out at the French nuclear research laboratories at Saclay in 1950 using neutron diffraction—the first time the technique was applied in France.[181] The study was carried out on the yttrium garnet, since the sublattice of yttrium ions, which were not magnetic, would not complicate the neutron analysis. It was less important to determine the structure of this sublattice than that of the ferric ions, the object of Néel's boldest hypothesis. The spectrum obtained showed that the coupling between the two iron sublattices is indeed antiparallel, as Néel had predicted.

The discovery of the garnets was particulary fruitful because these materials contain only trivalent ions and therefore are better insulators than the ferrites. They are better in all ways when one needs a magnetic compound for high-frequency electronic components. The laboratory at Grenoble did not patent them and gained nothing directly from the profits they brought to the electronics industry.

The discovery marks the end of the period when the bulk of magnetism research could be the responsibility of one man or a very small team. At this time, Néel ceased to concern himself with interpreting magnetism within the little group that he had until then inspired. Techniques belonging to numerous specialties were henceforth assembled around a given project. We have seen how, in the discovery of the garnets, chemists, crystallographers, and experimental and theoretical physicists worked in close collaboration. "We only found the garnets." Néel said, "because we had Bertaut, who was an excellent crystallographer."[182] Since the day in 1949 when Néel had shown Bertaut the article by Shull and Smart on the antiferromagnetism of manganese oxide, Bertaut knew the irreplaceable benefits to be obtained from neutron spectroscopy. To prepare for this, in 1953 he arranged an extended visit to the Brookhaven Laboratory in the United States.

To the technical and personal qualities of his co-workers, Néel added his vision of a strategy of research. He saw in the installation of a nuclear reactor the way to organize a research complex of "critical size." A recommendation from the French government to locate a new center of nuclear research in the provinces rather than in Paris led the French Atomic Energy Commission to choose Grenoble, where the reactor Mélusine was put in service in 1960. Néel was put in charge of the new center. He later remarked that this "did me a bad turn, because under these conditions I had to drop magnetism."[183] That was in 1956; the discovery of the garnets marked the end of his research career.

In conclusion we must emphasize that the work of Weiss, Néel, and their collaborators represents only one line of development in classical magnetism, although arguably the most important line. Here, as elsewhere in this book, in order to present a compact and unified story we have followed only selected developments, leaving aside much important research. We have even left it for future historians to describe important research in other areas of magnetism by Néel himself, as well as by many other physicists.[184]

6.2 The Genesis of a Modern Magnetics Community

Atomic Physics and Magnetic Properties

Modern theories of magnetism have roots in two distinct traditions of theoretical development. One of these traditions is represented by the phenomenological theories of Langevin and Weiss, described already. These theories were closely related to the bulk properties of magnetic materials, were able to account for known experimental results, and appeared self-consistent. At the same time, however, they contained what were initially speculative and even mysterious elements, such as the nature of the Weiss internal molecular field and the postulated existence of microscopically ordered domains. The key to understanding these obscure concepts was to come from quite independent developments in atomic physics where an understanding of magnetic phenomena at the atomic level was not aimed at the more limited quest to understand the properties of magnetic materials, but formed a crucial but integral aspect of the reworking of a new world view in physics.[185]

In 1913, Niels Bohr introduced the important concept of the quantization of angular momentum as part of his fundamental work on the structure of atoms. Although Bohr gave some thought to the implications of his general ideas for the theory of magnetism, he did not get as far as producing anything for publication.[186] Progress in this direction was left to others. Quantization of electron orbits implied the existence of an elementary magnetic moment, the "Bohr magneton." Sommerfeld and Debye extended Bohr's theory to the motion of electrons in space and showed that in accordance with quantum conditions, the elementary magnetic moments can assume only certain particular orientations in relation to the direction of the magnetic field.[187] Known as space quantization, this phenomenon was confirmed by Stern and Gerlach with a molecular beam experiment in 1921 and immediately provided some explanation for the Zeeman effect—the splitting of spectroscopic lines in magnetic fields.[188] Thus it was that during the early 1920s attempts to produce classification schemes for atomic structure from spectroscopic data were intimately linked to calculations of atomic magnetic moments. Furthermore, by knowing the energy distribution of possible states of the atom and with the resolved magnetic moments associated with them, it became possible, following a suggestion of Sommerfeld, to calculate values for magnetic susceptibility that could then be compared with observed experimental values.[189] The paramagnetic gases proved readily amenable to such a treatment, along with salts of the rare-earth and iron groups, which could be reasonably dealt with using a theory based on isolated atoms.

The subsequent development of a magnetic quantum theory of paramagnetism during the 1920s using first, with reasonable success, the Bohr-Sommerfeld scheme, and later, more satisfyingly, the new quantum mechanics, rested firmly within the context of atomic physics; for many physicists, a lifelong interest in the quantum theory of magnetism began with such problems. Prominent among them were the American physicists John H. Van Vleck and John Slater, Edmund Stoner in England, and the Dutch theoretician Hendrik Kramers. Stoner, for instance, in a retrospective account of this career, considered that his initial approach to magnetism was dominated by the idea that through the study of magnetic properties and behavior it might be possible to find out more about the structure of atoms.[190] Such a view is readily confirmed by his first book on magnetism, published in 1926.[191] In that year, the anonymous reviewer of work by Weiss and Foëx saw the study of the magnetic properties of materials leading "inevitably to the problem of the constitution of the atom."[192] Similarly, Peter Kapitza's interest in magnetism in the same period was motivated by a self-confessed belief that it would "open up a new way by which to approach the study of the atom," as spectroscopic information became exhausted.[193] There was also the double attraction that magnetism was a somewhat neglected theoretical field with a wealth of accumulated experimental data, although as Stoner later noted the situation was changing rapidly, and very soon much more was known about the structure of the atom than about the atomic structure of ordinary pieces of matter. Within a few years, the principal task became that of finding out more about the ordinary-scale magnetic properties of matter, and then trying to account for them in terms of the known properties of atoms.[194]

This reorientation from questions couched firmly within the framework of atomic physics toward a more general consideration of the bulk properties of matter, interpreted in terms of atomic structure, is well illustrated by comparing Stoner's first text on magnetism, previously referred to, with his book *Magnetism and Matter,* published in 1934. The latter book emphasizes the general problem of the magnetic properties of matter rather than the atomic structure per se, and a consideration of the properties of *solids* assumes a greater prominence. It is significant that this effective redefinition of a subject area had been anticipated from within an industrial context. In 1927, Karl Darrow, a senior scientist at Bell Labs, perceptively remarked:

> Some of the difficulties of ferromagnetism may be peculiar to it. Others, it is to be feared, are examples of the troubles which are reserved for scientists by the internal

properties of solid bodies generally, and which physicists will some day be forced to confront when the obvious problems of gases and free atoms are exhausted, if they are not sooner invited by curiosity or by the requirements of engineering.[195]

Ferromagnetism was proving a particularly intractable phenomenon to understand. Incidentally, Stoner's first interest in this branch of magnetism was for the same reason as his interest in paramagnetic salts—to attempt to calculate consistent values for atomic magnetic moments. Too little was understood about ferromagnetism to be able to draw any reliable conclusions, however. Although the old Bohr–Sommerfeld theory allowed a reasonable approach to paramagnetism, ferromagnetism remained somewhat mysterious at the atomic level at the time of Darrow's review. There were, however, certain clues. It had been suspected since the early 1920s that ferromagnetic behavior was linked to the concept of the spinning electron, implying that the electron possessed an intrinsic magnetic moment.[196] This view became fully acceptable when, in 1925, Uhlenbeck and Goudsmit demonstrated that the multiple structure of spectral lines could be explained on the assumption of electron spin.[197] Although this allowed a number of puzzling anomalies in atomic physics to be resolved, including gyromagnetic effects and the anomalous Zeeman effect, at face value the concept of the spinning electron offered little to the puzzle of ferromagnetism. Stoner, for instance, considered that the magnetic properties were "little elucidated" by the spinning electron "hypothesis."[198] His natural caution did not imply a rejection of the concept however. On the contrary, as he explained, "if the hypothesis of the spinning electron is necessary to account for atomic properties and if a new formal type of mechanics (i.e. quantum mechanics) is necessary to describe them briefly and coherently, both will be equally necessary in considering molecular and crystal properties."[199]

Such a project was already coming to fruition in Germany, and in 1928 Heisenberg showed how the previously obscure Weiss molecular field could be attributed to a quantum-mechanical exchange effect, in which forces of interaction between neighboring atoms give rise to a coupling between unpaired spinning electron.[200] The stable state on this view would correspond to the fact that at least a portion of the spins were parallel, representing spontaneous magnetization. Expressions he derived for the Curie temperature and the temperature variation of magnetization from a treatment of the exchange interaction agreed to a first approximation with the Weiss theory. At a more detailed level, there was more cause for concern. Although a quantitative explanation could be given for the ferromagnetism of iron, cobalt, and nickel, it was not immediately clear why other elements, such as manganese (which was manifestly not ferromagnetic), did not display ferromagnetic properties. It was not apparent which electrons in atoms were responsible for the exchange effect. Perhaps more significantly, for the vast majority of specialists working on magnetism, predominantly experimentalists, the quantum-mechanical foundations of the theory were as mysterious as the origins of the Weiss molecular field it sought to explain. For the day-to-day working practices of most of the experimentalists, it made not the slightest bit of difference that by the time of the Solvay Conference in 1930 the quantum theorists had transformed the picture of magnetic phenomena at the atomic level. What was more significant was that a difficult and, at times, mysterious subject had returned to the mainstream of interest in physics.

During a short period, covering roughly the years 1928 to 1933, it is possible to detect an unprecedented level of interest among physicists in general magnetic phenomena. This is indicated by a dramatic increase in scientific publications on various aspects of magnetism, and the occurrence of major meetings devoted to presenting results in the area. By studying the number of papers listed in *Physics Abstracts,* one can show that a steady increase of published papers on magnetism throughout the 1920s suddenly accelerated between 1928 and 1931, during which time the published output doubled, not matched by any increase in the general level of physics literature. Compared with only a handful of published papers appearing on magnetism each year in the early 1920s, *Physics Abstracts* for the year 1933 lists over 200 papers published on magnetism in the previous year. Although no country dominated this literature, the bulk of published material originated in France, Great Britain, the Netherlands, Germany, the United States, Japan, and the Soviet Union.

Perhaps the most intriguing aspect of this dramatic expansion is that it was not maintained, with the level of publication actually falling after 1933, although remaining in excess of the level of activity before 1928. Individual countries are more difficult to analyze on this basis. However, a count of papers indexed under the title "Magnetic Properties" in the *Physical Review,* a journal that should be a reasonable indicator of the level of scientific activity in the United States for the period under study, shows a pattern strikingly similar to that identified earlier. During the early 1920s, there appears to have been little basic research on magnetism in the United States. A gradual increase led to an identical doubling of published papers on magnetism between 1929 and 1931, reaching a peak in 1932, and thereafter declining for the remainder of the pre-war period.

The high relative incidence of publication in the field of magnetism in the early 1930s is further highlighted by the appearance of two major textbooks devoted to the subject, Stoner's *Magnetism and Matter* (1934) and Van Vleck's *Theory of Electric and Magnetic Susceptibilities* (1932).[201] Both are considered classic texts in modern magnetism. Evidence for an upsurge of interest in magnetism is further indicated by the occurrence of three major meetings having an international participation that met to discuss various aspects of magnetism. One was the Solvay Conference of 1930, and the other two were held respectively in London the same year and in Leipzig in 1932. Apparently no other gatherings of equivalent prestige met to discuss magnetism at any time before 1939.

At the Solvay Conference in Brussels in October 1930, the presentations reflected predominantly the recent achievements in atomic physics.[202] According to Mehra's retrospective observation, it was an opportune time to review a field that represented "a new frontier of research in physics," ripe for discussion.[203] Yet curiously, in view of this opinion, of the total of eight authors who presented papers at the two meetings held in 1930 from a quantum-theoretical perspective, only two would make any notable further contribution to the theory of magnetism. They were Stoner and Van Vleck. It is perhaps a more difficult question to ask why these two did continue to make fundamental contributions to magnetic theory than to ask why others turned to other problems in physics. Yet understanding this difficulty provides certain clues to the genesis of solid-state physics. One should first recall Stoner's remark, about how the situation was changing from one in which a knowledge of magnetic properties was useful in understanding atomic structure to one in which advances in atomic physics were coming to a stage where the reverse was true. For Stoner and Van Vleck, it seemed natural to follow this transition through; others were looking to new challenges and were even antagonistic to solid-state problems. Pauli is an often-quoted example. Pauli was invited to give a comprehensive review of the quantum theory of magnetism at the Solvay Conference. His own major contribution to magnetism was a single paper of fundamental importance on the temperature-independent paramagnetism of free electrons, essentially a contribution to ideas having a bearing on the quantum-mechanical treatment of the statistics of gases.[204] By discussing the application of Fermi-Dirac statistics to an electron gas, representing a rough model for the electrons in a metal, his primary object was not to discuss the properties of metals, but to construct a test case in quantum mechanics (Chapter 2). In fact, his contemporary Casimir has pointed out, the physics of solids was a subject for which Pauli had a profound dislike.[205]

Pauli, like many of his contemporaries in physics, was mainly interested in what really appeared as fundamental problems having a bearing on all of physics. In 1926, the year Pauli wrote his paper on paramagnetism, the theoretical framework of quantum theory was rapidly nearing completion, and as Felix Bloch recalled of the time, "almost any problem that had been tossed around years before could now be reopened and made amenable to a consistent treatment."[206] It was in such a context that magnetic phenomena held an attraction, albeit fleeting, for many. Their interest usually went no further than the qualitative demonstration that such phenomena were theoretically consistent and understandable within the new framework provided by quantum physics. As such, magnetism was not so much a new frontier for the quantum theorists as a staging post. At the 1930 Solvay Conference, Enrico Fermi was already on a new mount and indicating new directions for research. His discussion topic was the magnetic moments of nuclei. Understanding the nucleus would soon be the new challenge in theoretical physics, the new frontier, and the obvious subject for discussion at the next Solvay Conference. Magnetism held a more lasting attraction for only a few. Yet for those it was to be a route into solid-state physics.

From Atomic Physics to the Physics of Solids: The Examples of Van Vleck and Stoner

Although it may seem somewhat arbitrary to choose only two physicists to illustrate a shift in context from atomic physics to the physics of solids, in truth very few made such a shift in emphasis, and even fewer chose to specialize in the quantum theory of magnetism. From this standpoint, Van Vleck and Stoner were virtually unique. It is therefore of some interest to attempt to understand how both moved in this direction at a time when many of their contemporaries were attracted to theoretical nuclear physics. It also provides an excuse for a more detailed examination of two of the more prominent personalities of modern magnetism.

One is naturally curious to divine any elements of commonality to suggest why the two should have followed similar interests. Initially, however, more striking is the tremendous contrast between them, in both their family backgrounds and general personalities. Van Vleck was born into a family with a long and distinguished

American-based tradition.[207] In choosing a university career, he followed his father, who was a celebrated professor of mathematics, and befitting this continuity he carried a deep affection for traditional American college life, reveling in college football and songs and the easygoing atmosphere of academic work. He became noted for his convivial informality, was universally known as "Van" by his colleagues, and by the 1930s was a familiar face at some of the more important European conferences. His distinctive voice became equally familiar. Of a big international gathering at Strasbourg in 1939, Stoner privately wrote that "Van Vleck is the great talker of the Conference."[208]

Stoner was of a more reserved disposition, although this did not prevent him from voicing his own views strongly on subjects close to his interests.[209] For him, the achievement of a position of academic eminence did not come easily. Carpenters and small shopkeepers were more representative of his ancestry, and his father's work as a professional club cricketer brought inevitable financial insecurity for the Stoner family and an itinerant childhood. Academic advancement was made possible only by his ability to gain scholarships. As a student at the Cavendish Laboratory, he was plagued by ill health. Diabetes, in particular, seriously threatened to end his career in physics. The introduction of insulin treatment in 1927 brought crucial relief. Partly as a consequence of his diabetic condition, however, and the demands of controlling it, Stoner remained reluctant to travel and rarely ventured abroad.[210] It was perhaps also a consequence of his background that he developed a very deep sense of personal, even moral, responsibility that manifested itself in both his personal and professional relationships. He set high standards for himself and expected them from others, and could become emotionally distraught if he felt these standards were not being met. Temperamentally, he appears to have been very different from Van Vleck.

Yet despite such obvious difference, one senses certain similarities. One detects in both a certain conservatism, a respect for stability and tradition, and a fascination for complex yet ordered systems. Van Vleck had a passion for railways and delighted in railway timetables. Stoner's writings, scientific and administrative, public and private, show a particular emphasis on attention to detail, the compilation of lists and tables, and manual computation.

It is when the early careers of both these physicists are examined that the greatest similarities appear. Compared with most physicists who came to be associated with a special interest in the physics of solids, Van Vleck and Stoner were slightly older than the rest. For both, research began in the early 1920s. Van Vleck carried for life the intense pride of being one of the first "all-American" theoretical physicists.[211] In contrast, Stoner came to specialize as a theorist only after beginning his career as an experimentalist, having had no formal mathematical training. Both worked in an environment less than supportive to theoretical physics as a specialized vocation. Both the United States and England at the time were dominated by strong experimental traditions. Rutherford's Cavendish was not a natural location for a young scientist to develop an interest in theoretical physics, and for Stoner the push in that more specialized direction came purely by chance as a result of his poor health. In 1922, as a postgraduate student, he began work on an experimental study of x-ray absorption, stimulated by recent theoretical studies of atomic structure. A period of forced hospitalization, however, soon intervened; prevented from continuing with his experimental work, he began a more contemplative study of the physics literature.[212] In doing so, he became particularly familiar with the nomenclature of x-ray and optical levels and their quantum specifications in a way that would not have been possible but for this illness.

One evening, in May 1924, a particular distribution scheme for the electrons in atoms suddenly occurred to him as a possibility. He could simply relate the number of electrons occupying full energy levels to the quantum numbers specifying them, making a significant improvement on the rather arbitrary and unsatisfactory distribution schemes previously suggested. With the encouragement of R. H. Fowler, Stoner published his results in a paper that provided an important incentive for Pauli to publish a distribution scheme the following year in a more axiomatic and general form.[213] Stoner had essentially used inductive methods to correlate experimental data into a convincing theoretical scheme. In doing so, he made explicit reference to what would be more famously known as the Pauli exclusion principle.

Stoner's interest in atomic structure then led in quite a natural way to an interest in magnetic phenomena. For among the evidence directly relevant to the distribution of electrons in the atom was a sequence of values for the magnetic moments of ions of elements of the first transition series, as derived from the paramagnetic susceptibilities of their salts. By way of this introduction to the literature on magnetism, he realized that a vast field for the exploration of the relations between the magnetic properties of ordinary pieces of matter and atomic structure lay open to him. Kapitza, a close colleague at the Cavendish with similiar interests, shared and encouraged Stoner's growing concern for general magnetic properties. When Stoner moved to Leeds in 1924 as a newly appointed lecturer, he was resolved to write a text that could correlate a diverse literature on magnetism with ideas then current in atomic physics. The fruits of this work were published in 1926 as *Magnetism and Atomic Structure* and had a warm reception. Publication unfortunately coincided with the emergence of quantum mechanics, thus calling for an immediate revision of the book in the light of the new theoretical ideas. In the meantime, Stoner began interpreting the diamagnetic susceptibility of atoms and the paramagnetism of ions using a quantum-mechanical approach.[214] By the time Stoner turned to reworking his book, general interest rested less on atomic structure and more on the use of fundamental concepts to explore the bulk properties of matter. This was reflected even by the title of his new book, *Magnetism and Matter*, one of the first books that was pertinent, if not wholly devoted, to aspects of the solid state of matter. By this time, Stoner's own research interests had moved on to the ferromagnetism of metals and alloys, an area that would dominate the rest of his research career.

Van Vleck was by then also well established, primarily through his application of quantum mechanics to the theory of paramagnetic susceptibility. Compared with Stoner, who from the beginning of his work on magnetism was conscious of a more ambitious design, Van Vleck became a specialist in the subject almost incidentally, as a self-confessed pragmatist and opportunist.[215] Like Stoner, he began his theoretical work in the early 1920s, when as a student at Harvard he began research under the supervision of E. C. Kemble. Theoretical physics labored under the unsatisfactory scheme provided by the old quantum theory, and with many of his generation in physics Van Vleck experienced frustration, near misses, and missed opportunities. As he later observed, "We didn't know the rules of the game, so about all anybody could do was flounder if they wanted to write a paper and be a theoretical physicist."[216]

For a while, indeed, Van Vleck appears to have floundered, until quantum mechanics provided a new rule book and more promising opportunities. It was as a newly co-opted referee for the *Physical Review* that Van Vleck recognized such an opportunity when he received advanced notice of a paper by David Dennison. He realized that a computational technique that Dennison had used could be applied to derive the dielectric constant for a simple diatomic molecule and restore the classical value of ⅓ in the expression for susceptibility, which had mysteriously and nonsensically disappeared using the old quantum theory. Van Vleck's exhilaration at this success turned to frustration when, following a delay over publication, he found himself in practically a tie in publishing the result with three other papers that appeared in the summer of 1926.[217] Understandably disappointed, during the following winter he began to consider applying the same method to magnetism and succeeded in explaining the observed paramagnetic susceptibility of nitric oxide, another outstanding puzzle. Suitably encouraged, he pressed on to work on the general structure of the expressions for electric and magnetic susceptibilities, in terms of quantum mechanics, culminating in the publication of his widely acclaimed text on the subject in 1932.[218]

Stoner was full of admiration for Van Vleck's new book.[219] Since Van Vleck was one of the few people working in a field of his own interest, perhaps this was not surprising. More significantly, Van Vleck appeared to share his own methodological approach, and he was not alone in recognizing this similarity. As another reviewer pointed out, Van Vleck's book had the same aim as Stoner's work—to correlate and explain the great amount of experimental results in magnetism from the standpoint of the quantum theory.[220] If there was one obvious difference between the two, it was that Stoner repeatedly referred to his views on the nature of physical theory, whereas Van Vleck simply pressed on, seemingly less preoccupied with the general basis for his approach to theoretical physics.[221] Later in life, however, he was more than once to offer the view that his approach may have been unconsciously influenced by the operational philosophy of Percy Bridgman.[222]

Bridgman's views on science were expressed in his published works.[223] During his student days at Harvard, Van Vleck had been impressed by Bridgman's lectures, impressed enough to recall that the underlying approach might have had a lasting influence on his own research.[224] Incidentally, Van Vleck's first work on magnetism was as an undergraduate under Bridgman's supervision. Certainly his doctoral supervisor, Kemble, was a self-confessed advocate of the Bridgman philosophy.[225] It was a view, empirical in characters, in which theory found meaning only in terms of physical operations. As Bridgman himself noted, his method was suggested by the clear recognition that "the ultimately important thing about any theory is what it actually does."[226] Theory on this view becomes in compact and manageable form what is empirically known. In other words, theory organizes and makes understandable diverse operational data.

Stoner would have had a deep sympathy for such a view. His stated interest "was and remained very largely in the theoretical interpretation of experimental findings."[227] The publication of Van Vleck's book and his own review of it gave him an opportunity to dwell on the nature and role of physical theory, a preoccupation that, as already noted, recurs in his publications. He approvingly quoted Van Vleck's comment that "a theory is most 'physical' when it permits the calculation of a large number of experimentally observable quantities in terms of a few fundamental postulates."[228]

Stoner's attitude was that theorists had almost a moral responsibility to relate their work to experimental findings. Van Vleck's book had further attributes that appealed to him, in particular "a welcome absence of dogmatism and of that suggestion of finality which is often irritatingly conveyed in theoretical work."[229] What exactly Stoner had in mind here we do not know. Certainly, quantum mechanics was being seen by many as a fait accompli. As Dirac had suggested in 1929, with perhaps a hint of smug satisfaction,

> The general theory of quantum mechanics is now almost complete. . . . the underlying physical laws necessary for the mathematical theory of a large part of physics and the whole of chemistry are thus completely known, and the difficulty is only that the exact application of these laws leads to equations much too complicated to be soluble.[230]

This was, as many physicists have since discovered, by no means a trivial difficulty. The complexities of the n-body problem, of which ferromagnetism was one aspect, were so great that for only the very simplest molecule was it possible to integrate the Schrödinger equation with any real quantitative accuracy. It was little wonder that Pauli wanted nothing to do with the messy problems of the solid state.

Progress was dependent on the use of approximations and what Stoner liked to call "imaginative leaps." He thought of theory building as akin to erecting scaffolding, a temporary structure giving way later to a more permanent edifice.[231] Van Vleck shared such a view. Referring in this particular instance to the quantum theory of valence, he criticized the "pessimist" who demanded rigorous postulational theory and calculations, "devoid of any questionable approximations or of empirical appeal to known facts."[232] The "optimist," by contrast, was satisfied with approximate solutions and an appeal to experiment, thus providing "an excellent 'steer' and a very good idea of 'how things go,' permitting the systematization and understanding of what would otherwise be a maze of experimental data codified by purely empirical . . . rules."[233]

It was this concern for the systematization of experimental data that brought both Stoner and Van Vleck inevitably to take account of the properties of real solids, since they sought to demonstrate the general applicability of atomic physics to the bulk properties of matter. The simple problems dealt with gases. At the start of his work on magnetic susceptibilities, Van Vleck had intended to consider only gaseous media. Because the number of paramagnetic gases was so limited, he was inevitably drawn to the conclusion that any treatment of magnetism not applicable to solids was unfruitful.[234] Darrow's prophecy that physicists would have to confront the problems of solids after they had dealt with the similar cases was becoming a reality. Both Stoner and Van Vleck were drawn eventually to the problem of ferromagnetism, a subject on which there was to be serious disagreement between them regarding the nature of the magnetic carriers—whether they were localized or itinerant electrons. By the late 1940s, the formative context for their work on magnetism had been left far behind. Indeed, Stoner suggested that the theory of ferromagnetism on which he was then working should be seen "as part of the general theory of the solid state,"[235] an implicit recognition of the identifiable status of solid-state physics as a coherent area of theoretical research. During the 1930s, however, both Stoner and Van Vleck were working on problems that only with hindsight can be identified as part of present-day solid-state physics. For solids were not their only concern. Rather, an interest in them formed an integral part of an interest in applying atomic physics to the general properties of matter, although their work became more and more directed to matter in the solid state.

Both Van Vleck and Stoner established reputations internationally as authorities on the theory of magnetism in matter. One of Van Vleck's biographers has even gone as far as to call him the "father of modern magnetism."[236] In a field of such diverse characteristics, however, establishing paternity can be problematic. In the decade before the Second World War, Pierre Weiss was arguably the "godfather" of a field still dominated by experimentalists, many of whom, as Van Vleck became all too aware, had little understanding of and even less sympathy for quantum mechanics. The continuing influence of Weiss and a strong experimental tradition is well illustrated by the struggle to oppose the acceptance of the Weiss magneton, described earlier. Van Vleck found this a frustrating experience. To be invited to the 1930 Solvay Conference as the only American in attendance was flattering for the young theorist. Yet as he lamented years later, "Considerable time there was spent by the experimentalists in discussing evidence for the Weiss Magneton which of course the theorists knew was spurious. However, in deference they courteously refrained from making strongly critical remarks."[237]

The proceedings of the conference show that Van Vleck was highly involved in the discussion of the papers presented. Significantly, his memories of the event were dominated by the more ceremonial aspects of the occasion: the gala social event at which Dirac warned him against practicing his French on the king of Belgium; Einstein and the queen playing a violin duet after dinner. He had no memory of any important scientific exchange at the conference. Yet, in contrast, he had vivid recollections of his discussion with his contemporary, the Dutch theorist Kramers, on the sand dunes of Holland that same year.[238] The work of the young Hans Bethe on group theory was obviously a more interesting topic of conversation.

Bethe himself had a fleeting interest in magnetism, stimulated in part by his contact with Kapitza during a study period in Cambridge in the autumn of 1930, as well as by his general interest in the electron theory of metals. Indeed, electronic structure was so crucial to developing a quantum theory of ferromagnetism that the interests of a number of other theorists, concerned more with the general theory of solids, and the electron theory of metals in particular, became linked with ferromagnetism during the 1930s. They included Nevill Mott, John Slater, and the Soviet theorist Sergei Vonsovskii.[239] This was an important overlap. First because it encouraged theorists with more general interests to consider magnetic phenomena, and second because it brought magnetism into the contextual framework of the quantum theory of solids and thus into the general cognitive framework of a formative solid-state physics. This important relationship between the electron theory of metals and ferromagnetism guided Stoner's own ideas on developing a more satisfactory theory to explain ferromagnetic behavior in metals and alloys, and from 1930 onward his work was predominantly concerned with various aspects of ferromagnetism in metals. For Van Vleck, however, his work on paramagnetism provided the guiding light for his future research activities. This brings us back to his conversation with Kramers on the Dutch sand dunes. The background to this needs a brief explanation.

Usually one would not attempt to apply to solids a physical theory developed primarily for gases. Yet in 1925, using just such a theory, and a few months before the advent of quantum mechanics, the German physicist Hund showed that, except for two cases, he could successfully explain the observed susceptibilities of the rare-earth ions.[240] Four years later, Van Vleck and Amelia Frank demonstrated how the two recalcitrant ions, samarium and europium, could be brought into line. Marked deviations from the predicted behavior still existed, however, and the salts of the iron group also proved difficult to explain using this theory. Stoner had suggested that interaction effects between adjacent ions needed to be considered to improve the theory, a possibility that occurred to others independently, including Van Vleck.[241] What was required was some computational method for dealing with isolated atoms that could take account of electrostatic influences from surrounding ions. Bethe's paper on group theory, developed further by Kramers, provided such a technique, and using it Van Vleck was able to compute the effects of interaction on magnetic susceptibility. In collaboration with his students, Van Vleck was thus able to play a leading role in the application of what became known as ligand field theory. The significance of this little episode is not simply to show how Van Vleck's general interest in crystalline solids developed in the context of the theory of paramagnetism, for it has wider significance in the context of the emerging shape of a general magnetics community. The theory of paramagnetism held a strategic position with respect to two important areas of research, both essentially part of the framework of low-temperature physics that emerged in the 1930s: magnetic relaxation studies, and the introduction of magnetic cooling methods for producing very low temperatures.

The association of work on paramagnetism with low-temperature studies was not in itself a new development. Other workers, most notably Foëx at Strasbourg and Jackson in Leiden, had previously investigated the temperature dependence of paramagnetic susceptibility over ranges extending into the lower regions of temperature.[242] Using temperatures down to liquid hydrogen, they had noted deviations from the theoretical predictions then current. At even lower temperatures, Kamerlingh Onnes and Stet Woltjer had investigated the saturation effects predicted by Langevin in gadolinium sulfate.[243] Despite these precedents, it was the work of Van Vleck that provided the motivation for a rejuvenation of work on paramagnetism at low temperatures, led by the Dutch experimentalist C. J. Gorter. In the late 1920s at Leiden, Gorter began a study of the paramagnetic salts of the iron group and the rare earths.[244] Following the acceptance of his dissertation in 1932, at the suggestion of Ehrenfest he began to look for relaxation effects—that is, the mechanisms involved in the reestablishment of equilibrium when magnetization is altered by a changing magnetic field. As early as 1920, Ehrenfest had urged that this phenomenon would be best demonstrated at very low temperatures using rapidly alternating fields. Although Kamerlingh Onnes was unsuccessful in detecting such effects, Gorter took up the challenge and became the first to convincingly demonstrate the existence of paramagnetic relaxation.

In turn, this experimental work gave rise to a flowering of new theoretical work in this area, particularly in Holland itself, and in general it opened up a whole field of research devoted to the numerous frequency-dependent effects associated with the magnetic properties of solids.[245] In the longer term, these relaxation studies provided a crucial background to the discovery and application of resonance phenomena after the war. In turn, research techniques based on these phenomena would provide fundamental information on the behavior of solids at the atomic level. In the shorter term, during the 1930s the subject represented yet another new aspect of interest in magnetic phenomena.

Another fundamental development in low-temperature research in the 1930s should also be mentioned. In 1933 the chemist William Giauque, with the aid of a student, D. P. Macdougall, successfully demonstrated at the low-temperature laboratory at Berkeley the magnetic cooling method known as adiabatic demagnetization, an effect that Giauque had predicted.[246] Shortly after, the method was independently demonstrated by de Haas and Wiersma at Leiden, and was extensively developed at the Clarendon Laboratory at Oxford.[247] It brought within reach previously unexplored parts of the temperature scale. With the method based on the demagnetization of paramagnetic salts, any ultimate limit of cooling was dependent on the magnetic properties of the coolant. Theoretical and experimental studies of paramagnetism suddenly had a whole new context. Any physical theory that does not move forward becomes stagnant, and interest in it fades. Van Vleck had intermittent fears that this might happen in the theory of magnetism.[248] In the field of paramagnetism at least, the impetus provided by the developments in low-temperature physics ensured that this did not happen.

The continuing interest of theorists, including Van Vleck, Stoner, Kramers, Mott, and Slater, in various aspects of magnetic phenomena and the important context of low-temperature work provides some evidence for the growth of interest in magnetism in the period under study. It by no means gives a complete explanation, however. The low-temperature context was primarily relevant to paramagnetism. Moreover, the interest shown in magnetism by the quantum theorists is insufficient to explain the growth in work on ferromagnetism reflected in the literature. It also does not provide a convincing explanation for the rise and relative decline of interest in the early 1930s. To do this calls for a review of an area of magnetism that broadly falls under a category involved with the technical aspects of the magnetization curve. As we shall see, this area of study had its own renaissance in the late 1920s.

The Technological Dimension

Despite the undoubted advance in understanding magnetic phenomena on the atomic scale that came with the application of quantum mechanics, in practical terms quantum mechanics offered nothing immediate for the technologist interested in the magnetic properties of materials for commercial development and application. The fabrication of new materials remained an essentially empirical exercise, metallurgically oriented toward the production of materials with certain magnetic characteristics, resulting from the variation of the relative proportion of the constituent elements and the treatments involved in the production process.

The continual expansion and diversification of the market for magnetic materials, however, provides a crucial background for the development of ideas in magnetism, a context that we will now consider in greater detail.

The growth in the commercial importance of magnetic materials was closely associated initially with the rise and diversification of the electrical industries, beginning with power generation and distribution, although in more recent history applications in information storage and transmission have become equally, or even more, significant.[249] The initial utilization of dynamo-electrical machinery in the nineteenth century provided little incentive for improving the performance of magnetic materials. Soft iron with the necessary properties of high permeability (i.e., easily magnetized and demagnetized), suitable for such applications, was readily available and caused little trouble in a technology where mechanical inefficiency dominated any other sources of energy losses, including those associated with the magnetization process. Engineering design predominated. Incentives to improve on existing materials came with the introduction of alternating current power transmission, which provided a market for high-grade magnetic materials. The crucial factor behind this incentive for improvement was the efficiency of transformers. Because a transformer is essentially a closed magnetic system with no moving parts, its efficiency depends crucially on the magnetic properties of the materials used in its construction.

The most significant commercial development before 1930 involving the utilization of new materials based on scientific research was the introduction of silicon–iron during the first decade of the present century. If national claims for priority in commercial application of silicon–iron are uncertain, what is beyond question is the critical role of the preceding metallurgical researches carried out independently by Robert Hadfield, at his father's steelworks in Sheffield, and by Ernst Gumlich, at the Physikalisch-Technische Reichsanstalt in Germany.[250] The economic benefits were immediate and considerable. One estimate valued the savings on electrical energy that resulted in the United States from the utilization of the new alloy at $10 million for the year 1910 alone, with savings of $340 million worldwide between 1903 and 1920.[251]

Incremental improvements came as a result of further research by Gumlich, and independently by Yensen at Westinghouse, on the effects of impurities. Not until 1935 did any further major technical breakthrough of commercial importance relevant to silicon–iron occur. In that year, it was discovered that a combination of cold rolling and annealing resulted in much enhanced improvements in commercially available silicon–iron. Norman Goss, an American industrial metallurgist, is credited with this discovery, although he completely misinterpreted his x-ray data and provided an erroneous explanation for the phenomenon.[252] A correct interpretation was made possible only by the then recent developments in understanding the relationships between magnetization phenomena and crystal structure. Thus the observed improvement could be traced to a reorientation of the crystal grains aligning along an "easy" direction of rolling. Such anisotropic behavior (i.e., the existence of preferred directions of magnetization) had been confirmed by experiments using large single crystals, which became available only in the 1920s, particularly through the work of Honda and Kaya in Japan.

If the market for soft magnetic materials was dominated by silicon–iron, it also left room for more specialized applications in which particularly enhanced characteristics such as very high permeability were desirable. It was this potential economic role for specialized alloys that provided the stimulus for wide-ranging research on different alloy systems, research that was at least initially dominated by a metallurgical approach. One notable area of research concerned the study of high-permeability alloys based primarily on the iron-nickel system. Although these alloys had been extensively studied in the nineteenth century, it was Gustav Elmen at the laboratories of Western Electric, precursor to Bell Labs, who established their commercial potential for both power applications and the loading coils used in long-distance telephonic transmission.[253]

The research successes of one laboratory often stimulated further efforts elsewhere, either to evade existing patents or to modify such materials for alternative uses. At Westinghouse, for instance, Yensen's investigations led to a variant of Elmen's Permalloy, marketed under the name Hypernik, a material with lower permeability but higher saturation and resistivity, and more suited for power applications. Significant improvements came through the introduction of such materials. In transformers, the use of these iron–nickel alloys decreased their size by a factor of about two-thirds while considerably increasing performance. The use of Permalloy in the continuously loaded transatlantic cable laid in 1924 increased capacity from 250 to 1900 letters per minute.[254]

If some of the new materials did not create any immediate commercial demand, their unusual magnetic properties often stimulated wide scientific interest. A good example was the Heusler alloys, ferromagnetic alloys consisting of a particular structural arrangement of nonferromagnetic elements, which attracted interest simply because they were unusual. Such novelty focused attention on the link between internal structure and magnetic behavior. Intellectual puzzles, generated by the quest to relate magnetic properties to atomic and crystal structure, were not entirely divorced from the longer-term promise of establishing materials fabrication on a more detailed understanding of the physics of solids, particularly given the increasing propensity of industry to support basic research studies in magnetism. At Western Electric, the incentive to widen magnetic researches and undertake more basic studies came directly from the commercial success of Permalloy, and had the support of senior research administrators, including Harold Arnold, the first director of Bell Telephone Laboratories.[255] A small permanent research group was formed to pursue problems initiated by the work of Elmen and Arnold on Permalloy,[256] with an approach firmly based on the study of the relationship between structure and magnetic behavior. L. W. McKeehan, an x-ray diffraction specialist, was given charge of the group, and staff were recruited from other areas, including P. P. Cioffi, who had hitherto been working on vacuum tubes. When McKeehan left a few years later to take up a professorship at Yale, his place was taken by another convert to basic research in magnetism, Richard Bozorth, like McKeehan trained as an x-ray specialist. Elmen meanwhile continued to lead a larger group concentrating on the development of new materials.

By the late 1920s, Bell Labs was not alone in promotiong basic research, as other industrial laboratories sought to supplement their metallurgically oriented research

Richard M. Bozorth (1896–1981) and John
H. Van Vleck (1899–1980). (AIP Niels Bohr
Library, Physics Today Collection)

with more fundamental studies of ferromagnetism. Prominent among them in the
United States were General Electric and Westinghouse. For all three of these major
industrial companies, the period from 1928 to 1933 was outstandingly productive
for their scientific contributions to the field of magnetism. Consider, for example,
Francis Bitter, an ambitious young physicist appointed to work for Westinghouse
in 1930, specifically to research the field of ferromagnetism. In the two years before
he joined the company, he had published four principal papers dealing with magnetic susceptibility. In the following three years with Westinghouse, he published
16 papers dealing with a wide variety of problems related to his research on magnetic materials.[257] Prominent among them was the publication of his introduction
of the powder pattern method for visualizing magnetic domain structure.[258] The
same period saw the publication of internationally acclaimed work by Bozorth on
the Barkhausen effect, and a series of papers by two General Electric researchers,
Sixtus and Tonks, based on their study of large regions of reversed magnetizations.[259]

The general strength of industrially based magnetics research in the United States
during the early 1930s is illustrated by a major symposium on the subject of ferromagnetism held at Schenectady, New York, home of the General Electric
Research Laboratories, in 1931.[260] The organizing committee for this gathering
consisted of Bitter, Bozorth, Tonks, and McKeehan, with industrial scientists overwhelmingly represented in presentations given by Bozorth, Cioffi, Sixtus and
Tonks, Bitter, and Yensen. Academic contributions came from McKeehan, then
at Yale, and from Shirley Quimby, professor of physics at Columbia.

This remarkable flowering of industrially based research in magnetism came, significantly, at a time of intense industrial depression. Industrial laboratories were
certainly not immune to the depressed state of industry. The prevailing economic
conditions may, however, have actually stimulated basic research in some of the
laboratories or at least increased the propensity of industrial scientists to publish in
scientific journals. Laboratories were reluctant to lose their more senior research
staff, and, as has been suggested elsewhere, the recurrent temporary layoffs offered

more time for private study. At the same time, the threat of loss of employment stimulated publication as a means of defending scientific reputations and thereby increasing personal competitiveness and the prospects of staying in a post.[261] For others, active publication was a means of attracting academic attention, thus increasing their chances of university posts. Bitter is a case in point. Disallusioned by deteriorating working conditions at Westinghouse, Bitter actively attempted to bring academic attention to his scientific work.[262] As part of this, he initiated the organization for the Schenectady symposium with the express hope that the resulting visibility might result in an academic appointment. His strategy worked. An offer came from Karl Compton to move to MIT. Following a year supported by a Guggenheim Fellowship in England, Bitter took up his new academic post in 1934 and thereafter concentrated his efforts on the development of a new field—the design and building of big magnets for physical research.

At Bell Labs the Depression curtailed any expansion of research staff, and the layoffs and salary cuts began to bite. Bozorth took up the clarinet to occupy his enforced leisure time.[263] He also began writing more scientific papers. In 1932, when the layoffs were at their worst, he published more papers than he had in the five years from 1925 to 1930. Of these new papers, two related to the tail end of his work on the Barkhausen effect, and two others related to research unconnected to magnetism, using research results as much as six years old. As evidence of his increased leisure time and the curtailment of the experimental program, another paper was a lengthy theoretical treatment of anisotropy. In 1933, Bozorth did not publish anything. Not until 1938 did the budget of Bell Labs return to its 1930 level.[264] By this time a new research director, Mervin Kelly, was instituting changes that would have a far-reaching effect on the future contribution of Bell Labs to magnetism. In an effort to bring a new dimension to work on solids generally, young theorists proficient in quantum mechanics were recruited.[265] The impact of this on the magnetics research of the laboratory came later, but it was indicative of the arrival of a new generation. At the same time, the previously separate groups under Bozorth and Elmen were organizationally merged, with Bozorth taking the leading role in the new group's direction, as Elmen settled into more of a backseat position as his career drew to a close. As Bozorth later put it, "He'd usually sit there and smoke his cigar with his back to the door, and that's about all for a long time. Just worn out."[266]

If such personal reasons lay behind the reorganization, it was nevertheless symbolic of the way in which the previous metallurgical emphasis on the development of new magnetic materials was giving way before a more general context that acknowledged the role of basic physics. It was Bozorth who correctly interpreted the behavior of the so-called oriented-silicon steels invented by Goss, and who had an important role in instigating the introduction of magnetic annealing in the manufacture of magnetic materials in general.[267]

Concurrent with Bitter's departure from Westinghouse, the fruitful collaboration between Sixtus and Tonks at General Electric also came to an end. Tonks turned to his original interest in vacuum-tube technology, and although he retained an interest in magnetism, Sixtus returned to his native Germany in 1938 to work for the electrical firm AEG, one of several European-based firms that began to follow the American lead by hiring physicists to do basic research in magnetism. Siemens

Halske, for instance, decided to set up a magnetics laboratory. Here, amid modest facilities, a nucleus for research was formed by two physicists already established in magnetics research. One, Franz Preisach, was well known for his work in producing rectangular hysteresis loops from strained metal wires at Barkhausen's laboratory in Dresden. The other, Martin Kersten, was introduced to magnetism through Richard Becker's newfound interest in the subject at the Technische Hochschule in Berlin.[268] In the same period, a decision made by the senior research management at Philips Laboratories in Eindhoven initiated research on fundamental aspects of the solid state. Magnetism, a field in which the firm had strong commercial interests, was an obvious subject for study, and work began on the search for new magnetic materials.[269]

The demand for new and improved magnetic materials was a direct result of the relentless growth and diversification of the electrical industries and an ever-widening array of specialized applications. Permanent magnets in particular, a relatively neglected research field before 1930, were becoming commercially important. During the initial period of growth of the electrical industries, permanent magnets served few useful purposes, were expensive to produce, and had poor performance characteristics. This was no longer true by the 1930s. In Britain alone, the annual consumption of complete magnets and magnet parts for permanent-magnet moving-coil loudspeakers was estimated to be 1 million units and rising.[270] This was not the only application. They were extensively used in electric meters, telephone and communication apparatus, various indicating instruments, and magnetos, all areas of rapid commercial growth. Dramatic improvements in performance had been obtained principally through the efforts of a well-established research tradition in Japan, based on the use of alloying additions to carbon steels.[271] Such materials were commercially dominant until around 1934, when, primarily as a result of a discovery of Mishima in Japan three years previously, the whole outlook on permanent-magnet materials changed.

As with many of the early advances in magnetic materials, the circumstances surrounding Mishima's discovery were somewhat fortuitious, embedded in a wholly empirical approach to studying the properties of metals.[272] Mishima was experimenting with three-component (ternary) alloys using nickel and iron with additions of a third variable element. When he used aluminum in certain compositions, he found that his test pieces displayed abnormally high mechanical hardness and brittleness. Because his study included an approximate examination of magnetic properties, he noticed that these alloys possessed an exceptionally large value of coercive force, an important property for permanent magnets. Stimulated by Mishima's discovery and related work elsewhere, others were soon alive to the merits of carbon-free alloys. In 1933, Honda and his collaborators produced an alloy containing titanium, which showed the most outstanding technical characteristics of any permanent-magnet material yet known. Technical factors alone, however, did not guarantee commercial success.

The most successful permanent-magnet materials were developed from the addition of cobalt and copper to the Mishima ternary alloy, producing a generic series of commercial alloys known as the Alnicos. Commercial variants were patented almost simultaneously around 1934 in the United States (W. E. Ruder at General Electric), Britain (G.D.L. Horseborough and F. W. Tetley at Swift-Levick, Sheffield), and Germany (at Krupp). The technical improvement of these alloys came at a fast pace. At Jessop's, a Sheffield steel firm, D. A. Oliver and J. W. Sheddon, acting on a typically intuitive suggestion from Lawrence Bragg, attempted to assess the effect of magnetic annealing on Alnico. Until then, this method of improving the technical characteristics of magnetic materials had been limited to high-permeability alloys, such as Permalloy. The results on Alnico showed a disappointingly small improvement, and, with little commercial promise, Oliver and Sheddon published their results, noting in passing the "possible technological value."[273]

Oliver and Sheddon's results were picked up by researchers at Philips, working as part of the new magnetics research program. By increasing the cobalt content, the Philips scientists found that subsequent magnetic annealing produced a material far in advance of the performance of existing Alnico-type materials. The resulting commercial product, marketed under the trade name Ticonal, more than justified the research effort behind it and threatened to take the complete market. Philips's European competitors were under an immediate threat to come up with an alternative that could compete with Ticonal. The Sheffield manufacturers, acting together through the Permanent Magnet Association, a body set up initially as a price cartel, responded by attempting to develop a similar alloy outside the Philips patent specification. The immediate result of this initiative was a material called Alcomax, in competition with Ticonal; Alcomax was the single driving force for the Permanent Magnet Association to initiate moves during the early years of the Second World War to set up a central research laboratory in Sheffield. The Alcomax problem demonstrated convincingly that with so many variables now involved in the commercial production of magnetic materials, the days of struggling along with what one physicist referred to as "cookery book metallurgy" were over.[274]

From the physicist's point of view, the problem of improving the performance of magnetic materials was no less daunting. As Arthur Tyndall, professor of physics at Bristol, confided to his erstwhile colleague, the new professor of physics at Sheffield, Willie Sucksmith: "The Alcomax problem sounds dreadful. . . . I should visualize Jessop's getting impatient with lack of results after 12 months and yet except for a lucky chance the problem might take years to solve."[275]

The research program at the new Permanent Magnet Association's laboratory did eventually produce a satisfactory successor to the original Alcomax. The conclusion to this story will not concern us here. We are more interested in demonstrating how a succession of new magnet alloys, produced in a competitive economic environment, was making the traditional empirical approach to new alloy development less and less satisfactory. The complexity of the various factors contributing to the properties of these alloys demanded a more fundamental approach. Yet the physicists were still stuggling to fully understand the new materials. As Kurt Hoselitz, the young refugee physicist appointed in 1941 to tackle the Alcomax problem, later recalled: "We did not even understand what was the cause of the high coercive force in these alloys, nor could we in any way understand why the magnetic field acting for a comparatively short time during cooling had such a profound effect. The materials were a real puzzle."[276]

In general, by the end of the 1930s any understanding of the processes occurring during magnetization was fairly simple and accepted as crude and imperfect. Nevertheless, there were grounds for some optimism. Considerable progress had been made during the preceding decade toward at least a qualitative understanding, and the subject of ferromagnetism was showing all the indications of maturity. As Bozorth summed up in evidence for this:

> For the first time a plausible story can be told concerning the ultimate magnetic particles, the essential nature of the atom of a ferromagnetic substance, the kind of forces which determine the properties of magnetic crystals, the effect of strain on magnetic materials, and the manner in which these various phenomena combine to determine the properties of commercial materials. It is true that the story is largely qualitative and that there are still many points that are uncertain or missing entirely, but nevertheless it is possible to describe the major features with some confidence.[277]

By the time of Bozorth's assessment, the concept of magnetic domains was accepted, although the theory had yet to be fully formed. The intrinsic magnetization of metals and alloys was broadly understood. The magnetization curve had been extensively studied, particularly by Becker's school in Berlin, resulting in a fairly basic model based on the consideration of internal strains and stresses. A powerful stimulant to a continuing research program and supported by various experimental results, this model was still an unsatisfactory basis for explaining the properties of the new high-coercive alloys along with the complicated properties of the Permalloy group.

If not fully convincing, the work of Becker and many of his colleagues and compatriots in Germany provided a crucial intellectual boost to the study of ferromagnetism generally during the 1930s. Links between such properties as magnetostriction, anisotropy, and stress and the effects on the behavior of materials were revealed by a variety of experiments, beginning in the 1920s, made possible by the recent availability of large single-metal crystals. Perhaps most influential were Preisach's doctoral studies in Dresden, during which he succeeded in producing rectangular hysteresis loops when magnetizing metal wires subjected to stress.[278] This was a convincing demonstration of irreversible discontinuities in magnetization, produced by induced internal strains, and it caught the imagination of physicists. This work, including subsequent discussions with Preisach, appears to be the major cause of Becker's sudden new interest in magnetism in the late 1920s.[279] It was also the incentive for the work of Sixtus and Tonks at General Electric, where Langmuir suspected some kind of magnetic nuclei to be acting as an origin for the large Backhausen effects produced by Preisach.[280]

In what can be described as an intellectual avalanche came a rush of new ideas and models to explain the internal mechanisms giving rise to the shape of various part of magnetization curves, particularly concerning the process by which magnetization is reversed. Prominent on the theoretical side were Heisenberg, Becker, and Richard Gans at Königsberg, as well as the Soviet theorist Nikolai Akulov. Bloch introduced the concept of elementary magnetic excitations known as spin waves, which, in turn, led him to introduce the idea of domain boundaries, often referred to as Bloch walls. Bloch's work came as a direct development of Heisenberg's quantum-mechanical treatment of ferromagnetism and was also deeply influenced by the current discussions in Germany concerning general aspects of magnetization.[281] For a brief period before political events began to intervene, there was an intense social interaction between the theorists to develop ideas in magnetism.

During the same period, for the experimentalists it was a case of making hay while the sun shone, as is indicated by the activity of Gerlach. Gerlach retained an interest in magnetism throughout his research career, and during the 1920s he published nine papers on the subject.[282] In 1930 alone, however, he published seven papers on magnetism, with around 30 papers appearing in the decade before the war, dealing mainly with the physics of magnetization curves. Another important site for experimental work was Berlin, where Becker promoted a succession of young experimentalists working on magnetism.

From about 1929 to 1933, the force of numbers working on magnetism in Germany stimulated and maintained a rich social interaction. Meetings for discussions, small and large, formal and informal, were commonplace. Becker's student Kersten recalls, for instance, the unusual setting for his first meeting with Bloch. Acting in the junior capacity of helmsman in Becker's small jollyboat on the lakes west of Berlin, he kept the craft stable as Becker and Bloch got caught up in intensive discussion.[283] The year was 1932, and Bloch was again invited to explain his ideas in yet another unusual setting.[284] This time, a larger gathering was formed in response to a light-hearted priority conflict between Becker and Heisenberg. The matter was to be settled by a water-polo match at a rustic intermediary setting near Bad Duben. Becker arrived by car, supported by the Siemens trio of Kersten, Preisach, and Schottky. Heisenberg arrived with 15 cycling colleagues from Leipzig. With swimming limited to a muddy marshy pool, scientific discussion dominated the meeting. Schottky improvised coffee for the thirsty throng in an old enamel can found at a local farm.

Not all meetings on magnetism in Germany were held in such primitive conditions. In Febraury 1933 a more formal meeting convened in Leipzig to discuss magnetism at what was to be the last of the Debye colloquia. For many of the participants, it turned out to be the last opportunity to discuss the subject in Germany. Within months, the worsening political situation brought traumatic upheaval for many of the country's scientists. Bloch was one of the refugee physicists who left Germany, never to return. Preisach, a Hungarian Jew, lost his job at Seimens. Other casualties followed as German physics became a political battleground. In 1938 Gans, noted for his theoretical work in magnetism, was dismissed from his post at Königsberg to find only temporary employment with AEG. Others were affected more indirectly by the political situation. Both Gerlach and Becker were drawn into the protracted and sensitive negotiations that were involved in new academic appointments.[285] Of course, Germany was not the only country in which politics made wounds in the scientific community. When the Fascists took power in Spain, Cabrera, influential in developing physics in general and magnetics research in particular in that country, sought refuge in France. Kapitza, a great stimulus to English interest in magnetism, was another famous victim of political intervention when he was recalled to the Soviet Union and kept there. In Germany, research in magnetism did continue, although on the theoretical side it was virtually destroyed. Most were aware that a golden age had passed.

For a brief interlude, the initiative in the search for a fundamental understanding of the behavior of magnetic materials had moved from France to Germany. By the end of the 1930s, this was no longer true, and the future for many specialists in magnetism, in common with a great mass of humanity, looked decidedly uncertain. At a time when political tensions were stretched to the limit and beginning to break like the bonds of a shearing metal, magnetism as a field of intellectual study was assuming a state of maturity. With the benefit of hindsight, it is clear that despite dramatic developments in various areas of magnetics research that came in the postwar period, none of the developments provided a significant contradiction to the fundamental concepts already established as a result of prewar efforts. One contemporary indication of a field assuming a degree of maturity was a sudden rush of textbooks in all aspects of magnetism, which came in the late 1930s.[286] Finally, and as it turned out deeply symbolically, there was a major conference in Strasbourg.

Réunion sur le Magnétisme: Strasbourg, May 1939

Strasbourg, the traditional commercial and intellectual focus for the Alsace region, lies picturesquely at the junction of the Ill and Bruche rivers some 2 miles west of the Rhine. In 1939 it was, not for the first time in its history, under threat from an expansionist Germany. Bismark had once referred to the cities of Metz and Strasbourg as "the keys of the German house," both annexed at the end of the Franco-Prussian War by the German empire. At that time, Strasbourg was predominantly French in culture, although as recently as the French Revolution its university had been suppressed as a stronghold of German sentiment. At the end of the First World War, Strasbourg once more reverted to France, and, as described earlier, Pierre Weiss returned to his native province to create and direct a major physics institute at the reopened university. Under his guidance and with the aid of several associates brought from among the best of his former staff in Zurich, the new laboratory had soon surpassed even that which he had built up during his years in Switzerland.

It was perfectly natural that the first of what would later become a series of international conferences on magnetism should be held close to what most physicists would have regarded as the world's premier laboratory in the subject. Between 21 and 24 May 1939, over 30 invited participants gathered, along with the many researchers from Weiss's own institute, at the University of Strasbourg, for what one participant later described as a "landmark in magnetism."[287] International meetings on such a scale were at the time still rare and, for those participants new to such gatherings, impressive. Around a nearly closed ring table in the Salle Pasteur were seated the more senior physicists, with wider concentric circles of *observateurs* and a general audience.[288] One focal point was Weiss, "quiet but charming," as chairman; another was the centrally placed stenotypist, unhurriedly and efficiently recording the proceedings for posterity. Langevin had been invited to open the conference with a broad historical celebratory survey of magnetism. Because of his enforced absence, this task fell to Bauer. After lunch came more serious business. Van Vleck provided a survey of paramagnetism, and the theory established by Van Vleck, Kramers, and Stoner was a cause for great satisfaction. One reporter of the conference noted approvingly that "paramagnetism stands in rather the same relation to quantum theory as astronomy does to Newtonian mechanics.[289] Metallic paramagnetism remained problematic, however, and for similar reasons the quantum theory of ferromagnetism was in a far from satisfactory state.

The meeting marked the last time the Weiss magneton was seriously and hotly debated. E. C. Stoner objected to the straight-line segments, which, in his view, had been sought "over-enthusiastically." Bauer denied that the bends between segments indicated genuine changes from one magnetic state to another, since no thermal phenomena accompanied them. But Foëx defended the whole system by protesting that it was "easy to deny experiments that one has not done." Weiss appealed to scientific methodology, saying that "one does not have to doubt the existence of a phenomenon merely because it is not understood." Néel diplomatically split the difference, saying that one should draw "a general average curve through all the straight lines under discussion, and interpret this curve. Only then should one try to explain the small secondary straight lines."[290]

Fresh winds were blowing across the field of magnetism, with a younger generation beginning to take the lead. This was never more apparent than in France, where Weiss was clearly at the end of an outstanding career. Louis Néel was already proving to be an impressive successor to the Strasbourg chair, relinquished by Weiss in 1936. This was recognized, for instance, by Stoner at the Strasbourg conference. The subject came up as he left his hotel one morning to cross the Place Kleber with his British colleague Leslie Bates. As Bates recalled, "he told me that he felt that a new force, a new spirit had arisen in magnetism circles in France and that it was centered around that young man Néel and his companions."[291]

On 24 May at precisely 11:30 A.M., the conference participants gathered for the traditional group photograph. Captured in that split second of exposure was the core, with a few notable exceptions, of a modern magnetics community. Some would never meet again. With the formalities concluded, Néel entertained the English trio Stoner, Mott, and Sucksmith with a drive in his car to the wooded

First Conference on Magnetism, Strasbourg, May 1939. (AIP Niels Bohr Library)

Vosges.[292] The weather was beautiful, and the air was fresh with the smell of spring flowers and pines. A cuckoo was heard in the distance. It was almost possible to forget the ominous threat that hung like a pall over the whole of Europe. The evening's conversation turned to politics, and Gerlach spoke out in defense of Germany. Mrs. Becker was "rather more obstreperous. . . . they both want Strasbourg . . . a German city."[293] As Bates later recalled, "Most of us left Strasbourg with heavy foreboding."[294]

As the summer of 1939 advanced, preparations for war continued throughout Europe, and "the attitudes of diplomatists, the speeches of politicans and the wishes of mankind counted each day for less."[295] Less than three months after the magneticians had gone, Winston Churchill visited Strasbourg for very different reasons. All along the Rhine he noted that the temporary bridges across the river had been moved to one side or the other. Remaining permanent crossings were heavily guarded and mined, ready to be blown apart at a moment's notice. The riverside quarter of the city had been evacuated.[296]

For six weeks during May and June 1940, the battle for France was fought and lost. Thousands had left Strasbourg before Alsace was annexed by the invading German forces. Weiss, old and frail, fled to Lyons to meet up with his best friend, Jean Perrin. Others from his former institute rallied around Weiss's principal collaborator, Gabriel Foëx, in Clermont-Ferrand. Néel had left Strasbourg earlier to investigate the problems brought by magnetic mines. On the capitulation of France, he returned briefly to Clermont-Ferrand and then, with several of his old collaborators, including his own assistant Weil, began to build a new group at Grenoble. Not everyone got away, however. Two of the Strasbourg group were arrested and deported to Germany for forced labor.[297] Most of the equipment left behind was stolen. An epoch in the history of magnetism came to a tragic end. In October 1940, Pierre Weiss died of cancer. His old ally in the great debates over the magneton did not survive him long: Blas Cabrera died in exile in Mexico in 1945. Another close friend and colleague, Henri Abraham, died in Auschwitz. Preisach, the talented Hungarian experimentalist, was reportedly murdered by the SS.[298]

The War and Its Aftermath

When the war clouds over Europe finally cleared, scientists began to drift back to their personal research interests and attempt to reestablish old contacts. They found the context for fundamental research much changed. By popular acclaim, physicists had played a major role in the winning of the war, and were now set to have an influence in peacetime. Many among them were keen to apply new techniques developed in the context of wartime technology, and facilities were being more readily funded than ever before. All sectors of the physics community began to expand. Magnetism was ideally suited to benefit from the burst of enthusiasm for physics and for the benefits of basic research. It offered numerous fascinating research problems and had clear technological importance. At Bell Labs, for instance, several newly hired theoretical physicists began to work on problems in magnetism.[299]

A new generation was being introduced to the subject in industry and university departments, at a time when many of the senior hierachy of the subject had recently died or were no longer actively involved in magnetics research. Intellectural leadership had passed to the likes of Van Vleck, Néel, Stoner, Gorter, and Bozorth. The strong empirical tradition in metals research established in Japan, primarily through Honda's school, was to blossom in the postwar period together with the emergence of equally outstanding theoretical work in magnetism. In effect, the period of the Second World War interjected a marked generational shift. A younger generation is always likely to perceive the nature of a field, its problems, subject boundaries, and relationships with other areas of research, in a different way. The younger they were, the more likely they were to perceive magnetism as part of solid-state physics. Various significant wartime developments also influenced the perception of the subject.

For the most part, magnetics research closed up shop for the duration of the war. A small amount of highly influential research did take place, however. In France and the Netherlands, some magnetics research continued in complete isolation from the rest of the physics community. Following the German occupation of the Netherlands, the factories and laboratories of the Philips Company came under German supervision. In the intensely difficult circumstances that prevailed, the policy of the firm became that of keeping up appearance by seeming to perform an adequate amount of war work and sabotaging as much as possible.[300] Laboratory staff secretly provided radios for the Dutch resistance, and the factory became an important refuge. This could not prevent the incarceration and murder of many Jewish workers, and few survived the German concentration camps. The research program of the laboratory continued relatively unmolested, including the search for a suitable core material for filter coils based on ferrites. The success of this program became clear when, after the end of the war, Philips brought ferrites into commercial use. Not only was it obvious that ferrites would be immensely valuable for many commercial applications, but the Philips research had opened up an area of great scientific interest.

Further technical development and application of ferrites went hand in hand with the theoretical and experimental study of their properties, elucidated with particular success by Néel and others in France (see the section "Magnetism as seen by Néel").[301] The scientific interest in ferrites throughout the world was shown in the program for the second international conference on magnetism, held at Grenoble in 1950, where a large proportion of the presented papers was devoted to properties of ferrites.[302] The successful commercial development and scientific interpretation of ferrites marked an important shift in the scientific context of magnetic materials. Ferrites had little part in metallurgical tradition, being ceramic, manufactured by grinding and firing in ovens. Moreover, improvements in the performance of ferrites was totally dependent on, rather than incidental to, a fundamental understanding of their properties, and relied on the contributions of both physicists and chemists. As such, ferrite development represented one of the first truly interdisciplinary solid-state technologies, although the significance of this was overshadowed by the discovery of the transistor and the publicity devoted in general to semiconductor technology. In fact, more income has probably been generated through the manufacture of ferrites than through the manufacture of transistors.

Magnetics research also flourished during the war in France, where Néel made outstanding contributions to domain theory.[303] He became involved with developing an understanding of fine-particle magnetic materials, following a suggestion made in the 1930s that small magnetic particles could behave as single domains. Néel published his work on the theory of single-domain ferromagnetism after the end of the war, although not before it had been developed independently by Charles Kittel in the United States, and in a more general and comprehensive form by Stoner and his assistant, Peter Wohlfarth, in England.[304] The resulting theory was not simply something of academic interest, but it had significance in the development of magnetic tapes for sound recording and for computer memory systems. The properties of permanent magnets could now also be interpreted. These two examples, ferrites and fine-particle materials, were typical of the new realms of technology opening up, dependent in part on the development of magnetic materials through scientific understanding rather than empirical trial and error.

Another aspect of Néel's wartime work extended and criticized earlier contributions from Kondorsky and Kersten, and in doing so predicted the formation of "dagger blade" domains around cavities and imperfections using existing theory.[305] His prediction was confirmed after the war by H. J. Williams at Bell Labs, using more sophisticated powder pattern techniques than had been available to prewar researchers. These improvements in the powder pattern technique of visualizing domain structures were decisive in confirming the domain theory of ferromagnetism in all its essential aspects. Although Elmore in particular had considerably improved Bitter's crude initial demonstrations in the 1930s, it was the Bell researchers who exploited the technique with great success after the war and stimulated further research elsewhere. On the theoretical side, quantitative domain calculations became possible based on the previously rather neglected model introduced by the Soviet physicists Landau and Lifshitz in 1935.[306]

If the domain theory of ferromagnetism was taking its place among established and successful theories in the physics literature,[307] the quantum theory of ferromagnetism was proving a far more controversial subject of study, with the proponents of two apparently irreconcilable approaches locked in vigorous conflict. There are still two main approaches to the theoretical interpretation of the fundamental phenomena of ferromagnetism. One derives from the Heitler–London treatment of molecules, starting from atomic wave functions. The other stems from an electron energy-band treatment of metals and has a close analogy to the Hund–Mulliken treatment of molecular orbitals. The first approach considers localized electrons in direct quantum exchange interaction; the second applies quantum statistics to intinerant electrons in unfilled energy bands. The intellectual roots of these two approaches have been examined earlier (Chapter 3). In 1935 a model proposed by Mott of overlapping s and d bands provided a significant incentive for the practical application of an energy-band approach, first by Slater and then in two important prewar papers from Stoner, which set the scene for an intensive postwar debate.[308]

In its basic form, Stoner's model consisted of an assembly of electrons subject to Fermi–Dirac statistics and distributed in a parabolic energy band.[309] The latter, however, corresponds to the distribution of free electrons and was a deliberate simplifying assumption made by Stoner to facilitate the calculations, even then complex, that had to be carried out on a mechanical calculating machine. Although never considered by Stoner as ultimately essential, this assumption, together with the absence of other features, fueled objections. Opinion in the early postwar years was fiercely polarized concerning the respective merits of a localized or an itinerant approach, with the majority against the itinerant model. The localized model was more easily understood, and perhaps because of that alone became the dominant view in the scientific literature. This was despite the fact that in its detailed application it was inadequate for explaining a wide range of experimental results pertaining to metallic ferromagnetism. In contrast, as even one of its main detractors, Van Vleck, admitted, the conclusions of Stoner's model were consistent with experiment at least qualitatively.[310] This alone was not considered sufficient evidence in favor of the model.

What Stoner called the collective electron approach was perceived by its critics as having fundamental theoretical weaknesses. Many of these Stoner did not dispute. The debate thus rested on the degree of simplification that was acceptable and necessary to explain the essential features of ferromagnetism consistent with experimental evidence. Stoner never considered his model as more than "scaffolding with cross-links and planking" used to highlight the essential features of the problem and to coordinate a wide array of experimental phenomena.[311] Nevertheless, he remained, quite defensive. A major objection remaining was the inability of the itinerant approach, as developed by Stoner and extended with the assistance of Wohlfarth, to be consistent with the concept of spin waves introduced by Bloch and thoroughly established in nearly everyone's mind by the postwar efforts of Charles Kittle and Conyers Herring. Perhaps more significant for the reception and temporary fate of the itinerant approach was an ongoing debate in the late 1940s and early 1950s on the general validity in solid-state theory of different kinds of approximations, which broke out into open confrontation at a meeting in Washington in 1952. Acrimonious discussion, described later as "lengthy but lively," followed the presentation of papers.[312] One casualty was the itinerant approach to ferromagnetism. Its convalescence lasted about a decade.

Various factors contributed to its recuperation, including theoretical and experimental developments, most of which came from fields away from the mainstream of magnetics research. These included improved techniques for handling the complex process of interactions between electrons giving rise to ferromagnetic behavior. On the computational side, energy-band structures could be determined in a way not possible in earlier years. Experimental evidence on metallic ferromagnets became overwhelmingly supportive of the itinerant approach. This included Fermi surface studies and solid-state spectroscopy. A major publication from Herring in 1966 provided a great incentive to the reestablishment and further development of the itinerant approach to ferromagnetism while demonstrating that this approach and the localized model were not mutually exclusive.[313] (Herring, incidentially, is only one of the theorists recruited by Bell Labs who made important contributions to the theory of magnetism. Others included Kittel and Philip Anderson.)

Many of the experimental techniques so important to the development of magnetism in the postwar years were developed from wartime technology. Experience in microwave technology gained from radar development paved the way for the discovery of magnetic resonance and its subsequent application to the study of atomic and nuclear structure. Nuclear pile technology opened another avenue, pro-

Charles Kittel (b. 1916). (AIP Niels Bohr Library)

viding powerful sources of neutrons with which to probe the complexities of magnetic structure and consolidate theoretical development.[314] Neutron-scattering techniques developed in tandem with nuclear experimental facilities, with the United States giving the lead from its Oak Ridge and Brookhaven installations to emergent teams at Harwell in England, Chalk River in Canada, and Saclay in France. By the early 1950s, the Oak Ridge group led by Clifford Schull had successfully confirmed the ferri- and antiferromagnetic structures predicted by Néel. New magnetic structures were discovered, and particularly in France and the Soviet Union an entirely new theoretical field of magnetic space groups opened up.

Philip W. Anderson (b. 1923). (AIP Meggers Gallery of Nobel Laureates)

By the end of the 1950s, magnetism no longer had the small and relatively cohesive community represented at Strasbourg in 1939. Like many other areas in science by 1960, the lead tended to come from the United States, although the subject was till vigorously and successfully studied in other traditional centers. Newer national representations were evident, such as Israel. Most evident, however, was the growth of interest in Japan, where a strong theoretical school of magnetism began to develop to complement its more empirical traditions in the subject. Japan shows signs of having become the leading influence in magnetism. The scientific literature in magnetism grew steadily year by year after the end of the war, and the strategic importance of magnetic materials for various electronic applications placed magnetism at the center of the explosion in the published scientific literature signaled in 1956 by the launching of *Sputnik*. Magnetism conferences became vast affairs with myriad represented interests. Whereas in the 1930s the number of active researchers worldwide in all aspects of the subject was probably fewer than 100, by the 1960s this figure had increased, according to one estimate by a factor of 46.[315] In 1956 a major international conference in Moscow devoted to magnetism was attended by around 800 Soviet scientists alone.[316] Five years later, an equal number of Japanese physicists attended a similar event in Kyoto.[317] To give another example, in 1951 Richard Bozorth was the sole author of an encyclopedic handbook on magnetic materials; a project to update Bozorth's efforts more recently has taxed the efforts of some 40 authors.[318] One end of the subject may be totally isolated from another. Most researchers in all the diverse areas of magnetism, however, would identify their work as forming part of solid-state physics.

Notes

1. S. Evershed, "Permanent Magnets in Theory and Practice," *Journal of the Institution of Electrical Engineers* 58 (1920): 780–835, p. 780.

2. S. Evershed, "Permanent Magnets in Theory and Practice—Part 2," *Journal of the Institution of Electrical Engineers* 63 (1925): 725–821, p. 818.

3. E. C. Stoner, "Magnetism in Retrospect and Prospect" (Thirty-ninth Guthrie Lecture), in *Yearbook of the Physical Society—1955* (London: Physical Society, 1955), 23–43.

4. See Chapter 2.

5. A. Ewing, "Ferromagnetism and Hysteresis," *Proceedings of the Physical Society* 42 (1930): 355–357, p. 355.

6. S. Evershed, Reply to Ewing (n. 5) paper, *Proceedings of the Physical Society* 42 (1930): 454–455.

7. A. Ferguson, "The Physical Society's Discussion on Magnetism," *Nature* 125 (1930): 874–875, p. 875.

8. F. Bitter, "On the Interpretation of Some Ferromagnetic Phenomena," *Phys. Rev* 39 (1932): 337–345, p. 337.

9. C. S. Smith, "The Pre-history of Solid-State Physics," *Physics Today* 18, no. 12 (1965): 18–30.

10. E. C. Stoner, "Magnetism in the Twentieth Century," *Proceedings of the Physical Society* 42 (1930): 358–371, p. 370.

11. E. P. Wohlfarth, "Iron, Cobalt and Nickel," in *Ferromagnetic Materials. I*, ed. E. P. Wohlfarth (Amsterdam: North-Holland, 1980), 1–70, p. 3.

12. P. Curie, "Propriétés magnétiques des corps à diverses températures," *Annales de chimie et de physique*, 7th ser., vol. 5 (1895): 289–405.

13. Ibid., 290.

14. Ibid., 294.

15. Ibid., 340.

16. Ibid., 404.

17. P. Curie, *Notice sur les travaux scientifiques de M. P. Curie* (Paris: Gauthier-Villars, 1902), 17 (copy in P. Curie file, archives of the Academy of Sciences, Paris).

18. P. Langevin, "Magnétisme et théorie des électrons," *Annales de chimie et de physique,* 8th ser., vol. 5 (1905): 70–127.

19. Ibid., 78.

20. Ibid., 89.

21. Ibid., 87.

22. Ibid., 121.

23. Ibid.

24. Ibid., 122.

25. A. Herpin, *Théorie du magnétisme* (Paris: Presses Universitaires de France, 1968), 101.

26. Curie (n. 12).

27. Langevin (n. 18), 124.

28. M. Brillouin, "Rapport sur la candidature de P. Weiss" (prepared for the Comité Secret meeting of 31 May 1926), 29 May 1926, pp. 4–5, P. Weiss file, archives of the Academy of Sciences, Paris.

29. E. Bauer, "Le Magnétisme depuis cinquante ans," in International Congress on Magnetism, *Le Magnétisme* (Paris: Institut International de Coopération Intellectuelle, 1940), 1:xxvi.

30. W. Thomson, "On the Theory of Magnetic Induction in Crystalline and Non-crystalline Substances," *Phil. Mag.,* 4th ser., vol. 1 (1851): 177–186, p. 183.

31. Brillouin (n. 28), 4.

32. Information in the archives of the Escola de Minas de Ouro Prêto, Minas Gerais, Brazil, where Da Costa Sena was director from 1900 to 1919.

33. P. Weiss, "Sur l'aimantation plane de la pyrrhotine," *C.R.* 126 (1898): 1099–1101.

34. P. Weiss, "Les Propriétés magnétiques de pyrrhotine," *Journal de physique,* 4th ser., vol. 4 (1905): 469–508, 829–846, p. 829.

35. Ibid., 489.

36. Ibid., 829.

37. Langevin (n. 18), 121.

38. P. Weiss, "L'Hypothèse du champ moléculaire et la propriété ferromagnétique," *Journal de physique,* 4th ser., vol. 6 (1907): 661–690.

39. Ibid., 668.

40. Ibid., 662.

41. P. Weiss and H. Kamerlingh Onnes, "Recherches sur l'aimantation aux très basses températures," *Journal de physique,* 4th ser., vol. 9 (1910): 555–584, p. 555.

42. P. Weiss, "Curriculum vitae, extrait de la liste des publications et notice sur les travaux scientifiques," 26 November 1908, p. 3, Weiss file, archives of the Academy of Sciences, Paris.

43. P. Weiss, *Notice sur les travaux scientifiques de M. P. Weiss* (Strasbourg: Imprimerie Alsacienne, 1926), 14 (copy in Weiss file, archives of the Academy of Sciences, Paris).

44. A. Ewing, "Contributions to the Molecular Theory of Induced Magnetism," *Proc. Roy. Soc.* 48 (1890): 342–359, p. 354.

45. L. Néel, "Le Champ moléculaire de Weiss et le champ moléculaire local," in *Actes du Colloque National de Magnétisme tenu à Strasbourg du 8 au 10 juillet 1957* (Paris: Centre National de la Recherche Scientifique, 1958), 1–24, p. 7.

46. P. Weiss and P.-N. Beck, "Chaleur spécifique et champ moléculaire des substances ferromagnétiques," *Journal de physique,* 4th ser., vol. 7 (1908): 249–264.

47. J. Pionchon, "Recherches calorimétriques sur les chaleurs spécifiques et les changements d'état aux températures élevées," *Annales de chimie et de physique,* 6th ser., vol. 11 (1887): 33–111.

48. P. Weiss and A. Picard, "Sur un nouveau phénomène magnétocalorique," *C.R.* 166 (1918): 352–354.

49. Langevin (n. 18), 123.

50. W. F. Giauque, "A Thermodynamic Treatment of Certain Magnetic Effects. A Proposed

Method of Producing Temperature Considerably Below 1° Absolute," *Journal of the American Chemical Society* 49 (1927): 1864–1870.

51. Weiss and Picard (n. 48), 352.

52. Ibid., 354.

53. A. Perrier and G. Balachowsky, "Sur la dépendance entre l'aimantation rémanente, l'aimantation spontanée et la température," *Archives des sciences physiques et naturelles* (Geneva), 5th ser., vol. 2 (1920): 5–29.

54. P. Weiss and R. Forrer, "Aimantation et phénomène magnétocalorique du nickel," *Annales de physique,* 10th ser., vol. 5 (1926): 153–213.

55. Weiss (n. 38), 663.

56. Ibid., 667.

57. Ibid., 670.

58. Weiss and Picard (n. 48), 353.

59. P. Weiss, "Sur la rationalité des rapports des moments magnétiques moléculaires et la magnéton," *Journal de physique,* 5th ser., vol. 1 (1911): 900–912, 965–988, p. 905.

60. H. Barkhausen, "Zwei mit hilfe der neuen Verstärker entdeckte Erscheinungen," *Phys. Zs.* 20 (1919): 401–403.

61. B. van der Pol, "Magnetisation Discontinuities Revealed by the Valve Amplifier," *Verhandelingen der Koninklijke Akademie van Wetenschappen te Amsterdam,* 23, no. 4 (1921): 637–643, 980–988.

62. P. Weiss and G. Ribaud, "Sur les discontinuités de l'aimantation," *Journal de physique et le radium,* 6th ser., vol. 3 (1922): 74–80.

63. Ibid., 75.

64. Ibid., 79.

65. W. Gerlach and P. Lertes, "Über magneto-elastische Effekte," *Zs. Phys.* 4 (1921): 383–392.

66. E.P.T. Tyndall, "The Barkhausen Effect," *Phys. Rev.* 24 (1924): 439–451.

67. Weiss and Forrer (n. 54), 156.

68. Tyndall (n. 66), 451.

69. F. Bitter, "On Inhomogeneities in the Magnetisation of Ferromagnetic Materials," *Phys. Rev.* 38 (1931): 1903–1905.

70. J. Frenkel and J. Dorfman, "Spontaneous and Induced Magnetization in Ferromagnetic Bodies," *Nature* 126 (1930): 274–275.

71. K. J. Sixtus and L. Tonks, "Propagation of Large Barkhausen Discontinuities," *Physical Reveiw* 37 (1931): 930–958.

72. F. Bloch, "Zur Theorie des Ferromagnetismus," *Zs. Phys.* 61 (1930): 206–219.

73. L. Landau and E. Lifshitz, "On the Theory of the Dispersion of Magnetic Permeability in Ferromagnetic Bodies," *Physikalische Zeitschrift Sowjetunion* 8 (1935): 153–169.

74. Weiss (n. 59), 903.

75. P. Langevin, "La Théorie cinétique du magnétisme et les magnétons," in *La Théorie du rayonnement et las quanta* (Paris: Gauthier-Villars, 1912), 393–404, p. 396.

76. P. Weiss, "Sur une propriété du ferromagnétisme," *C. R.* 167 (1918): 74–76.

77. P. Weiss, "L'Hypothèse du champ moléculaire," *Annales du physique,* 10th ser., vol. 17 (1932): 97–136.

78. P. Weiss, "Sur l'équation caractéristique des fluides," *C. R.* 167 (1918): 232–235.

79. P. Villard in a report on the titles and work of P. Weiss, 2 June 1919, p. 4, P. Weiss file, archives of the Academy of Sciences, Paris.

80. P. Weiss, "Sur le champ moléculaire et l'action magnétisante de Maurain," *C. R.* 158 (1914): 29–32.

81. C. Maurain, "Sur une action magnétisante de contact et son rayon d'activité," *Journal de physique,* 4th ser., vol. 1 (1902): 90–100.

82. P. Debye, "Einige Resultate einer kinetischen Theorie der Isolatoren," *Phys. Zs.* 13 (1912): 97–100.

83. P. Weiss, "Sur la nature de champ moléculaire," *Annales de physique,* 9th ser., vol. 1 (1914): 134–162, p. 162.

84. P. Weiss, "Un Argument en faveur de la nature électrostatique de champ moléculaire," *C. R.* 178 (1924): 739–742.

85. K. Honda, "Thermomagnetic Properties of the Elements," *Annalen der Physik* 32 (1910): 1027–1063.

86. W. Ritz, "Sur l'origine des spectres en séries," *C. R.* 145 (1907): 178–180.

87. Weiss and Kamerlingh Onnes (n. 41), p. 560.

88. Weiss (n. 59).

89. Ibid., 982, 979.

90. Ibid., 986ff.

91. P. Weiss to Aimé Cotton, 27 February 1911, in possession of Eugène Cotton, whom I thank for kindly giving me access.

92. Langevin (n. 75).

93. J. L. Heilbron and T. S. Kuhn, "The Genesis of the Bohr Atom," *Historical Studies in the Physical Sciences* 1 (1969): 211–290, pp. 230ff.

94. B. Cabrera, "Paramagnetismo de los elementos de la tierras raras y el magnetón de Weiss," *Trabajos del Laboratorio de Investigaciones físicas* (Madrid: Imprenta clásica española) 75 (1926): 289–320, p. 289.

95. H. Kamerlingh Onnes and E. Oosterhuis, "Magnetic Researches. VIII. On the Susceptibility of Gaseous Oxygen at Low Temperature," *Communications from the Physical Laboratory of the University of Leiden* 134d (1914): 31–33.

96. W. Kopp, *Der thermische Verlauf des Paramagnetisimus bei Magnetit, Platin und Palladium* (Saint Gall: Buchdruckerei H. Tschudy, 1919), 73ff.

97. G. Foëx, "Recherches sur le paramagnétisme," *Annales de physique,* 9th ser., vol. 16 (1921): 174–305.

98. B. Cabrera and A. Duperier, "Sur le propriétés paramagnétiques des terres rares," *C. R.* 188 (1929): 1640–1642.

99. J. H. Van Vleck, "$\chi = -k + (c/T + \Delta)$," the Most Overworked Formula in the History of Paramagnetism," *Physica* 69 (1973): 177–192.

100. Foëx (n. 97).

101. H. Kamerlingh Onnes, "Further Experiments with Liquid Helium; Appearance of Beginning Paramagnetism Saturation," *Communications from the Physical Laboratory of the University of Leiden* 140d (1914): 27–32.

102. W. Pauli, "Quantentheorie und Magneton," *Phys. Zs.* 21 (1920): 616–617.

103. W. F. Giauque, "Paramagnetism and the Third Law of Thermodynamics, Interpretation of the Low-Temperature Magnetic Susceptibility of Gadolinium Sulfate," *Journal of the American Chemical Society* 49 (1927): 1870–1877.

104. F. Hund, "Atomtheoretische Deutung des Magnetismus der seltenen Erden," *Zs. Phy.* 33 (1925): 855–859.

105. Foëx (n. 97), 305.

106. B. Cabrera, "Les Terres rares et la question du magnéton," *C. R.* 180 (1925): 688–671.

107. S. Meyer, "Magnetisierungzaalen seltener Erden," *Phys. Zs.* 26 (1925): 51–54.

108. J. H. Van Vleck and A. Frank, "The Effect of Second Order Zeeman Terms on Magnetic Susceptibilities in the Rare Earth and Iron Group," *Physical Review* 34 (1929): 1494–1496, 1625.

109. Cabrera and Duperier (n. 98).

110. Kopp (n. 96).

111. A. Serres, "Recherches sur les moments atomiques," *Annales de physique* 17 (1932): 5–95, pp. 57ff.

112. H. Forestier and G. Chaudron, "Caractères ferromagnétiques du sesquioxyde de fer," *C. R.* 183 (1926): 787–789.

113. Serres (n. 111), 90.

114. C. Sadron, "Les Moments ferromagnétiques des éléments et le système périodique," *Annales de physique,* 10th ser., vol. 17 (1932): 371–452.

115. Ibid., 447.

116. P. Weiss and A. Piccard, "Sur l'aimantation de l'oxyde azotique et le magnéton," *C. R.* 157 (1913): 916–918.

117. P. Weiss, E. Bauer, and A. Piccard, "Sur les coefficients d'aimantation de l'oxygène, de l'oxyde azotique et le théorie du magnéton," *C. R.* 167 (1918): 484–487.

118. Ibid., 487.

119. B. Cabrera, "La Théorie du magnétisme," *Journal de physique et le radium,* 6th ser., vol. 8 (1927): 257–275, p. 250; Cabrera, "L'Étude expérimentale du paramagnétisme; le magnéton," *Le Magnétisme* (Paris: Gauthier-Villars, 1932), 81–159.

120. P. Weiss, discussion of B. Cabrera's report, in *Le Magnétisme* (n. 119), 161–174, p. 163.

121. Cabrera (1932) (n. 119),

122. J. Dorfman, in *Le Magnétisme* (n. 119), 170.

123. C. J. Gorter, "Die Suszeptibilitäten paramagnetischer Lösungen," *Phys. Zs.* 34 (1933): 462–464.

124. A. Lallemand, "Influence de l'état physique sur les propriétés magnétiques de quelques sels de la famille du fer," *Annales de physique,* 11th ser., vol. 3 (1935): 97–180.

125. L. Néel to P. Quédec, 10 August 1983.

126. P. Weiss and G. Foëx, *Le Magnétisme,* 4th ed. (Paris: Armand Colin, 1931, 1951), 198.

127. G. Foëx to L. de Broglie (permanent secretary of the Academy of Sciences), 6 March 1962, Weiss file, archives of the Academy of Sciences, Paris.

128. Weiss (n. 38).

129. Ibid., 688.

130. Ibid., 689.

131. L. Néel, "Influence des fluctuations du champ moléculaire sur les propriétés magnétiques des corps," *Annales de physique* 18 (1932): 5–105, p. 84.

132. R. Forrer, "Le problème des deux points de Curie," *Journal de physique et le radium* 1 (1930): 49–64.

133. W. Heisenberg, "Zur Theorie des Ferromagnetismus," *Zs. Phys.* 49 (1928): 619–636.

134. L. Néel, interview with P. Quédec, 7 October 1983, archives of the Musée de la Villette, Paris.

135. Weiss (n. 80); Maurain (n. 81).

136. Néel interview (n. 134).

137. Néel (n. 131).

138. See the section "Consequences of Molecular Field Theory."

139. E. Lapp, "Etude de la chaleur spécifique vraie du nickel," *Annales de physique* 12 (1929): 442–521.

140. L. Néel, "Application au nickel d'une nouvelle méthode de mesure des chaleurs spécifiques vraies," *C.R.* 207 (1938): 1384–1385.

141. We have used the term *fluctuations* in the sense Néel used it. Today one would speak of a calculation of the mean magnetization in the canonical ensemble. Néel's "fluctuations" characterize the physical system's exploration of the different configurations accessible to it.

142. Néel (n. 131), 57.

143. L. Néel, "Sur l'interprétation des propriétés magnétiques des alliages," *C.R.* 198 (1934): 1311–1313.

144. Néel (n. 131), 100–101.

145. L. Néel, "Propriétés magnétiques du manganèse et du chrome en solution solide étendue," *Journal de physique et la radium* 3 (1932): 160–171.

146. Néel interview (n. 134).

147. L. Néel, "L'Equation d'état et le porteur élémentaire de magnétisme du nickel," *Journal de physique et le radium* 5 (1934): 104–120, p. 116.

148. L. Néel, "Théorie du paramagnétisme constant; application au manganèse," *C.R.* 203 (1936): 304–306.

149. H. Bizette, C. F. Squire, and B. Tsai, "Le Point de transition de la susceptibilité du protoxyde de manganèse (MnO)," *C.R.* 207 (1938): 449–450.

150. F. Bitter, "A Generalization of the Theory of Ferromagnetism," *Phys. Rev.* 54 (1938): 79–86.

151. C. G. Shull and J. S. Smart, "Detection of Antiferromagnetism by Neutron Diffraction," *Phys. Rev.* 76 (1949): 1256–1257.

152. L. Néel, "Propriétés magnétiques de l'état métallique et énergie d'interaction entre atomes magnétiques," *Annales de physique* 5 (1936): 232–279.

153. N. J. Poulis and G.E.G. Hardeman, "Nuclear Magnetic Resonance in an Antiferromag-

netic Single Crystal," *Physica* 18 (1952): 201–220, 315–328; Poulis and Hardeman, "Behaviour of a Single Crystal of CuCl₂, 2H₂O near the Néel Temperature," *Physica* 18 (1952): 429–432.

154. Serres (n. 111).

155. J. L. Snoek, "Magnetic and Electrical Properties of the Binary System MOFe₂O₃," *Physica* 3 (1936): 463–483; Snoek, *New Developments in Ferromagnetic Materials,* 2nd ed. (New York: Elsevier, 1949), 68ff.

156. E.J.W. Verwey and E. L. Heilmann, "Physical Properties and Cation Arrangements of Oxides with Spinel Structures," *Journal of Chemical Physics* 15 (1947): 174–180.

157. L. Néel, "Propriétés magnétiques des ferrites. Ferrimagnétisme et antiferromagnétisme," *Annales de physique* 3 (1948): 137–198.

158. Néel (n. 143).

159. C. G. Shull, E. O. Wollan, and W. A. Strauser, "Magnetic Structure of Magnetite and Its Use in Studying the Neutron Magnetic Interaction," *Phys. Rev.* 81 (1951): 483–484.

160. C. Starr, F. Bitter, and A. R. Kaufmann, "The Magnetic Properties of the Iron Group Anhydrous Chlorides at Low Temperatures," *Phys. Rev.* 58 (1940): 984–992; J. Becquerel and J. Van Den Handel, "Le Métamagnétisme," *Journal de physique et le radium* 10 (1939): 10–13.

161. J. Becquerel file, archives of the Academy of Sciences, Paris.

162. L. Néel, "Antiferromagnétisme et métamagnétisme," in Institut International de Physique Solvay, *Les Électrons dans les métaux* (Brussels: Stoops, 1955). The meeting was held 13–17 September 1954.

163. H. Forestier, "Transformations magnétiques du sesquioxyde de fer, de ses solutions solides et de ses combinaisons ferromagnétiques," *Annales de chimie,* 10th ser., vol. 9 (1928): 316–401, pp. 389ff.

164. H. Forestier and G. Guiot-Guillain, "Une Nouvelle série de corps ferromagnétiques: les ferrites de terres rares," *C.R.* 230 (1950): 1844–1845.

165. Forestier and Chaudron (n. 112).

166. G. Guiot-Guillain, "Structure cristalline des ferrites de lanthane et de praséodyme," *C.R.* 232 (1951): 1832–1833.

167. H. Forestier and G. Guiot-Guillain, "Ferrites de terres rares à double point de Curie ferromagnétique," *C.R.* 235 (1952): 48–50.

168. G. Guiot-Guillain, "Influence des diamètres ioniques des terres rares sur les propriétés ferromagnétiques de leurs ferrites," *C.R.* 237 (1953): 1654–1656.

169. J. C. Slater, "Atomic Shielding Constants," *Phys. Rev.* 36 (1930): 57–64.

170. Néel (nn. 152, 143).

171. Néel (n. 157).

172. Y. Yafet and C. Kittel, "Antiferromagnetic Arrangements in Ferrites," *Phys. Rev.* 87 (1952): 290–294.

173. R. Pauthenet and P. Blum, "Etude thermomagnétique du ferrite de gadolinium," *C.R.* 239 (1954): 33–37.

174. L. Néel and R. Pauthenet, "Etude thermomagnétique d'un monocristal de Fe₂O₃ α," *C.R.* 234 (1952): 2172–2174.

175. L. Néel, "Sur l'interprétation des propriétés magnétiques des ferrites de terres rares," *C.R.* 239 (1954): 8–11.

176. Néel (n. 157). Here Néel showed that the spontaneous magnetizations of each of the two sublattices of a ferrimagnetic structure could decrease with temperature in a different fashion. If the most intense magnetization decreased faster than the other, one could expect to find an "inversion temperature," where the magnetization resulting from the two partial and opposed magnetizations changes direction. This is in effect what was first observed by E. W. Gorter and J. A. Schulkes at Eindhoven in 1953: "Reversal of Spontaneous Magnetization as a Function of Temperature in LiFeCr Spinels," *Phys. Rev.* 90 (1953): 487–488.

177. G. Guiot-Guillain, R. Pauthenet, and H. Forestier, "Etude thermomagnétique des ferrites de dysprosium et d'erbium," *C.R.* 239 (1954): 155–157.

178. F. Bertaut and F. Forrat, "Structure des ferrites ferrimagnétiques de terres rares," *C.R.* 242 (1956): 382–384.

179. F. Bertaut, interview with P. Quédec, 28 February 1983, archives of the Musée de la Villette, Paris.

180. R. Pauthenet, "Propriétés magnétiques des ferrites de gadolinium," *C.R.* 242 (1956): 1859–1862.

181. F. Bertaut, F. Forrat, A. Herpin, and P. Mériel, "Etude par diffraction de neutrons de grenat ferrimagnétique $Y_3Fe_5O_{12}$," *C.R.* 243 (1956): 898–901.

182. Néel interview (n. 134).

183. Ibid.

184. See L. Néel, *Oeuvres scientifiques* (Paris: CNRS, 1978).

185. For general accounts of the development of ideas concerning magnetic phenomena in the context of atomic physics, see E. C. Stoner, *Magnetism and Matter* (London: Methuen, 1934); J. H. Van Vleck, "Landmarks in the Theory of Magnetism," *American Journal of Physics* 18 (1950): 495–509; Stoner (nn. 3, 10).

186. For Bohr's unpublished work in magnetism, see Ulrich Hoyer, ed., *Niels Bohr Collected Works,* vol. 2: *Work on Atomic Physics (1912–1917)* (Amsterdam: North-Holland, 1981), 256–265.

187. Stoner (n. 10).

188. W. Gerlach and O. Stern, "Der Experimentelle Nachweis der RichtungsQuantelung im Magnetfeld," *Zs. Phys.* 9 (1922): 349–355.

189. Stoner (n. 10).

190. Stoner (n. 3), 32.

191. E. C. Stoner, *Magnetism and Atomic Structure* (London: Methuen, 1926).

192. Unsigned review of Weiss and Foëx (n. 126), *Nature* 118 (1926): 584.

193. P. Kapitza, "The Future of Magnetism," *Nature* 119 (1927): 809–810.

194. Stoner (n. 3), 32; see also Stoner (n. 185), preface.

195. K. Darrow, "Ferromagnetism," *Bell System Technical Journal* 6 (1927): 295–366, p. 296.

196. A. H. Compton, "The Magnetic Electron," *Journal of the Franklin Institute* 192 (1921): 145–155; E. H. Kennard, "Moment of Momentum of Magnetic Electrons," *Phys. Rev.* 19 (1922): 420.

197. See G. Uhlenbeck and S. Goudsmit, "Spinning Electrons and the Structure of Spectra," *Nature* 117 (1925): 264–265.

198. E. C. Stoner, "Recent Developments in Magnetism," *Science Progress* 21 (1927): 600–620, p. 619.

199. E. C. Stoner, "Magnetism and Molecular Structure," *Phil. Mag.* 3 (1927): 336–356, p. 355.

200. W. Heisenberg, "Zur Theorie des Ferromagnetismus," *Zs. Phys.* 49 (1928): 619–636; see Chapter 2.

201. Stoner (n. 185); J. H. Van Vleck, *Theory of Electric and Magnetic Susceptibilities* (Oxford: Oxford University Press, 1932).

202. See A. Langevin, "Paul Langevin et les Congrès de Physique Solvay," *Pensée,* no. 129 (1966): 3–32, 17ff; J. Mehra, *The Solvay Conferences: Aspects of the Development of Physics Since 1911* (Dordrecht: Reidel, 1975), 182ff.

203. Mehra (n. 202), 182.

204. W. Pauli, "Über Gasentartung und Paramagetismus," *Zs. Phys.* 41 (1927): 81–102.

205. H.B.G. Casimir, "Pauli and the Theory of the Solid State," in *Theoretical Physics in the Twentieth Century,* ed. M. Fierz and V. F. Weisskopf (New York: Wiley Interscience, 1960), 137–139.

206. F. Bloch, "Heisenberg and the Early Days of Quantum Mechanics," *Physics Today* 29, no. 12 (1976), 23–27, p. 26.

207. See B. Bleaney, "John Hasbrouck Van Vleck," *Biographical Memoirs of Fellows of the Royal Society* 28 (1982): 627–665.

208. Stoner's diary, entries for 21–24 May 1939, Stoner Papers, Brotherton Library, University of Leeds, Leeds, England.

209. For an account of Stoner's life, see L. F. Bates, "Edmund Clifton Stoner," *Biographical Memoirs of Fellows of the Royal Society* 15 (1969): 201–237; an autobiographical account of his early life as a child and student at Cambridge, "Family Background and Early Years," is in the Stoner Papers (n. 208).

210. Throughout his life he turned down numerous invitations, including one in the 1930s from Dorfman inviting him to visit the Soviet Union.

211. See, for instance, J. H. Van Vleck, "Reminiscences of the First Decade of Quantum Mechanics," *International Journal of Quantum Chemistry,* no. 5 (1971): 3–20; Van Vleck, "American Physics Comes of Age," *Physics Today* 17, no. 6 (1964): 21–26.

212. This episode is described in Stoner (n. 209).

213. E. C. Stoner, "The Distribution of Electrons Among Energy Levels," *Phil. Mag.* 48 (1924): 719–736; W. Pauli, "Über der Zusammen des Abschlusses der Elektronengruppen in Atom mit der Komplexstrukter der Spektren," *Zs. Phys.* 31 (1925): 765–783.

214. E. C. Stoner, "Diamagnetism and Space Charge Distribution of Atoms and Ions," *Proceedings of the Leeds Philosphical Society* 1 (1929): 484–490; Stoner, "Ionic Magnetic Moments," *Phil. Mag.* 8 (1929): 250–266.

215. J. H. Van Vleck, interview with C. Weiner, 19 January 1973, AIP.

216. J. H. Van Vleck, interview with T. S. Kuhn, 4 October 1963, AIP.

217. Ibid.; Van Vleck (1971) (n. 211), 7ff.

218. Van Vleck (n. 201).

219. See Stoner's review, "Magnetism and Quantum Mechanics," *Nature* 130 (1932): 490–491.

220. G. Uhlenbeck, Review of *Theory of Electric and Magnetic Susceptibilities* by J. H. Van Vleck, *Phys. Rev.* 42 (1932): 737–740, p. 738. Van Vleck was careful in his book to acknowledge Stoner's book and stressed that his own text was meant to complement it rather than duplicate it in any way.

221. See, for instance, Stoner (nn. 3, 10, 219); E. C. Stoner, "Magnetism in Theory and Practice," *Journal of the Institution of Electrical Engineers* 91 (1944): 340–349.

222. Van Vleck (1971) (n. 211); J. H. Van Vleck, "Biographical Information" (unpublished and undated account prepared for the AIP project on the history of recent physics, AIP).

223. P. W. Bridgman, *The Logic of Modern Physics* (London: Macmillan, 1927); Bridgman, *The Nature of Physical Theory* (New York: Dover, 1936).

224. Van Vleck (1971) (n. 211), 4.

225. E. C. Kemble, *The Fundamental Principles of Quantum Mechanics* (New York: McGraw-Hill, 1937), preface; Kemble cites his "consistent emphasis on the operational point of view."

226. Bridgman (n. 223), 5.

227. Stoner (n. 185), 27.

228. Stoner (n. 219), 490.

229. Ibid., 491.

230. P.A.M. Dirac, "Quantum Mechanics of Many-Electron Systems," *Proc. Roy. Soc.* A123 (1929): 714–733, p. 714.

231. Stoner (n. 185), 40; E. C. Stoner, "Ferromagnetism," *Reports on Progress in Physics* 2 (1946–1947): 43–112, p. 44.

232. J. H. Van Vleck and A. Sherman, "The Quantum Theory of Valance," *Reviews of Modern Physics* 7 (1935): 167–228, p. 169.

233. Ibid.

234. Van Vleck (n. 201), preface.

235. Stoner (n. 231), 44.

236. Bleaney (n. 207), 627.

237. Van Vleck (1971) (n. 211), 13.

238. Ibid., 12.

239. H. Bethe, "Recollections of Solid State Theory, 1926–33," in *Beginnings,* 49–51; Bethe, interview with C. Weiner, 17 November 1967, AIP. Mott used the collective electron treatment to describe metals such as nickel and palladium in which s and d bands were thought to overlap, a model accepted by Stoner. See N. F. Mott, *Proceedings of the Physical Society (London)* 47 (1935): 571–588; N. F. Mott and H. Jones, *Theory of the Properties of Metals and Alloys* (Oxford: Oxford University Press, 1936), 191. Mott retained a belief in this model during a period when it was unfashionable, see N. F. Mott, "Electrons in Transition Metals," *Advances in Physics* 13 (1964): 325–422.

240. These developments are reviewed in Van Vleck (n. 185).

241. Stoner (n. 214).

242. See L. C. Jackson, "Investigations on Ferromagnetism at Low Temperatures," *Philosphical Transactions of the Royal Society* 224 (1923): 1–48.

243. For the background to this work, see K. Mendelssohn, *The Quest for Absolute Zero,* 2nd ed. (London: Taylor and Francis, 1977).

244. See J. Van Den Handel, preface to *Physica* 69 (1973); Van Vleck (n. 99); R. Kronig, "Magnetic Relaxation in Retrospect," *Physica* 69 (1973): 1–4; for reminiscences of Leiden in this period, see H.B.G. Casimir, *Haphazard Reality* (New York: Harper & Row, 1983).

245. See K. H. Chang, *Evaluation and Survey of a Subfield of Physics: Magnetic Resonance and Relaxation Studies in the Netherlands* (Utrecht: Stichting F.O.M., 1975).

246. Debye predicted the effect independently. Giauque was primarily interested in entropy and successfully confirmed the third law of thermodynamics.

247. For further background, see N. Kurti, "Cryomagnetic Research at the Clarendon Laboratory," in *Search and Research,* ed. M. Ryle et al. (London: Mullard, 1971), 56–92; K. Mendelssohn, "The Clarendon Laboratory, Oxford," *Cryogenics* 6 (1966): 129–140.

248. Van Vleck (n. 185), 506.

249. For the general technological background, see B. Bowers, "The Generation, Distribution and Utilization of Electricity," in *A History of Technology,* ed. T. I. Williams (Oxford: Clarendon Press, 1978), 6:1068–1090; D. G. Tucker, "Electrical Communication," in *History of Technology,* 1220–1267. For accounts of the development of magnetic materials and their application, see G. R. Polgreen, *New Applications of Modern Magnets* (London: MacDonald, 1966); D. A. Oliver, "Permanent Magnets," in *Physics in Industry: Magnetism* (London: Institute of Physics, 1938), 7: 71–88; U. Enz, "Magnetism and Magnetic Materials: Historical Developments and Recent Role in Industry and Technology," in *Ferromagnetic Materials,* ed. E. P. Wohlfarth (Amsterdam: North-Holland, 1982), 3:1–36; J. H. Bechtold and G. W. Wiener, "The History of Soft Magnetic Materials," in *The Sorby Centennial Symposium on the History of Metallugy,* ed. C. S. Smith (New York: Gordon and Breach, 1965), 501–518; T. D. Yensen, "The Development of Magnetic Materials," *Electric Journal* (March 1921): 93–95; I. S. Jacobs, "The Role of Magnetism in Technology," *Journal of Applied Physics* 40 (1969): 917–928; Jacobs, "Magnetic Materials, A Quarter-Century Overview," *Journal of Applied Physics* 50 (1979): 7294–7306.

250. Bechtold and Wiener (n. 249).

251. Yensen (n. 249), 94.

252. Bechtold and Wiener (n. 249).

253. See M. D. Fagan, *A History of Engineering and Science in the Bell System: The Early Years 1875–1925* (New York: Bell Telephone Laboratories, 1975), 801ff.

254. Ibid., 802.

255. For the Bell system, see L. Hoddeson, "The Emergence of Basic Research in the Bell Telephone System, 1875–1915," *Technology and Culture* 22 (1981): 512–514; more specific to research in magnetism are Fagan (n. 253), 977ff; R. Bozorth, interview with L. Hoddeson, 28 July 1975, AIP; Bozorth, interview with M. D. Fagan, 23 May 1967, Bell Telephone Archives.

256. H. D. Arnold and G. W. Elmen, "Permalloy, an Alloy of Remarkable Magnetic Properties," *Journal of the Franklin Institute* 195 (1923): 621–632.

257. T. Erber, "Francis Bitter: A Biographical Sketch," in *Francis Bitter, Selected Papers and Commentaries,* ed., T. Erber and C. M. Fowler (Cambridge, Mass.: MIT Press, 1969), 3–19; W. F. Brown, "Ferromagnetism Studies 1930–1936," in *Francis Bitter,* 67–75; F. Bitter, *Magnets: The Education of a Physicist* (London: Heinemann, 1960).

258. F. Bitter, "On Inhomogeneities in the Magnetization of Ferromagnetic Materials," *Phys. Rev.* 38 (1931): 1903–1905.

259. R. M. Bozorth, "Barkhausen Effect in Iron, Nickel, and Permalloy. I. Measurement of Discontinuous Change in Magnetization," *Phys. Rev.* 34 (1929): 772–784; R. M. Bozorth and J. F. Dillinger, "Barkhausen Effect. II. Determination of the Average Size of the Discontinuities in Magnetization," *Phys. Rev.* 35 (1930): 733–752; R. M. Bozorth, "Barkhausen Effect: Orientation of Magnetization in Elementary Domains," *Phys. Rev.* 39 (1932): 353–356; K. J. Sixtus and L. Tonks, "Propagation of Large Barkhausen Discontinuities," *Phys. Rev.* 37 (1931): 930–958; Sixtus and Tonks, "Propagation of Large Barkhausen Discontinuities. II," *Phys. Rev.* 42 (1932): 419–435, and others.

260. The proceedings of this conference were published in *Phys. Rev.* 39 (1932): 337–377.

261. L. H. Hoddeson, "The Entry of the Quantum Theory of Solids into the Bell Telephone Laboratories, 1925–40: A Case Study of the Application of Fundamental Science," *Minerva* 18

(1980), 422–477, p. 440. For a more general view of the impact of the Depression on the American physics establishment, see S. Weart, "The Physics Business in America, 1919–1940: A Statistical Reconnaissance," in *The Sciences in the American Context: New Perspectives,* ed. N. Reingold (Washington, D.C.: Smithsonian Institution Press, 1979), 295–358, pp. 305ff.

262. Bitter (n. 257), 97ff.

263. Bozorth interview (1975) (n. 255).

264. Hoddeson (n. 261), 440.

265. Ibid.

266. Bozorth interview (1975) (n. 255). Elmen retired from his post at Bell Labs in 1941. For the rest of his life, he remained a consultant to the newly established magnetics laboratory at the Naval Ordnance Laboratory.

267. R. M. Bozorth, "The Orientation of Crystals in Silicon-Iron," *Transactions of the American Society of Metals* 23 (1935): 1107–1111; Bozorth, "Directional Ferromagnetic Properties of Metals," *Journal of Applied Physics* 8 (1937): 575–588.

268. M. Kersten, interview with M. Eckert, S. T. Keith, and H. Schubert, May 1982, DM.

269. H.B.G. Casimir, interview with S. T. Keith, 25 February 1981.

270. Oliver (n. 249), 84.

271. Kotaro Honda was the central and guiding figure for Japanese work in magnetism. See N. Kawamiya, "Kotaro Honda: Founder of the Science of Metals in Japan," *Japanese Studies in the History of Science,* no. 15 (1976): 145–158; S. Chikazumi, "Evolution of Research in Magnetism in Japan," *Journal of Applied Physics* 53 (1982): 7631–7636.

272. Oliver (n. 249), 77–78.

273. D. A. Oliver and J. W. Sheddon, "Cooling of Permanent Magnet Alloys in a Constant Magnetic Field," *Nature* 142 (1938): 209; K. Hoselitz, interview with S. T. Keith, 23 July 1981.

274. A. M. Tyndall to W. Sucksmith, 13 October 1941, papers and correspondence of the Wills Physics Laboratory, University of Bristol (kindly made available by Professor Norman Thompson).

275. Ibid.

276. K. Hoselitz, "30 Years of Magnetism in Britain" (Paper presented at the Second Conference on Advances in Magnetic Materials and Their Applications, 1–3 September 1976) (copy provided by the author).

277. R. M. Bozorth, "The Physical Basis of Ferromagnetism," *Bell Systems Technical Journal* 19 (1940): 1–39, p. 2.

278. F. Preisach, "Untersuchungen über den Barkhauseneffekt," *Ann. Phys.* 3 (1929): 737–799.

279. Kersten interview (n. 268).

280. I am grateful for the opinion of George Wise, historian at General Electric, on this matter. Wise considers that Sixtus, Tonks, or Langmuir initially read the Preisach paper, and in a discussion of the results Langmuir proposed an interpretation.

281. F. Bloch to S. T. Keith, 14 April 1982.

282. "Walter Gerlach Zum Neunzigsten," *Physikalische Blätter* 35 (1979): 370–374; a list of Gerlach's publications was kindly provided by Michael Eckert.

283. M. Kersten, "Some Personal Reminiscences of the History of Bloch Walls," *Journal of Magnetism and Magnetic Materials* 19 (1980): 1–5, p. 2.

284. Ibid.

285. Munich was a focus for the bitter struggle for the future of German physics; Gerlach's role is extensively discussed in A. D. Beyerchen, *Scientists under Hitler* (New Haven, Conn.: Yale University Press, 1977). Becker was rejected as a candidate to fill Sommerfeld's former chair in Munich in 1937, before his appointment to the chair of theoretical physics at Göttingen, left vacant by Born's departure. Becker's chair in Berlin was then immediately abolished.

286. L. F. Bates, *Modern Magnetism* (Cambridge: Cambridge University Press, 1939); R. Becker and W. Döring, *Ferromagnetismus* (Berlin: Springer, 1939); F. Bitter, *Introduction to Ferromagnetism* (New York: McGraw-Hill, 1937); R. Becker, ed., *Problem der Technischen Magnetisierungskurve* (Berlin: Springer, 1938); H.B.G. Casimir, *Magnetism and Very Low Temperatures* (Cambridge: Cambridge University Press, 1940).

287. L. F. Bates, "A Link Between Past and Present in British Magnetism," *Contemporary Physics* 13 (1972): 601–614, p. 610.

288. Stoner describes the proceedings in his diary entries for 21–24 May 1939, Stoner Papers (n. 208).

289. Proceedings of the conference were published as *Le Magnétisme* (n. 29). General reports of the conference included "Modern Views on magnetism," *Nature* 143 (1939): 1032–1033; S. J. Barnett, "International Conference on Magnetism, Strasbourg, 21–24 May 1939," *Science* 104 (1946): 70–73. Quote is from "Modern Views on Magnetism," 1033.

290. "Discussion générale sur le paramagnétisme," in *Le Magnétisme* (n. 29), 3:286–319, pp. 308ff. This paragraph was provided by P. Quédec.

291. Bates (n. 287), 610.

292. This episode is described in Stoner's diary entry for 24 May 1939, Stoner Papers (n. 208).

293. Ibid.

294. Bates (n. 287), 611.

295. Winston S. Churchill, *The Second World War,* 3rd ed. (London: Cassell, 1950), 1:342.

296. Ibid.

297. Barnett (n. 289), 73.

298. Kersten (n. 283), 3.

299. P. W. Anderson, "Some Memories of Developments in the Theory of Magnetism," *Journal of Applied Physics* 50 (1979): 7281–7284.

300. On Philips during the war, see J. G. Crowther, *Science in Liberated Europe* (London: Pilot Press, 1949), 217ff; Casimir (n. 244), 204ff.

301. On ferrites, see J. L. Snoek, *New Developments in Ferromagnetic Materials,* 2nd ed. (New York: Elsevier, 1949); *Interactions of Science and Technology in the Innovative Process: Some Case Studies* (Columbus, Ohio: Battelle, 1973), II-1ff; *Technology in Retrospect and Critical Events in Science (TRACES)* (Chicago: IIT Research Institute, 1969), 2:11ff.

302. "Colloque International de Ferromagnétisme et d'Antiferromagnétisme de Grenoble," *Journal de physique et le radium* 12 (1951): 149–508.

303. L. Néel, interview with A. Guinier and L. Hoddeson, 29 May 1981; L. Néel to E. C. Stoner, 10 August 1947, Stoner Papers (n. 208).

304. L. Néel, "Propriétés d'un ferromagnetique cubique en grains fins," *C.R.* 224 (1947): 1488–1490; Néel, "Le Champ coercif d'une poudre ferromagnetique cubique à grains anisotropes," *C.R.* 224 (1947): 1550–1551; E. C. Stoner and E. P. Wohlfarth, "Interpretation of High Coercivity in Ferromagnetic Materials," *Nature* 160 (1947): 650–651; Stoner and Wohlfarth, "A Mechanism of Magnetic Hysteresis in Heterogeneous Alloys," *Philosphical Transactions of the Royal Society* 240 (1948): 599–642; C. Kittel, "Theory of the Structure of Ferromagnetic Domains in Films and Small Particles," *Phys. Rev.* 70 (1946): 965–971. Kittel was certainly the first to publish, Néel the first to begin serious calculations, and Stoner probably the first to think about it.

305. L. Néel, "Effet des cavités et des inclusions sur le champ coercitif," *Cahiers de physique* 25 (1944): 21–44.

306. H. J. Williams, "Directions of Domain Patterns on Single Crystals of Silicon Iron," *Phys. Rev.* 75 (1949): 155–178, W. F. Brown, Jr., *Micromagnetics* (New York: Wiley Interscience, 1963), 11–23.

307. For a comprehensive review of the theory of magnetic domains, see C. Kittel, "Physical Theory of Ferromagnetic Domains," *Reviews of Modern Physics* 21 (1949): 541–583.

308. Mott (n. 239); J. C. Slater, "The Ferromagnetism of Nickel," *Phys. Rev.* 49 (1936): 537–545, 931–937; E. C. Stoner, "Collective Electron Ferromagnetism," *Proc. Roy. Soc.* A165 (1938): 372–414; Stoner, "Collective Electron Ferromagnetism II. Energy and Specific Heat," *Proc. Roy. Soc.* A169 (1939): 339–371. For a summary of prewar contributions, see J. H. Van Vleck, "A Survey of the Theory of Ferromagnetism," *Reviews of Modern Physics* 17 (1945): 27–47.

309. E. P. Wohlfarth, "History of Stoner Theory and Its Developments," *Institute of Physics Conference Series,* no. 55 (1981): 161–172.

310. J. H. Van Vleck, "The Present Status of the Theory of Ferromagnetism," *Physica* 15 (1949): 197–206, p. 202.

311. See Stoner (n. 3), 40; E. C. Stoner, "Collective Electron Ferromagnetism in Metals and Alloys," *Journal de physique et le radium* 12 (1951): 372–388.

312. For the conference proceedings, see *Reviews of Modern Physics* 25 (1953): 1–392.

313. C. Herring, "Exchange Interactions Among Itinerant Electrons," vol. 4 of G. Rado and H. Suhl, *Magnetism* (New York: Academic Press, 1966), 118ff.

314. On resonance, see Chang (n. 245); on neutron diffraction, see G. E. Bacon, "Neutron Beams for the Study of Condensed Matter: A View of the First Half Century," *Contemporary Physics* 23 (1982): 541–552; Bacon, ed., *Fifty Years of Neutron Diffraction. The Advent of Neutron Scattering* (Bristol: Adam Hilger, 1987); P. Schofield, ed., *The Neutron and Its Application, 1982* (London: Institute of Physics, 1982).

315. J. E. Goldman et al., "Magnetism and Magnetic Materials," in *Perspectives in Materials Research*, ed. L. Himel, J. J. Hasweed, and W. J. Harns (Washington, D.C.: Office of Naval Research, 1963), 75–154, p. 77.

316. Bates (n. 287), 613.

317. Ibid.

318. E. P. Wohlfarth, ed., *Ferromagnetic Materials,* 4 vols. (Amsterdam: North-Holland, 1980–1983).

7 | *Selected Topics from the History of Semiconductor Physics and Its Applications*

ERNEST BRAUN

7.1 **The Early Years**
7.2 **Semiconductors and the Second World War**
7.3 **The Postwar Period**
7.4 **Conclusion**

So vast is the field of semiconductor physics and so great the variety of semiconductor devices that the restricted treatment of this chapter has to be emphasized even at its beginning and in its cumbersome title. To treat the history of semiconductors with anything resembling completeness would exceed all reasonable bounds of a single chapter in this volume. A severe selection had to be made, and this was determined by many considerations. First was the requirement to show the emergence of a basic body of knowledge that could be identified as semiconductor physics. Second was the attempt to show the relationship between semiconductor physics and the most significant technological devices and innovations associated with semiconductors. The dominance of early semiconductor materials over later ones is a historical necessity and will, hopefully, point the way toward a sequel to this volume. Finally, the choice of episodes was, to some extent, determined by availability of material and by personal choice, knowledge, and interest. The treatment of some episodes may be more extensive than the balance of historical interest would warrant, but the author hopes that these sections convey something of the spirit of the history and reveal some not generally known facets of the development. To all those who find my choice too idiosyncratic, I offer my sincere apologies.

7.1 The Early Years

As with most things electrical, the beginnings of knowledge about semiconductors can be traced back to Michael Faraday. As early as 1833 he discovered that the resistance of silver sulfide decreases with temperature, whereas that of metals was known to increase.[1] Faraday soon added a series of other substances to the list of those with negative temperature coefficients of resistance and with conductivities well below those for the common metals. These facts remained curious isolated episodes, however, and were not immediately absorbed into the main body of theoretical knowledge.[2]

In 1874, Ferdinand Braun observed that the conductivity of some metallic sulfides depended on the polarity of the applied potential difference. In a series of papers published between 1877 and 1883, Braun showed that this effect was most pronounced if one of the electrodes consisted of a sharp wire tip pressed into one face of the mineral. Thus Braun had discovered the rectifying contact and had laid the foundation for what became known as the cat's whisker detector for radio waves, several years before the waves themselves were discovered.[3] Braun's discovery was of fundamental scientific importance and later gained considerable importance in radio and microwave technology. For the rectification of larger currents and power, the discovery by A. Schuster, in 1874, of the rectifying properties of the contact between copper and its oxide proved of even greater significance because it served both as a foundation for technical devices and as a model for the theoretical understanding of metal–semiconductor contacts.[4]

The previous year, Willoughby Smith discovered that the resistivity of selenium decreased when it was illuminated and thus established the phenomenon of photoconductivity. Although doubts were expressed about this effect and it was thought by some to be caused by heat rather than by light, the painstaking work of L. Sale and Werner von Siemens proved the photoconducting properties of selenium beyond doubt.[5]

With the discovery by Edwin Herbert Hall, in 1878, of the deflection of charge carriers in a solid by a magnetic field, an effect that has been called after its discoverer, the foundations for a proper understanding of the properties of semiconductors had been laid.[6] Although these substances were not thought of as a separate class in a systematic way as yet, it was clear that some materials exhibited mediocre electrical conductivity which increased with temperature, were photoconductive, could make rectifying contacts, and exhibited the Hall effect. The last fact in itself does not distinguish semiconductors from metals, but detailed measurements of Hall coefficients give detailed information about the sign and density of charge carriers, which allows a very clear distinction between semiconductors and metals to be made. One further characteristic of semiconductors—as we may call these substances with the benefit of hindsight—is the photovoltaic effect, discovered in 1876 by Adams and Day in England and, independently, by Charles Fritts in New York.[7] Siemens received some samples of Fritts's selenium specimens and reported to the Berlin Academy of Sciences in 1877 "an entirely new physical phenomenon, . . . whereby the direct conversion of light energy into electrical energy becomes possible."[8]

Thus the basic properties of semiconductors had been discovered by about 1880, but the term *semiconductor (Halbleiter)* was apparently introduced in 1911 by Königsberger and Weiss.[9] Two events of some significance for the practical development of semiconductor science occurred in the first decade of the twentieth century. The American G. W. Pickard was the first to show that silicon was a good detector material for radio waves, which by then had attained some practical importance, and the German K. Bädeker discovered that the conductivity of a layer of copper iodide could be altered by exposing it to iodine vapor.[10] Thus the first steps had been taken toward establishing silicon as a useful semiconductor material and, of equal significance, toward understanding the role of the precise chemical composition for the properties of a semiconductor.

Oddly, arguments about whether silicon was a metal or a semiconductor were to continue until the 1940s. The German professor B. Gudden, a close collaborator of R. Pohl, believed silicon to be a metal, presumably because of the experimental difficulty of obtaining samples that were good enough to establish its true nature with any certainty. So powerful was Gudden's authority that when Karl Seiler measured the conductivity of silicon, in about 1943, and found an intrinsic and extrinsic range, he dared not publish this result for fear of contradicting Gudden.[11]

From the early days of this century—that is, right from their infancy—semiconductors began to play a role in technology. The first application was as radio wave detectors. Disregarding the ingenious but distinctly quaint and somewhat impractical coherer, the cat's whisker semiconductor detector was the only useful device for the detection of signals in the early days of radio.[12] The cat's whisker radio detector, although extensively used in early applications and much beloved by radio hams even in the 1930s, never reached the stage of a technically mature product. It was soon displaced by the valve detector as part of a move toward electronically processed and amplified signals.[13]

Although the success of semiconductors as high-frequency electronic components was initially short-lived—with a spectacular future as yet unforeseen—success as rectifiers and photoelectric cells became more firmly established.

In 1925, E. Presser developed a selenium rectifier, and one year later L. O. Grondahl announced a copper oxide rectifier.[14] All major electrical firms took up further development of these devices, and in the ensuing 30 years or so many were produced all over the world. One of the first applications was in measuring instruments, but by 1939 the technology of copper oxide rectifiers was sufficiently developed to produce systems that could supply 30,000 Amp at 6 V for electrolytic coating or 1 Amp at 5000 V for radio transmitters.[15]

Selenium rectifiers proved of even greater practical significance. Up to and during the Second World War, the device was extensively used in battery chargers and portable radio receivers.[16] During the war, the selenium rectifier became vital to the German military machine because it was used in all communication and radar equipment. Major production problems arose in 1942, when the so-called thallium catastrophe hit production, an episode illustrative of the difficulties of taming these wayward substances. By the end of the 1920s, it was common knowledge that trace impurities influenced the behavior of semiconductor rectifiers both in their resistance to forward flow of current and in their ability to withstand reverse voltages. Details of the mechanisms were not understood, however, and manufacture was based on almost pure empiricism. The "thallium catastrophe," when devices with very poor forward characteristics came off the production line, was the result of using solder from a different source than had provided ealier solder. It was soon discovered that thallium was present as an impurity in the new solder. Although this impurity had beneficial effects on the reverse characteristic, its effect on the forward conductance was catastrophic. No fundamental progress was made, however, because no truly major research effort was mounted during the war in Germany. A change of solder brought production back to scratch, and nobody was any the wiser.[17]

The manufacture of the copper–cuprous oxide and the selenium rectifiers, although carried out in very large numbers throughout the world, never left the realm of empiricism. The processes employed are described in some detail by Heinz

Henisch, in an early attempt to bring together the theory and practice of semiconductor rectifiers.[18]

While practical semiconductor devices made gradual progress by empirical development, with only few results of general validity, theoretical progress was made in two major directions: band theory of solids, including semiconductors, and the theory of rectifying contacts. Soon after the pioneering efforts that led to the understanding of the behavior of electrons in metals in terms of energy bands (Chapter 2), A. H. Wilson proposed that the distinction among metals, insulators, and semiconductors could be understood by considering the band structure. Wilson belonged to the generation of Cambridge students whose theoretical understanding was strongly influenced by R. H. Fowler. Indeed, he was Fowler's research student and tells one of the best anecdotes about the great man. When he was given a rather intractable problem to solve for his doctoral dissertation and returned to Fowler for advice, the only comfort he was offered was the somewhat discouraging "if I had been able to solve the problem, I would not have given it to you."[19]

The great breakthrough in Wilson's thinking about band theory came while he was in Leipzig as a postdoctoral research fellow, working in Heisenberg's group. With hindsight, the basic idea seems simplicity itself. By 1930 it was becoming established that the Pauli principle, as applied to periodic structures, required that the electrons occupy energy bands instead of single energy levels and that the electrical and thermal properties of metals could be explained by assuming a large number of free electrons obeying Fermi–Dirac statistics. Wilson's great contribution was the consideration that if electrons in a crystalline solid are located in energy bands, then the puzzling differences among the electrical conductivity of metals, of semiconductors, and of insulators could be explained on a single assumption. If the highest occupied energy band is not quite full, as is the case in metals, then an applied electric field can accelerate electrons in the direction of the field and an electric current results. If the band is full, then an applied field cannot impart energy to the electrons and the crystal behaves as an insulator. Semiconductors are insulators with a small forbidden energy gap between the last full band (the valence band) and the first empty one (the conduction band). Thus at elevated temperatures, some electrons leave the full band and populate the nearby almost empty band. These electrons can participate in a conduction process, and the semiconductor becomes increasingly conductive as the temperature is raised. This is the mechanism of intrinsic conduction.[20]

Each electron raised into the conduction band leaves behind an unoccupied state in the valence band. The collective motion of the electrons in the valence band can then be described by the motion of the vacant state, which became known as a hole. The hole has properties equivalent to those of a positive charge carrier. The elucidation of the concept of the hole, and with it the explanation of the previously puzzling observation of a positive Hall effect in some materials, was achieved by Rudolf Peierls. Peierls had been a student of Sommerfeld, and joined Heisenberg in Leipzig in the spring of 1928. At that time Felix Bloch, also in Leipzig, had just completed his celebrated work on electrons in a periodic potential, which represented a considerable advance over the previous Sommerfeld theory. In the autumn of 1928, Heisenberg suggested that Peierls look into the problem of the positive Hall effect. Peierls treated the problem by considering the mean velocity of electrons in a peri-

Fig. 7.1 The energy levels in a semiconductor. (From A. H. Wilson, *Theory of Metals* [Cambridge: Cambridge University Press, 1954], 114)

odic field to be given by dE/dK and the action of a magnetic field to be represented by the Lorentz force.[21]

> Thus for an electron near the band edge an electric field could cause a decrease, rather than an increase, in the velocity in the field direction. One's first shock on seeing this result is the fear that it might lead to a negative conductivity. One soon realizes, however, that for an ensemble of electrons in statistical equilibrium the positive acceleration of the electrons near the bottom of the band outweighs the negative acceleration of those near the top, until for a full band the current just vanishes.[22]

If the temperature is low enough for the conduction band of a semiconductor to be empty, then a pure semiconductor behaves exactly as an insulator. If, however, impurities and imperfections exist—as indeed they must—these may give rise to localized electronic states within the forbidden gap (Fig. 7.1). Wilson recognized this fact in his second 1931 paper and saw clearly that the localized states can become ionized at temperatures low enough for intrinsic conduction to play no role.[23] Thus an extrinsic semiconductor is one in which free charge carriers have become available by ionization of localized states. A state that is normally neutral and becomes positively charged by donating an electron to the conduction band is called a donor state; one that becomes negatively charged by accepting an electron from the valence band is called an acceptor state. The trapping of a valence electron in a localized state causes the creation of a hole.

The crucial role of impurities in determining the properties of semiconductors, long recognized by Bädeker, Pohl and his school, and others (Chapter 4), which had caused so much puzzlement and had given solid-state physics a bad name, suddenly became clear and almost blindingly obvious. Generally, theoreticians mistrusted phenomena that depended on minute impurities and tended to regard such matters as "dirty physics." As Peierls remarked, "I was, of course, aware of the work by Pohl and others on the origin of luminescence and other optical phenomena, but since these results depended strongly on the state of purity of the substance they seemed too complicated for a basic theoretical approach.[11] [24]

Wilson, who had come to Leipzig after Peierls had left for Zurich, had not only found a way to generalize the effect of impurities for a certain case, but also tried to expand this success into neighboring fields. The invasion, as it were, of neighboring fields by a method once used successfully is a most common way of scientific and technical progress. Wilson tackled the very problems Peierls regarded as unsuitable and wrote a paper on photoconductivity in alkali halides, a domain dominated by Pohl's experimental school.[25]

Obviously, the simple picture just described does not adequately explain the full intricacies of the band structure of metals, insulators, and semiconductors. Nevertheless, a vital first step was made in connecting diverse experimental observations with a new fundamental theory. It was shown that band theory, which was a consequence of quantum theory, could explain the extremely diverse behavior of different classes of solids, and thus the understanding of solids was considerably deepened.

Wilson's theory was an immediate success, and the distinction of metals, semiconductors, and insulators in terms of band theory became part of basic accepted knowledge. This is not to say that doubts about the proper classification of individual materials did not persist, but one of the great aims of scientific theory was fulfilled—a single principle was enunciated that could explain a wide range of experimental phenomena.

The great breakthrough in semiconductor physics had been achieved by the logical and consistent use of Bloch's band theory, which, in turn, had been the result of quantum mechanics and the exclusion principle. Thus we might speak of cross-pollination in that theories were transferred from other fields of knowledge, but the theory of semiconductors soon acquired so much of its own flavor that it achieved its own status as a theory in its own right. Perhaps Peierls's treatment of the hole was the first step in this direction, for the behavior of charge carriers in a nearly empty band is an essential feature of semiconductors and of semiconductors only.

We now turn to another puzzle that exercised the minds of physicists: the theoretical explanation of rectifying contacts. As shown earlier, the copper oxide rectifier, consisting of a contact between copper and cuprous oxide, was widely used as a technical device. A theoretical understanding of the rectifying process was not available, however, and empirical technical practice was, as is not uncommon, far ahead of theoretical knowledge.

The successful theory of rectifying contacts is connected with three people who worked on it more or less independently and simultaneously in the late 1930s and early 1940s: W. Schottky, N. F. Mott, and B. Davydov, although they were not the only significant contributors. As with most successful theories, the theory of rectifying contacts had several precursors—early attempts that failed to gain general acceptance. Several of these are described by Henisch.[26] One of the earliest was proposed in 1930 and assumed that the application of a voltage caused the piezoelectric contraction or expansion of the rectifier and thus affected the true contact area between copper and cuprous oxide. Plausible as the theory is, it failed when it became clear that neither copper nor cuprous oxide exhibited any piezoelectric effect. Other theories sought explanations in terms of ionic conduction within the oxide layer and were based on a mix of hypothesis and microscopic examination of the structure of the contact region. These theories never reached a stage where a generally valid principle could be distilled from detailed descriptions. Similarly, theories that described large-area contacts as a series of point contacts did not represent much of a breakthrough because they only shifted the lack of understanding from large-area contacts to similarly puzzling point contacts.

Perhaps the most immediate forerunner of the modern theory of rectification was a suggestion put forward by the main inventor of the copper rectifier, Grondahl.[27] He suggested that a cloud of electrons existed in the oxide layer and could be swept away if a forward voltage was applied, while in the reverse direction the cloud accumulated and the space charge prevented the flow of current. The theory could not explain details of the process, but it established the principle of a space

Walter Schottky (b. 1886). (AIP Niels Bohr
Library, Brattain Collection)

charge as an important feature of rectifying contacts, and this principle stood the
test of time.

The development of the modern theory of rectification, with particular reference
to the work of Schottky, has been described by Spenke, and this description is essentially followed here.[28] The model rectifier was the copper–cuprous oxide system
because the selenium rectifier and point contact detectors appeared messier and
more intractable.

Walter Schottky was, undoubtedly, one of the key figures in the rectifier story
and, thus, in the history of semiconductor physics. He spent most of his working
life in one industrial firm (Siemens, in Germany), and his scientific interests were
invariably stimulated by technical problems, the rectifier being a case in point. Yet
his work was always of fundamental theoretical character and of great generality.
Schottky's interest in thermionic emission, in connection with thermionic valves,
led to fundamental considerations of the thermodynamics of charged particles and
to the development of the so-called electrochemical potential (equal to the Fermi
energy) as a determinant of thermodynamic equilibrium.[29]

This was an extension of the basic treatment of equilibrium by J. W. Gibbs. In
essence, the equilibrium of a system demands that charged particles be in equilibrium both in their roles as particles of a chemical system and as charge carriers; that
is, the electrochemical potential throughout the system should be equal and free of
gradients.

One of the results of Schottky's preoccupation with thermodynamics was the
realization that lattice vacancies and other singularities were subject to statistical
treatment. The equilibrium density of lattice defects can thus be calculated.[30] One
result of this work is the name given to lattice vacancies: the Schottky defect. The
occupancy of localized electronic states is similarly determined by their position in
relation to the electrochemical potential and by the temperature.[31]

Schottky's interests in thermionic emission and in crystal defects combined in
his work on rectification. Here was a true scientific puzzle brought to Schottky's
attention by industrial practice: Why do metal–semiconductor contacts have rectifying properties? The approach adopted by him was determined by his previous
interests, and this proved to be a case of extraordinarily successful cross-pollination.

Schottky's first contribution to the theory of metal–semiconductor contacts goes back to 1923.[32] At that stage, he thought that the effect could be explained by the transition of electrons from a good emitter to a poor emitter through an intervening thin insulating layer. The theory became untenable when measurements carried out by his co-workers on copper–cuprous oxide rectifiers showed that the capacitance associated with the contacts was too low and hence the barrier had to be rather thick. The nearly full theory of barrier layer rectification was published in 1939.[33]

Apparently, Schottky had set up the equations for a thick barrier layer, assuming it to be a region depleted of charge carriers as early as 1932, but did not then publish his results for two reasons. For one, the theory gave the wrong sign for the rectification effect of a copper–cuprous oxide layer, and, for another, it was hard to explain why the oxide layer needed special treatment in the practical manufacture of rectifiers.[34] The first difficulty was resolved when it turned out that cuprous oxide was, in fact, a *p*-type semiconductor, and previous Hall-effect measurements, which had classed it as an *n*-type conductor, had been erroneous.[35] This is a good example of a correct theory being rejected because of a wrong experimental result. Schottky does not explain how the second difficulty was removed, and it seems to be a case of a problem simply going away. The explanatory power of Schottky's theory was great, too great to worry about the fact that there were complexities of oxide layers which the theory could not explain. Deviations of practical devices from ideal behavior no longer surprised anybody. It was realized that the interface between a metal and its oxide was not in reality the abrupt transition postulated by theory. Similarly, it was appreciated that contacts between metals and semiconductors other than oxides were similarly bedeviled by practical problems. It is understandable, however, that Schottky worried about ascribing the rectifier effect to an intrinsic property of the physical system, when in practice the system had to be chemically manipulated to function. The theory became to be regarded as essentially correct, even if not all practical cases could be fully explained by it. Thus theories are rejected only if objections weigh heavily against them, while minor blemishes are tolerated. What is regarded as a minor blemish depends on a balance between the positive powers of explanation of the theory and the gravity of the objections. The weight of argument is a matter of judgment for both individual scientists and the scientific community in the formation of a consensus view.[36]

Back in 1932, one of Schottky's correspondents on problems of rectification was Rudolf Peierls. In March 1932, Peierls wrote: "It is plausible to assume that in practical detectors the density of electrons in the boundary layer is smaller than the equilibrium density and thus it becomes a layer of high resistivity."[37] This helped Schottky in the final realization that a depletion layer in the semiconductor could arise as a result of a contact potential difference between the metal and the semiconductor. Schottky's great contribution was an exact calculation of the shape and width of the depletion layer and a very detailed description of the equilibrium conditions that led to the barrier layer. Schottky thus was the main exponent of the so-called physical barrier theory. This assumes that the barrier layer is caused by equilibrium requirements in the region of the semiconductor close to the metal.

The crux of Schottky's work is the realization that the concentration of electrons in the bulk of the material is determined by factors quite distinct from those that determine the electron concentration in the surface layer of the semiconductor

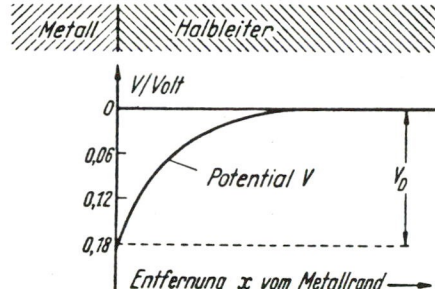

Fig. 7.2 Metal–semiconductor rectifying contact, showing a depletion layer. (From E. Spenke, *Elektronische Halbleiter* [Berlin: Springer-Verlag, 1955], 75)

close to the adjacent metal. The density of free electrons in the bulk of the semi-conductor, n_B, is determined by the density n_D of donor states that have been thermally ionized, or, in other words, have become locally positively charged by donating an electron to the conduction band. In the simplest case, $n_B = n_D$. Near the contact with the metal, the density of electrons is not determined by the density of donors, but is dominated by the ease with which electrons can leave the metal to enter the semiconductor—in other words, by the contact potential difference between the two substances. Electrons will enter or leave the semiconductor until the potential difference between the two materials disappears. Because the density of electrons in the semiconductor is low, the field caused by the potential difference extends over a substantial distance. Figure 7.2 shows the case in which the field is such as to cause a depletion layer in the semiconductor (i.e., a layer in which the density of electrons is lower than that in the bulk). The depletion layer may have a resistance substantially in excess of that of the bulk.

If an external voltage is applied such as to deplete the layer further, no current will flow; if the external voltage lowers the barrier V_D (*Diffusions-Spannung*, according to Schottky), then current can flow through the device. Thus the rectifying properties of the contact are explained. The essential feature is that the positively charged donor centers in the barrier layer give rise to the field that compensates the contact potential difference between metal and semiconductor. The depletion layer becomes wider when a reverse voltage is applied and narrower with the application of a forward voltage. The metal must have positive polarity in the forward direction. Schottky pointed out that measurements of the capacity of the barrier layer would yield information on the concentration of donor centers, as is indeed the case.

The 1939 *Zeitschrift für Physik* paper uses a great deal of plausibility arguments, while the more quantitative aspects were published in a paper with Spenke in the same year.[38] In a final paper on the barrier layer theory of rectification, in 1942, Schottky brought the whole edifice to virtual perfection.[39]

Schottky's theory was not, however, unrivaled. One of the competing theories assumed some form of intervening chemical layer interposed between metal and semiconductor. The theory was developed by N. F. Mott in 1939, and a preprint of this paper was sent to Schottky and is mentioned by him.[40] For the copper–cuprous oxide rectifer the chemical barrier was thought to consist of a layer of semiconductor that was free of impurity centers and thus represents a depletion layer on chemical rather than on physical grounds. The theory of the potential in such a structure is then developed both for equilibrium and for current flow in the forward and reverse directions. Although the shape and, more important, the cause of the barrier in Mott's theory were not eventually accepted, his treatment of the flow of current as a function of applied voltage was accepted and proved identical with the independent treatment by Schottky and Spenke. The essence of the treatment is that the current across the barrier layer is composed of a drift current (charge carriers moving in an electric field [i.e., a gradient in electrostatic potential]) and a diffusion current (charge carriers moving in a concentration gradient [i.e., a gradient in chemical potential]). Good results are obtained in this way, although some corrections have to be applied for certain cases that need not detain us here.

Having done Mott some injustice by treating his contribution too briefly, we shall compound the felony by meting out the same treatment to Davydov. Davydov proposed his theory as an alternative to the tunneling theory (see later), in response to new knowledge on the excessive thickness of the barrier layer and the sign of the majority carriers in copper–cuprous oxide rectifiers. He considered a contact between an n-type and a p-type semiconductor and explained the rectifier action by a "blocking layer" with a surplus of "electropositive atoms or ions." This view seems to be close to the view of a "chemical" rather than an inherent physical layer. Although Davydov drew an abrupt potential transition in the region of the $p–n$ junction, he did realize that the equilibrium concentration of carriers in the barrier region is altered by the applied potential and said that these "must lead to the generation of concentration electromotive forces." Davydov's rather complex calculations yield an equation that gives the correct result over a very limited range of applied potential.[41] Although Davydov's treatment was essentially correct, it lacked "Schottky's thoroughness and the felicity of his concept of electrochemical potential."[42] In a lecture, Herring noted that

> his [Davydov's] equations, though not entirely correct, contain the essential idea of minority carrier injection or depletion, and the existence of non-equilibrium concentrations within a diffusion length from the junction. But the physical picture of what was happening was not made clear, and it seems that even within Russia neither theorists nor experimenters acquired from this paper a clear enough understanding of the important junction phenomena to make further advances which would correct and verify the theory.[43]

The victorious diffusion current theory was not always unchallenged. In fact, two competing theories were proposed, and each had famous progenitors. While the diffusion theory assumes that the barrier is much wider than the mean free path of charge carriers, the diode theory, proposed by Bethe, does not require this condition.[44] The results of the diode theory are not much different from those of the diffusion theory, but the latter is applicable to a wider range of situations. For germanium and silicon, the diode theory is quite valid.

The second rival theory was proposed by Wilson, by Nordheim, and by Frenkel and Ioffé.[45] In this theory, the electrons were supposed to tunnel through the potential barrier at the metal–semiconductor interface. The theory is not applicable to the thick barriers found in practice and predicts rectification in the wrong direction. Although this theory never found acceptance, it represents an interesting unsuccessful attempt to transfer knowledge from one application to another. The quantum-mechanical tunnel effect was then very new, and it was worth the effort to try to apply the theory to a problem that it might conceivably fit. Especially for Lothar Nordheim, who had worked out the theory of field emission based on the tunnel effect in collaboration with Fowler, the attempt to transfer his knowledge to the puzzling problems of rectification is very understandable.

The work of Schottky was, as stated earlier, done in close association with experimental and development work carried out by Siemens. Clearly, other firms were not idle in the matter of solid-state rectifiers. As one example, Philips in the Netherlands not only produced such rectifiers, but also made contributions to their theory. These contributions are described by J. Schopman.[46] In the main, W. C. van Geel supported the chemical barrier theory in combination with the theory of field-effect emission, which is essentially similar to the tunnel theory described previously.

It would appear that no harm was suffered by Philips from the fact that van Geel bet on the wrong horse. The coupling between theory and practice was still weak enough to enable a firm to manufacture perfectly good rectifiers, despite its faith in a theory that was eventually rejected.

If the copper–cuprous oxide system proved difficult to understand, the selenium rectifier proved infinitely more resistant to theoretical analysis. In the immediate postwar years in Germany, much effort was put into reconstituting a viable electronic industry, including the manufacture of selenium rectifiers. Siemens decided to open a small laboratory in Pretzfeld, with E. Spenke in charge, where research and pilot plant production of selenium rectifers went hand in hand. Initially, chemical analysis had to be improved in order to establish the effects of various impurities on the performance of the rectifiers. At the same time, Siemens began the development of evaporated selenium layers, a process that earlier had given Allgemeine Elektrizitäts Gesellschaft (AEG) a technical lead.[47] A long series of experimental investigations eventually led to the conclusion that the selenium rectifier consisted of a contact between p-type selenium and n-type CdSe. The selenide was formed chemically by a reaction between the selenium layer and a top electrode consisting of a CdZn eutectic. This work was greatly facilitated by the theoretical understanding of barrier layers.

The technical success of selenium rectifiers was fairly spectacular, and they virtually displaced the mercury-vapor rectifiers in high-power applications, although they were also used extensively for small current purposes in communications, radio, and so on. Millions of units were produced each month.

At this postwar stage, theory was certainly helping and guiding technology, while in earlier years of rectifier development, technology had been way ahead of theory and had paid little attention to it. According to Henisch, "When theories change very rapidly and each apparently incorporates an equal measure of truth without being of any immediate assistance to the development engineer, there is a tendency to regard all theories as more or less false."[48]

Although many scientific questions concerning selenium rectifiers remained unresolved, the selenium work was stopped in Pretzfeld in 1952 and all further effort was devoted to the development of germanium and silicon devices for power applications. This decision was, of course, taken in the light of the first years of (mainly American) experience with germanium transistors and diodes. History has proved the wisdom of this decision. We shall follow these developments in a later section and return now to a more chronological description. Before turning our attention to the dark years of the Second World War, we shall briefly describe two discoveries that illustrate the convoluted ways in which understanding grows.

It is characteristic of semiconductor physics that a great deal of experimental evidence is accumulated in an almost random exploratory process before a clear picture emerges. Inevitably, numerous experimental observations have to be made before a theory can be put together, and then more results are accumulated to test and refine the theory. Because solids have the unhappy property of almost individual behavior of different specimens, this process of trial and error is extremely laborious. We briefly mention only two items of this apparently random collection of facts, both of which are connected, with hindsight, to the motion of minority carriers. The work is also connected with the Russian school of solid-state physics, which is rather underrepresented in our narrative.

The first of these is the so-called Dember effect, discovered in 1931.[49] A sample of Cu_2O was illuminated and an emf was observed between the illuminated face and the back of the specimen. Dember explained this on the basis of strong optical absorption and the creation of internal photoelectrons. The emf results from a gradient of free electrons along the path of the light. We now know that we are dealing with the creation and diffusion of minority carriers, but the matter gave rise to a great deal of discussion in the pre-transistor days. Mott and Gurney describe the discussion in some detail and pointed out that W. W. Coblentz had discovered the Dember effect some 10 years before Dember.[50]

The second discovery was made by I. K. Kikoin and M. M. Noskov in 1933 and has become known as the photoelectromagnetic (PEM) effect.[51] The authors themselves produced a somewhat cumbersome explanation, involving the external photoelectric effect and a surface space charge. Frenkel explained the experimental observations in a very much simpler theory, assuming that holes and electrons have comparable mobilities.[52] The excess carriers produced by light diffuse away from the illuminated face, and the magnetic field, perpendicular to the diffusion current, causes a voltage to appear as in the Hall effect. The effect can be used to measure lifetimes of carriers and recombination velocities.[53] The PEM effect was later studied extensively and may also be regarded as an example of the diffusion of minority carriers. At the time, however, no understanding of the behavior of minority carriers was available, and the two discoveries remained relatively isolated and somewhat puzzling effects. Herring remarked that "although these two effects produced quite a flurry in the literature at the time, they did not succeed in imprinting permanently on the minds of research workers the possibilities of non-equilibrium minority carriers as a research tool."[54]

7.2   Semiconductors and the Second World War

Accepted wisdom has it that the Second World War considerably enhanced our knowledge of semiconductors and accelerated their widespread technical application. Although the proposition is not as well founded as its wide acceptance would imply, it certainly is part of the distortions of war that some aspects of science receive intensive attention while others are forced to lie dormant. It so happened that semiconductors showed great utility as radar detectors and were therefore regarded as vital to the war effort on the Allied side. The Germans discovered this utility much later, and their research effort never equaled the Allied, particularly the United States, scale.

Radar (i.e., the use of the echo of high-frequency electromagnetic waves for the detection of objects) was a prewar invention, and early developments were undertaken by several military establishments, which could all see the utility of detecting ships or aircraft at considerable distance in conditions of poor or zero visibility. The trouble with early developments was the very low power of electromagnetic waves available, particularly at the desirable high frequency. The other problem was that amplification of high-frequency signals was not possible, and the signal had to be detected and "mixed" at available signal strength. "Since no satisfactory amplifier for microwaves exists, the conversion to intermediate frequency must be made at the low level of received signal powers. . . . The most satisfactory device that has been found is the rectifying contact between a metallic point and a crystal of silicon."[55]

Undoubtedly, the British effort on radar was the most successful. By a combination of a high-power magnetron, which far surpassed any source of microwaves available elsewhere, silicon detectors, and skillful engineering, the British were able to build truly useful radar sets well before anybody else. It is arguable that these made an important contribution to the Battle of Britain.

The earliest radar mixer crystals were made by British Thomson-Houston (BTH) and used commercially available silicon of about 98% purity. It was common practice to remelt the commercial silicon with some addition of aluminum, thus obtaining more homogeneous and uniformly p-type ingots.[56] The "crystals" were, of course, coarse-grained polycrystalline specimens and exhibited sensitive and less sensitive spots that had to be found by trial and error. Their sensitivity varied considerably from lot to lot. Somewhat later, the General Electric Company developed the so-called red dot detector, which could withstand fairly large currents and was altogether superior. In radar parlance, it was a "high-burn-out" detector and thus had a longer service life.[57]

The two keys to success were thus the magnetron and the silicon detector. Although both the U.S. Army and the U.S. Navy had developed their own radar sets in the late 1930s, these sets were fairly rudimentary efforts and had involved little basic science. When it became clear that the Americans would not avoid involvement in the war for very long, they began to give serious thought to the further development of radar. The first visible sign of this thinking was the founding of a special research center for radar development, the Radiation Laboratory, in association with the Massachusetts Institute of Technology. The Rad Lab, as it became commonly known, undertook and coordinated massive research-and-development work on all aspects of radar, including work on semiconductor detectors.[58]

In June 1940, President Roosevelt authorized the establishment of the National Defense Research Committee (NDRC) and appointed Vannevar Bush, at that time president of the Carnegie Institution and of the National Advisory Committee on Aeronautics, as chairman. One of the eight members of the committee was Karl Compton, a prominent physicist and president of MIT. Compton headed Division

D, which was to look after most radar development, in particular detection, controls, and instruments. Compton selected his close friend, the wealthy Wall Street lawyer and amateur physicist Alfred L. Loomis, as his vice chairman and head of Section D-1. Loomis was not only a true "insider," but also an anachronism in that he kept a private research laboratory in which he worked side by side with professional physicists on problems of microwaves. In fact, Loomis had, in cooperation with MIT, produced an 8.6-cm radar set in early 1940, even before the Rad Lab was in operation.

Loomis's committee, which became known as the Microwave Committee, included high-ranking industrial as well as academic scientists, and they believed their task was "to organize and coordinate research, invention, development, design and manufacture so as to obtain the maximum number of effective military applications of microwaves in the minimum time."

Loomis and his committee gathered around them many experts and much information, but the weak point of early centimeter-wave radar sets remained the power source. This situation changed dramatically when a British scientific and technical mission, headed by Sir Henry Tizard, arrived in the United States in September 1940. In their luggage, the mission brought a most precious gift, a cavity magnetron capable of delivering 10 KW at a 10-cm wavelength, a performance 1000 times better than anything produced at that time in the United States. Some members of the Tizard mission presented the magnetron in Loomis's laboratory in Tuxedo Park, and the idea of founding a central U.S. microwave radar laboratory was born at the very same meeting.

The British mission described successful attempts to involve civilian scientists in the war effort—for example, in the Air Ministry Research Establishment at Swanage. The British report encouraged the Americans in their decision to set up a laboratory staffed and directed by civilians. In view of Loomis's close association with MIT, it is not surprising that the Rad Lab was eventually set up there, although not without tricky negotiations about rival claims. Everything moved very quickly, and Lee DuBridge, a physics professor from the University of Rochester, was appointed director of the laboratory in mid-October. Intensive recruitment brought some very fine physicists to the lab, among whom was I. I. Rabi from Columbia University, who immediately left on a recruitment trip. "I would go to some place and say: Now look, we have a laboratory at Cambridge. I can't really tell you what it's about. It isn't a very good laboratory, but hell, I'd like to see you there in about two weeks. And they'd come."[59] By mid-December about 40 scientists had joined the Rad Lab, and they became known as the First Forty. The organization into groups was functional and resembled a hierarchy, but all boundaries were permeable and the atmosphere was informal. At the peak of its development, the Rad Lab employed some 1200 physicists and nearly 4000 people in all.

The army and navy each continued work on longer-wave sets, but liaison between the civilian Rad Lab and the military existed at all times, although with different degrees of ease and success. Military advice was always available at various levels, and eventually radar sets designed by the Rad Lab were produced in large numbers by industrial contractors for the armed forces. Constant liaison with the British research effort was also maintained, and a high-level British scientist was always located at the laboratory. Rabi noted, "Everything we learned at the beginning about radar we learned from them [the British]. . . . We came to know a great deal, not because we were smarter, but because we were smart enough to listen to the British, who had had real operational experience."[60]

The Microwave Committee and the Rad Lab parceled out research-and-development contracts to outside organizations and coordinated their efforts. The main beneficiaries were industrial concerns that were, of course, also manufacturing contractors. Some contracts went to universities, such as MIT, Columbia, Purdue, and Pennsylvania. As far as semiconductors are concerned, the most important contractors were Purdue on the properties of germanium, the University of Pennsylvania on the properties of silicon, and Bell Telephone Laboratories and Sperry Gyroscope on the design and manufacture of detectors. Work on these aspects was coordinated by H. C. Torrey.

Frederick Seitz moved to the University of Pennsylvania early in 1939 and was busy building up a mixed experimental and theoretical solid-state physics group when he became involved in Rad Lab research. He wrote:

> In 1940 when the rectifier problem became acute, some of my friends at the Radiation Laboratory approached me and asked if we would consider taking a contract to work on silicon rectifiers at the University of Pennsylvania. We were glad to do this and I was joined by colleagues there, particularly Andrew Lawson . . . and Park Miller. . . . Still later we were joined by others such as Robert Maurer, Leonhard Schiff, Simon Pasternack, Bernard Serin and some graduate students. We rapidly set up equipment for melting silicon and for measuring both conductivity and the Hall coefficient.[61]

One of the first tasks tackled by the new silicon group was an attempt to improve the raw material. Lawson did a lot of re-melting and doping and eventually settled on boron as a doping agent. Du Pont was placed under contract to produce pure silicon, and the firms's effort was, by the standards of the day, highly successful.

The various groups working on similar problems under the aegis of the Rad Lab soon formed into little scientific communities. "I should make it very clear that something in the nature of a scientific club was quickly formed by all individuals and organizations involved in the work on silicon." Thus Seitz, although in later passages of his letter it becomes very clear that he should have said "silicon and germanium." The social nature of modern science, even in secret war work, stands out as one of its most characteristic features. There can be no science without the interaction of scientists with their peers, as Seitz acknowledged:

> We met regularly, perhaps every six weeks or so, to review results, trading information about new experiments and related concepts. Since the band theory of solids was well established, we quickly recognized that silicon and germanium were band gap semiconductors and that the doping agents provided extra donor or receptor levels in the gaps. We also noted the transition from intrinsic to impurity conductivity. Interestingly enough, the first such analysis was made by our group at Pennsylvania.[62]

Obviously, the recognition of germanium and silicon as band-gap semiconductors was not at all obvious at the time. The experimental distinction between intrinsic and extrinsic conduction in germanium was independently discovered several times over, and the Pennsylvania group was among the small band of original discoverers of this fundamental phenomenon.

Somewhat later than Seitz and his group, the Department of Physics at Purdue University, under the leadership of Karl Lark-Horovitz, also became a contractor to the Rad Lab. Because of the efforts of some individuals, the Purdue war work was particularly well documented, and thus some historical injustice has been done to other, equally deserving, groups. A list of problems to be worked on outside the Rad Lab was prepared in January 1942. In February, Lark-Horovitz visited various laboratories to learn the state-of-the-art on radar detector development. Among the places visited was the Sperry Gyroscope Corporation, which had apparently been the first to introduce a germanium detector, while in most applications the silicon detector, based on British developments, was used. Lark-Horovitz was allowed to join the "club" and obtained all the knowledge and information available at the time. This proved a considerable starting and continuing help.

The brief given to the Purdue group was to "investigate the problem of the crystal detector experimentally and theoretically, so as to develop a compact, sensitive and shock-proof crystal unit for ultra-high frequencies." This is clearly the sort of brief which is part of the success of U.S. science, a combination of fundamental theoretical and experimental investigations with a very clear, albeit elusive, technical goal in mind.

In March 1942 the germanium research group was set up at Purdue and started work on purification of germanium. By May 1942, when a meeting was held at MIT, Lark-Horovitz was able to report that both p- and n-type germanium had been produced, using boron, aluminum, gallium, indium, arsenic, and bismuth as doping agents. Like all other solid-state experimentalists, the Purdue group soon found that the behavior of commercially available germanium was quite erratic. Hence they were forced to develop purification procedures to enable them to obtain reasonably clean germanium from the dioxide bought from a commercial supplier.

In August 1942 a meeting was held at Columbia University at which Lark-Horovitz expressed his conviction that both germanium and silicon were true intrinsic semiconductors. This means that in measurements of resistivity against temperature, an intrinsic range could be separated from an impurity range. Numerous plots of resistivity versus inverse temperature were produced during March and April 1942 for germanium. Thus the Purdue group confirmed what had earlier been found at the University of Pennsylvania.

Only a little later and quite independently, a young student in Pohl's laboratory in Göttingen also found the intrinsic conductivity of germanium. J. Stuke measured the resistance of germanium as a function of temperature from early 1944 to the end of the war. When he attempted to submit his results for a doctoral dissertation, Pohl refused to accept it. He lacked faith in elementary semiconductors, and it took two more years and some American publications before Pohl not only accepted the thesis but humbly acknowledged his error of judgment.[63]

The Purdue group worked on four experimental aspects of the germanium problem: (1) measurement of electrical parameters, such as Hall effect, mobility of carriers, lifetime; (2) purification and preparation of materials; (3) burnout and high back-voltage rectifiers; and (4) radio frequency properties of semiconductors. The work was supported by at least one theorist in addition to the direction by Lark-Horovitz.

Hans Bethe was one of the consultants for the detector project, and his theory of the rectifying contact was apparently used throughout this work.[64] His was the diode theory, which applies to barrier layers that are thinner than or comparable with the mean free path of carriers. Although his theory was later found to be less universally applicable than that of Schottky, it proved correct for this case and served well in guiding the development efforts. Lark-Horovitz was aware of Schottky's work in 1939, but clearly had no access to the work published in 1942. The theoretician Bethe was keenly interested in the practical performance of radar detectors and suggested, for example, that the excellent performance of British "red dot" crystals was owing to the knife-edge metal contact on a polished crystal surface.

The wartime contractual work at Purdue stretched from March 1942 to November 1945 and was described in a final report by Lark-Horivitz. The summary of this report is quoted here, because it is better able to give the flavor of the work than any transcription could.

1. Germanium of high purity has been prepared by reduction from pure oxide. By using various impurities in a varying range of concentration (.001% to 17%), it has been shown that both N-type (excess conductor) and P-type (defect conductor) semi-conductors can be produced. B, Al, Ga, In, all produce P-type germanium semi-conductors. N, P, As, Sb, Bi and Sn and other elements produce N-type germanium semi-conductors.

2. Hall effect and thermo-electric power measurements and sign of rectification determine the sign of the carrier. Both Hall effect and thermo-electric power become negative at high temperature for all samples, indicating that at these temperatures current carriers are released from the intrinsic levels and the mobility of the "holes" (P-type carrier) is smaller than the mobility of the electrons. The fact that the high temperature slope for ρ is parallel and identical at the high temperature region for most samples has been interpreted as an indication of an intrinsic level. (Band width ΔB equal .76 eV.)

3. Measurements of resistivity as a function of temperature indicate that the resistivity of germanium alloys is caused not only by the scattering of the electrons by the lattice but also by the ionized impurity centers. The computation of resistivity, calculated from lattice scattering (ρ_L) and impurity scattering (ρ_i), allows a theoretical prediction of the resistivity curve throughout the temperature range from liquid air up to the melting point.

4. Using the complete expressions for the chemical potential of the electrons in the semi-conductor, the thermo-electric power curve as a function of temperature can be calculated from Hall effect measurements.

5. Structure investigation of the alloys shows that the lattice constants are the same for P-type, N-type, and pure germanium. All of the samples investigated show the existence of large and perfect crystallites indicating that lattice distortions and foreign enclosures are not primarily responsible for the electrical properties of these semi-conductors.

6. The investigation of the rectifier model has shown:
   a) In germanium and silicon crystal rectifiers, the barrier layer is small or comparable with the mean free path (diode behavior as compared to diffusion in cuprous oxide).
   b) The D.C. characteristic has a slope smaller than that predicted from theory,

   a fact which can be explained by the assumption of a distribution of contact
   potentials at the surface of the semi-conductors (multi-contact theory).

   c) The behavior of spreading resistance and contact capacity in germanium
      crystals indicates that the present model used to calculate high frequency
      performance is oversimplified and cannot explain the measurements.

   d) Experiments with germanium rectifiers in a high vacuum or in controlled
      atmospheres show that while rectification exists in a high vacuum, a change
      in atmosphere produces adsorbed layers which may greatly influence the
      shape of the D.C. characteristics.

7. Germanium semi-conductors containing P or Sb can be used in microwave
   mixer crystals, comparing well in performance to silicon crystals.

8. Sn or N added to germanium produces crystal detectors which withstand a high
   voltage in the order up to 250 V or more in the back direction, while passing
   adequate currents in the forward direction. Depending on power treatment
   and preparation of the surface, the back resistance is of the order of megaohms,
   while a forward resistance from the slope at 1 V can be calculated as equal to
   about 50 to 200 ohms.

9. Various types of photo effects have been observed with a sensitivity maximum
   at about 1.3μ, and a threshold of about 1.5μ.

10. Various methods for the rapid determination of gain and noise in crystal rec-
    tifiers have been developed and checked by comparison with standard meth-
    ods.[65]

When Lark-Horovitz first applied for the Rad Lab contract, he had in mind to
work on galena cat's whisker detectors, of which he had experience as a communications officer in the Austrian army during the First World War. His mind was
turned to germanium mainly by discussion with scientists from the Sperry Corporation and, later, by finding germanium point contact rectifiers described in the literature.[66]

One of the Purdue wartime results that had some influence on fundamental
semiconductor theory was the finding that the mobility of free electrons varied as
$T^{3/2}$ with temperature. Lark-Horovitz came to the conclusion that this indicated
impurity rather than phonon scattering. The matter was later taken up by Esther
Conwell and Victor Weisskopf.[67]

Ralph Bray, a young graduate student, joined the germanium effort at Purdue in
November 1943 and was given the tricky task of measuring the spreading resistance
at the metal–semiconductor contact. In essence, the question is one of the relationship between the current that spreads from the metallic point contact into the bulk
of the semiconductor and the resistivity of the semiconductor. Bray found a great
many anomalies, such as internal high-resistivity barriers in some samples of germanium. The most curious phenomenon was the exceptionally low resistance
observed when voltage pulses were applied. This effect remained a mystery because
nobody realized, until 1948, that Bray had observed minority carrier injection—
the effect that was identified at Bell Labs and made the transistor a reality. Bray
wrote,

   That was the one aspect that we missed, but even had we understood the idea of
   minority carrier injection . . . we would have said, "Oh, this explains our effects."
   We might not necessarily have gone ahead and said, "Let's start making transis-

tors," open up a factory and sell them. . . . At that time the important device was the high back voltage rectifier.[68]

This is a classical example which shows that scientists rarely see what they do not expect to see and that mysteries can get in the way of a busy research schedule and get brushed aside. There is no time to follow up each mystery, and most of them lead up blind alleys; yet some prove the key to unexpected riches of discovery. The difference between using the key or ignoring it often resides in the perception of whether the solution of this particular mystery is relevant to the goal pursued, be it basic research or the search for a practical device. Since it is practically impossible to follow up every question, a selection is made on the basis of the expected contribution of the solution to the ultimate goal of the research.[69]

Another Purdue scientist, Randall Whaley, put the matter in a nutshell:

> The irony of the whole thing is that two or three of us were occasionally around lunch asking ourselves, "Why can't we put a grid on this and make a triode of it to control the electrons?" But in the press of getting degrees and putting detectors together for MIT, we didn't take the next step and try this.[70]

Bray put his finger on yet another feature of the scientific community that kept him back. Scientists, like anybody else, do not like talking about their weaknesses; they like to shine. Thus in most Purdue presentations, their considerable successes were paraded while they kept the unresolved problems to themselves. Bray's comment was: "The spreading resistance was sort of a mystery and no one understood it and so I wasn't pushed forward, pushed in front to talk about this very much."[71]

The greatest practical success of the Purdue group was in developing high back-voltage diodes. Seymour Benzer made a first step toward stable diodes by accidentally welding the metal whisker to the germanium. Benzer and Whaley obtained higher and higher back-voltages with germanium containing tin as an impurity by a combination of better processing of the material and better surface treatment. Other suitable impurities were found to be nickel, copper, bismuth, and barium. The high back-voltage diodes were reported to the Rad Lab some time in 1943, and further development and testing work were put in hand. This work occupied much of the time at Purdue until the device was handed over to Bell Labs and Western Electric for mass production in late summer of 1944. Thus germanium detectors became a real product, and the type of material used for their production was also used for the first transistors to emerge from Bell Labs three years later.

There was considerable transfer of knowledge between Bell Labs and all the participants in the Rad Lab projects throughout the period, and undoubtedly knowledge flowed both ways. The last batch of knowledge, as it were, was exchanged during a visit by Stanley Morgan and Willian Schockley of Bell Labs when they included Purdue in a fact-finding tour in the autumn of 1945, in preparation of their postwar program of semiconductor research.

As mentioned earlier, Bell Laboratories also worked on radar detectors during the war and had worked on semiconductor devices in prewar days. There was a degree of continuity between prewar work on copper oxide rectifiers and prewar attempts to produce solid-state amplifiers, on the one hand, and the postwar work leading to the invention of the transistor, on the other. The wartime effort on radar detectors provided this link in some sense; in another sense, warwork retarded progress. Who can tell when the transistor would have emerged but for the war? Bardeen himself commented: "I think the pressure was to try to develop an amplifying device using a semiconductor. I don't think the war had any direct effect on that. It probably slowed that down if anything because the people were all working on other subjects during the war."[72]

In fact, Bardeen had himself worked for the navy on problems of magnetic fields of ships and the firing of torpedoes. Another prominent semiconductor scientist, Gerald L. Pearson, who spent the war years at Bell Labs working on radar and on infrared bombsights, expressed sentiments about this period closely similar to those of Bardeen: "The work during the war was a diversion from my earlier interest in fundamental semiconductor research."[73]

One of the people who did provide some continuity was Walter Brattain, who worked on copper oxide rectifiers at Bell Labs as early as 1933 or 1934.[74] Brattain was also among those many physicists who attempted, from time to time, to produce solid-state amplifiers before the war. He wrote, "If one knew how to put the third electrode in the cold rectifier—like the grid in the vacuum tube—one would have an amplifier."[75]

Although the ultimate emergence from Bell Labs of the solid-state amplifier—in the shape of the point contact transistor—contained, as we shall see, an element of serendipity, it is true to say that it was the deserved reward for years of painstaking, knowledgeable, and conscious effort. In the late 1930s, William Shockley suggested a design for a field-effect solid-state amplifier and Brattain carried out the experiments, but, like all other attempts of this ilk at that time, it did not work. In fact, the number of people who tried to produce solid-state amplifiers grows steadily as one probes further. Another Bell man, Russell S. Ohl, recalls: "In fact, I took out a patent in which I controlled (on paper) the electron current in a copper oxide rectifier. This was 1927. . . . many people had the same thought."[76]

Ohl was also one of the pioneers of the crystal detector and certainly played an important part in the transistor story. As early as 1922, he got rather disgusted when he had burned out all the thermocouples at his disposal, using them as detectors for short radio waves. He then changed to silicon crystal point contact detectors. Around 1926, Ohl demonstrated that copper oxide detectors could not be operated at high frequencies, while silicon point contact detectors could. It is not clear where or how Ohl obtained these silicon detectors, but they were apparently generally available and G. C. Southworth bought some in lower Manhattan even in the late 1930s. Apparently because of his earlier experience, Ohl remained convinced that silicon was the proper semiconductor to use and tried to push this line in the radar detector work and in the early years of the transistor.

Russell Ohl and Jack Scaff were the first to find a grown *p–n* junction in silicon and had understood the effect of impurities in causing *n*-type and *p*-type silicon, although in the early days they had called them *x*-type and *y*-type. Ohl held patents on *p–n* junctions and was also probably the first to find photovoltaic effects in silicon. He was trained as an electrochemist and proved a prolific and profound practical physicist, yet had little use for quantum mechanics: "I had to work in my own terms." At an internal conference held at Bell Labs to discuss progress on semiconductors, Ohl explained "how we had learned to process the silicon, how we had learned to make a back contact, how we had learned to solder it fast, and how we had learned to polish the surfaces and fabricate it in quantity so we could manufacture it."[77] "It" was a silicon point contact detector, and the very repetitiveness of Ohl's statement shows vividly how laborious it all was. There were thousands of little tricks to be learned the hard way, by trial and error, endlessly repeated for each and every step. There is a long way from even a theoretically sound idea to its practical realization. According to Bardeen,

> While in retrospect progress seems rapid, there were times when it seemed painfully slow. Why did it take ten years to develop diffusion technology when the basic ideas (e.g. double diffusion) were well known? The first diffused device was the silicon solar cell by G. L. Pearson and C. S. Fuller. Why did it take 15 years to learn to make a silicon–silicon oxide interface free of surface states so as to make MOS transistors practical? Why did it take so long to develop ion implantation as a method of doping, although the concept was noted in Shockley's earliest patents?[78]

During the war, Ohl was able to purchase what was, by the standards of the day, hyperpure silicon at the expense of the Rad Lab program. Earlier, Bell had refused to pay $250,000 for about 25 lb of this material! Ohl suggested that attempts should be made to grow single crystals of silicon for the detector work. The required effort was not made available. Who is to say, even with the benefit of hindsight, whether this was a wise decision.

Not all work on semiconductors during the war was associated with microwave detectors. Another quite important aspect was the detection of infrared radiation, mainly for bomb and gunsights. There are dozens of examples of this work. Gerald L. Pearson, who had joined Bell in 1929 on the motto "If you wanted to be somebody, you wanted to work at Bell Labs," had done some work on "thermistors" before the war. The change of conductivity of semiconductors with temperature was used, among other things, for the rather mundane purpose of controlling the temperature of buried cables. During the war, this experience proved of benefit for the production of infrared bombsights, using mainly thin films of oxide semiconductors.

The Germans also used semiconductor infrared sights, and some of them were found in German planes shot down over Britain and taken to the Telecommunications Research Establishment at Malvern. As a result of analyzing these sights, apparently some improvements could be made to the British designs.[79]

The counterpart to this episode happened with Allied war planes shot down over Germany. The meager German efforts at semiconductor radar detectors received a considerable fillip when British or American detectors were found and analyzed by German scientists.[80]

There can be little doubt that German radar developments lagged considerably behind the British and American efforts throughout the war. Although the weakest link presumably was the power source, a component on which the Allies had a dramatic advantage with the cavity magnetron developed by Randall and Wilkins in Birmingham, undoubtedly German detectors were also inferior to Allied technology. It would appear that initially only valve detectors of various kinds were in actual use in radar instruments, and they could operate only at undesirably low frequencies and, thus, long wavelengths. Experimentally, the use of composite point contact detectors of the older type, such as the veteran lead sulfide and the newer pyrite, was known. Such detectors proved extremely troublesome in practical use because the material under the point contact tended to decompose as the device heated up by the flow of current. Germanium and silicon detectors did not come into use until quite late in the war.

The development of germanium detectors in Germany can be traced mainly to Heinrich Welker, an *Assistent* to the famous Arnold Sommerfeld in Munich. Welker was a theoretical physicist who had worked on a variety of problems, among them questions of superconductivity. Being in Sommerfeld's ambience and participating in various conferences, he became acquainted with a number of theoreticians of the day, including Walter Schottky. By early 1940, Welker had become restless in Munich. Sommerfeld was retiring, and his successor was to be a political appointee. Welker's own path to a teaching career was made thorny by political circumstances, and after years of pure theory, he felt the desire to do something practical (and patriotic?).[81] Welker left the University of Munich and moved to the Wireless Telegraphy Research Station in Gräfelfing in April 1940 to work on radar. After making several contributions to problems of paraboloid aerials, he concentrated on the critical issue of microwave detectors. He knew Schottky and was among the few who had read and understood Schottky's early publications on contact barrier layers. He did some calculations and decided that pyrite had suitable qualities as a detector material for microwaves, but so did germanium, which had the advantage of being an elementary semiconductor and was thus not prone to thermal decomposition.

The management did not think much of Welker's idea, since they regarded point contact detectors as obsolete devices from the infancy of radio. But the reputation of Sommerfeld and his co-workers was the highest, and thus Welker was allowed, although not encouraged, to follow his hunch. It became necessary to produce purified germanium, and this could not be undertaken in the "wooden huts" of the Wireless Telegraphy Research Station. Instead, in 1941, Welker went back to the University of Munich to work in the Institute of Physical Chemistry under Klaus Clusius. The almost immediate result of this move was a German patent of October 1942. Interestingly, the patent states explicitly that the preparation of pure silicon for detectors is impossible because of its high melting point. The invention is concerned with the use of germanium as a detector of radio or centimeter waves. The purified germanium of about 1 ohm-cm either is pulled in a capillary tube or is evaporated onto a substrate. In either case, it should be as nearly monocrystalline as possible. The surface is etched either chemically or electrolytically, and the point contact consists of platinum. The nonrectifying back contact can be hard-soldered to the germanium crystal.[82]

The germanium detectors thus invented were developed and manufactured by Siemens, although in the early stages the technicians in the Institute of Physical Chemistry in Munich produced such detectors in the laboratory. The material was prepared from the chloride by reaction with zinc or, more particularly, aluminum.[83] Only the capillary method was used, and the crystal was hard-soldered on one side and cut, polished, and etched on the other. The contact consisted of molybdenum, not of platinum, as suggested in the patent. Obviously, great problems were encountered in securing the necessary materials. On 10 February 1943, R. Rompe of

Osram in Berlin wrote to Welker: "Unfortunately, I have to disappoint you in the matter of tungsten capillaries. It would be difficult, at the moment, to attempt such 'tricky feats.'"[84] Industrial development and production of the germanium diodes started late in 1942 at the Siemens works in Berlin. The work was interrupted by the destruction of the plant and its laboratory in an air raid in September 1943, and was transferred to Vienna. Production there started in April 1944, reached 650 detectors a month in May, 1000 in August, and never reached the goal of 20,000 detectors a month. There were four variants, mostly useful only for wavelengths between 10 and 20 cm, although the last design could just about be used down to 9 cm. In the last model at least, the contacts were "formed" by a current pulse, as had been common practice in England and the United States.[85]

There can be no doubt that Schottky was the spiritual father of the modern German point contact detector. Welker was in constant correspondence with Schottky, although the latter's interest concentrated on the theoretical side. Not that Schottky was not interested in practical matters; indeed, he invented many practical devices, but in semiconductors his theoretical interest prevailed. Obviously, he was not altogether happy with the use made of detectors in the war, for on 23 May 1943, he wrote to Welker: "I am glad that your detector work has led to production which, hopefully, will one day acquire importance in peaceful uses."[86]

Earlier, on 20 April 1940, Schottky had written to Welker:

> I think that for relatively slow processes the matter (theory of crystal rectifiers) is now substantially clear. . . . At very high frequencies new questions arise; apart from questions of capacity, which Dr. Spenke is studying even now during the war, there are questions of reaction kinetics in reaching an equilibrium distribution of defects.[87]

Curiously, this seems to have been one of those problems that later simply "went away."

German work on silicon detectors started much later and by quite a different path. Germanium detectors were associated with Siemens, but the silicon work was carried out by Telefunken. While germanium was produced in the form of small, nearly monocrystalline rods, silicon was deposited on a carbon substrate. While germanium developments started with Schottky's and Welker's theoretical considerations, silicon was picked up mainly as a result of finding silicon detectors in crashed Allied aircraft.

One of the people who were brought back from the war to develop radar detectors was Karl Seiler. According to Seiler, the Allies used 9.5-cm radar in 1943, whereas the Germans had nothing much below 20 cm. When he analyzed a device from a downed American plane, Seiler discovered a silicon point contact detector. Using a method developed by Gunter and Rebentisch for depositing silicon on graphite rods, Seiler soon developed a useful point contact detector. One of the first uses made of this device was for the detection of Allied radar by German U-boats.[88]

There can be little or no doubt that German work on semiconductors during the war was not nearly as well coordinated, as fundamental, or as extensive as Allied, particularly American, work. Although several meetings on detectors were held in Jena and Berlin under the auspices of the Air Ministry, there was no extensive contact among the few small groups working on these problems. The Germans were most reluctant to release scientists from active military service to work in research, and some of their meager effort was wasted on anachronistic developments of sonic aircraft detectors.[89] Only relatively few measurements of basic properties of semiconductor materials were made, and apart from Schottky, Spenke, and Welker, nobody seems to have done any theoretical work on devices. The practical work, on the other hand, was severely disrupted by bombing raids.

7.3   The Postwar Period

While the period up to the end of the Second World War produced a great deal of progress in the theoretical understanding and practical knowledge of semiconductors, it remained a time of limited effort and scope. Thus it is possible, at least in principle, for the historian to trace the major developments of this period, although we do not claim to have done this. The postwar years, on the other hand, particularly after the invention of the transistor, are characterized by explosive growth, and the only way the historian can deal with it is by severe, probably idiosyncratic, selection. The writing of history and the award of prizes have a disturbing feature in common: by emphasizing and praising the one deserving actor, they relegate the many, perhaps equally deserving, to obscurity. The distillation process of history brings to the fore the one fraction favored, by design or by accident, by the historian; in the selection process of prize giving, the jurors do much the same. In one case, an often unwarranted slant is the result; in the other, for every happy winner there are a multitude of embittered losers.

The analogy is not fortuitous. To the contrary, the history of semiconductors serves as a case in point. For undoubtedly the most important development that has emerged from semiconductor physics is the transistor as the ancestor of microelectronics, and the credit for this feat of science has gone to three deserving scientists with the award of the Nobel Prize. John Bardeen, Walter Brattain, and William Shockley were awarded this coveted prize on 10 December 1956 for "their research on semiconductors and their discovery of the transistor effect." Without wishing to detract from the achievements of these three men, it is obvious that their efforts were part of a much greater enterprise in which they played leading, but not exclusive, roles. This section will attempt to show the great achievement of the Nobel Prize winners in a slightly broader context. The narrative will be kept very brief, since numerous accounts of the development of microelectronics have been published in recent years.[90]

After the war, scientists began drifting back to civilian employment, and industrial laboratories began to forge their plans for a big surge of peacetime scientific and engineering progress. Bell Labs decided to undertake a major effort in research on semiconducting elements, which meant, in practice, germanium and silicon. The idea was, of course, the one that had been around in so many minds for so long—to produce a solid-state amplifier. Although, no doubt, a solid-state switch and improvements to semiconductor diodes developed during the war would be welcome, the real prize was the amplifier. This was not, however, a development program. It was a program of basic scientific research into the properties of ger-

Walter Houser Brattain (1902–1987), John Bardeen (b. 1908), and William Shockley (1910–1989). (AIP Niels Bohr Library, Brattain Collection)

manium and silicon, albeit undertaken in the hope of obtaining devices useful to the Bell System. The work was to be based on knowledge gained during the war at Bell Labs and elsewhere, and was to establish continuity with prewar Bell activities in the field of semiconductors. The main instigator of the new Solid State Division was M. J. Kelly, with William Shockley as co-head of the division and head of the semiconductor research group.

The first idea to be tried out more thoroughly was an old one—that is, the modulation of the conductivity of a thin layer of semiconductor material by the application of an electric field perpendicular to the flow of current. This "field effect" had been suggested and tried in numerous unworkable inventions before, but somehow the idea was so plausible that another effort seemed worthwhile. The result, alas, was the same as in previous attempts: the simple, plausible, infallible device failed to work. When all doubt had been dispelled about the experimental result, it became necessary to reexamine the theory. This really was a case where the prevailing theory had been falsified, and some revision or addition became necessary. It is not, of course, a case of the downfall of a major theory—something that happens extremely rarely—but an everyday event of small failures and consequent modification of theory, usually for a special case.

The required modification of theory was undertaken by John Bardeen. His postulate was that the carriers induced in the semiconductor by the electrostatic field (the field effect) failed to modulate its conductivity because they became trapped and immobilized in surface states. Although Bardeen's suggestion proved of decisive value and had some important novel features, it was not without precursors.

As early as 1932, I. Tamm calculated that the termination of a one-dimensional Kronig–Penney potential led to a surface state, and this calculation can be generalized to the effect that many atomic states lead to a surface state when a periodic structure is terminated. In 1939, Shockley postulated surface states arising if two electronic bands within a finite periodic structure intersect. This calculation would allow the existence of a surface conduction band. Thus quantum theory allowed both localized surface states and a continuum of such states.[91]

A further indication that all was not well with theories which assumed that the band structure was unaffected by the surface came from detailed measurements on rectifying contacts. In theory, the rectifying properties of a metal–semiconductor contact should depend on the difference in work function between the two materials. For this reason, Welker suggested in his patent that platinum contacts should be used. We have seen that in practice molybdenum served as the metallic contact, and many experiments on the dependence of rectifier characteristics on the nature of the metal yielded inconclusive results. There was reason to believe that the contact potential difference was not determined by the work functions of the clean surfaces, but that each surface was covered by adsorbed material that formed an electric double layer. Further support for this assumption was derived from the fact that practical metal–semiconductor contacts had to be formed by passing a current pulse through them, which presumably meant some form of punching through the adsorbed layers. To cut a long story short, there were good theoretical reasons to believe in localized electronic surface states and good practical reasons to believe that adsorbed layers were associated with electrical double layers, equivalent to localized charges. Putting these items of scattered circumstantial evidence together, Bardeen concluded that the failure of the field-effect amplifier must be caused by the fact that the charge carriers induced by the applied field became immobilized in surface states, whether on a clean or a dirty surface, and could not contribute to the conductivity. "The novel feature was not the idea of surface states . . . ., but to apply the idea to understand the real surface of a semiconductor."[92]

Bardeen considered the surface states to be continuously distributed through the forbidden gap and the surface as a whole to be neutral when the states are filled up to the Fermi level. If the density of surface states is high, then a large shift in surface potential can be accommodated without an appreciable change in the free-carrier density. A density of 10^{13} states/cm^2 would render the rectifying properties of a metal–semiconductor contact insensitive to the work function of the metal. This condition is readily fulfilled on contaminated surfaces. The first measurements of surface state densities in a field-effect experiment were reported by Shockley and Pearson, and indeed a density of 5×10^{13} states/cm^2 was found on germanium.[93]

Bardeen's theory and this first field-effect experiment gave rise to a long series of experimental and theoretical investigations on surface states, carried out in several industrial and university research groups all over the world. The topic remained fashionable until the early 1960s, because it seemed to provide fundamental insights and was of considerable practical importance in the then-current designs of transistors. As far as the transistor is concerned, surface problems were not so much solved as bypassed. With the introduction of the planar technique and oxidized silicon surfaces, the earlier problems of surface recombination and deterioration virtually disappeared, and interest shifted toward questions concerned with the silicon–silicon dioxide interface. The fundamental problems of surface physics did not disappear; to the contrary, surface science became a major field of study with implications for catalysis, surface finishes, fatigue, and so on.

The main results of a dozen years of semiconductor surface studies were the distinction between surface states on clean surfaces, which reached equilibrium with the bulk semiconductor very rapidly—the so-called fast states—and surface states associated with adsorbed layers. The latter have very long time constants in reaching equilibrium and became known as slow states. In attempts to clarify the highly complex and confused situation, much effort was put into ultrahigh vacuum work. At a time when commercial systems commonly worked at 10^{-6} Torr, surface physicists were achieving 10^{-12} Torr and undoubtedly influenced commercial developments in this direction. Ingenious methods for cleaving specimens in vacuum were developed and, more important, methods for producing clean surfaces by sputtering and heat treatment. These methods have remained important and are widely used.

After this brief excursion into the world of surface physics, we must return to Bell Labs in 1947. Bardeen's theory of surface states led to a shift of emphasis in semiconductor work onto surfaces. The investigation of surface states afforded the possibility of gaining fundamental insights as well as the possibility of achieving the elusive goal of an amplifier if the surface states could somehow be neutralized. A large number of experiments were performed by many people. The first results promising success toward amplification were obtained in field-effect experiments by Brattain and R. B. Gibney on germanium immersed in a liquid electrolyte. Bardeen wrote,

> The idea was to get the effect of a thin film in bulk material by modulating the flow in an inversion layer of opposite conductivity type. This is the principle of present day MOS transistors. It was known that part of the difficulty with earlier attempts to make a field effect transistor with evaporated thin films was the poor mobility of carriers in the films. I was credited for this invention and was issued a little known patent. . . .[94]

After several further stages of experimentation, mainly by Brattain and Bardeen, aided by the chemist Gibney, a germanium surface was prepared by anodizing it and evaporating gold spots onto the oxide. Transistor action was first observed on 15 December 1947, when "it was found that current flowing in the forward direction from one contact influenced the current flowing in the reverse direction in a neighboring contact in such a way as to produce voltage amplification." The germanium block was *n*-type, and obviously holes were being injected by a contact biased in the forward direction toward a contact biased in the reverse direction. In the memorandum by W. S. Gorton, from which this quotation is taken, 12 people are mentioned as directly involved in the invention of the transistor in the Bell Laboratory. Undoubtedly, many more, inside and outside Bell Labs, were important indirect contributors.[95]

On 15 December, Brattain used a gold spot cut into three sections. In his own parlance, he used area A of the spot as the plate and area B as the grid, in complete analogy with electron-tube notation. When the points were "very close together got voltage amp about 2 but not power amp. This voltage amplification was independent of frequency 10 to 10,000 cycles."[96]

On 16 December, Brattain had a gadget made of polystyrene for putting two gold leaf contacts very close together on the germanium surface. He wrote:

> Using this double point contact, contact was made to a germanium surface that had been anodized to 90 volts, electrolyte washed off in H_2O and then had some gold spots evaporated on it. The gold contacts were pressed down on the bare surface. Both gold contacts to the surface rectified nicely. . . .
>
> The separation between points was about 4×10^{-3} cm. One point was used as a grid and the other point as a plate. The bias (D.C.) on the grid had to be positive to get amplification. . . .
>
> power gain 1.3 voltage gain 15 at a plate bias of about 15 volts[97]

It turned out that the oxide was not necessary for the process, and the gold spots were eventually replaced by other metal points. When the electrodes were more closely spaced, it was found that power amplification of 18 could be obtained. On 23 December 1947, a speech amplifier of this gain was demonstrated by Brattain and H. R. Moore to several of their colleagues and managers.

J. R. Pierce gave the device the name transistor because the concept of transresistance suggested itself to him as an analogy with transconductance in vacuum tubes.[98] Not in their wildest dreams could they have imagined that they had laid the foundation stone to the edifice of modern microelectronics. According to Bardeen:

> The initial experiments with the gold spot suggested immediately that holes were being introduced into the germanium block, increasing the concentration of holes near the surface. The names emitter and collector were chosen to describe this phenomenon. The only question was how the charge of the added holes was compensated. Our first thought was that the charge was compensated by surface states. Shockley later suggested that the charge was compensated by electrons in the bulk and suggested the junction transistor geometry. A little later [February 1948] John Shive showed definitely that the latter could occur by placing emitter and collector on opposite faces of a thin slab of germanium. Later experiments carried out by Brattain and me showed that very likely both occur in the usual point-contact transistor. The patent issued to Brattain and me has always been considered basic to bipolar transistors, including the junction transistor.[99]

The invention of the point contact transistor was the result of what Shockley has termed the "creative failure methodology."[100] When the proposed field-effect amplifier failed to work, a theory for this failure was proposed. While investigating this theory, the researchers never lost sight of the objective of creating a solid-state amplifier. When an unexpected amplification mechanism was finally found, this was not regarded as an anomaly and a nuisance, but was followed up both in theory and in practice. Thus the theory of minority carrier injection was created, and the point contact transistor invented.

This procedure is by no means self-evident. To the contrary, it is very common to see only what one is looking for and to disregard as anomalies effects that do not fit expectations. Bell had the supreme advantage of knowing what it was looking for and having a team of about a dozen superb scientists to discuss results and carry out new measurements to test ideas very quickly. Flexibility and speed of response, so vital in industrial technological innovation, proved important in this invention based on science.

Once the Bell Labs management was convinced that the lab was dealing with a major invention, feverish activities started. Patents had to be filed; more devices had to be prepared and tested; the processes had to be better understood to be better controlled. Strict secrecy was observed, and nobody who did not need to know knew much about the Surface States Project, as the work was called.

Some extracts from a conference report may serve as an illustration of the multifaceted activities and the involvement of numerous scientists during this hectic period. The strictly internal conference on the Surface States Project was held at Bell Labs on 28 April 1948, and was attended by "about 30 or 40 people." The conference was chaired by Ralph Bown, director of research, and the speakers were Bardeen, Brattain, Shockley, Pfann, Shive, Koch, Black, and Bown.

> Bardeen discussed mathematical relationships between input and output currents and voltages of the surface states triode. . . .
>
> Brattain discussed measurements of γ (from $I_1 = f(V_1 + \gamma I_2)$ . . . and mentioned temperature data on Pfann unit. . . .
>
> Shockley briefly explained a theory he had derived to account for α's > 1 at very low collector voltage. . . .
>
> Koch discussed exploratory work on circuit applications. . . .
>
> H. S. Black introduced a demonstration of a 7-amplifier radio made entirely with s.s. triodes, plus a selenium rectifier and a crystal detector. . . . Of the 7 s.s. units 5 were Pfann units (plane type) and 2 were Shive units (coaxial type). The latter were used in the audio section because they handled more power. The former were used in the RF and IF sections because of their higher cut-off frequency, and possibly their somewhat lower noise.[101]

Clearly, the Surface States Project had the attention of top management. Equally clearly, several people were engaged in designing and producing surface state triodes, as point contact transistors were then called. Pfann and Shive had produced different types, each with its own advantages and disadvantages. While Bardeen and Shockley were busy interpreting results and guiding new experiments, others were already engaged in using the new devices in circuits. The complete spectrum of sensible activities was thus immediately tackled, despite conditions of secrecy and great urgency.

The news of the transistor was sprung on an unsuspecting and disinterested world on 30 June 1948, at a major press conference. Although the world at large took quite some time to realize that something important was happening, scientists in the electronics industry, followed by university scientists all over the world, sat up and listened. Bell Labs, after some internal discussion, decided to make its knowledge widely available. This meant not only that Bell assiduously and skillfully avoided classification of its knowledge as a military secret, but also that industrial proprietary rights were not guarded as zealously as they might have been. According to Bell Labs' own story, the then vice president of Bell Laboratories, J. W. McRae, explained his advocacy of openness by the argument "that the Nation and the Bell System had more to gain by rapid development of the transistor than by conceal-ment and restriction of the production techniques."[102] A further discussion of this issue and its significance for the relatively rapid spread of transistor technology may be found in Braun and Macdonald.[103]

Within the Bell organization, three strands of further development seem the most significant in the infant years of the newly born transistor. The first line of work was progress on theory, both of semiconductors as materials and of transistors as circuit elements. Much of the early theory is incorporated in Shockley's book *Electrons and Holes in Semiconductors,* first published in November 1950. The book arose out of a series of lectures that Shockley gave at Bell Laboratories "in connection with the growth of the transistor program."[104] The book, with its very detailed the-ory, proved a milestone in the understanding of semiconductor physics.

The injection of minority carriers is the key feature of transistor action. Injection of majority carriers would quickly lead to the neutralization of their space charge and thus to thermal equilibrium. Injection of minority carriers is also followed by neutralization, but in this case with increased carrier density. This feature distin-guishes semiconductors uniquely from metals.

> In a semiconductor containing substantially only one type of current carrier, it is impossible to increase the total carrier concentration by injecting carriers of the same type; however, such increases can be produced by injecting the opposite type since the space charge of the latter can be neutralized by an increased concentration of the type normally present.[105]

The second strand is associated with the fact that the first point contact transistor, as all its successors in the rapidly evolving field of solid-state electronics, was obso-lescent even as it was first produced. By the time the first transistors were being shipped, Shockley and his colleagues knew that there was a better way of producing transistor action. Alas, this better way, using two *p–n* junctions in close proximity, remained a paper design for some time. Point contact transistors remained in pro-duction for about 10 years, although it certainly did not take nearly that long to get junction transistors into production.

Shockley derived the equation for current flow through a *p–n* junction. If a volt-age *V* is applied across the junction, then the current is given by

$$I = I_s \left[\exp\left(\frac{eV}{kT}\right) - 1 \right]$$

where I_s is determined by the concentrations of carriers and their mobilities. This is a rectifier equation, based on a similar equation derived previously, as we have seen, in the 1930s. He then derived an energy level and charge density model for *p–n* junctions, although, curiously, he did not show the position of the Fermi level in his initial diagrams. When a second junction is placed in close proximity to the first, current flowing through the one junction will influence that flowing through the other (Fig. 7.3). Shockley explained the action of this transistor in analogy with the triode vacuum tube—an analogy that was very powerful in many a mind. The new and unfamiliar inevitably seeks orientation from the old and familiar.[106] "In the transistor structure the application of voltage between emitter and base has an effect

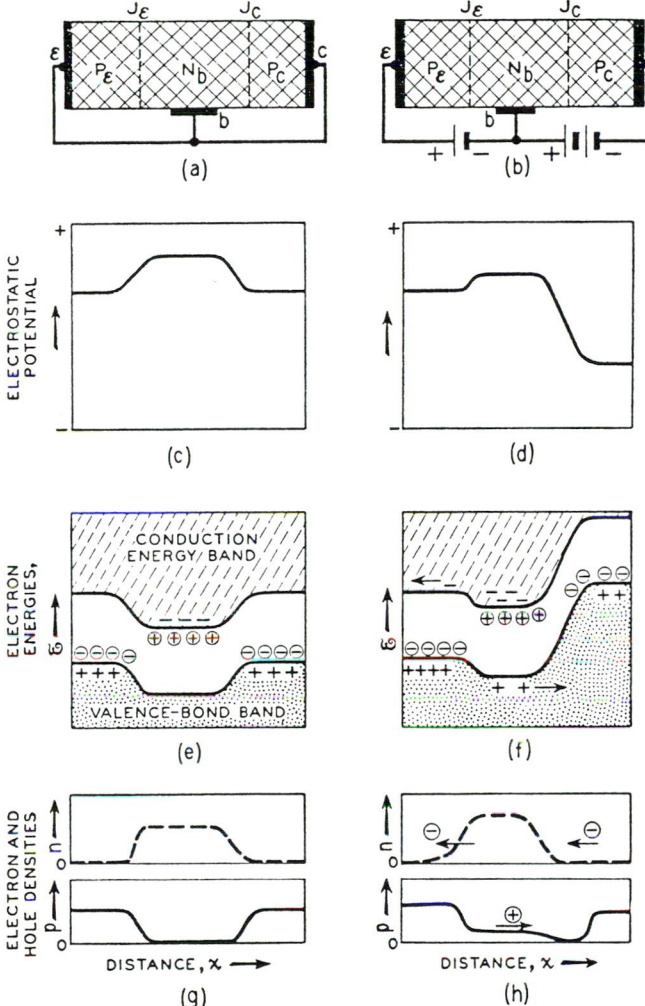

Fig. 7.3 A *p-n-p* transistor compared with a vacuum-tube triode. (From W. Shockley, *Electrons and Holes in Semiconductors* [New York: Van Nostrand, 1950], 92)

similar to the application of voltage between grid and cathode in the vacuum tube."[107] This and similar analogies greatly helped understanding.

Although Ohl had found a *p–n* junction in silicon in the prewar days, this was a far cry from being able to fabricate *p–n–p* transistors at will. In fact, it soon became apparent that no kind of transistor could be reliably and uniformly manufactured unless mastery was gained over the properties of the basic material, which, in those early days, was germanium. Despite heated debates about whether it was feasible to build industrial production on a raw material that had to be so very strictly controlled and grown into single crystals, the view that this was both necessary and possible prevailed.

The father of clean, controllable, single-crystal semiconductor material, William Pfann, was one of the many Bell scientists busily making point contact transistors immediately after their discovery. At the same time, however, Pfann, too, was thinking of a better transistor, an *n–p–n* device, and of ways to make it. On 18 February 1948, Pfann wrote in his notebook:

> It is known that addition of aluminum makes HBV Ge P-type and that melting in water vapor makes it N-type. Hence, begin with a slab of high-back-voltage germanium, deposit a layer of aluminum on its surface; heat it to cause diffusion inward (heat in air or some inert medium—Al diffuses at quite low T), thus producing a P-layer of thickness, t, at the surface. Next heat the slab in water vapor at a suitably high temperature, transforming a thin surface layer, of thickness less than t, to N-Ge. . . .

> From present knowledge of effects of impurities & treatments on Ge & Si obtained by Theuerer & Scaff, the above example, and also the opposite case, P-N-P, seem quite likely of attainment in either Ge or Si. The possibilities of such materials as circuit elements and as photoelectric devices seem to have many aspects. The layers could in various ways be made available for contact by control electrodes. For example, polishing the surface at a very slight angle.[108]

These were prophetic words, indeed, although it was to take many more years of painstaking development work before diffused junction transistors became a mass-produced technical reality.

This takes us to the third category of major effort—the preparation of pure single crystals of germanium. The crystals were grown by pulling a seed crystal from the melt, a method known after its originator, Czochralski.[109] The main proponent of single-crystal production at Bell Labs was Gordon Teal, who was to play a significant role in making silicon available for practical applications. Although the growth of large single crystals of germanium—and by large were meant in those days diameters of 10 to 20 mm and lengths of perhaps 100 to 150 mm—proved rather difficult, it was accomplished. The crystals had to have uniformly high purity, had to be doped to the right level, and had to have reasonably low densities of dislocations. New methods had to be developed at each stage: measurement of resistivities by the four-probe method, cutting of the crystals by diamond wheels, etching for purposes of cleaning the surfaces as well as revealing dislocations, and many, many more.

The history of semiconductor physics is not one of grand heroic theories, but one of painstaking intelligent labor. Not strokes of genius producing lofty edifices, but great ingenuity and endless undulation of hope and despair. Not sweeping generalizations, but careful judgment of the border between perseverance and obstinancy. Thus the history of solid-state physics in general, and of semiconductors in particular, is not so much about great men and women and their glorious deeds, as about the unsung heroes of thousands of clever ideas and skillful experiments— progress of a purposeful centipede rather than a sleek thoroughbred, and thus a reflection of an age of organization rather than of individuality. This is not to detract from individual achievement, but only to stress that what has been done was done by virtue of superb efforts by great teams and many individuals within them.

A precondition for obtaining materials of required purity was to find a new method of purification. Here the methods of zone leveling and zone refining, invented by W. G. Pfann, proved crucial. Zone refining "is one of a general family of techniques known as zone melting. . . . a short liquid (or molten) zone travels through a relatively long charge (or ingot) of solid, carrying with it a portion of the soluble impurities."[110]

As with almost all inventions, it could be proved after the event that a predecessor had existed. The multitude of theories, techniques, and inventions that get nowhere because they are conceived under an inauspicious constellation of circumstances is so great that it is not really surprising when almost all successful inventions find, with hindsight, unknown and unwanted precursors. Pfann himself found, years after his own paper, one written by P. Kapitza, in which he had described a one-pass zone-melting operation. "He was concerned with crystal growth, however, and apparently did not recognize the potential of a molten zone as a distributor of solutes."[111]

Pfann himself came to zone melting via the route of studying slip bands in crystals of lead containing a small fraction of antimony. He tried to prevent segregation of antimony by passing a molten zone along the ingot, a process now known as zone leveling. At that time, in about 1939, he was unsuccessful and did not consider his attempt as anything more than an unremarkable commonsense approach. Pfann did not return to these questions until after the war, when the need for uniform purity in germanium arose. He noted, "Suddenly the lightning of zone refining struck: a molten zone moving through an ingot could do more than level out impurities—it could remove them."[112]

Zone refining quickly developed into an indispensable technique. It was first applied to germanium, then to silicon, and later to all kinds of materials used in laboratory experiments. For silicon, it has become a regular manufacturing technique used on a very large industrial scale. Technically, the availability of eddy current heaters, pure gases, vacuum apparatus, and fused silica or pure graphite vessels helped greatly. Physically, the process depends on different solubilities of impurities in the liquid and solid phases and can be repeated at will. The purities that have to be, and are, achieved are an almost unbelievable one foreign atom to 10^{10} germanium atoms. Pfann wrote, "The 'Secret' of zone refining, which now seems altogether obvious, is to pass a molten zone through an ingot *repeatedly*."[113]

As all these techniques of crystal preparation developed, the first attempts at changing the doping agent during crystal growth were made and thus grown p–n junctions, and from these, junction transistors were obtained. Small batches of such grown junction transistors were fabricated before the law of obsolescence hit this rather clumsy product. The next generation of junction transistors was fabricated by the so-called alloying process. In this, a small slab of germanium is cut out of the crystal, and impurities are alloyed into both sides of the slab by rapid heating in a process reminiscent of hard soldering. The alloyed germanium junction transistor became an important product, although it was later replaced by the diffused germanium transistor, the silicon transistor, and many more developments until the planar transistor and, from it, the integrated circuit became the state-of-the-art. It is tempting to describe all these matters here, but this would mean an unwarranted repetition and the interested reader is referred to the literature.[114]

The "open policy," referred to earlier, found expression in several symposia that proved of key importance. They were organized by Bell Labs and consisted of a school on transistor physics and technology provided jointly with Western Electric staff for military, university, and industrial scientists and engineers who wished to build up knowledge of transistor technology. The first of these key events was held during the week of 17 September 1951 at Murray Hill and was attended by 121 military personnel, 41 university people, and 139 industrial scientists. The proceedings came to 792 pages, and 5500 copies were distributed.[115]

The symposium clearly mirrored the great interest of the military in this new technology. Indeed, military applications of transistors were among the earliest ones, and the military was briefed on the emerging technology on 23 June 1948, a few days before the press conference. Presumably, clearance had to be obtained for public disclosure. Military funding of Bell Labs transistor research began in June 1949. Between 1949 and 1958, it amounted to $8.5 million, representing about 25% of total Bell Labs expenditure on device and material development during this period.[116]

So rapid was progress and so great was interest, that a second transistor technology symposium was arranged on 21–29 April 1952. This was, apparently, an entirely industrial affair, and 26 domestic and 14 foreign firms were represented. The information was given to the companies on the basis of confidentiality. "The transmission or the revelation of this information in any manner to any unauthorized person is prohibited by law."[117] Although the second symposium was mainly a civilian affair, the two volumes of *Transistor Technology,* published in 1952, were not declassified until the end of 1953. The preface includes the passage, curious for so international a gathering, "It is hoped also that, through the efforts of those receiving these volumes, transistor technology will advance rapidly and will bring enduring benefits to our national security and economy."[118]

Whatever one may think of the advantages or disadvantages of contemporary electronics, certainly nobody would doubt that actual events must have exceeded even the wildest expectations of the sponsors. Whether the fairy who granted these wishes was a good one or one with mixed motives, it certainly was most powerful. The open policy of Bell paid off handsomely and undoubtedly contributed to the amazingly fast growth of the emerging semiconductor industry. This fact also found expression in two further Bell symposia, held in 1956, at a time when diffusion technology was beginning to mature.[119]

So far, the story of the transistor has been a story of Bell Laboratories. Whether because of the open policy or because of the enormous demand for semiconductor experts, which was largely filled by an exodus from Bell Labs, some of the action soon shifted elsewhere. Many factors contributed to the mushrooming of a new industry. We shall not repeat here what has been described elsewhere,[120] but shall try to fill in at least one of the many gaps left in the narrative of Braun and Macdonald. Thus we shift our focus from Bell and the United States to the battered scene of postwar Germany and the shaken giant Siemens.

By 1951, economic recovery was sufficient to allow some scientific activity in Germany. Siemens was engaged in the manufacture of the old-fashioned solid-state rectifiers, had had considerable experience with germanium detectors, and was a Bell licensee for germanium transistors. Communication with American scientists was reestablished, and Siemens gradually reentered the world scientific–industrial community.

It was obvious to all that for high-power applications at least, silicon would be a much more suitable rectifier material than germanium and indeed was the only material that could replace the wayward selenium then in use. The main reason for preferring silicon was that the intrinsic conductivity becomes significant in silicon only at about 220°C, whereas germanium becomes an intrinsic semiconductor at about 70°C. The further advantage of silicon, the possibility of forming a superior oxide, was not considered important until later. In December 1952 the semiconductor division of Siemens decided to embark on a silicon development program.[121]

There were several known methods available for producing silicon powder or small crystallites. The classical, Du Pont, method consisted of a reaction of silicon tetrachloride with zinc, magnesium, or aluminum. One of the difficulties of this procedure was the further processing of the silicon into a single crystal, because molten silicon reacts with every conceivable crucible material and absorbs impurities from it.

A second method was tried, in which gaseous silicochloroform was reduced in a hydrogen atmosphere in a gas discharge. A polycrystalline rod of silicon grew between the carbon electrodes. The method, developed in cooperation with the Institute for Inorganic Chemistry at the University of Munich, did not yield the purity hoped for. The initial results of the cooperation proved, however, sufficiently promising to warrant further development, and, indeed, improvements in the apparatus and method yielded silicon of 2.5×10^{-9} (10^2 ohm-cm) purity.

In the next stage, it was attempted to deposit further silicon onto the thin polycrystalline silicon rods prepared by the foregoing method. The rods were resistance heated, and gaseous $SiCl_4$ was reduced by hydrogen and deposited. Later, this method was used alone, instead of the discharge, and became known as the Siemens C-process. The classical method was known at Siemens as the A-process.

The rods produced in the C-process had to be further purified and converted to single crystals. This was done by the process of floating zone melting invented not only at Bell Labs and the U.S. Signal Corps Laboratory, but simultaneously in two Siemens laboratories. The zone-refining process, as used for germanium, had to be modified somewhat for silicon, because the latter, when molten, reacted with all available containers. Henry C. Theuerer of Bell Labs hit on the idea of a "floating zone," although P. H. Keck and M.J.E. Golay of the U.S. Army Signal Corps Laboratories and R. Emeis, K. Siebertz, and H. Henker, of two Siemens laboratories, all independently invented the same process. The simple point is that the material to be refined is positioned vertically rather than horizontally. Both the A- and C-processes remained in use, and continual improvements were made so that by 1957 the amazing resistivity of 5000 ohm-cm—that is, a purity of 5×10^{-11}—was achieved. The lifetime of carriers reached 500 μsec.

Improvements in materials technology went hand in hand with improvements in measuring techniques and with greater theoretical understanding. Resistivity and Hall coefficient measurements had to be refined. A combination of theoretical analysis and measurement of temperature dependence of carrier concentrations, mobilities, and lifetimes yielded results on donor and acceptor concentrations as well as on scattering and recombination processes. It soon became apparent that high resistivity alone was no guarantee of purity, for if donors and acceptors are present in roughly equal numbers, the resistivity is the same as in an intrinsic semiconductor, yet the quality of the material is by no means the same.

Difficulties arise because carriers can spend considerable time in deep or shallow traps, and because mobilities can be affected by surface effects and by linear crystal imperfections, such as grain subboundaries and dislocations. The elucidation of all these matters was a mammoth task, let alone the elimination of undesirable imperfections.

Two problems stand out as bedeviling silicon development. One was contamination by boron, and the other by "donor X," where X eventually turned out to be phosphorus. Boron cannot be eliminated by zone refining because its solubility in the solid is slightly higher than in the liquid. The boron could be removed in the Siemens C-process, and in the Du Pont A-process, when hydrogen reduction is employed, because boron appears as a contaminant of $SiCl_4$ or $SiHCl_3$ and is removed in hydrogen reduction but not in Zn reduction. Phosphorus was removed by detective work reminiscent of that of Sherlock Holmes and involving 10 scientists in the painstaking elimination of all conceivable and inconceivable sources of contamination.

The careful preparation of silicon paid off handsomely. The entire semiconductor industry is wholly dependent on the quality of its raw material, and the history of semiconductor device development runs closely in parallel with material development. Better materials make better devices possible; better devices demand better material—a chicken-and-egg relationship of demand pull and ability push.

In the case of Siemens, superior mastery of silicon technology made possible the design and manufacture of successful power rectifiers and controlled rectifiers. The design of rectifiers demands a compromise between the requirement of a high reverse voltage and a high forward conductance. The former demands low doping levels, while the latter demands high doping. This dilemma was resolved by R. N. Hall and W. C. Dunlap, who suggested a *p-i-n* or *p-s-n* structure, where *i* stands for an intrinsic and *s* for a weakly doped region inserted between the *p* and *n* regions of the rectifier. The reverse voltage is determined by the *s–p* barrier, while in the forward direction the *s* region is flooded with carriers and adds little resistance to the forward flow of current.

The silicon power rectifier soon became firmly established and displaced virtually all other power rectifiers. There was one further step to take, however—the solid-state replacement of the thyristor or controlled rectifier. This was achieved by a *p-n-p-n* structure suggested by Moll and his colleagues. This product, too, achieved a high degree of perfection and captured large markets. One effect, just as an example, is that electric locomotives can now carry rectifiers on board and the transmitted current can be alternating, instead of using the old method of transmitting direct current.

Probably because of the impact of the invention of the transistor, but also because the first generation of postwar students came "on stream," the year 1948 marked the beginning of a veritable explosion in the number of papers published annually on the physics of semiconductors. Figure 7.4 gives a rough idea of this development. To pick from this wealth of information what is germane to our chapter is a well-nigh-impossible task, but I shall attempt to sketch (and sketch only) a few developments that were recommended to me by prominent and wise members of the semiconductor physics community.

The first development is concerned with a substantial deepening of theoretical understanding. In the early years of band theory and semiconductor physics, it was

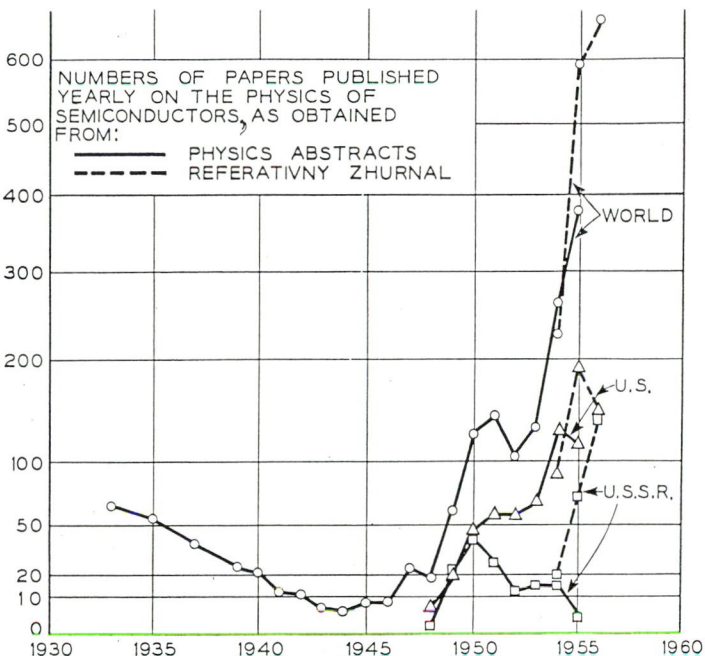

Fig. 7.4From a lecture given by Conyers Herring at Bell Laboratories, 27 March 1957.

simply assumed that the bands were spherically symmetrical in k-space, which means that the momentum of carriers is uniquely determined by their energy and does not depend on their direction of motion. The attitude to band theory up to the late 1940s has been described by Herring as "we don't know the band structure, so we might as well assume a simple isotropic band with some unknown effective mass."[122] Two trends began to countervail this complacency. First, it became possible to calculate band structures from first principles by methods initially developed for metals (Chapter 3). Second, some experimental results on magnetoresistance could not be interpreted without assuming more complex band-edge states in silicon and germanium. One of the "first principles" methods was the orthogonalized plane-wave (OPW) approach, developed by Herring and first applied to beryllium in collaboration with A. G. Hill. Further progress was delayed when this work was interrupted by the war, but Herring's interest remained. Some years later, he met Frank Herman, who had used the OPW method for diamond, and encouraged him to calculate the band structure of germanium. Herman, who had meanwhile joined the RCA Laboratories, performed such calculations in collaboration with J. Callaway. The calculated band structure certainly was not isotropic and showed several minima.[123]

The magnetoresistive effect provides an interesting example of the way progress was made in semiconductor physics. It is the wont of science to test the effect of temperature and of various fields on virtually every measurable parameter of a substance. Thus it soon became known that the application of a magnetic field to a semiconductor increases its resistivity and decreases the Hall coefficient. Measurements by Estermann and Foner at the Carnegie Institute of Technology in 1950 were carried out on polycrystalline specimens, and a little later G. L. Pearson and H. Suhl of Bell Laboratories specifically measured the effect of crystal orientation on magnetoresistance in germanium.[124] Unfortunately, theory failed to explain the magnitude of the effect. Although Gans had treated the problem of conductivity in magnetic fields in 1906, there was no adequate theory to deal with the question in semiconductors. F. Seitz set up a fairly general theory, but an attempt by Johnson and Whitesell to explain the magnetoresistive effect in detail failed yet again. In the same year, Shockley set up a very general theory in a mere letter, but a lot of further work was required to apply this to real experimental results.[125] The first real success in this direction was achieved by B. Abeles and S. Meiboom, at that time at the Weizmann Institute in Israel.[126] They took the energy of electrons in k-space to have several extrema and assumed the energy function near the extrema to be quadratic. The theory gave excellent agreement with the results obtained by Pearson and Suhl for n-type germanium. By this successful interpretation of experimental results, it became highly plausible to assume that indeed the image of a single minimum in the energy function was too simple and that the band structure was as complex as the calculation by Herman and Callaway had shown.

The anisotropy of magnetoresistance in silicon, obtained experimentally by Pearson, confirmed the band anisotropy for silicon. The multivalleyed band structure for silicon, as inferred from magnetoresistance measurements, was obtained by Herring and Pearson at Bell Labs in 1954.[127] The connection between the work of Abeles and Meiboom, on the one hand, and that of Herring and Pearson, on the other, was not fortuitous. Herring formed the link between most workers in this field and is acknowledged by nearly everyone as an important adviser and discussant.

Further experimental confirmation and absolute values for the anisotropic effective mass of carriers were obtained in the next few years by the new method of cyclotron resonance. The theory of cyclotron resonance absorption goes back to Drude, Voigt, and Lorentz (i.e., to the classical school of German physics). It achieved some importance in the propagation of radio waves in the upper atmosphere. A possible application of the theory to semiconductors was discussed by R. B. Dingle in 1952 and by Shockley in the already mentioned letter in the *Physical Review* in 1953. The first actual observations were made by Dresselhaus, Kip, and Kittel on germanium.[128] Herman reviewed the state-of-the-art on the band structure of germanium and silicon in 1955, and Figure 7.5 shows an example of such a structure from this review.[129]

The fact that the conduction bands of silicon and germanium near the band edges have constant energy surfaces that are either eight half-ellipsoids or six ellipsoids of revolution, corresponding to four and six energy valleys, respectively, has many consequences. For example, the charge transport and scattering processes are entirely different from those for a single-valley conduction band. These matters are discussed in detail by Karlheinz Seeger.[130] It is not possible to cover the history of this complex topic. We mention only one consequence. The fact that electrons have different mobilities in different valleys and that transitions within the band are possible leads to a region of negative resistance. Among the first to discuss this possibility was Cyril Hilsum.[131] His model materials were gallium arsenide and anti-

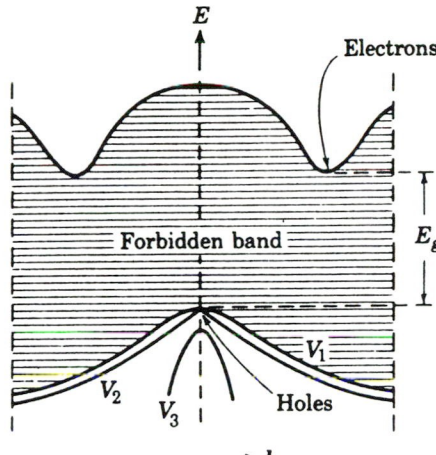

Fig. 7.5 Schematic representation of the energy band structure in Si along a ⟨100⟩ axis. (Based on F. Herman, "The Electronic Energy Band Structure of Silicon and Germanium," *Proceedings of the Institute of Radio Engineers* 43 [1955]:1703–1732)

monide (intermetallic semiconductors to be discussed next). Hilsum did not have sufficiently perfect material to demonstrate the effect, and it remained for Ian Gunn to show that negative resistance devices could be used as amplifiers or oscillators for high frequencies. We now briefly turn to intermetallic semiconductor compounds.

Intermetallic semiconductors imitate elementary materials, such as germanium and silicon, by mixing elements from group III with those of group V, or, more rarely, group II with group VI. Thus tetrahedral bonds are formed with four valence electrons, as in the group IV elements germanium and silicon. The inventor of the III–V compounds, Heinrich Welker, described his own train of thought in a retrospective paper of 1976.[132] Apparently, the idea of stable tetrahedral bonding goes back to H. G. Grimm and A. Sommerfeld in 1926, and the theoretical possibility of III–V compounds was discussed by V. M. Goldschmidt in 1927. Welker came to the conclusion that tetrahedral crystals with strongly covalent bonding should provide semiconductors with high carrier mobilities, as shown by the examples of germanium and silicon. It might be possible that III–V compounds with even higher mobilities than germanium or silicon could be obtained, and, in the wake of the transistor discovery, it seemed worthwhile to explore the properties of materials such as GaAs, AlSb, and InP. Although Welker made some progress in establishing which compounds should be metals and which insulators by using Pauling's theory of the chemical bond, the question was only finally resolved by Heywang and Seraphin in 1956.

In 1951, Welker produced his first ingot of indium antimonide and was delighted to find that it was a semiconductor. At about that time, he joined the newly established Siemens research laboratory in Erlangen as head of the solid-state research division. Indium antimonide and aluminum antimonide were chosen as the most promising of the III–V compounds for intensive investigation. Rectifying contacts were successfully made, and thus the semiconducting nature of the materials was readily confirmed. Welker had predicted from theory that the electron mobility in InSb should be as high as in germanium. Galvanomagnetic (Hall effect and magnetoresistance) measurements at first gave disappointing values for the mobility, but as the purity of the samples increased, the mobility crept up. Eventually, a value of 70,000 cm^2/V-sec was reached, which is nearly 20 times higher than in germanium. This was the decisive breakthrough. Several new devices were developed immediately, among them the so-called Hall generator, a device that utilizes the Hall effect for the measurement of magnetic fields.[133]

Welker's first paper on the topic was published in the autumn of 1952;[134] 10 years later, over 1000 papers on compound semiconductors had been published by researchers all over the world. So large is the world scientific–technical community and such is its lust for truly new openings that any scent is instantly picked up by the pack. The prize to be fought for seemed high, for it appeared possible that compound semiconductors might replace germanium and silicon altogether. This was not to be, since the combination of silicon and the planar process proved a truly winning combination that ushered in the age of microelectronics. On the other hand, the very high mobilities and certain radiative recombination processes give materials such as gallium arsenide advantages that secure a position of some importance for them in the marketplace.

From today's vantage point, two applications of compound semiconductors stand out: their use as light-emitting diodes (LEDs) or semiconductor lasers, and their use in very-high-speed integrated circuits. In LEDs the fact that conduction electrons accelerated through a forward-biased p–n junction return to the valence band with the emission of radiation is the crucial point. The complex band structure of these semiconductors not only makes such radiative transitions possible, but also allows a choice of emitted wavelength. A recent edition of Hans Landolt-Börnstein's handbook lists eight III–V compounds as technically useful as LEDs, operating either as lasers or as incoherent emitters.[135]

Very-high-speed integrated circuits, made possible by the extremely high mobility of charge carriers in materials such as GaAs, give these compounds a small but extremely important slice of the total microelectronics market.[136]

7.4   Conclusion

If we attempt a final bird's-eye glimpse of the development of semiconductor physics, we can discern several historical periods, each associated with the elucidation of some of the characteristic features and concepts of semiconductors.

The first period is the random collection of unconnected facts, without a clear recognition of semiconductors as a separate class of materials. We could call it the preparadigmatic period and note that it is related to a constant search for new facts. If these facts do not fit current theories, they may lead to their falsification, but are more likely to lead to small modifications and extensions of current theories or, as is the case in the preparadigmatic period, unaccountable results may simply be observed, noted, documented, and accumulated in the hope of some future use or understanding. Many of these are forgotten and rot away in the unread pages of scientific journals; some are picked up by later developments.

During the preparadigmatic period, technical applications are perfectly possible and, in the case of semiconductors, became widespread. Technology can often use effects without any need to understand them either in great depth or in a broad context. Semiconductors, which even in their infancy were used extensively as rectifiers and detectors, provide a good example of technology being able to dispense with deep physical insights. Semiconductors also provide the extreme counterexample, when a deeper understanding of the physics and chemistry of the materials enabled technology to develop new and better applications that were not possible without such scientific spadework.

Because of the early and continuing interest of technologists in the properties of semiconductors, much of the progress made in their scientific understanding was made in industrial research laboratories. The development of semiconductor applications coincided with and contributed to the rise of the industrial laboratory.

The next historical period, in our case the 1930s, is that of the emergence of a first general concept or paradigm. It was the understanding of the nature of the semiconductor in terms of band theory and the elucidation of the three most fundamental properties of semiconductors: the possibility of affecting the conductivity by impurities, the nature of positive charge carriers (the hole), and the fundamental significance of minority carriers. It was perhaps one of the special features of semiconductors as technically important materials that part of the elucidation of the role of minority carriers coincided with the explanation of their technically most significant feature, the rectifying contact.

The next historical period must be that of first consolidation. In the light of the newly gained theoretical insight, new measurements are carried out to test and develop the theory. In a sense, the period of first consolidation is also one of second accumulation, for new results often do go beyond the theoretical framework and form the raw material for the next period of theoretical refinement. In our case, the period of first consolidation is the decade 1940 to 1950. Apart from the general and historically common feature of consolidation, it has three special features: (1) the war and the single-minded, though one-sided, effort to develop certain semiconductor applications; (2) the great need for purer and better materials and the confusion created by structure-dependent properties; and (3) the intimate relationship between physical understanding and technical application. The second and third properties led to the need for interdisciplinary work, while the third property became decisive for the entire development of postwar research-and-development activities.

The final period treated here, roughly the 1950s, is one of refinement and extension. The first theories begin to show their deficiencies, and their simplifying assumptions suddenly appear naive. This is seen here in the example of the refinement of band theory, although many other examples could have been treated. Not only are theories refined, but the scope of the subject is also widened. In our case, we exemplified this by briefly mentioning compound semiconductors.

In modern science, the periods of consolidation and of refinement are also marked by explosive growth. The initial handful of papers grows into thousands. To maintain any sort of grip on developments, conferences, organizations, meetings, and abstracts become necessary, and we see the coincidence of these historical periods with periods of growth of social organization.

Because of the unique feature of semiconductor physics as the progenitor of the microelectronics revolution, all the foregoing developments have occurred in an almost grotesquely exaggerated manner. Thus the physics of semiconductors has been absorbed into a vast technological enterprise and owes its present stature to huge industrial organizations.

Notes

It remains to express my very sincere thanks to all those who have given me advice and information and especially those who have so wisely criticized the nearly final draft of this chapter. I have profited greatly from their criticism and hope that my readers will share in the profit.

1. See E. Braun und E. Macdonald, *Revolution in Miniature,* 2nd ed. (Cambridge: Cambridge University Press, 1982); H. Goetzeler, "Aus den Anfängen der Halbleiter," *Technikgeschichte* 3a (1972): 31–50; G. L. Pearson and W. H. Brattain, "History of Semiconductor Research," *Proceedings of Institute of Radio Engineers* 43 (1955): 1794–1806.

2. E. Braun, "Science and Technological Innovation," *Physics Education* 14 (1979): 353.

3. F. Braun, "Über die Stromleitung, Schwefel-metalle." *Annalen der Physik und Chemie* 229 (1874): 556–563; Goetzeler (n. 1).

4. A. Schuster, "On Unilateral Conductivity," *Phil. Mag.* 48 (1874): 251–257.

5. Goetzeler (n. 1).

6. E. H. Hall, "On a New Action of the Magnet on Electric Currents," *American Journal of Mathematics* 2 (1879): 287–292.

7. W. G. Adams and R. E. Day, "Action of Light on Selenium," *Proc. Roy. Soc.* 25 (1876): 113–117; C. Fritts, "A New Form of Selenium Cell," *Americal Journal of Science* 26 (1883): 465–472.

8. W. von Siemens, *Wissenschaftliche und technische Arbeiten* (Berlin, 1889), 1: 311.

9. J. Koenigsberger and J. Weiss, "Uber die thermoelektrischen Effekte und die Wärmeleitung in einigen Elementen und Verbindungen und Über die experimentelle Prüfung der Elektronentheorien," *Ann. Phys.* 35 (1911): 1–46.

10. G. W. Pickard, U.S. patent 836,531 (1906); K. Bädeker and E. Pauli, "Electrical Conductivity of Solid Copper Iodides," *Phys. Zs.* 9 (1908): 431. See also W. Kaiser, "Karl Baedekers Beitrag zur Halbleiterforschung," *Centaurus* 22 (1978): 187–200.

11. E. Spenke and W. Heywang, "Twenty Five Years of Semiconductor-Grade Silicon," *Physica Status Solidi* 64 (1981): 11–44, p. 14; B. Gudden, "Conduction Elecrons and Photoelectrons in Insulators and Semi-Conductors," *Phys. Zs.* 32 (1931): 825–833; K. Seiler, interview with E. Braun and J. Teichmann, 2 June 1982.

12. See, for example, J. A. Fleming, *The Principles of Electric Wave Telegraphy* (London: Longman, 1908).

13. See Braun and Macdonald (n. 1), chap. 2.

14. E. Presser, *Funkbastler* (1925): 558; L. O. Grondahl, "A New Type of Contact Rectifier," *Phys. Rev.* 27 (1926): 813.

15. E. Spenke, "Beitrag zur Frühgeschichte der Halbleiter-Elektronik und der Kupferoxydul und Selen-Gleichrichterentwicklung vornehmlich im Hause Siemens," unpublished report, p. 9. I am much indebted to Dr. Spenke for providing me with this report, which served as the basis for the treatment of this topic.

16. Ibid., 18.

17. Ibid., 18–21.

18. H. K. Henisch, *Metal Rectifiers* (Oxford: Clarendon Press, 1949), 4–13.

19. A. Wilson, interview with S. Keith and E. Braun, 1980.

20. A. H. Wilson, "The Theory of Electronic Semi-conductors," *Proc. Roy. Soc.* A133 (1931): 458–491.

21. R. E. Peierls, "Recollections of Early Solid State Physics," in *Beginnings,* 28–38.

22. Ibid, 30; For the full results of this work, see R. Peierls, "Zur Theorie der galvanomagnetischen Effekte," *Zs. Phys.* 53 (1929): 255–266; Peierls, "Zur Theorie des Hall-Effekts," *Phys. Zs.* 30 (1929): 273–274. For comprehensive treatments, see A. H. Wilson, *The Theory of Metals* (Cambridge: Cambridge University Press, 1936); W. Shockley, *Electrons and Holes in Semiconductors* (New York: Van Nostrand, 1950).

23. A. H. Wilson, "The Theory of Electronic Semi-conductors, II," *Proc. Roy. Soc.* A134 (1931): 277–287.

24. Peierls (n. 21), 34–35.

25. A. H. Wilson, "A Note on the Theory of Rectification," *Proc. Roy. Soc.* A136 (1932): 487–498.

26. Henisch (n. 18), 109–114.

27. L. O. Grondahl, "The Copper–Cuprous Oxide Rectifier and Photoelectric Cell," *Review of Modern Physics* 5 (1933): 141–168.

28. Spenke (n. 15), 11–17.

29. H. Rothe, E. Spenke, and C. Wagner, "Zum 65. Geburtstag von Walter Schottky," *Archive der elektrischen Übertragung* 5 (1951): 306–313; W. Schottky, "Zur Halbleitertheorie der Sperrschicht und Spitzengleichrichter," *Zs. Phys.* 113 (1939): 367–414.

30. W. Schottky and C. Wagner, "Theorie der geordneten Mischphasen," *Zeitschrift für physikalische Chemie* 11 (1930): 163–210.

31. For a summary of Schottky's views, see W. Schottky, "Statistics and Thermodynamics of the States of Disorder in Crystals, Especially Those with a Slight Degree of Disorder," *Zeitschrift für Elektrochemie* 45 (1939): 33–72.

32. W. Schottky, "Cold and Hot Electron Discharges," *Zs. Phys.* 14 (1923): 63–106.

33. W. Schottky, "Zur Halbleitertheorie der Sperrschicht und Spitzengleichrichter," *Zs. Phys.* 113 (1939): 367–414, p. 369.

34. Ibid., 410.

35. Ibid., 370, quoting O. Fritsch, "Electrical and Optical Properties of Semi-conductors. Electrical Measurements on ZnO," *Ann. Phys.* 22 (1935): 375–401.

36. For a detailed argument of the formation of consensus in scientific theory, see J. Ziman, *Public Knowledge* (Cambridge: Cambridge University Press, 1968).

37. Schottky (n. 33), 408.

38. W. Schottky and E. Spenke, "Quantitative Aspekte der Theorie der Randschichten und Kristall-Detektoren," *Wissenschaftliche Veröffentlichungen aus den Siemens-Werken* 18 (1939): 225–291.

39. W. Schottky, "Vereinfachte und erweiterte Theorie der Randschicht-gleichrichter," *Zs. Phys.* 118 (1942): 539–592.

40. N. F. Mott, "The Theory of Crystal Rectifiers," *Proc. Roy. Soc.* 171 (1939): 27–144. See also N. F. Mott and R. W. Gurney, *Electronic Processes in Ionic Crystals,* 2nd ed. (Oxford: Clarendon Press, 1948), 174–185.

41. B. Davydov, "The Rectifying Action of Semi-Conductors," *Journal of Technical Physics of the USSR* 5 (1938): 87–95.

42. C. Herring to E. Braun, 30 January 1986.

43. C. Herring, in a lecture given at Bell Laboratories, 27 March 1957.

44. H. Bethe, "Theory of the Boundary Layer of Crystal Rectifiers," *MIT Radiation Laboratory Report* 43 (1942).

45. Wilson (n. 25); J. Frenkel and A. Joffe, "Electric and Photoelectric Properties of Contacts between Metals and Semi-conductors," *Physikalische Zeitschrift der Sowjetunion* 1 (1932): 60–87; L. Nordheim, "Theory of Detector Action," *Zs. Phys.* 75 (1932): 434–441.

46. J. Schopman, "The Dutch Contribution to Barrier-Layer Semiconductors in the Pre-Germanium Era," *Janus* 69 (1982): 1–28.

47. Spenke (n. 15). See also S. Poganski, "Geschichtlicher Überblick über die Entwicklungs- und Forschungsarbeiten an Selengleichrichtern bei der Allgemeinen Elektricitäts-Gesellschaft (AEG)," unpublished report.

48. Henisch (n. 18), 51–52.

49. H. Dember, "Über eine photoelektromotorische Kraft in Kupferoxydul-Kristallen," *Phys. Zs.* 32 (1931): 554–556.

50. Mott and Gurney (n. 40), 192–196.

51. I. K. Kikoin and M. M. Noskov, "A New type of Photoelectric Effect in Cuprous Oxide in a Magnetic Field," *Nature* 31 (1933): 725–726; a much more detailed description was published by Kikoin and Noskov in "New Photoelectric Effect in Cuprous Oxide. Part I.," *Physikalische Zeitschrift der Sowjetunion* 5 (1934): 586–596.

52. J. Frenkel, "On the Explanation of the Photoelectromagnetic Effect in Semi-conductors," *Physikalische Zeitschrift der Sowjetunion* 5 (1934): 597–598.

53. N. Cusack, *The Electrical and Magnetic Properties of Solids* (London: Longman, 1958), 240.

54. Herring (n. 43).

55. L. N. Ridenour, *Radar Systems Engineering* (New York: McGraw-Hill, 1947), 412.

56. F. Seitz to E. Braun, 3 February 1986.

57. H. C. Torrey and C. A. Whitmer, *Crystal Rectifiers* (New York: McGraw-Hill, 1948), 8.

58. The following description of the development of the Radiation Laboratory is based on an unpublished draft report by P. Henriksen, "The Rad Lab and Solid State Physics," University of Illinois, January 1984. I am much indebted to him for supplying me with this valuable information.

59. J. Bernstein, "Profiles—I. I. Rabi II," *New Yorker,* 20 October 1975, 48ff; I. I. Rabi, Karl T. Compton lecture, 8 March 1962, tape in Institute Archives and Special Collection, MIT Record Group, MC-109.

60. Bernstein (n. 59), 49.

61. Seitz letter (n. 56).

62. Ibid.

63. J. Stuke, interview with E. Braun, 5 July 1984.

64. H. Bethe, "Theory of HF Rectification by Silicon Crystals," *MIT Radiation Laboratory Report,* 43 (1942).

65. K. Lark-Horovitz, "Preparation of Semiconductors and Development of Crystal Rectifiers," NDRC Report 14-585 (1946).

66. P. W. Henriksen, "Solid State Physics Research at Purdue," *Osiris* 3, ser. 2 (1987): 237–260; C. Benedicks, "Elektrischer Widerstand Einiger Seltener Metalle: Thermokraft und Gleichrichterwirkung des germaniums," *Internationale Zeitschrift für Metallographie* 7 (1915): 225–238; E. Merritt, "Contact Rectification by Metallic Germanium," *Proceedings of the National Academy of Sciences of the United States* 11 (1925): 743–748.

67. Henriksen (n. 66); E. M. Conwell and V. F. Weisskopf, "Theory of Impurity Scattering in Semicoductors," *Phys. Rev.* 77 (1950): 388–390.

68. R. Bray, interview with P. Henriksen, 14 May 1982, AIP.

69. For a discussion of this issue, see also Braun and Macdonald (n. 1), 38–39.

70. R. Whaley, interview with P. Henriksen, 29 October 1982, AIP.

71. Bray interview (n. 68).

72. J. Bardeen, interview with L. Hoddeson, 1 December 1977, AIP.

73. G. L. Pearson, interview with L. Hoddeson, 23 August 1976, AIP.

74. W. Brattain, interview with A. N. Holden and W. J. King, May 1974, AIP.

75. Ibid.; see also Braun and Macdonald (n. 1), 37.

76. R. S. Ohl, interview with L. Hoddeson, August 1976, AIP.

77. Ibid.

78. J. Bardeen to E. Braun, 20 February 1986.

79. E. H. Putley and D. H. Parkinson, interview with E. Braun, P. Hoch, and S. Keith.

80. K. Seiler, interview with E. Braun and J. Teichmann, 2 June 1982.

81. H. Welker, interview with M. Eckert, J. Teichmann, and G. Torkar, 4 December 1981, AIP (shortly before his death on Christmas Day, 1981).

82. K. Clusius, E. Holz, and H. Welker, Patent 966,387, "Elektrische Gleichrichteranordnung mit Germanium als Halbleiter und Verfahren zur Herstellung von Germanium fur eine solche Gleichrichteranordnung." This FRG patent was transferred from the 1942 patent of the German Reich and is dated August 1957.

83. G. Rosenberger, interview with E. Braun, January 1984; see also K. Seiler, G. Goubau, and J. Zenneck, eds., *Naturforschung und Medizin in Deutschland, 1939–1946,* vol. 15, pt. 1, *Elektronenemission* (Wiesbaden: Dieterichsche Verlagsbuchhandlung, n.d.).

84. From a collection of Welker's papers in DM.

85. Gaudlitz, "Historischer Rückblick auf die Anfänge der Halbleiterentwicklung bei Siemens & Halske bis zum Sommer 1950," internal Siemens memorandum, Munich, April 1962. See also

E. Hofmeister, "Hundert Jahre Kristallgleichrichter," in *Proceedings of Nürnberg Conference, 23–27 September 1974,* ed. Wernheim (n.p.: Physik Verlag, n.d.), 421–452.

86. Welker papers (n. 84).

87. Ibid.

88. Seiler interview (n. 80).

89. H. Pick, interview with J. Teichmann and G. Torkar, 2 October 1981, AIP.

90. See Braun and Macdonald (n. 1); L. Hoddeson, "The Discovery of the Point-Contact Transistor," *Historical Studies in the Physical Sciences* 12 (1981): 41–76; C. Weiner, "How the Transistor Emerged," *IEEE Spectrum* 10 (1973), 24–33.

91. E. Braun, "Semiconductor Surface Physics," research laboratory report, AEI Ltd., March 1961; I. Tamm, "Über eine mögliche Art der Elektronenbindung an Kristalloberflächen," *Physikalische Zeitschrift der Sowjetunion* (1932): 733–746; W. Shockley, "On the Surface States Associated with a Periodic Potential," *Phys. Rev.* 56 (1939): 317–323. See also J. N. Zemel, ed., *Semiconductor Surface Physics* (Oxford: Pergamon Press, 1960); this constitutes the proceedings of the second conference on this topic held in December 1959.

92. Bardeen letter (n. 78); J. Bardeen, "Surface States and Rectification at a Metal Semi-Conductor Contact," *Phys. Rev.* 71 (1947): 717–727.

93. W. Shockley and G. L. Pearson, "Modulation of Conductance of Thin Films of Semi-conductors by Surface Charges," *Phys. Rev.* 74 (1948): 232–233.

94. Bardeen, letter (n. 78). The patent referred to is No. 2,524,033, 3 October 1950: J. Bardeen, "Three-Electrode Circuit Element Utilizing Semiconductive Materials," filed 26 February 1948.

95. W. S. Gorton, "Genesis of the Transistor," written in December 1949 and intended for volume 3 of *A History of Engineering and Science in the Bell System.* My thanks are due to Bell Labs archives for allowing me to see this and much other material.

96. W. H. Brattain, entry of 15 December 1947, laboratory notebook, case 38139-7, Bell Laboratories archives. Reproduced here by kind permission.

97. Brattain, entry of 16 December 1947 (ibid.).

98. This is mentioned by Brattain in an appendix to Gorton (n. 95), and was mentioned by J. R. Pierce, interviews with S. Macdonald and E. Braun.

99. Bardeen letter (n. 78).

100. W. Shockley, "The Invention of the Transistor: An Example of Creative-Failure Methodology," in *Solid State Devices* (London: Institute of Physics, 1972), 55–75.

101. Minutes of conference on the Surface States Project, W. G. Pfann, entry of 28 April 1948, laboratory notebook, case 38235-5, Bell Laboratories archives. At this stage, some transistor nomenclature (e.g., "collector") had been introduced.

102. "Development History of the Transistor in the Bell Laboratories and Western Electric (1947–1975), Section I: General Narrative," manuscript, August 1980, 10–12 (kindly made available by Bell Labs).

103. Braun and Macdonald (n. 1), 46–49.

104. Shockley (n. 22), x.

105. Ibid., 59.

106. S. Macdonald and E. Braun, "The Transistor and Attitude to Change," *American Journal of Physics* 45 (1977): 1061–1065.

107. Shockley (n. 22), 94.

108. Pfann, entry of 18 February 1948, laboratory notebook (n. 101). "HBV" stands for high back voltage and is a reference to the type of material used in the manufacture of radar detectors.

109. G. K. Teal and J. B. Little, "Growth of Germanium Single Crystals," *Phys. Rev.* 78 (1950): 647.

110. W. G. Pfann, "Zone Refining," *Scientific American* 217 (1967): 63.

111. Ibid.; the original scientific paper by Pfann was published in "Principles of Zone Melting," *Journal of Metals* 4 (1952): 747–753.

112. Pfann (n. 110), 63.

113. Ibid.

114. See Braun and Macdonald (n. 1).

115. M. D. Fagen, ed., *A History of Engineering and Science in the Bell System, National Service*

in War and Peace (1925–1975) (Murray Hill, N.J.: Bell Telephone Laboratories, 1978), 622. The publication resulting from the symposium was *The Transistor* (Murray Hill, N.J.: Bell Telephone Laboratories, 1951).

116. Fagen (n. 115), 621.

117. "Development History of the Transistor" (n. 102), Appendix II.

118. Ibid., 24–25. The proceedings of the symposium were published as *Transistor Technology* 2 vols. (New York: Western Electric, 1952; New York: Van Nostrand, 1958).

119. Fagen (n. 115), 622. It would appear that separate symposia were held for Bell licensees and the military.

120. Braun and Macdonald (n. 1).

121. Spenke and Heywang (n. 11); H. Fischer, J. Pfaffenberger, K. Siebertz, and E. Spenke, "Historische Übersicht über die Silizium Entwicklung im H Siemens," unpublished 1957. I am much indebted to Dr. F. Spenke for letting me have this material and much advice. See also F. Spenke, "History and Future, Notes on Silicon Technology," in *Semiconductor Silicon,* ed. R. R. Haberecht and E. L. Kern (New York: Electrochemical Society, 1969), 1–35.

122. Herring letter (n. 42).

123. C. Herring, "Recollections," in *Beginnings,* 68–71.

124. I. Estermann and A. Foner, "Magneto-resistance of Germanium Samples Between 20° and 300° K," *Phys. Rev.* 39 (1950): 365–372; G. L. Pearson and H. Suhl, "The Magneto-Resistance Effect in Oriented Single Crystals of Germanium," *Phys. Rev.* 83 (1951): 768–776.

125. F. Seitz, "Note on the Theory of Resistance of a Cubic Semiconductor in a Magnetic Field," *Phys. Rev.* 79 (1950): 372–375; W. Shockley, "Cyclotron Resonances, Magnetoresistance, and Brillouin Zones in Semiconductors," *Phys. Rev* 90 (1953): 491; V. A. Johnson and W. J. Whitesell, "Theory of the Magnetoresistive Effect in Semiconductors," *Phys. Rev.* 89 (1953): 941–947.

126. B. Abeles and S. Meiboom, "Theory of the Galvanomagnetic Effects in Germanium," *Phys. Rev.* 95 (1954): 31–37.

127. F. Herman and J. Callaway, "Electronic Structure of Germanium Crystals," *Phys. Rev.* 89 (1953): 518; G. L. Pearson and C. Herring, "Magneto-Resistance Effect and the Band Structure of Single Crystal Silicon," *Physica* 20 (1954): 975–978.

128. G. Dresselhaus, A. F. Kip, and C. Kittel, "Observation of Cyclotron Resonance in Germanium Crystals," *Phys. Rev.* 92 (1953): 827.

129. F. Herman, "The Electronic Energy Band Structure of Silicon and Germanium," *Proceedings of Institute of Radio Engineers* 43 (1955): 1703–1732.

130. K. Seeger, *Semiconductor Physics* (New York: Springer, 1973), chap. 7.

131. C. Hilsum, "Transferred Electron Amplifiers and Oscillators," *Proceedings of Institute of Radio Engineers* 50 (1962): 185–189.

132. H. Welker, "Discovery and Development of III–V Compounds," *IEEE Transactions on Electron Devices* ED 23 (1976): 664–674.

133. O. Madelung, "Die III–V Verbindungen und ihre Bedeutung für die Halbleiterphysik" (lecture given at the University of Munich in memory of H. Welker, 20 December 1982); Madelung, interview with E. Braun, 5 July 1984.

134. H. Welker, "Über neue halbleitende Verbindungen," *Zeitschrift für Naturforschung* 7a (1952): 744–749.

135. H. H. Landolt-Börnstein, *Numerical Data and Functional Relationships in Science and Technology,* vol. 17: *Semiconductors,* subvol. d: *Technology of III–V, II–VI and Non-Tetrahedrally Bonded Compounds* (Berlin: Springer, 1984), 68ff.

136. W. Heywang, "III–V Verbindungen," *Umschau* 78 (1978): 771–777; Heywang, interview with E. Braun and J. Teichmann.

8 | *Collective Phenomena*

LILLIAN HODDESON, HELMUT SCHUBERT, STEVE J. HEIMS, AND GORDON BAYM

In the 1930s and 1940s, the quantum theory of solids dealt primarily with single-electron phenomena. Although surprisingly many problems were solvable within this framework, some were not—for example, the explanation from first principles of superconductivity and the calculation of the cohesive energy of metals. A related problem, whose connection to superconductivity was recognized almost immediately, was the superfluidity of ^{4}He, discovered experimentally in 1937. More generally, phase transitions—ordinary liquid–gas and melting, as well as those to the superconducting, superfluid, and ferromagnetic states—lacked adequate theory. These various phenomena, unexplainable within a single-particle framework, depend strongly on the interactions between particles in the systems; today they are recognized as "collective" or "cooperative" phenomena.

The range of validity of the one-electron theory, which had proved successful in so many instances, appears not to have been seriously examined in the pre–Second World War period. Yet since solids contain not simply single electrons, but more than 10^{23} interacting particles, the question of why this theory worked for any solid-state problem at all lurked in the background. Even in the 1930s, a few leading physicists suspected that the insoluble problems, such as superconductivity, had in common the need for proper treatment of the interactions between electrons (Chapter 2).[1] But despite these intuitions, the basis for significant advance was not available; methods suitable for dealing with the interactions in a many-body system—"many-body theory"—would not be invented until the 1950s. These techniques were then developed into a powerful new framework for the theory of solids, within which one could show rigorously why the one-electron theory worked where it did, as well as treat cases where the earlier theory had failed. The story of the emergence of the collective framework is the theme of this chapter.[2]

Like the larger area of solid-state physics as a whole (Chapter 9), the study of many-particle systems developed through a process of consolidation of previously independent or loosely interacting streams of research. Eventually, these streams would be seen as the subfields of the subject of collective phenomena, areas sharing many general themes and concepts, such as elementary excitations (e.g., phonons, spin waves, quasiparticles), macroscopic wave functions, order parameters, and broken symmetries.[3] We examine four streams of collective phenomena: the electron gas, superconductivity, superfluid helium, and critical phenomena. Several other important streams (e.g., ferromagnetism and ferroelectricity) are not treated, either for lack of space or because it was originally planned to discuss them elsewhere in this volume. Since the Second World War served as an influential interlude in the emergence of collective phenomena research, we divide the chapter into two sections, the first covering the prewar and the second the postwar history up to the early 1960s.

8.1   Pre–Second World War Roots

The Interacting Electron Gas

Four developments during the 1920s and 1930s helped to set the stage for the remarkable progress in the 1950s on the theory of the interacting electron gas: (1) work on "screening" in classical plasmas in Zurich by Peter Debye and his *Assistent* Erich Hückel in the winter of 1922; (2) vacuum-tube investigations by Irving Langmuir and collaborators at the General Electric Company (GE) in 1925 to 1929, resulting in their theory of classical plasmas; (3) theoretical work in Budapest and Princeton in 1934 and 1938 by Eugene Wigner on the correlation energy of the alkali metals; and (4) studies by John Bardeen at Harvard in 1936 to 1938 on screening and the specific heat of the electron gas.

The Debye–Hückel study of screening arose in the context of attempts to understand electrolytes.[4] As Debye recalled in 1965,[5] the physical chemist E. Bauer, who was interested in electrical fuel cells, learned from W. Nernst in Berlin, "a kind of pope in Physical Chemistry," of a new theory of electrolytes put forth by an Indian, J. C. Ghosh.[6] The problem was that strong electrolytes, such as sodium chloride solution, did not follow Ostwald's dilution law relating the conductivity of electrolytes to the concentration, based on the then-accepted theory of J. van't Hoff and S. Arrhenius. Ghosh attempted to describe the concentration dependence of the electrical properties of electrolytes by conceiving of them as expanded crystals of sodium and chlorine, in which the distance between ions is determined by the concentration. Bauer presented this theory at a colloquium of physical chemists held in one of the guildhouses in Zurich. In a heated discussion, Debye objected to the idea that the sodium and chlorine could be treated as a crystal in a solution, because of ionic Browian motion. Debye recalls that in response to Bauer's remark, "If you want to make criticism . . . do something better!", he "went home and just that same evening I wrote out how you should do it . . . with this atmosphere of ions around each central ion of opposite charge."[7]

Using the Boltzmann and Poisson equations to develop a self-consistent solution, Debye formulated the notion of screening, which was to have great importance in the subsequent history of the interacting electron gas. By the winter of 1921, he had found that the potential created by a charge in a plasma falls off exponentially with a characteristic "Debye screening length" $L_D = \sqrt{k_b T/4\pi n e^2}$. Debye and Hückel continued work on the electrolyte problem over the next year, deriving in particular the result that the mobility of carriers should vary as the square root of the concentration.[8]

The pioneering work on plasmas by Langmuir and his group grew out of studies in the early 1920s of electron velocity distributions in hot vacuum-tube cathode discharges at low pressures, following up the interest of the GE management in the development of mercury arc rectifiers. In the course of exploring such rectifiers, Langmuir noticed that the uniformly glowing ionized gas in the arc discharges assumed its own form and shape—hence Langmuir's word *plasma*—"a kind of crystalline gas," having about equal numbers of ions and electrons, and thus a very small resultant space charge.[9] With Harold Mott-Smith, his chief assistant from 1921 to 1928, Langmuir conducted a series of scattering experiments to determine the velocities of the electrons; he found that certain electrons have much greater and certain others much smaller velocity than that corresponding to the voltage drop across the tube. Langmuir's theory of classical plasmas grew out of his idea that these variations might be caused by oscillations of the electronic current—"compressional waves somewhat analogous to sound waves." Tonks and Langmuir then clarified the picture of these collective oscillations of plasmas in 1929, explaining[10] that in a plasma "when the electrons oscillate, the positive ions behave like a rigid jelly with uniform density of positive charge *ne*. Embedded in this jelly and free to move there is an initially uniform electron distribution of charge $-ne$."[11] Here *n* is the density of electrons, and *e* the charge on the electron.

Another separate root of the problem of an interacting electron gas was Eugene Wigner's perturbation-variational approach to the cohesive energy of metals, developed in 1934 and extended in 1938.[12] This work, written in Budapest during the summer and sent to Frederick Seitz at Princeton for typing, grew out of Wigner's attempts to improve on the cohesive energy predicted by his innovative calculations with Seitz in 1933 and 1934 of the band structure of sodium (Chapter 3).[13] The Hartree treatment of the cohesive energy of metals, based on an independent electron model, in which the effect of the other electrons is represented by a self-consistent potential, failed to explain why the alkalis are bound. Hartree–Fock theory led to cohesion, since the exclusion principle keeps electrons of parallel spin apart, but it was still quantitatively inadequate. In his paper, Wigner attempted to include the repulsive correlation of electrons having antiparallel spin, an effect that reduces the repulsion between such electrons. The additional binding energy from this effect was named by Wigner the "correlation energy." Pointing out that at very small kinetic energies (i.e., sufficiently low densities) the electrons form a lattice, he was able to estimate the correlation energy for the limits of high and low density.[14] Then, as he showed in his 1938 paper, an estimate for the ground-state energy could be obtained by interpolating between the two limits.[15] This estimate was reasonably good, but the work attracted little attention at the time.[16]

Although going from the Hartree to the Hartree–Fock approach led to qualitative improvements in the calculation of the cohesive energy, quite the opposite was true for the electronic specific heat. Not only was it surprising that the primitive

Hartree method worked so well in calculating specific heats, since electron correlations ought to play a role, but improving the Hartree treatment by using the Hartree–Fock equation worsened agreement with experiment! The first hint of troubles was Dirac's 1930 demonstration that the exchange energy leads to a logarithmic singularity in the electron density of states at the Fermi surface.[17] Then Bardeen, Wigner's former graduate student at Princeton and a junior fellow at Harvard at the time, showed in 1936 that the Hartree–Fock approximation with an unscreened interaction between the electrons leads to a logarithmic singularity in the specific heat, which was not observed. Bardeen published only an abstract of a talk on this subject[18] because he felt he had not found a real solution to the question of why exchange corrections do not appear empirically.[19] The following year, Bardeen considered the role of screening of the ionic charges by the electrons, which shows up in the electron–lattice interaction as a term due to the polarization of electrons by the ionic motion and drastically alters the interaction of the conduction electrons in metals with long-wavelength ionic motion.[20]

The Debye and Langmuir studies established the correct long-wavelength phenomenology of interacting electron systems, but the cohesive energy and specific heat studies underlined the need to take these phenomena into account systematically in the treatment of electron–electron interactions. However, it would not be until the postwar period that these separate strands would intertwine. Another area in which electron–electron interactions were in fact playing a crucial role was superconductivity, and indeed, only after the development of the techniques to handle the normal interacting electron gas would the microscopic problem of superconductivity be amenable to solution. We turn now to the prewar understanding of this problem.

Superconductivity

The discovery of the remarkable phenomenon of superconductivity, the loss of certain metals and alloys of any measurable electrical resistance at temperatures close to absolute zero, was made in Leiden in the laboratory of Heike Kamerlingh Onnes only several years after the capability of conducting experiments at very low temperatures was first achieved, also in Leiden.[21] Liquid helium was first produced in 1908 by using a liquefier built with the help of Master G. J. Flim. Kamerlingh Onnes's technician (for the Leiden laboratory, see "Low-Temperature Methods" section in Chapter 1, and for helium see the next section). At the low temperatures made accessible by the liquefier, down to 1.2 K, research initially centered on measurement of the temperature dependence of the electrical resistance of metals. This problem arose in connection with the determination of the purity of metals using Matthiessen's rule (a relation between the purity and the residual resistance at low temperatures). Electrical resistance had also been used for the exact determination of low temperatures (e.g., by the platinum thermometer). The properties of helium would soon become a second focus of low-temperature research.

At liquid hydrogen temperatures (about 20 K), Kamerlingh Onnes and his assistant J. Clay found,[22] in agreement with measurements by Dewar and Fleming,[23] deviations of the resistance curve of metals from linearity. In early measurements in the new temperature regime, however, Kamerlingh Onnes found that below 4.5

Flim and Dana preparing for an experiment. (AIP Niels Bohr Library)

K the resistivity approached a constant value, which was lower if the metal was more pure, in accord with Matthiessen's rule. Since there was no indication of another rise in the resistance, Kamerlingh Onnes rejected Kelvin's dissociation theory of the resistivity. He put forth instead an explanation based on the Einstein–Planck theory of specific heat:

> It seems that the free electrons in the main remain free, and it seems to be the movable parts of the vibrators that are now bound, their motion at ordinary temperature forming the obstacles to conduction; these disappear when the temperature is lowered sufficiently as the vibrators become practically immovable.[24]

Identifying these vibrators with those in Einstein's theory of specific heat, Kamerlingh Onnes was able to obtain a vanishingly small resistance at helium temperatures, in qualitative agreement with experiment.

   This result appeared to point toward the disappearance of electrical resistance at absolute zero. Using oscillator energy data from elasticity and specific heat measurements, Kamerlingh Onnes estimated the expected resistivities of various metals and found that only for pure mercury would the resistance in the neighborhood of 4 K lie outside experimental uncertainty. The resistance of gold, for example, came immeasurably close to zero already at 13 K.[25] The first measurements on pure mercury at helium temperatures showed that at 4.3 K, the boiling point of helium, the resistance was actually lower than at hydrogen temperatures and still temperature-dependent. At the lowest temperature of 3 K, it fell below the limit of sensitivity of the instruments. Yet these values appeared to confirm the predictions of the simple quantum expression, "that a pure metal can be brought to such a condition that its electrical resistance becomes zero, or at least differs inappreciably from that value."[26]

This result, published in April 1911 "with all reserve," was "repeated without delay"; only a month later the next report came out, entitled "The Disappearance of the Resistance of Mercury."[27] Gilles Holst, a student under Kamerlingh Onnes responsible for the technique of exact measurement of resistance, during this time had increased the sensitivity of his method some 3 orders of magnitude, and was able to demonstrate that the resistance of the mercury sample at 3 K fell to a value 10^{-7} that at 0°C! With rising temperature, the resistance at "slightly more than 4.2 K took on a measureable value." At only 4.3 K, the boiling point of helium, the resistance was already 300 times greater. This rapid increase was incomprehensible according to the simple quantum-mechanical formula. "A point of inflection which does not appear in the formula . . . seems to occur between the melting point of hydrogen and the boiling point of helium in the curve which represents the resistance as a function of temperature."[28] According to Flim's recollection, Holst believed at first that a short circuit had developed in the apparatus, and Holst and Flim searched to find it. But even when the wires were switched, they found that above 4.2 K the "short" would always disappear, whereas below this temperature it returned.[29]

Kamerlingh Onnes presented the finding of sudden loss of resistance at the first Solvay Conference, held from 30 October to 3 November 1911 in Brussels on the theme of the theory of radiation and quanta. It was by no means clear that mercury had entered into a new state. In the discussion following Kamerlingh Onnes's talk, Paul Langevin asked whether the strikingly rapid change that the conductivity of mercury showed near 4 K might correspond to a sudden contraction rather than to a change of state. Kamerlingh Onnes preferred to attribute the rise in conductivity to a sudden change in the period of Einstein oscillators.[30] On 25 November, in a talk at the University of Leiden, Kamerlingh Onnes reported that the rise in the resistance curve still agreed with the expected theoretical formula based on quantum oscillators, down to 4.21 K, but that then the resistance suddenly disappeared in an interval of only 0.02 K. He concluded by describing experiments planned at higher current intensities, which aimed to find whether, in the temperature region of the seemingly vanishing resistance, there existed a microresistance obeying Ohm's law. He concluded that

> as long as the contrary is not experimentally proved, we shall however adhere to this law, because we have first to try to refer the phenomena as much as possible to already known ones and so far under appropriate suppositions from the domain of known phenomena the results obtained did not seem incompatible with Ohm's law.[31]

Additional experiments in October 1911 had yielded sufficiently unexpected results to cause Kamerlingh Onnes to withhold publishing further announcements until 1913. At a certain "threshold" (now called "critical") current, which increased with decreasing temperature, the resistance suddenly rose to a measurable value. Moreover, the resistance curve was more level at large than at small currents.[32] Since above a threshold current as well as above a critical temperature, the resistance rose, he tried to attribute both phenomena to the same cause. Believing that strong currents would heat the mercury, he investigated Joule heating, the Peltier effect, and external heat leaks. Even the mercury wire itself was considered; there might be resistance from "bad places" in the sample (e.g., impurities, dislocations, poor contacts, and places with different crystal structure) that might produce local ohmic heating at high currents. However, later experiments with added traces of gold and cadmium in the mercury indicated that the impurities were unlikely to be "bad places."[33]

Kamerlingh Onnes's next experiments, in December 1912, showed that although gold and platinum did not become superconducting at accessible temperatures, tin and lead did. He observed the resistance of a pure tin wire to disappear at 3.78 K, but since lead had already turned superconducting at boiling helium temperatures, he could not at first determine its transition temperature.[34] With lead and tin available, samples of varying shapes could be easily prepared. When long wires were drawn and wound in coils, to investigate the "micro-residual resistance," the critical current was found to depend on the shape of the sample. For example, although 8 A could flow through a straight lead wire before it became normal, only 1 A destroyed the superconductivity in the coiled wire. Comparing various geometries, Kamerlingh Onnes felt that his old hypothesis was confirmed, that bad places in the wire were the likely explanation:

> if we may therefore be confident that they can be removed . . . and if moreover the magnetic field of the coil itself does not produce any disturbance then this miniature coil may be the prototype of magnetic coils without iron by which in future much stronger magnetic fields may be realised than are at present reached in . . . the strongest electromagnets.[35]

In a paper delivered for him at the Third International Congress of Refrigeration, held in Washington, D.C., and Chicago in September 1913, Kamerlingh Onnes speculated about the economic advantages of using superconducting materials to produce extremely high magnetic fields (100,000 Gauss). But at this point, he already suspected a limitation imposed by the magnetic field. By the time his speech was published, his suspicion had been confirmed; in lead, a field of a few hundred gauss destroyed the superconducting state. In a footnote he remarked: "An unforeseen difficulty is now found in our way but this is well counterbalanced by the discovery of the curious property which is the cause of it."[36]

Although experiments at high currents carried out to demonstrate the "micro-residual resistance" fell short of their goal, they led to the knowledge that the cause of the disappearance of superconductivity lay not in heating by bad places, but in the effect of the magnetic field itself. At the same time, they suggested a new way of determining the resistance in the superconducting state more precisely. During the last measurements he made before the onset of the First World War, Kamerlingh Onnes joined the ends of a lead coil and observed the decay of a current with time. A magnetic needle showed that the current flow persisted for hours and first broke down when the coil became warmed above the critical temperature. "The experiment shows that it is possible in a conductor without electromotive force or leads from outside to maintain a current permanently and then approximately to imitate a permanent magnet or better a molecular current as imagined by Ampere."[37] The upper limit of a "micro-residual resistance" was pushed down by this experiment

Willem Hendrik Keesom (1876–1956). (AIP Niels Bohr Library, Francis Simon Collection)

to 0.2×10^{-10} of the resistance at O°C. Superconductors were thereby distinguished in their conductivity from normal conductors just as strongly as were insulators.

At this point, the First World War intervened. The number of publications in the *Communications* of the University of Leiden, 56 in 1913 and 1914, fell to a total of 50 during the following seven years, and most of these reported work at higher temperatures.

In the United States, where the war's effects on research were less disruptive, F. B. Silsbee, a physics assistant at the National Bureau of Standards, studied the published data from Leiden. In 1916 he formed a hypothesis on the relation of the critical current and critical magnetic fields of superconductors: "the threshold value of the current is equal to that value at which the magnetic field which caused the current is equal to the magnetic threshold value."[38] After the resumption of work in Leiden, this hypothesis could be tested and verified.

Willem H. Keesom reported the first new results obtained after the war at the Fourth Solvay Conference in April 1924, representing Kamerlingh Onnes, who had taken ill. The "micro-residual resistance" had failed to appear, although it was sought using more sensitive techniques. Several experiments on the current-carrying state were carried out by Kamerlingh Onnes and Tuyn. In one of these, continuous currents were induced in two concentric rings, the inner one suspended so that it could move freely. Since the inner ring responded like a permanent magnet to a constant torque during the course of the experiment, it was concluded that the magnitude of the current in the superconductor remained constant. Even more astonishing was a variation of the experiment, in which a hollow ball in place of the inner ring similarly behaved like a permanent magnet.[39] The apparent explanation of this phenomenon—that with the cooling of a superconductor in a magnetic field, the magnetic flux in the interior of the superconductor becomes frozen[40]—appeared to reveal a nonuniqueness of the superconducting state, since depending on its previous history, the superconductor should remain field-free or be permeated by a frozen-in field.[41] Under this assumption, a thermodynmaic treatment of the transition from the normal to the superconducting state was out of the question.

During the discussion of Keesom's talk, Percy Bridgman said that he in no way considered this conclusion as binding. He suggested that one could approach the problem of superconductivity in terms of a reversible thermodynamic cycle, analogous to that employed to derive the Clausius–Clapeyron relation between pressure and temperature, and that one could characterize the transformation from one state into another by a discontinuous change of the magnetic permeability.[42] But the authority of the Leiden physicists was so great that this suggestion fell by the wayside, and not until a decade later did the Meissner–Oschenfeld experiment (discussed later) put an end to the apparent asymmetry of the transition from the normal to the superconducting state and revive the thermodynamic approach.

Experiments on the influence of magnetic fields on superconductivity remained central in Leiden. In January 1925, W. J. de Haas, who together with Keesom succeeded Kamerlingh Onnes, and G. J. Sizoo noted that the transition curve from the normal to the superconducting state assumes a different shape when the field is raised or lowered, a phenomenon analogous, as they realized, to hysteresis in ferromagnetism.[43] A second significant experiment was carried out a few years later by de Haas and J. Voogd, who found that if the lines of force of a magnetic field are perpendicular to a superconducting wire, the field needed to destroy superconductivity is much smaller than when the field is parallel to the wire.[44]

In the mid-1920s, two new centers for superconductivity research emerged: first in 1923 in Toronto under J. C. McLennan,[45] and then in 1925 at the Physikalisch Technische Reichsanstalt (PTR) in Berlin, under Walther Meissner. Corresponding actively, the three low-temperature laboratories compiled a rich body of data.

James Franck (1882–1964), Max Born (1882–1970), and Walther Meissner (1882–1974) in Como, 1927. (AIP Niels Bohr Library)

Meissner formulated his research program on superconductivity thus: "The first problem is whether all metals become superconducting . . . or whether there is only a certain group of superconducting metals. The second problem is, what the nature of superconductivity consists in, which laws apply . . . to the currents,"[46] a problem closely connected with the magnetic properties of a superconductor. Like Kamerlingh Onnes, Meissner placed priority on measurements. By 1928 he had already investigated 40 metallic conductors for superconductivity when he discovered tantalum as the first superconductor from the fifth group of the periodic table.[47] Then followed titanium, thorium, and niobium, which at 8.4 K had the highest transition temperature. After all available elements had been tested, Meissner and his co-workers performed measurements on chemical compounds and later on alloys. The case of copper sulfate showed that a material made up of a nonconductor and a metal that remained normal down to the lowest temperatures could become superconducting; consequently, superconductivity could not be a purely atomic property.[48]

The Leiden discovery of hysteresis raised the question of the difference between the current distributions in normal and superconducting states. In a 1932 review on superconductivity in the *Ergebnissen der exakten Naturwissenschaften,* Meissner characterized the issue: "There are practically no experiments under consideration on the very important question of the nature of the superconducting current—the distribution of the current over the cross-section."[49] In a letter in December 1932 to Ralph Kronig, he conjectured: "It could very well be that if one starts out from the non-superconducting state, in the superconducting state a uniform distribution remains preserved in the cross-section. But this is not certain. I am busy with preparations to resolve this question experimentally."[50]

Max von Laue, who since 1925 had served at the PTR as scientific adviser for special projects, talking one half-day per week with the experimenters, proposed the decisive experiment.[51] His idea was to measure the magnetic field between two long, single tin crystals arranged in parallel in a small interval, by means of a tiny sample coil, since this allowed direct conclusions to be drawn on the distribution of current in the two rods. Von Laue and his assistant Friedrich Möglich calculated the field distribution for the case where supercurrent flows on the surface, finding that in contrast to flow in bulk, the average line of force of a current should be displaced from the center in the direction of the space between the two cylinders.[52]

Since Meissner was working mainly on producing low temperatures by compressing helium,[53] he chose Robert Ochsenfeld, then in the magnet laboratory of the PTR, to carry out the main work of the experiment.[54] Meissner and Ochsenfeld's preliminary experiments were aimed at detecting changes in the magnetic field strength between two cylindrical single crystals of tin, in an external magnetic field, as they became superconducting.[55] They found the unexpected result that even when the cylinders carried no current, the magnetic field between the cylinders suddenly increased when the superconducting transition temperature was reached. The magnetic field, which had penetrated the conducting cylinders when normal, appeared to have been driven out of the cylinders.[56] At this point, Meissner and Ochsenfeld decided to postpone trying to determine the current distributions in the superconducting cylinders in order to investigate this unexpected behavior more thoroughly .

In a letter of 24 July 1933 to Karl Scheel, secretary of the German Physical Society, Meissner suggested a talk at the forthcoming ninth *Physiker Tag,* in Würzburg, on current distribution in superconductors. As the measurements leading to Meissner and Ochsenfeld's communication[57] to *Naturwissenschaften* in October 1933 proceeded, Meissner asked Scheel, on 29 July, to change the title to "Change of Current Distribution and Susceptibility in Superconductors at the Onset of Superconductivity. According to Measurements with R. Ochsenfeld."[58]

The Meissner–Ochsenfeld effect was a turning point in the history of superconductivity. It suggested that the basic idea that had guided all theoretical attempts to formulate a theory of superconductivity up to that time—that zero resistance is the characteristic feature of a superconductor, and that magnetic flux lines are frozen into superconductors—was not valid. The more fundamental property now appeared to be perfect diamagnetism. Moreover, by demonstrating that the transition to superconductivity is reversible, rather than irreversible (as was commonly believed), the effect gave new stimulus to a surge of thermodynamic considerations.

As described in Chapter 2, the few theoretical attempts to understand superconductivity microscopically before the development of quantum mechanics foundered immediately on the failure to understand even normal electrical conductivity. The development of the quantum theory of metals between 1926 and 1933 led to at least a dozen serious but ultimately unsuccessful attempts to formulate a microscopic quantum theory of superconductivity. Although unsuccessful, these early quantum theories of superconductivity show the rich imagination of physicists in that period as well as their optimism. And if none of these early quantum theories of superconductivity worked, they at least helped to pose the problem solved some 30 years later by Bardeen, J. Robert Schrieffer, and Leon Cooper, as described in Section 8.2. Felix Bloch, who had spent much time trying to develop a quantum theory of superconductivity between 1929 and 1931, summed up the frustration of those who had tried to formulate a quantum theory of superconductivity in the 1930s in what was to become a much quoted (but never published) theorem: all theories of superconductivity can be disproved![59] However, as indicated in Chapter 2, the failure to explain superconductivity does not appear to have called into question the quantum theory of solids. As Bethe wrote in his widely read 1933 *Handbuch* article with Sommerfeld, "Despite failures up to now, one may assert that superconductivity must be explainable on the basis of our present quantum mechanical understanding."[60]

The difficulties of the microscopic description encouraged phenomenological studies of superconductivity. Initial attempts focused on its possible thermodynamics.[61] Keesom and J. A. Kok had observed in 1932 a discontinuity in the specific heat at the transition temperature of superconducting lead.[62] Noting the similarity of this behavior to the phase transition of liquid helium at the critical temperature (the λ-point), Paul Ehrenfest, in a paper published in February 1933, suggested that the concept of phase transformations might also apply to superconductors. The suggestion was followed up by his student A. J. Rutgers.[63] Cornelius J. Gorter, who had recently completed his degree with de Haas, then took up the problem and "suggested Rutgers join our efforts and that we write a paper together, but he was of the opinion that the time was not ripe and so I published my calculations."[64] In his article,[65] Gorter restricted himself to transitions in those superconductors with

Huygens Club astronomers: Kuiper, [?], Bok, and Gorter. (AIP Niels Bohr Library, Uhlenbeck Collection)

zero magnetic induction B. But immediately after Meissner and Ochsenfeld's discovery, he wrote in a note in *Nature* that $B = 0$ is not a restriction, but in fact the general case.[66] He sent a copy of his work to Meissner, commenting, "It is remarkable, apropos of my thermodynamic considerations, that Dr. Casimir as well as I had, before your measurements, conjectured that $B = 0$ and had inquired of Dr. Voogd whether that would be impossible!"[67]

Hendrik Casimir, who had taken his degree in 1931 with Ehrenfest and worked with Pauli in Zurich, returned to Leiden in 1933 at Ehrenfest's instigation, and entered into a collaboration in 1934 with Gorter on superconductivity. Their fundamental studies established the connection between the critical magnetic field $H_c(T)$ and the free-energy difference between the normal and superconducting states—that this difference equals $(H_c(T)^2 - H^2)/8\pi$, where H is the external field.[68] In particular, this relation implied the formula Rutgers had established earlier between the specific heat discontinuity at the critical temperature T_c and the critical field, and showed that the transition in zero field between the normal and superconducting states should be second order. Gorter and Casimir also showed that it became thermodynamically favorable in certain geometries—for example, a sphere—for an external field below the critical value to penetrate a superconductor. They thus were the first to recognize the "intermediate state" of superconductors in magnetic fields near the critical field, in which the field threads the sample; "part of the superconductor will be perforated or reduced to pieces," causing a domain structure in which certain regions become normal while other regions remain superconducting.

To account for the transport properties, Gorter and Casimir developed the first of many subsequent "two-fluid" models in which the normal and superconducting

A. J. Rutgers and Hendrik B. G. Casimir. (AIP Niels Bohr Library, Goudsmit Collection)

electrons were considered to be two separate interpenetrating fluids. At zero temperature, all the electrons are superfluid, but the superfluid component decreases with increasing temperature and goes to zero at T_c. Thus they laid down the concept, already discussed by Bloch in 1932 in the ferromagnetic context[69] and common to all subsequent phenomenological theories of superconductivity, that one may assign to the system a parameter that measures the degree of ordering of the system, the "order parameter." Here the order parameter describes the fraction of the total electrons in the superconducting state, varying from 0 at $T = T_c$ to 1 at $T = 0$.

At about the same time that Gorter and Casimir were developing their thermodynamical theory of superconductivity at Leiden, Fritz and Heinz London, two recent refugees from Nazi Germany, began work at the Clarendon Laboratory in Oxford on an alternative phenomenological theory. This was oriented toward explaining electromagnetic properties. Heinz, the younger of the London brothers, had worked on his dissertation since 1932 with Franz E. Simon (later Sir Francis Simon), a student of Nernst, in Breslau. In the summer of 1933, when Simon saw that as a Jew he would have no future in Germany, he had accepted the invitation of another Nernst student, F. A. Lindemann, to go to Oxford. He was joined at Oxford by his Jewish co-workers Kurt Mendelssohn, Nicholas Kurti, and Heinz London. Mendelssohn aptly described the situation: "As I was leaving the University library after reading Keesom's paper, I had to duck a few bullets which were flying through the streets of Breslau, heralding the approach of Nazi rule, and it became clear that the cooling experiment would have to be deferred a little."[71] Simon's group helped to initiate the low-temperature physics program at the Clarendon Laboratory, with funds from the Rockefeller Foundation; the first British helium liquefier was installed by Mendelssohn. Fritz London also had to leave Ger-

Heinz London, Mr. and Mrs. Laub, Lady Charlotte Simon, and Sir Francis Simon. (AIP Niels Bohr Library, Francis Simon Collection)

many; he had taken a degree in philosophy in Munich and had distinguished himself in work with Heitler that applied quantum mechanics to the description of chemical bonds. Between 1933 and 1936, he worked under an Imperial Chemical Industries fellowship at the Clarendon Laboratory in Oxford, and during this period his brother's interests drew him into research in superconductivity.

In an attempt to explain his experimental observations on high-frequency electromagnetic loss in superconductors, Heinz London had developed a phenomenological theory of superconductivity, containing the important idea that supercurrents flow in a small but finite penetration depth.[72] Only later was it realized that the penetration depth formula had already been published in 1925 by Mrs. de Haas-Lorentz.[73] Her starting point was the question of whether a magnetic field, held completely outside a superconductor by screening currents, exerts an influence on the superconductor. The work, being in Dutch, had failed to attract wide attention. R. Becker, G. Heller, and F. Sauter had also independently shown in 1933 that in the specific problem of the current distribution in a superconducting sphere in a magnetic field, the field penetrates somewhat into the superconductor to a depth of order 10^{-5} cm.[74]

Living together in Oxford, in 1934 the London brothers began to work out Heinz's earlier theory in detail. The work of de Haas-Lorentz and of Becker, Heller, and Sauter had tried to account for the infinite direct current conductivity of a superconductor by assuming that metals are perfect conductors in which the electrons are freely accelerated by an electric field. However, this line of argument led to a dependence of the magnetic field on its history, a result untenable following

Meissner and Ochsenfeld's results. The Londons therefore made the crucial assertion that the supercurrent was directly related to the magnetic field H by the equation curl $(\lambda J_s) = -H/c$, where J_s is the superconducting current density and λ is a new material constant, characteristic of the particular superconductor and proportional to the square of the magnetic field penetration depth. "Through the equation the superconductor becomes characterized as a single large diamagnetic atom."[75] The vanishing resistance followed from the "acceleration equation" $d(\lambda J_s)/dt = E$, where E is the electric field.[76] The solution of the combined set of London and Maxwell equations implied that the magnetic field exponentially decays inward from the surface of a superconductor with a characteristic "penetration" depth c ($\lambda/4\pi)^{1/2}$, typically about 10^{-6} to 10^{-5} cm. The current flowing in the "penetration layer" near the surface serves to shield the interior from the external magnetic field, thus explaining the Meissner effect.

On the other hand, to explain the response of superconductors to electromagnetic waves, it was necessary to assume the existence of a normal current, J_n that satisfies Ohm's law but is short-circuited by the superconducting current under steady-state conditions. In decomposing the total current into a superconducting and a normal part, the Londons, like Gorter and Casimir, constructed a phenomenological two-fluid model that would account for the observation "that not all conduction electrons become superconducting as soon as superconductivity is established."

The Londons then turned to the quantum theory basis of their simple phenomenological equations for the electrodynamics of superconductors. They were able to interpret their results in terms of a very simple statement of the "rigidity" of the ground-state wave function in the presence of a magnetic field: "as long as the magnetic field is sufficiently weak there should not be more than a negligible disturbance of the eigenfunctions"; that is, to a first approximation, the wave function of the superconducting state remains independent of an impressed magnetic field. As a consequence, the current density in the presence of a field described by a vector potential A is given by $J = -A/\lambda c$, which yields immediately the two London equations.[77]

The question was now opened, Why should the wave function have this rigid behavior? The Londons' answer was the existence of an energy gap between the ground state and the low-lying excited states. At one of the first topical conferences in low-temperature research, held at the Royal Society in London in May 1935, Fritz London suggested in his paper "Discussion of Superconductivity and Other Low Temperature Phenomena" that the electrons were

> coupled by some form of interaction in such a way that the lowest state may be separated by a finite interval from the excited ones. Then the disturbing influence of the field on the eigenfunctions can only be considerable if it is of the same order of magnitude as the coupling force.[78]

Other than a hint by Léon Brillouin in a work on the self-consistent field method, this appears to have been the first suggestion of an energy gap for excitations from the ground state.[79]

London's notion of a superconductor as a macroscopically long-range ordered crystal—a single large diamagnetic atom with a rigid wave function for the ground

Fritz London (1900–1954). (AIP Niels Bohr Library, Francis Simon Collection)

state—may have grown out of earlier chemical interests. Reflecting on the early 1930s, Lothar Nordheim recalls that London used to

> talk about macromolecules, about how long chains would know what was happening at a far end. . . . Thus there started the growth of an idea which was to be the leitmotif for all his later work . . . the conception of quantum mechanics of macroscopic scale . . . of wave functions of macroscopic dimensions, influenced by the geometry of the sample, but nevertheless withdrawn from the disorder of thermal agitation in the same manner as is the electronic motion within atoms and molecules.[80]

The program of the meeting at the Royal Society at which Fritz London talked on the superconducting state reflected an institutional change in low-temperature research: the formation of a community of low-temperature physicists. This was accompanied by the rise of new low-temperature centers, made possible by the development of helium liquefiers that did not require a laboratory equipped on the Leiden scale. Besides London's contribution on the macroscopic interpretation of superconductivity, the meeting included talks on Peter Kapitza's technique of liquefaction of helium (by John Cockcroft), magnetic properties of superconductors at low temperatures (by David Shoenberg), lambda phenomena in liquid helium (by Keesom), magnetic effects occurring in the transition to the superconducting state (by Meissner), liquid-helium studies at low temperature (by McLennan), the propagation of electromagnetic waves in metallic conductors and the bearing of this phenomenon on superconductivity (by Kronig), difficulties of interpretation of superconductivity, including Bloch's theorem (by Brillouin), the specific heat of iron ammonium alum below 1 K and the thermodynamic scale of temperature (by

Kurti and Simon), and the induction and energy content in various superconductors (by Mendelssohn). Others present at the meeting included Nevill Mott, M. Blackman, J. D. Bernal, and Rudolf Peierls.

Although the Londons' theory, together with thermodynamics, provided a rudimentary understanding of the properties of superconductors, particularly the penetration depth and its temperature dependence, the experimental situation was far more complicated than these simple theories could begin to account for; the properties observed depended strongly on the particular materials studied. Looking back at the situation then, Mendelssohn recalled that one saw

> a whole spectrum of behavior, from a complete Meissner effect in pure mercury to a complete freezing in of flux in alloys. In between, there were all the intermediate stages, showing clearly that the presence of even a small proportion of a second constituent caused radical departure from the ideal behavior. Moreover, we found that, in those cases which differed from ideal behavior, there were two critical fields instead of one . . . the field at which the electrical resistance became normal, and for which we retained the term "threshold field" (now H_{c2}) and a much lower value at which magnetic flux first began to penetrate the sample. This we call the "penetration field" (now H_{c1}).[81]

The underlying complication was the lack of recognition at this time that there were in fact two types of superconductor, with different reactions to external magnetic fields. In what came to be called "type I" superconductors, the magnetic field completely penetrates the sample on reaching the critical value H_c, whereas in "type II" superconductors (as designated by L. Schubnikov)—Mendelsohn's nonideal cases—superconductivity is preserved even above H_c, although field lines begin to penetrate below this critical strength. To add to the confusion, an external magnetic field below the critical value could, as Gorter and Casimir showed, also penetrate a (type I) superconductor. The clear distinction between the two types was not formulated until 1950 by Vitaly Ginzburg and Lev Landau, as described in Section 8.2.

Around 1935, the intermediate state, the state first discussed by Gorter and Casimir, in which the material divides into small superconducting domains, became a principal focus in superconductivity research. Experiments were carried out by the low-temperature groups at both Oxford and Cambridge.[82] The Londons' theory, when examined more carefully, led to the incorrect conclusion that rather than exhibiting a Meissner effect, a superconductor in a magnetic field would always go into an intermediate state. To avoid this conclusion, Heinz London in 1935 introduced ad hoc the existence of a positive surface energy between a normal and a superconducting region, to inhibit the formation of the domain structure in weak magnetic fields.[83] Two independent theoretical studies of the magnetic behavior of the intermediate state on the level of a macroscopic average over the microstructure were carried out in 1936, by Fritz London[84] and by Peierls,[85] who was then in Cambridge. They both pointed out that the presence of a surface energy prevents the partition of the sample from being arbitrarily fine. Peierls interacted strongly with Shoenberg, one of the experimentalists in Cambridge, who recalls that Peierls "had all sort of extraordinarily helpful ideas" that guided him in his choice of experiments—for example, the suggestion to use alternating fields to study the intermediate state.[86]

Research on the intermediate state was also strong in Moscow and Kharkov, although the experimental program in Kharkov ran aground in 1937 when Shubnikov, who lead the research there, was arrested, never to return. Shoenberg worked further on the intermediate state in Moscow with Kapitza and Landau, whom Kapitza, after his return to the Soviet Union, had persuaded in 1936 to leave Kharkov and join him. In 1937, Landau developed a theory of the intermediate state, showing that for a range of average magnetic fields, a superconductor of any shape (except a long cylinder with axis parallel to the field) breaks up into a laminar structure of alternating superconducting and normal layers.[87] Later experiments carried out in Moscow verified Landau's theory and confirmed that flux is carried in the normal regions, in which the magnetic field is essentially the thermodynamic critical field H_c. However, direct evidence for the intermediate state did not come until the immediate postwar period, with experiments by A. I. Shalnikov and A. Meshkovsky.[88]

Nearly all the low-temperature laboratories also investigated the peculiar properties of superconducting (type II) alloys. The history of these dated from 1930, when de Haas and Voogd observed that certain lead–bismuth alloys remained superconducting in fields of nearly 20,000 Gauss.[89] Indeed, at such high field strengths the material did not behave like an ideal diamagnet, and a discontinuity in the specific heat, expected according to Rutger's formula, could not be observed.[90] In contrast to the intermediate state of type I superconductors, which was a consequence of the geometry of the sample, in these alloys the superconducting state, when penetrated by magnetic fields of strength between the two critical fields H_{c1} and H_{c2} (designated the "mixed" state), was a property of the materials. But the nature of this superconducting mixed state was completely mysterious. Mendelssohn suggested viewing the alloys as a multiply-branched net of fine superconducting threads, a system of rings with continuous currents that locked in the lines of force like a superconducting sponge. But as Heinz London cautioned, "There is too vast a field open for arbitrary assumptions for it to be worth while going into details before more experimental data are available."[91] Heinz London had speculated, on surface energy grounds, that the alloys studied by Mendelssohn might have a negative total surface energy.[92] But it would not be until the end of the 1950s that A. A. Abrikosov extended the Ginzburg–Landau theory to a phenomenological description of type II superconductors.[93]

In the late 1930s, experimental research on superconductivity slowed down. Gorter reminisced:

> the number of research workers in the field was small . . . some of them almost simultaneously left it. Shubnikov disappeared. Mendelssohn concentrated a large part of his attention on the superfluid properties of helium II, while I returned to magnetism. As to the properties of alloys, I feel that the lack of matallurgical facilities and experience also weighed heavily.[94]

One significant theoretical advance in this period was an insight by Heinrich Welker, one of Sommerfeld's last assistants, into the origin of the energy gap that underlay the Londons' electrodynamics. In presenting an account of the London theory in his seminars at the end of 1937, Welker considered the role of the magnetic exchange interaction of antiparallel conduction electrons, which if sufficiently strong gives rise to a state in which an energy gap appears above the Fermi energy. In this way, he was able to predict the order of magnitude of the critical temperature.[95] Welker's full development of this theory was interrupted in 1940 when he took up war work in wireless telegraphy.

A major experimental advance came out of German war research into metals with high melting points suitable for filaments. Eduard Justi, a successor of Meissner at the PTR, and co-workers found that the heavy-metal compound niobium nitride has a remarkably high transition temperature of 16.1 K, reachable with boiling hydrogen.[96] However, large-scale applications of superconductivity—in magnets, power transmission, switching, and measurement techniques—were not made until the postwar period; then large industrial laboratories, including Bell Labs, and universities, such as MIT, turned to developing superconducting materials with high critical magnetic field strengths and high transition temperatures.

Helium

At the beginning of the twentieth century, three research groups—those of Kamerlingh Onnes at the University of Leiden, James Dewar at the Royal Institution in London, and Karol Olszewsky and Zygmunt Wroblewsky at the University of Cracow—entered a race to liquefy the last perfect gas to resist such a change of phase, helium.[97] Two approaches were available, neither of which seemed promising: an expansion method and the Joule–Thomson method. In the expansion

Kamerlingh Onnes laboratory, Leiden, 1926. (AIP Niels Bohr Library, Uhlenbeck Collection)

method, highly compressed helium gas had to be cooled below its critical temperature, which Olszewsky had experimentally estimated at 2 K and Dewar at 8.2 K. But even using liquid hydrogen, the lowest temperature that could be reached was 15 K. On the other hand, the Joule–Thomson method, which had enabled the liquefaction of air and hydrogen, required that a very large quantity of the gas be liquefied, and helium was both expensive and rare. All three groups settled on the first method.

By January 1908 the problem appeared close to solution in Leiden. Kamerlingh Onnes, through painstaking measurements of the isotherms of helium, had determined the critical temperature to be 5 K. By expansion of helium initially confined by a pressure of over 100 atmospheres, he had observed the formation of a fog.[98] Unfortunately, further experiments showed this fog to be condensed droplets of hydrogen, not helium (which had not been carefully enough purified).[99] On realizing his mistake, Kamerlingh Onnes mentioned that liquefaction should be possible by the Joule-Thomson effect. At this point, he had 7 L of helium at his disposal, and a fortunate circumstance soon helped him obtain even more. His brother, the director of the Office of Commercial Intelligence in Amsterdam, obtained some monazite sand, a thorium-bearing mineral in which a small amount of helium is trapped.[100] In the early summer of 1908, the Leiden experimenters produced enough helium from this substance to carry out successfully the first Joule–Thomson experiment.

This first experiment almost failed. It began in the afternoon of 10 July 1908 after 75 L of liquid air and 20 L of liquid hydrogen had been produced for precooling the helium. In the evening, as the last reserve bottle of hydrogen was connected, the temperature of the helium fell to 6 K, and no liquid was seen. But the temperature continued to fall slowly until, at somewhat less than 5 K, during the pressure increase part of the cycle, the temperature unexpectedly rose slightly and then fell off to a constant value. A visitor to the laboratory, Schreinemakers, attributed this behavior to the fact that the thermometer was already immersed in a liquid.[101] In fact, the experimenters had missed the point of influx of the liquid helium. With sufficient illumination from underneath, the liquid surface "stood out sharply defined like the edge of a knife against the glass wall."[102] The dream was fulfilled.

Firmly convinced of the validity of the law of corresponding states, Kamerlingh Onnes immediately tried to solidify the helium. The vapor pressure over the liquid was lowered, but the helium remained liquid even at the lowest pressure, at a temperature of 1.72 K.[103] Further experiments during 1922 reached 0.83 K (through improvement of the thermal insulation of the cryostat and through the insertion of diffusion pumps), but still failed to produce solidification. Looking on the bright side, Kamerlingh Onnes pointed out, "Each time that it is found with further lowering of the vapor pressure that the helium remains liquid, this failure represents a gain: a new region of temperature has proved accessible to us, which is extremely important on account of its extreme situation." Ultimately, he had to accept the fact that the behavior of helium deviates from the classical laws of physics: "We cannot escape the question whether helium will not remain perhaps liquid even if it is cooled down to absolute zero."[104]

In the 14 years from 1908 to 1922 that Kamerlingh Onnes had been experimenting with liquid helium, he had found several anomalies, but he did not realize that helium was a substance whose properties would keep researchers in as much suspense as superconductivity. It was not his style to speculate, and he hesitated to publish results that he could not confirm through repeated experiments. His work on superconductivity was different. The determination of electrical resistances at low temperatures had been part of his research program for years, and the anomaly that the galvanometer showed was investigated and reinvestigated until it was certain. But in the case of liquid helium, quantities like vapor pressure, boiling point, and density, which had to be measured through study of deviations from the law of corresponding states, showed no such strongly anomalous relationship, nothing that would lead conclusively to the discovery of the strange properties of superfluid helium.

For example, in 1911 when Kamerlingh Onnes reported on a maximum of pressure at 2.2 K (later learned to be the "lambda point," the onset of the superfluid phase), he did not mention that the height of the liquid column simultaneously passed through a minimum, although he should have observed this.[105] Although other low-temperature laboratories in the 1920s must also have observed this unexpected behavior of helium, it was first reported in 1932 by McLennan:

> At 4.2 K it was impossible to keep the column of liquid free from bubbles that moved rapidly to the surface.... When a pressure of 38 mm was reached, the appearance of the liquid underwent a marked change and the rapid ebullition ceased instantly. The liquid became very quiet and the curvature at the edge of the miniscus appeared to be almost negligible.[106]

But McLennan did not go beyond mere description, and this key to the "super" properties of helium escaped notice.

On another occasion, Kamerlingh Onnes's caution prevented him from publishing an anomaly of liquid helium. In 1925 he and Leo Dana, a Sheldon fellow from Harvard University, measured the specific heat at a temperature above 2.5 K.[107] They had also taken measurements at lower temperatures, but since the results were not linear, as at higher temperatures, and were also more uncertain, they were put aside.[108] In the same issue of the Leiden *Communications,* they mentioned in a footnote an anomaly in the measurement of the latent heat; like the density it took on an extreme value at 2.2 K, pointing to a discontinuity at the same temperature:

> Further there is a noteworthy anomaly at the temperature of maximum density. It is possible that only the variation of the state of convection at surpassing this temperature causes a change in a systematic error [*sic*]. The accuracy of the results, however, being sufficient with regard to the deviation with near temperatures [*sic*], it is remarkable that these results would indicate that near the maximum density something happens to the helium, which within a small temperature range is perhaps even discontinuous. The change of density of the liquid also indicates something of the same kind.[109]

As Dana had to return to America and Kamerlingh Onnes died, this point was not pursued further.

Seven years elapsed before Keesom and the visiting German Klaus Clusius published measurements of the specific heat of liquid helium that indicated a sudden rise at a temperature of 2.19 K.[110] Since the curve resembled the Greek letter λ (lambda), following Ehrenfest's suggestion they designated the anomaly a lambda

(rear) Samuel A. Goudsmit, J. Tinbergen, R. Kronig; *(front)* Gerhard Dieke, Paul Ehrenfest, Enrico Fermi in Leiden, 1924. (AIP Niels Bohr Library)

curve, and the temperature at the anomaly the lambda point. Keesom and M. Wolfke had earlier found a jump in the temperature dependence of the dielectric constant at the same temperature as the density maximum; they interpreted this as a transition to another phase. They called the two states helium I above the critical temperature and helium II beneath it.[111] Keesom and Clusius naturally interpreted the specific heat discontinuity as a phase transition from He I to He II, but this transition, not accompanied by a latent heat, differed from an ordinary phase change.[112]

By 1933, the thermodynamic concept of phase transitions had become widely discussed in Leiden. Thus far, only "first-order" transitions exhibiting a discontinuity in the entropy (the first derivative of the free energy) had been encountered. Ehrenfest, whom Keesom brought into the discussion, suggested an extension of the concept. Perhaps transitions could exist with discontinuities only in higher derivatives—that is, with continuous entropy (and thus no latent heat) but discontinuous specific heat. The transition between the phases of liquid helium would be an example of such a "second-order" phase transition. Ehrenfest cautiously worded his 1933 paper: "I would very much like to be able better to formulate and understand this characteristic difference in relation to the 'usual' phase transformation."[113]

Ehrenfest considered his formalism suitable as well for describing superconductors at the transition temperature and ferromagnets at the Curie temperature. Inferences about helium were drawn from the analogy to known behaviors of solids. For example, in ferromagnetism, the specific heat anomaly was attributed to an order–disorder transition, hinting that the mysterious helium phase below 2.2 K might also be connected with a spatial order.[114] Explanations based on this idea were excluded in 1938 by Keesom and K. W. Taconis, who found no difference in Debye–Scherrer photographs above and below the lambda point.[115] Nevertheless, the analogy suggested that during the transition, helium becomes more ordered. In 1935, Keesom and his daughter, in analyzing the asymmetrical form of the lambda curve and the influence of thermal fluctuations, inferred that the lambda transition could not be an atomic phenomenon, but must depend on the interaction between many neighboring atoms, "a kind of correlation effect."[116]

During their measurements, the Keesoms noticed that the heat exchange in the calorimeter at temperatures below the lambda point took place very rapidly, while at higher temperatures this exchange was much slower, indicating a substantial increase in the thermal conductivity at the transition from He I to He II. Their first preliminary results the following year on thermal conductivity were meaningful only above the lambda point, since the apparatus was constructed for a much poorer thermal conductor than helium II proved to be. With the help of a meter-long capillary, the Keesoms estimated that the thermal conductivity rose at the transition by a factor of 3×10^6 and was at least 200 times higher than that of the best thermal conductors, copper and silver. They proposed that helium II be designated as "supra-heat-conducting."[117]

Keesom and his daughter were not without competition in their experiments on thermal conduction. By the mid-1930s, the superproperties of helium were a subject of research at more than 10 low-temperature physics laboratories. The expansion of low-temperature facilities began in 1922, when McLennan built the world's second helium liquefier, in Toronto. The 14-year interval that had elapsed since the original liquefaction of helium was not due to lack of interest, but to the great expenditure of personnel and equipment required and to the limited availability of helium. Shortly before the First World War, the loss of men and equipment in exploding hydrogen dirigibles led to interest in Canada in filling them with helium from Canada's natural-gas deposits. McLennan, who was commissioned to investigate the helium content of the gas deposits, obtained funding for scientific research on helium from the Council for Scientific and Industrial Research of Canada, the University of Toronto, the Carnegie Foundation for Research, and the Air Ministry of Great Britain. At the time the Toronto liquefaction plant was planned, however, available helium was being used for dirigibles.[118]

Liquid helium entered many laboratories only when Simon, in Breslau, constructed a very simple liquefier suitable for producing small quantities of liquid helium.[119] Simon brought the technology with him when he established a low-temperature group at the Clarendon Laboratory in 1933. A new type of liquifier, which bypassed the step of precooling by liquid hydrogen, and supplied the relatively large quantity of some 3 L of liquid helium per hour, was developed at the Royal Society Mond Laboratory in Cambridge by Kapitza in 1934, shortly before he returned to the Soviet Union. Cockcroft, Kapitza's successor, planned to build the Mond Laboratory into a cryogenics center matching Leiden.

Most of the startling flow properties of helium were discovered in 1937 and 1938. In 1938 alone, no fewer than 14 publications on liquid helium appeared in *Nature*. (To establish priority, results had to be published as quickly as possible, and letters to the editors of journals like *Nature* became an increasingly popular form for first

Sir Francis E. Simon (1893–1933) and Paul Ehrenfest (1880–1933). (AIP Niels Bohr Library, Francis Simon Collection)

reports.) The series of flow experiments had its origin in 1937 when Jack F. Allen, Peierls, and H. Z. Uddin of the Mond Laboratories developed a method for carrying out thermal conductivity measurements with greater accuracy.[120] In a thermally isolated glass container, helium could be heated with the help of a filament. The helium could communicate with the surrounding helium bath through the bottom of the vessel, which was constricted to a capillary. The differing vapor pressures over the helium in the container and in the bath led to a difference in the heights of the liquid surface, which provided a direct measure of the temperature difference of the liquids. Unlike the normal thermal conductivity process, the heat flux produced here turned out not to be proportional to the temperature difference; but the smaller the difference, the greater the conductivity.

The authors certainly questioned their results, since when the temperature difference was made very small, the helium surface inside the flask suddenly rose higher than that outside. The vapor pressure difference indicated the impossibility that the heated liquid in the flask had cooled down.[121] Allen and Harry Jones then repeated the experiment with a container open on top, so that the same pressure existed over both liquid columns. The capillaries were replaced in this experiment with fine emery powder, illuminated with a flashlight in order to warm the helium. In the course of warming, the liquid in the flask rose again. With the top side of the flask formed into a nozzle, Allen and Jones quite unexpectedly observed that as long as the helium was being heated, a fountain of helium squirted out of the container—the "fountain effect."[122]

This thermomechanical effect suggested to Mendelssohn and his student John Daunt, in the Clarendon Laboratory at Oxford, that the reverse process—the mechanocaloric effect—might also be observable. An experiment in which helium was pressed out of a reservoir through canals of fine powder actually showed a warming of the reservior; cooling occurred when the helium was pushed back into the container. These observations suggested that the heat content of the liquid that flowed through the finest canals was less than that of the remaining liquids.[123] But how could helium below the lambda point flow through fine pores that were completely impermeable to helium above this temperature?

Indications of this phenomenon had in fact shown up in several laboratories, where researchers had observed that cryostats leakproof above the lambda point would suddenly leak on further cooling.[124] The Keesoms had conjectured in 1932 that the cause could be a decrease in viscosity.[125] In 1935, J. O. Wilhelm, A. D. Misener, and A. P. Clark, at Toronto, carried out the first direct measurements of viscosity by observing the damping of a rotating cylinder immersed in the liquid.[126] Three years later, Keesom and G. E. MacWood, a fellow from Columbia University, also investigated the viscosity, but used rotating disks instead of a cylinder because this method was easier to analyze.[127] Despite their quantitative differences, both measurements led to the conclusion that the viscosity fell off sharply below the lambda point. But in neither case was the magnitude of the decrease comparable with the enormous rise of the thermal conductivity below the lambda point, nor was it sufficient to explain the observed leak rate.

The superfluidity of He II was finally discovered in 1938 by Kapitza, now head of the Institute for Physical Problems of the Academy of Sciences in Moscow. In the January 1938 issue of *Nature*, letters to the editors from Kapitza, as well as from Allen and Misener of the Mond Laboratory, referred to the difficulty that the helium viscosity could be correctly measured only for laminar flow.[128] Due to the small viscosity, this assumption appeared not to be satisfied, for the smaller the viscosity, the smaller the vessel had to be in order to avoid turbulence. Kapitza even proposed explaining the Keesoms' heat conduction as convection. In order to eliminate turbulence, Kapitza had permitted the helium to flow only through a narrow fissure half a micron wide, formed by two plane parallel glass plates lying on top of each other. Although impermeable to normal liquids like water, this fissure did not hinder helium II. Kapitza estimated an upper limit for the viscosity 1500 times smaller than that of helium I and some 10,000 times smaller than that of gaseous hydrogen. In analogy with superconductivity, he suggested that this new phase of helium be called superfluid.

A similar result was reported by Allen and Misener for an experiment with thin capillaries. However, they rejected Kapitza's proposed explanation of the thermal conductivity via convection; to produce such enormous thermal transport, the liquid would have to move with the velocity of a bullet. The velocity of the helium flow was much lower than this, and since it scarcely varied with the pressure difference at the ends of the capillaries, they concluded that the flow could be explained neither by laminar nor by turbulent flow: "It may be possible that the liquid helium II slips over the surface of the tube." Thus was cautiously suggested a fundamental property of helium II, film flow.

It turned out that in 1922 Kamerlingh Onnes had already observed, but falsely interpreted, film flow. At a joint meeting in October of the Faraday Society and the British Cold Storage and Ice Association, on the topic "The Generation and Utilization of Cold," he had reported the remarkable observation that if two concentric Dewar flasks were filled unequally high with helium, the difference in level equalized with striking speed. He tried to account for this phenomenon in terms of distillation and the attendant condensation, but before making final judgment he awaited further planned measurements of the thermal conductivity of glass, helium vapor, and liquid helium.[129] However, the problem was not pursued further until 1936, when Rollin determined that the rate of vaporization of helium in his cryostat was higher below the lambda point than expected from the known influx of heat into the cryostat. Since at this point the extremely high thermal conductivity of helium II had already been discovered, it was natural for Rollin to infer that the heat input due to thermal conductivity resulted in a film that had condensed on the inner walls of the cryostat.[130]

Mendelssohn and Daunt were motivated by Rollin's allusion to repeat Kamerlingh Onnes's 1922 experiment, and found that an empty beaker immersed in liquid helium filled itself even when the helium surface did not reach up to the edge of the beaker. When pulled out of the helium, the beaker emptied by itself, the liquid on the outside of the base forming droplets that fell into the bath. There was therefore no distillation present: the helium flowed in a thin film along the solid surface.[131] The publication of these results in the famous volume 141 of *Nature* was preceded by a letter from A. V. Kikoin and B. G. Lasarev of the Ukrainian Physico-Technical Institute of Kharkov confirming Mendelssohn and Daunt's "film flow" and estimating the film thickness to be of order 10^{-5} cm.[132] Their further measurements lowered the estimate to only about 100 atom layers. This raised another question: if helium could flow in a film that was even thinner than the fissure in Kapitza's experiment, then the maximum value of the viscosity also had to be lower than that assumed by Kapitza.[133] Negligible viscosity made film flow plausible; the fluid moistened the container walls with a thin imperceptible film, and only because of its superfluidity did helium flow so quickly out of the container and across the surface.

Microscopic understanding of any of the "super" properties of helium was still completely lacking. The first attempt at a microscopic explanation in terms of a higher ordering of the superfluid phase of helium was based on comparison with crystals. Thus Fröhlich in 1937 compared the lambda transition of helium to transitions in binary alloys. Helium was thought to form a body-centered cubic lattice, partially occupied with helium atoms; in the ordered state, both the helium atoms and the vacancies were taken to form diamond structure sublattices. Fritz London (who took a professorship at the Sorbonne after his successful work on superconductivity) had argued that a static spatial model of liquid helium with regular structure was not energetically possible.[134]

In his publication in volume 141 of *Nature,* London broke new ground in understanding helium.[135] From spectroscopic evidence he realized that helium atoms should obey Bose–Einstein statistics rather than the Fermi–Dirac statistics followed by electrons in a metal. As he later put it, he therefore "advanced, in 1938, the hypothesis that the strange change of state in liquid helium at 2.19 K, even though it occurs in the liquid and not in the gaseous state, is due to the condensation mechanism of the Bose-Einstein gas."[136] Einstein had shown in 1925, in his work on gas degeneracy, that an ideal gas obeying Bose–Einstein statistics would undergo a transition at a certain temperature, with a macroscopic fraction of the particles in the system condensing into the lowest single-particle energy state.[137] This condensation represents a higher order, but does not manifest itself through crystallization, since the condensation takes place not in position but in momentum space. London computed that the specific heat resulting from this condensation had a maximum and a discontinuity in its derivative, indicating a phase change of third order, a difference he attributed to the fact that the gas model for liquid helium represented only a crude approximation. "Though actually the λ-point of helium resembles rather a phase transition of second order, it seems difficult not to imagine a connection with the condensation phenomenon of the Bose-Einstein statistics."

London discussed his ideas with Laszlo Tisza, a young Hungarian theoretician then working in the Laboratory of Experimental Physics at the Collège de France in Paris. Tisza applied-London's suggestions to the transport phenomena of helium, developing a picture in which he described helium II as a degenerate Bose gas of n_0 helium atoms/cm^3, consisting of two independent fluids. The first, the normal component, is made up of the n atoms occupying excited states; the second, the superfluid component, is made up of the remaining $n_0 - n$ particles condensed in the ground state. Since the condensed atoms transfer momentum neither to atoms of the wall nor to excited atoms, and therefore should not be able to interact with them, the particles in the different states should behave like two independent mutually penetrating fluids—indeed, fluids in which the number of excited atoms and the relative proportion depend only on the temperature.[138] This two-fluid model could explain contradictory thermal measurements involving flow. The conventional method of measuring took into account only the viscosity of the normal component. However, the superfluid component, having zero viscosity, would flow through very thin capillaries. If a temperature gradient is sustained in such a capillary, a pressure gradient would be produced and the superfluid component would flow to the warmer end. The normal component, however, cannot flow in the opposite direction because of its finite viscosity. Eventually, the increased pressure at the warmer end would result in the fountain effect.

London, initially not pleased that Tisza used his suggestion so quickly, considered it out of the question that two such different fluids could arise from a single group of indistinguishable atoms.[139] On the other hand, despite its seemingly untenable and controversial assumptions, the model was quite successful—predicting, for example, the inverse of the fountain effect, the mechanocaloric effect, prior to its experimental demonstration. A second prediction pointed to a completely new type of wave propagation: Tisza conceived the idea that a local disturbance of the concentration ratios of the two components of the fluid would not equalize through diffusion, but through wave propagation, in analogy to the density in the case of normal sound. Since the concentration ratio is directly related to the temperature, a temperature wave must propagate in He II simultaneously with a density concentration wave.[140] Not until 1944 did Vasilii Peshkov observe propagation of this "second sound."[141]

The burst of theoretical work was published in 1938 when the Second World War was imminent, and as scientists turned to applied research and helium again became scarce, fundamental research on helium slowed. Kapitza, in the Soviet Union, was at the time continuing his experimental investigations of helium's high

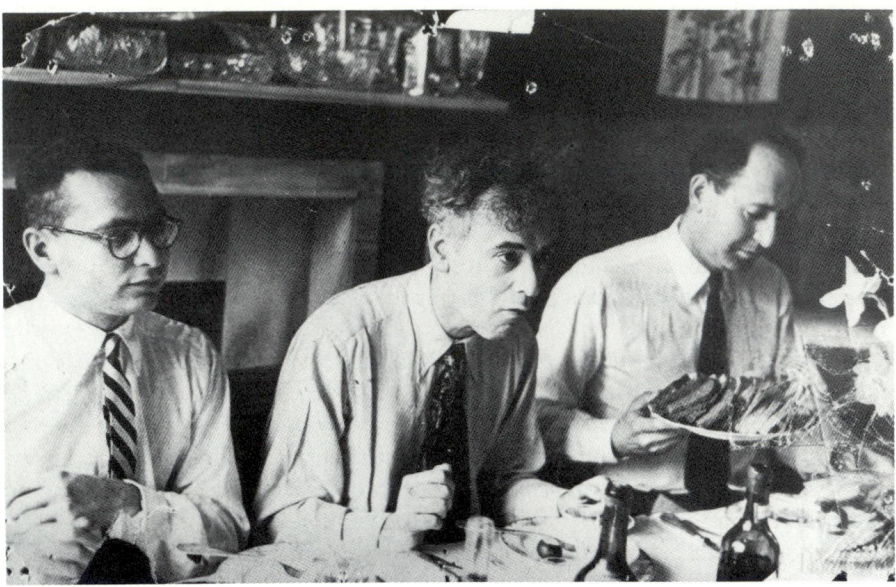

Murray Gell-Mann (b. 1929), Lev Landau (1908–1968), and Robert Marshak (b. 1916) at Moscow conference, 1956. (AIP Niels Bohr Library)

thermal conductivity. In attempting to understand this phenomenon, Kapitza, "in the absence of any guiding thought . . . examined the heat transfer of helium II as a function of pressure, gravity, time, etc."[142] After a year's efforts, he noticed that the random pressure variation in the input of helium strongly influenced the thermal conductivity, suggesting an important role of flow in heat conduction. Directly performing measurements on flowing helium, he found a thermal conductivity about 1000 times smaller than in the stationary case.

The connection between the flow and the thermal conductivity supplied the key to further research. Kapitza measured the flow by observing the diversion of a leaf attached at one end of a capillary; the second end, formed into a glass container, held a filament. During warming, a strongly directed stream of helium flowed out of the capillary, and, astonishingly, the vessel did not become empty as long as the flow of liquid continued.[143] Landau called this observation "a biblical experiment like the bush that burned with fire and was not consumed."[144] Helium had to flow back through the capillary by some other way. Kapitza thought that film flow across the small surface was the only possibility. The correct explanation in terms of two counterflowing components, the normal and the superfluid, was given in 1941 by Landau as part of an epochal quantum-theoretical treatment of helium.[145]

Since 1937, Landau had been head of the Section of Theoretical Physics, which had been organized for him at the Institute of Physical Problems at Kapitza's instigation. In this position, he was in close contact with Kapitza's helium program, a contact well reflected in Landau's paper on the theory of superfluidity of helium II. He began this paper with a criticism of Tisza's two-fluid model as being completely unsatisfactory for explaining superfluidity. "Nothing could prevent atoms in the ground state from colliding with excited atoms, i.e., when moving through the liq-

uid they would experience a friction and there would be no superfluidity at all."
The key to the explanation of the properties of the quantum liquid helium, Landau
recognized, was that the low-lying excitations of helium had nothing to do with
individual atomic excitations (the idea underlying Tisza's picture), but the states
near the ground state could be described in terms of a spectrum of weakly interacting collective "elementary excitations" above the ground state. The simplest excitations in helium, Landau proposed, were analogous to a longitudinal potential
(curl-free) flow of a normal fluid, and were quantized sound waves, or "phonons,"
in the helium fluid, closely related to the phonon excitations of the lattice in a solid.
The energy ϵ of the individual phonons was simply the sound velocity c times their
momentum p.

Besides these longitudinal vibrations, Landau postulated that since vortex
motions can occur in any fluid, there should be a corresponding second type of elementary excitation, the "roton" (the name was given by I. Tamm, to suggest the
connection to rotational motion). In this first paper of Landau's, the roton was
assumed to be a quantized collective motion in which the entire aggregate of helium
atoms participates in a vortex motion. Since the vortex motions have a nonvanishing curl, they must be separated in energy from the ground state by a minimum gap
Δ, and so he assumed that the rotons had an energy spectrum of the form $\epsilon = \Delta +$
$p^2/2\mu$, where μ is the "effective mass" of a roton. Landau would later go on to realize
that the second branch of excitations, which were needed to understand the thermodynamics, were not in fact related to rotational motion but were continuously
joined to the phonon spectrum; nevertheless, the name roton stuck.

Although the number of atoms in the fluid remains constant, the number of elementary excitations vanishes at absolute zero and increases with rising temperature. At liquid-helium temperatures, so few quasiparticles exist that their influence
on one another can be neglected and one can treat them as gas particles. The phonon gas clearly obeyed Bose statistics, and Landau assumed that the roton gas
would too. In terms of this picture, Landau explained that the observed heat capacity as a sum of a T^3 contribution from the phonons together with a roton contribution which, owing to the roton energy gap Δ, vanishes at $T = 0$.

In the same paper, Landau went on to argue that when helium flows through a
capillary at absolute zero, an interaction between the wall and the liquid can be due
only to the creation of elementary excitations. Since at low-flow velocities it is
impossible, on the basis of energy and momentum conservation, to create either
phonons or rotons, the flow must proceed unhindered; therefore, helium is superfluid. At finite temperature, helium behaves similarly to a two-fluid system, with
the normal fluid composed of the gas of excitations. But Landau cautioned,

> when we look upon helium as a mixture of two liquids it is no more than a method
> of expression, convenient for describing phenomena which take place in helium II.
> Like every description of quantum phenomena in classical terms it is not quite ade-
> quate. . . . We particularly emphasize that there is no division of the real particles
> of the liquid into "superfluid" and "normal" ones here.[146]

Kapitza's observations were also simply explained by Landau's theory. If the liquid helium in the glass container becomes heated, phonons and rotons are thermally produced. The excitations carry the heat and flow from the hot to the cold
end. As they leave the capillary, they divert the leaf in Kapitza's experiment. The

counterflow of the superfluid component guarantees that no macroscopic flow exists in the helium, but being potential flow, it can neither exert pressure on the leaf nor exchange momentum with the outflowing quasiparticles. Contrary to Kapitza's conception, the flow takes place over the entire cross-sectional area of the capillary, which obviates the need for a film flow with impossibly high velocity.

Intrinsic in Landau's theory is the deep assumption that the superfluid flow velocity is an independent thermodynamic variable of the system, added to the usual variables of pressure and temperature. In an equilibrium state of helium, the excitation gas would remain in thermodynamic equilibrium with the walls of the container but would still have superfluid flow. On this basis, Landau calculated the superfluid mass density as a function of temperature—a prediction later confirmed by E. Andronikashvilii, a student of Kapitza, in a measurement of the moment of inertia of a stack of very close metal disks rotating in a container of helium II.[147]

A further major step in Landau's paper was the development of the macroscopic hydrodynamic equations of superfluid helium, in which the superfluid velocity is assumed to be irrotational. From these equations followed the prediction of two modes of sound propagation, normal sound and temperature waves, the latter qualitatively similar to the mode predicted by Tisza.

The paper closed with a remarkable discussion of the similarities between superfluidity and superconductivity. Superfluidity arises in helium, according to Landau's theory, from the kinematic impossibility that superflow can generate excitations when the flow velocity is low. Referring to recent ideas of an energy gap in superconductors,[148] he remarked that superconductivity "can also be explained by . . . an energy gap in the spectrum of the 'electron liquid' in a metal"; here "one must suppose that there is an energy interval between the ground state of the electron liquid and the beginning of the continuous spectrum of the states with an inner motion."[149] He then went on to introduce a description of the current-carrying states of the superconductor in terms of a single-electron phase χ. (This would play a fundamental role as the phase of the order parameter in the later Ginzburg–Landau phenomenological theory of superconductivity, described in Section 8.2.) At nonzero temperatures "one can in a certain sense speak of the division of the whole charge density into 'superconducting' and 'normal' parts, but not forgetting all the reservations mentioned for . . . liquid helium." He concluded that like helium superflow, supercurrents cannot transfer heat, which was borne out by the lack of observation at that time of thermoelectric phenomena in superconductors.

Landau's paper was a culmination of prewar research on helium. It provided the understanding of the essential phenomenology of liquid helium; it introduced several basic tools that would be taken up in the postwar development of the quantum many-body problem (e.g., elementary excitations and effective hydrodynamics), and it explicitly connected the phenomena of superconductivity and superfluidity. The microscopic origin of Landau's picture would begin to be developed after the war.

Phase Transitions and Critical Phenomena

The fourth root in the study of collective phenomena that we consider in this chapter is the field of phase transitions and critical phenomena—for example, long familiar phase changes such as melting and ferromagnetism. Like superconductivity and superfluidity, these are phenomena associated with numerous identical interacting units.

The subject of phase transitions has a long prehistory.[150] At least three relevant features had been prominent before the twentieth century: the change of symmetry or order accompanying some phase transitions, the sharp boiling and freezing points of common substances, and the latent heat of first-order transitions.

The modern history of phase transitions and critical phenomena starts with Johannes Diderik van der Waals's doctoral dissertation at Leiden in 1873.[151] This work was intended to enhance understanding of intermolecular forces and to identify experimental methods for obtaining data for the construction of kinetic-molecular descriptions of matter. Drawing on Laplace's much earlier consideration of a short-range attractive force of molecules and on Clausius's kinetic theory of gases,[152] van der Waals sought a description of a continuous transition between the gaseous and liquid states of matter.[153] In his own words, he

> strictly speaking desired to prove more; that is the identity of the two states of aggregation. For if the supposition which is partly established, that in the liquid state the molecules do not merge into each other to form greater atomic complexes—if this supposition should be fully confirmed—there would then only be a difference of greater or smaller density in the two states, and thus only a quantitative difference.[154]

Van der Waals's ideas were developed at a time when the kinetic-molecular theory of heat was still highly controversial. Although at that time he was among the relatively few believers in the reality of molecules, even he later admitted to wondering at times "whether in the final analysis a molecule is a figment of the imagination."[155] His equation of state—which takes into account the existence of short-range intermolecular forces—successfully described both the gaseous and liquid states of a fluid. In particular, it yielded a "critical point" at a certain critical pressure, temperature, and density, a point at which the vapor pressure curve separating the gaseous and liquid phases terminates. Beyond the critical point, the equation describes continuous transitions between gas and liquid. First developed in a time of controversy between energeticists and proponents of molecular-kinetic theories, the van der Waals equation provided a major first step and stimulus to the development of the modern theory of phase transitions and critical phenomena.

Van der Waals's assertion that continuity from liquid to gas "was a fact" was based on Thomas Andrews's earlier investigation of the continuous change of carbon dioxide from the gaseous to the liquid state above a "critical point"; this was the first experiment to find such a point.[156] The van der Waals equation served as a qualitative theoretical description of the Andrews experiments. These experiments and van der Waals's dissertation stimulated many theoretical and experimental studies of the phase transitions and the equation of state of fluids.[157]

The van der Waals equation was primarily an extension of the molecular-kinetic theory from gases to liquids, following simple thermodynamic considerations of stability. To obtain his equation, van der Waals had applied the Clausius virial theorem to show that the attractive forces between pairs of molecules in a gas gives rise to an internal pressure, $-a/V^2$, in the equation of state, $(P + a/V^2)(V - b) = RT$, where V is the volume, b is the "excluded" volume due to the finite size of the molecules, and R is the gas constant. At low temperatures on a given isotherm, the equation describes coexistence between phases of different density; the thermodynamically stable coexistence point could be found, as Maxwell showed, by a simple "straight-line" geometric construction.[158]

The van der Waals equation of state, when written in units of the critical temperature, pressure, and volume instead of the constants a, b, and R, took a universal form—the "law of corresponding states," which indicated an underlying similarity of all fluids. As remarked earlier, Kamerlingh Onnes's strong belief in this "scaling law," and his surprise at its failure for liquid helium, would guide his early helium research.[159]

Quantitative agreement between the van der Waals equation and the carbon dioxide experiments of Andrews was within 4% for the critical volume and temperature, but the discrepancy for the critical pressure was a large 13%. When the equation was tested experimentally for numerous other fluids, it was found that quantitative discrepancies of that order of magnitude were typical. Leading theoretical physicists—including Maxwell, Lorentz, Boltzmann, and Clausius—tried their hand at improving the van der Waals equation, with little success. Max Planck concerned himself with computational schemes to help comparison with experiment.[160]

The other major nineteenth-century development in the theory of phase transitions was Gibbs's geometric representation of thermodynamic functions and his systematic statement of the conditions for stable equilibrium of coexisting phases, even for substances containing diverse chemical compounds. His phase rule was among the results.[161] The science of thermodynamics was then relatively new. Only in 1865 had Clausius identified the entropy function as sufficiently important to deserve a name.[162] Gibbs used the van der Waals equation as well as the Andrews experiments to illustrate his own theories,[163] and, in turn, some European scientists, especially Maxwell,[164] quickly learned and utilized Gibbs's methods. The rigor of thermodynamic reasoning, the absence of special hypotheses and mechanical models, and the applicability to phase equilibria of all kinds gave Gibbs's theory unique status. It must be viewed as the first general theory of phase transitions.

The mechanical models underlying the molecular-kinetic theories could be made plausible on the basis of numerous macroscopic experiments, but in the late nineteenth century the legitimacy of any molecular model whatsoever was under severe attack, especially from the German "energeticists" Wilhelm Ostwald and Georg Helm.[165] Yet calculation of an equation of state, and of such things as transport coefficients, specific heats, and density fluctuations, was outside the range of thermodynamics, and these quantities could be determined from a molecular-kinetic theory. To mention one important example, Kamerlingh Onnes generalized the van der Waals equation by introducing a systematic "virial expansion,"[166] a calculation of the equation of state of an imperfect gas as a power series in the particle density. The lowest-order term yielded the equation of state of an ideal gas, while the second virial gave corrections to the equation of state at relatively low densities. The van der Waals constants a and b could be systematically derived from the second virial coefficient, calculated from the intermolecular potential and fitted to low-density data.

Thermodynamics and molecular-kinetic models complemented each other. A unified and comprehensive theoretical method more complete than thermodynamics became available in principle with Boltzmann's and, especially, Gibbs's statistical thermodynamics and statistical mechanics. Gibbs noted that the large fluctuations of observable quantities when two phases coexist provide an exception to the rule that the expectation values calculated for macroscopic physical systems by his methods have negligible dispersion.[167] He thus implicitly raised the question whether computation of mean values of macroscopic variables by statistical mechanics is useful near the critical point. But he did not attempt to apply his statistical theory to the description of phase transitions.

The modern theory of critical phenomena emerged with the joining of several streams, both experimental and theoretical, among them van der Waals's work,[168] Kamerlingh Onnes's virial expansion, Lindemann's theory of melting,[169] studies of opalescence and fluctuations at the critical point, ferromagnetism, and Bose–Einstein condensation, as well as the streams of superconductivity and superfluidity already discussed. Critical opalescence—the cloudiness of a liquid at its critical point—had been noted by Andrews, and was systematically investigated in the 1870s.[170] Smoluchowski in 1908 and Einstein in 1910, using kinetic theory and thermodynamic arguments, showed it to be a consequence of large density fluctuations.[171] The 1914 Ornstein–Zernike theory of critical fluctuations in terms of density correlations would play a major role in the development of theories of phase transitions.[172] Meanwhile, in the 1890s the existence of a new phase transition, that between the ferromagnetic and unmagnetized states at the Curie point, was discovered (Chapter 1).[173] Indeed, Weiss's 1907 theory of ferromagnetism was an early instance of viewing a phase transition as a cooperative phenomenon, although the origin of his "molecular field" proportional to the average magnetization was to remain a source of puzzlement until Heisenberg's theory of ferromagnetism (Chapters 2 and 6).[174]

The period between the First World War and the early 1930s, during which the development of quantum mechanics was the principal focus of physics, saw relatively little work related to critical phenomena. Yet there were a few developments that would prove of great importance later. Among them was Einstein's statistical-mechanical extension in 1924 and 1925 of the Bose photon gas theory to a quantum gas of particles with mass, which led to the theory of Bose–Einstein condensation.[175] This theory would be seen as a possible guide for evaluating the partition function for ordinary gas–liquid phase transitions,[176] and would become the underlying model for the theory of superfluid helium.

In 1920, Wilhelm Lenz introduced the model of ferromagnetism that is now named after his student Ernst Ising, who in 1925 showed that the model would not lead to a phase transition in one dimension; he also mistakenly declared that it would not lead to phase transitions in two or three dimensions either.[177] The model, while grossly oversimplifying the interatomic interactions, provided an example of the full complexity of the statistical features of critical phenomena. But since the details of the interactions were believed to be of primary importance, and Ising had argued that the model would not lead to a phase transition, the model did not then receive a great deal of attention in relation to critical phenomena. Some work also proceeded in this period in rigorously calculating the equation of state of gases from the virial expansion near liquid densities. However, such an approach required the calculation of many higher-order terms (corresponding to a molecule interacting with 2, 3, or *n* other molecules at the same time), an apparently intractable problem.[178]

The 1930s brought new studies in the quantum theory of ferromagnetism, the description of the condensation of gases, the recognition of other types of phase transitions, and a great deal of comparison of theory with experiment. Statistical mechanics was now accepted as the appropriate theory, and methods of calculation were improved, but physicists understood that even the best of these could not be trusted to give reliable results in the immediate neighborhood of critical points. The 1930s also brought an important shift in outlook: increasingly, such disparate physical phenomena as boiling, ferromagnetism, and orderly structure in binary alloys came to be seen as belonging to a common field of study. The idea of a statistical-mechanical theory of "cooperative phenomena" began to be used to describe all types of phase transitions, in spite of a lingering uncertainty about the inherent ability of statistical mechanics to yield the singularities associated with critical behavior.

Four lines of research in phase transitions were now evident: (1) general theories based on thermodynamic or combinatorial arguments; (2) attempts to construct an exact theory of condensation; (3) elucidation of the mechanism of ferromagnetism; and (4) development of successive approximation schemes for the thermodynamic functions, and application to types of phase transitions other than condensation and ferromagnetism.[179]

Prominent among the general theories of phase transitions was Ehrenfest's thermodynamic work on the second-order phase transitions, discussed earlier, which extended the concept beyond the conventional fluid transitions to include the transitions in liquid helium, superconductors, and ferromagnets. The following year, Peierls suggested that one can examine theoretically whether long-range order occurs naturally in a system in equilibrium by testing the stability of an assumed perfectly ordered state. He also called attention to dimensionality as an element determining whether a system can sustain long-range order and exhibit a sharply defined transition temperature. To illustrate, he gave a general argument for the nonexistence of a sharply defined melting point in a one-dimensional crystal, in contrast to the case in a three-dimensional crystal. The argument was based on the growth of positional fluctuations with distance in a one-dimensional crystal.[180] Subsequently, he examined the simple Ising model–type of interaction in terms of its dimensionality, showing that the Ising model does lead to a phase transition in two and three dimensions.[181] Peierls's method, however, provided no information about the nature of the singularity at the transition point.

In 1935, Landau, in his first paper on phase transitions, introduced the general concept of an order parameter to describe the degree of order of a substance; the order parameter is zero in the disordered phase and positive in the ordered phase.[182] As discussed in detail in Chapter 2, the notion of an order parameter had first been introduced by Bloch in 1932 in his postdoctoral dissertation *(Habilitationsschrift)*, where he used it to describe spatially varying ferromagnetic states.[183] Shortly thereafter, Landau, in his 1933 paper on superconductivity, used the saturation current as an order parameter for the superconducting state, expanding the free energy near the transition temperature in terms of the saturation current; in the same year, he also gave a theory along similar lines of the antiferromagnetic phase transition.[184] In his general paper of 1935, Landau contended that the onset of order character- izes a large class of phase transitions; carrying out a similar expansion of the free energy as a polynomial in the order parameter, he again concluded that the specific heat must grow as $(T_c - T)^{-1/2}$ as the critical point is approached from below.

In 1937, Landau published, in Russian and German, two classic papers on the thermodynamics of phase transitions.[185] The first is noteworthy for introducing the question of changes of symmetry in phase transitions. Discussing the effect of the symmetry of the high- and low-temperature phases on the possible transitions, Landau observed that "elements of symmetry are either present or absent; no inter- mediate case is possible." This implies a sharp discontinuity in the state of a system undergoing a transformation from crystal to liquid, or between different crystal structures, or from magnetized to unmagnetized. Similarly to Ehrenfest, Landau distinguished the two types of transitions, those with and those without a discon- tinuity in energy. From the symmetry argument he concluded that, unlike the behavior of liquid and gas above the critical point, no continuous change of state between fluid and crystal is possible. He derived from the expansion of thermody- namic potentials around the Curie point or critical point the nature of the discon- tinuities in specific heat and the latent heats.

The second paper proved the famous Landau–Peierls theorem: long-range order is possible only if the order parameter varies in all three spatial directions. This drew on Peierls's earlier arguments for the lack of a phase transition in a one-dimensional crystal.

Ehrenfest's taxonomy, and especially Landau's theory, served to call attention to the possiblity of a general theory of phase transitions—a theory that would apply to such disparate phenomena as ferromagnetism, melting, vaporization, order–dis- order transitions, superconductivity, and so on. The theory put equations of state into the background and identified discontinuities, singularities, and order param- eters as the main items of interest. Despite the relative simplicity of Landau's approach, the question of the rigorous evaluation of the partition function (the only point, apart from experiment, from which one could appraise the legitimacy of Landau's expansion near the critical point) remained an important question.

The matter was of particular interest in Holland, where van der Waals's influence was strong. A November 1937 Amsterdam conference commemorating the hun- dredth anniversary of van der Waals's birth provides insight into some of the topics and issues in the approach to phase transitions at that time.[186] Although van der Waals was seen primarily as a pioneer in research on intermolecular potentials and other diverse topics in physical chemistry, and the theory of phase transitions and critical phenomena was not yet very active, some 20% of the papers at the confer- ence dealt with this subject.[187]

The American Joseph Mayer, who did not attend the meeting, had just published a paper entitled "The Statistical Mechanics of Condensing Systems," which excited the interest of the theorists.[188] In a step toward evaluating the classical configura- tional partition function, which Ursell and others had expanded to give successive virial coefficients—an expansion valid at low density—Mayer rearranged its terms into a "cluster expansion." This expansion appeared to hold for any density, and could thus be expected to provide a description of the gas-to-liquid phase transition. Max Born, speaking of Mayer's theory at the conference, said, "I consider this work a most important contribution to the development of van der Waals' theory . . . in spite of the fact that Mayer's methods are rather difficult to understand and his results not completely satisfactory." In particular, the resulting equation of state gave metastable loops in the van der Waals isotherm between coexisting gas and liquid, rather than the correct equilibrium Maxwell construction. But most seriously, as Born stated in a "note added in proof," was the use of a formula for the partition function that is invalid for densities higher than the critical density.[189]

Boris Kahn and George Uhlenbeck found Mayer's theory promising and noted that at low temperatures, Mayer's expansions for particle density and pressure were analogous to a pair of series describing the ideal Bose–Einstein gas.[190] The analogy suggested that the Mayer expansion would also lead to condensation. If the theory were to replicate the Maxwell construction, statistical mechanics would have a mathematical form that could generate singularities—a central point whose consideration would be a long-remembered highlight of the conference. J. de Boer recalled "a very lively discussion about whether one might ever expect to obtain with Gibbs' ensemble theory and the partition function an experimental isotherm with a section in which the pressure is independent of the density, or whether one would obtain isotherms with loops of the Van der Waals type." The issue was that the flat section implied nonanalyticity in the partition function. He went on, "it was Kramers who pointed out that we are really interested not in the partition function iteslf, but in the thermodynamic limit of the free energy per particle, and that this limiting process may be able to introduce a non-analytic behavior at certain densities and temperatures."[191]

Kramer's concept of the thermodynamic limit in which the particle number N and the volume V go to infinity, with the density N/V fixed, would help to focus attention on the discontinuities and singularities, rather than on the equation of state. It defined a research program for future statistical mechanics of phase transitions and critical phenomena. Although probably the most significant event at the Amsterdam conference, it was recorded nowhere in the proceedings. Uhlenbeck recalls there was a morning-long debate, which ended with a vote on the issue "Does the partition function contain the information necessary to describe a sharp phase transition?"—the conferees splitting about evenly on the question.[192]

Mayer's theory was foremost among attempts to derive an exact theory of phase transitions by rigorous evaluation of the partition function. As subsequent work of Mayer and co-workers showed, the theory with a realistic intermolecular potential led to a singularity in the free energy for all T below a certain T_c, which one could associate with the critical temperature.[193] In a tour de force, they derived from the theory an equation of state that exhibited condensation.[194]

Within a few years, the political and social circumstances within which the exact approach was attempted would lead to a sudden break in the development. Born had been the professor and director of the Institute of Theoretical Physics in Göttingen during the heroic years of the quantum theory, the time of the Weimar Republic in Germany. But being a Jew, he found himself suddenly unemployed in 1933, and emigrated first to England and then to Scotland, where in 1936 he obtained a professorship in Edinburgh. His co-worker Klaus Fuchs likewise emi-

Paul Ehrenfest and George Uhlenbeck. (AIP Niels Bohr Library, Uhlenbeck Collection)

grated, eventually working on the Los Alamos atomic bomb project and some years later gaining notoriety as a spy for the Soviet Union. Uhlenbeck, who had been a student of Ehrenfest in Leiden, came to the United States in 1927 to join the University of Michigan's physics department; he was hired as part of a concerted effort in the United States at the time to bring European theoretical physicists to America.[195] However, in 1935 he returned to Holland to be the physics professor at Utrecht. Kahn, whose doctoral dissertation in 1938 had been an impressive study of the Mayer theory and its extension to quantum statistics, was murdered during the Second World War in a concentration camp.

A third line in the study of phase transitions was ferromagnetism, whose development has been described in Chapters 2 and 6. In the 1930s interest in ferromagnetism was focused primarily on elucidating the quantum-mechanical and electronic mechanisms responsible for the phenomenon in solids, rather than on their detailed behavior in the immediate neighborhood of the Curie point. The qualitative origins of ferromagnetic phase transitions were understood in terms of the quantum theory, and unlike superconductivity, the phenomenon no longer seemed mysterious. Theoretical models were used to calculate expectation values of the magnetic moment as a function of temperature and magnetic field strength, primarily because such calculations provided a useful means to test the assumed mechanisms for ferromagnetism.

But measurements of bulk magnetization were of limited precision, given the combined effects of the formation of ferromagnetic domains, dependence on sample shape, the complexity of interpreting data from polycrystalline samples, anisotropies, impurities in the sample, and the curvature of the magnetic field. In view of the complexity of the electronic structure of actual metals and alloys, any quantum-mechanical theory entailed simplifications and assumptions difficult to assess. Moreover, even within the Heisenberg model, detailed evaluation of the partition function required additional approximations that would alter considerably the detailed behavior of the magnetization, susceptibility, and specific heat close to the Curie point.

One of the more productive developments in the 1930s in the theory of phase transitions was the application of approximate methods, particularly to cooperative phenomena; modern statistical-mechanical theories of lattices have their roots in this work. For the second, 1936, edition of his influential 1929 book *Statistical Mechanics,* R. H. Fowler added a chapter that opened by calling attention to a new subject matter:

> The most striking recent advances in the application of statistical mechanics have been made in the study of cooperative phenomena. These are phenomena which cannot be interpreted in terms of the properties of an assembly of distinct systems only slightly linked to each other, but can only be interpreted in terms of the states of the assembly as a whole, because the states of any distinct system are fundamentally influenced by which states of the other systems are occupied. The classical example is ferromagnetism.[196]

Fowler went on to discuss three "recently recognized cooperative phenomena": order-disorder transitions in alloys, "rotational melting," and critical adsorption isotherms.

Approaches to cooperative phenomena included both the study of kinetics, which proved useful in forming physical pictures of the processes associated with cooperative effects, and equilibrium statistical-mechanical theories. W. L. Bragg and E. J. Williams employed the kinetic method to examine the "effects of thermal treatment on the arrangement of atoms among available sites" in a binary alloy.[197] Paying particular attention to the relaxation times for reaching thermal equilibrium, they showed how rapid cooling (quenching) and slow cooling (annealing) lead to different stable configurations at room temperature, and were thus able to address the question of whether samples were in thermal equilibrium. Frenkel similarly studied condensation and freezing in terms of kinetics.[198] This permitted him to consider especially near transition points, the metastable or stable equilibrium between, for example, a liquid and small gas bubbles or a liquid and small crystals forming within it.

Unlike the study of kinetics, statistical-mechanical theories were restricted to thermal equilibrium. Among the various types of theories for cooperative effects, the Weiss theory of ferromagnetism was especially attractive because of its mathematical simplicity; in effect, it replaced an extremely complex many-particle problem with that of a single particle in an average field. The resulting magnetization, which was found by solving a nonlinear equation graphically, was not accurate in detail, but it did exhibit the phase transition and permitted one to interpret the critical temperature in terms of the average field. Similarly, in the van der Waals model of the liquid–gas transition, the critical temperature, density, and pressure were expressed in terms of average molecular fields contained in the "internal pressure" and "excluded volume" coefficients *a* and *b*. Other types of phase transitions presumably also resulted from the cooperation of a large number of interacting (usually submicroscopic) units.

It was recognized that an extremely useful step in describing such phase transitions consisted of formulating a model in terms of the interactions among the elementary units, then solving it by replacing the many-body interactions with an average interaction analogous to Weiss's molecular field. For example, although traditionally the adsorption of a monolayer of molecules on a solid surface had been studied in terms of its kinetics, Fowler formulated the problem in terms of statistical mechanics and applied a molecular field approximation to its solution.[199] Bragg and Williams, in addition to their analysis of the kinetics of the order–disorder problem in alloys, set up a relatively simple scheme for the possible energies of that system; it turned out to be nearly equivalent to the Ising model, involving, however, two sublattices as in antiferromagnetism.[200] To obtain the free-energy and other macroscopic variables for the model, they used the molecular field approximation, according to which the field tending to produce order is proportional to the degree of order itself. In agreement with experiment, the model led to a phase transition at a sharply defined temperature.

J. E. Lennard-Jones and A. F. Devonshire[201] adapted the Bragg–Williams model for the order–disorder transition in alloys to make it serviceable for a description of melting as well as condensation; they said they did this also to bring these transitions into the category that Fowler called "cooperative phenomena."[202] They assumed an intermolecular potential with a short-range repulsion of order r^{-12} and a long-range attraction of order r^{-6}, fitted to the equation of state at low densities. Then they computed critical temperatures by a carefully devised variant of molecular field theory, reminiscent of the Wigner–Seitz approach in which the molecules of the liquid were pictured as moving in individual cells that constituted a lattice. Although physically the cell model of a liquid seemed rather artificial and partly begged the question of how crystalline structures form as an amorphous liquid is cooled, the model was easy to handle computationally where the exact theories of Mayer and others had failed.[203] As Lennard-Jones reported at the van der Waals centennial, they obtained expressions for the pressure in the critical region that differed considerably from the van der Waals equation but led to much better numerical agreement with observations.[204]

Theories of the molecular field type suffered from ignoring short-range correlations between the particles (or spins). Could one perhaps improve on the molecular field approximation by treating the immediate neighbors of an atom exactly, replacing only the more distant ones by an average field? Hans Bethe, at the University of Bristol in 1934, considered the Bragg–Williams model for a binary alloy—a model that consisted of two interlocking sublattices with nearest-neighbor interaction of the Ising type, whose magnitude depended on whether two neighbors are atoms of the same or different kind.[205] Bethe defined a long-range order, which was strictly zero above the transition point, and an "order of neighbors," equal to the difference of the probabilities of finding a pair of neighboring atoms to be of the same or different type. The latter type of order was neglected in the molecular field approximation. In his first approximation, Bethe considered the "order-of-neighbors" for a central atom and its nearest neighbors only; in the second approximation, he included all the nearest neighbors of the nearest neighbors, and so on. Bethe's results led not only to corrections to the Bragg–Williams value for the Curie point, but also to qualitative differences. For example, in the molecular field approximation the specific heat vanished above T_c, whereas it remained finite in Bethe's calculation. Bethe's improvement on mean-field theories led to application of his method in problems where previously only molecular field theory had been used.[206]

Alternatives to Bethe's method were developed by Fowler and Guggenheim (their "quasi-chemical method") and by Kirkwood. Although based on the theory of liquid solutions, Fowler and Guggenheim's method was in fact equivalent to Bethe's first approximation.[207] Kirkwood expanded the configurational partition function—for a fixed value of the long-range order, but a function of the local order parameter—in terms of the lowest-order terms representing the mean interaction energies between nearest neighbors and the fluctuations about the mean.[208] Comparison of the Bethe and Kirkwood methods showed that they corresponded closely but not completely.[209] In both Bethe's and Kirkwood's methods, the higher-order terms were very complicated, and convergence of the series was slow near the critical temperature. Only an exact solution could be relied on to give the correct behavior of specific heats and other quantities in the critical region, but the first exact solution was still in the future.

A decisive event in the history of the theory of critical phenomena was the solution by Lars Onsager,[210] published in 1944, of the two-dimensional Ising model,[211] the first rigorous evaluation of a partition function demonstrating explicitly the existence of a phase transition and associated critical behavior in the thermodynamic limit. This work, both a permanent monument and a beacon lighting the way for subsequent research, was a virtuoso performance in applied mathematics, which once and for all put to rest doubts about the suitability of equilibrium statistical mechanics for describing phase transitions and critical phenomena. The solution provided the first instance of a rigorous analytic derivation of the critical behavior of substances in thermal equilibrium. It demonstrated that all the approximation methods and series expansions used hitherto gave qualitatively wrong results in the critical region. It furthermore brought to life a branch of the theory of phase transitions that made higher mathematical demands than any of the earlier analyses.[212]

Onsager was already a major figure in the physical sciences because of his earlier work on electrolytes and on the theory of irreversible processes. Three other researchers in particular, Hendrik A. Kramers, Gregory Wannier, and the younger mathematical physicist Elliot Montroll, who had been Wannier's student, had prepared the way in 1941 for Onsager's work on the Ising model.[213] But to them a rigorous evaluation of the partition function in closed form seemed too formidable even to attempt.[214] Kramers, Wannier, Montroll, and Onsager were part of an informal network of researchers; quite independently of them and outside the network, Edwin N. Lassettre and John P. Howe had pursued a similar line of research on the partition function for a binary solid. This was also published in 1941, and Onsager had learned of it before he wrote his landmark paper.[215] Lassettre was then in the Chemistry Department at Ohio State University, and Howe was at Brown University, but both would become involved in the atom bomb project soon after their collaboration. In their work, Lassettre and Howe introduced the transfer matrix formalism that would form the basis of the Onsager solution, and showed a connection between a phase transition and the degeneracy of the largest eigenvalue of the matrix. Still further removed from the network was Ryogo Kubo in wartime Japan, who also formulated the problem in terms of the eigenvalues of the transfer matrix.[216]

Kramers, the oldest of the three who directly prepared the way, had worked from 1916 to 1919 with Niels Bohr using the old quantum theory to study the interaction of electrons and radiation; in the late 1930s and early 1940s, he worked actively on a modern formulation of quantum electrodymamics and had developed the method of mass renormalization. More directly important for the theory of phase transitions, however, was the fact that he had been a pupil of Ehrenfest in Leiden, while Ehrenfest had in turn been initiated "into both the substance and the spirit of theoretical physics," including statistical mechanics, by Boltzmann.[217] More than a decade after Kramer's death in 1952, a methodological convergence, especially in the use of renormalization techniques, brought the two seemingly disparate areas of his interest—quantum field theory and the theory of phase transitions—into close proximity.

Kramers had returned to the Netherlands as professor at Utrecht following his work with Bohr, and after Ehrenfest's death he became professor of physics at Leiden, remaining there throughout the Second World War. The Netherlands was invaded in May 1940 and for the next five years was a victim of German military conquest and occupation.

The Swiss-born Wannier had become familiar with the topic of phase transitions while studying with R. H. Fowler in Cambridge. In 1937, after a year of research in Wigner's group at Princeton, where he worked on solid-state theory, Wannier became an instructor at the University of Pittsburgh; when his contract there was not renewed, he moved to the University of Texas. Visiting Leiden, he and Kramers together analyzed the partition function of the Ising model.[218] Their motivation was to "gain sound statistical information about some model of a ferromagnet. The Ising model [for a two-dimensional square net] has been chosen because its extreme simplicity makes it particularly suitable for such a purpose."

When Wannier came to Pittsburgh, Montroll, a mathematically inclined student there, eagerly talked with and learned statistical mechanics from him. Wannier started his student on a study of Mayer's new cluster expansions, and from this a doctoral dissertation grew. Montroll soon recognized that the problem of evaluating certain integrals, the "cluster integrals," in the Mayer theory was equivalent to the problem of finding the eigenvalues of an integral equation. He also showed that a certain set of terms in the cluster expansions (the "ring diagrams") could be summed to all orders.[219] Montroll's work represents an early instance of evaluating functions approximately by carrying out an expansion in which the terms could be represented by diagrams, and then summing a particular class of diagrams to all orders. Although his calculation did not particularly help in finding the complete partition function, it is noteworthy that the method preceded the invention of Feynman diagrams and also the use of "summation of diagrams" in solid-state theory, which would be highly important in the quantum-mechanical many-body problem in the 1950s and 1960s. After completing his dissertation, Montroll spent a year working directly with Mayer, then at Columbia. When Wannier returned from Leiden, Montroll recalls, he visited Montroll and the two of them spent "a couple of days discussing things," especially the Kramers–Wannier theory of the Ising model.[220]

An abstract[221] by Wannier of his work with Kramers entitled "Statistics of the Two-Dimensional Ferromagnet," published in February 1941, was followed by two full-length papers in August.[222] One is reminded of wartime conditions in a footnote explaining that "owing to communication difficulties, one of the authors (G. H. W.) is entirely responsible for the printed text"; a footnote referring to these papers further stated that "unfortunately the present state of world affairs has delayed publication of this work."[223]

A crucial mathematical device in the Kramers–Wannier paper was the conversion of the problem of evaluating the partition function into that of finding the largest eigenvalue of a matrix, the "transfer matrix." More precisely, they found a sequence of finite matrices whose largest eigenvalue becomes equal to the partition function as the size of the system, and hence the matrix, becomes infinite. The authors referred to borrowing the "elegant form of procedure" first used by Montroll in connection with the theory of molecular chains. In turn, Montroll credited discussion with Wannier as a source of his inspiration.

Montroll outlined a general mathematical procedure for evaluating the partition function of any system with nearest-neighbor interactions by means of solving linear homogeneous operator equations, which he illustrated by applying it to the two-dimensional Ising net.[224] He noted that an "abstractly equivalent" process had been used by him and Mayer to evaluate cluster integrals. Kramers and Wannier showed that a certain transformation connects high-temperature to low-temperature solutions and reveals a rigorous symmetry between the two limits. Although they did not obtain an exact solution for the partition function of the two-dimensional Ising model in the critical region, they were able to use the symmetry property to obtain several results. The computed the Curie temperature exactly and correctly concluded that, contrary to earlier calculations (the Bragg–Williams and Bethe approximations), the specific heat is infinite at T_c. Quite incidentally, Montroll showed from his formalism that a necessary condition for a phase transition in a crystal is that it be infinite in at least two directions. He called attention to the twofold degeneracy of the eigenvalue of the matrix at the transition temperature, but only Lassettre and Howe had recognized that the degeneracy applied for all $T < T_c$.

In 1941, Onsager was a professor at the Sterling Chemistry Laboratory at Yale University, funded by the Sterling Foundation. Originally from Norway and trained in chemical engineering, Onsager was invited by Yale in 1933 as a Sterling fellow in Chemistry, on the strength of his important work on the transport properties of electrolytes. As T. Shedlovsky and Montroll recounted,

> The Chemistry Department was surprised and somewhat embarrassed after awarding this postdoctoral fellowship when they discovered that he had never received his Ph.D. They were even more shocked by his dissertation of over 100 pages on properties of Mathieu functions, hardly a suitable topic for a chemistry thesis. The Yale mathematicians came to the rescue by agreeing that the dissertation was an outstanding contribution to the subject.[225]

Maria Goeppert-Mayer (1906–1972), Joseph Mayer (1904–1983), Robert d'Escourt Atkinson (b. 1898), Paul Ehrenfest, and Lars Onsager (1903–1976) at the University of Michigan summer school, Ann Arbor, 1930. (AIP Niels Bohr Library, Goudsmit Collection)

At Yale, Onsager was reputed to be a singular individual, extraordinarily insightful but also taciturn and enigmatic in conversations about his own area of physical chemistry.

> However, those who know him will witness the fact that he is clarity itself, and often responds at great length if the questions presented to him refer to Norse mythology, gardening, the more subtle aspects of "Kriegspiel" (a form of blindfold chess involving two opponents and a referee), and even encyclopedic facts of organic chemistry.[226]

Onsager had done significant work on various topics, which he had left unfinished or at least never wrote up for publication.[227] When Montroll, after his stint with Mayer, became a Sterling research fellow at Yale, he believed that the university expected that he would function as an assistant to Onsager—an assistant who might exhume some of the latter's incomplete work, fill it in where needed, and write it up for publication, to the greater glory of Yale University and scientific progress.[228]

At the time, Montroll recalled, Onsager was interested in calculating the entropy of ice. But hearing about Wannier's and Montroll's ideas on the Ising problem, he became interested and made various suggestions that furthered Montroll's study.[229] Onsager and Montroll shared an office at Yale, and they talked a good deal about the Ising model. "Onsager's technique was first to do the one-dimensional problem, then he did a line of three rows, then he did four and five and started to notice the systematics and literally guessed the answer for a line of n and for the two-dimensional problem."[230] But the discovery preceded the proof of the answer; the complete detailed analysis of the problem took a long time. Beyond the confidence that the problem could be solved, the successful derivation required bringing to bear little-known tools and devices from diverse branches of mathematics, and Onsager was the rare physical scientist who had the requisite broad mathematical knowledge.

At a meeting of the New York Academy of Sciences on 28 February 1942, Onsager wrote on the blackboard, as a remark during informal discussion, the exact result for the partition function for the two-dimensional rectangular Ising net.[231] In his paper describing the derivation, which appeared two years later, he used the approach of Kramers, Wannier, and Montroll to evaluate the partition function by finding the largest eigenvalue of the transfer matrix.[232] To find this eigenvalue, Onsager decomposed the matrix into a direct product of 2×2 matrices, although he emphasized the abstract properties of the operators of which the Kramers–Wannier matrices are an explicit representation. Onsager's tour de force consisted of carrying through the complicated algebra to obtain the largest eigenvalue of the matrix, and thus the partition function.

He furthermore generalized Kramer's and Wannier's connection between the low- and high-temperature limits to lattices other than the square net by constructing a "dual lattice" to a given lattice structure, and proved that the symmetry property connects the low-temperature solution of one lattice to the higher-temperature solution of its dual. He noted that although he was calculating only the case of the rectangular net, a similar calculation is possible for hexagonal and triangular nets, duals of each other, if an additional symmetry transformation, the "star–triangle" transformation, is also used. The star–triangle transformation connects the low-temperature behavior of a triangular lattice to that of the hexagonal one, and together with the dual-transformation it links the high- and low-temperature values of the partition function for the same lattice;[233] in particular, it can be utilized to locate the critical temperature for either of these lattice structures. Onsager's exact result for the specific heat, expressed in terms of elliptic integrals, yielded a singularity of the form $-\log |T - T_c|$ at the critical temperature, whether approached from above or below. The internal energy, however, was continuous across the critical point. The shape of the specific heat curve as a function of temperature was qualitatively different from that obtained by all previous approximate calculations, for the previous calculations gave a finite discontinuity instead of a logarithmic divergence symmetrical about T_c.

The discovery of the matrix formulation of the Ising partition function in 1941 stirred renewed interest in the theory of phase transitions. Onsager's announcement of his formula for the Ising partition function, in February 1942, primarily reached people in the New York area. Since most American theoretical physicists were already diverted or soon to be diverted to war-related problems, physics ideas unrelated to military objectives were sequestered until after the war.

An Onsager paper of 1944, totally unrelated to military projects, was an exception. Although Onsager was interested in the militarily important isotope separation problem and had some unpublished papers and calculations on the subject, he did not become heavily involved in war work.[234] At Columbia University, goal-oriented military projects notwithstanding, the Wednesday night theoretical physics colloquium met regularly, and Montroll (who himself became involved in war projects) recalls lecturing on the unpublished Onsager solution, generating an interest in the problem among Willis Lamb, Arnold Nordsieck, and others. Out of this interest grew a Ph.D. dissertation by Julius Ashkin at Columbia, which contained an analysis, published by Ashkin and Lamb, of the low-temperature Ising model pair correlation function using the matrix approach.[235] A paper by Ashkin and Edward Teller at the same time referred to the unpublished Onsager solution, saying that Onsager's "ingenious methods appear to lend themselves readily to generalizations," although the paper merely extended the solution to a four-component alloy.[236]

The extent and the limitations on generalizations of the Onsager approach would occupy and frustrate later investigators. Especially frustrated were those unsuccessfully attempting a three-dimensional Ising model. Wannier, reviewing the matrix-eigenvalue method in 1945, conjectured,

> It would be surprising if the specific heat remained finite at the Curie point for a three-dimensional Ising model. It is conceivable, on the other hand, that we may find a curve which is more asymmetric about the singularity. . . . It is to be hoped that a three-dimensional calculation will, before long, furnish the answer to these questions.[237]

8.2 Post–Second World War Development

The onset of the Second World War disrupted traditional research in collective phenomena.[238] In countries that mobilized for war work, both manpower and resources were reallocated.[239] As scientists turned to problems of weaponry, materials used for experiments, such as helium, became scarce. In only a few exceptional cases was significant nonmilitary research carried out on collective phenomena—for example, the development by Justi and co-workers in Germany of the superconductor niobium nitride, with its high transition temperature; Kaptiza's work in the Soviet Union on superfluid helium, which continued into the wartime period;[240] and independent theoretical advances by Onsager in the United States and Kubo in Japan on the Ising model.

However, the framework for further progress in the field changed dramatically during the course of the war. By the 1950s, the enormous investments made in wartime research and development would lead to a remarkable burst of advances.[241] Technological developments stimulated by military concerns, the contact of scientists with new and practical problems, and social and institutional changes would all have major impact. New technologies useful in collective phenomena research included microwave generation and detection, neutron diffraction, and the Collins helium liquefier. Such developments, beyond the capabilities of individuals or small groups in the relatively short wartime period, for the most part grew out of large-scale war efforts. These new experimental tools also stimulated the invention of theoretical techniques. Leading examples are the immediate postwar studies by Lamb and Robert Retherford of the shift of the levels of the hydrogen atom caused by interaction of the electron with the radiation field, and by Polykarp Kusch on the magnetic moment of the electron; these experiments, in turn, contributed to the development (by Bethe, Victor Weisskopf, Julian Schwinger, Richard Feynman, and others) of renormalized quantum electrodynamics.[242] Wartime research made new materials readily available—for example, pure isotopes and ^{3}He in bulk quantity—which made possible the discovery and analysis of new aspects of collective phenomena.

The framework was also shaped by social and institutional changes brought about by the wartime circumstances. The most striking social change was the movement of German refugee physicists to Great Britain and the United States, an upheaval affecting all areas of modern physics. Furthermore, in Great Britain and the United States many science students who had been diverted into war work found themselves working side by side in the large radar or bomb projects with leading scientists; this experience motivated many to continue in research and to make up hastily after the war for the time lost. Furthermore, the wartime experience provided a high level of technological training, enabling such students to "really put on the steam and develop things very rapidly."[243] The large American wartime laboratories, such as the MIT Radiation Laboratory, the Chicago Metallurgical Laboratory, and the Los Alamos Laboratory, also provided new institutional models for large-scale, goal-oriented scientific research programs; such programs were now established at a number of institutions both inside and outside academia. The large wartime laboratories served also as a source of trained personnel for these new programs.[244] Finally, since the war drove home the lesson that science could be effective, new government funding became available for physics, particularly in the United States, where new agencies, such as the Office of Naval Research and later the National Science Foundation, supported solid-state research.[245] Thus with a multitude of new theoretical and experimental tools on hand, the postwar generation of researchers was in an unusually favorable position to break new ground rapidly.

Finally, in a few important cases, particular problems studied for wartime purposes turned out to be relevant to the postwar development of the theory of collective phenomena. One important example, to which we will turn in the next section, was David Bohm's wartime study of the behavior of plasmas, which led directly to a many-body theory. Overall, the advance of collective phenomena studies made possible by the war effort greatly outweighed the curtailment of progress caused by the interruption of ordinary research activities.

The Quantum Electron Gas

The first step in the formation of a coherent microscopic picture of collective phenomena in many-particle systems was the development of the theory of the electron gas. The electron gas was a prototypal many-body system for which one would be able to construct a systematic analysis, at least at high densities. The theory, by enabling a description of the effective electron–electron interactions in metals, would play a crucial role in the development of the microscopic theory of superconductivity. Interestingly, the emergence of the theory of the electron gas was strongly affected by wartime research, particularly on plasmas and microwaves.

Electromagnetic separation of isotopes in plasmas, a technique pioneered by Ernest Lawrence in Berkeley, was being developed around 1943 as a method for

David Pines (b. 1924). (AIP Niels Bohr Library)

obtaining uranium-235 for the Manhattan Project. The need to understand plasmas engaged David Bohm, a research associate at Berkeley, who in 1943 had taken his physics doctorate under J. Robert Oppenheimer. Bohm found plasmas "captivating"—an "autonomous medium that determined its own conditions and had its own movements which were self-determined"; furthermore, "all the individuals contributed to the collective movement and at the same time had their own autonomy."[246] After Oppenheimer left Berkeley in March 1943 to head the Los Alamos bomb project, Bohm worked on synchrotron radiation and on the behavior of plasmas in magnetic fields.[247] Among the questions he considered in 1944 was turbulence in plasmas. Suspecting that instabilities were the cause of plasma turbulence, he worked out intuitively, essentially on dimensional grounds, a description of the phenomenon now known as "Bohm diffusion." The microscopic theory for this phenomenon would not be worked out for two more decades. David Pines recently reflected that "during the years of the early work on the thermonuclear project, people were just fascinated by how Bohm could have figured this out, but the experiments showed that this is right."[248] While the interest at Berkeley in plasmas sharply waned in 1946, Bohm began to publish and lecture on the subject. At this time, Bohm also worked on other problems, including superconductivity. For example, in 1949 he extended Bloch's theorem to many-particle systems; but he did not, at this point, expect that his work on plasmas would eventually contribute to solving the superconductivity problem.[249] Among those who attended Bohm's lectures was Pines, then a Berkeley graduate student.[250] Out of Bohm's studies of plasmas would emerge the pivotal Bohm–Pines theory of the electron gas.

At Princeton, where Bohm moved in 1946, he and his student Eugene Gross began systematic study of the oscillations in classical plasmas, extending the work done in the 1920s by Debye and Hückel and by Langmuir and his collaborators.[251]

The concept of ordering—a key idea in many of the collective phenomena problems of this period—was a chief concern in the Bohm–Gross work: "that each particle because of long-range interaction affected all the others."[252]

Schwinger's recently developed version of renormalized quantum electrodynamics, also rooted in war work, as mentioned, played a historical role in the next stage of Bohm's work. For Bohm realized that in dealing with a quantum plasma he would have to extend Debye's concept of static screening to a dynamical concept of a deformable cloud moving with the particles. This notion, he recognized, "is basically the same as renormalization in elementary particle theory—to say the actual charge on an electron is infinite according to that theory but surrounds itself with a cloud that makes it finite."[253] Bohm recalls that he came to this idea on hearing Schwinger's famous eight-hour lecture at the March 1948 Pocono Manor Conference.[254] Another idea Bohm borrowed from Schwinger was to separate the Coulomb interaction through a series of canonical transformations.

At the time Pines arrived at Princeton in 1947, people in the Physics Department there were excited by reports on the June 1947 Shelter Island Conference on the development of a full theory of quantum electrodynamics. Pines and Bohm drew on these studies in exploring the consequences of interactions in a quantum plasma, "possibly with the idea that if we could really understand it, maybe we could have some sense of explaining superconductivity."[255] No one had yet dealt satisfactorily with the long-range Coulomb interaction. Although work of Bardeen and others had shown that the long-range Coulomb interaction gave trouble when electron–electron interactions were taken into account, calculations of the ground-state energy from first principles using a screened Coulomb interaction disagreed with experiment. Pines recently summed up the situation: "In retrospect, certainly all of the formalism was there . . . the problem could have been solved in the early 1930s once one had sorted out the quantum mechanics of an electron gas." But no one had yet worked out how to apply this formalism to the problems at hand. He reflects:

> One had the classicial situation in physics where you find when you do the problem a little bit better you're led to all kinds of divergences. . . . And when you fix up one part of the problem, you're led to deep problems with the other. . . . That I think is one of the conditions which require a wholly new approach.[256]

Bohm and Pines realized that to solve the long-range Coulomb force problem, it was necessary to clarify the relationship between the collective modes—the plasma oscillations corresponding to the long-range part of the Coulomb interaction—and the single-particle modes. Although theoretical arguments for plasma oscillations in metals seemed strong, at first Bohm and Pines were not aware of experimental evidence for them. Then in the spring of 1950, Conyers Herring called their attention to pioneering experiments by Ruthemann in 1941 and 1942 and by Lang in 1948; these experiments showed peaks in the scattering of a kilovolt beam of electrons by thin metallic films of beryllium and aluminum, corresponding in fact to the creation of quantized plasma oscillations by the high-energy electron beam.[257] Since the electrons in a metal form a dense gas, Bohm suspected an analogy with the vacuum in quantum electrodynamics, and suggested that Pines try out in this context Schwinger's canonical transformation approach.

In the first of their four classic papers on the collective description of electron interactions, Bohm and Pines tried, using collective coordinates, to decouple the longitudinal collective modes from the individual electron motions.[258] In order to avoid the complications of the longitudinal mode, they first treated only the magnetic interaction, which allowed them to work out a trial problem by a perturbative approach in which the variables were already known and simply had to be renormalized. They also explored the "equation of motion" approach for a time-dependent Hartree–Fock treatment using second quantization. The concept of screening came out of the calculations naturally, and they showed that one derived the same plasma oscillations as in the canonical transformation approach. Pines, in this work, developed a method of calculating the plasma modes to higher order in the wave vector k, relying on a "random-phase approximation" (now called the RPA). This neglected the coupling between the electron–plasmon and electron–electron interactions, keeping only that component of the Coulomb interaction involving the momentum transfer k.

Two subsequent papers, the first using classical equations of motion and the second using quantum-mechanical canonical transformations, treated the longitudinal Coulomb field by separating out the long-range part of the interaction. That left a short-range screen Coulomb interaction that could be handled by Hartree–Fock perturbation theory. Out of this work emerged the picture that the basic degrees of freedom of the plasma are the long-wavelength quantized plasma oscillations (the collective modes in which the electrons interact via long-range Coulomb fields) and a system of screened "quasiparticles," electrons interacting via a short-range, screened Coulomb interaction.[259] Having established the existence of quantized plasma oscillations, and recognizing that essentially all the energy associated with long-wavelength energy transfers goes into these collective modes, Bohm and Pines could now explain the energy loss of fast electrons passing through a plasma, studied experimentally by Ruthemann and Lang.[260]

The Bohm–Pines papers were written at the height of the Cold War. Bohm, having been indicted for contempt of Congress as a result of his refusal to testify about his past associations before Senator Eugene McCarthy's House Un-American Activities Committee, was prohibited from appearing at Princeton University. He was permitted to come to the Institute for Advanced Study, which was organizationally and physically separate from the university, but he mainly worked at home.[261] The third Bohm–Pines paper was written by correspondence, since by that time Bohm, as an "unfriendly witness" to the McCarthy committee, had lost his position at Princeton; with no American institution willing to hire him, he was working in Brazil. Pines had moved first to the University of Pennsylvania and then to the University of Illinois to work as a research assistant professor with John Bardeen.

In the fourth paper of the series,[262] as well as in his dissertation,[263] Pines turned to the calculation of the ground state of an interacting electron gas. By separating the problem into single-particle excitations and collective modes, he was able to calculate a correlation energy close to the earlier estimate by Wigner, and to obtain the quite substantial enhancement of the Pauli spin susceptibility, which results from electron–electron interaction, in good agreement with experiment. Wigner was then at Princeton, but Bohm and Pines did not discuss plasmas with him, since his rigorous approach differed sufficiently from their intuitive one to make communication difficult. Furthermore, for Bohm and Pines the interaction energy Wigner had calculated was not the main point; they were more interested in the general theory of plasmas. Somewhat later, after Pines, who obtained his Ph.D at Princeton in 1950, had been working at Pennsylvania for over a year, Bohm and Pines realized that they could significantly improve on Wigner's calculation using their collective-coordinate formulation.[264]

Herring, Bardeen, Seitz, Charles Kittel, Nevill Mott, Harry Jones, and Dennis Gabor were among the physicists interested in the Bohm-Pines work in its initial stages. Gabor had independently explored energy loss and excitation of the collective modes in thin films. At the 1954 Solvay meeting (which Pines opened with a summary of his work on the electron gas and on his work with Bardeen on the combined influence of electron–electron and electron–phonon interaction on metals), Mott suggested a semiclassical approach for treating plasma oscillations as a polarization wave. He pointed out how one could use the Thomas-Reiche-Kuhn sum rule to explain why plasma oscillations should have a well-defined frequency, not only for nearly free electron models, but also for solids.[265]

Parallel work on exchange and interaction correlations of electrons in metals was carried out independently in the same period by Peter Landsberg and by E. P. Wohlfarth in England.[266] In 1949, Landsberg had used an empirical screen Coulomb potential to estimate correlation forces, with an eye toward understanding soft x-ray anomalies,[267] an approach suggested by Wigner in 1938. Wohlfarth learned from Jones of Landsberg's work, and having found "disturbing" Heinz Koppe's 1947 calculation of the influence of electron exchange on the specific heat of the electron gas,[268] which failed to agree with Bardeen's 1936 calculation of the same effect,[269] he proceeded to employ a similar screened Coulomb interaction to calculate the specific heat, arriving at a result in better agreement with experiment.[270] Both Wohlfarth's and Landsberg's works were cited as background in Bohm and Pines's first paper. The importance of screening in calculating the cohesive energy in metals would also be shown subsequently.

An alternative to the Bohm–Pines approach to the ground-state properties of the electron gas grew out of consulting work done by Murray Gell-Mann and Keith Brueckner in 1956 at the RAND Corporation, a military-oriented think tank in California.[271] The correlation energy of the electron gas in the high-density region was of practical importance in equation of state calculations, relevant to thermonuclear fusion problems; these calculations had been developed during the war at Los Alamos, based on the Fermi–Thomas method.[272] Since a good low-density calculation was available—the Wigner lattice—a rigorous high-density calculation would enable an interpolation between the two regions to obtain a reasonable estimate for the medium-density range, for which Wigner's earlier variational calculation was insufficient. The principal difficulty in the calculation was that because of the long range of the Coulomb interaction, each term in an expansion of the correlation energy in powers of e^2 (where e is the charge on the electron) was divergent.

Gell-Mann was attempting to produce a finite result by summing the singularities when Brueckner visited RAND and made several suggestions, pointing to the solution in terms of expanding the various contributions in the perturbation series in Feynman diagrams.[273] Gell-Mann then devised a diagrammatic method of summing the divergent contributions in the perturbation expansion. Essentially, the method showed which terms diverged, and since these formed a geometric series under the integral sign, a finite summation was straightforward. The Gell-Mann–Brueckner approach yielded an exact result for the specific heat of the electron gas in the high-density limit, a result that included Bardeen's logarithmic term, as well as a rigorous method for determining the corrections to the correlation energy and specific heat.[274] Following a routine clearance procedure (for the work had been conducted under conditions of secrecy), Gell-Mann sent a preprint to Pines. The equivalence of the Gell-Mann–Brueckner and Bohm–Pines approaches was soon apparent to the workers in the field, and would be formally discussed by John Hubbard.

But there remained a mystery: plasma oscillations were not apparent in the Gell-Mann–Brueckner work. For this reason Brueckner, who in the fall of 1956 was at the University of Pennsylvania, and a Japanese postdoctoral student, K. Sawada, began to examine more carefully the work with Gell-Mann. Brueckner and Sawada soon found the oscillations in the formal structure of the theory.[275] Gell-Mann also resumed work on the problem with W. Karzas at RAND and exhibited the plasmon contributions using analyticity properties. Subsequently, a graduate student of Gell-Mann's at Caltech, Donald DuBois, took up the problem of computing the next set of corrections using Feynman diagrams. He developed a field-theoretical description of the plasmons, in which one could systematically examine corrections to the plasmon dispersion relation arising from interactions between particles. At the same time, in Paris, Philippe Nozières together with Pines showed how the Gell-Mann–Brueckner results for the ground-state energy of the electron gas could be derived from the Bohm–Pines theory. They also formulated the problem in terms of the frequency- and wave-number–dependent dielectric constant of the system. Nozières had originally come from Paris as a student to work with Pines at Princeton, to which Pines returned as an assistant professor in the spring of 1955.[276]

In two papers in 1957, Hubbard derived a theory equivalent to that of Bohm–Pines, using a many-body diagrammatic technique based on an infinite perturbation series expansion along with the linked-cluster technique of Brueckner, Bethe, and Goldstone. By this means, Hubbard was able to demonstrate the relationship between the Gell-Mann–Brueckner and Bohm–Pines theories, showing that in the long-range limit they give identical results for the ground-state energy, with both giving rise to plasmons as well as single-particle modes.[277] Hubbard also showed how the random-phase approximation breaks down for short-range correlations, and presented approximate calculations of the correlation energy of the electron gas at lower densities. By this time, the formalism for treating the electron gas was sufficiently well developed that one could confidently hope to understand the role of electron–electron interactions in more complex problems, including superconductivity.

It was in 1957 that Landau put forth his fundamental theory of interacting "Fermi liquids," showing that the elementary excitations of a low-temperature system of fermions would be interacting quasiparticles. (We treat the early history of the theory in the "Helium," section, since historically Landau developed it for application in that context.) The theory also provided the fundamental basis for

Rudolf Peierls and Joaquim Luttinger. (AIP Niels Bohr Library, Rudolf Peierls Collection)

understanding the one-electron picture of metals, developed in the 1920s and 1930s, and it led to considerable fruitful work in the late 1950s and 1960s by a new generation of solid-state theorists in applying field-theory methods to the electron gas problem.

Among the achievements in the application to electrons in normal metals was justification from first principles of the important notion of the Fermi surface (Chapter 3) by Walter Kohn and Joaquim Luttinger, and by Arkady Migdal, and the theorem by Luttinger that the volume of the Fermi surface is unchanged by electron interactions. Expanding their theoretical arsenal with many-body techniques in the early 1960s, Luttinger and Nozières were able to derive Landau's Fermi liquid theory from many-body perturbation theory. Another milestone of the field-theory approach was the solution, by Migdal, of the problem of electron–phonon interactions, even when they are not weak, through a careful analysis of the relative sizes of terms in the perturbation expansion.[278]

Three factors were crucial to the enormous success of the many-body field-theoreticians: they had quantum mechanics and field theory "in their blood," as Luttinger put it; they tended to work in close contact with experimentalists; and experiments themselves became in this period far more precise and comprehensive, providing theorists with clean results on which to test their work. Luttinger recalls that earlier,

> most of the people who had worked in solid state physics . . . tended to be close to phenomenology. People like Walter [Kohn] and I and Phil Anderson . . . a whole generation . . . came up after the war and knew quantum mechanics. . . . We had seen it in a million contexts and were able to really think in those terms.

Significantly, Luttinger, Kohn, and Anderson all worked during the 1950s at Bell Laboratories. After the discovery of the transistor in 1947 (Chapter 5), Bell invested heavily in solid-state research, not only supporting in-house experiment and theory, but also inviting outside physicists to work or consult at the laboratory, particularly during summers. Such visitors interacted vigorously with Bell's in-house theoretical staff (which in the 1950s included Charles Kittel, Melvyn Lax, John Bardeen, William Shockley, Gregory Wannier, Conyers Herring, and Philip Anderson) and experimentalists (such as Berndt Matthias and George Feher). Luttinger, who gave lectures to Bell experimentalists, remarked,

> Bell was special because there were so many people, and there were such good experiments going on. You could walk down the hall and talk to anybody and find out what they were doing. There was a high concentration of people who were specialists in various aspects. . . . If anything came up . . . there was somebody who knew about it or knew where to look.[279]

The solid-state field-theory community was international in character, with strong contingents in the United States, the Soviet Union (including Migdal, A. A. Abrikosov, L. P. Gor'kov, I. E. Dzyaloshinskii, S. Beliaev, and G. M. Eliashberg, surrounding Landau in Moscow, and Bogoliubov's group), Japan (e.g., Takeo Matsubara and Ryogo Kubo), and France (e.g., Roger Balian, Jacques Friedel, and Cyrano de Dominicis).

The application of field theory to solid-state physics represents a turning point in the electron theory of metals, at which the problem joined with others in diverse areas of solids, such as magnetism and Fermi surface studies, and other areas of physics, including low-temperature physics, critical phenomena, nuclear physics, and quantum electrodynamics. While the full history of the field-theory approach to collective phenomena, including quantum transport theory, is beyond the scope of this chapter, we touch on major developments in subsequent sections on superconductivity, helium, and critical phenomena.

Superconductivity

After a relatively slow and steady development following the war, in which useful, as well as blind, alleys were explored and considerable information gathered,[280] a burst of progress in superconductivity began about 1950. This activity brought confidence in the Londons' ideas of long-range order, insight into the mechanism behind superconductivity, and new theoretical techniques for dealing with the many-body problem of superconductivity. Several new ideas entered the discussion, including the coherence length, the order parameter, the role of the electron–phonon interaction, and development of the many-body formalism. Much work focused on determining and understanding the energy gap, which had been suggested in the 1930s by both Fritz London and Welker. Then, over a remarkable period of a year and a half, starting in the fall of 1955, Bardeen, Leon Cooper, and J. Robert Schrieffer wove all the threads together to create the BCS theory, solving the problem that had stymied physicists for almost half a century.

Major attempts to develop microscopic theories of superconductivity during the first five postwar years were made by Heisenberg and his student H. Koppe at the

Max Planck Institute in Göttingen, and by Max Born and Kai Chia Cheng at the University of Edinburgh. Heisenberg's work on superconductivity during 1947 and 1948 was based on the idea that at sufficiently low temperatures, the long-wavelength part of the Coulomb interaction causes localized asymmetric fluctuations in the electron density, fluctuations that thus carry current. The direction of the currents is taken to vary from domain to domain in the metal, reminiscent of Landau's 1933 attempt at the problem. Heisenberg described "the condensed phase [in position space] as a lattice of . . . wavepackets," which indirectly produces an energy gap.[281] Like Kronig over a decade earlier (Chapter 2), Heisenberg attributed the rearrangement of the electron distribution to the Coulomb force. Curiously, he persisted in regarding perfect conductivity, rather than diamagnetism, as the primary feature of superconductivity. Koppe developed further certain aspects of Heisenberg's theory, including the specific heat and the size of the energy gap. However, the theory was short on new predictions and was not thermodynamically consistent. As Shoenberg remarked in 1952, "Perhaps the main value of the theory will prove to have been the revival, which it has undoubtedly stimulated, of active discussion of the problem."[282]

Born and Cheng developed a similar spontaneous current theory in the spring of 1948, arguing on the basis of the Brillouin zone structure that in certain cases Coulomb interactions cause the ground-state electron distribution to be asymmetric and to carry a current.[283] They were thus able to offer an explanation for why certain columns of the periodic table are more favorable than others for superconductivity. Like Heisenberg's theory, Born and Cheng's violated Bloch's theorem that the state of lowest free energy should carry no current.

More productive than attempts to develop a theory of superconductivity on first principles were phenomenological approaches anchored in experiment. Between 1950 and 1953, five crucial developments pointed toward the present understanding of superconductivity: (1) verification of the Londons' ideas of rigidity and long-range order, through microwave studies of the surface impedance and penetration depths of superconductors; (2) development of a phenomenological theory by Vitaly Ginzburg and Landau that could explain the surface energy at the boundary between normal and superconducting states; (3) establishment of the existence of the energy gap suggested in 1935 by Fritz London and further developed by Welker before the war; (4) identification of the basic interaction underlying superconductivity, through the discovery of the isotope effect; and (5) development of proper theoretical tools for dealing with the many-body problem.

The London theory predicted the depth to which magnetic fields penetrate into superconductors, and it was therefore of great interest to measure this penetration depth precisely. The first determinations were based on extrapolations, using thermodynamic arguments, from measurements on thin films. For example, in 1938 von Laue made such an estimate from measurements of the critical field of thin lead wires, measurements made in 1937 by Rex Pontius, a research student from Idaho working at Oxford.[284] At the same time, Shoenberg determined the penetration depth of superconductors by carrying out a series of measurements of the magnetic susceptibilities of very small samples at different temperatures, inferring the penetration depth from the size dependence of the susceptibility.[285]

In 1940, Heinz London at Bristol carried out pioneering determinations of the penetration depth by the direct method of measuring the surface resistance of superconductors at finite frequency by calorimetric techniques.[286] Examining superconducting tin at 1460 Mc/sec (now commonly designated MHz; we use Mc/sec as used in the period), using a method that extended earlier work by McLennan and co-workers using electromagnetic waves,[287] he discovered the "anomalous skin effect": the normal state resistance of metals such as tin is higher at microwave frequencies than expected theoretically—that is, when the electron mean free path is longer than the skin depth. Furthermore, London found that for superconductors the surface resistance at microwave frequencies does not suddenly jump to zero at the critical temperature but falls continuously as the temperature is decreased. Attributing this result to energy lost by normal electrons in the superconducting penetration depth, he concluded that this "anomalous resistance" at high frequencies and low temperatures is probably due to the electron mean free path being "considerably larger than the penetration depth." This was an experimental indication of the long-range order that he and his brother, Fritz, had suggested some five years earlier. However, Heinz London dropped these studies when the war began, and full exploitation of the effect would await the introduction of wartime microwave techniques.

The problem of explaining the penetration depth and anomalous drop of resistance below the transition temperature at high frequencies was taken up in the fall of 1945 by A. Brian Pippard, who joined the Mond Laboratory in Cambridge as a graduate student under Shoenberg. Pippard had spent the previous four years in microwave development at the British Royal Radar Establishment at Malvern. "The war," he reflected recently, "turned me into a physicist rather than a chemist . . . it taught me about microwave techniques, waveguides and aerials and I came back from the Army with first class techniques at my fingertips."[288] Shoenberg suggested that Pippard use this experience to reexamine London's work on the response of superconductors to microwave radiation. Since superconductors did not dissipate energy at zero frequency and dissipated normally at optical frequencies, studying an intermediate microwave range of frequencies might offer a clue to the mechanism of superconductivity. In 1946 and 1947, Pippard studied the surface impedance of normal and superconducting tin at 1200 Mc/sec, and in 1948 went up to 9400 Mc/sec.[289]

Parallel studies were made in the United States by Emanuel Maxwell, Paul Marcus, and John Slater at MIT and by William Fairbank at Yale—all physicists who, like Pippard, had worked on radar during the war. Slater, then chairman of the MIT Physics Department, through his war work on the magnetron had sensed the possible application of microwave techniques and was very eager to bring these to bear on solid-state physcis.[290] He also wanted to use another wartime spin-off, the Collins cryostat, a device

much simpler and more compact than any previously developed cryostat. We became very interested in this development, which promised to revolutionize low temperature research, a field in which before the war very few institutions had taken part on account of the very expensive and massive equipment which was then necessary, and yet a field of very great interest in the study of the solid state.[291]

This compact helium liquefier, completed in 1946 by Samuel Collins of the MIT Mechanical Engineering Department, drew on technology he had developed from 1940 onward, under the auspices of the National Defense Research Committee, for portable oxygen generators for submarines and airplanes.[292] The Collins cryostat would enable experiments to be conveniently carried out over long periods at temperatures as low as 2° above absolute zero. It was a principal reason why low-temperature physics was conducted on a large scale after the war.

Together with Maxwell, a graduate student during the war at the MIT Rad Lab, and Marcus, a theoretician who also worked there, Slater began studying the resistance of superconducting tin and lead at the very high frequency of 24,000 Mc/sec.[293] Fairbank, a Yale graduate student who had also been at the MIT Rad Lab during the war, took up the same problem for his dissertation; he determined the surface resistance of superconducting tin over a wide temperature range, like Pippard at 9400 Mc/sec.[294] But Pippard, in British string-and-sealing-wax tradition, using only waveguides, transmission lines, a crystal rectifier, and a galvanometer, arrived at results faster than either of these American efforts. The Americans, he later remarked, lost up to a year setting up elaborate electronic instrumentation.[295] All these studies, and their continuation over the next several years, would help to verify the Londons' ideas about the rigidity of the superconducting wave function and macroscopic long-range order.

In a 1950 series of microwave measurements of the surface resistance of impure, or "dirty," superconductors, Pippard found that the penetration depth, for magnetic fields between zero and the critical value, changes by less than 3%, a result indicating wave function rigidity.[296] Furthermore, an extensive series of penetration depth measurements in dilute alloys showed that the range over which effects were transmitted coherently was about 10^{-4} cm. This was long, comparable with the penetration depth in the London theory, and, suggestively, was consistent with the range of the ordering in a semiphenomenological theory of superconductivity just proposed by Ginzburg and Landau, which we will discuss shortly.[297] Pippard recalls that "it was from that that the coherence idea came into my mind."[298] As he described the idea, the sudden nature of the transition to superconductivity and the fact that the transition is second-order in idealized form "suggest that the co-operation of [a] very large number of electrons is involved, in such a way as to reduce enormously those local fluctuations (e.g., persistence of local order above the transition temperature) which usually play an important role in the neighborhood of a lambda point."[299]

Pippard's microwave studies of superconducting and normal metals brought him to develop an important nonlocal reformulation of the London relation between the current and magnetic vector potential in a superconductor, which explained the increase of the penetration depth with increasing impurity of the sample. He had recently worked on the anomalous skin-effect problem, whose solution lay in rewriting Ohm's law to say that the current at a given point depended on the electric field over a finite region whose size was determined by the electron mean free path about the point.[300] Now Pippard proposed a similar nonlocal relation in a superconductor: that the current at a given point depended on the vector potential within a distance, the "coherence length," around the point. A visit in 1949 to Heisenberg in Göttingen gave Pippard the "incentive and the formalism which was

Alfred Pippard, Vitaly Ginzburg, I. M. Lifshitz, A. H. Cook, and Peter Kapitza in Moscow, around 1955. (AIP Niels Bohr Library)

needed" to take this step. Pippard was troubled by Heisenberg's use of thermody-namics in his theory of superconductivity, and recalls telling Heisenberg that "he couldn't get away with thermodynamics like that, because what he'd managed to prove was not that superconductors obeyed the London equation but that all metals were superconductors!" Heisenberg then modified his theory, which Pippard used to derive "a very nice little integral equation for the supercurrent which had the penetration depth increasing as impurity was added.... I didn't for a moment believe that Heisenberg's explanation was right but it was a very good model to lead one to the equation which made sense."[301]

The microwave work led to greater understanding of two of the building blocks of the modern theory of superconductivity: the Londons' ideas of long-range order, and rigidity of the superconducting wave function. A further important step in the phenomenological understanding of superconductivity was the Ginzburg–Landau theory of 1950, which fused the Londons' electrodynamics with Landau's earlier theory of phase transitions. Ginzburg, in the Soviet Union, was at this time inter-ested in understanding the surface energy at the boundary between normal and superconducting phases. Having completed a study on superfluidity, he began to work on this problem during the war. Unlike in America and Great Britain, where most physicists were mobilized by the government to work on war research, in the Soviet Union it was not recognized clearly that physicists could be helpful in the war effort. Therefore, many Soviet physicists simply continued their own research, although sometimes the emphasis shifted to applied issues.[302]

In 1944, in his first scientific paper, Ginzburg postulated a quasiparticle spectrum and, on the basis of the Londons' phenomenological theory, discussed the relation between the experimental penetration depth and the surface energy of thin films.[303] The failure of this approach to lead to a consistent understanding of the surface energy played, in Ginzburg's words, "a role of catalyst" for the 1950 theory. About 1947, drawing heavily on the framework of Landau's earlier theory of second-order phase transitions, Ginzburg worked particularly on achieving a positive surface energy in the calculation. Through frequent consultations with Ginzburg on this exciting problem, Landau joined in the work as a collaborator.[304]

The issue, as Ginzburg and Landau noted in their paper, was that "the present phenomenological theory of superconductivity . . . does not permit the determination of the surface energy at the boundary between the normal phase and the superconducting phase and does not give the possibility of correctly describing the destruction of superconductivity by magnetic fields and currents." To generalize the London theory, they drew on Landau's theory of second-order phase transitions to introduce a complex order parameter ψ of the superconducting state.[305] The absolute value of ψ^2 was interpreted as the density of superconducting electrons. However, it would not be until the development of the BCS theory that the detailed meaning of the superconducting "wave function" in the Ginzburg–Landau theory would become clear. The value of ψ in the superconducting state was determined by minimizing the free energy; ψ vanished in the normal state. A critical feature introduced here was the possibility of spatial variation of the order parameter;[306] the superconducting electron wave function in the Ginzburg–Landau theory was not perfectly rigid (as was its equivalent in the London theory). Generally, ψ was determined by a Schrödinger-like equation with terms nonlinear in ψ. Magnetic fields were included, as in the Schrödinger equation, in a gauge invariant fashion. With their generalization, Ginzburg and Landau were able to treat the boundary between normal and superconducting regions, as may be found in the intermediate state. The first experimental verification of the predictions of the Ginzburg–Landau equations of the temperature and thickness dependence of the critical fields and currents of thin films came from the experiments of N. V. Zavaritski.[307]

One day Zavaritski altered his technique of preparing samples, causing the atoms to form an amorphous structure for which the Ginzburg–Landau relationship between critical field and thickness appeared not to hold. This discovery led Alexei Abrikosov, a young theoretician in Landau's group, to propose in 1952 the existence of a new class of "superconductors of the second group," or type II superconductors. These were characterized by a *negative* surface energy, and consequently had quite different behavior in magnetic fields.[308] (Ordinary, or type I, superconductors have a positive surface energy.) In his paper, Abrikosov used the Ginzburg–Landau theory to study the onset of superconductivity with decreasing applied magnetic field in these systems, a regime where nonlinear effects could be neglected.

Abrikosov went on, between 1952 and 1957, to develop the theory of type II superconductivity for general magnetic field strength.[309] Extending the solution to below the critical field, he found that the system was described by a periodic Ginzburg–Landau wave function and magnetic field—the "mixed state." He recalls that when he told his theory to Landau in 1953, Landau "did not agree' with the solution because the magnetic field did not obey the London field equation with zero current in the bulk of the system. Neither appears to have recognized at the time that the solution described quantized vortex lines, and indeed Landau as late as 1955 published a paper with Lifshitz attempting to explain the properties of rotating liquid helium in terms of a laminar intermediate state structure. The theory was resurrected only after Feynman's article on vortices in liquid helium arrived in Moscow in 1955, when it became clear that Abrikosov's solution contained quantized vortices.[310]

Up to the time of Abrikosov's theory, all theories of superconductivity were designed for pure materials, which with the exception of niobium were type I. Yet experimental superconductors were, as a consequence of impurities, quite generally type II. Type II superconductors have practical importance because they have high critical fields, and superconducting alloys can therefore be used to make very strong permanent magnets.[311] One of the theory's initial successes was the interpretation of certain of Shubnikov's experiments carried out before the war.[312] Initially, however, this important theory was overshadowed by the appearance of the BCS theory. That began to change in 1961 when Bruce Goodman in Cambridge rediscovered that alloys with high critical fields have negative surface energy; in 1962 he calculated their properties using the Abrikosov theory and noted that the experimental fit to the theory was good.[313] Goodman's article brought recognition to the Abrikosov theory.

The Ginzburg–Landau theory was not received well by the Pippard–Shoenberg group. They felt that the theory was "highly speculative and introduced quite arbitrarily a wave function whose meaning they did not know." More generally, Pippard comments, the Ginzburg–Landau theory had been "very much more to the taste of the American physicists than to us much more empirical practical experimentalists, and I think we've always had more interest in genuinely dirty, inhomogeneous, mucky superconductors than in homogeneous pure things, which are a theorist's delight."[314]

At the same time as the work of Pippard and of Ginzburg and Landau, the existence of the energy gap in the electron energy spectrum was established through various experiments. An early example came in 1945, when Daunt and Mendelssohn invoked such a gap to explain their experimental finding that supercurrents have no Thompson heat and therefore transport no entropy.[315] As in semiconductors, a gap in the energy spectrum of superconductors implies that the number of quasiparticles at finite temperature should vary as $\exp(-W/kT)$, where W is the gap energy. Goodman found the first clear evidence of this dependence in 1953 in his measurements of the thermal conductivity of tin.[316] He also showed that the Heisenberg–Koppe theory could be interpreted in terms of a temperature-dependent energy gap of order kT_c in the spectrum of electron energy levels.[317]

Experimental evidence for the gap in superconductors continued to mount—for example, in measurements of the temperature dependence of the electronic specific heat. The earliest measurements, by Keesom, Kok, and others in Leiden, agreed with the T^3 law predicted by the Gorter–Casimir theory. But later measurements, starting with those in 1953 by A. Brown, M. Zemansky, and H. Boorse on niobium, showed an exponential behavior as T approaches zero, giving evidence for a gap.[318] W. Corak and co-workers at Westinghouse found a similar exponential decrease in vanadium in 1954, and in tin in 1956.[319] Measurements of ultrasonic attenuation, nuclear spin relaxation, and, later, tunneling gave further information about the gap. The first definitive measurements of the $T = 0$ gap were made in 1956 and 1957 by M. Tinkham and his student R. E. Glover, using transmission and reflection of electromagnetic absorption in the far-infrared region.[320]

Meanwhile, another major stream was developing in the study of superconductivity. The atomic weapons program in both Britain and the United States made large numbers of stable isotopes of elements available for research. Such isotopes of mercury, with mass numbers between 198 and 202, were used by Maxwell, then at the National Bureau of Standards, and by Bernard Serin and collaborators at Rutgers, in a series of decisive experiments in the winter of 1950. As we will see, these experiments demonstrated an unexpected but fundamental feature of superconductivity: the interaction responsible for superconductivity arises from coupling between the conduction electrons and the lattice.[321] Although Pauli had casually suggested this possibility several years earlier, the idea was not taken seriously by theoreticians.[322] As Maxwell remarked recently, it was then felt that superconductivity was "something in which the electrons suddenly become free of the lattice."[323]

When Maxwell arrived at the Bureau of Standards in June 1948, he learned that the spectroscopy department was using as a standard mercury isotopes obtained from neutron bombardment of gold at Los Alamos. Hearing of Karl Hertzfeld's suggestion that one search for a relation between superconductivity and the lattice using uranium isotopes, Maxwell thought of using the mercury isotopes on hand to examine whether there was any relation between the transition temperatures and the isotopic mass of atoms in the lattice.[324] Aware that two earlier searches for this mass effect, in 1922 and 1941, using natural lead isotopes from radioactive decay, had failed, the experiment was, he recalls, a "shot in the dark."

Maxwell initially planned a resistance measurement with the rare isotope ^{198}Hg, but he eventually changed to a simple induction technique, in which the sample was connected in series with a rheostat placed inside a pickup coil, and then connected to a ballistic galvanometer with a shunt. Starting at a given temperature, he slowly raised the field until he saw a kick in the galvanometer. He found to his surprise that the "lighter the mass [of isotopes making the lattice] the higher the transition temperature."[325]

Meanwhile, parallel work was under way in the Rutgers group under Serin, who had joined the faculty in 1947. Serin conceived his version of the experiment in discussion with Michael Garfunkel, his senior graduate student, and Charles Reynolds, a young faculty member. Trying to think of tests that would reveal factors on which superconductivity depends, they looked at a 1947 article by Jules de Launay and Richard Dolocek of the Naval Research Laboratory, in which the transition temperatures of the 17 known superconductors were plotted as a function of their Debye temperatures.[326] Garfunkel recalls that Serin, noticing that the curve showed a very rapid change of superconducting temperature near the element mercury, suggested an experiment focused on that element. Appropriate isotope samples, separated by a mass spectrometer, were ordered from Oak Ridge.[327] Later Lloyd Nesbitt joined the experiment as a technician, and Wilbur Wright as a graduate student. Carrying out the first experiment soon after the isotopes arrived, they observed the transition using a slight alternating current "tickling field" superimposed on the direct current field, a technique made possible by wartime advances in electronics. Like Maxwell, they were surprised to find "a systematic decrease of transition temperature with increasing mass." Henry Torrey recalls "Serin coming to me when he got the results. . . . He was very excited, because they seemed to show the proportional relation to the mass. This immediately suggested to Serin the relationship to the electron–phonon interaction."[328] Further confirmations came with later experiments extended to low temperatures and to other materials including tin and thallium.

The experiments were announced for the first time at a meeting held at the Georgia Institute of Technology in Atlanta on 20 and 21 March 1950, sponsored by the Office of Naval Research, the major United States funding agency for solid-state physics just after the war. Although there had been no previous communication between Serin and Maxwell about their isotope work, they agreed at this meeting to submit simultaneous letters on their results to the *Physical Review;* the letters were submitted on 24 March and appeared on 15 May. Not long afterward, Mendelssohn's group in Oxford and Shoenberg and Pippard's group in Cambridge started work on the isotope effect in tin. Confirming the results of Serin and Maxwell, they demonstrated a square root dependence on mass with high precision. The two groups published their results together.[329]

While Maxwell and Serin were carrying out their experiments, Herbert Fröhlich, spending the spring 1950 semester at Purdue University on leave from Liverpool, was studying the electron–phonon interaction, beginning with electrons in polar crystals.[330] Fröhlich realized that the field-theory approach used in this problem might also be the key to dealing with superconductivity, since it described processes in which lattice displacements follow changes in the electron density and react back on the electrons. In a paper submitted to the *Physical Review,* received the day after the Maxwell–Serin results were published, he set forth a theory of the superconducting ground state.[331]

Arguing for the importance of the "electron–phonon" in producing superconductivity, Fröhlich pointed out that it is not "accidental that very good conductors do not become superconductors, for the required relatively strong interaction between electrons and lattice vibrations gives rise to large normal resistivity." He started by deriving, using second-order perturbation theory, that the electrons, as a consequence of their coupling to the lattice vibrations, experience an effective interaction that, under certain circumstances, is attractive. He then argued that the attractive interaction can make the usual electron Fermi distribution in momentum space unstable close to the Fermi surface; rather, it yields "a new distribution which—subject to later confirmation—will be identified with the superconducting state." This turned out to be unsuccessful as a theory of superconductivity, yet Fröhlich's paper would have a major impact on future developments, by formulating the problem in terms of the effects of the attractive phonon-induced interaction–the "Fröhlich Hamiltonian."

Fröhlich's calculation implicitly contained the isotope effect, as he found that the energy difference between the normal and superconducting states (expressed in terms of the square of the critical magnetic field) should be proportional to $1/M$, where M is the ionic mass. The relation arose in his work because in heavier materials the lattice vibrations are slower and therefore interact less strongly with the electrons. He did not make here an explicit connection with the effect on the critical temperature. Fröhlich learned of the experimental isotope shift results a day or two after they appeared in the *Physical Review,* while spending several days at Princeton. "There, at my breakfast table, [I] found the two letters reporting the isotope effect. On checking I found my M-dependence confirmed and on 19 May 1950 sent a letter[332] to claim confirmation of the basic idea, electron-phonon interaction."[333] In his letter he showed that his calculation of the energy leads directly to the $M^{-1/2}$ dependence of the critical temperature. He also added a note in proof to his *Physical Review* paper on the isotope effect.

Bardeen, at Bell Laboratories, heard of the experimental results on 15 May through a telephone call from Serin. He wrote a note to himself on the following day: "These results indicate that electron-lattice interactions are important in determining superconductivity.... It is important to include their effect on the free energy of the electrons."[334] He immediately revived work that he had done on superconductivity before the war at the University of Minnesota.[335] In this uncompleted work, he had attempted to account for the energy gap in superconductors in terms of a one-electron model in which a small static lattice distortion introduced gaps at the Fermi surface and lowered the energy of electrons just below the Fermi surface. This work was interrupted by the war, and in the first postwar years Bardeen was immersed in semiconductor research leading to the invention of the transistor (Chapter 7).

Bearing the Serin and Maxwell results in mind, Bardeen drafted in the next few days a new theory of superconductivity. Here the lowering of the energy of superconductors came not from a periodic lattice distortion, as in his earlier theory, but from "the zero-point fluctuations of the lattice which lower the electron energies."[336] The letter was completed and sent to the *Physical Review* on 22 May.[337] Whereas Fröhlich had calculated the interaction energy by second-order perturbation theory, Bardeen used a variational method to show that the zero-point energy of superconducting electrons is lowered as a result of the electron–phonon interaction. He worked further on the theory during the subsequent weeks, attempting to determine the superconducting wave function, the energy of the quasibound states and other relevant quantities such as the effective mass of the electrons.[338]

Approximately a week after Bardeen submitted his letter on superconductivity, Fröhlich visited Bell Laboratories, giving the two an opportunity to converse about their new theories of superconductivity. Bardeen later recalled, "Although there were mathematical difficulties in both his method and mine, we were convinced that at last we were on the road to an explanation of superconductivity."[339] Part of the difficulties lay in the fact that simple perturbation theory was not valid in this situation. As was understood only much later, the two theories explained the isotope effect but failed to explain superconductivity because they focused on the single-electron self-energies, rather than on the full effects of the interaction between electrons. Furthermore, these theories did not show that the attractive interaction between the electrons resulting from the electron–phonon interaction actually dominates the repulsive screened Coulomb interaction.

In a letter to Peierls in mid-July 1951,[340] Bardeen reflected on the limitations of various methods for treating the electron–phonon interaction in superconductivity, including Fröhlich's method, his own, and others. All indicated "some sort of instability of the electron distribution when Fröhlich's criterion for superconductivity is satisfied," but the nature of the instability was not understood, and "it is not certain yet whether the best 'one-electron' picture will give a shell distribution or whether the electrons near the Fermi surface are simply lowered in energy by $\sim kT_c$ in the superconducting state." Bardeen added, presciently, "In either case, I believe that the explanation of the superconducting properties is to be found along the lines suggested by F. London: The wave functions for the electrons are not altered very much by a magnetic field."

At this point, however, with the establishment of the importance of the electron–phonon interaction, theorists began to mount a full-scale attack on the problem of superconductivity. In 1954, Fröhlich, who worked at reformulating his theory of superconductivity in terms of a renormalized field theory,[341] had a crucial insight into the mathematical structure of the theory. Using a self-consistent method for a one-dimensional model of the electron gas interacting with the lattice vibrations, he showed that for sufficiently strong electron–phonon coupling, the system developed an instability leading to an energy gap at the Fermi surface. He computed that the energy of electrons just below the surface is lowered by an amount proportional to $\exp(-1/g)$, where g is proportional to the electron–phonon coupling constant. We know now that this mathematical "essential singularity" in the interaction parameter is characteristic of the superconductivity problem, and since the exponential function cannot be expanded in powers of g, it explained why any approach based on a perturbation expansion in λ had to fail.[342]

This difficulty had in fact been alluded to earlier by M. R. Schafroth. In 1951, at Pauli's suggestion, Schafroth had studied the interaction of a superconductor with an electromagnetic field and showed that one cannot derive the Meissner effect in any order of perturbation theory starting with the Fröhlich Hamiltonian. This result did not nullify the basic idea that the electron–phonon interaction is the source of superconductivity, but it showed clearly the inapplicability of perturbation theory.[343]

In 1954, Schafroth suggested examining the correlations between pairs of electrons, and presented a model based on analogy with the hydrogen molecule.[344] In this model, a pair of electrons, above but very near the top of the Fermi surface and having opposite spin, in the presence of an attractive interaction could form resonant states of pairs, "pseudomolecules," obeying Bose–Einstein statistics. Superconductivity, like superfluidity in helium, was viewed as a Bose–Einstein condensation; in order to lose energy, electrons would have to enter energy levels below the Fermi surface, which was prohibited by the Pauli exclusion principle since such states are occupied. Schafroth later developed this picture more completely with the collaboration of S. T. Butler and John M. Blatt.[345] Although Schafroth correctly suspected that pairing of electrons would play a crucial role in the theory of superconductivity, the bosonic nature of his pairs was fundamentally different from the actual pairing that would emerge in 1957 in the BCS theory.[346]

After the discovery of the isotope effect, Bardeen devoted himself full-time to solving the problem of superconductivity. As he sketched out in his notes of 23 October 1951, the "problems in superconductivity" included

> (1) Derivation of London equations for multiply connected body. (2) Proof of current and effective mass theories for one-electron wave functions. (3) Analysis of

diamagnetic properties of gas of electrons with small effective mass. (4) Extension of 3 to include high frequencies and scattering of electrons (effect of electric field). (5) Boundary energy for thin films. (6) Better calculation of interaction energies. (7) Specific heat and other thermal properties.[347]

Bardeen learned about the Bohm-Pines theory during an extended visit to Princeton during the spring 1950 semester, when he was studying the role of the electron–phonon interaction in superconductivity. Suspecting that the approach was useful for dealing with the electron–phonon as well as the electron–electron interaction, he and Frederick Seitz in 1952 invited Pines, who was then at the University of Pennsylvania, to join the research staff of the University of Illinois at Urbana, with support from an Army Research Office grant to Bardeen. When Pines arrived at Illinois in July 1952, Bardeen suggested that Pines examine the polaron problem, recently studied by Fröhlich—the problem of the motion of an electron that is strongly coupled to the optical mode of lattice vibrations—in a polar crystal.[348]

Tsung-Dao Lee was also spending that summer visiting Illinois. Lee and Pines recognized that the intermediate coupling method of quantum field theory, on which Lee had been working, could be adapted to deal with polarons, and they developed a solution valid for the coupling region found in polar crystals. Subsequently, with Francis Low, they used a canonical transformation to arrive at an especially transparent formulation of the Lee–Pines solution, a formulation that would significantly influence Schrieffer in his invention of the microscopic pair wave function of BCS theory.[349]

Pines began work on a generalization of the Bohm–Pines theory to treat the combined influence on the motion of electrons in a metal of their interactions among themselves and with the phonons, and in 1954 derived the net effective electron–electron interaction. He and Bohm found that for interactions in which the energy transfer is small, the phonon-induced interaction between electrons studied by Fröhlich is stronger than the repulsive Coulomb interaction, so that for pairs of electrons close to the Fermi surface (within a Debye energy), the net electron–electron interaction is attractive. The Bardeen–Pines effective interaction served as the jumping-off point for subsequent work at Illinois on the microscopic theory of superconductivity.[350]

By 1953 the field of low-temperature physics seemed ripe for systematic review, and the editor of the *Handbuch der Physik,* S. Flügge, arranged for two major reviews on superconductivity for the 1956 volume: one by Serin on the experimental situation,[351] and the other by Bardeen on theory.[352] In his review, written in 1954, Bardeen emphasized the concepts that he felt would be crucial to solving the microscopic problem. He particularly stressed the energy gap, which could not then be derived from a microscopic theory. He showed how to develop from this concept the electrodynamic properties and a generalization of the London equations similar to Pippard's nonlocal superconductor electrodynamics.[353] Bardeen argued for London's approach based on the notion that all supercurrents are in origin diamagnetic. He also emphasized the following points: Fritz London's notion of superconductivity as an "ordered phase in which quantum effects extend over large distances in space . . . of the order of 10^{-4} cm in pure metals"; the second-order phase transition between the normal and superconducting states, with the "superconductor probably characterized by some sort of order parameter which goes to zero at the transition point" (admitting, however, that "we do not have any understanding at all of what the order parameter represents in physical terms"); the pivotal role of the electron–phonon interaction; and the importance of considering the electrons as screened in treating the combined electron–electron and electron–phonon systems. Bardeen commented on the promise of using recently developed field-theory techniques to break the impasse in developing the microscopic theory of superconductivity. He noted, for example, Tomonaga's strong-coupling approach to the interacting one-dimensional Fermi gas (the Lee, Low, and Pines intermediate coupling polaron solution) and the methodology developed by Bohm and Pines in the collective approach to the quantum plasma. In conclusion, he wrote: "A framework for an adequate theory of superconductivity exists, but the problem is an exceedingly difficult one. Some radically new ideas are required, particularly to get a really good physical picture of the superconducting state and the nature of the order parameter, if one exists."

To Bardeen, "it was becoming clear that field theory might be useful in solving the many-body problem of a Fermi gas with attractive interactions between the particles."[354] He recognized that Fröhlich's application of the field-theory techniques of canonical transformations and second quantization was only the beginning, and that the complete machinery of the new quantum electrodynamics as used in particle physics had not yet been brought to bear on the problem. So in the spring of 1955 he telephoned C. N. Yang at the Institute for Advanced Study at Princeton, asking him to send to Urbana someone "versed in field theory who might be willing to work on superconductivity." Yang recommended Leon Cooper, who was then spending a year at Princeton as a postdoctoral fellow, having recently taken his Ph.D. at Columbia in nuclear theory. Cooper joined the Illinois department in September 1955, not long after Pines left to go to Princeton.[355]

The final member of the team that would develop the microscopic theory of superconductivity was J. Robert Schrieffer, a graduate student at Illinois. Schrieffer had decided to go to graduate school at Urbana, having "heard outstanding things about Professor Bardeen." Soon after arriving there, he began to work with Bardeen, and when it came time, he chose superconductivity for his dissertation, as it "looked like the most exciting thing." Schrieffer was a second-year graduate student when, during 1954 to 1955, he learned about superconductivity by proofreading Bardeen's *Handbuch* article.[356]

After Cooper's arrival in Urbana, Bardeen asked him to give a series of informal seminars on field theory. His lectures included discussion of Feynman diagrams applied to virtual excitations of a Fermi gas. But Cooper was doubtful that these perturbative methods would be useful in dealing with the superconducting ground state.[357]

Bardeen orchestrated the research effort and played an important role in educating, guiding, and communicating both his taste in physics and his approach to problems—to break up larger ones into small soluble pieces and "use the smallest weapon in your arsenal to kill a monster." A lack of space in the Urbana Physics Department increased intellectual contact; Bardeen and Cooper shared an office (as Bardeen and Pines had done from 1952 to 1955), and, as Schrieffer recalls, they "could wheel around their chairs and join in the discussion when he came to speak with either one." As a graduate student, Schrieffer himself was based elsewhere in a large room shared by graduate students, affectionately called the Institute for Retarded Study.[358]

Bardeen kept emphasizing certain concepts of Fritz London's program for developing a microscopic theory of superconductivity, outlined in his 1950 book.[359] In particular, as London put it: (1) "in an isolated, simply connected superconductor and for a given applied magnetic field, there is just one stable current distribution." (2) In thermal equilibrium, there is no "permanent current in an isolated superconductor (in agreement with Bloch's theorem) except in the presence of an applied magnetic field, and there is no conservation of these currents; they differ for every variation of the strength or direction of the applied field." (3) There is a "long-range order of the (average) momentum . . . due to the wide extension in space of the wave functions representing the same momentum distribution throughout the whole metal in the presence as well as the absence of a magnetic field." Bardeen discussed the long-range order in terms of "a phase coherence," the coherence length, or typical size of the correlations, of the order of a micron (as Pippard had suggested).[360] (4) There is a rigidity of the wave function with respect to magnetic perturbations caused by the long-range order of the wave function, making the superconducting state "a quantum structure on a macroscopic scale associated with . . . a kind of solidification or condensation of the average momentum distribution." This rigidity causes the energy gap in the electronic excitation spectrum.

Another principle Bardeen worked with was that the superconducting states should be in a one-to-one correspondence with the normal ones, in the sense that their wave functions should be expressible as a linear combination of low-lying normal configurations defined by quasiparticle and phonon-state occupation numbers. Thus one could concentrate on the small energy differences between normal and superconducting states.[361]

Bardeen encouraged the group to explore the interaction he and Pines had developed as a generalization of Fröhlich's work, being quite certain that it was responsible for superconductivity.[362] During 1955 to 1956, Schrieffer examined Brueckner's "t-matrix" methods for nuclei—a nucleus, too, was a system of physically close fermion particles—which Bardeen felt might be applicable.[363] Cooper began to consider, in the context of general theories of quantum mechanics, why there is an energy gap. In one approach, he studied the class of matrices and determinants having the property that the lowest eigenvalue is split from the rest by a finite amount independent of the size of the matrix.[364]

In September 1956, Cooper made an important advance through his detailed study of the Bardeen–Pines interaction. At Bardeen's suggestion, Cooper studied whether the interaction could give rise to an energy gap by examining a simplified model of just two electrons outside the Fermi surface, which interact via an attractive interaction provided their energies lie within a Debye energy range of each other. The remaining electrons were assumed to occupy all the states up to the Fermi energy, thus making those states unavailable to the two electrons being studied. He found that if the net force is attractive, no matter how weakly, the two electrons form a "bound" state, separated by an energy gap (essentially singular in the interaction strength) below the continuum states. It was immediately clear that this result—the "Cooper pair"—was an important step toward explaining superconductivity.[365]

Cooper's theory suggested that if one formed the superconducting ground state out of normal excitations, the matrix elements of the attractive interaction would contribute negatively to the energy (i.e., would lower the energy below that of the normal state) if the electrons were associated in pairs.[366] The relevant part of the Hamiltonian appeared from Cooper's work to be the part that coupled together pairs; now the outstanding problem was how to write down a many-body wave function taking the pairing into account.

A first problem was that if all the electrons near the Fermi surface in a superconductor were paired in the way Cooper described, the pairs would strongly overlap—the problem was more difficult than that of a collection of "molecular" pairs. Furthermore, as Bardeen had already recognized in his *Handbuch* article, the correlations of the superconducting state were extremely subtle. The energy change in the transition from the normal to the superconducting phase is of the order of 10^{-8} eV per electron, far smaller than the accuracy with which one could hope to calculate the energy of either the normal or the superconducting state.[367] The group was rightly concerned that in working with only a piece of the Hamiltonian, the "pairing Hamiltonian," they might have left out a part so important that the whole problem would be treated incorrectly.[368] By the fall of 1956, Schrieffer, already a fourth-year graduate student, was so discouraged that he told Bardeen he might want to change his dissertation topic. In December, just before leaving for Stockholm to receive the 1956 Nobel Prize for the discovery of the transistor, Bardeen suggested that Schrieffer give the problem another month or so.[369]

Indeed, during a meeting on the many-body problem, held on 28 and 29 January 1957 at the Stevens Institute of Technology in Hoboken, New Jersey, Schrieffer took the important step of writing down a variational form for the wave function of the ground state that took the pairing into account.[370] Drawing on the polaron wave function of Lee, Low, and Pines, he wrote down the many-body wave function, expressed as a product of creation operators acting on the vacuum; when expanded, this gave a series of terms with varying total number of pairs. Schrieffer was excited to find that he could readily carry out the calculation of the corresponding ground-state energy in terms of a variational "gap parameter" in the wave function. Soon after Bardeen returned to Urbana, Schrieffer and Cooper discussed this progress with him, and he immediately set out to compute the energy gap in the excitation spectrum. Within a few days, he had derived the gap and found it to be the same as the gap parameter in the ground-state energy.[371]

Now all members of the team began to work quickly. They suspected that Feynman—who had spoken in September on superfluidity and superconductivity at the International Theoretical Physics Conference in Seattle, and who was able to draw on complicated new field-theory methods that the team knew little about—was also hot on the trail.[372] Several days later, Bardeen calculated the condensation energy both in terms of the observed critical field (the Gorter–Casimir relation) and in terms of the energy gap, obtaining a relationship between these two experimentally determined quantities.[373] They now divided tasks: Schrieffer would work on ther-

John Bardeen (b. 1908), Leon N. Cooper (b. 1930), and John Robert Schrieffer (b. 1931). (AIP Niels Bohr Library)

modynamic properties; Cooper, on the Meissner effect and other electrodynamic properties; and Bardeen, on transport and nonequilibrium properties. At last, the pieces of the puzzle, which Bardeen had been patiently collecting since 1939, began to fit together.

After several weeks of unsuccessful attempts to derive the second-order phase transition at the critical temperature, Bardeen decided not to let this failure hold up publication of their fundamental result any longer. The historic letter on the BCS theory was sent to the *Physical Review* on 15 February.[374] Along with it went a covering note from Bardeen to the editor, Samuel Goudsmit:

> I know that you object to letters, but we feel that this work represents a major breakthrough in the theory of superconductivity and this warrants special handling. We believe, further, that the essential elements of the theory are adequately described in the letter. It will probably be some time before we write a complete paper on the subject.[375]

Their paper explained that the foundation of superconductivity is the attractive interaction between electrons which results from their coupling to phonons. In the presence of an attractive interaction between the electrons, the system forms a coherent superconducting ground state, characterized by occupation of the "individual particle states in pairs, such that if one of the pair is occupied, the other is also." In the ground state, the pairs have opposite spin and momenta, and displacing the whole distribution in momentum space would produce a state with a net supercurrent flow. Calculating in a simplified model, they found the energy difference between the normal and the superconducting phase at zero temperature to be proportional to the square of the "number of electrons, n_c, virtually excited in coherent pairs above the Fermi surface." This number of electrons is proportional to the Debye (or mean phonon) energy times the same, essentially singular, function of the interaction strength that occurs in Cooper pairing. The electron–hole spectrum, they showed, contains a gap proportional to n_c.

Bardeen, Cooper, and Schrieffer summarized the "advantages" of their theory:

> (1) It leads to an energy-gap model of the sort that may be expected to account for the electromagnetic properties. (2) It gives the isotope effect. (3) An order parameter, which might be taken as the fraction of electrons above the Fermi surface in virtual pair states, comes in a natural way. (4) An exponential factor in the energy may account for the fact that kT is very much smaller than $\hbar\omega$ [the Debye energy] (5) The theory is simple enough so that it should be possible to make calculations of thermal, transport, and electromagnetic properties of the superconducting state.[376]

Just after the BCS letter was sent in, Bardeen succeeded in computing the second-order phase transition.

The BCS theory had three immediate major experimental confirmations. In early February 1957, at the time the finite temperature properties were being worked out, Charles Slichter and his student Charles Hebel at Illinois, using nuclear magnetic resonance (NMR), measured the nuclear spin relaxation rate in aluminum as a function of temperature.[377] They found that as the aluminum underwent the transition from normal to superconducting, the relaxation rate increased to values more than twice that of the normal state instead of decreasing, but as the temperature was further reduced below the transition temperature, the rate began to decrease. The effect was contrary to the expectations of the two-fluid model, which focused on only the "number" of normal electrons. It was explained by Bardeen, Cooper, and Schrieffer (with "considerable help in working out the details of the calculation of matrix elements" from Hebel and Slichter) as due effectively to the increased density of states below the transition temperature.

Experiments by R. W. Morse and H. V. Bohm at Brown University on ultrasonic attenuation, which showed a rapid decrease of the attenuation in very pure superconductors as the temperature fell below T_c, provided additional confirmation.[378] Still further confirmation came from Glover and Tinkham's experiment in Berkeley on far-infrared transmission through thin superconducting films, which gave direct evidence for an energy gap corresponding to the frequency at which absorption sets in.[379]

The theory was announced officially at the American Physical Society meeting in Philadelphia, 21 to 23 March 1957. Two postdeadline papers were arranged. Bardeen, concerned that Cooper and Schrieffer receive due credit, did not attend. But Schrieffer got word too late to attend the meeting, and so Cooper delivered both papers.[380] (The discovery letter appeared one week later.) The theory was further described at the annual meeting of the National Academy of Sciences in Washington, D.C., on 23 and 24 April.

A full article was sent to the *Physical Review* in July 1957, fulfilling the promises of the earlier letter.[381] The paper opened with an outline of its contents:

> The main facts which a theory of superconductivity must explain are (1) a second-order phase transition at the critical temperature, T_c, (2) an electronic specific heat varying as $\exp(-T_0/T)$ near $T = 0°$ K and other evidence for an energy gap for

individual particle-like excitations, (3) the Meissner-Ochsenfeld effect ($B = 0$), (4) effects associated with infinite conductivity ($E = 0$), and (5) the dependence of T_c on isotopic mass, $T_c \sqrt{M} = $ const. We present here a theory which accounts for all of these, and in addition gives good quantitative agreement for specific heats and penetration depths and their variation with temperature when evaluated from experimentally determined parameters of the theory.

How the theory can give rise to persistent current flow was explained here for the first time. In the ground state, Bloch states are occupied in pairs of opposite momentum and spin. A displacement of the whole distribution in momentum space, in which the electrons are paired not to zero momentum, but to a common net momentum, produces a supercurrent-carrying state, about which "nearly all fluctuations will increase the free energy." Only those fluctuations that can change the common net momentum of the pairs can decrease the free energy, and these "are presumably extremely rare, so that the metastable current carrying state can persist indefinitely."

As in their letter, in this paper Bardeen, Cooper, and Schrieffer worked in terms of the "reduced Hamiltonian," which takes into account only interactions of electron pairs of zero net momentum. Now they constructed, in addition to the ground state, the elementary quasiparticle excitations and the finite temperature states of the system, including the equation for the temperature dependence of the energy gap. This technology enabled them to give an extensive account of the electric and thermal properties for low-frequency or static fields, the second-order phase transition, the Meissner effect, absorption of ultrasonic and electromagnetic waves, and nuclear spin relaxation.

It took remarkably little time for most experimentalists, especially those not previously in the superconductivity field, to accept the basic tests given in the first long BCS paper and in its immediate aftermath. But it was much longer before the theorists' skepticism was quelled. Eventually, the BCS theory achieved the ultimate stamp of approval—it won the authors the 1972 Nobel Prize for Physics. Meanwhile, the processes by which most of the skepticism was dispelled led to healthy additions and changes in the BCS theory. The story of the immediate aftermath of BCS was well told by Philip W. Anderson at the March 1987 meeting of the American Physical Society in New York City.[382]

The response of the community was summarized in a letter Bardeen sent to Cooper in September 1957, describing recent meetings at Madison, Wisconsin, and Banff: "While very few had had a chance to study the theory carefully, the reaction was generally favorable. Experimentalists were particularly enthusiastic about the theory. Objections were raised mainly by those with preconceived notions of what the theory should be like."[383] Pippard later remarked that many earlier theorists had thought of the electron–phonon interaction as crucial to superconductivity and rejected it. Thus "anyone who had travelled that road and decided it led nowhere was understandably irritated by the suggestion that he had missed discovering the theory of superconductivity; and it was easy enough for the expert to discover flaws in the working out that enabled him to overlook any possible merits." Pippard claims his initial disappointment "sprang from regret for the end of an era, when superconductivity as an unsolved mystery posed the sort of problem that keeps an experimental physicist happy."[384]

Bardeen, in his September 1957 letter to Cooper, mentioned Schafroth's criticisms of the BCS derivation of electrodynamic properties, and his assertion (which Bardeen did not believe) that "there is no way to define a current with our interaction." Bardeen remarked that "some of the English (Kuper and others) took the viewpoint that the check of our theory did not mean much because almost any energy gap model would lead to equivalent results. . . . I think the check goes beyond that." He also added in the same letter that "Pippard wrote recently with several objections. . . . The most significant was on our electrodynamics, and particularly on the temperature variation [of the coherence length]."[385]

Schrieffer communicated to Bardeen from Birmingham, England, in November with "comments on the theory from the 'grand tour'" he was conducting of British universities. At Bristol, he wrote that Bohm quoted Bohr as feeling that

> the many body aspects are much more complicated than we have assumed. . . . Bohm agrees with Bohr on the role of many-body correlations of other than Pauli origin . . . [and] thinks we have the essential answer but the understanding of what the "pairs" really are and why other terms are unimportant is completely obscure.[386]

Schrieffer recently recalled Bohr commenting, "It's an interesting idea but nature isn't that simple."[387]

Blatt, Butler, and Schafroth rejected BCS for a number of years, offering an alternative in the summer of 1957. This "quasi-chemical approach" expanded on Schafroth's earlier theory of superconductivity as a Bose–Einstein condensation of pairs of electrons.[388] Although the BCS paper emphasized the difference between their picture and this one—"Our pairs are not localized . . . , and out transition is not analogous to a Bose-Einstein condensation"[389]—the issue of the equivalence of the two theories remained. In defense of BCS, Bardeen responded to Freeman Dyson, one of those who felt that the Blatt-Butler-Schafroth theory was equivalent to BCS:

> We believe that there is no relation between actual superconductors and the superconducting properties of a perfect Bose-Einstein gas. The key point in our theory is that the virtual pairs all have the same net momentum. The reason is not Bose-Einstein statistics, but comes from the exclusion principle. . . . Our wave functions indicate that there is no localized pairing of electrons into "pseudo-molecules" . . . the correlation occurs over such large distances that the electrons cannot be grouped into localized pairs.[390]

The objections of many theorists to BCS focused initially on its apparent lack of gauge invariance—the fundamental freedom to make certain formal changes in the electromagnetic potentials without changing the electromagnetic fields. Any valid theory should be independent of such changes; the issue had been in the air just prior to the BCS paper, when Buckingham criticized Bardeen's derivation of the Meissner effect in terms of an energy gap.[391] In response, Bardeen had stressed the advantages of working in the particular gauge (with vanishing divergence of the vector potential) in which the London electrodynamic equation is valid. He pointed out in addition that the electromagnetic response of a superconductor can be divided into two parts: first, a gauge-invariant transverse part, determined by individual particle excitations, and second, a gauge-dependent longitudinal part, which comes from collective excitations. He remarked, "On physical grounds . . . one would not expect a magnetic field to excite collective modes."[392] Indeed, the BCS calculation was correct in the gauge with vanishing divergence of the vector potential.

Problems of gauge invariance in BCS theory were raised, among others, by Gregor Wentzel of the University of Chicago, by Blatt and T. Matsubara, and by Schafroth,[393] as well as by the University of Chicago field theorist Yoichiro Nambu. Although a note added in proof to the BCS paper discussed the need to include collective excitations when calculating in a general gauge, the argument would be carried through explicitly by Philip Anderson at Bell Laboratories.[394] In particular, Anderson pointed out that a neutral BCS system would have a branch of longitudinal acoustic-like excitations (a form of zero sound); in a realistic system of charged particles, this "Anderson mode" is raised in frequency to become the longitudinal plasma oscillation, which when included guarantees gauge invariance of the electromagnetic properties of the superconductor. (This analysis would later have an important influence on theories of elementary particles, where the phenomenon is known as the "Higgs mechanism.") Similar arguments appeared independently in a rapid series of papers by N. N. Bogoliubov and co-workers at the Mathematical Institute of the Soviet Academy of Sciences in Moscow, and in a technically very important paper by Nambu in 1959.[395]

The history of the response to BCS by Soviet physicists was complicated by the fact that it was not possible at the time to mail preprints of the full BCS article, completed in June 1957, to the Soviet Union. However, detailed news of the dis-

Paul A. M. Dirac (1902–1984), Nikolai N. Bogoliubov (b. 1900). Victor Weisskopf (b. 1908), and D. D. Ivanenko (b. 1904) at the Max Planck Centenary Jubilee Conference, Berlin, 1958. (AIP Niels Bohr Library, Physics Today Collection)

covery and one preprint of the full article arrived there in August; the preprint was delivered by G. Picus, who had received it from Seitz at the August 1957 summer study at Varenna. The published paper arrived in the Soviet Union in November. In the meantime, starting from the first BCS letter, the groups of both Landau and Bogoliubov independently rederived much of the theory.[396]

The solution of the gauge invariance problem relied on what we now call the "broken symmetry" of the BCS state, a notion not explicit in the original BCS theory. In the BCS state, the order parameter (a complex number) selects a given phase, thus breaking the gauge invariance of the theory. The situation is analogous to the way the magnetization in the ground state of a ferromagnet selects out a direction in space, apparently breaking the rotational symmetry. The gauge invariance of the theory is recovered when one includes changes of the phase of the order parameter when changing the electromagnetic potentials, and takes into account the collective (Anderson) mode associated with these changes in phase. This explanation of the gauge problem was rapidly accepted by the physics community.

The controversy over gauge invariance had an important by-product. Having become aware of the structure of BCS, Nambu was able to introduce the concept of broken symmetry into particle physics.[397] This concept would be responsible for many parts of the present "standard model"—for example, in electroweak theory and current ideas of grand unification. If solid-state physicists had borrowed ideas and techniques from field theory, the loan was now repaid. The particle physicists' borrowing of broken symmetry then had further repercussions on solid-state physicists, who were forced to refine and conceptualize their vague notions of broken symmetry, which had in fact been in the air (e.g., in ferro- and antiferromagnetism).

A second objection to BCS, the main one from the point of view of solid-state physics—championed by Berndt Matthias with the help of Fröhlich, Felix Bloch, and others—addressed the phonon mechanism, and the lack of quantitative energetics and ability to predict which materials would become superconductors. At first, this opposition focused on the isotope effect, which soon was measured well enough to bring out real or imagined deviations from the rule that the transition temperature should vary as the inverse of the square root of the isotopic mass.[398]

Initially, James Swihart, and then Pierre Morel and Anderson, set out to make roughly realistic calculations of T_c's from first principles.[399] Casting the BCS equation as an integral equation in time enabled Morel and Anderson to take into account the more complete time-retarded interaction between electrons due to phonon exchange. This work was able to predict the general run of T_c's and isotope effects (albeit only within broad limits).

The eventual quantitative theory of energy gaps had, as an important beginning, an informal collaboration between Schrieffer and Anderson. During a Utrecht many-body theory meeting in June 1960, they conversed while on a long bus tour of the polders in Holland on a rainy day. Anderson told Schrieffer about his work with Morel on integration in the time domain; Schrieffer told of his study of the Green's function formalism developed by Russians, especially Eliashberg, which was the correct way to do this integration and to express the tunneling current.[400] By the summer of 1962, the lines of approach taken by Anderson, who with Morel became concerned with systematics, and Schrieffer, who had become concerned with computer integrations of the gap equation, began to converge. That summer, Bardeen and Anderson visited Schrieffer at the University of Pennsylvania, bringing along data on lead and a suggested rough model. The result was twin letters to the *Physical Review* by Schrieffer, Douglas Scalapino, and John Wilkins and by John Rowell, Anderson, and D. E. Thomas. These letters founded the quantitative theory of T_c and the coupling parameters, a theory later carried to great precision by William McMillan and Rowell.[401]

The third objection, best expressed by Bloch and Casimir, questioned whether the basic phenomenon of superconductivity had really been explained. Bloch kept arguing that BCS theory did not explain the most fundamental observation, that of persistent currents, which he himself had shown could not be present in an equilibrium state (Chapter 2). Casimir asked the related question: How could the voltage be exactly zero along a mile of dirty lead wire?[402]

The answer to these questions, which depended on the fact that a persistent current in a ring carried with it a quantized magnetic flux through the ring, had already been given in principle, almost simultaneously with the answer to gauge invariance. In 1958, Lev Gor'kov derived the Ginzburg–Landau theory from BCS theory reexpressed in Green's function language.[403]

If a logical historical scenario had been followed, Gor'kov would have noticed that his derivation of the Ginzburg–Landau theory from BCS implied that the charge parameter e^* in the Ginzburg–Landau theory was twice the charge of the electron. Then Abrikosov would have noticed that his paper of 1956 on type II superconductors contained an actual calculation of the magnitude of the flux, $hc/2e$, in a quantized flux line.[404] Then everyone would have noticed that to stop a persistent current required passing flux quanta through the superconductor, and that to do so would require overcoming an energy barrier of order electron volts per Angstrom of flux line—a number easily derived from the Ginzburg–Landau theory. None of this happened. Landau speculated that e^* in Ginzburg–Landau had to be an integer, but Abrikosov's dimensionless numbers concealed the value of the flux quantum. And no one went back and read Fritz London's arguments on persistent currents, set forth in 1950.[405]

Meanwhile, a race to do the flux quantum experiment, involving at least five laboratories, was won early in 1961 in a dead heat by Doll and Näbauer, on the one hand, and Fairbank and Deaver, on the other.[406] Anderson recalls that at Utrecht, Fairbank, while planning the experiment, asked various theorists what value he would get, if any, but received no unequivocal answers. The theory of the effect within the framework of BCS was given by Nina Byers and C. N. Yang in a letter that appeared in the *Physical Review* sandwiched between the two experimental letters. The theorists pointed out that "the quantization of flux is an indication of the pairing of the electrons in the superconductor," and also noted the connection with the Aharonov–Bohm effect of a vector potential on the quantum mechanics of charged particles.[407]

A deeper understanding of the persistence of currents eventually came from Brian Josephson. This line began with work in 1960 by Ivar Giaever at the General Electric Research Laboratory in Schenectady. He used electron tunneling between a superconductor and a normal metal, or between two superconductors, as a powerful means for studying superconductors.[408] Giaever, who originally came from Norway and was educated as a mechanical engineer, had come up with the idea that the energy gap could be observed directly by tunneling, and in the spring of 1960 he was able to demonstrate its effect in tunneling between two aluminum superconductors.

Josephson, Pippard's graduate student at Cambridge, attending Philip Anderson's lectures there in 1961 to 1962, became fascinated by the concept of the phase of the BCS-Ginzburg-Landau order parameter as a manifestation of the quantum theory on a macroscopic scale. Playing with the theory of Giaever tunneling, Josephson found a phase-dependent term in the current; he then worked out all the consequences in a series of papers, private letters, and a privately circulated fellowship thesis. In particular, Josephson predicted that a direct current should flow, without any applied voltage, between two superconductors separated by a thin insulating layer. This current would come as a consequence of the tunneling of electron pairs between the superconductors, and the current would be proportional to the sine of the phase difference between the superconductors. At a finite applied voltage V, an alternating supercurrent of frequency $2eV/h$ should flow between the superconductors.[409] Josephson's work established the phase as a fundamental variable in superconductivity.

By June 1962, Rowell noted that some junctions might be showing the Josephson effect, and by December, Anderson and Rowell had firm experimental evidence for it.[410] Josephson's relation that the time rate of change of the phase of the order parameter is proportional to the voltage—which, as Josephson pointed out, is implicit in Gor'kov's theory[411]—then provided the final answer to Casimir's question about the "mile of dirty lead wire." As long as the wave function is time-independent, there can be no voltage drop.

Bloch's question about persistent currents had to wait until they were actually shown not to be persistent. Goodman had awakened physicists in the West to Abrikosov's paper, enabling understanding of why large critical fields were possible in the type II flux quantum array state, but not why they could carry large currents.[412] In the eventful fall of 1962, Young Kim and co-workers at Bell Labs demonstrated that critical currents did in fact decay, and Anderson showed that this was a consequence of the pinning of flux lines, their slow release, and Josephson's equation.[413] How and under what conditions currents persist was now clear. Shortly thereafter, Kim demonstrated the phenomenon of flux flow: large diagmagnetism, a Meissner effect, but large voltage and no persistent current, underlining that Casimir and Bloch were correct in separating these phenomena out as independent from the original BCS theory.[414]

By early 1963 all rational objections to BCS had been quelled, to the satisfaction of the answerers, at least, if not the objectors. Until 1987, when high T_c superconductivity was announced, the phenomenon was no longer the outstanding mystery of solid-state physics.

Soon after its discovery, the BCS theory was applied to other systems. During the summer of 1957, Pines, Aage Bohr, and Ben Mottelson treated the pairing energy in nuclei using BCS, applying for the first time the concept of superfluidity to nuclear matter.[415] As pointed out by A. B. Migdal in 1959, and then in 1964 by

Ginzburg and D. A. Kirzhnits, the interiors of neutron stars were likely to contain nuclear matter in BCS-like superfluid states.[416] Extensions of the theory have been particularly successful in describing the superfluidity of ^{3}He.[417]

Helium

The Second World War would have an important influence on liquid-helium research, as it did for superconductivity, by providing such new tools as the Collins cryostat (discussed earlier), nuclear magnetic resonance, and low-energy neutrons, as well as by making available ^{3}He in bulk. But the immediate concern in the postwar period centered on the prewar question of understanding in microscopic terms the two-fluid behavior of liquid ^{4}He at very low temperatures.

Peshkov's January and April 1945 experiments on the velocity of second sound provided a crucial stimulus to clarifying the microscopic picture while fueling the debate between Landau and Tisza over their theories.[418] Peshkov's experiment confirmed both theories qualitatively. But as he pointed out, "the microscopic theory agrees with the experiments less satisfactorily." To fit the temperature dependence of the sound velocity would require a "marked" temperature dependence of the roton gap Δ and mass μ in Landau's elementary excitation model. Regarding Tisza's two-fluid model, Peshkov wrote that "although Tisza explains some of the facts qualitatively satisfactorily, . . . quantitatively the theory yields incorrect results."

Peshkov's measurements showed a small but clear decline in the second sound velocity with falling temperatures below 1.6 K, ending in a minimum or at least in a constant value at 1.2 K. Tisza, who had moved to MIT during the war, in a 1947 quasi-thermodynamic theory of liquid helium[419] showed his theoretical curve passing through Peshkov's points, while Landau's theory, as calculated by Evgenii Lifshitz,[420] was shown as passing some 5 m/sec above the data.[421]

In order to improve the fit to Peshkov's measured velocity (e.g., 19 m/sec at 1.6 K, compared with his calculated value of 25 m/sec), Landau soon modified his two-branch model of the elementary excitations of helium II, the phonons and the rotons. Instead, he presented—what has since become well established—a single branch spectrum consisting of phonons at low wave vector and rotons at high wave vectors. "With such a spectrum it is of course impossible to speak strictly of rotons and phonons as of qualitatively different types of elementary excitations."[422] The roton region now was specified by three parameters: the energy gap, the effective mass, and the momentum of the roton minimum.

Tisza, on seeing Landau's paper after his own had been completed, added a footnote attacking Landau's excitation spectrum:

> Apart from the unsatisfactory nature of this procedure [with three arbitrary constants] it tends to modify the theory in the wrong direction. Originally Landau failed to notice that every vortex element can be associated with a definite mass contained in the volume in which the vorticity is different from zero. . . . In contrast to this situation, phonons are associated with the liquid as a whole. Only elementary excitations associated with definite masses can make it understandable that the liquid breaks up into two components. . . . Thus Landau has ignored an important difference between phonons and rotons. In his latest paper this difference is even more blurred.[423]

Landau replied in a short paper written to express his "opinion on some of the statements which have been put forward, especially in connection with the last paper by L. Tisza."[424] Before his discussion, Landau remarked in a footnote:

> I am glad to use this occasion to pay tribute to L. Tisza for introducing, as early as 1938, the conception of the macroscopical description of helium II by dividing its density into two parts and introducing, correspondingly, two velocity fields. This made it possible for him to predict two kinds of sound waves in helium II. (Tisza's detailed paper . . . was not available in the U.S.S.R. until 1943 owing to war conditions, and I regret having missed his previous short letter. . . .) However, his entire quantitative theory (microscopic as well as thermodynamic-hydrodynamic) is in my opinion entirely incorrect.

Thus after rebutting Tisza's criticisms, he took Tisza's own theory to task, commenting, for example, "every consideration of the motion of individual atoms in a system of strongly interacting particles is in contradiction with the first principles of quantum mechanics." Landau drew support from a recent paper of N. N. Bogoliubov, who "has succeeded recently, by an ingeneous application of second quantization, in determining the general form of the energy spectrum of a Bose-Einstein gas with a weak interaction between the particles."

Bogoliubov's theory played a very important role in two regards.[425] It was the first calculation of liquid helium to take the forces between the particles into account quantum mechanically, and would form the basis for later microscopic work in the 1950s and 1960s. Second, the techniques of the paper would later appear in Bogoliubov's theory of superconductivity.[426] Bogoliubov showed that the interaction had the effect of depleting the condensate. Unlike in an ideal Bose–Einstein gas, only a finite fraction of the particles populate the ground state even at $T = 0$ K; other momentum states are occupied, in combinations (e.g., pairs) contributing zero net momentum.

Bogoliubov, assuming that for low densities the most important part of the interaction of noncondensed particles was with particles in the condensate, derived an effective Hamiltonian that could be readily diagonalized in terms of new quasiparticle degrees of freedom. The long-wavelength spectrum emerged as linear in momentum—the phonon form predicted by Landau. The spectrum of elementary excitations had a minimum velocity; thus, according to Landau's criterion, the system would be superfluid.

With Landau's new spectrum and Bogoliubov's calculation, the nature of the excitation spectrum of helium II underlying its two-fluid behavior became established theoretically. Measurements of the second sound velocity undertaken by Peshkov in 1948, and a year later by John Pellam and Russell Scott at the U.S. National Bureau of Standards, clearly showed a strong rise of the velocity of second sound with falling temperature, as predicted by Landau.[427] The predicted behavior was further confirmed when Kenneth Atkins and Donald Osborne at the Mond Laboratory extended the measurement of second sound down to 0.1 K, finding a velocity there about 150 m/sec, admirably close to the theoretical zero-temperature value of 157 m/sec.[428]

A second issue in the postwar period was the rotational properties of helium II. In his prewar paper, Landau had assumed that flow of the superfluid component would be derivable from a velocity potential, and thus would have no vorticity; the superfluid would be incapable of rotating uniformly.[429] What, then, would happen when a bucket of helium below the lambda point was rotated? Following a suggestion of Peierls, Osborne undertook a rotation experiment.[430] No deviation was found: the helium rotated with its entire mass. The problem was that in this experiment, the helium was rotated too fast. Osborne's negative result, also obtained independently by Andronikaskvilii in his doctoral dissertation work, was not totally unexpected.[431] Heinz London, at an international conference on fundamental particles and low temperatures in Cambridge in 1946, suggested that in analogy to the behavior of a superconductor in a magnetic field, in a rotating container filled with liquid helium, "there may be some rotational movement in an intermediate state, in which certain angular sections perform vortex-free circulations with velocities inversely proportional to their distance from the axis, and with vortex sheets between the sections."[432] Landau and Lifschitz would attempt in 1955 to explain the rotating experiments from this point of view.[433]

In fact, the explanation of the apparent rotation of the fluid as a whole turned out to be that rapid rotation of helium II would generate vortices in the superfluid. Prior to Osborne's experiment, at the May 1949 International Union of Pure and Applied Physics meeting in Florence, Onsager commented that superluid flow need only be free of vorticity if the sample was simply connected. If the region was multiply connected, then the only restriction was that the circulation—the integral of the velocity around a closed path, which vanishes for potential flow—would be quantized in integral multiples of Planck's constant divided by the atomic mass of helium, h/m.

> If we admit the existence of quantized vortices [Onsager said], then a superfluid is able to rotate, but the distribution of the vorticity is discrete rather than continuous. . . . Finally we can have vortex rings in the liquid, and the thermal excitation of helium II, apart from the phonons, is presumably due to vortex rings of molecular size.[434]

Five years later, Feynman independently reached the same conclusions, and proposed that rotating helium would contain an array of quantized vortex lines.[435]

The first experiment to explore successfully the internal structure of rotating helium was done by William Vinen and Henry Hall in the Mond Laboratory in 1956. They found that the damping of second sound in rotating helium was proportional to the angular velocity of rotation, and could be described in terms of an anisotropic "mutual friction" between the normal and superfluid components. The detailed anisotropy did not agree with the Landau-Lifshitz model of vortex sheets; only Feynman's model led to results in qualitative agreement with their experiment.[436] Unfortunately, Hall and Vinen's experiments provided only indirect evidence for the quantization of superfluid circulation. However, in 1960, Vinen, by measuring the influence of the circulation of liquid helium II on the transverse vibrational modes of a fine wire within rotating helium, was able to verify the quantization of superfluid circulation in units of h/m.[437]

A new field of helium research was opened in the late 1940s by the increasing availability of bulk quantities of the lighter helium isotope, ^{3}He, resulting from the radioactive decay of tritium (^{3}H) produced for thermonuclear fusion bombs.

Although by the mid-1930s, ^{3}He had been known to exist, its natural abundance, of order one part in 10^7, was too small to permit cryogenic experiments. High fluxes of neutrons in reactors made it possible to produce tritium from the interaction of neutrons with ^{6}Li, resulting in tritium and ^{4}He. Tritium beta-decays into ^{3}He with a half-life of about 12.5 years, and can be "milked off" by diffusion methods. The first usable tritium samples were produced by Herbert Anderson, Fermi's long-time collaborator. Having learned that Bloch and others had recently discovered nuclear magnetic resonance, Anderson became interested in measuring the magnetic moments of both tritium and ^{3}He by such techniques. In the fall of 1945, while still at Los Alamos, he conceived of making tritium by sending cans of lithium carbonate to the Hanford reactor for neutron irradiation, and then extracting the tritium using a method developed in collaboration with Aaron Novick of the University of Chicago.[438]

Important questions about ^{3}He, a nucleus with half-integer spin obeying Fermi–Dirac statistics, included whether it would turn into a liquid under normal pressure, and whether like ^{4}He it would become a superfluid. Both Fritz London and Tisza believed that the zero-point motion of the very light ^{3}He atoms would prevent a ^{3}He gas from becoming solid, except under applied pressure.[439] By late 1948, a sufficient quantity of ^{3}He was assembled at Los Alamos to enable the first "condensed matter" experiment on this substance. Three members of the low-temperature group there, Edward Hammel, Edward Grilly, and Stephen Sydoriak, succeeded in condensing pure ^{3}He, and found its boiling point to be 3.2 K, only 1° below that of ^{4}He.[440] An experiment in 1949 at Argonne, the other laboratory capable of carrying out ^{3}He experiments in the early years, looked for superfluid flow of ^{3}He through a narrow pipe, but failed to detect superfluidity at temperatures as low as 1.05 K.[441]

The possibility that ^{3}He would become superfluid strongly stimulated investigation of its properties at lower and lower temperatures. While failure to observe a superfluid transition in ^{3}He supported London's contention that the superfluidity of ^{4}He was a manifestation of Bose–Einstein, the BCS theory provided strong support for the possibility that Fermi particles can become superfluid.[442] In the following years, a series of many-body approaches was taken to ^{3}He, based mainly on the BCS theory. The studies strongly suggested that a superfluid phase should exist, as we will discuss presently. By 1957, sizable quantities of ^{3}He were available for research, enabling study of the role of statistics for superfluidity, but it was not until 1971 that the superfluid phases were discovered experimentally.[443]

Further evidence came from nuclear magnetic resonance—a technique that grew out of the wartime microwave radar program—in which the energy transfer from a radiofrequency circuit to a system of nuclear spins yields a sharp absorption line.[444] This became useful for investigating ^{3}He, a nucleus with half-integer spin and thus a magnetic moment. In 1954, at Duke University, William Fairbank with his graduate student Geoffrey Walters and postdoctoral student W. B. Ard measured the nuclear magnetic resonance absorption signal of ^{3}He down to 0.23 K, using a 20-cc gas sample obtained from the Stable Isotope Division of the Oak Ridge National Laboratory.[445] The experiment showed a tendency for parallel alignment of the spins, an effect understood a few years later within the framework of the Landau theory of Fermi liquids.[446]

A second technique arising out of wartime research, thermal neutron scattering, also became useful for helium research, initially for ^{4}He and later for ^{3}He. Low-energy neutrons became available with the development of nuclear piles. Unlike x-ray scattering from materials, for which the energy change is always small compared with the incident energy, from slow neutron bombardment the energy absorbed can be comparable to the energy of the incident neutron, thus enabling one to probe the excitation energies of the material. Furthermore, neutrons possess a magnetic moment and are sensitive to the magnetic properties of the scatterer. David Kleinman, at Brookhaven, realized in 1952 that the vibrational spectrum of solids could be determined by measuring the angular and energy distribution of inelastically scattered neutrons. The technique was soon discussed theoretically by George Placzek and Leon Van Hove.[447] After an unsuccessful experiment in 1952, Bertram Brockhouse and A. T. Stewart at Chalk River, the Canadian nuclear-reactor center, in the summer of 1955 managed to obtain the dispersion spectrum of phonons in aluminum.[448] The earliest experiments on liquid helium were those of D. G. Henshaw and D. G. Hurst at Chalk River, who measured the static structure function of normal and superfluid by elastic neutron diffraction; they did not detect a difference in the scattering distributions from normal and superfluid helium.[449]

An important theoretical analysis by Feynman and his graduate student Michael Cohen in 1957 demonstrated the feasibility of measuring the ^{4}He excitation spectrum by neutron scattering.[450] As they showed, the creation of single excitations, signaled by sharp lines in the neutron energy loss spectrum, should dominate the continuous background at low temperatures in the superfluid phase. This work stimulated low-temperature reactor experiments to observe the excitation spectrum of superfluid helium. The first group to detect the elementary excitations of the superfluid phase was at the Aktiebolaget Atomenergi (Atomic Energy Corporation) in Stockholm; Harry Palevsky, a visitor in the group from Brookhaven National Laboratory, had previously discussed the neutron scattering on helium with Cohen and Feynman. Below the lambda point, the group observed the shift in the energy distribution of scattered neutrons caused by emission of rotons in the superfluid. Later, the Los Alamos group of John Yarnell detected both roton and phonon points, which provided direct evidence for the forms of the excitation curve predicted by Landau. In 1960, Henshaw and A. David Woods carried out measurements at Chalk River similar to those of Yarnell and co-workers, but with higher resolution. They not only confirmed the Landau energy spectrum, but were able to observe the termination of the spectrum at high-momentum transfers. The behavior observed there was consistent with the decay of single high-momentum rotons into pairs of rotons, a possibility already predicted theoretically by Lev Pitaevskii.[451]

The increasing experimental research on the quantum liquids ^{4}He and ^{3}He in turn greatly stimulated theoretical work on the many-body problem. The theories of liquid helium in the 1950s were not developed by specialists in low-temperature physics; the ideas used grew out of various subjects, including the electron gas, nuclear matter, and quantum field theory. In the preface to his 1953 volume on superfluids, Fritz London wrote,

> There can be little doubt at present that the superfluids are essentially quantum mechanisms . . . on a macroscopic scale. Indeed, it is just this aspect that is drawing

the attention of the theoretical physicists to these fascinating low temperature phenomena. In superfluid helium the play of quantum mechanics can be seen, so to speak, with the naked eye, inasmuch as one can literally "tap ground state" from the liquid.[452]

In his conclusion, he sketched what he felt would be a way to proceed:

We have recognized the importance, for liquid helium, of the combined effect of small molecular mass, of unusually weak intermolecular forces, and of Bose-Einstein degeneracy. Still it seems that we shall have to invent a more drastic approach to the techniques of the quantum-mechanical many-body problem. We should think that a fluid at absolute zero ought to be accessible to a simplifying idealization, as simple and as workable as that used for the solid state. . . . For the superfluid state the concept of an order in momentum space . . . seems to point in a promising direction by putting it into its place as a kind of complementary counterpart to the ordinary space lattice order of the crystals. Yet this so tantalizingly simple idea is a conjecture which will have to be substantiated or else reduced to its true bounds by a workable analysis of the quantum-mechanical many-body problem before we can say that we have penetrated the mystery of the superfluids.[453]

The spectrum of the elementary excitations was postulated by Landau intuitively, not derived from first principles. The first formulation of a fully microscopic theory of interacting Bose particles, by Bogoliubov in 1947, did not draw on any ad hoc assumptions concerning the structure of the energy spectrum, but it was restricted to a weakly interacting system. The justification of the Landau spectrum on the basis of a microscopic theory, as well as the microscopic description of the ground state, remained an open problem. In 1953 to 1954, Feynman undertook to formulate a microscopic theory of ^{4}He. By means of highly elegant physical arguments, he developed wave functions for the low-lying excitations above the ground state; from these, he derived a Landau-type energy spectrum, and was able to show how phononlike excitations evolved into roton excitations at large momenta. In this initial work, the excitation energies in the neighborhood of the roton minimum were too high compared with neutron-scattering measurements. To overcome this difficulty, Feynman and Cohen improved the wavefunctions by including another physical phenomenon, the "backflow" of other atoms about a given atom as it moved along. Now they were able to derive an energy spectrum in quantitative agreement with experiment. The concept of backflow was a key contribution of this work to the many-body problem more generally.[454]

A separate line of inquiry into the microscopic properties of liquid helium was begun by Oliver Penrose and Onsager, who introduced the concept of long-range order in the system. As they showed, the criteria for Bose–Einstein condensation in an interacting system is that the single-particle density matrix (the quantity that measures the amplitude for adding a particle at one point in the system and removing it at another point, thus returning to the same state) approaches a nonzero value as the two points become far separated. In an approximate calculation of the single-particle density matrix, they found that interactions strongly deplete the condensate of zero-momentum particles, finding roughly only 8% of the particles condensed.[455]

The generalization of Bogoliubov's calculation to more strongly interacting systems, by means of many-body perturbation theory, was undertaken first by Brueckner and Sawada at the University of Pennsylvania in 1956. They used a Green's function method that followed Brueckner's theory of the nuclear many-body system. In the case of a hard-sphere interaction, they derived an approximate excitation spectrum containing some of the essential features of the Landau phonon-roton spectrum. T. D. Lee, Kerson Huang, and C. N. Yang, calculating the ground state and the low-lying excited states of a low-density gas with a hard-sphere interaction described by a pseudopotential, found, at the same time, results similar to those of Brueckner and Sawada. Spartak Beliaev, in the Soviet Union, generalized Bogoliubov's treatment of the weakly interacting low-density gas by means of quantum field-theory techniques. His method, a diagrammatic perturbation expansion in terms of Green's functions—a generalization of earlier techniques applied to the electron gas—was valid for all strengths of interactions, and took into account depletion of the condensate. Numerically reliable microscopic calculations of the ground-state properties of liquid helium would not begin until the work of William McMillan, in his 1964 dissertation, who drew on Monte Carlo methods developed by Nicholas Metropolis and co-workers to calculate physical quantities with a variational ground-state wave function.[456]

In the work of Beliaev and of Hugenholtz and Pines, as in Bogoliubov's earlier work, the amplitude describing the occupation of the ground state, the condensate, was thought of simply as a numerical parameter in the theory. The notion that this amplitude was in fact the more generally spatially varying order parameter would not be introduced until the work of Pitaevskii and Eugene Gross, in a direct analogy with the Landau–Ginzburg theory of superconductivity.[457]

In 1960, Pines who had by then joined the faculty at Illinois, and a postdoctoral researcher, Allen Miller, turned to the problem of how to understand backflow in terms of the interaction of the elementary excitations, which Pines and Nozières had begun work on in 1958 during Pines's year at the Ecole Normale Supérieure in Paris. Together with Nozières, they showed that backflow arose from the interaction between "a Feynman excitation and higher configurations involving several elementary excitations." This paper also introduced an analysis of the spectrum in terms of the exact sum rules obeyed by the neutron scattering cross sections of the system, which enabled more precise analysis of the experimental data.[458]

Just as Landau had formulated the ^{4}He problem in terms of the elementary excitations of the system, so in 1957 he presented a general theory of the elementary excitations of an interacting normal (i.e., not superfluid) "Fermi liquid."[459] The basic assumption of the theory was that the states of the interacting Fermi liquid are in a one-to-one correspondence with the states of the ideal Fermi gas, as one can see by gradually turning on the interactions between the particles. The elementary excitations near the ground states would be quasiparticles associated with the motion of a real particle in the self-consistent field created by its interactions with its neighbors; the excitation spectrum would be similar to that of noninteracting fermions, with a modified effective mass. A very subtle feature pointed out by Landau was that in near-equilibrium states, one can keep track of properties of the system, such as the particle and momentum density and currents, in terms of the quasiparticles.

In its region of validity, at low temperatures where the lifetimes of the quasiparticles are sufficiently long, Landau's theory predicted the properties of liquid ^{3}He such as the specific heat and magnetic susceptibility, in terms of a few parameters describing the interactions of the quasiparticles. It also gave a framework for calculating transport coefficients such as spin diffusion, thermal conductivity, and viscosity in terms of the quasiparticle interactions, and predicted exactly their temperature dependence at low temperatures.

Landau, in addition, predicted a new collective mode in Fermi liquids: "zero sound." At very low temperatures, collisions become too infrequent for the system to have a well-defined ordinary sound mode, and instead a collective oscillation driven by the self-consistent interaction fields of the quasiparticles, similar to that observed in the vibrational states of atomic nuclei or in plasma oscillations, would occur. This oscillation was seen experimentally in liquid ^{3}He in the mid-1960s.[460]

Landau's Fermi-liquid theory, a triumph of theoretical work on this problem, was semiphenomenological in that it made no predictions for the interactions between the quasiparticles. The derivation of the structure of the theory from microscopic first principles would be a major challenge to many-particle theorists, as would the microscopic calculation of the parameters of the theory.[461]

An early fully microscopic approach to the ground-state energy and quasiparticle excitations of a fermion system, by diagrammatic field-theory methods, was carried out in 1957 by V. M. Galitskii, in Moscow, for a low-density gas.[462] He derived the energies of quasiparticles near the Fermi surface and, as predicted by Landau, found a phonon collective mode corresponding to zero sound in the long-wavelength limit. To go beyond this low-density calculation, which was not applicable to ^{3}He, Brueckner and J. L. Gammel applied the Brueckner nuclear many-body theory, summing a selected set of higher-order diagrams to obtain a coupled set of equations. These were solved iteratively with the aid of a fast electronic computer (IBM 704) at Los Alamos. The computed binding energy turned out to be extremely sensitive to the potential used; small changes could improve the fit to experiment. In this way, they calculated the compressibility, the temperature and pressure dependence of the specific heat, and the enhancement of the magnetic susceptibility with pressure, in qualitative agreement with experimental numbers.[463]

None of these theories of a Fermi liquid touched on the question of possible superfluid behavior for ^{3}He. In the first microscopic discussion of this question, Cooper, together with Robert Mills and Andrew Sessler, in 1958 applied the BCS theory to a system of neutral fermions, assuming, as in BCS, that the pairing would be in s-wave states.[464] They demonstrated that in the absence of a repulsive hard core, an attractive two-body interaction between the atoms would lead to a paired superfluid state, and that a core of the interatomic interaction would not necessarily forbid such a state. Examining a number of trial superfluid ground-state wave functions with a model interaction between the ^{3}He, they found that while a transition to a superfluid did not appear likely (within their assumptions), they could not definitely exclude it. The crucial point missed in this calculation was that pairing in ^{3}He would take place not in s waves, but in a higher angular-momentum state, in fact p wave with total spin of the pair equal to unity.

The possibility of pairing in p and d states, with consequent anistropy of the superfluid, was proposed soon after by Brueckner, Soda, Anderson, and Morel, who estimated that the transition would be to a d-wave superfluid at a critical temperature of 0.1 K. A similar conclusion was reached by Victor Emery and Sessler.

Anderson and Morel subsequently examined the physical consequences of the anisotropic superfluidity, including the expected thermodynamic and flow properties, presaging many properties of the ^{3}He-A phase. They refined their estimate of the transition temperature, finding onset of d-wave pairing at the still lower temperature of 0.07 K. At this stage, however, it remained difficult to predict accurately the subtle interplay of the repulsive and attractive parts of the interatomic potential in producing pairing; as they remark, the d-state ($l = 2$) pairing "is a consequence of the strongly repulsive short-range part of the He^3 = He^3 interaction potential . . . preventing a condensation in the $l = 0$ or l configurations." In an attempt to avoid the "physically implausible" anisotropy of the Anderson–Morel state, Roger Balian and Richard Werthamer showed that by including a superposition of all three projections of the unit spin of the pair in the pair wave function, one could produce an isotropic p-wave paired state. Shortly after the discovery by Douglas Osheroff, David Lee, and Robert Richardson in 1972 of two very-low-temperature phase transitions in liquid ^{3}He along the melting curve, Anthony Leggett as well as Anderson showed that the new phases were anisotropic superfluids; one phase corresponded to that discussed by Anderson and Morel, and the other, the B phase, to the Balian–Werthamer state.[465]

Critical Phenomena

Save for the disruption of work in the Netherlands and the shift toward applied work by American researchers in the field, the Second World War appears to have had little immediate influence on the highly theoretical course of research on phase transitions and critical phenomena, in contrast with research on the electron gas, superconductivity, and liquid helium. The problem that stimulated primary interest after the war remained the Ising model, particularly the extension of Onsager's method to three dimensions. Pauli, Joseph Jauch, and Montroll were among those who worked on it. John Maddox of Kings' College in London was, according to Montroll, reputed to have solved the three-dimensional Ising problem by Onsager's method, but comparison by Montroll and others of the solution with the high-temperature expansion showed that it did not work.[466]

Kramers's student R.M.F. Houtappel and Nordsieck's student Gordon Newell, among others, were assigned the three-dimensional model as their dissertation problem. Both ended up changing their problem to the problem of extending the Onsager method to other two-dimensional lattices, a possibility foreseen by Onsager in his 1944 article.[467] Extensions worked out in detail in the early 1950s included triangular, hexagonal, honeycomb, and "Kagome" lattice structures. The problem of extension to various lattice structures particularly caught the fancy of a number of Japanese investigators.

Work on the two-dimensional Ising model was, however, by no means exhausted by the Onsager paper and its generalization to other planar lattices. Could the derivation be clarified or simplified? Could one derive it in other ways more readily generalizable? The Onsager evaluation of the partition function applies to the symmetric case of zero external field: Could one calculate the magnetization and susceptibility in the case where a magnetic field, however weak, selects one direction for the spins as preferable to the opposite direction? The perspective of those working on the theory of the Ising model differed considerably from that of the students of phase transitions in van der Waals's day. The forces between molecules were no longer mysterious. Work on phase transitions covered a spectrum from highly mathematical investigations at one end to the study of microscopic mechanisms and phenomenological descriptions at the other. Underlying the interest in the Ising model was a desire to compute correctly the statistical features of phase transitions, even if actual physical interactions between the "atoms" were replaced by a highly simplified model. The Ising model presented a problem of considerable physical interest, and was particularly attractive to some because it challenged one's mathematical perspicacity and prowess.

Bruria Kaufman, who had been a graduate student in physics at Columbia University, was familiar with the theory of group representations when she came to work with Onsager. She found a shorter, mathematically simpler, and more elegant derivation of Onsager's result. To evaluate the required matrix eigenvalue, she made use of sets of anticommuting matrices, whose general properties had been worked out by Richard Brauer and Hermann Weyl.[468]

The pair correlation (sometimes called the "short-range order" of a crystal) in an Ising net played an important role in all improvements over the mean-field theories in which it was completely neglected, such as the theories of Ornstein and Zernicke, and the Bethe and the Kirkwood approximations. Knowledge of the behavior of the pair correlation function could provide insight into the mechanism of the phase transition, and was of special interest also because it could be measured directly in x-ray scattering experiments. The "long-range order," the correlation in the limit of large separation, was closely related to the magnetization. Kaufman and Onsager worked out, by a difficult mathematical route making use of anticommuting matrices, the pair correlation as a function of temperature and distance between sites for the plane Ising lattice; they found no long-range order.[469] Yet even without magnetization or long-range order, the regions above and below T_c were mathematically distinguishable.

Kaufman moved on to the Institute for Advanced Study in Princeton, where she later became Einstein's last assistant and worked on unified field theories. Some of the group at the institute in the early 1950s—J. M. Luttinger, T. D. Lee, Yang, Van Hove—were interested in the theory of phase transitions. At the American Physical Society meetings in the late 1940s, the prominent theoretical topics were quantum electrodynamics and the resonances of nuclear physics. Only a small group of theoreticians—who called themselves "the lunatic fringe"—would participate in little separate sessions on phase transitions; among them were Lamb, Onsager, Montroll, Yang, Lee, Kaufman, Luttinger, Mark Kac, and Arnold Siegert.[470]

In 1948, on the occasion of two separate conferences, Onsager added piquancy to the mathematical challenge of the Ising model by presenting a surprisingly simple closed-form expression for the magnetization of a rectangular Ising net. The calculation of the magnetization in the limit of vanishing field, or the equivalent calculation of the long-range order, had clearly been one of the outstanding problems ever since the partition function in the absence of any field had been known. However, he gave no hint about the derivation.[471] As Montroll and co-workers put it, "In the days of Kepler and Galileo it was fashionable to announce a new scientific result through the circulation of a cryptogram which gave the author priority and his colleagues headaches. Onsager is one of the few moderns who operates in this tradition."[472]

A derivation of this result, a mathematical work as difficult as Onsager's for the zero-field partition function, was carried through for the square lattice by Yang, using a generalization of Kaufman's method.[473] He thanked Kaufman for stimulating discussions and for showing him her notes on Onsager's work on the long-range order, thus reinforcing the picture of a network, closely linking most of the major contributors to the Ising problem. Just as the logarithmic behavior of the specific heat at the Curie point, so also the mathematical form of the magnetization (involving a one-eighth power behavior) had no counterpart in earlier theories. Molecular field theories of magnetism, as well as the Landau theory of phase transitions, would have predicted a square root behavior for the magnetization.

In 1947, George Uhlenbeck introduced the Onsager solution of the Ising problem to Mark Kac, a Polish mathematician, who had immigrated to the United States in 1936 and joined the Cornell Mathematics Department. Before Onsager's solution appeared, it had been known that the partition function for the Ising net could be written as a sum, in which the coefficients were derived from "graphs" on the lattice.[474] But the counting of graphs was itself problematic, and in the spirit of using the exact solution as a guide to the proper method of counting, Kac and John Ward rederived the Onsager result for the partition function by purely combinatorial methods.[475]

The exact solution of the Ising model opened up the possibility of a rigorous formulation of other phase transition problems. While the Ising model was originally constructed to describe a ferromagnet, its physical applicability was extended later to entirely different physical systems. The 1930s models such as that of Bragg and Williams for order–disorder transitions in alloys, and the Lennard-Jones and Devonshire model of melting and condensation, bore considerable resemblance to the Ising model. In contrast to Kramers and Wannier, who in their 1941 paper refer to the "Ising model of ferromagnetism," Montroll was immediately concerned with treating mathematically all kinds of "nearest neighbor systems" including not only ferromagnets but also binary alloys and hindered rotations. Ashkin and Lamb in 1943 likewise thought in terms of binary alloys. Onsager in his 1944 paper observed that while the Ising model was originally intended to describe ferromagnetism, it is "more properly representative of condensation phenomena in the two-dimensional systems formed by the adsorption of gases on the surfaces of crystals." The formal analogies among lattice models of fluids, antiferromagnetic systems, and liquid–solid systems would soon be developed.[476]

Frustration with the three-dimensional Ising model encouraged an interest in simplified models for which the partition function could be evaluated exactly in three dimensions. Since the Ising model seemed intractable there, Kac proposed a "spherical model" in which the spins, rather than taking on only the values ± 1, are allowed continuous values, with the only restriction being that the sum of the squares of the spins must equal the total number of lattice sites. For the spherical model, the partition function, rather than being an extremely difficult sum, was an integral over a many-dimensional Gaussian function. Theodore Berlin, an Uhlenbeck student at Ann Arbor, realized that this function could readily be evaluated by the method of steepest descent.[477] Unlike the Ising model, the spherical model had no phase transition in two dimensions, but it did have one in three dimensions, as well as long-range order below the critical temperature.

In the 1937 Amsterdam Conference (see Section 8.1), the question had been debated whether the partition function of a fluid could and would in the thermodynamic limit mathematically yield a first-order phase transition—that is, that segment of the isotherms corresponding to the Maxwell construction for which pressure is independent of density. The Onsager solution was a full example of how the partition function could and did lead to singularities at the critical temperature. But some theoreticians sought mathematical proof for a model of a fluid in three dimensions without actually evaluating the partition function. Not that anyone after Onsager's solution seriously expressed doubts about the matter, but the formal mathematical investigation was looked to for mathematical insight into how smooth thermodynamic functions can develop sudden discontinuities and singularities.

Along these lines, Van Hove in 1949 assumed a potential with a hard core and an attractive portion or arbitrary shape but finite range, and proved from the Gibbs canonical distribution that the free energy per particle approaches a definite value in the thermodynamic limit, and that along an isotherm the pressure is a monotonic nondecreasing function of the volume per particle. Several years later, Lee and Yang utilized the grand canonical ensemble and found results in complete agreement with those of Van Hove. In their proof, they introduced the elegant approach of studying the analytic structure of the grand partition function as a function of a complex "fugacity" variable $\exp(\mu/T)$, where μ is the chemical potential and T is the temperature. The distribution of zeroes, as they showed, provided a mathematical language for characterizing phase transitions, a formulation they illustrated for the Ising model.[478]

In social terms, by the late 1940s the group of physicists and mathematicians working on rigorous theories of phase transitions was no longer centered in Holland and other European countries, as in the 1930s. Rather, it was based almost entirely in America. This small network of theoreticians, whose topic and methods fell outside the main thrust of theoretical physics, seemed to work more cooperatively than competitively—with some esprit de corps—on the mathematical edge of physics. The fact that their subject had relatively little concern or contact with experimental data, or with other branches of physics, was perhaps part of the reason why, unlike most other areas of physics, this one seemed little affected by the war. And while the number of physicists greatly increased in the 1950s, the membership of this small network remained almost the same. (The one exception appears to be the group of Japanese physicists mentioned earlier.) Moreover, the very completeness of the work of Onsager and others on the Ising model would make subsequent work anticlimactic, once the attempt to work out the three-dimensional Ising case began to be felt as hopeless. Not many in the younger generation were attracted to this austere area of rigorous theoretical physics.

The state of the field at the end of the 1940s is reflected in two major conferences: the first, "Phase Transitions in Solids,"[479] held in August 1948 at Cornell; and the second, "Statistical Mechanics," held on 17 to 20 May 1949 in Florence, Italy.[480] Both the content and the character of the Cornell conference are of historical interest. Montroll recalls that Onsager's discussion remark (not included in the published proceedings) giving the magnetization formula for the two-dimensional Ising model was the highlight for some participants. Otherwise, discussion of exact mathematical solutions relating to the Ising model was all but absent; the emphasis was empirical and phenomenological. The meeting, while retaining its connection to university research, was interdisciplinary, with invitations extended not only to physicists but also to chemists, metallurgists, ceramicists, crystallographers, glass experts, and specialists. One major aspect of wartime research had been collaboration among researchers across disciplines to fulfill goal-oriented missions; the Cornell conference had been organized by the National Research Council within this model. The meeting was sponsored by the Office of Naval Research through a contract with the National Academy of Sciences. The spirit of collaboration at this conference was for some scientists transitional between the extensive interdisciplinary teamwork characteristic of wartime research and the practice at typical universities of conducting research within narrow departmental structures.[481]

In one of the few theoretical statistical-mechanics papers at the Cornell conference, Kirkwood applied the molecular distribution function approach to the problem of fusion and cystallization, extending his earlier work with Monroe.[482] The starting point of this research was the formulation of the problem of crystallization in terms of how the density and pair correlation functions, with their high degree of symmetry in a fluid, change over to the corresponding functions with lower symmetry that describe the crystal structure. Kirkwood did not attempt a general solution to the system of equations he presented, but contented himself with the tractable problem of studying the stability of the liquid phase under small perturbations.

Also at this conference, Mayer[483] called attention to the considerable work done by Max Born with collaborators and students[484] in developing a theory of melting, and added a new wrinkle, dealing again with method more than results, to incorporate correction terms to the potential energy arising from crystal imperfections.

Tisza presented a paper in which, reconsidering Gibbs's thermodynamic theory of phase transitions, he observed that neither Ehrenfest's classification of phase transitions nor Landau's power series expansion of the free energy could accommodate a logarithmic singularity in the specific heat. Underlying these theories was the assumption that the singularity at the ferromagnetic, superconducting, and superfluid transitions led to a finite discontinuity in the specific heat. Tisza addressed this difficulty by formulating a general thermodynamic theory of phase transitions; he applied this to various physical systems—for example, critical points in fluids and ferroelectric Curie points, and the lambda points in substitutional alloys, molecular crystals, and liquid helium.[485] Wannier, Anderson, and J. M. Richardson, then working at the Bell Telephone Laboratories, pointed out that Tisza's formulation needed to be extended to accommodate fully the two-dimensional Ising model.

The Florence conference had been the first major meeting on statistical mechanics since the start of the Second World War. While it and the Cornell meeting overlapped in subject matter, they had quite different but complementary emphases; at Cornell the main point was thermodynamic approaches to the description of phase transitions and empirical studies of phase transition of industrial interest, while at Florence the emphasis was on the conceptual and mathematical problems of a subspecialty within theoretical physics. Whereas the Cornell conference had emphasized the crossing of disciplinary boundaries and was sponsored by agencies based in the United States, the Florence conference emphasized crossing international boundaries. The ideal of international scientific cooperation was beginning to reassert itself after the war. One of the conference sponsors was UNESCO, the chief United Nations agency promoting international cultural contacts. Another sponsoring agency was the International Union of Pure and Applied Physics (IUPAP). Locally the conference was organized by the Italian Physical Society and the University of Florence in Arcetri. Notwithstanding the international ideal, all the speakers as well as all those who commented informally were based in either North America or Western Europe (Sweden, France, Great Britain, Belgium, the Netherlands, Switzerland, Italy, but no one from Germany or Austria). None attended from the Soviet Union or from any of the Eastern European countries.

Kramers, then president of IUPAP, opened the Florence conference by listing the outstanding problems of statistical mechanics. Among these he included problems of foundations, the extension to irreversible processes, and the development of methods to evaluate the partition functions for systems in which many-particle correlations are prominent. In that connection, he referred to Weiss's treatment of the ferromagnetic phase transition as the seminal first step. Returning to the present, Kramers spoke hopefully of the recent "development of new methods."[486] It was a time of searching for mathematical techniques and formulations of the phase transition problem that would make at least some aspects of it tractable. The conference mirrored all the major issues in statistical mechanics of the time. Some of the controversies are preserved in the published discussion remarks following each paper.

The Florence conference hinted at further approaches to the theory of phase transitions: thermodynamics and phenomenological theories, field-theory techniques, time-dependent correlation functions, and series expansions and other approximation methods, which would later become the fabric of the field. On the theory of phase transitions, highlights of the meeting included papers by E. A. Guggenheim, Mayer, G. S. Rushbrooke, and Montroll. Also noteworthy were three short sentences uttered by Onsager following Rushbrooke's paper, in which he stated the unpublished Kaufman–Onsager result (which we quoted in Section 8.1) for the degree of order, or magnetization, of a two-dimensional rectangular Ising net. Another discussion remark by Onsager at this conference described quantization of circulation in a superfluid. Pauli, in the discussion following a paper by Oskar Klein, remarked that metastable macroscopic states can be more adequately described by statistical mechanics than by thermodynamics.[487]

Guggenheim introduced a heuristic quasi-thermodynamic concept, the "cooperative free energy" for, say, an atom located in a crystal in the average field of the remainder of the cooperative system. He used this concept to explore systems including liquid–gas, regular solutions, order–disorder in alloys, and thermionic emission.[488] Mayer's quite formal paper,[489] following earlier approaches of Kirkwood and others,[490] focused on formulating coupled integral equations for the many-particle distribution functions and obtaining limited information on the equation of state in the critical region.

Rushbrooke reported on an innovative and eventually very important approach to the problem of critical behavior, carried out with his research student A. J. Wakefield, using high- and low-temperature series expansions.[491] He first of all developed the mathematical analogy between an Ising ferromagnet and a "regular solution" containing two molecular species, A and B, with species-dependent interaction energies between neighbors. He then considered the partition function of the three-dimensional simple cubic Ising lattice by means of low-temperature and high-temperature expansions, which were laboriously carried out by combinatorial methods to a large number of terms. With the aid of extrapolation, this calculation permitted an estimate of the critical mixing temperature for a solution with equal number of A and B molecules. The result differed from other approximate calculations (Bragg–Williams, Bethe–Guggenheim) in the same direction as the exact Wannier–Onsager result differed in the two-dimensional case. Rushbrooke found this an indication that he was on the right track, and rightly offered the series expansions as a generally promising approach to the study of critical behavior where exact solutions were unavailable. The technology of such series expansions would soon become increasingly refined and fruitful.

In the discussion period following, Kirkwood asked Rushbrooke whether he had tested the method in the two-dimensional case where the exact solutions were known; he had not yet done so. In fact, C. Domb had independently obtained in 1949 a low-temperature series for the two-dimensional Ising net, thereby not only replicating the low-temperature end of the Onsager solution in the absence of a magnetic field, but also deriving new results for both magnetization and specific heat in the case in which the magnetic field is present.[492] Domb also obtained several terms in the high-temperature expansion for the Ising net with and without a field.

The discussion continued with DeBoer, Onsager, and Guggenheim assessing measurements on the critical point of pure liquids, where fluctuations are very large and relaxation times extremely long. Neither Guggenheim, who remarked that "it is extremely difficult to determine experimentally with certainty what happens in the immediate neighborhood of the critical point," nor Onsager, who felt that "it is very difficult to reach true equilibrium near the critical point," was willing to accept such measurements at face value. In short, it was conveniently permissible for the theoreticians to pay no heed to experimental data close to the critical point!

In the next paper, Montroll defined a generalized model of cooperative phenomena that reduced to the Ising model in one limit and to the Kac–Berlin spherical model in another. Having come to the conclusion that "it will be exceedingly difficult if not impossible to extend Onsager's algebraic solution of the two-dimensional Ising problem to three dimensions," Montroll solved the spherical model by mathematical approaches different from those of Kac and Berlin, in the hope that his methods might lend themselves more readily to generalization.[493]

With Onsager's solution of the two-dimensional Ising model demonstrating that the methods of statistical mechanics were appropriate to the problem of phase transitions, thermodynamic methods for describing phase transistion—classical nineteenth-century techniques—were less in vogue and less interesting to most physicists than statistical mechanics. A few physicists, however, continued to take a strong interest in the thermodynamic approach, especially Landau and Lifshitz, Pippard, and Tisza.

Although the second English edition of Landau and Lifshitz's *Statistical Physics,* which appeared in 1958, was a quite new text going far beyond its 1938 predecessor,[494] the thermodynamic theory of phase transitions in the 1958 edition was based largely on Landau's earlier work of 1935 and 1937. Their method, appealing for its mathematical simplicity and directness, was buoyed by the success of the Landau–Ginzburg theory in describing the region of the phase transition in superconductors. It was the only theory to do so, after the Weiss mean-field theory (which did not work very close to the critical temperature). The Landau–Lifshitz method, however, entailed particular assumptions that certain thermodynamic derivatives behave at the critical point in the same way as in mean-field theories, such as the Weiss model for ferromagnetism and the Bragg–Williams model for the order–disorder transition in beta-brass. In particular, second-order phase transitions, including the ferromagnetic, superconducting, and superfluid transitions, as well as the order–disorder transition for a binary alloy, were characterized in the Landau–Lifshitz theory by the appearance of a nonvanishing long-range order signaling a change of symmetry; the order parameter "vanishes continuously, without a discontinuous jump," at the transition. Without justification, they boldly assumed that the Gibbs free energy could be expanded as a power series in terms of the order parameter near the transition point, but they proceeded aware that "a complete description of the nature of the behavior of the thermodynamic potential at the transition point is extremely difficult and it has not yet been done."[495]

But long before 1958, the Onsager solution for the Ising model had yielded results entirely different from those of Landau and Lifshitz, an event that would at first sight seem to have required major revision, if not dismissal, of the thermodynamic theory. But Landau and Lifshitz were undaunted. In their book, they merely stated that their theory (with, for example, a finite jump in the specific heat) applied to transitions in ordinary three-dimensional physical systems, whereas the Onsager solution (with a logarithmic singularity in the specific heat at the transition) applied to plane lattice structures only.[496] Experimental data for the three-dimensional case were at that time not sufficiently precise, and the statistical-mechanical theory was too crude, either to corroborate or to refuse Landau and Lifshitz unambiguously. However, it seemed implausible that the three-dimensional counterpart of the two-dimensional problem solved by Onsager would bear so little resemblance to the latter. Moreover, experimental observations were strongly suggestive of critical behavior in disagreement with the Landau theory.

When Pippard in 1957 surveyed the experimental data and gave an overview of the phenomenology of phase transitions, he supplemented Ehrenfest's and Landau's categories by the addition of the lambda transition:

> three distinct types of thermal behavior occurring along lines separating different phases or modifications of a substance, the normal transition with latent heat, the λ transition without latent heat but a very high (perhaps infinite) peak of specific heat, and the so-called second-order transition in which there is no latent heat and a finite discontinuity in specific heat.[497]

He further noted that "it is probably true to say that of physically interesting systems there is only one class, the superconducting transition, which bears any resemblance to an ideal second-order transition." The Landau theory had indeed proved most fruitful for superconductivity, resulting in the Landau–Ginzburg theory. Among lambda transitions, Pippard listed liquid helium, the antiferromagnetic transition in $MnBr_2$, and order–disorder transformations in beta-brass, ammonium salts, crystalline quartz, and many other solids. He suggested tentatively that the Curie point of many ferromagnets may also represent a lambda transition, but stated that the experimental evidence for the magnetic materials is too unclear to permit reliable classification. In 1949, Tisza had given a more general thermodynamic theory of phase transitions that included lambda transitions, in which he concluded that the nature of the singularities cannot be established by thermodynamic arguments alone but requires statistical physics.[498]

Another approach to phase transitions in this period centered about pair correlation functions and scattering. The density–density correlation function, which measures how the values of the density at a given point is correlated with its values at another point, had been introduced by L. S. Ornstein and H. Zernike in 1914 to discuss critical opalescence.[499] They expressed the fluctuation in the number of particles in a macroscopic volume of fluid, a number proportional to the isothermal compressibility, as an integral over the correlation function. In particular, they were able by this means to describe the anomalously large forward scattering of light near the critical point in terms of the correlation function. Zernike himself measured the scattering and found good agreement between their theory and experiment. The theory was an improvement over the earlier theories of "critical opalescence" by Einstein and by Smoluchowski, who had considered fluctuations but not spatial correlations.[500]

In a subsequent paper Zernike had gone further and showed that the correlation function $g(r)$ of a fluid decays exponentially at large distances with a characteristic "correlation length" $1/\kappa$—that is, $g(r) = A \exp(-\kappa r)/r$.[501] Here both the factor A and κ are functions of temperature, and κ vanishes at the critical point; that is, the correlation length becomes infinite at the critical point. Then in 1926, Zernike and J. A. Prins extended the earlier theories of optical scattering to the scattering of monochromatic x rays in terms of a pair correlation function.[502] It was also found possible to compute the correlation function from an experimental angular scattering distribution.[503]

Although Einstein had recognized in 1910 that if the density fluctuations were Fourier-analyzed, the optical scattering from a particular mode must satisfy a principle of conservation of wave vector, the full physical understanding of the scattering took several decades. Brillouin in France and Mandel'shtam in the Soviet Union described the fluctuations in fluids as a superposition of sound waves, and on that basis predicted a shift in the frequency between the incident and the scattered radiation, resulting in a doublet.[504]

A more complete theory of optical scattering by a liquid was worked out by Landau and George Placzek in a "hydrodynamic" approximation, which assumed that the incident wavelength is large compared with molecular mean free paths. In principle, as they found, a good deal of information about the liquid could be gleaned from optical scattering.[505] However, optical resolution was for a long time inadequate for measuring accurately the linewidths of the sidebands and similar fine details of the spectrum. Only gas laser techniques and the associated high-resolution interferometry developed in the early 1960s would permit such precise measurements. New experimental optical techniques would, in turn, stimulate further theoretical developments.[506]

Two different methods were tried to derive the theory of critical fluctuations and scattering. One was the Ornstein–Zernike approach, which began by relating the direct intermolecular interaction to the pair correlation function through an integral equation; the other was the semiphenomenological approach, used by Landau and his co-workers.[507] The latter approach to a certain extent paralleled Onsager's consideration of the time correlations of fluctuating variables that led to his famous reciprocal relations for the response coefficients in linear irreversible processes.[508] Onsager's work, as well as work on the theory of stochastic processes,[509] had called attention to the use of correlations in time, rather than space, for statistical mechanics. The great usefulness of the space–time correlation function in connection with critical phenomena was cogently demonstrated only in the 1950s, when Van Hove applied it to the scattering of slow neutrons from a system of interacting particles.

Van Hove, after working at the Institute for Advanced Study in Princeton with Placzek on inelastic neutron scattering from crystalline solids,[510] gave a general formulation, in Born approximation, of the scattering of neutrons by systems of interacting particles.[511] In his paper, Van Hove introduced "the time-dependent generalization of the familiar pair distribution function to indicate its interest from the standpoint of statistical mechanics, and to establish its role in scattering theory, showing at the same time how slow neutron scattering makes it experimentally accessible."[512] To evaluate the time-dependent correlation function for a nonmagnetic crystalline solid, he assumed purely harmonic forces between atoms. And to obtain explicit results for the scattering from liquids or dense gases, he required the assumption (first introduced by Onsager)[513] that spontaneous fluctuations of macroscopic variables decay to equilibrium according to the same laws as artificially created deviations from thermal equilibrium, as in linear heat conduction. Van Hove found that for neutrons, the intensity of scattering is exceptionally large in the neighborhood of the critical point of the fluid—which was what would be expected from the long-familiar case of electromagnetic scattering.

Van Hove soon submitted another article dealing with neutron scattering from ferromagnetic crystals, described by the Heisenberg model for a ferromagnet.[514] The scattering cross section was essentially the Fourier transform of the space- and time-dependent spin–spin correlation function. The large spontaneous fluctuations in magnetization density in the neighborhood of the Curie point, he found, lead to large "critical scattering," which he was able to calculate, at any rate just above the Curie temperature. His result was consistent with contemporary experimental work, most of which had not yet been described in print.[515] As a result of atomic bomb and reactor development during the Second World War, neutron beams had become readily available for experimental studies.

Within a few years, more detailed calculations were made of the critical spin–spin correlation function for a Heisenberg ferromagnet and of the scattering of x rays or neutrons from binary alloys described by the Ising model.[516] These were promptly compared with the growing fund of experimental information. Band-theoretical calculations for the ferromagnetic correlation function and critical scattering of transition metals of the iron group were not carried through until the 1960s.[517]

The highly mathematical renormalization group techniques later developed by Leo Kadanoff, Ben Widom, Kenneth Wilson, and others would eventually provide new physical insights into the partition function and correlation functions near critical points, and into the nature of critical scattering.[518]

In the 1950s the technology of high- and low-temperature series expansions continued to be developed, particularly in England, and was applied to the Heisenberg as well as the Ising model.[519] At low temperatures, the series expansion technique resembled the earlier spin-wave approximations for the Heinsenberg model. The expansions of Domb, Rushbrooke, and co-workers were distinguished from other approximation methods in that the procedure was systematic and based on exact power series expansions of the free energy. Although at first sight the series expansions were a numerical method offering no physical insight, the combinatorial problem in calculating coefficients in the series did entail physical considerations suggestive of the processes entailed in critical behavior. The series expansions permitted estimation with considerable accuracy of the behavior of thermodynamic functions close to the critical point. Furthermore, they applied to three as well as two dimensions, and to the Heisenberg and other models as well as to the Ising case. Beginning in 1961, the technology would be supplemented by bringing to bear the tool of Padé approximants, which provided a means for continuing the series results to the critical point.

One of the most important developments in statistical mechanics in the postwar period was the introduction of field-theory techniques, modeled initially on quantum electrodynamics. We have earlier traced their development in the context of the electron gas and have noted their use in the problems of superconductivity and liquid helium. Eventually, the field theory approach would become a common underpinning for nearly all the various areas of the many-body problem. By 1961, Pines could appropriately say that "the recent developments in the many-body problem . . . have tended to change it from a quiet corner of theoretical physics to a major crossroad."[520]

The application of field theory to statistical mechanics had analogies with earlier cluster expansions and with the use of coupled integral equations for classical *n*-particle distribution functions. In the 1940s the classical coupled integral equations for the distribution functions had been developed by Born and Green, Kirkwood, and Bogoliubov.[521]

The use of diagrams in the finite temperature quantum many-body problem may be traced in large measure to the pioneering work of Takeo Matsubara at the Research Institute for Fundamental Physics in Kyoto in 1955.[522] Matsubara approached the calculation of the partition function of many-body systems with interactions by introducing "explicitly the quantized field of particles and utiliz[ing] the various techniques of operator calculus in quantum field theory as far as possible in evaluating the quantum-statistical average of the field quantities." As he stated in his abstract,

> the grand partition function, which is a trace of the density matrix expressed in terms of field operators, can be evaluated in a way almost parallel with the evaluation of the vacuum expectation values of the *S*-matrix in quantum field theory, provided that appropriate modifications in notation and definition are made.

Some physicists developed perturbation theories for the many-body problem at finite temperatures, and time-dependent Green's functions (generalization of those introduced by Matsubara) were variously defined, the differential equations they satisfy derived, and theorems about them proved. Crucial to making thermal Green's functions into a practical calculational tool was the discovery by Ryogo Kubo of the boundary conditions that they obey. The boundary conditions, a form of periodicity in imaginary time, exploited the similarity, first recognized by Felix Bloch in his 1932 *Habilitationsschrift* (Chapter 2), of the thermal density matrix and the Heisenberg time-development operator for imaginary times.[523] Primarily, interest lay in understanding normal properties of metals, superconductivity, and superfluidity. Field-theory methods and language were eventually brought to bear on critical phenomena, and by the early 1970s, Wilson could rightly state that "the efforts of many years to apply Feynman diagram methods to critical phenomena finally succeeded."[524]

How far had physicists come by the 1960s in understanding phase transitions? The question concerns underlying scientific objectives—how different researchers would define an "adequate understanding" of phase transitions and critical phenomena. Ernst Mach might have considered it sufficient for an elaborate numerical calculation based on the canonical ensemble to yield the experimental curves for thermodynamic functions in the critical region. Ehrenfest might have insisted rather that physical insight into the nature of the process is of central importance. Philosophical attitudes, rarely articulated in the published articles, were muted by the positivism dominating physics in the United States and Western Europe since the Second World War. The inclinations of leading physicists of an earlier generation—Bohr, Einstein, Planck, Schrödinger—had been intensely philosphical; Bloch and Peierls, by contrast, expressed a sentiment characteristic of the younger generation when they complained of Bohr's preoccupation with fundamental issues as a distraction from the challenging task of physics to calculate answers.[525] By the end of the Second World War, the antiphilosophical bias had hardened among physicists.[526]

The appeal of a theory of phase transitions was that it could potentially encompass the most diverse physical systems. To the extent that the primary objective of such a theory was to provide a simple and lucid description of the physical and mathematical origin and nature of the discontinuities and singularities of otherwise smooth functions describing macroscopic matter, the Weiss field and early kinetic theory models, such as that of van der Waals, had been major steps. But adequate mathematical description required exact calculations so complex as to constitute a subject of study in their own right; the primary objective might appear, at least temporarily, forgotten. Increasingly after the war the focus was on the exact nature of the singularities and the precise "critical exponents."

The discrepancy between the logarithmic singularity of the Onsager solution and all previous results had highlighted the issue. Once identified as crucial, the question could and did engender a variety of approaches and a fairly well-defined general research program to calculate and measure those exponents. The laborious series expansions were a natural and fruitful part of such a research program. The thermodynamic theories reflected the natural requirement that phase transitions and critical phenomena had somehow to fit into general thermodynamics, and the description in terms of correlation functions, so closely related to scattering experiments, fluctuations, and the theory of steady-state irreversible phenomena, linked critical phenomena to other portions of physics. Through subsequent progress by renormalization group techniques, coupled with extensive computer simulation, the theory of critical phenomena came to fruition.

8.3   Conclusion

A number of common themes and elements unify the subfields of collective phenomena. Among them are the concepts of long-range order, changes of symmetry in phase transitions, collective modes, low-lying excitations above the ground state, Bose–Einstein condensation, pairing, broken symmetry, order parameters, and macroscopic quantum phenomena. Although certain of these unities were recognized well before the Second World War, even though often phrased in different language, recognition alone did not weld the subfields together. And although individuals did occasionally cross over among subfields in their work (e.g., Landau on helium, superconductivity, and phase transitions; Fritz London on superconductivity and helium), the communities of researchers working on different problems of collective phenomena did not generally see themselves as working under a common umbrella. Only very gradually, in the 1950s and 1960s, did these subcommunities and subfields coalesce around the unifying strands.

As an illustration of the intellectual unification of the subfields, consider the development of the concept of a spatially varying order parameter to describe the broken-symmetry states of superconductivity and superfluidity. The notion of an order parameter distinguishing a condensed state can be traced back to the local magnetization in the Weiss theory of ferromagnetism (Chapters 2 and 6). Bloch, in his *Habilitationsschrift* on ferromagnetic domain walls, then generalized the notion to spatially varying situations and showed how to calculate the energy in terms of the order parameter.[528] The idea was soon picked up by Landau in his 1935 theory of phase transitions,[529] and was eventually introduced into superconductivity in the 1950 Ginzburg–Landau theory.[530] With Gor'kov's 1957 formulation of the BCS theory of superconductivity in terms of diagrammatic field theory and his subsequent derivation of the Ginzburg–Landau theory from BCS, the role of a spatially varying order parameter characterizing the superconducting state became established.[531]

The recognition of an order parameter to describe superfluidity in helium came through a basically independent route. The idea was implicit in London's 1938 picture of helium undergoing Bose–Einstein condensation into a single macroscopic quantum state,[532] and was also probably in the back of Onsager's mind when he argued, in 1949, that the vorticity had to be quantized. (He certainly understood its role in superconductivity by the time of the flux quantization experiments.) As early as 1951, O. Penrose suggested how the single-particle correlation function of a Bose fluid would exhibit long-range order in terms of what we now recognize as the order parameter.[533] But an order parameter that was possibly spatially varying did not enter into any discussion of the microscopic behavior of helium II until the work of Pitaevskii and Gross in 1961.[534] Aside from quantization of circulation, the concept would become useful only when Josephson and Anderson, in the context of tunneling between superconductors, demonstrated the importance of the phase of the order parameter.[535] The commonality of the order parameters used to describe the broken symmetry in both superconductivity and helium superfluidity was recognized by workers in the field by the early 1960s, and was codified in Yang's review of "off-diagonal" long-range order.[536] By this point, the two phenomena were seen as similar manifestations of quantum mechanics on a macroscopic scale.

The growing application in the late 1950s of diagrammatic field-theory methods (Feynman diagrams and Green's functions) led to a general theoretical framework for collective phenomena. Although the first diagrammatic technqiues were introduced into problems of collective phenomena in a classical context—the cluster expansions—the modern techniques had their root in work on quantum electrodynamics after the Second World War. As we have seen, these techniques grew independently in the study of the electron gas and of nuclear matter, were extended to finite temperature by Matsubara and others,[537] were first applied to neutral Fermi systems by Migdal and Galitskii,[538] and were then applied to superconductivity by Gor'kov and to helium by Beliaev and by Hugenholtz and Pines.[539]

The war aided this development by providing experimental technologies that contributed to the development of quantum electrodynamics,[540] by increasing the funding for solid-state physics (Chapter 9), and by making it fashionable for those talented in physics and mathematics to enter solid state and work on applied problems. By the 1950s, solid-state theory had clearly developed from an area in which practitioners worked on rather abstract problems of ideal solids (Chapter 2) to one in which physicists could deal with the physics of solids.

The connection of the many-body problem to field theory helped as well to fertilize high-energy physics. For example, the concept of broken symmetry and the accompanying Anderson–Higgs mechanism has played a major role in theories of elementary particles, such as the unified theory of electroweak interactions. Just as solid-state physics had been a proving ground for quantum mechanics in the late 1920s and early 1930s (Chapter 2), collective phenomena problems would eventually serve as a source of ideas as well as a testing ground for quantum field theory.

By the late 1950s, approaches to collective phenomena in different physical contexts—the electron gas, superconductivity, helium, and nuclear matter—had converged sufficiently to define a field of "many-body physics." An international community working on various aspects of the problem had formed and begun to meet regularly at specialized conferences and to train students. The first major symposium on the many-body problem was held in January 1957, at the Stevens Institute of Technology, where relationships between theoretical work in these various areas were examined and put into perspective. The Les Houches summer school of 1958, entitled "The Many-body Problem" and organized by Nozières, included courses by Schrieffer on the theory of superconductivity; Huang on the hard-sphere boson gas and the binary collision approach of Lee and Yang; Brueckner on the application of many-body theory to nuclear matter and other problems; Bohm on collective coordinates; Pines on electrons, plasmons, and phonons; Beliaev on the Bogoliubov canonical transformation method; and Hugenholtz on perturbation theory of many-fermion systems.[541] The second major meeting on the many-body problem, held at Utrecht in June 1960 and attended by several leading Soviet physicists, added phase changes, a problem of continuing interest in the Netherlands, to the topics covered at the Stevens meeting.[542]

Until the 1960s, phase transitions remained somewhat apart from the streams that converged into many-body physics. However, as superconductivity and superfluidity became better understood, the attention of many-body physicists began to turn to applications of their proven methods to problems of critical phenomena. Important examples were the studies of the critical behavior of superfluid ^{4}He by Patashinskii and Pokrovskii, and by Josephson.[543] With the development of renormalization group techniques starting in the mid-1960s, this area too came under the umbrella, and the study of collective phenomena had come together as a single area of the physics of condensed matter.

Notes

S. Heims is the major author of the sections on phase transitions and critical phenomena, and H. Schubert is responsible for the sections on liquid helium and the prewar history of superconductivity. The discussion of the post-BCS period in superconductivity is based in large part on an unpublished manuscript of P. W. Anderson, "It's Not Over till the Fat Lady Sings," presented at the March 1987 meeting of the American Physical Society, New York City. We would like to thank Professor Anderson for making this manuscript available to us for use in this chapter.

1. See, for example, F. Bloch, *Die Elektronentheorie der Metalle, Handbuch der Radiologie,* 2nd ed. (Leipzig: Akademische Verlagsgesellschaff, 1933), 6: pt. 1, 226–278, 238–239, 278; Bloch, "Some Remarks on the Theory of Superconductivity," *Physics Today* 19, no. 5 (1966): 27–36; L. Brillouin, *Die Quantenstatistik und ihre Anwendug auf die Elektronentheorie der Metalle,* vol. 13 of *Struktur der Materie in Einzeldarstellungen* (Berlin: Springer, 1931), intro., 45, 413–415; H. G. Smith and J. O. Wilhelm, "Superconductivity," *Reviews of Modern Physics* 7 (1935): 237–271, p. 270.

2. We would like to acknowledge very helpful discussions with John Bardeen, Anthony Leggett, Christopher Pethick, and David Pines in developing this history, as well as all those, named explicitly in the notes, whom we interviewed. All interviews referred to in this chapter have been deposited at AIP unless otherwise indicated. For their critical reading of the entire chapter, and numerous corrections and additions, we are particularly grateful to Spencer Weart, David Pines, and Philip Anderson. These readers also made many valuable suggestions of areas in which the material deserves further expansion; we regret that project deadlines and space limited the extent to which we were able to add such material in appropriate detail. Scientific areas that have strongly influenced many-body theory and certainly warrant expanded treatment include quantum transport in solids and the collective theory of magnetism; it was unfortunately not possible to include the latter in Chapter 6 either. Valuable historical sources in these area are Anderson's interviews with Hoddeson on 13 July 1987 and 10 May 1988, as well as his letter to Hoddeson of 11 February 1988 (all on file at the Center for History of Physics at the AIP), and his article "Some Memories of Developments in the Theory of Magnetism," *Journal of Applied Physics* 50, no. 11 (1979): 7281–7284. We also recognize that the present history does not, for lack of access to primary documents, adequately treat Soviet or Japanese contributions to the modern development of the theory of collective phenomena; we look forward to more complete material from Soviet and Japanese historians of solid-state physics.

3. The principal elementary excitations in the theory of collective phenomena are the following: the phonon, which is a quantum of lattice vibration above the ground state in a crystal, or a long-wavelength-density oscillation in liquid ^{4}He; the quantized spin wave in a ferromagnet, or magnon (Chapters 2 and 6); the exciton, or bound electron–hole in an insulator or a semiconductor (Chapter 4); the polaron, an electron moving through a polar crystal lattice "dressed" by the comoving

cloud of phonons formed by the interactions of the electron and the ions (Chapter 4); the roton, which is the elementary excitation at short wave numbers in superfluid ^{4}He; the quasiparticle, a dressed particle excitation above the Fermi sea, clothed by interactions with other excitations—for example, an excited electron in normal or superconducting metals, or a ^{3}He atom in liquid ^{3}He; and the plasmon, the quantum associated with longitudinal (Langmuir) oscillations of an interacting plasma. An account of the discovery and naming of the elementary excitations in condensed matter is given in C. T. Walker and G. A. Slack, "Who Named the -On's?" *American Journal of Physics* 38 (1970): 1380–1389.

4. P. Debye and E. Hückel, "Zur Theorie der Elektrolyte. I. Gefrier punktserniedrigung und verwandte Erscheinungen," *Phys. Zs.* 24 (1923): 185–208; Debye and Hückel "Zur Theorie der Elektrolyte. II. Das Grenzgesetz für die elektrische Leitfähigkeit," *Phys. Zs.* 24 (1923): 305–328.

5. See P. Debye, interview with D. M. Kerr, Jr., and L. P. Williams, 22 December 1965, from which most of the background material given here has been abstracted, Cornell University Archives. Also see Debye and Hückel ("Gefrier punktserniedrigung") (n. 4); P. Debye, interview with T. S. Kuhn and G. Uhlenbeck, 3 May 1962. pp. 17–18, Archives for the History of Quantum Physics, American Philosophical Society, Philadelphia, copies at AIP and elsewhere.

6. J. Ghosh, "Abnormality of Strong Electrolytes. I. Electrical Conductivity of Aqueous Salt Solutions," *Chemical Society Journal, Transactions* 113 (1918): 449–458, Ghosh, "Electrical Conductivity of Acids and Bases in Aqueous Solutions," *Chemical Society Journal, Transactions* 113 (1918): 790–799; Ghosh, "The Abnormality of Strong Electrolytes. III. The Osmotic Pressure of Salt Solutions and Equilibrium between Electrolytes," *Chemical Society Journal, Transactions* 113 (1918): 707–715; Ghosh, "The Abnormality of Strong Electrolytes. Part II. The Electrical Conductivity of Non-aqueous Solutions," *Chemical Society Journal, Transactions* 113 (1918): 627–638; Ghosh, "Eine allgemeine Theorie der Elektrolösungen," *Zeitschrift für Physikalische Chemie* A98 (1921): 211–238.

7. Debye interview (1965) (n. 5).

8. Debye and Hückel (n. 4).

9. I. Langmuir, "Scattering of Electrons in Ionized Gases," *Phys. Rev.* 26 (1925): 583–613; Langmuir, "Oscillations in Ionized Gases," *Proceedings of the National Academy of Sciences* 14 (1928): 627–637, p. 628. Other industries and universities did not, however, use the term *plasma,* regarding it as G. E. jargon; Bohm's work circa 1950 appears to have reintroduced it.

10. Actually, such waves had first been detected in 1920 by H. Barkhausen and K. Kurz, "Die kürzesten mit Vakuumröhren herstellbaren Wellen," *Phys. Zs.* 21 (1920): 1–6, and discussed theoretically in 1922 by E.W.B. Gill and J. H. Morrel, "Short Electric Waves Obtained by Valves," *Phil. Mag.* 44 (1922): 161–178. L. Tonks and I. Langmuir, "Oscillations in Ionized Gases," *Phys. Rev.* 33 (1929): 195–210.

11. The general picture of this research in the 1920s is given by H. Mott-Smith, interview with G. Wise, 1 and 2 March 1977, Archives of G. E. Corporation, Schenectady, N.Y. See also G. Wise, *General Electric and the Origins of U.S. Industrial Research* (New York: Columbia University Press, 1985); M. R. Andrews, "Reports of Progress in Lamp and Tube Work at the Research Laboratory of the General Electric Company," 4 June 1929, pp. 13–14, and 11 February 1929, pp. 9–10, mimeograph. We thank George Wise at G. E. for kindly supplying this material.

12. E. Wigner, "On the Interaction of Electrons in Metals," *Phys. Rev.* 46 (1934): 1002–1011; Wigner "Effects of the Electron Interaction on the Energy Levels of Electrons in Metals," *Transactions of the Faraday Society* 34 (1938): 678–685.

13. E. Wigner, interview with L. Hoddeson, G. Baym, and F. Seitz, January 1981.

14. The next order contribution to the ground-state energy of the electron gas, in the Rayleigh–Schrödinger perturbation expansion in the electron charge, is in fact logarithmically divergent, a divergence missed by Wigner in his approximate calculation. According to David Pines, this discouraging divergence, arising from the long range of the Coulomb interaction "set back the systematic application of perturbation theory to the electron gas for some 20 years." See D. Pines, *Elementary Excitation in Solids* (New York: Benjamin, 1964), 89.

15. In his 1938 paper, Wigner gives both an interpolation formula for the correlation energy (p. 680) based on his numerically incorrect low density limit for the correlation energy of 1934, and a corrected version in a footnote (p. 684). See ibid., 91–95.

16. Wigner interview (n. 13).

17. P.A.M. Dirac, "Note on Exchange Phenomena in the Thomas Atom," *Proceedings of the Cambridge Philosophical Society* 26 (1930): 376–385.

18. J. Bardeen, "Electron Exchange in the Theory of Metals," *Phys. Rev.* 50 (1936): 1098–1099.

19. J. Bardeen, interview with L. Hoddeson, 12 and 16 May 1977.

20. J. Bardeen, "Conductivity of Monovalent Metals," *Phys. Rev.* 52 (1937): 688–697.

21. The discovery and early experimental work on superconductivity have been treated in extensive detail in two outstanding articles by P. Dahl: "Kamerlingh Onnes and the Discovery of Superconductivity: The Leyden Years, 1911–1914," *Historical Studies in the Physical Sciences* 15, no. 1 (1984): 1–37, and "Superconductivity after World War I and Circumstances Surrounding Discovery of a State $B = O$," *Historical Studies in the Physical Sciences* 16, no. 1 (1986): 1–58. These articles appeared after the initial drafting of this section, and we present here only a brief summary of the important events. Other useful sources of the history of superconductivity include D. Shoenberg, *Superconductivity*, 2nd ed. (Cambridge: Cambridge University Press, 1952); Shoenberg, "Forty Odd Years in the Cold," *Physics Bulletin* 29 (1978): 16–19; J. Bardeen, "History of Superconductivity Research," in *Impact of Basic Research on Technology*, ed. B. Kursunoglu and A. Perlmutter (New York: Plenum, 1973); R. de Bruyn Ouboter, "Superconductivity: Discoveries During the Early Years of Low Temperature Research at Leiden, 1908–1914," *IEEE Transactions Magnetics* MAG-23, no. 2 (1987): 355–370; C. J. Gorter, "History and Future Applications of Superconductivity," in *Physics in the Sixties*, ed. S. K. Runcorn (London: Oliver and Boyd, 1963), 79–95; B. Serin, "Superconductivity, Experimental Part," in *Handbuch der Physik*, ed. S. Flügge (Berlin: Springer, 1956), 15: 211–273; M. von Laue, "Geschichtliches über Supraleitung," *Zeitschrift für Forschungen und Fortschritte* 25 (1949): 287–289.

22. H. Kamerlingh Onnes and J. Clay, "On the Change of the Resistance of the Metals at Very Low Temperatures and the Influence Exerted on It by Small Amounts of Admixtures," *Communications from the Physical Laboratory of the University of Leiden* 99c (1908): 17–26.

23. J. Dewar and J. A. Fleming, "The Electrical Resistance of Metals and Alloys at Temperatures Approaching the Absolute Zero," *Phil. Mag.* 36 (1893): 271–299.

24. H. Kamerlingh Onnes, "The Resistance of Platinum at Helium Temperatures," *Communications from the Physical Laboratory of the University of Leiden* 119 (1911): 18–26; p. 22.

25. Ibid.

26. H. Kamerlingh Onnes, "The Resistance of Pure Mercury at Helium Temperatures," *Communications from the Physical Laboratory of the University of Leiden* 120b (1911): 3–5, p. 4.

27. H. Kamerlingh Onnes, "The Disappearance of the Resistance of Mercury," *Communications from the Physical Laboratory of the University of Leiden* 122b (1911): 13–15.

28. Ibid., 14. After the decisive experiments on mercury, Kamerlingh Onnes remarked that he had overlooked the 1910 quantum theory of electrical resistance of Lindemann, and having been unacquainted with the work, it was a fortunate circumstance that the unfavorable predictions of the Lindemann formula might have led to different experimental directions. Footnote in H. Kamerlingh Onnes, "Remarks on the Preceding Communications," *Communications from the Physical Laboratory of the University of Leiden* 123 (1911); 3–9.

29. See H. Kamerlingh Onnes, in Nobel Institute, *Nobel Lectures, Physics, 1, 1901–1921* (Amsterdam: Elsevier, 1967), 306–336; "Meester Flim," student interview with G. J. Flim, ca. 1965, Bardeen files, University of Illinois.

30. H. Kamerlingh Onnes, "Über den elektrischen Widerstand," *Abhandlungen der Deutschen Bunsen-Gesellschaft für Angewandte Physikalische Chemie* 7 (1913): 245–250.

31. H. Kamerlingh Onnes, "Further Experiments with Liquid Helium. H. On the Electrical Resistance of Pure Metals, etc. VII. The Potential Difference Necessary for the Electrical Current Through Mercury Below 4.19 K," *Communications from the Physical Laboratory of the University of Leiden* 133a (1913): 3–26, p. 13.

32. Ibid.

33. H. Kamerlingh Onnes, "Further Experiments with Liquid Helium. H. On Electrical Resistance of Pure Metals etc. VII. The Potential Difference Necessary for the Electrical Current Through Mercury Below 4.19 K. (Continuation)," *Communications from the Physical Laboratory of the University of Leiden* 133b (1913): 29–32, p. 30.

34. H. Kamerlingh Onnes, "Further Experiments with Liquid Helium. H. On the Electrical

Resistance etc. (continued). VIII. The Sudden Disappearance of the Ordinary Resistance of Tin, and the Super-conductive State of Lead," *Communications from the Physical Laboratory of the University of Leiden* 133d (1913): 51–68.

35. Ibid., 64. According to C. J. Gorter, who joined the Kamerlingh Onnes low-temperature group in 1927 as a graduate student, Onnes "immediately" had the idea of applying superconductivity in building high-field superconducting electromagnets that would not dissipate any Joule heat. C. J. Gorter, "Superconductivity until 1940 in Leiden as Seen from There," *Reviews of Modern Physics* 36 (1964): 3–7. See also L.S.J.M. Henkens and P. F. de Châtel, "Physics of Condensed Matter," mimeograph, sec. 74, p. 233, AIP.

36. H. Kamerlingh Onnes, "Report on Research Made in the Leiden Cryogenic Laboratory Between the Second and Third International Congress of Refrigeration," *Communications from the Physical Laboratory of the University of Leiden* suppl. 34b (1913): 35–70, p. 65.

37. H. Kamerlingh Onnes, "The Imitation of an Ampere Molecular Current or of a Permanent Magnet by Means of a Supra-conductor," *Communications from the Physical Laboratory of the University of Leiden* 140b (1914): 9–18, p. 17.

38. F. B. Silsbee, "Note on Electrical Conduction in Metals at Low Temperatures," U.S. Bureau of Standards, *Scientific Papers* 14, no. 307 (1917): 301–306, p. 302.

39. H. Kamerlingh Onnes, "Les Supraconducteurs," in Instituts Solvay, *Conductibilité électrique des métaux, rapports et discussions du quatrième Conseil de Physique* (Paris: Gauthier-Villars, 1927), 251–301.

40. Ibid.

41. Gabriel Lippman had already derived this result from the Maxwell theory. See G. Lippmann, "Sur les propriétés des circuits électriques dénués de résistance," *C.R.* 168 (1919): 73–78.

42. See Kamerlingh Onnes (n. 39), 282–289. In the discussion, Keesom did set up such formulas in accord with Bridgman's suggestion, but he as well as A. H. Lorentz ruled out a reversible cycle.

43. G. J. Sizoo, W. J. de Haas, and H. Kamerlingh Onnes, "Measurements on the Magnetic Disturbance of the Supraconductivity with Tin. I. Influence of Elastic Deformation. Hysteresis Phenomena," *Communications from the Physical Laboratory of the University of Leiden* 180c (1925): 29–53.

44. W. J. De Haas and J. Voogd, "The Influences of Magnetic Fields on Supraconductors," *Communications from the Physical Laboratory of the University of Leiden* 208b (1930): 9–10.

45. See Chapter 2 for additional discussion of the work of McLennan's group and, in particular, on its influence on Bohr's 1932 theory of superconductivity.

46. W. Meissner, "Über die Heliumverflüssigungsanlage der PTR und einige Messungen mit Hilfe von flüssigen Helium," *Phys. Zs.* 26 (1925): 689–694, p. 691.

47. W. Meissner, "Supraleitfähigkeit von Tantal," *Phys. Zs.* 29 (1928): 897–904.

48. W. Meissner, "Supraleitfähigkeit von Kupfersulfid," *Zs. Phys.* 58 (1929): 570–572.

49. W. Meissner, "Supraleitfähigkeit," *Ergebnisse der exacten Naturwissenschaften* 11 (1932): 219–263, p. 248.

50. W. Meissner to R. Kronig, 10 December 1932, DM.

51. E. Justi, interview with H. Schubert, 22 July 1982, DM. R. Ochsenfeld, interview with H. Schubert, 22 October 1982, DM.

52. M. von Laue and F. Möglich, "Über das magnetische Feld in der Umgebung von Supraleitern," *Berliner Berichte* (1933): 544–565.

53. See W. Meissner to F. Simon, 6 April 1933; W. Meissner to K. Clusius, 8 April 1933, DM.

54. W. Meissner to R. Gans, 9 March and 26 May 1933, DM.

55. Ochsenfeld interview (n. 51).

56. For a detailed description of the Meissner–Ochsenfeld experiments, as well as Meissner's laboratory notebook, see Dahl (1986) (n. 21).

57. W. Meissner and R. Ochsenfeld, "Ein neuer Effekt bei Eintritt der Supraleitfähigkeit," *Die Naturwissenchaften* 21 (1933): 787–788. The authors chose this journal because it guaranteed quick publication, an apparent necessity in order to publish before the Leiden group, which was also investigating the influence of the magnetic field during cooling. See W. J. de Haas, "Supraleiter im Magnetfeld," in *Leipziger Vorträge, 1933: Magnetismus* (Leipzig: Hirzel, 1933), 59–73.

58. W. Meissner to K. Scheel, 24 July 1933, DM.

59. Bloch's "theorem" was a consequence of his assumption that superconductivity corre-

sponds to a minimum energy state, and of his many calculations that showed that all such states carry zero current. See references in Chapter 2. Much later, in 1949, Bohm extended Bloch's theorem to many-body systems, showing "that even when interelectronic interactions are taken into account, the state of lowest electron free energy corresponds to a zero net current." See D. Bohm, "Note on a Theorem of Bloch Concerning Possible Causes of Superconductivity," *Phys. Rev.* 75 (1949): 502–504.

60. A. Sommerfeld and H. Bethe, "Elektronentheorie der Metalle," *Handbuch der Physik* (Berlin: Springer, 1933), 24: pt. 2, 333–622, p. 558. Similarly, Peierls notes that "further development in several different directions" in the quantum theory of solids is needed to explain superconductivity. R. Peierls, "Elektronentheorie der Metalle," *Ergebnisse der Exakten Naturwissenschaften* 11 (1932): 264–321, p. 320.

61. An accessible description of the thermodynamics of superconductors, with a historical perspective, is given in F. London, *Superfluids,* vol. 1: *Macroscopic Theory of Superconductivity* (New York: Wiley, 1950).

62. W. H. Keesom and J. A. Kok, "On the Change of the Specific Heat of Tin When Becoming Superconductive," *Communications from the Physical Laboratory of the University of Leiden* 221c (1932): 27–32. In fact, the transition in the presence of a magnetic field is first order, but reduces at zero field to a second-order transition.

63. P. Ehrenfest, "Phasenumwandlungen in üblichen und erweiterten Sinn, classifiziert nach den entsprechenden Singularitäten des thermodynamischen Potentiales," *Communications from the Physical Laboratory of the University of Leiden,* suppl. 75b (1933): 8–13. A. J. Rutgers, "Notes on Supraconductivity," *Physica* 1 (1934): 1055–1058; Rutgers, "Bemerkung zur Anwendung der Thermodynamik auf die Supraleitung," *Physica* 3 (1936): 999–1005.

64. Gorter (n. 35), 3.

65. C. J. Gorter, "Some Remarks on the Thermodynamics of Superconductors," *Archives Musée Teyler* 7 (1933): 378–386.

66. C. J. Gorter, "Theory of Superconductivity," *Nature* 132 (1933): 931.

67. C. J. Gorter to W. Meissner, 27 November 1933, DM.

68. C. J. Gorter and H. Casimir, "Zur Thermodynamik des supraleitenden Zustandes," *Zs. Phys.* 35 (1934): 963–966; Gorter and Casimir, "Zur Thermodynamik des supraleitenden Zustandes," *Zeitschrift für technische Physik* 15 (1934): 539–542; Gorter and Casimir, "On Superconductivity," *Physica* 1 (1934): 306–320.

69. F. Bloch, "Zur Theorie des Austauschproblems und der Remanenzerscheinung der Ferromagnetika," *Zs. Phys.* 74 (1932): 295–335; Bloch, interview with L. Hoddeson, December 1981.

70. C.W.F. Everitt and W. M. Fairbank, "London, Fritz," in *Dictionary of Scientific Biography* (New York: Scribner, 1973), 8: 473–479. See also L. W. Nordheim, "Fritz London," *Physics Today* 7 no. 7 (1954): 16–17; D. Shoenberg, "Heinz London, 1907–1970," *Biographical Memoirs of Fellows of the Royal Society* 17 (1971): 441–461.

71. K. Mendelssohn, "Prewar Work on Superconductivity as Seen from Oxford," *Reviews of Modern Physics* 36 (1964): 7–12.

72. H. London, "Über die Möglichkeit des Auftretens eines Hochfrequenz-Restwiderstandes bei Supraleitern" (Ph.D. diss., Berlin University, 1934).

73. R. Becker, G. Heller, and F. Sauter, "Über die Stromverteilung in einer supraleitenden Kugel," *Zs. Phys.* 85 (1933): 772–787.

74. G. L. de Haas-Lorentz, "Iets over het Mechanisme van Inductieverschijnselen," *Physica* 5 (1925): 384–388.

75. F. London and H. London, "Supraleitung und Diamagnetismus," *Physica* 2 (1935): 341–354, p. 348.

76. The Londons credit this equation to Becker, Heller, and Sauter, although Heinz London had derived but not published it several years earlier. See Shoenberg (n. 70).

77. F. London, "Macroscopical Interpretation of Supraconductivity," *Proc. Roy. Soc.* A152 (1935): 24–34.

78. Ibid., 31.

79. L. Brillouin, "Les Bases de la theéorie électronique des métaux et la méthode des champs self-consistents," *Helvetica Physika Acta* 7 (1934): 33–46.

80. See Nordheim (n. 70). Much later, London would make the concept of the long-range

ordered state in a superconductor (as well as in superfluid ⁴He) fully explicit: "a kind of condensed state in momentum space implying a long-range order of the momentum vector in ordinary space," a concept essential to the subsequent development of the BCS theory. F. London, "On the Problem of the Molecular Theory of Superconductivity," *Phys. Rev.* 74 (1948): 562–573.

81. Mendelssohn (n. 71).

82. For an overview see Shoenberg (1952) (n. 21), 95–137; Mendelssohn (n. 71).

83. H. London, "Phase-Equilibrium of Supraconductors in a Magnetic Field," *Proc. Roy. Soc.* A152 (1935): 650–663.

84. F. London, "Zur Theorie Magnetischer Felder im Supraleiter," *Physica* 3 (1936): 450–462.

85. R. Peierls, "Magnetic Transition Curves of Supraconductors," *Proc. Roy. Soc.* A155 (1936): 613–628.

86. D. Shoenberg, interview with P. Hoch, 10 February 1981. Shoenberg's interest in superconductivity began when he was a second-year graduate student at Cambridge, through Peter Kapitza and the informal seminar series, the Kapitza Club. As Shoenberg recalls, Kapitza asked him to lecture to the club on superconductivity; among the new work he spoke on was the recently discovered Meissner effect.

87. L. Landau, "Zur theorie der Superleitfähigkeit," *Physikalische Zeitschrift der Sowjetunion* 11 (1937): 129–140. See also Landau, "On the Theory of the Intermediate State of Superconductors," *Journal of Physics (Moscow)* 7 (1943): 99.

88. A. I. Shalnikov, *Journal of Physics (Moscow)* 9 (1945): 202; A. Meshkovsky and A. Shalnikov, "The Structure of the Superconductors in the Intermediate State. II," *Journal of Physics (Moscow)* 11 (1947): 1–15; Meshkovsky and Shalnikov, *Zhurnal Eksperimental noi i Teoreticheskoi Fiziki* 17 (1947): 851.

89. De Haas and Voogd (n. 44).

90. K. Mendelssohn and J. R. Moore, "Magneto-Caloric Effect in Supraconducting Tin," *Nature* 133 (1934): 413.

91. H. London (n. 83).

92. Ibid.

93. A. A. Abrikosov, "On the Magnetic Properties of Superconductors of the Second Group," *Zhurnal Eksperimental'noi i Teoreticheskoi Fiziki* 32 (1957): 1442–1452 (trans. in *Soviet Physics JETP* 5 [1957]: 1174–1182).

94. Gorter (n. 35), 7.

95. H. Welker, "Über ein elektronentheoretisches Modell des Supraleiters," *Phys. Zs.* 39 (1938): 920–925; Welker, "Supraleitung und magnetische Austauschwechselwirkung," *Zs. Phys.* 114 (1939): 525–551; Welker, interview with J. Teichmann, M. Eckert, and G. Torkar, 4 December 1981, DM; Welker, "Impact of Sommerfeld's Work in Solid State Research and Technology," in *Physics of the One- and Two-Electron Atoms. Proceedings of Arnold Sommerfeld Centennial Meeting, 1968* (Amsterdam: North-Holland, 1969), 32–43. A contradiction in Welker's argument was pointed out much later by M. J. Buckingham, "A Note on the Energy Gap Model of Superconductivity," *Nuovo Cimento* 5 (1957): 1763–1765. See also H. Fröhlich, "The Theory of the Superconducting State," *Reports on Progress in Physics* 24 (1961): 1–23, p. 6.

96. G. Aschermann, E. Friederich, E. Justi, and J. Kramer, "Supraleitfähige Verbindungen mit extrem hohen Sprungtemperaturen," *Phys. Zs.,* 42 (1941): 349–360. Another German military application was a superconducting bolometer, whose pronounced change of resistance at small temperature changes permitted sensitive measurement of heat radiation at the transition temperature. A. Goetz, "The Possible Use of Superconductivity for Radiometric Purposes," *Phys. Rev.* 55 (1939): 1270–1271.

97. For the prehistory of the liquefaction of helium see Chapter 1.

98. H. Kamerlingh Onnes, "Die Cascade zur Erhaltung tiefer Temperaturen bis −259°C," *Communications from the Physical Laboratory of the University of Leiden* supp. 18a (1907): 3–8.

99. H. Kamerlingh Onnes, "Experiments on the Condensation of Helium by Expansion," *Communications from the Physical Laboratory of the University of Leiden* 105 (1908): 3–6.

100. Onnes (n. 29).

101. Ibid.

102. H. Kamerlingh Onnes, "The Liquefaction of Helium," *Communications from the Physical Laboratory of the University of Leiden* 108 (1908): 3–23.

103. This temperature was recomputed by W. H. Keesom, "Experiments to Decrease the Limit of the Temperatures Obtained," *Communications from the Physical Laboratory of the University of Leiden* 219a (1932): 1–9.

104. H. Kamerlingh Onnes, "Further Experiments with Liquid Helium. P. On the Lowest Temperature Yet Obtained," *Communications from the Physical Laboratory of the University of Leiden* 159 (1922): 3–36.

105. H. Kamerlingh Onnes, "Further Experiments with Liquid Helium. A. Isotherms of Monoatomic Gases, etc. VIII. Thermal Properties of Helium." *Communications from the Physical Laboratory of the University of Leiden* 119 (1911): 3–19.

106. J. C. McLennan, H. D. Smith, and J. O. Wilhelm, "The Scattering of Light by Liquid Helium," *Phil. Mag.* 14 (1932): 161–167.

107. L. I. Dana and H. Kamerlingh Onnes, "Further Experiments with Liquid Helium. B. B. Preliminary Determinations of the Specific Heat of Liquid Helium," *Communications from the Physical Laboratory of the University of Leiden* 179d (1926): 35–45.

108. See R. J. Donnelly and A. W. Francis, eds., *Cryogenic Science and Technology: Contributions by Leo I. Dana* (Danbury: Union Carbide, 1985); W. H. Keesom, *Helium* (Amsterdam: Elsevier, 1942).

109. L. I. Dana and H. Kamerlingh Onnes, "Further experiments with Liquid Helium. B. A. Preliminary Determinations of the Latent Heat of Vaporization of Liquid Helium," *Communications from the Physical Laboratory of the University of Leiden* 179c (1926): 21–34.

110. W. H. Keesom and K. Clusius, "Über die spezifische Wärme des flüssigen Heliums," *Communications from the Physical Laboratory of the University of Leiden* 219e (1932): 42–58.

111. W. H. Keesom and M. Wolfke, "Two Different Liquid States of Helium," *Communications from the Physical Laboratory of the University of Leiden* 190b (1927): 15–22.

112. Keesom and Clusius (n. 110), 55.

113. Ehrenfest (n. 63).

114. See Keesom and Wolfke (n. 111), 22. Crystal-based theories of liquid helium were presented by H. Fröhlich, "Zur Theorie Des λ-punktes Des Heliums," *Physica* 4 (1937): 639–644, F. London, "On Condensed Helium at Absolute Zero," *Proc. Roy. Soc.* A153 (1936): 576–583.

115. W. H. Keesom and K. W. Taconis, "Debye-Scherrer Exposures of Liquid Helium," *Physica* 5 (1938): 270–280.

116. W. H. Keesom and A. P. Keesom, "New Measurements on the Specific Heat of Liquid Helium," *Physica* 2 (1935): 557–569.

117. W. H. Keesom and A. P. Keesom, "On the Heat Conductivity of Liquid Helium," *Physica* 3 (1936): 359–360.

118. J. C. McLennan, "The Cryogenic Laboratory of the University of Toronto," *Nature* 112 (1923): 135–139. Kamerlingh Onnes welcomed the emergence of new low-temperature centers and placed detailed plans of his equipment at McLennan's disposal.

119. F. Simon, "Heliumverflüssigung mit Arbeitsleistung," *Zeitschrift für die Gesamte Kälte-Industrie* 39 (1932): 89–90.

120. J. F. Allen, R. Peierls, and M. Zaki Uddin, "Heat Conduction in Liquid Helium," *Nature* 140 (1937): 62–63.

121. See K. Mendelssohn, *The Quest for Absolute Zero* (New York: McGraw-Hill, 1966).

122. J. F. Allen and H. Jones, "New Phenomena Connected with Heat Flow in Helium II," *Nature* 141 (1938): 243–244.

123. J. G. Daunt and K. Mendelssohn, "Surface Transport in Liquid Helium," *Nature* 143 (1939): 719–720.

124. See Mendelssohn (n. 121).

125. W. H. Keesom and A. P. Keesom, "On the Anomaly in the Specific Heat of Liquid Helium," *Communications from the Physical Laboratory of the University of Leiden* 221d (1932): 19–26.

126. J. O. Wilhelm, A. D. Misener, and A. R. Clark, "The Viscosity of Liquid Helium," *Proc. Roy. Soc.* A151 (1935): 342–347.

127. W. H. Keesom and G. E. MacWood, "The Viscosity of Liquid Helium," *Physica* 5 (1938): 737–744.

128. P. Kapitza, "Viscosity of Liquid Helium Below the λ-point," *Nature* 141 (1938): 74; J. F. Allen and A. D. Misener, "Flow of Liquid Helium II," *Nature* 141 (1938): 75.

129. Kamerlingh Onnes (n. 104).

130. J. G. Daunt and K. Mendelssohn, "Transfer of Helium II on Glass," *Nature* 141 (1938): 911–912 (see reference 1 to Rollin [p. 911]).

131. Ibid.

132. A. K. Kikoin and B. G. Laserew, "Experiments with Liquid Helium II," *Nature* 141 (1938): 912–913.

133. A. K. Kikoin and B. G. Laserew, "Further Experiments on Liquid Helium II," *Nature* 142 (1938): 289–290.

134. Fröhlich (n. 114); London (n. 114).

135. F. London, "On the Bose-Einstein Condensation," *Phys. Rev.* 54 (1938): 947–954; London, "The Lambda-Phenomenon of Liquid Helium and the Bose-Einstein Degeneracy," *Nature* 141 (1938): 643–644.

136. London (n. 61), 5.

137. A. Einstein, "Quantentheorie des Einatomigen Idealen Gases," *Sitzungberichten der Preussischen Akademie der Wissenschaften zu Berlin* 18–25 (1924): 261–267, 1–2 (1925): 3–14, 3–5 (1925): 18–25; S. N. Bose, "Plancks Gesetz und Lichtquanten-hypothese," *Zs. Phys.* 26 (1924): 178–181.

138. L. Tisza, "Transport Phenomena in Helium II," *Nature* 141 (1938): 913.

139. See Mendelssohn (n. 121).

140. L. Tisza, "Thermodynamique statistique—sur la supraconductibilité thermique de l'helium II liquide et la statistique de Bose-Einstein," *C.R.* 207 (1938): 1035–1137; Tisza, "Thermodynamique statistique—la viscosité de l'helium liquide et la statistique de Bose-Einstein," *C.R.* 207 (1938): 1186–1189.

141. V. Peshkov, " 'Second Sound' in Helium II," *Journal of Physics (Moscow)* 8 (1944): 381.

142. P. L. Kapitza, *Experiment, Theory, Practice,* vol. 46 of *Boston Studies in the Philosophy of Science* (Dordrecht: Reidel, 1980), 24.

143. P. L. Kapitza, "The Study of Heat Transfer in Helium," *Journal of Physics (Moscow)* 4 (1941): 181–210; Kapitza, "Heat Transfer and Superfluidity of Helium II," *Journal of Physics (Moscow)* 5 (1941): 59–69.

144. A. Livanova, *Landau: A Great Physicist and Teacher,* trans. J. B. Sykes (New York: Pergamon Press, 1980), 149.

145. L. D. Landau, "The Theory of Superfluidity of Helium II," *Journal of Physics (Moscow)* 5 (1941): 71–90; Landau, "The Theory of Superfluidity of Helium II," *Phys. Rev.* 60 (1941): 356–358. This work was the formal basis of the award of the 1962 Nobel Prize to Landau.

146. Landau (n. 145).

147. E. L. Andronikashvilii, "A Direct Observation of Two Kinds of Motion in Helium II," *Journal of Physics (Moscow)* 10 (1946): 201–206.

148. Referring to Welker (1939) (n. 95).

149. Although the gap is a sufficient condition, it would not prove to be necessary. Were Landau's criterion correct, then neither dilute solutions of ^{3}He and ^{4}He nor ^{3}He = A would be superfluid. A complete understanding of superfluidity would require the concept of borken symmetry.

150. Historical overviews include C. S. Smith, *A History of Metallography* (Chicago: University of Chicago Press, 1960); Smith, *A Search for Structure* (Cambridge, Mass.: MIT Press, 1981); D. Roller, *The Early Development of the Concepts of Temperature and Heat* (Cambridge, Mass.: Harvard University Press, 1950); J. Needham and L. Gwei-Djen, "The Earliest Snow Crystal Observations," *Weather* 16 (1961): 319–327. Significant historical references include V. Biringuiccio, *Pirotechnia* (1540; reprint, Cambridge, Mass.: MIT Press, 1942); G. della Porta, *Magiae Naturalis,* trans. as *Natural Magick* (1589; reprint New York: Basic Books, 1958); R. Boyle, *Essay About the Origins and Virtue of Gems* (London, 1672); W. Gilbert, *De Magnete* (1600; reprint, New York: Basic Books, 1958); J. Keill, *Introductio ad Veram Physicam* (London: T. Bennet, 1702); E. Swedenborg, *Principia rerum naturalium* (Dresden, 1734). See also E. Mach, *Die Principien der Wärmelehre,* 3rd ed. (Leipzig: Barth, 1919).

151. J. D. van der Waals, *Over de continuieteit van den gas—en vloeistoftoestand* (Leiden: Sijthoff, 1873); van der Waals, *Die Continuität des gasförmigen und flüssigen Zustandes* (Leipzig: Barth, 1881). That the theory of phase transitions arose as a by-product of the effort to learn something about the attractive forces between molecules appears ironic from today's perspective, for, as we now understand, the nature of phase transitions is relatively independent of detailed features of intermolecular force laws.

152. R. Clausius, "On the Nature of the Motion Which We Call Heat," *Phil. Mag.* 14 (1857): 108–127.

153. van der Waals (n. 151); the existence of a continuous transition had been previously asserted by G. A. Hirn, "Mémoire sur la thermodynamique," *Annales de chimie et de physique*, 4th ser., vol. 11 (1867); 5–112, on the basis of theoretical arguments close to those of van der Waals. See also J. P. Kuenen, "Die Eigenschafte der Gase," in *Handbuch der Allgemeinen Chemie*, vol. 3, ed. W. Ostwald and C. Drucker (Leipzig, 1919).

154. From author's preface to J. D. van der Waals, "The Continuity of the Liquid and Gaseous States," trans. in *Physical Memoirs*, vol. 1 (London: Taylor and Francis, 1888).

155. J. D. van der Waals, "The Equation of State for Gases and Liquids," in *Nobel Lectures* (n. 29), 254–265. For a discussion of van der Waal's focus of interest, see M. J. Klein, "The Historical Origins of the van der Waals Equation," *Physica* 73 (1974): 28–47.

156. T. Andrews, "On the Continuity of the Gaseous and Liquid States of Matter," *Philosophical Transactions of the Royal Society (London)* 159 (1869): 575–590 (reprinted in *Cooperative Phenomena Near Phase Transitions*, ed. H. E. Stanley [Cambridge, Mass.: MIT Press, 1973]).

157. Kuenen (n. 153).

158. J. C. Maxwell, "On the Dynamical Evidence of the Molecular Constitution of Bodies," *Nature* 11 (1875): 357–359.

159. H. Kamerlingh Onnes, "Algemeene Theorie der Vloeistiffen," *Verhandelingen der Konenlijke Akademie van Wetenschappen te Amsterdam* 21 (1881): 1–24. See also J. de Boer, "Van der Waals in His Time and the Present Revival," *Physica* 73 (1974): 1–27.

160. For a brief description of all of these contributions, see Kuenen (n. 153).

161. J. W. Gibbs, "On the Equilibrium of Heterogeneous Substances," *Transactions of the Connecticut Academy of Sciences* 3 (1876): 108–248, 343–524; Gibbs, *The Scientific Papers of J. Willard Gibbs*, vol. 1 (London: Longmans, Green, 1906; reprint, New York: Dover, 1961).

162. Clausius's memoirs on thermodynamics are collected in R.J.E. Clausius, *The Mechanical Theory of Heat with Its Applications to the Steam Engine and to the Physical Properties of Bodies*, ed. T. A. Hirst (London: John Van Voorst, 1867).

163. See, for example, E. B. Wilson, "Papers I and II as Illustrated by Gibbs' Lectures on Thermodynamics," in *Commentary on the Scientific Writings of J. Willard Gibbs*, pt. 1, ed. R. G. Donnan (New York: Arno Press, 1980), 19–59, describing Gibbs's 1899–1900 lectures on thermodynamics.

164. See E. Garber, "James Clerk Maxwell and Thermodynamics," *American Journal of Physics* 37 (1969): 146–155.

165. See, for example, E. Hiebert, "The Energetics Controversy and the New Thermodynamics," in *Perspectives in the History of Science and Technology*, ed. Duane Roller (Norman: University of Oklahoma Press, 1971), 67–86.

166. See Kuenen (n. 153); H. Kamerlingh Onnes, *Communications from the Physical Laboratory of the University of Leiden* 71, 72 (1901); Kamerlingh Onnes, *Archives Neerlandaises des Sciences Exactes et Naturelles*, 2d ser., vol. 6 (1901): 874.

167. J. W. Gibbs, *Elementary Principles in Statistical Mechanics* (London: Edward Arnold, 1902), 203.

168. Whereas in 1873 van der Waals had been led to study critical phenomena only to elucidate the force between molecules, his focus in the early twentieth century had shifted to the description of phase transitions and critical phenomena per se. See, for example, the textbook based on his Amsterdam lectures "Thermodynamics in Its Application to the Equilibrium of Systems with Liquid and Gas Phases," in *Lehrbuch der Thermodynamic in ihrer Anwendung auf das Gleichgewicht von Systemen mit Gasförmig-Flüssigen Phasen, Based on Lectures by J. D. van der Waals*, 2 vols., comp. P. Kohnstamm (Amsterdam: Maas and Van Suchtelen, 1908).

169. F. A. Lindemann, "Molecular Frequencies," *Phys. Zs.* 11 (1910): 609–612. This theory, based on Einstein's theory of specific heats of solids, discussed melting in terms of the atomic vibrations becoming sufficiently large to disrupt the lattice structure.

170. Andrews (n. 156).

171. A. Einstein, "Theorie der Opaleszenz von homogenen Flüssigkeiten und Flüssigkeitsgemischen in der Nähe des kritischen Zustands," *Ann. Phys.* 33 (1910): 1275–1298; M. von Smoluchowski, "Molekular-kinetische Theorie der Opaleszenz von Gasen im kritischen Zustande, sowie einiger verwandter Erscheinungen," *Ann. Phys.* 25 (1908): 205–226; Smoluchowski, "On Opalescence of Gases in the Critical State," *Phil. Mag.* 23 (1912): 165–173.

172. L. S. Ornstein and F. Zernike, "Accidental Deviations of Density and Opalescence at the Critical Point of a Single Substance," *Verhandelingen der Konenlijke Akademie van Wetenschappen te Amsterdam* 17 (1914): 793–806; Ornstein and Zernike, "The Linear Dimensions of Density Variations," *Phys. Zs.* 19 (1918): 134–137; Ornstein and Zernike, "Bemerkung zur Arbeit von Herrn K. C. Kar: Die Molecularzerstreuung des Lichtes beim kritischen Zustande," *Phys. Zs.* 27 (1926): 761–763.

173. J. Hopkinson, "Magnetic Properties of Alloys of Nickel and Iron," *Proc. Roy. Soc.* 48 (1890): 1–13; P. Curie, "Propiétés magnétiques des corps à diverses temperatures," *Annales de chimie et physique* 5 (1895): 289–405.

174. P. Weiss, "L'Hypothèse du champ moléculaire et la propriété ferromagnétique," *Journal de physique et le radium* 6 (1907): 661–690; P. Weiss and R. Forrer, "Aimantation et phénomène magnétocalorique du nickel," *Ann. Phys.* 5 (1926): 153–213.

175. Einstein (n. 137); Bose (n. 137).

176. G. E. Uhlenbeck, "Over Statistische Methoden in de Theorie de Quanta" (Ph.D. diss., Nyhoff, The Hague, 1927); also later publications by B. Kahn and Uhlenbeck, discussed subsequently.

177. W. Lentz, "Beitrag zum Verständnis der magnetischen Erscheinungen in fastkörpern," *Phys. Zs.* 21 (1920): 613–615; E. Ising, "Betrag zur Theorie des Ferromagnetismus," *Zs. Phys.* 31 (1925): 253–258. See also S. Brush, "History of the Lenz-Ising Model," *Reviews of Modern Physics* 39 (1967): 883–893. See Chapter 2 for a discussion of Heisenberg and the Ising model.

178. The most rigorous and influential such calculation was by H. D. Ursell, "Evaluation of Gibbs' Phase Integral for Imperfect Gases," *Proceedings of the Cambridge Philosophical Society* 23 (1927): 685–697.

179. Although work in this period dealt primarily with abstracted models, the more concrete issue of structural phase transitions in solids also had its roots during this time. Noteworthy were the papers by H. Jones, "The Theory of Alloys in the Gamma-Phase," *Proc. Roy. Soc.* A144 (1934): 225–234, and "Applications of the Bloch Theory to the Study of Alloys and of the Properties of Bismuth," *Proc. Roy. Soc.* A147 (1934): 396–417, which made the first connection between structure and the electron theory of metals. These papers explain how the electron energies induce a distortion of the lattice in bismuth, a predecessor of the modern "Peierls transition," in one-dimensional structures. R. Peierls, *Quantum Theory of Solids* (London: Oxford University Press, 1955), chap. V. Such transitions are now understood in terms of a "softening" to zero frequency of certain vibrational modes of the lattice.

180. R. Peierls, "Bemerkungen über Umwandlungstemperaturen," *Helvetica Physica Acta* 7, suppl. 2 (1934): 81–83. Peierls mentions that the subject of this article emerged from conversations with H. A. Bethe. See also H. Bethe, "Statistical Theory of Superlattices," *Proc. Roy. Soc.* A150 (1935): 552–575, who thanks Peierls for valuable suggestions. The actual calculations of the long-range thermal fluctuations of one- and three-dimensional lattices were given in R. Peierls, "Quelques propriétés typiques des corps solides," *Annales d'Institut Henri Poincaré* 5 (1935): 177–222.

181. R. Peierls, "Statistical Theory of Adsorption with Interaction Between the Absorbed Atoms," *Proceedings of the Cambridge Philosophical Society* 32 (1936): 471–475; Peierls, "Ising's Model of Ferromagnetism," *Proceedings of the Cambridge Philosophical Society* 32 (1936): 477–481. For a modern critique of Peierls's arguments, see R. B. Griffiths, "Peierls' Proof of Spontaneous Magnetization in a Two-Dimensional Ising Ferromagnet," *Phys. Rev.* 136A (1964): 437–439.

182. L. Landau, "Zur Theorie der Anomalien der Spezifischen Wärme," *Physikalische Zeitschrift der Sowjetunion* 8 (1935): 113–118.

183. F. Bloch, "Zur Theorie des Austauschproblems und der Remanenzerscheinung der Ferromagnetika," *Zs. Phys.* 74 (1932): 295–335.

184. L. Landau, "Zur Theorie der Supraleitfähigkeit. I," *Physikalische Zeitschrift der Sowjetunion* 4 (1933): 43–49; Landau, "Eine mögliche Erklärung der Feld abhängigkeit der Suszeptibilität bei niedrigen Temperaturen," *Physikalische Zeitschrift der Sowjetunion* 4 (1933): 675–678. Bethe also introduced at this time order parameters, called the "long range order" and "order of neighbors" in the binary alloy problem. See H. A. Bethe, "Statistical Theory of Superlattices," *Proc. Roy. Soc.* A150 (1935): 552–575.

185. L. Landau, "Zur Theorie der Phasenumwandlungen. I, II," *Physikalische Zeitschrift der Sowjetunion* 11 (1937): 26–47, 545–555. Both papers were also published in Russian in the *Zhurnal Eksperimental'noi i Teoreticheskoi Fiziki* 7 (1937): 19, 627.

186. Proceedings of the Amsterdam Conference, *Physica* 4 (1937): 915–1180.

187. Several papers dealt with the theory of fluids, particularly J. E. Lennard-Jones, "The Equation of State of Gases and Critical Phenomena," *Physica* 4 (1937): 941–956 (discussed later, a paper on the calculation of the second virial coefficient); E. Beth and G. Uhlenbeck, "The Quantum Theory of the Non-ideal Gas. II. Behavior at Low Temperatures," *Physica* 4 (1937): 915–924; L. Waldmann, "Über eine Verallgemeinerung der Boltzmann'schen Abzählungsmethode auf das van der Walls'sche Gas," *Physica* 4 (1937): 1117–1132.

188. J. E. Mayer, "The Statistical Mechanics of Condensing Systems. I," *Journal of Chemical Physics* 5 (1937): 67–73; J. E. Mayer and P. G. Ackermann, "The Statistical Mechanics of Condensing Systems, II." *Journal of Chemical Physics* 5 (1937): 74–83.

189. M. Born, "The Statistical Mechanics of Condensing Systems," *Physica* 4 (1937): 1034–1044.

190. B. Kahn and G. E. Uhlenbeck, "On the Theory of Condensation," *Physica* 4 (1937): 1155–1156; despite the inspirational role of Bose–Einstein historically, Uhlenbeck recorded a quarter century later that he now believed "there is no real connection between the Bose-Einstein condensation and ordinary condensation phenomena." M. Kac, G. E. Uhlenbeck, and P. C. Hemmer, "On the van der Waals Theory of the Vapor-Liquid Equilibrium. I. Discussion of a One-Dimensional Model," *Journal of Mathematical Physics* 4 (1963): 216–228; Kac et al., "On the van der Waals Theory of the Vapor-Liquid Equilibrium. II. Discussion of the Distribution Functions," *Journal of Mathematical Physics* 4 (1963): 229–247.

191. J. de Boer, "Van der Waals in His Time and the Present Revival," *Physica* 73 (1974): 1–27. See also B. Kahn, "On the Theory of the Equation of State" (Ph.D. diss., Utrecht, 1938). See also M. Dresden, *H. A. Kramers: Between Theory and Revolution* (New York: Springer, 1987).

192. A. Pais, *Subtle Is the Lord* (New York: Oxford University Press, 1982), 432–433.

193. J. E. Mayer and S. F. Harrison, "Statistical Mechanics of Condensing Systems. III, IV," *Journal of Chemical Physics* 6 (1938): 87–100, 101–104; J. E. Mayer and M. Goeppert-Mayer, *Statistical Mechanics* (New York: Wiley, 1940), chap. 14.

194. Unfortunately, the equation of state was valid only in a limited range of parameters. In particular, it failed to describe the liquid in a pressure-volume diagram. A major source of the difficulty, as Born and Klaus Fuchs, on the one hand, and Kahn and Uhlenbeck, on the other, agreed, lay in the details of the taking of the infinite volume limit. M. Born and K. Fuchs, "The Statistical Mechanics of Condensing Systems," *Proc. Roy. Soc.* A166 (1938): 391–414; B. Kahn and G. E. Uhlenbeck, "On Theory of Condensation," *Physica* 5 (1938): 399–416.

195. C. Weiner, "A New Site for the Seminar," in *The Intellectual Migration: Europe and America, 1930–1960*, ed. Donald Fleming and Bernard Bailyn (Cambridge, Mass.: Harvard University Press, 1969), 190–234.

196. R. H. Fowler, *Statistical Mechanics* (New York: Cambridge University Press, 1936), 289; F. C. Nix and W. Shockley, "Order-Disorder Transformations in Alloys," *Reviews of Modern Physics* 10 (1938): 1–71, spoke of "cooperational phenomena," but the earlier designation "cooperative" stuck.

197. W. L. Bragg and E. J. Williams, "The Effect of Thermal Agitation on Atomic Arrangement in Alloys," *Proc. Roy. Soc.* A145 (1934): 699–730; Bragg and Williams, "The Effect of Thermal Agitation on Atomic Arrangement in Alloys. II," *Proc. Roy. Soc.* A151 (1935): 540–566.

198. J. Frenkel, "A General Theory of Heterophase Fluctuations and Pre-transition phenomena," *Journal of Physics (Moscow)* 1 (1939): 315–324.

199. R. H. Fowler, "Adsorption Isotherms. Critical Conditions," *Proceedings of the Cambridge Philosophical Society* 32 (1936): 144–151.

200. Bragg and Williams (n. 197). They refer to previous work by W. Gorsky, "Röntgenographische Untersuchung von Umwandlungen in der Legierung Cu Au," *Zs. Phys.* 50 (1928): 64–81; U. Dehlinger, "Stetiger Übergang and kritischer Punkt zwischen zwei festen Phasen," *Zeitschrift für Physikalische Chemie* B26 (1934): 343–352; and G. Borelius, "Zur Theorie der Umwandlungen von metallischen Mischphasen," *Ann. Phys.* 20 (1934): 57–74. See also E. J. Williams, "The Effect of Thermal Agitation on Atomic Arrangement in Alloys. III.'" *Proc. Roy. Soc.* A152 (1935): 231–252.

201. J. E. Lennard-Jones and A. F. Devonshire, "Critical Phenomena in Gases, I," *Proc. Roy. Soc.* A163 (1937): 53–70; Lennard-Jones and Devonshire, "Critical Phenomena in Gases, II," *Proc. Roy. Soc.* A165 (1938): 1–11; Lennard-Jones and Devonshire, "Critical Phenomena in Gases, III," *Proc. Roy. Soc.* A169 (1939): 317–338; Lennard-Jones and Devonshire, "Critical Phenomena in Gases, IV, " *Proc. Roy. Soc.* A170 (1939): 464–484.

202. Lennard-Jones and Devonshire (1938) (n. 201).

203. This model for a fluid seemed far removed from attempts at rigorous evaluation of the partition function; however, it was later shown to correspond to a well-defined approximation. J. G. Kirkwood, "Critique of the Free Volume Theory of the Liquid State," *Journal of Chemical Physics* 18 (1950): 380–382.

204. Lennard-Jones (n. 187).

205. Bethe (n. 184).

206. R. Peierls, "Statistical Theory of Superlattices with Unequal Concentrations of the Components," *Proc. Roy. Soc.* A154 (1936): 207–222; Williams (n. 200); E. Lennard-Jones and A. F. Devonshire, "Critical and Co-operative Phenomena," *Proc. Roy. Soc.* A169 (1939): 317–338; J. R. Lacher, "Statistics of Hydrogen-Palladium System," *Proceedings of the Cambridge Philosophical Society* 33 (1937): 518–523; G. S. Rushbrooke, "A Note on Guggenheim's Theory of Strictly Regular Binary Liquid Mixtures," *Proc. Roy. Soc.* A166 (1938): 296–315; J. Wang, "Properties of Adsorbed Films with Repulsive Interaction Between the Adsorbed Atoms," *Proc. Roy. Soc.* A161 (1937): 127–140; E. A. Guggenheim, "The Statistical Mechanics of Cooperative Assemblies," *Proc. Roy. Soc.* A169 (1938): 134–148.

207. R. H. Fowler and E. A. Guggenheim, *Statistical Thermodynamics* (Cambridge: Cambridge University Press, 1939), chap. 13; T. S. Chang, "The Number of Configurations of an Assembly with Long-Distance Order," *Proc. Roy. Soc.* A173 (1939): 48–58.

208. J. Kirkwood, "Order and Disorder in Binary Solid Solutions," *Journal of Chemical Physics* 6 (1938): 70–74.

209. H. Bethe and J. Kirkwood, "Critical Behavior of Solid Solutions in the Order-Disorder Transformation," *Journal of Chemical Physics* 7 (1939): 578–582.

210. L. Onsager, "Crystal Statistics. I. A Two-Dimensional Model with an Order-Disorder Transition," *Phys. Rev.* 65 (1944): 117–149.

211. See also Brush (n. 177).

212. According to Elliott Montroll, in 1945 Pauli, in response to an inquiry from H. G. B. Casimir in Holland about new developments in theoretical physics in the United States during the war, wrote a letter stating that the only interesting event was the rigorous solution by L. Onsager. E. Montroll, interview with S. J. Heims, 26 September 1983.

213. H. A. Kramers and G. H. Wannier, "Statistics of the Two-Dimensional Ferromagnet. I, II," *Phys. Rev.* 60 (1941): 252–262, 262–276. E. Montroll, "Statistical Mechanics of Nearest Neighbor Systems," *Journal of Chemical Physics* 9 (1941): 706–721.

214. Montroll interview (n. 212).

215. E. N. Lassettre and J. P. Howe, "Thermodynamic Properties of Binary Solid Solutions on the Basis of the Nearest-Neighbor Approximation," *Journal of Chemical Physics* 9 (1941): 747–754; Lassettre and Howe, "Thermodynamic Properties of Binary Solid Solutions. Phase Separations," *Journal of Chemical Physics* 9 (1941): 801–806.

216. R. Kubo, *Busseiron Kenkyu* 1 (1943). Translation by University of California Radiation Laboratory, 1030 (L).

217. M. Klein, *Paul Ehrenfest* (Amsterdam: North-Holland, 1970), 36–38, 94–140.

218. Kramers and Wannier (n. 213).

219. J. E. Mayer and E. W. Montroll, "Statistical Mechanics of Imperfect Gases," *Journal of Chemical Physics* 9 (1941): 626–637.

220. Montroll interview (n. 212).

221. G. H. Wannier, "Statistics of the Two-Dimensional Ferromagnet," *Phys. Rev.* 59A (1941): 683.

222. Montrol (n. 213).

223. Ibid.

224. Ibid.; E. Montroll, "Statistical Mechanics of Nearest Neighbor Systems. II," *Journal of Chemical Physics* 10 (1942): 61–77.

225. T. Shedlovsky and E. Montroll, "Proceedings of Conference on Irreversible Thermodynamics and the Statistical Mechanics of Phase Transitions," *Journal of Mathematical Physics* 4 (1963), two pages of introductory information at beginning of volume and two unnumbered pages before p. 147 and after p. 146.

226. Ibid.

227. The Onsager papers are preserved in microfilm form at the Yale University Archives.

228. Montroll interview (n. 212).

229. See acknowledgement at end of Montroll (n. 213).

230. Montroll interview (n. 212).

231. Shedlovsky and Montroll (n. 225).

232. Onsager (n. 210).

233. The motivation for the various steps of the Onsager derivation are elucidated by G. Newell and E. Montroll, "On the Theory of the Ising Model of Ferromagnetism," *Reviews of Modern Physics* 25 (1953): 159–160; for discussion of dual and star-triangle transformations, see G. H. Wannier, "The Statistical Problem in Cooperative Phenomena," *Reviews of Modern Physics* 17 (1945): 50–60.

234. Montroll interview (n. 212).

235. J. Ashkin and W. E. Lamb, "The Propagation of Order in Crystal Lattices," *Phys. Rev.* 64 (1943): 159–178.

236. J. Ashkin and E. Teller, "Statistics of Two-Dimensional Lattices with Four Components," *Phys. Rev.* 64 (1943): 179–184.

237. Wannier (n. 233).

238. For example, research at the Kamerlingh Onnes Laboratory ground to a near halt in May 1940 when Germany invaded the Netherlands; in 1942 most of the staff resigned. H. G. B. Casimir, *Haphazard Reality* (New York: Harper & Row, 1983), 204.

239. British and American universities continued their teaching programs, but with greatly reduced staffs. See, for example, N. F. Mott, "Notes from Abroad," *Physics Today* 1, no. 5 (1948): 9.

240. A. B. Migdal, private communication to L. Hoddeson and G. Baym. Soviet low-temperature physicists were generally less handicapped by the war.

241. For further material, see K. Szymborski, "The Effect of World War II on Solid-State Physics," and references therein (in prep.).

242. V. F. Weisskopf, "Growing Up with Field Theory: The Development of Quantum Electrodynamics," in *The Birth of Particle Physics,* ed. L. Brown and L. Hoddeson (New York: Cambridge University Press, 1983), 56–81; J. Schwinger, "Renormalization Theory of Quantum Electrodynamics: An Individual View," in ibid., 329–353. Interestingly, Sin-itiro Tomonaga, who independently developed renormalized quantum electrodynamics, also worked on microwaves in Japan during the war. See also S. Schweber, "Some Chapters for a History of Quantum Field Theory 1938–1952," in *Relativity, Groups and Topology II,* ed. B. S. De Witt and R. Slora (Amsterdam: North-Holland, 1984), 37–220; Schweber, "Shelter Island, Pocono and Oldstone: The Emergence of American Quantum Electrodynamics after World War II," *Osiris* 2 (1986): 265–302; Schweber, "Some Reflections on the History of Particle Physics in the 1950s," in *Pions to Quarks: Particle Physics in the 1950s,* ed. L. Brown, M. Dresden, and L. Hoddeson (New York: Cambridge University Press, 1989); L. M. Brown, R. Kawabe, M. Konuma, and Z. Maki, *Elementary Particle*

Theory in Japan, 1935–1960 (Kyoto: Research Institute for Fundamental Physics, Kyoto University, 1988).

243. E. Maxwell, interview with G. Baym, 4 March 1983.

244. Maxwell recalls that the MIT Radiation Laboratory, where he worked on radar, became on closing "a mecca for industry, universities, for everybody to come and hire people." Maxwell interview (n. 242). See also L. Hoddeson, P. Henriksen, R. Meade, and C. Westfell, *Critical Assembly: A Technical History of Los Alamos During the Oppenheimer Years, 1943–45* (New York: Cambridge University Press, in press).

245. See, for example, D. Kevles, *The Physicists* (New York: Vintage Books, 1979), 300–301, 356–366.

246. D. Bohm, interview with L. Hoddeson, 8 May 1981.

247. Ibid.

248. D. Pines, interview with L. Hoddeson, 13 and 16 April 1981.

249. Bohm's first paper, "Excitation of Plasma Oscillations," *Phys. Rev.* 70 (1946): 448, explored the conditions determining stability of plasma waves excited by means of fast beams. Soviet work on the microscopic theory of plasmas in this period includes L. Landau's fundamental paper, "On the Vibrations of the Electronic Plasma," *Journal of Physics (Moscow)* 10 (1946): 25–40, discussing plasma oscillations and their damping in terms of the plasma kinetic equation formulated by A. A. Vlasov in his 1938 dissertation, "On the Vibrational Properties of the Electron Gas," *Zhurnal Eksperimental'noi i Teoreticheskoi Fiziki* 8 (1938): 291–318; see also Vlasov, "On the Kinetic Theory of an Assembly of Particles with Collective Interaction," *Journal of Physics (Moscow)* 9 (1945): 25–40.

250. Pines interview (n. 248); Bohm interview (n. 246).

251. D. Bohm and E. P. Gross, "Plasma Oscillations as a Cause of Acceleration of Cosmic-Ray Particles," *Phys. Rev.* 74 (1948): 624; Bohm and Gross, "Excitation and Damping of Plasma Oscillations," *Phys. Rev.* 75 (1949): 1323; Bohm and Gross, "Theory of Plasma Oscillations. A. Origin of Medium-Like Behavior," *Phys. Rev.* 75 (1949): 1851–1864; Bohm and Gross, "Theory of Plasma Oscillations. B. Excitation and Damping of Oscillations," *Phys. Rev.* 75 (1949): 1864–1876; Bohm and Gross, "Effects of Plasma Boundaries and Oscillations," *Phys. Rev.* 79 (1950): 992–1001. How unfamiliar most physicists then were with plasmas is revealed by the fact that Bohm and Gross felt it useful to include in their papers a definition of the word *plasma!*

252. Bohm interview (n. 246).

253. Ibid.

254. Ibid. See also Schwinger (n. 242).

255. Pines interview (n. 248).

256. Ibid.

257. G. Ruthemann, "Diskrete Energieverluste mittelschneller Elektronen beim Durchgang durch dünne Folien," *Ann. Phys.* 2 (1948): 113–134; Ruthemann, "Elektronenbremsung an Röntgenniveaus," *Ann. Phys.* 2 (1948): 135–146; W. Lang, "Geschwindigkeitsverluste mittelschneller Elektronen beim Duchgang durch dünne Metallfolien," *Optik* 3 (1948): 233–246.

258. D. Bohm and D. Pines, "A Collective Description of Electron Interactions. I. Magnetic Interactions," *Phys. Rev.* 82 (1951): 625–634. For a complete description of the Bohm–Pines theory, and later connections with helium, see D. Pines, "The Collective Description of Particle Interactions: From Plasmas to the Helium Liquids," in *Quantum Implications: Essays in Honour of David Bohm*, ed. B. Hiley and F. D. Peat (Henley-on-Thames: Routledge, 1987), 66–84.

259. A similar description of a one-dimensional Fermi gas in terms of bosonic collective coordinates was studied at the same time by S. Tomonaga, "Remarks on Bloch's Method of Sound Waves Applied to Many-Fermion Problems," *Progress in Theoretical Physics* 5 (1950): 544–569, who was also at the Institute for Advanced Study at Princeton.

260. The "stopping power" problem played a role in the development of the theory of collective phenomena analogous to that played by the hydrogen atom in the development of quantum mechanics. The stopping power had been of theoretical interest since the discussion by N. Bohr, "Decrease of Speed of Electrified Particles on Passing Through Matter," *Phil. Mag.* 25 (1913): 10–31, in terms of electrons bound elastically to their equilibrium positions. As was found, taking unscreened Coulomb forces into account leads to a divergent result; see, for example, H. Bethe,

"Zur theorie des Durchganges schneller Korpuskularstrahlen durch Materie," *Ann. Phys.* 5 (1930): 325–400; C. F. von Weizsäcker, "Durchgang schneller Korpuskularstrahlen durch ein Ferromagnetikum," *Ann. Phys.* 17 (1933): 869–896. The problem was reapproached during the war by R. Kronig and J. Korringa, "Zur Theorie der Bremsung schneller geladener Teilchen in metallischen Leitern," *Physica* 10 (1943): 406–418, 800; also R. Kronig, "A Note on the Stopping of Fast Charged Particles in Metallic Conductors," *Physica* 15 (1949): 667–670, who attempted to show that the stopping power becomes finite when electron–electron interactions are accounted for, here approximately in terms of an artificial internal friction. H. A. Kramers, "The Stopping Power of a Metal for Alpha Particles," *Physica,* 13 (1947): 401–412, and A. Bohr, "Atomic Interaction in Penetration Phenomena," *Koninklijke Danske Videnskabernes Selskab, Matematisk-Fysiske Meddelelser* 24, no. 19 (1948), then introduced screening, which cut off the divergences in the stopping power. Although Kramers and A. Bohr had correctly treated the short-wavelength or high-momentum transfer part of the Coulomb interaction with a charged particle, neither had allowed for the possibility of exciting collective plasma oscillations. Pines considered this possibility early in 1951, obtaining a more accurate expression for the stopping power. D. Pines, "The Stopping Power of a Metal for Charged Particles," *Phys. Rev.* 85 (1952): 931.

261. Bohm interview (n. 246).

262. D. Pines, "A Collective Description of Electron Interactions: IV. Electron Interaction in Metals," *Phys. Rev.* 92 (1953): 626–636.

263. D. Pines, "The Role of Plasma Oscillations in Electron Interactions" (Ph.D. diss., Princeton University, 1950).

264. Bohm interview (n. 246); Pines interview (n. 248).

265. See comment by Mott, pp. 67–68, on paper of D. Pines, "The Collective Description of Electron Interaction in Metals," in Institut International de Physique Solvay, *Les Electrons dans les métaux, Dixième Conseil de Physique, Bruxelles, 1954,* ed. R. Stoops (Brussels: Coudenberg, 1955), 9–70.

266. We would like to thank E. P. Wohlfarth for sending us historical material on this work, including a memorandum of 4 February 1983 and copies of letters by J. Bardeen to A. Lidiard, 12 April 1950, and H. Koppe to Wohlfarth, 18 August 1950.

267. P. T. Landsberg, "A Contribution to the Theory of Soft X-ray Emission Bands of Sodium," *Proceedings of the Physical Society (London)* A62 (1949): 806–816.

268. H. Koppe, "Der Einfluss der Austauschenergie auf die spezifische Wärme des Elektronengases," *Zeitschrift für Naturforschung* 2a (1947): 429–432.

269. Bardeen (n. 18). Bardeen wrote to Lidiard on 12 April 1950: "There seems to be no empirical evidence that the exchange terms are important. Wigner believes that correlation effects compensate for the effects of exchange on the density of energy levels" (n. 266).

270. E. P. Wohlfarth, "The Influence of Exchange and Correlation Forces on the Specific Heat of Free Electrons in Metals," *Phil. Mag.* 41 (1950): 534–542. See also W. Macke, "Über die Wechselwirkungen in Fermi Gas; Polarisations erscheinungen, Correlationsenergie, Elektronenkondensation," *Zeitschrift für Naturforschung* 5a (1950): 192–208. Wohlfarth's general interest in these phenomena grew out of his attempts in 1949 to develop in a quantum-mechanical context Stoner's ideas of ferromagnetism. For example, E. C. Stoner, "Collective Electron Ferromagnetism," *Proc. Roy. Soc.* A165 (1938): 372–414; Stoner, "Collective Electron Ferromagnetism. II. Energy and Specific Heat," *Proc. Roy Soc.,* A169 (1939): 339–371; reviewed in E. P. Wohlfarth, "The Theoretical and Experimental Status of the Collective Electron Theory of Ferromagnetism," *Reviews of Modern Physics* 25 (1953): 211–219. See also chapter 6.

271. M. Gell-Mann, interview with L. Hoddeson, July 1982.

272. Discussed in Hoddeson et al. (n. 244).

273. Application of Feynman diagram techniques in the many-body problem was in its infancy at this stage. Brueckner had recently formulated a theory of nuclear matter in terms of a "linked-cluster" expansion. K. A. Brueckner and C. A. Levinson, "Approximate Reduction of the Many-Body Problem for Strongly Interacting Particles to a Problem of Self-Consistent Fields," *Phys. Rev.* 97 (1955): 1344–1352; K. A. Brueckner, "Two-Body Forces and Nuclear Saturation. III. Details of the Structure of the Nucleus." *Phys. Rev.* 97 (1955): 1353–1366, for which J. Goldstone and Bethe and, independently, N. Hugenholtz provided direct proofs: J. Goldstone, "Derivation of

Brueckner Many-Body Theory," *Proc. Roy. Soc.* A239 (1957): 267–279; N. M. Hugenholtz, "Perturbation Theory of Large Quantum Systems," *Physica* 23 (1957): 481–532. See also H. Bethe, "Nuclear Many-Body Problems," *Phys. Rev.* 103 (1956): 1353–1390; R. J. Eden, "The Brueckner Theory of Nuclear Structure," *Proc. Roy. Soc.* A235 (1956): 408–418. The work of Hubbard on the electron gas is discussed later.

274. M. Gell-Mann and K. A. Brueckner, "Correlation Energy of an Electron Gas at High Density," *Phys. Rev.* 106 (1957): 364–368; M. Gell-Mann, "Specific Heat of a Degenerate Electron Gas at High Density," *Phys. Rev.* 106 (1957): 369–372.

275. K. Sawada, "Correlation Energy of an Electron Gas at High Density," *Phys. Rev.* 106 (1957): 372–383; K. Sawada, K. A. Brueckner, N. Fukuda, and R. Brout, "Correlation Energy of an Electron Gas at High Density: Plasma Oscillations," *Phys. Rev.* 108 (1957): 507–514.

276. D. F. DuBois, "Part I. Field Theory of a Degenerate Electron Gas," *Annals of Physics (N.Y.)* 7 (1959): 174–237; DuBois, "Part II. Properties of a Dense Electron Gas," *Annals of Physics (N.Y.)* 8 (1959): 24–77; P. Nozières and D. Pines, "Correlation Energy of a Free Electron Gas," *Phys. Rev.* 111 (1958): 442–454; Nozières and Pines, "A Dielectric Formulation of the Many-Body Problem: Application to the Free Electron Gas," *Nuovo Cimento* 9 (1958): 470–490.

277. J. Hubbard, "The Description of Collective Motions in Terms of Many-Body Perturbation Theory," *Proc. Roy. Soc.* A240 (1957): 539–560; Hubbard, "The Description of Collective Motions in Terms of Many-Body Perturbation Theory; II. The Correlation Energy of a Free-Electron Gas," *Proc. Roy. Soc.* A243 (1957): 336–352.

278. J. M. Luttinger and W. Kohn, "Quantum Theory of Cyclotron Resonance in Semiconductors," *Phys. Rev.* 96 (1954): 529–530; Luttinger and Kohn, "Hyperfine Splitting of Donor States in Silicon," *Phys. Rev.* 96 (1954): 802–803; Luttinger and Kohn, "Hyperfine Splitting of Donor States in Silicon," *Phys. Rev.* 97 (1955): 883–888; Luttinger and Kohn, "Theory of Donor Levels in Silicon," *Phys. Rev.* 97 (1955): 1721; Luttinger and Kohn, "Theory of Donor States in Silicon," *Phys. Rev.* 98 (1955): 915–922; Luttinger and Kohn, "Motion of Electrons and Holes in Perturbed Periodic Fields," *Phys. Rev.* 97 (1955): 869–883; J. M. Luttinger, "Quantum Theory of Electrical Transport Phenomena. I," *Phys. Rev.* 108 (1957): 590–611; Luttinger and Kohn, "Quantum Theory of Electrical Transport Phenomena. II," *Phys. Rev.* 109 (1958): 1892–1909; Luttinger and Kohn, "Ground State Energy of a Many-Fermion System," *Phys. Rev.* 118 (1959): 41–45; J. M. Luttinger and J. Ward, "Ground State Energy of a Many-Fermion System. II," *Phys. Rev.* 118 (1960): 1417–1427; Luttinger, "Fermi Surface and Some Simple Equilibrium Properties of a System of Interacting Fermions," *Phys. Rev.* 119 (1960): 1153–1163; J. M. Luttinger and P. Nozières, "Derivation of the Landau Theory of Fermi Liquids. I. Formal Preliminaries," *Phys. Rev.* 127 (1962): 1423–1431; Luttinger and Nozières, "Derivation of the Landau Theory of Fermi Liquids. II. Equilibrium Properties and Transport Equation," *Phys. Rev.* 127 (1962): 1431–1440; A. B. Migdal, "The Momentum Distribution of Interacting Fermi Particles," *Zhurnal Eksperimental'noi i Teoreticheskoi Fiziki* 32 (1957): 399–400 (trans. *Soviet Physics JETP* 5 [1957]: 333–334); Migdal, "Interaction Between Electrons and Lattice Vibrations in a Normal Metal," *Zhurnal Eksperimental'noi i Teoreticheskoi Fiziki* 34 (1958): 1438–1446 (trans. *Soviet Physics JETP* 7 [1958]: 996–1001). See also P. W. Anderson, interview with L. Hoddeson, 10 May 1988.

279. J. M. Luttinger, interview with L. Hoddeson, 9 May 1988; Anderson interview (n. 278).

280. A good review of superconductivity through the early postwar period is Shoenberg (1952) (n. 21). The theory of superconductivity through 1952 is reviewed in V. L. Ginzburg, "The Present State of the Theory of Superconductivity," *Uspekhi Fizicheskikh Nauk* 48 (1952): 25ff (German trans., "Der gegenwärtige Stand der Theorie der Superleitung," *Fortschritte der Physik* 1 [1953]: 101–163).

281. W. Heisenberg, "Zur Theorie der Supraleitung," *Zeitschrift für Naturforschung* 2a (1947): 185–201; Heisenberg, "Das elektrodynamische Verhalten der Supraleiter," *Zeitschrift für Naturwissenschaften* 3a (1948): 65–75; Heisenberg, "Thermodynamische Betrachtungen zum Problem der Supraleitung," *Ann. Phys.* 3 (1948): 289–96; Heisenberg, "Electron Theory of Superconductivity," in *Two Lectures* (Cambridge: Cambridge University Press, 1949), 27–51.

282. H. Koppe, "Die spezifische Wärme der Supraleiter nach der Theorie von W. Heisenberg," *Annalen der Physik (Leipzig)* 1 (1947): 405–414; Koppe, "Zur theorie der Supraleitung II. Die Berechnung der Sprungtemperatur," *Zeitschrift für Naturforschung* 3a (1948): 115; Koppe, "Theo-

rie der Supraleitung," *Ergebnisse der exakten Naturwissenschaften* 23 (1950): 283–358; Koppe, "Zur phänomenologischen Theorie der Supraleitung," *Zeitschrift für Naturwissenschaften* 6a (1951): 284–287; Shoenberg (1952) (n. 21), 218.

283. M. Born and K. C. Cheng, "Theory of Superconductivity," *Nature* 161 (1948): 968–969, 1017–1018; *Nature* 163 (1949): 247; Born and Cheng, "Sur la théorie de la supraconductivité," *Journal de physique et le radium* 9 (1948): 249–252.

284. M. von Laue, "Zur Thermodynamik der Superleitung," *Ann. Physik* 32 (1938): 71–84; Laue, "Der magnetische Schwellenwert für Supraleitung," *Ann. Phys.* 32 (1938): 253–258; R. B. Pontius, "Threshold Values of Supraconductors of Small Dimensions," *Phil. Mag.* 24 (1937): 787–797. Penetration in thin films was also studied in 1938 by Shalnikov in Kapitza's Institute and Appleyard at the Mond Laboratory, among others.

285. D. Shoenberg, "Superconducting Colloidal Mercury," *Nature* 143 (1939): 434–435; Shoenberg, "Properties of Superconducting Colloids and Emulsions," *Proc. Roy. Soc.* A175 (1940): 49–70. Shoenberg's results were consistent with the Londons' theory if he assumed that the concentration of superconducting electrons varies with temperature, as predicted by the Gorter–Casimir theory. Shoenberg (n. 21) recalls that this work originated in a suggestion by Landau.

286. H. London, "The High-Frequency Resistance of Superconducting Tin," *Proc. Roy. Soc.* A176 (1940): 522–533.

287. J. C. McLennan, A. C. Burton, A. Pitt, and J. O. Wilhelm, "The Phenomena of Superconductivity with Alternating Currents of High Frequency," *Proc. Roy. Soc.* 136 (1932): 52–76; McLennan et al., "Further Experiments on Superconductivity with Alternating Currents of High Frequency," *Proc. Roy. Soc.* 138 (1932): 245–258.

288. A. B. Pippard, interview with P. Hoch and E. Braun, 22 January 1981.

289. For example, A. B. Pippard, "The Surface Impedance of Superconductors and Normal Metals at High Frequencies," *Proc. Roy. Soc.* A191 (1947): 385–415.

290. Maxwell interview (n. 243); J. C. Slater, *Solid-State and Molecular Theory: A Scientific Biography* (New York: Wiley, 1975), 217–225.

291. J. Slater, "History of the MIT Physics Department," John Clarke Slater Papers, American Philosophical Society Library, Philadelphia, copy at AIP; J. Burchard, *MIT in World War II* (New York: Wiley, 1948), 180; Maxwell interview (n. 243).

292. S. C. Collins, "Helium Liquifier," *Science* 116 (1952): 289–294; S. C. Collins and R. L. Cannaday, *Expansion Machines for Low Temperature Processes* (London: Oxford University Press, 1958); Collins, "A Helium Cryostat," *Review of Scientific Instruments* 18 (1947): 157–167.

293. E. Maxwell, P. M. Marcus, and J. C. Slater, "Surface Impedance of Normal and Superconductors at 24,000 Megacycles per Second," *Phys. Rev.* 76 (1949): 1332–1347.

294. W. M. Fairbank, "High Frequency Surface Resistivity of Tin in the Normal and Superconducting States," *Phys. Rev.* 76 (1949): 167, 1106–1111. T. G. Berlincourt, "Low Temperature Physics at Yale" (Talk at Symposium on Low Temperature Physics, held at Yale in honor of Cecil T. Lane, 13 April 1973), Bardeen Papers, University of Illinois, Urbana.

295. A. B. Pippard, interview with L. Hoddeson and G. Baym, 14 September 1982.

296. A. B. Pippard, unpublished notes, Bardeen Papers (n. 294); Pippard, "Theory of Boundary Effects in Superconductors," manuscript, Bardeen Papers (n. 294); J. Bardeen, "Change in Superconducting Penetration Depth Field," *Bulletin of the American Physical Society* 27, no. 3 (1952): 16.

297. A. B. Pippard, "Metallic Conduction at High Frequencies and Low Temperatures," *Advances in Electronics and Electron Physics* 6 (1954): 1–45, and references therein.

298. Pippard interview (n. 295).

299. A. B. Pippard, "The Coherence Concept in Superconductivity," *Physica* 19 (1953): 765–774, pp. 766–768.

300. G.E.H. Reuter and E. H. Sondheimer, "The Theory of the Anomalous Skin Effect in Metals," *Proc. Roy. Soc.* A195 (1948): 336–364; A. B. Pippard, G.E.H. Reuter, and E. H. Sondheimer, "The Conductivity of Metals at Microwave Frequencies," *Phys. Rev.* 73 (1948): 920–921; R. G. Chambers, "The Kinetic Formulation of Conduction Problems," *Proceedings of the Physical Society (London)* A65 (1952): 458–459. The Reuter–Sondheimer work led to a powerful new method of determining the Fermi surface of metals; see Chapter 3.

301. Pippard interview (n. 295).

302. V. L. Ginzburg and L. D. Landau, "On the Theory of Superconductivity," *Zhurnal Eksperimental'noi i Teoreticheskoi Fiziki* 20 (1950): 1064–1082; V. Ginzburg, interview with G. Baym, 17 November 1987.

303. V. Ginzburg, "On the Surface Energy and the Behavior of Supraconductors of Small Dimensions," *Journal of Physics (Moscow)* 9 (1945): 305–311. Also Ginzburg, *Sverhprovodimost* (Superconductivity) (Moscow: Academy of Sciences of USSR, 1946); Ginzburg, "On the Surface Energy and the Behaviour of Supraconductors of Small Dimensions," *Zhurnal Eksperimental'noi i Teoreticheskoi Fiziki* 16 (1946): 87–96.

304. V. L. Ginzburg to L. Hoddeson, 28 December 1982; Ginzburg interview (n. 302).

305. Landau (n. 185). An order parameter for superconductivity, the fraction of electrons that are superconducting, had earlier appeared in the discussion of the stability of the superconducting state by Gorter and Casimir (n. 68).

306. In his discussion of domain walls in ferromagnetism, Bloch (n. 183) had already introduced a spatially varying order parameter; indeed, the Ginzburg–Landau equation in the absence of magnetic fields or supercurrents is essentially equivalent to Bloch's equation for the order parameter.

307. N. V. Zavaritski, "Investigation of Superconducting Properties of Thallium and Tin Films Condensed at Low Temperatures," *Doklady Akademii Nauk SSSR* 86 (1952): 501–504.

308. A. A. Abrikosov, "My Years with Landau," *Physics Today* 26, no. 1 (1973): 56–60; Abrikosov, "Influence of Dimensions on the Critical Field of Superconductors of the Second Group," *Doklady Akademii Nauk SSSR* 86 (1952): 489–492.

309. A. A. Abrikosov, "On the Magnetic Properties of Superconductors of the Second Group," *Zhurnal Eksperimental'noi i Teoreticheskoi Fiziki* 32 (1957): 1442–1452 (trans. *Soviet Physics JETP* 6 [1957]: 1174–1182).

310. Sources for this somewhat controversial history are Abrikosov (1973) (n. 308); E. Lifshitz to J. Bardeen, 26 June 1978; N. Zavaritski to A. B. Pippard, 10 October 1978; and A. Abrikosov to J. Bardeen, 31 October 1978, Bardeen Papers (n. 294); L. Landau and E. M. Lifshitz, "On the Rotation of Liquid Helium," *Doklady Akademii Nauk SSSR* 100 (1955): 669, in *Collected Papers of L. D. Landau*, ed. D. Ter Haar (New York: Gordon and Breach, 1965), 650–654; R. P. Feynman, "Application of Quantum Mechanics to Liquid Helium," in *Progress in Low Temperature Physics* ed. C. J. Gorter (1955), 1: 17–53.

311. The creation of such alloys and original construction of superconducting magnets is due largely to the American group at Bell Labs, headed by J. E. Kunzler. After 1957 a large experimental program to study dirty superconductors was mounted at Bell Labs, Westinghouse, and General Electric.

312. L. V. Shubnikov, W. I. Khotkevich, D. Shepiliov, and J. N. Rabinin, "Magnetic Properties of Superconducting Metals and Alloys," *Zhurnal Eksperimental'noi i Teoreticheskoi Fiziki* 7 (1937): 221–237.

313. B. Goodman, "The Magnetic Behavior of Superconductors of Negative Surface Energy," *IBM Journal of Research and Development* 6 (1962): 63–67.

314. Pippard interview (n. 295).

315. J. G. Daunt and K. Mendelssohn, "An Experiment on the Mechanism of Superconductivity," *Proc. Roy. Soc.* A185 (1946): 225–239.

316. B. B. Goodman, "The Thermal Conductivity of Superconducting Tin below 1°K," *Proc. Roy. Soc.* A66 (1953): 217–227.

317. See B. B. Goodman, "Applications of Superconductivity," in *Trends in Physics, Volume of Plenary Lectures at 2nd General Conference of European Physical Society, 3–6 October 1972, Weisbaden* (Geneva: European Physical Society, 1973), 67–94.

318. A. Brown, M. W. Zemansky, and H. A. Boorse, "The Superconducting and Normal Heat Capacities of Niobium," *Phys. Rev.* 92 (1953): 52–58.

319. W. S. Corak, B. B. Goodman, C. B. Satterthwaite, and A. Wexler, "Exponential Temperature Dependence of the Electronic Specific Heat of Superconducting Vanadium," *Phys. Rev.* 96 (1954): 1442–1444; Corak et al., "Atomic Heats of Normal and Superconducting Vanadium," *Phys. Rev.* 102 (1956): 656–661; W. S. Corak and C. B. Satterthwaite, "Atomic Heats of Normal and Superconducting Tin Between 1.2 Degrees and 4.5 Degrees K," *Phys. Rev.* 102 (1956): 662–

666. See also the review by M. A. Biondi, A. T. Forrester, M. B. Garfunkel, and C. B. Satterthwaite, "Experimental Evidence for an Energy Gap in Superconductors," *Reviews of Modern Physics* 30 (1958): 1109–1136.

320. R. E. Glover III and M. Tinkham, "Transmission of Superconducting Films at Millimeter-Microwave and Far Infrared Frequencies," *Phys. Rev.* 104 (1956): 844–845; Glover and Tinkham, "Conductivity of Superconducting Films for Photon Energies Between 0.3 and $40kT_c$," *Phys. Rev.* 108 (1957): 243–256.

321. C. A. Reynolds, B. Serin, W. H. Wright, and L. B. Nesbitt, "Superconductivity of Isotopes of Mercury," *Phys. Rev.* 78 (1950): 487; E. Maxwell, "Isotope Effect in the Superconductivity of Mercury," *Phys. Rev.* 78 (1950): 477.

322. Maxwell interview (n. 243).

323. Ibid.

324. E. Maxwell notebook entry for 12 November 1948. We would like to thank Maxwell for making available to us relevant sections of his notebook from the period.

325. E. Maxwell, "Isotope Effect in the Superconductivity of Mercury," *Phys. Rev. Letters* 78 (1950): 477.

326. J. de Launay and R. L. Dolocek, "Superconductivity and the Debye Characteristic Temperature," *Phys. Rev.* 72 (1947): 141–431.

327. M. Garfunkel, telephone interview with L. Hoddeson, 20 January 1983.

328. H. Torrey, telephone interview with L. Hoddeson, 19 January 1983.

329. W. D. Allen, R. H. Dawton, J. M. Lock, A. B. Pippard, and D. Shoenberg, "Superconductivity of Tin Isotopes," *Nature* 166 (1950): 1071; W. D. Allen, R. H. Dawton, M. Bär, K. Mendelssohn, and J. L. Olsen, "Superconductivity of Tin Isotopes," *Nature* 166 (1950): 1071–1072. Both groups used isotopes prepared by Allen and Dawton at Harwell (Pippard interview [n. 295]), and both carried out essentially the same measurements by different techniques.

330. H. Fröhlich, H. Pelzer, and S. Zienau, "Properties of Slow Electrons in Polar Materials," *Phil. Mag.* 41 (1950): 221–242.

331. H. Fröhlich, "Theory of the Superconducting State. I. The Ground State at the Absolute Zero of Temperature," *Phys. Rev.* 79 (1950): 845–856.

332. H. Fröhlich, "Isotope Effect in Superconductivity" (letter), *Proceedings of the Physical Society (London)* A63 (1950): 778.

333. H. Fröhlich, "History of the Theory of Superconductivity," manuscript. We are grateful to Paul Hoch for a copy of this manuscript.

334. J. Bardeen, handwritten note, 16 May 1950, Bardeen Papers (n. 294); Bardeen, "History of Superconductivity Research," in *Impact of Basic Research on Technology*, ed. B. Kursonoglu and A. Perlmutter (New York: Plenum, 1973), 15–57.

335. J. Bardeen, manuscript, Bardeen Papers (n. 294); only an abstract was published: J. Bardeen, "Theory of Superconductivity," *Phys. Rev.* 59 (1941): 928.

336. Bardeen notes, 15–18 May 1950, Bardeen Papers (n. 294).

337. J. Bardeen, "Zero-Point Vibrations and Superconductivity," *Phys. Rev.* 79 (1950): 167–168.

338. J. Bardeen, "Wave Functions for Superconducting Electrons," *Phys. Rev.* 80 (1950): 567–574.

339. Bardeen (1973) (n. 334), 31.

340. Bardeen to R. Peierls, 17 July 1951, Peierls Papers, Oxford.

341. H. Fröhlich, "Interaction of Electrons with Lattice Vibrations," *Proc. Roy. Soc.* A215 (1952): 291–298; Fröhlich, "Superconductivity and Lattice Vibrations," *Physica* 19 (1953): 755–764.

342. H. Fröhlich, "On the Theory of Superconductivity: The One-Dimensional Case," *Proc. Roy. Soc.* A223 (1954): 296–305. In fact, the instability Fröhlich found is associated with the formation of charge density waves in a one-dimensional conductor.

343. M. R. Schafroth, "Bemerkungen zur Fröhlichschen Theorie der Supraleitung," *Helvetica Physica Acta* 24 (1951): 645–662.

344. In 1958, Arkady Migdal solved the Fröhlich Hamiltonian for normal metals and showed that no instabilities similar to those found by Fröhlich in the one-dimensional case appear. Migdal (1958) (n. 278).

345. M. R. Schafroth, "Theory of Superconductivity," *Phys. Rev.* 96 (1954): 1442; Schafroth, "Superconductivity of a Charged Boson Gas," *Phys. Rev.* 96 (1954): 1149; Schafroth, "Superconductivity of a Charged Ideal Bose Gas," *Phys. Rev.* 100 (1955): 463–475. Apparently, R. A. Ogg was the first to suggest that superconductivity could be explained in terms of Bose–Einstein condensation of pairs of electrons, in a paper on purported observations of extremely high-temperature superconductivity, "Bose-Einstein Condensation of Trapped Electron Pairs. Phase Separation and Superconductivity of Metal-Ammonia Solutions," *Phys. Rev.* 69 (1946): 243–244. The proposal was also made by Fröhlich (1953) (n. 341), by Ginzburg (n. 280), and by R. E. Peierls, *Quantum Theory of Solids* (Oxford: Clarendon Press, 1955), 221.

346. M. R. Schafroth, S. T. Butler, and J. M. Blatt, "Quasichemical Equilibrium Approach to Superconductivity," *Helvetica Physica Acta* 30 (1957): 93–134.

347. J. Bardeen, L. N. Cooper, and J. R. Schrieffer, "Theory of Superconductivity," *Phys. Rev.* 108 (1957): 1175–1204. For a more technical account of BCS and some of its prehistory, see J. Bardeen and J. R. Schrieffer, "Recent Developments in Superconductivity," in *Progress in Low Temperature Physics,* ed. C. J. Gorter (New York: Interscience, 1961), 3: 170–287. A clear popular introduction to the discovery of BCS is Joan N. Warnow, ed., "Moments of Discovery: Superconductivity," unpublished audio tape, 1976, AIP.

348. J. Bardeen, handwritten notes, 23 October 1951, Bardeen Papers (n. 294).

349. H. Fröhlich, "Interaction of Electrons with Lattic Vibrations," *Proc. Roy. Soc.* A215 (1952): 291–298; T. D. Lee, F. E. Low, and D. Pines, "The Motion of Slow Electrons in a Polar Crystal," *Phys. Rev.* 90 (1953): 297–302; T. D. Lee and D. Pines, "Interaction of a Nonrelativistic Particle with a Scalar Field with Application to Slow Electrons in Polar Crystals," *Phys. Rev.* 92 (1953): 883–889; F. E. Low and D. Pines, "Mobility of Slow Electrons in Polar Crystals," *Phys. Rev.* 98 (1955): 414–418.

350. J. Bardeen and D. Pines, "Electron-Phonon Interaction in Metals," *Phys. Rev.* 99 (1955): 1140–1150.

351. B. Serin, "Superconductivity. Experimental Part," in *Handbuch der Physik* (Berlin: Springer, 1956), 15: 210–273.

352. J. Bardeen, "Theory of Superconductivity. Theoretical Part," in *Handbuch der Physik* (n. 351), 15: 274–369.

353. See J. Bardeen, "Theory of the Meissner Effect in Superconductors," *Phys. Rev.* 97 (1955): 1724–1725. During the 1950s, there was increasing evidence for an energy gap.

354. J. Bardeen, "Talk on Superconductivity," unpublished notes, Bardeen Papers (n. 294).

355. Bardeen (1973) (n. 334).

356. J. R. Schrieffer, interview with J. Warnow and R. M. Williams, 26 September 1974.

357. Ibid.

358. Ibid.

359. London (n. 61), 142–155.

360. Bardeen et al. (n. 347), 1177 n. 18.

361. See, for example, Bardeen (n. 338), 567.

362. Schrieffer interview (n. 356).

363. Brueckner and Levinson (n. 273); Brueckner (n. 273).

364. Schrieffer interview (n.356).

365. L. Cooper, "Bound Electron Pairs in a Degenerate Fermi Gas," *Phys. Rev.* 104 (1956): 1189–1190.

366. J. Bardeen, private communication to L. Hoddeson.

367. Bardeen (n. 352), 276.

368. Schrieffer interview (n. 356).

369. Ibid.

370. Ibid.

371. Ibid.

372. R. P. Feynman, "Superfluidity and Superconductivity," *Reviews of Modern Physics* 29 (1957): 205–212. Schrieffer assisted in preparing the manuscript of this talk for publication.

373. Glover and Tinkham (n. 320) were just then measuring the gap for lead and tin and were thus able to verify this prediction of theory.

374. J. Bardeen, L. N. Cooper, and J. R. Schrieffer, "Microscopic Theory of Superconductivity," *Phys. Rev.* 106 (1957): 162–164.

375. J. Bardeen to S. A. Goudsmit, 15 February 1957, Bardeen Papers (n. 294).

376. Bardeen et al. (n. 374).

377. L. C. Hebel and C. P. Slichter, "Nuclear Relaxation in Superconducting Aluminum," *Phys. Rev.* 107 (1957): 901–902. Schrieffer interview (n. 356); C. P. Slichter, interview with L. Hoddeson, 1976.

378. R. W. Morse and H. V. Bohm, "Superconducting Energy Gap from Ultrasonic Attenuation Measurements," *Phys. Rev.* 108 (1957): 1094–1096; R. W. Morse to J. Bardeen, 24 September 1957, and 14 March 1958, Bardeen Papers (n. 294).

379. Glover and Tinkham (n. 320).

380. Schrieffer interview (n. 356).

381. Bardeen et al. (n. 347).

382. P. W. Anderson, "It's Not Over Till the Fat Lady Sings," manuscript. We are grateful to Anderson for allowing us to incorporate the material of his talk into the rest of this section. See also Anderson, "Superconductivity in the Past and the Future," in *Superconductivity,* ed. R. D. Parks (New York: Marcel Dekker, 1969), 2: 1343–1358.

383. J. Bardeen to L. N. Cooper, 19 September 1957, Bardeen Papers (n. 294).

384. A. B. Pippard, "The Historical Context of Josephson's Discovery (Opening talk at NATO Advanced Study Institute on Small-Scale Superconducting Devices), in *Superconducting Applications: SQUIDS and Machines (Proceedings of the NATO Advanced Study Institute on Small-Scale Applications of Superconductivity),* ed B. B. Schwartz and S. Foner (New York: Plenum, 1977), 1–20.

385. Bardeen to Cooper (n. 383); A. P. Pippard to J. Bardeen, 11 September 1957, Bardeen Papers (n. 294).

386. J. R. Schrieffer to J. Bardeen, 19 November 1957, Bardeen Papers (n. 294).

387. Schrieffer interview (n. 356).

388. Schafroth et al. (n. 346); Schafroth (n. 345).

389. Bardeen et al. (n. 347), 1177 n. 18.

390. J. Bardeen to F. Dyson, 23 July 1957, Bardeen Papers (n. 294). Blatt later showed that their theory could be transformed to BCS. See J. M. Blatt, *Theory of Superconductivity* (New York: Academic Press, 1964), 181–185.

391. Bardeen (n. 353); M. J. Buckingham, "A Note on the Energy Gap Model of Superconductivity," *Nuovo Cimento* 5 (1957): 1763–1765.

392. J. Bardeen, "Gauge Invariance and the Energy Gap Model of Superconductivity," *Nuovo Cimento* 5 (1957): 1766–1768.

393. G. Wentzel, "Meissner Effect," *Phys. Rev.* 111 (1958): 1488–1492; M. R. Schafroth, "Remarks on the Meissner Effect," *Phys. Rev.* 111 (1958): 72–74; J. Blatt and T. Matsubara, "The Meissner-Ochsenfeld Effect in the Bogoliubov Theory," *Progress in Theoretical Physics* 20 (1958): 781–783. Notably, the last paper fails to reference Bardeen, Cooper, and Schrieffer!

394. P. W. Anderson, "Coherent Excited States in the Theory of Superconductivity: Gauge Invariance and the Meissner Effect," *Phys. Rev.* 110 (1958): 827–835; Anderson, "Random-Phase Approximation in the Theory of Superconductivity," *Phys. Rev.* 112 (1958): 1900–1916. See also D. Pines and J. R. Schrieffer, "Gauge Invariance in the Theory of Superconductivity," *Nuovo Cimento* 10 (1958): 496–504. A fuller demonstration of the gauge invariance of the BCS theory when collective modes are accounted for was given by G. Rickayzen, "Collective Excitations in the Theory of Superconductivity," *Phys. Rev.* 115 (1959); 795–808.

395. N. N. Bogoliubov, "On a New Method in the Theory of Superconductivity," *Nuovo Cimento* 7 (1958): 794–805; N. N. Bogoliubov et al., "A New Method in the Theory of Superconductivity. I," *Zhurnal Eksperimental'noi i Teoreticheskoi Fiziki* 34 (1958): 58–65 (trans. *Soviet Physics JETP* 7 [1958]: 41–46); V. V. Tolmachev and S. V. Tiablikov, "A New Method in the Theory of Superconductivity, II," *Zhurnal Eksperimental'noi i Teoreticheskoi Fiziki* 34 (1958): 66–72 (trans. *Soviet Physics JETP* 7 [1958]: 46–50); N. N. Bogoliubov, "A New Method in the Theory of Superconductivity, III," *Zhurnal Eksperimental'noi i Teoreticheskoi Fiziki* 34 (1958): 73–79 (trans. *Soviet Physics JETP* 7 [1958]: 51–55); N. N. Bogoliubov, D. N. Zubarev, and Iu. A. Tserkovnikov, "The Compensation Principle and the Self-Consistent Field Method," *Uspekhi*

Fizicheskikh Nauk 67 (1959): 549–580 (trans. *Soviet Physics Uspekhi* 2 [1959]: 236–254). These and other papers appeared as a book: N. N. Bogoliubov, V. V. Tolmachev, and D. V. Shirkov, *A New Method in the Theory of Superconductivity* (Moscow: Academy of Sciences USSR, 1958). In his derivation of the ground state and excitation spectrum (in agreement with the original BCS theory), Bogoliubov introduced a clever quasiparticle transformation to a new set of fermion particles, which are linear superpositions of the original particles and holes; the transformation was independently derived by Valatin in Birmingham. J. G. Valatin, "Comments on the Theory of Superconductivity," *Nuovo Cimento* 7 (1958): 843–857; Y. Nambu, "Quasiparticles and Gauge Invariance in the Theory of Superconductivity," *Phys. Rev.* 117 (1960): 648–663.

396. Bardeen (1973) (n. 334), 41–43; D. Pines, personal communication, 21 December 1987.

397. Y. Nambu and G. Jona-Lasinio, "Dynamical Model of Elementary Particles Based on an Analogy with Superconductivity. I," *Phys. Rev.* 122 (1961): 345–358.

398. Matthias, in exploring the isotope effect with John Hulm at Bell Labs, found that certain substances behaved differently from the BCS predictions; he had started with Hulm at Chicago in 1950, at Fermi's suggestion, an experimental program to determine the empirical rules governing the occurrence of superconductivity, by applying principles of structural chemistry. Developing empirical rules for estimating critical temperatures based on electron-to-atom ratios and atomic volumes, Matthias was able to discover a vast number of new superconductors. See B. T. Matthias, T. H. Geballe, and V. B. Compton, "Superconductivity," *Reviews of Modern Physics* 35 (1963): 1–22; J. K. Hulm, J. E. Kunzler, and B. J. Matthias, "The Road to Superconducting Materials," *Physics Today* 34, no. 1 (1981): 34–43.

399. J. C. Swihart, "Solutions of the BCS Integral Equation and Deviations from the Law of Corresponding States," *IBM Journal of Research and Development* 6 (1962): 14–23; P. Morel and P. W. Anderson, "Calculation of the Superconducting State Parameters with Retarded Electron-Phonon Interaction," *Phys. Rev.* 125 (1962): 1263–1271.

400. G. M. Eliashberg, "Interaction Between Electrons and Lattice Vibrations in a Superconductor," *Zhurnal Eksperimental'noi i Teoreticheskoi Fiziki* 38 (1960): 996–976 (trans. *Soviet Physics JETP* 11 [1960]: 696–702).

401. J. M. Rowell, P. W. Anderson, and D. E. Thomas, "Image of the Phonon Spectrum in the Tunneling Characteristic Between Superconductors," *Physical Review Letters* 10 (1963): 334–336; J. R. Schrieffer, D. J. Scalapino, and J. W. Wilkins, "Effective Tunneling Density of States in Superconductors," *Physical Review Letters* 10 (1963): 336–339; W. McMillan and J. M. Rowell, "Lead Phonon Spectrum Calculated from Superconducting Density of States," *Physical Review Letters* 14 (1965): 108–112.

402. H. B. G. Casimir, "On the Theory of Superconductivity," in *Niels Bohr and the Development of Physics,* ed. W. Pauli (New York: Pergamon Press, 1955), 118–133.

403. L. P. Gor'kov, "Microscopic Derivation of the Ginzburg-Landau Equations in the Theory of Superconductivity," *Zhurnal Eksperimental'noi i Teoreticheskoi Fiziki* 36 (1959): 1918–1923 (trans. *Soviet Physics JETP* 9 [1959]: 1364–1367). Anderson recalls Landau remarking in December 1958, after Gor'kov had derived Ginzburg–Landau from BCS, that since Ginzburg–Landau was gauge invariant there was no problem. Gor'kov's work followed his formulation of the BCS theory in diagrammatic perturbation theory in "On the Energy Spectrum of Superconductors," *Zhurnal Eksperimental'noi i Teoreticheskoi Fiziki* 34 (1958): 735–739 (trans. *Soviet Physics JETP* 7 [1958]: 505–508).

404. Abrikosov (n. 309).

405. Quantization of flux was predicted—although only in units of hc/e rather than the observed $hc/2e$—by both London (n. 61), 152, and L. Onsager, discussion remarks after paper by Ichimura, "Statistical Mechanics of Electron-Lattice System," in *Proceedings of the International Conference on Theoretical Physics, Kyoto and Tokyo, September 1953* (Tokyo: Science Council of Japan, 1954), 935–936; Onsager, "Magnetic Flux Through a Superconducting Ring," *Physical Review Letters* 7 (1961): 50.

406. B. S. Deaver, Jr., and W. M. Fairbank, "Experimental Evidence for Quantized Flux in Superconducting Cylinders," *Physical Review* 7 (1961): 43–46; R. Doll and M. Näbauer, "Experimental Proof of Magnetic Flux Quantization in a Superconducting Ring," *Physical Review Letters* 7 (1961): 51–52.

407. N. Byers and C. N. Yang, "Theoretical Considerations Concerning Quantized Magnetic Flux in Superconducting Cylinders," *Physical Review Letters* 7 (1961): 46–49.

408. I. Giaever, "Energy Gap in Superconductors Measured by Electron Tunneling," *Physical Review Letters* 5 (1960): 147–148. For a lucid account of the discovery of tunneling in superconductors, see I. Giaever, "Discovery of Electron-Tunneling into Superconductors: One Researcher's Personal Account," *Adventures in Experimental Physics* (Princeton, N.J.: World Science Communications, 1974), 4: 133–142.

409. B. D. Josephson, "Possible New Effects in Superconductive Tunneling," *Physics Letters* 1 (1962): 251–253. See also Pippard (n. 384). According to Pippard, Anderson suggested a sin ϕ dependence of the current. Anderson ("It's Not Over") (n. 382); P. W. Anderson, "How Josephson Discovered His Effect," *Physics Today* 23, no. 11 (1970): 23–29.

410. P. W. Anderson and J. M. Rowell, "Probable Observation of the Josephson Superconducting Tunneling Effect," *Physical Review Letters* 10 (1963): 230–232. The alternating current Josephson effect was observed first by S. Shapiro, "Josephson Currents in Superconducting Tunneling: The Effect of Microwaves and Other Observations," *Physical Review Letters* 11 (1963): 80–82.

411. Also in V. Ambegaokar and L. P. Kadanoff, "Electromagnetic Properties of Superconductors," *Nuovo Cimento* 22 (1961): 914–935.

412. Goodman (n. 313).

413. Y. Kim, C. F. Hempstead, and A. R. Strnad, "Critical Persistent Currents in Hard Superconductors," *Physical Review Letters* 9 (1962): 306–309; P. W. Anderson, "Theory of Flux Creep in Hard Superconductors," *Physical Review Letters* 9 (1962): 309–311.

414. P. W. Anderson and Y. B. Kim, "Hard Superconductivity: Theory of the Motion of Abrikosov Flux Lines," *Reviews of Modern Physics* 36 (1964): 39–43.

415. A. Bohr, B. Mottelsen, and D. Pines, "Possible Analogy Between the Excitation Spectrum of Nuclei and Those of the Superconducting Metallic State," *Phys. Rev.* 110 (1958): 936–938.

416. A. B. Migdal, "Superfluidity and the Moments of Inertia of Nuclei," *Nuclear Physics* 13 (1957): 655–674; V. L. Ginzburg and D. A. Kirzhnits, "On the Superfluidity of Neutron Stars," *Zhurnal Eksperimental'noi i Teoreticheskoi Fiziki* 47 (1964): 2006–2007 (trans. *Soviet Physics JETP* 20 [1965]: 1346–1348); G. Baym, C. Pethick, and D. Pines, "Superfluidity in Neutron Stars," *Nature* 224 (1969): 673–674.

417. See, for example, A. J. Leggett, "Theoretical Description of the New Phases of Liquid ³He," *Reviews of Modern Physics* 47 (1975): 331–414; J. C. Wheatley, "Experimental Properties of Superfluid ³He," *Reviews of Modern Physics* 47 (1975): 415–470.

418. V. Peshkov, "Determination of the Velocity of Propagation of the Second Sound in Helium II," *Journal of Physics (Moscow)* 10 (1946): 389–398; Peshkov (n. 141).

419. L. Tisza, "The Theory of Liquid Helium," *Phys. Rev.* 72 (1947): 838–854.

420. E. Lifshitz, "Radiation of Sound in Helium II," *Journal of Physics (Moscow)* 8 (1944): 110–114.

421. Curiously, London, in his book on superfluidity, showed the data clearly confirming the decline predicted by Tisza, but omitted the points at the lowest temperatures that show the flattening of the velocity predicted by Landau. F. London, *Superfluids II* (New York: Dover, 1954), *80*.

422. L. Landau, "On the Theory of Superfluidity of Helium II," *Journal of Physics (Moscow)* 11 (1947): 91ff., in *Collected Papers* (n. 310), 466–468.

423. Tisza (n. 419), 852.

424. L. Landau, "On the Theory of Superfluidity," *Doklady Akademii Nauk SSSR* 61 (1948): 253–256, in *Collected Papers* (n. 310), 474–477; Landau, "On the Theory of Superfluidity," *Phys. Rev.* 75 (1949): 884–885.

425. N. N. Bogoliubov, "On the Theory of Superfluidity," *Journal of Physics (Moscow)* 11 (1947): 23–41.

426. See n. 395.

427. V. Peshkov, "Velocity of Second Sound from 1.3 to 1.03°K," *Zhurnal Eksperimental'noi i Teoreticheskoi Fiziki* 18 (1948): 951–952; J. R. Pellam and R. B. Scott, "Second Sound Velocity in Paramagnetically Cooled Liquid Helium II," *Phys. Rev.* 76 (1949): 869–870.

428. K. R. Atkins and D. V. Osborne, "The Velocity of Second Sound below 1°K," *Phil. Mag.* 41 (1950): 1078–1081. The extension of measurements on helium to below 1°K was made possible

by the development of the technique of cooling by adiabatic demagnetization of paramagnetic salts, which dated back to independent suggestions of Debye and Giaque in 1926. P. Debye, "Einige Bemerkungen zur Magnetisierung bei tiefer Temperatur," *Ann. Phys.* 81 (1926): 1154–1160; W. F. Giaque, "A Thermodynamic Treatment of Certain Magnetic Effects. A Proposed Method of Producing Temperatures Considerably Below 1° Absolute," *Journal of the American Chemical Society* 49 (1927): 1864–1870; Giaque, "Paramagnetism and the Third Law of Thermodynamics. Interpretation of the Low-Temperature Magnetic Susceptibility of Gadolinium Sulfate," *Journal of the American Chemical Society* 49 (1927): 1870–1877. By the mid-1950s, more than a dozen laboratories throughout the world were employing this method. D. Ambler and R. P. Hudson, "Magnetic Cooling," *Reports on Progress in Physics* 18 (1955): 251–303.

429. Landau (n. 145).

430. D. V. Osborne, "The Rotation of Liquid Helium II," *Proceedings of the Physical Society (London)* 63 (1950): 909–912. The basis of the experiment was that whereas in an ordinary rotating liquid the meniscus of the free surface is a paraboloid independent of the liquid density (because both the centrifugal and gravitational forces of an element of the fluid are proportional to the density), in liquid helium, where the centrifugal force should act only on the normal part, a deviation from a paraboloid should occur.

431. E. L. Andronikaskvilii, "Investigations of Hydrodynamic Superfluidity" (Ph.D. diss., 1948), quoted in L. Landau and E. Lifshitz, "On the Rotation of Liquid Helium," *Doklady Akademii Nauk SSSR* 100 (1955): 669, in *Collected Papers* (n. 310), 650–654.

432. H. London, "The Superfluid Flow of Liquid Helium II," *Report of the Physical Society of Cambridge Conference* 2 (1946): 48–52.

433. Landau and Lifshitz (n. 431). See also London (n. 421), 151–155.

434. L. Onsager, in a contribution to the discussion following C. J. Gorter, "The Two Fluid Model for Helium II," *Nuovo Cimento* 6, suppl. 2 (1949): 245–250, as well as an unpublished remark at the 1948 Low Temperature Physics Conference at Shelter Island. See also V. L. Ginzburg, "Theory of Superfluidity and Critical Velocity of Helium II," *Doklady Akademii Nauk SSSR* 69 (1949): 161–164.

435. Feynman (n. 310).

436. H. E. Hall and W. F. Vinen, "The Rotation of Liquid Helium II. I. Experiments on the Propagation of Second Sound in Uniformly Rotating Helium II," *Proc. Roy. Soc.* A238 (1956): 204–214; Hall and Vinen, "The Rotation of Liquid Helium II. The Theory of Mutual Friction in Uniformly Rotating Helium II," *Proc. Roy. Soc* A238 (1956): 215–234.

437. W. F. Vinen, "The Detection of Single Quanta of Circulation in Liquid Helium II," *Proc. Roy. Soc.* A260 (1961): 218–234. The evidence for quantization of the circulation was not fully accepted by "many who were familiar with the caprices of He II"—W. E. Keller, *Helium-3 and Helium-4* (New York: Plenum, 1969), 281–282—until the repetition and extension of the experiment some five years later by S. C. Whitmore and W. Zimmermann, Jr., "Observation of Superfluid Circulation in Liquid-Helium at the Level of One, Two, and Three Quantum Units," *Physical Review Letters* 15 (1965): 389–392. The existence of quantized vortex rings, as suggested by Onsager, was proved only in 1964 by experiments showing that ions in helium II can create charge-carrying structures having the properties of vortex rings of quite large size, up to 500 Å and more. G. W. Rayfield and F. Reif, "Quantized Vortex Rings in Superfluid Helium," *Phys. Rev.* 136 (1964): 123–137.

438. H. Anderson et al., "Proposal for Tritium Production Using the Hanford Piles," 1 September 1945; E. R. Jette to N. E. Bradbury, "Notes and Proposals Based on Discussion at a Meeting Held on November 28, 1945, Dealing with Tritium Production," 29 November 1945; L. R. Groves to N. Bradbury, 17 December 1945; F. Daniels to N. E. Bradbury, 24 January 1945; Bradbury to Daniels, 2 February 1946; Daniels to Bradbury, 13 February 1946; all in Los Alamos National Laboratories Archives and Record Center, Los Alamos, New Mexico; H. Anderson to L. Hoddeson and G. Baym, private communication, January 1981.

439. London (n. 421), 165; L. Tisza, "Helium, the Unruly Liquid," *Physics Today* 1, no. 8 (1948): 4–8, 26.

440. S. G. Sydoriak, E. R. Grilly, and E. F. Hammel, "Condensation of Pure Helium 3 and Its Vapor Pressures Between 1.2 Degrees K and Its Critical Point," *Phys. Rev.* 75 (1949): 303–305.

441. D. W. Osborne, B. Weinstock, and B. M. Abraham, "Comparison of the Flow of Isotopically Pure Liquid He³ and He⁴," *Phys. Rev.* 75 (1949): 988.

442. L. D. Landau, in his first paper on the Fermi liquid theory, after pointing out that Bose fluids necessarily become superfluid, remarked that "the converse theorem that a liquid consisting of Fermi particles cannot be superfluid . . . is in general form not true" ("The Theory of a Fermi Liquid," *Zhurnal Eksperimental'noi i Teoreticheskoi Fiziki* 30 [1956]: 1058–1064 [trans. *Soviet Physics JETP* 3 (1957): 920–925]).

443. D. D. Osheroff, R. C. Richardson, and D. M. Lee, "Evidence for a New Phase of Solid He³," *Physical Review Letters* 28 (1972): 885–888.

444. E. M. Purcell, H. C. Torrey, and R. V. Pound, "Resonance Absorption by Nuclear Magnetic Moments in a Solid," *Phys. Rev.* 69 (1946): 37–38; F. Bloch, W. W. Hansen, and M. Packard, "Nuclear Induction," *Phys. Rev.* 69 (1946): 127.

445. W. M. Fairbank, W. B. Ard, and G. K. Walters, "Fermi Dirac Degeneracy in Liquid ³He below 1 deg K," *Phys. Rev.* 92 (1954): 566–568.

446. Landau (n. 442).

447. D. A. Kleinman, "The Diffuse Scattering of Neutrons and X-rays," *Phys. Rev.* 86 (1952): 622; G. Placzek, B.R.A. Nijboer, and L. Van Hove, "Effect of Short Wavelenth Interference on Neutron Scattering by Dense Systems of Heavy Nuclei," *Phys. Rev.* 82 (1952): 392–403; G. Placzek, "Crystal Dimensions and Inelastic Scattering of Neutrons," *Phys. Rev.* 93 (1954): 1207–1214; G. Placzek and L. Van Hove, "Crystal Dynamics and Inelastic Scattering of Neutrons," *Phys. Rev.* 93 (1954): 1207–1214.

448. B. N. Brockhouse and A. T. Stewart, "Scattering of Neutrons by Phonons in an Aluminum Single Crystal," *Phys. Rev.* 100 (1955): 756–757.

449. D. G. Henshaw and D. G. Hurst, "Neutron Diffraction by Liquid Helium," *Phys. Rev.* 91 (1953): 1222; D. G. Hurst and D. G. Henshaw, "Atomic Distribution in Liquid Helium by Neutron Diffraction," *Phys. Rev.* 100 (1955): 994–1002.

450. M. Cohen and R. P. Feynman, "Theory of Inelastic Scattering of Cold Neutrons from Liquid Helium," *Phys. Rev.* 107 (1957): 13–24.

451. H. Palevsky, K. Otnes, K. E. Larson, R. Pauli, and L. Stedman, "Excitations of Rotons in He II by Cold Neutrons," *Phys. Rev.* 108 (1957): 1346–1347; J. L. Yarnell, G. P. Arnold, P. J. Bendt, and E. C. Kerr, "Energy versus Momentum Relation for the Excitations in Liquid Helium," *Physical Review Letters* 1 (1958): 9–11; D. G. Henshaw and A.D.B. Woods, "Modes of Atomic Motions in Liquid Helium by Inelastic Scattering of Neutrons," *Phys. Rev.* 121 (1961): 1266–74; L. P. Pitaevskii, "Properties of the Spectrum of Elementary Excitations Near the Disintegration Threshold of the Excitations," *Zhurnal Eksperimental'noi i Teoreticheskoi Fiziki* 36 (1959): 1168–1178 (trans. *Soviet Physics JETP* 9 [1959]: 830–837).

452. London (n. 421), xi, 201.

453. Ibid.

454. R. P. Feynman, "Atomic Theory of Liquid Helium Near Absolute Zero," *Phys. Rev.* 91 (1953): 1291–1295, 1301–1308; Feynman, "Atomic Theory of the Two Fluid Model of Liquid Helium," *Phys. Rev.* 94 (1954): 262–277. Feynman's relation between the elementary excitation spectrum and the static structure function was derived in 1940 by Bijl in a paper that did not attract adequate attention for many years. A. Bijl, "The Lowest Wave Function of the Symmetrical Many Particles System," *Physica* 7 (1940): 869–886. R. P. Feynman and M. Cohen, "Energy Spectrum of the Excitation in Liquid Helium," *Phys. Rev.* 102 (1956): 1189–1204.

455. O. Penrose and L. Onsager, "Bose-Einstein Condensation and Liquid Helium," *Phys. Rev.* 104 (1956): 576–584.

456. K. A. Brueckner and K. Sawada, "Bose-Einstein Gas with Repulsive Interactions: General Theory," *Phys. Rev.* 106 (1957): 1117–1127; Brueckner and Sawada, "Hard Spheres at High Density," *Phys. Rev.* 106 (1957): 1128–1135; T. D. Lee, K. Huang, and C. N. Yang, "Eigenvalues and Eigenfunctions of a Bose System of Hard Spheres and Its Low Temperature Properties," *Phys. Rev.* 106 (1957): 1135–1145; S. T. Beliaev, "Application of the Methods of Quantum Field Theory to a System of Bosons," *Zhurnal Eksperimental'noi i Teorieticheskoi Fiziki* 34 (1958): 417–432 (trans. *Soviet Physics JETP* 34 [1958]: 289–299); S. T. Beliaev, "Energy Spectrum of a Non-ideal Bose Gas," *Zhurnal Eksperimental'noi i Teoreticheskoi Fiziki* 34 (1958): 433–446 (trans. *Soviet Physics JETP* 34 [1958]: 299–307). Beliaev acknowledged Migdal's communicating to him the

applicability of Green's function techniques in the many-body problem. W. L. McMillan, "The Growth State of Liquid Helium-4" (Ph.D. diss., University of Illinois, 1964); McMillan, "Ground State of Liquid He⁴," *Phys. Rev.* 138 (1965): A442–451.

457. D. Pines, interview with L. Hoddeson and H. Schubert, 11 February 1985; N. M. Hugenholtz and D. Pines, "Ground-State Energy and Excitation Spectrum of a System of Interacting Bosons," *Phys. Rev.* 116 (1959): 489–506; L. P. Pitaevskii, "Vortex Lines in an Imperfect Bose Gas," *Zhurnal Eksperimental'noi i Teoreticheskoi Fiziki* 40 (1961): 646–651 (trans. *Soviet Physics JETP* 13 [1961]: 451–454); E. P. Gross, "Structure of a Quantized Vortex in Boson Systems," *Nuovo Cimento* 20 (1961): 454–477.

458. A. Miller, D. Pines, and P. Nozières, "Elementary Excitations in Liquid Helium," *Phys. Rev.* 127 (1962): 1452–1464.

459. L. D. Landau, "The Theory of a Fermi Liquid," *Zhurnal Eksperimental'noi i Teoreticheskoi Fiziki* 30 (1956): 1058–1064 (trans. *Soviet Physics JETP* 3 [1957]: 920–925); Landau, "Oscillations of a Fermi Liquid," *Zhurnal Eksperimental'noi i Teoreticheskoi Fiziki* 32 (1957): 59–62 (trans. *Soviet Physics JETP* 5 [1957]: 101–108); Landau, "On the Theory of the Fermi Liquid," *Zhurnal Eksperimental'noi i Teoreticheskoi Fiziki* 35 (1958): 97–103 (trans. *Soviet Physics JETP* 8 [1959]: 70–74).

460. B. E. Keen, P. W. Matthews, and J. Wilks, "The Acoustic Impedance of Liquid He³ and Zero Sound," *Physics Letters* 5 (1963): 5–6; Keen et al., "The Acoustic Impedance of Liquid Helium-3," *Proc. Roy. Soc.* A284 (1965): 125–136; W. R. Abel, A. C. Anderson, and J. C. Wheatley, "Propagation of Zero Sound in Liquid He³ at Low Temperatures," *Physical Review Letters* 17 (1966): 74–78.

461. J. Luttinger and J. C. Ward, "Ground-State Energy of a Many-Fermion System," *Phys. Rev.* 118 (1960): 1417–1427, and other references in P. Nozières, *Theory of Interacting Fermi Systems* (New York: Benjamin, 1964).

462. V. M. Galitskii, "The Energy Spectrum of a Non-ideal Fermi-Gas," *Zhurnal Eksperimental'noi i Teoreticheskoi Fiziki* 34 (1958): 151–162 (trans. *Soviet Physics JETP* 7 [1958]: 104–112). The microscopic approach underlying this paper was developed in a preceding paper by V. M. Galitskii and A. B. Migdal, "Application of Quantum Field Theory Methods to the Many Body Problem," *Zhurnal Eksperimental'noi i Teoreticheskoi Fiziki* 34 (1958): 139–150 (trans. *Soviet Physics JETP* 7 [1958]: 96–104).

463. K. A. Brueckner and J. L. Gammel, "Properties of Liquid He-3 at Low Temperatures," *Phys. Rev.* 109 (1958): 1040–1046; Brueckner and Gammel, "Properties of Nuclear Matter," *Phys. Rev.* 109 (1958): 1023–1039. See n. 278.

464. L. N. Cooper, R. L. Mills, and A. M. Sessler, "Possible Superfluidity of a System of Strongly Interacting Fermions," *Phys. Rev.* 114 (1959): 1377–1382.

465. K. A. Brueckner, T. Soda, P. W. Anderson, and P. Morel, "Level Structure of Nuclear Matter and Liquid He³," *Phys. Rev.* 118 (1960): 1442–1446; V. J. Emery and A. M. Sessler, "Possible Phase Transition in Liquid He³," *Phys. Rev.* 119 (1960): 43–49; P. W. Anderson and P. Morel, "Generalized Bardeen-Cooper-Schrieffer States and the Proposed Low-Temperature Phase of Liquid He³," *Phys. Rev.* 123 (1961): 1911–1934; R. Balian and N. R. Werthamer, "Superconductivity with Pairs in a Relative *p* Wave," *Phys. Rev.* 131 (1963): 1553–1564. The experimental discovery paper is Osheroff et al. (n. 443). A. J. Leggett, "Interpretation of Recent Results on He³ Below 3 mK: A New Liquid Phase?" *Physical Review Letters* 29 (1972): 1227–1230; P. W. Anderson, "Some Macroscopic Considerations on Motions of Anisotropic Superfluids," *Physical Review Letters* 30 (1973): 368–370; see also Leggett (n. 417); Wheatley (n. 417); Anderson interview (n. 278).

466. Montroll interview (n. 212).

467. R. M. F. Houtappel, "Order-Disorder in Hexagonal Lattices," *Physica* 16 (1950): 425–455; G. Newell, "Crystal Statistics of a Two-Dimensional Triangular Ising Lattice," *Phys. Rev.* 79 (1950): 876–882.

468. B. Kaufman, "Crystal Statistics. II. Partition Function Evaluated by Spinor Analysis," *Phys. Rev.* 76 (1949): 1232–1243; R. Brauer and H. Weyl, "Spinors in *n* Dimensions," *American Journal of Mathematics* 57 (1935): 425–449.

469. B. Kaufman and L. Onsager, "Crystal Statistics. III. Short Range Order in a Binary Ising Lattice," *Phys. Rev.* 76 (1949): 1244–1252.

470. Montroll interview (n. 212).

471. Onsager appears to have presented his answer in a discussion remark on 23 August 1948 at a conference on phase transitions at Cornell. For a report on the conference, see R. Smoluchowski, J. Mayer, and H. Weyl, eds., *Phase Transformations in Solids* (New York: Wiley, 1951). Also, description in E. W. Montroll, R. B. Potts, and J. Ward, "Correlations and Spontaneous Magnetization of the Two-Dimensional Ising Model," *Journal of Mathematical Physics* 4 (1963): 308–322. Later that year, he stated it at the IUPAP statistical mechanics conference in Florence. L. Onsager, two lines in "Discussione e Osservazioni," *Nuovo Cimento* 6, suppl. 2 (1949): 261; C. N. Yang, "The Spontaneous Magnetization of a Two-Dimensional Ising Model," *Phys. Rev.* 85 (1952): 808–813 n. 10.

472. Montroll et al. (n. 471).

473. Yang (n. 471). The extension to the rectangular lattice is due to C. H. Chang, "The Spontaneous Magnetization of a Two-Dimensional Rectangular Ising Model," *Phys. Rev.* 88 (1952): 1422.

474. B. L. van der Waerden, "Die lange Reichweite der regelmässigen Atomanordnung in Mischkristaller," *Zs. Phys.* 118 (1941): 472–488.

475. M. Kac and J. C. Ward, "A Combinatorial Solution of the Two-Dimensional Ising Model," *Phys. Rev.* 88 (1952): 1332–1337. The correlation function was similarly calculated by R. B. Potts and J. C. Ward, "The Combinatorial Method and the Two-Dimensional Ising Model," *Progress in Theoretical Physics* 13 (1955): 38–46.

476. Y. Muto, "On the Order-Disorder Transition in Solids. Part I, II," *Journal of Chemical Physics* 16 (1948): 519–523, 524–525; Muto, "Errata: On the Order-Disorder Transition in Solids. Parts I and II," *Journal of Chemical Physics* 16 (1948): 1176; T. D. Lee and C. N. Yang, "Statistical Theory of Equations of State and Phase Transitions. II. Lattice Gas and Ising Model," *Phys. Rev.* 87 (1952): 410–419; H.N.V. Temperley, "Application of the Mayer Method to the Melting Problem," *Proc. Roy. Soc.* 74 (1905): 183–195; Temperley, "On the Asymptomic Behaviour of the Mayer Cluster Series in the Antiferromagnetic Problem," *Proceedings of the Physical Society (London)* 74 (1959): 432–443; Temperley, "Can the 'Lattice' Model of a Gas Describe Both Liquefication and Solidification?" *Proceedings of the Physical Society (London)* 74 (1959): 444–448. For example, in the lattice model of a fluid the spin-up of the ferromagnet corresponds to an occupied lattice site and spin-down to an unoccupied site, whereas the magnetization has its analogue in the density of the fluid, and the magnetic field in the chemical potential.

477. T. H. Berlin and M. Kac, "The Spherical Model of a Ferromagnet," *Phys. Rev.* 86 (1952): 821–825; E. Montroll, "Continuum Models of Cooperative Phenomena," *Nuovo Cimento* 6, suppl. 2 (1949): 265–278; Montroll, "Observation," *Nuovo Cimento* 6, suppl. 2 (1949): 278. Since the model completely ignores the quantization of energy levels, it is not suprising that its low temperature behavior is unphysical and even violates the Nernst theorem.

478. L. Van Hove, "Quelques propriétés générales de l'intégrale de configuration d'un système de particules avec interaction," *Physica* 15 (1949): 951–961; C. N. Yang and T. D. Lee, "Statistical Theory of Equations of State and Phase Transformations. I. Theory of Condensation," *Phys. Rev.* 87 (1952): 404–409; Lee and Yang, "Statistical Theory of Equations of State and Phase Transitions. II. Lattice Gas and Ising Model," *Phys. Rev.* 87 (1952): 410–419. A good exposition of the theories was given by Uhlenbeck in his 1960 lectures at the University of Colorado, appropriately sponsored by the American Mathematical Society. G. Uhlenbeck and G. W. Ford, eds., *Lectures in Statistical Mechanics* (Providence, R.I.: American Mathematical Society, 1963).

479. The proceedings of the Cornell conference are in Smoluchowski et al. (n. 471).

480. The Florence conference proceedings are published in *Nuovo Cimento* 6, suppl. 2 (1949): 149–306. The following participated in the conference as speakers or through discussion remarks: H. A. Kramers, M. Born, O. Klein, J. G. Kirkwood, H.B.G. Casimir, J. DeBoer, I. Prigogine, S. R. DeGroot, W. Pauli, E. A. Guggenheim, C. J. Gorter, G. S. Rushbrooke, L. Onsager, J. Yvon, F. London, J. E. Mayer, E. W. Montroll, G. Wataghin, E. Bauer, G. Careri, and F. Perrin.

481. Smoluchowski et al. (n. 471), foreword.

482. J. G. Kirkwood, "Crystallization as a Cooperative Phenomenon," in Smoluchowski et al. (n. 471), 67–76; J. G. Kirkwood and E. Monroe, "Statistical Mechanics of Fusion," *Journal of Chemical Physics* 9 (1941): 514–526.

483. J. E. Mayer, "A General Method for Imperfect Crystals and Phase Transitions," in Smoluchowski et al. (n. 471), 38–66.

484. M. Born, "Thermodynamics of Crystals and Melting," *Journal of Chemical Physics* 7 (1939): 591–603; Born, "On the Stability of Crystal Lattices. IX. Covariant Theory of Lattice Deformations and the Stability of Some Hexagonal Lattices," *Proceedings of the Cambridge Philosophical Society* 38 (1942): 82–99, and references therein to earlier papers with R. D. Misra, R. Fürth, H. W. Peng, and S. C. Power; M. M. Gow, "The Thermodynamics of Crystal Lattices. IV. The Elastic Constants of a Face-Centered Cubic Lattice with Central Forces," *Proceedings of the Cambridge Philosophical Society* 40 (1944): 151–166, and references therein to earlier papers of Born and M. Bradburn.

485. L. Tisza, "On the General Theory of Phase Transitions," in Smoluchowski et al. (n. 471), 1–37. Tisza at that time was on the staff of the Research Laboratory of Electronics at MIT, which had evolved out of the wartime Radiation Laboratory.

486. Florence Conference Proceedings (n. 480), 158–159.

487. Ibid., 178.

488. G. S. Rushbrooke, "Cooperative Free Energy," *Nuovo Cimento* 6, suppl. 2 (1949): 181–186.

489. J. E. Mayer, "Distribution Functions and Integral Equation Methods," *Nuovo Cimento* 6, suppl. 2 (1949): 209–226.

490. Kirkwood and Monroe (n. 482); J. E. Mayer and E. Montroll, "Molecular Distribution," *Journal of Chemical Physics* 9 (1941): 2–16; J. Mayer, "Integral Equations between Distribution Functions of Molecules," *Journal of Chemical Physics* 15 (1947): 187–201; M. Born and H. S. Green, "A General Kinetic Theory of Liquids. IV. Quantum Mechanics of Fluids," *Proc. Roy. Soc.* A191 (1947): 168–181, and earlier references therein.

491. G. S. Rushbrooke, "On the Theory of Regular Solutions," *Nuovo Cimento* 6, suppl. 2 (1949): 251–260.

492. C. Domb, "Order-Disorder Statistics. II. A Two-Dimensional Model," *Proc. Roy. Soc.* A199 (1949): 199–221.

493. E. Montroll, "Continuum Models of Cooperative Phenomena," *Nuovo Cimento* 6 suppl. 2 (1949): 265–278.

494. L. D. Landau and E. M. Lifshitz, *Statistical Physics,* 2nd ed. (Reading, Mass.: Addison-Wesley, 1958), chaps. 8, 14.

495. Ibid., 435.

496. Ibid., 438–439.

497. A. B. Pippard, *Elements of Classical Thermodynamics for Advanced Students of Physics* (Cambridge: Cambridge University Press, 1957), 136.

498. Tisza (n. 485).

499. L. S. Ornstein and F. Zernike, "Accidental Deviations of Density and Opalescence at the Critical Point of a Single Substance," *Koninklijke Adademie van Wetenschappen te Amsterdam, Proceedings* 17 (1914): 793–806. This work was apparently unknown in Germany until the authors reiterated and reviewed their findings in 1918, in "Die Linearen Dimensionen der Dichteschwankungen," *Phys. Zs.* 19 (1918): 134–137. For example, M. von Laue wrote in "Temperatur und Dichteschwankungen," *Phys. Zs.* 18 (1917): 542–544, of the futile attempts to seek a characteristic length for density fluctuations in a field.

500. A. Einstein, "Theorie der Opaleszenz von Homogenen Flüssigkeiten und Flüssigkeitsgemischen in der Nähe des kritischen Zustandes," *Ann. Phys.* 33 (1910): 1275–1298; M. von Smoluchowski, "Molekular-kinetische Theorie der Opaleszenz von Gases im kritischen Zustande, Sowie einiger verwandter Erscheinungen," *Ann. Phys.* 25 (1908): 205–226; Smoluchowski, "On Opalescence of Gases in the Critical State," *Phil. Mag.* 23 (1912): 165–173.

501. F. Zernike, "The Clustering-Tendency of the Molecules in the Critical State and the Extinction of Light Caused Thereby," *Koninklijke Adademie van Wetenschappen te Amsterdam, Proceedings* 18 (1916): 1520–1527.

502. F. Zernike and J. A. Prins, "Die Beugung von Röntgenstrahlen in Flüssigkeiten als Effekt der Molekülanordnung," *Zs. Phys.* 41 (1927): 184–194; see also I. Waller, "Über eine verallgemeinerte Streüngsformel," *Zs. Phys.* 51 (1928): 213–231.

503. P. Debye and H. Menke, "Bestimming der inneren Struktur von Flüssigkeiten mit Röntgenstrahlen," *Phys. Zs.* 31 (1931): 797–978.

504. L. Brillouin, "Diffusion de la lumière et des rayons X par un corps transparent homogène

influencé de l'agitation thermique," *Annales de physique (Paris)* 17 (1922): 88–122; L. I. Mandel'shtam, *Zhurnal Russkago Fiziko-Khimicheskago Obshchestva* 58 (1926): 381.

505. L. Landau and G. Placzek, "Structure of the Undisplaced Scattering Line," *Physikalische Zeitschrift der Sowjetunion* 5 (1934): 172–173. The authoritative 1930s monographs in Western Europe relating electromagnetic scattering to distribution functions were J. Yvon, "La Théorie statistique des fluides et l'equation d'éte," in *Actualités scientifiques et industrielles,* no. 203 (Paris: Hermann, 1935); Yvon, "Fluctuations en densité," in *Actualités scientifiques et industrielles,* no. 542 (Paris: Hermann, 1935); Yvon, "La Propagation et al diffusion de la lumière," in *Actualités scientifiques et industrielles,* no. 543 (Paris: Hermann, 1937). The hydrodynamic density fluctuations can be divided into two types: pressure fluctuations at constant entropy that shift the light frequency and are responsible for the Brillouin–Mandel'shtam doublet, and entropy fluctuations at constant pressure that do not shift, but broaden the line. The frequency shift is a measure of sound velocity in the fluid, the ratio of intensity of the central line to the sidebands yields the ratio of specific heat at constant pressure to that at constant volume, and the linewidths provide measurements of combinations of the thermal conductivity and the shear and bulk viscosities of the fluid.

506. In surveying the literature on the electromagnetic scattering by fluids from the time of the Mandel'shtam paper in 1926 through the 1950s, one encounters the names of many Soviet scientists (Landau, Lifshitz, Ginzburg, Rytov, Fabelinskii, Leontovich, Molchanov, Motulevich, and Frenkel, to mention a few), indicating that the topic was an active field of research in the Soviet Union. Although the subject of the pair correlation function appeared in several presentations at the 1949 Florence conference, the question of critical scattering (i.e., near the critical point) did not. De Boer noted there that the equation of state of a fluid, as well as its energy, can be expressed without approximation in terms of the pair correlation function. J. de Boer, "The Caloric and Thermal Equation of States in Classical and in Quantum Statistical Mechanics," *Nuovo Cimento* 6, suppl. 2 (1949): 199–207. In "Molecular Distributions and Equations of State of Gases," *Reports on Progress in Physics* 12 (1949): 305–374, de Boer discussed the pair distribution more fully, including the scattering of light and x rays, but did not cite Soviet work.

507. M. Fisher, "Correlation Functions and the Critical Region of Simple Fluids," *Journal of Mathematical Physics* 5 (1964): 944–962. The former approach was followed in particular by Yvon (n. 505); the latter was used by, among others, Y. Rocard, "Theorie des fluctuations et opalescence critique," *Journal de physique et le radium* 4 (1933): 165–185 (whose work preceded the Landau–Placzek paper), and M. J. Klein and L. Tisza, "Theory of Critical Fluctuations," *Phys. Rev.* 76 (1949): 1861–1868.

508. L. Onsager, "Reciprocal Relations in Irreversible Processes. I.," *Phys. Rev.* 37 (1931): 405–426; L. Onsager, "Reciprocal Relations in Irreversible Processes. II.," *Phys. Rev.* 38 (1931): 2265–2279.

509. For example, G. E. Uhlenbeck and L. S. Ornstein, "On the Theory of Brownian Motion," *Phys. Rev.* 36 (1930): 823–841; M. C. Wang and G. E. Uhlenbeck, "On the Theory of Brownian Motion. II," *Reviews of Modern Physics* 17 (1945): 323–342.

510. See n. 465.

511. L. Van Hove, "Correlations in Space and Time and Born Approximation Scattering in Systems of Interacting Particles," *Phys. Rev.* 95 (1954): 249–262; Van Hove, "Temperature Variation of the Magnetic Inelastic Scattering of Slow Neutrons," *Phys. Rev.* 93 (1954): 268–269. Not only does Van Hove acknowledge Placzek "for many stimulating discussions and suggestions on the various aspects of the present work," but he also expresses his indebtedness to him "for illuminating discussions and communication of unpublished work on this subject." He also refers to "private communication" with Roy Glauber, whose work on the subject had not yet appeared in print.

512. Van Hove ("Correlations") (n. 511), 262.

513. Onsager (n. 508).

514. L. Van Hove, "Time-Dependent Correlations Between Spins and Neutron Scattering for Ferromagnetic Crystals," *Phys. Rev.* 95 (1954): 1374–1384.

515. L. Van Hove, "Temperature Variation of the Magnetic Inelastic Scattering of Slow Neutrons," *Phys. Rev.* 93 (1954): 268–269; G. L. Squires, "The Scattering of Slow Neutrons by Ferromagnetic Crystals," *Proceedings of the Physical Society (London)* A67 (1954): 248–253; C. G.

Shull and M. K. Wilkinson, "Neutron Diffraction Studies on Iron at High Temperature," *Phys. Rev.* 103 (1956): 516–524; A. W. McReynolds and T. Riste, "Magnetic Neutron Diffraction from Fe_3O_4," *Phys. Rev.* 95 (1954): 1161–1167.

516. R. J. Elliott and W. Marshall, "Theory of Critical Scattering," *Reviews of Modern Physics* 30 (1958): 75–89; P. G. de Gennes, "Theory of Neutron Scattering by Magnetic Crystals," in *Magnetism*, ed. G. T. Rado and H. Suhl (New York: Academic Press, 1963), 3:115–147. The state of the theory was briefly summarized in 1965 by M. E. Fisher, "Theory of Critical Fluctuations and Singularities," in *Critical Phenomena*, ed. M. S. Green and J. V. Sengers (Washington, D.C.: National Bureau of Standards, 1966), 108–115, and by W. Marshall, "Critical Scattering of Neutrons by Ferromagnets," in ibid., 135–142.

517. T. Izuyama, D. Kim, and R. Kubo, "Band Theoretical Intepretation of Neutron Diffraction in Ferromagnetic Metals," *Journal of the Physical Society of Japan* 18 (1963): 1025–1042.

518. See, for example, L. P. Kadanoff, "Scaling Laws for Ising Models near T_c," *Physics* 2 (1966): 263–272; B. Widom, "Surface Tension and Molecular Correlations near the Critical Point," *Journal of Chemical Physics* 43 (1965): 3892–3897; Widom, "Equation of State in the Neighborhood of the Critical Point," *Journal of Chemical Physics* 43 (1965): 3898–3905; K. G. Wilson, "The Renormalization Group," in *Phase Transitions and Critical Phenomena*, ed. C. Domb and M. S. Greene (London: Academic Press, 1976), 6: 1–5; Wilson, "The Renormalization Group and Block Spins," in *Statistical Physics*, ed. L. Pal and P. Szepfalusy (Amsterdam: North-Holland, 1976), 17–28; also L. P. Kadanoff, W. Götze, D. Hamblen, R. Hecht, E.A.S. Lewis, V. V. Palciauskas, M. Rayl, J. Swift, D. Aspnes, and J. Kane, "Static Phenomena near Critical Points: Theory and Experiment," *Reviews of Modern Physics* 39 (1967): 395–431. One of the origins of the renormalization group approach was a stimulating paper on the lambda transition in ^{4}He by A. Z. Patashinskii and V. L. Pokrovskii, "Second Order Phase Transitions in a Bose Fluid," *Zhurnal Eksperimental'noi i Teoreticheskoi Fiziki* 46 (1964): 994–1016 (trans. *Soviet Physics JETP* 19 [1964]: 677–691), in which they studied the behavior of length scales in the system near the phase transition. See also B. D. Josephson, "Relation Between the Superfluid Density and Order Parameter for Superfluid He near T_c," *Physics Letters* 21 (1965): 608–609.

519. For example, G. S. Rushbrooke and P. J. Wood, "On the High Temperature Susceptibility for the Heisenberg Model of a Ferromagnet," *Proceedings of the Physical Society (London)* A68 (1955): 1161–1169; Rushbrooke and Wood, "On the Curie Points and High Temperature Susceptibilities of Heisenberg Model Ferromagnets," *Molecular Physics* 1 (1958): 257–283; C. Domb and M. F. Sykes, "The Calculation of Lattice Constants in Crystal Statistics," *Phil. Mag.* 2 (1957): 733–749; Domb and Sykes, "On the Susceptibility of a Ferromagnetic above the Curie Point," *Proc. Roy. Soc.* A240 (1957): 214–228; Domb and Sykes, "Specific Heat of a Ferromagnetic Substance above the Curie Point," *Phys. Rev.* 108 (1957): 1415–1416; M. E. Fisher and M. F. Sykes, "Excluded-Volume Problem and the Ising Model of Ferromagnetism," *Phys. Rev.* 114 (1959): 45–58; see also C. Domb, "On the Theory of Cooperative Phenomena in Crystals," *Advances in Physics* 9 (1960): 149–361.

520. D. Pines, *The Many Body Problem* (New York: Benjamin, 1961), 1.

521. M. Born and H. S. Green, "A General Kinetic Theory of Liquids. I. The Molecular Distribution Functions," *Proc. Roy. Soc.* A188 (1946): 10–18; G. Kirkwood, "The Statistical Mechanical Theory of Transport Processes. I. General Theory," *Journal of Chemical Physics* 14 (1946): 180–201; N. Bogoliubov, "Problems of a Dynamical Theory in Statistical Physics," originally published in Russian about 1946 and translated into English by J. de Boer and G. Uhlenbeck, eds., *Studies in Statistical Mechanics* (Amsterdam: North-Holland, 1962), 1: 5–118.

522. T. Matsubara, "A New Approach to Quantum Statistical Mechanics," *Progress in Theoretical Physics* 14 (1955): 351–378.

523. R. Kubo, "Statistical-Mechanical Theory of Irreversible Processes. I. General Theory and Simple Applications to Magnetic and Conduction Problems," *Journal of the Physical Society of Japan* 12 (1957): 570–586. Significant works in the application of field theory to statistical mechanics also include P. C. Martin and J. Schwinger, "Many Body Problem with Quantum Statistics," *Bulletin of the American Physical Society* 3 (1958): 202; Martin and Schwinger, "Theory of Many-Particle Systems. I," *Phys. Rev.* 115 (1959): 1342–1373; E. W. Montroll and J. C. Ward, "Quantum Statistics of Interacting Particles; General Theory and Some Remarks on Properties of an Electron Gas," *Physics of Fluids* 1 (1958): 55–72; L. P. Kadanoff and G. Baym, *Quantum Statistical*

Mechanics (New York: Benjamin, 1962); A. A. Abrikosov, I. E. Dzyaloshinskii, and L. P. Gor'kov, *Methods of Quantum Field Theory in Statistical Physics* (Moscow: Fizmatgiz, 1961; New York: Prentice-Hall, 1963). Further references may be found in Pines (n. 520).

524. Wilson, in Domb and Green (n. 518).

525. F. Bloch, interview with T. Kuhn, 14 May 1964; R. Peierls, interview with J. L. Heilbron, 17 June 1963, Archives for the History of Quantum Physics, American Philosophical Society, Philadelphia, copies at AIP and elsewhere.

526. Schweber (1989) (n. 242).

527. For a descriptive summary, see D. Pines, "Elementary Excitations in Quantum Liquids," *Physics Today* 34, no. 11 (1981): 106–131.

528. Bloch (n. 183).

529. Landau (n. 182).

530. Ginzburg and Landau (n. 302).

531. Gor'kov (n. 403).

532. London (n. 135).

533. O. Penrose, "On the Quantum Mechanics of Helium II," *Phil. Mag.* 42 (1951): 1373–1377. See also Penrose and Onsager (n. 455).

534. Pitaevskii (n. 457); Gross (n. 457).

535. Josephson (n. 409); Pippard (n. 384); Anderson ("It's Not Over") (n. 382).

536. C. N. Yang, "Concept of Off-Diagonal Long-Range Order and the Quantum Phases of Liquid He and of Superconductors," *Reviews of Modern Physics* 34 (1962): 694–704.

537. See nn. 273, 274, 277, 522, 523.

538. Galitskii (n. 462); Galitskii and Migdal (n. 462).

539. Gor'kov (n. 403); Beliaev (n. 456).

540. P. Galison, "Impact of World War II on Postwar Particle Physics" (contribution to the Los Alamos Symposium on Transfer of Technology from World War II to Postwar Science, May 1987, ed. R. Seidel [to be published]); Schweber (n. 242).

541. *The Many Body Problem,* Proceedings of Les Houches summer school, 1958 (London: Methuen, 1959).

542. Early texts on the many-body problem include D. ter Haar, *Introduction to the Physics of Many-body Systems* (New York: Wiley Interscience, 1958); Pines (n. 520); Abrikosov et al. (n. 523); Kadanoff and Baym (n. 523).

543. Pathashinskii and Pokrovskii (n. 518); Josephson (n. 518).

9 | *The Solid Community*

SPENCER R. WEART

In 1930 solid-state physics did not exist. The very term was unknown, nor was there any intellectual or social entity to which the term could have applied. Certainly there flourished a number of specialties, such as the electron theory of metals and experimental studies of magnetic alloys, which would eventually fall within the field of solid-state physics, but there was no field as a whole, no sense that the physicists who studied solids should be distinguished as a group from those who studied gases, radioactivity, or whatever. By 1960 the situation had entirely changed. The term "solid-state physics" not only was familiar but could be attached to a number of institutions: university chairs, conferences, journals, research groups, funding mechanisms, even buildings entirely devoted to the subject. These institutions were knit together in a worldwide solid-state physics community. Like a mirror, this social entity reflected an intellectual entity, the study of the physics of solids. This chapter asks how social coherence came about, and what some of the deeper intellectual corollaries were.

What does it mean to say that a research field or a community has its own identity? One way to define it would be in terms of communications. If through their publications the scientists in a group interact more strongly among themselves than they interact with scientists in other groups, then we can call them members of an intellectual field and at the same time of a social community. It happens that this definition can be investigated in a quantitative and objective way. The method is the study of "co-citations," as developed by Henry Small and others at the Institute for Scientific Information.

Suppose a scientific paper cites in its footnotes two other papers X_1, X_2. Then we say X_1 and X_2 are co-cited, and we suspect they have something in common. If many papers all have both X_1 and X_2 among the works they cite, then the co-citation link is a strong one, and (as can be verified by checking the actual writings) X_1 and X_2 almost certainly deal with related scientific subjects. Continuing in this way, we can build up a set of papers X that are all joined by strong co-citation links. There may be another set of papers Y that are also joined tightly to one another, but that rarely appear in a set of footnotes alongside any of the Xs. In this case, we would expect that the Xs and the Ys constitute separate intellectual fields. Inspection of the writings themselves shows this nearly always to be the case.

The method has been applied to modern science as a whole with the aid of digital computers—solid-state physics coming to the aid of history. Analysis of the entire *Science Citation Index* for 1972 cleaves physics as neatly as a diamond struck with a chisel. Against a background of miscellaneous subjects, two tight clusters of strong co-citation links stand out: the study of individual particles, and solid-state physics. The study of particles, in turn, decomposes into high-energy physics, nuclear physics, and atomic physics; solid-state physics divides into tight clusters for the study of metals, of crystal lattices, and of semiconductors, along with other subjects that are more widely and loosely interlinked. In the broader context of all science, solid-state physics as a whole turns out to be more closely linked with chemistry than with the physics of individual particles. There is every reason, then, to call solid-state physics a distinct field.[1]

The co-citation pattern for the years 1920 to 1929 has also been analyzed; it was utterly different. Physics at that time did not appear to be separated into broad fields. There were certainly many distinct specialties, but each was linked with many others with no major zones of cleavage. At the center of physics in the 1920s were quantum mechanics, spectroscopy, and x rays, all closely connected with one another; most other research areas were connected with these subjects. In short, where physics in the early 1970s could be partially disentangled into separate skeins, in the 1920s all the strands ran through a central knot, mainly atomic physics.[2]

The difference between the 1920s and the 1970s represents a historic transformation of physics, with implications that extend far beyond solid-state research. But even if we ignore changes elsewhere in science, the consolidation of solid-state physics is a striking and unusual phenomenon.

Most discussions of the birth of an important specialty see it as like a plant growing from a seed. Almost nothing is visible until a few pioneering papers appear. Then the number of papers in the subject—radioactivity, radio astronomy, molecular biology, or whatever—grows rapidly at a rate far exceeding the growth of the general scientific literature. Something that was little studied before, perhaps something whose existence was not even suspected before, is now being understood. At length, maturity is reached when the rate of publication levels off or even declines.[3]

Solid-state physics displays an altogether different pattern. Communities of people who studied the properties of solids such as metals or ceramics existed in unbroken continuity from prehistoric times into the twentieth century. Meanwhile, the subjects of study became increasingly specialized, with metals, for example, not only separate from ceramics but divided into studies of pure crystals, industrial alloys, magnetism, and so forth; the separate communities did not combine within an overarching field. When we speak of the emergence of solid-state physics, then, we do not mean the creation of something *de novo*. We mean a grand rearrangement of an entire array of specialties, old and new, into a novel constellation.

The usual, vegetative model of specialization sees the process as an almost continuous activity, a ceaseless division and redivision, as when ivy splits into branches as it creeps up a wall. A look at history on a larger scale shows that this is not always the case: an entire set of specialties may rearrange themselves into fields at a single time when conditions reach some critical point. One such time came around the middle of the nineteenth century, when "physics" was propagated as a term, as an intellectual discipline, and as a community with institutions distinct from chemistry and allied fields. Nothing like that happened again until the middle of the twentieth century, when not only solid-state physics but also other fields of physics acquired, for the first time, identities within the world of science.

In the first decades of the twentieth century, it was common for a physicist to do original work in a wide variety of areas. Albert Einstein's 1905 set of four papers on statistical thermodynamics, Brownian motion, the photoelectric effect, and special relativity is only the most famous example. At a time when not only quantum mechanics and relativity, but even electromagnetic theory and statistical mechanics were just becoming established, there was no very extensive territory of well-known experimental results and mathematical techniques to slog through before reaching the frontier in any direction. Only in the specialty known as rational mechanics, which elaborated the implications of Newton's laws for such questions as the theory of the moon's motion and the tides, was specialization required to become a master—so much so that rational mechanics had become almost a branch of mathematics rather than of physics.

Experimentalists, of course, tended more than theorists to specialize. In the schools of Robert Pohl and Pierre Weiss, as previous chapters have shown, it was possible to spend the bulk of a career on a particular set of materials and techniques; similarly, the schools of Ernest Rutherford and the Curies devoted themselves to radioactivity to the exclusion of almost all else. Yet such specialization was not the rule early in the century. Pierre Curie, for example, had come to radioactivity after distinguished experimental and theoretical work on piezoelectricity, and shortly before his tragic death in a street accident he had decided to turn back to his old love, the study of crystals and their symmetries. Marie Curie had done her doctoral dissertation on magnetic properties of steels, but moved quickly into radioactivity research. Around the turn of the century, it was still possible for Wilhelm Röntgen, a physicist of ordinary talent, after spending decades on such subjects as crystals, to take up an entirely different subject such as cathode rays and almost immediately stumble on a major discovery.

The universe offers a limited stock of such discoveries. By the mid-twentieth century, significant experimental results usually required a heavy investment of funds, time, and, above all, specialized expertise. The first physics subject to show this effect clearly, and also to become more or less a recognized specialty, was radioactivity, or, as it began to be named in the 1930s, nuclear physics. Marie Curie's Radium Institute in Paris, Otto Hahn's Kaiser Wilhelm Institute for chemistry near Berlin, Ernest O. Lawrence's Radiation Laboratory in Berkeley, and others showed the separate existence of their field in the most visible way: entire buildings devoted to the study of the nucleus. Special conferences and a number of classic textbooks also showed that by the early 1930s, if not earlier, the field was internally cohesive.[4]

However, even in the 1930s nuclear physics lacked such marks of a fully separate specialty as its own journals and societies. All its institutions were either transitory, like the conferences and textbooks, or local, like the laboratories. Even the strongest nuclear research centers often swayed to the pull of neighboring fields, ranging from the radiation medicine that became one of the central activities of Lawrence's laboratory, to solid-state questions like radioactive tracer studies of crystallization and adsorption on surfaces, a specialty of Hahn's institute. The intermingling of fields was most evident in theoretical work. Bohr, Heisenberg, and others of their generation may have done nuclear physics theory, but in the mid-1930s few would have called them "nuclear physicists," as if that was all they would do.

Yet with the coming of quantum mechanics, as well as the elaboration of electromagnetic theory and statistical mechanics, the theorist too was under pressure to specialize in order to understand a subject fully. The dividing line can be seen clearly in solid-state physics. Lorentz and Drude at the turn of the century, and even Bethe and Wigner in the early 1930s, saw theory of solids as only a narrow example of universal atomic physics. Easily and quickly they entered the electron theory of metals; easily and quickly they left it for other realms. But Seitz, Bardeen, and many others of that generation, only a few years younger than Bethe and Wigner, spent their entire careers within the theory of solids. They never felt cramped, for by that time theory had become developed enough to allow an endless variety of work within the boundaries of the field.

The effects can be shown quantitatively. We have taken a sample of 50 people whose work in solid-state physics was particularly important, according to the earlier chapters of this book, and whose bibliographies are available.[5] The dates of birth of the people in the sample ranged from 1862 to 1931, with the median at 1905. We divided their published articles into those that had to do with the physics of solids, and those in other fields. Up to 1932, about three-quarters of our sample of physicists straddled the boundary, publishing both on solids and on other subjects. From 1933 to 1947, about half of the sample published research both on the physics of solids and on other topics. The proportion seems to have held steady throughout these years, probably because the Second World War made for diverse interests: some who had been confined to solid-state physics went to work on microwaves for radar, while others who had turned away from solids moved back, like Wigner with his work on the damage that neutrons do to graphite in nuclear reactors. Finally, after the papers describing wartime research were published—that is, after about 1948 (and up to about 1962, which we took as our end point)—only about a quarter of the sample published both in and out of solid-state physics.

These proportions—three-quarters overlap between solids and other fields before 1933, one-half from 1933 to 1947, and one-quarter after—held equally for the physicists born before 1905, and for those born after. That is, the effect depended simply on the continuing development of physics, and there was no narrowing of research interests because of age. The increased specialization after the war is even more marked than the statistics suggest, for some of those who still straddled the boundary, theorists such as Leon Cooper, David Pines, and Lev Landau, who worked on nuclei and even astrophysics as well as on solids, were in fact largely restricted within a specialty of their own—quantum many-body theory. This was far from the diversity commanded in earlier years by such a one as Fermi, who ranged from the neutrino to the Fermi–Dirac statistics, which laid the groundwork for the quantum theory of metals, with experiments along the way. Of course, even after the war there remained some who worked in truly diverse specialties, such as Francis Simon with his low-temperature experiments on a variety of subjects. And at any time there were others, such as Pohl with his work on the alkali halides and Vivian Johnson with her semiconductors, who preferred to mine a single rich vein of ore.

Now a word of caution. If we say that in the 1920s, for example, a certain person published papers on the physics of solids, we are applying an artificial criterion. We are retroactively defining as "solid-state physics" a set of specialties that were not connected in anyone's mind as parts of such an entity until later. Such use of hind-sight will be found throughout this chapter. A retroactive criterion would not be legitimate if we used it to pretend that the field of solid-state physics existed when it did not, but hindsight is a legitimate tool for studying exactly when and how the field did come into existence.

The effects seen in our sample of elite physicists can also be traced through the community as a whole. Jerome Rowley, aided by Kris Szymborski, checked through *Science Abstracts* in selected periods for papers on what would now be called solid-state physics. He found what other papers the same authors published in the same three-year period. In the 1921 to 1923 period, of a sample of 800 authors, 55% published not only in solid state but also in other topics (many of the remaining 45% had published only one paper, of course). By 1937 to 1939, the pro-portion combining solid state with other fields—notably, quantum theory, ther-modynamics, radioactivity, and spectroscopy—had fallen to 29% of a sample of 400 authors, and in 1955 to 1957 only 17% of 400 authors who published on solid state also published on other topics. Average physicists, then, narrowed down to solid-state physics even more sharply than the elite ones.

The other side of the coin was that within "solid-state physics" itself, people increasingly ranged among subfields, as defined by the subcategories established by *Science Abstracts.* In the 1921 to 1923 period, only 22% of the authors who pub-lished in one subfield of solid state also published in another subfield, whereas in 1937 to 1939 this had risen to 44% and in 1955 to 1957, to 65%; in this last period some 15% of the authors published in four or more solid-state subfields.

Statistics cannot verify that the tendency to stick within the overall field of solid state was related to the growth of theory, but we can at least confirm that theory was on the rise. In the 1921 to 1923 period, we found by inspecting abstracts that only 16% of solid-state papers included some new theory—restricting the term to work that not only had some ideas about structure, but expressed such ideas with the use of mathematics. In 1937 to 1939, after the coming of quantum mechanics, the pro-portion of papers with some theory had nearly doubled to 30%, and in 1955 to 1957 this had risen a bit more, to 36%. By the way, in the first period, two-thirds of the papers with new theoretical ideas included little or no new experimental informa-tion, whereas by the last period only one-quarter were so restricted. Thus it had become common practice to publish experimental work and theoretical explana-tion in one paper, a sign of the maturity of the field, and of the increased demand for expertise.

The discussion so far has treated specialization as an intellectual matter. But spe-cialization also had a demographic side. If there were only three physicists in the world and each chose a different research topic, we would not speak of a division of physics into fields. We could speak of that only when there were hundreds work-ing on each subject, and in particular when—as revealed, for example, in the co-citation analysis—their communications with one another took precedence over their communications with colleagues in neighboring fields. Fields in this sense could not separate from one another until there were, at least in some countries, many hundreds of physicists, enough so that each separate field was strong enough to stand on its own as a community of researchers. Indeed, once so many people were at work, an individual would not have time to read and grasp all the important papers in more than a few specialties. That line was crossed by physics around the middle of the twentieth century.

Is it a coincidence that the intellectual and demographic dividing lines were crossed at about the same time? Certainly, theoretical and experimental techniques could not have been swiftly elaborated in all directions unless many people were at work. At the same time, the intellectual and practical successes of physics did much to attract a rapidly increasing number of people to the field, and also to win the funds that sustained them. In sum, the division of physics into fields was no simple, continuous, vegetable process, but a severe change of state in a system with complex feedbacks.[6]

9.1   To 1940: Schools for Solids

Earlier chapters describe a variety of institutions where the study of solids flourished during the first 40 years of the twentieth century, and above all in the 1930s. These were the "schools" of Heisenberg, Ioffe, Mott, Pauli, Pohl, Slater, Sommerfeld, Weiss, Wigner, and so forth—university departments and institutes, at each of which a handful of students and fewer than a handful of professors worked on aspects of the physics of solids. The previous chapters have shown how all the schools were interlinked through visits and formal meetings. However, except within more restricted specialities like crystallography and physical metallurgy, there is little evidence for a special solid-state network independent of the rest of physics. In the 1930s almost all physicists of note, along with their brighter students, were on familiar terms with most of their colleagues in all research areas.

In a few cases, there was a clear institutional emphasis on solids, as with the group at Bristol University that was aided in 1930 when the British government gave a grant for "theoretical investigation of the physical properties of the solid state of matter," and the group at the University of Pennsylvania founded in 1939 when Frederick Seitz set out to build up both theory and experiment in that area of investigation. But such schools were no firm base for a field. Each depended on temporary circumstances; if only one person such as Mott had left, the Bristol theorists would not have seemed specially oriented toward solids, and when Seitz did leave the University of Pennsylvania in 1942, the school there faltered.

In scarcely any case did the boundaries of a school correspond to the boundaries of what we would now call solid-state physics. Usually the subjects studied were far too limited to bestride the variety of research that a retroactive definition of solid-state physics would require. For example, Sommerfeld and his school devoted themselves almost exclusively to the electron theory of ideal metals and crystals, shrugging their shoulders at the irregularities that are responsible for most real properties; meanwhile, Pohl and others like him studied effects of structural details almost without reference to quantum mechanics.

However, a few of these schools were moving in the direction of something larger. Around the mid-1930s, Pohl's group began to speak of the "physics of the solid state" *(Physik der Festkörper),* and sooner or later they would have to incorporate into their work all that the words implied (Chapter 4). Still farther advanced was Ioffe's Leningrad Physical-Technical Institute. Having raised funds from various Soviet government agencies by promising benefits for industry and national prestige, Ioffe set people to work on problems ranging from the strength of materials to ferroelectricity. A 1935 reorganization established the Department of Physics of the Solid State *(tverdykh tel).*[7]

A 1966 survey of the solid-state physics community suggested that a certain critical size was required before a group could properly enter the field. According to the experience of the 1960s, the report said, this critical size "is somewhere between seven and ten men at or above the assistant-professor level, plus a research staff, including postdoctorals, of similar size. This number would accommodate two or three theorists and five to eight experimenters."[8]

Probably the critical size would have been smaller in the 1930s, but even so, only a very few schools of the period could muster resources sufficient to span a field as broad as the study of solids.

However, there was more to the matter than limitations of size. Some groups were large enough to touch on a variety of research topics, but they rarely felt they had crossed a boundary if they left solids. For example, Slater and his companions at MIT and Harvard showed about as much interest in molecules as in solids. Ioffe's school in Leningrad studied almost everything in physics; besides the Department of Physics of the Solid State there were a number of others, all overlapping, including the Department of X-Rays and Electrical Phenomena, in which studies of positrons went on alongside studies of the photoelectric properties of metals.

At this point, we need to say more clearly just what should be included in a retrospective definition of solid-state physics. For if understanding of a research area demands investigation of the social community that pursued it, still more must investigation of the social community take into account the research program or paradigm—that is, the set of questions and techniques that they share. The preceding chapters have given this matter much attention and it need only be summarized here.

Solid-state physics stood on three pillars. The earliest to be erected was x-ray crystallography, as developed after 1912 in many laboratories. It was this technique and the study of crystal lattices in general that gave physicists the precise, geometric, atomic picture of solids they required before they could get anywhere at all. The second essential common element was, of course, quantum mechanics, which since the end of the 1920s was known to provide a complete theoretical foundation for the description of solids. The third common element, rather more subtle, was an appreciation for what Adolf Smekal called "structure-sensitive" properties. In the mid-1930s, as W. H. Bragg remarked in 1934, interest began to focus on the question of which properties of a solid depend on its idealized crystal pattern, and which on "accidents" of the surface or interior arrangements. We would consider an institution to be fully involved in modern solid-state physics only when it paid attention to all three elements.[9]

The pattern of co-citations in the physics literature of the 1920s casts a useful light on the matter. Henry Small of the institute for Scientific Information carried out a special study for us, a study in which most of the tangle of linkages was cut away to reveal some underlying patterns. We simply eliminated from the calculation all physics papers except those that fell into about 50 categories centered on the study of solids. As a check, we also included a half-dozen other categories of papers that dealt with radioactivity and gamma rays (but not x rays) and one set of general relativity papers. This time, with the overlay of intermediate linkages removed, physics did show lines of cleavage. All the solid-state papers were well connected with one another; radioactivity papers were connected with one another but hardly at all with solid-state papers; general relativity stood entirely apart. Thus even in the 1920s, physics could be divided into parts, including a proto-field of solid-state research, although this is revealed only by artificially stripping away papers that formed intermediate links between fields.

Still more interesting is the shape of the solid-state proto-field in the 1920s. All the specialties within it were interconnected, with the exception of papers on magnetism, which formed an interlinked cluster of their own off to the side. However, the linkages between specialties were still loose and indirect. For example, while papers on crystal deformation were woven together in a cluster, the members of this cluster were scarcely ever cited together with papers on photoeffects, which clustered together in their own specialty. What bound together all the specialties dealing with solids in the 1920s, including even to some extent magnetism, was less their linkages with one another, than their common tendency to be joined by co-citation with papers on x rays and with papers on quantum mechanics. These are precisely the first two essential common elements listed above; the third, imperfections, would not appear until the 1930s.

With these criteria in mind, let us look at the places where we might expect a field of science to show its existence: conferences, monographs, and the like. Its appearance in the classroom is of special interest. Many particular questions dealing with solids were treated together by some professors from an early date; the proto-field had long maintained a quasi-existence. For example, Voigt's *Lehrbuch der Kristallphysik* of 1910 already gave sophisticated and often mathematical treatment of topics ranging from thermal expansion to ferromagnetism, covering classical facts and calculations that formed the foundation beneath the subsequent growth of x-ray crystallography and allied subjects.[10]

Attempts to connect classical facts such as conductivity with quantum ideas, in the lecture room, were formalized already in the winter of 1913/1914 when Paul Langevin gave a course at the Collège de France on the electron theory of metals, extending Drude's and Lorentz's results along the lines indicated by Bohr's atom. After the First World War, electron theory of metals was taught in France in the Ecole Polytechnique, until a general tightening of the curriculum in 1924 made the material (along with other unsettled topics, such as quantum mechanics and relativity) "optional," which to the young *polytechniciens* meant in effect forbidden. As described in Chapter 2, in the summer of 1927 Sommerfeld took up the matter in his course, and like Langevin he simultaneously taught and worked on the connection between the quantum theory of metals and the facts given in handbooks. Meanwhile, from 1927 to 1930 Karl Weissenberg taught courses at Berlin on *Theorie der Festkörper*. By the mid-1930s, there were a number of such courses, along the lines of the spate of review articles such as the famous Sommerfeld–Bethe *Handbuch der Physik* monograph of 1933.[11]

None of this was yet solid-state physics. Prior to the work of Wigner and Seitz, theorists were not able to get deeply into the connections between gross mechanical or electrical properties and the structural details that the x-ray crystallographers were probing. Neither did quantum theorists get far with the complexities of structure-sensitive properties. In short, the quantum theory of metals was still a specialty only loosely connected with the other specialties that dealt with solids.

Still more important, when specialties were put into groups it was not common practice to set apart those that dealt with solids. Works like Smekal's 1933 *Handbuch der Physik* article "Zusammenhängende Materie" and Slater and Frank's 1933 general handbook of theoretical physics tended to interweave discussions of solids with discussions of molecules and chemical reactions.[12] This parallels what we have already seen for the "schools" at various universities: there was no particular entity intermediate between physics as a whole and its divers small specialties. Many institutions of the 1920s and 1930s showed a tendency to take as the basic subdivision some specialty rather than the entire study of solids. Conferences and similar gatherings in this vein (mentioned in earlier chapters) included the 1930 Leipziger Vorträge on metals, the 1935 Royal Society meeting on superconductivity, and the 1939 Bristol Conference on Internal Strains in Solids.

As a detailed example, consider the extent to which magnetism stood on its own, not only in the co-citation patterns, but as an intellectual and social entity. As mentioned in Chapter 1, there was a centuries-old tendency to study a fundamental subject such as magnetism by looking not only at solids but also at liquids, gases, and so forth, a tendency that often outweighed the reverse idea of using magnetism as a tool to study solids. This way of thinking was still alive in the Sixth Solvay Conference, "Magnetism," held in 1930, where papers on the magnetic properties of iron and other solids were overshadowed by papers on the magnetism of gas molecules, atoms, and nuclei. Solid and nonsolid topics were also mixed together in the Leipziger Vorträge on magnetism held in 1933. Less important than the mixture of topics, however, was the fact that such conferences held up magnetism as a separate specialty. The trend reached a new height with a 1939 international conference on magnetism held under Pierre Weiss in Strasbourg (Chapter 6). The specialty also had its own shelf full of textbooks. These tended either to pull toward the central, basic questions of physics by covering more than solids, or to pull away toward practical matters by specializing in ferromagnetic metallurgy.[13]

Magnetism had concrete, and expensive, institutions. The laboratory surrounding a giant electromagnet constructed in the 1920s by Aimé Cotton near Paris, the pulsed magnets used by Kapitza in Cambridge in the early 1930s, and the magnets that Francis Bitter began to build at the MIT Spectroscopy Laboratory in the late 1930s were "big physics" on a par with radioactivity laboratories and all but the largest particle accelerators. These laboratories were used, on the one hand, for fundamental atomic and nuclear studies and, on the other hand, for metallurgical work. As far as anyone could see in the 1930s, there was nothing here to indicate a specialty that was somehow part of a larger solid-state entity. The same could be said for other areas, such as low-temperature work, where large cryogenic facilities handled a variety of research areas.

Another example of a specialty that was taking on some identity of its own covered a subject even more fundamental to the new understanding of solids: x-ray

Early "big science": Aimé Cotton with electromagnet used to investigate solids and other subjects, Paris, 1920s. (AIP Niels Bohr Library)

crystallography. Its importance was recognized by researchers of the time and is reflected in almost every retrospective account by solid-state physicists, for x rays revealed the skeleton that theorists then fleshed out with quantum theory. Yet few quantum theorists were attracted to the mathematical puzzles of x-ray crystallography itself; the specialty was less like an integral part of their work than like a subject off to one side that only happened to provide useful facts.

For their part, even before x-ray diffraction was discovered, traditional crystallographers had been ambiguous about their relations with physics. Originally, most scientists had lumped crystallography with mineralogy, almost a branch of geology although related also to chemistry and metallurgy. A few objected to that categorization, as did a geophysicist who in 1919 refused to join the nascent Mineralogical Society of America because it would not name itself a Crystallographic Society; he explained that crystallography, as "the science which has to deal with matter in the solid state," was "on a par with physics and chemistry." Similarly, at the turn of the century in Russia one of the leaders of the field insisted that mathematical crystallography was a separate fundamental science "on the same level" with any other; however, another leader held that crystallography was only "a chapter in physics," a particular approach to studying solids. The question, then, was whether the revolutionary x-ray techniques developed in the 1920s would pull the field definitively into physics, or whether the x-ray crystallographers, many of whom had begun in physics, would be pulled away.[14]

The books of W. H. and W. L. Bragg showed that the varied interests of x-ray crystallographers, extending from rock salt to proteins, could fit together as a single subject. It was a subject that, since the discovery of x-ray diffraction, had grown rapidly in the classic fashion of a fresh specialty. The Braggs' 1915 text, *X-rays and Crystal Structures,* together with Max Born's *Dynamik der Kristallgitter* of the same year, were followed in 1923 and 1924 by textbooks by P. P. Ewald, Charles Mauguin, and Ralph Wyckoff, giving a firm base for the teaching of x-ray crystallography as a separate subject. Students and researchers were accordingly recruited, and communities began to develop. In Britain above all, "schools" of x-ray crystallography grew up around W. H. Bragg at Leeds and the Royal Institution in London and around W. L. Bragg at Manchester. Their methods and research questions were taken up at various British universities in the 1920s and 1930s by dozens of other crystallographers, nearly all of them students of the Braggs or at least students of their students, all forming a close-knit community.[15]

Another sign of institutionalization was international conferences. These began modestly in 1925 when Ewald invited a number of eminent x-ray crystallographers to meet for five congenial days in a large room at his mother's home, putting them up at an inn by a lake outside Munich. Other, more formal crystallography meetings followed at irregular intervals. One result was an international set of crystal structure tables, published in 1935. Crystallography also had its own well-established journal, the *Zeitschrift für Kristallographie und Mineralogie,* founded in 1877 by the dean of Munich mineralogists, Paul von Groth. In the early 1920s, after the old man had stepped down as editor, "Mineralogy" was dropped from the title. The new editors announced that the *Zeitschrift für Kristallographie* would become an international journal, and aspired to gather in "all those papers which attempt to elucidate the Solid State by means of a study of the Crystalline State."

That announcement suggested that the editors wanted x-ray crystallography to remain a specialty at the center of physics and, indeed, to be the core of the study of solids. The only major conference of the period devoted explicitly to "the solid state," held in 1934 as part of an international physics congress under the presidency of W. H. Bragg, was (except for a little molecular physics) largely crystallographic. Indeed, before the 1940s it was British scientists specializing in x-ray crystallography, W. L. Bragg and J. D. Bernal, who voiced the strongest demands for considering the "physics of the solid state" to be an independent "branch of physics." But as with magnetism, so with x-ray crystallography, ties with other approaches to the physics of solids seemed less strong than the tendency to develop independent institutions. In neither specialty had the tendency advanced so far as separate and permanent organizations such as societies and university chairs. In the 1930s it was impossible to say just where the final lines of division would fall.[16]

In sum, inspection of the institutions of physics through the 1930s, ranging from university departments to handbooks to journals, reveals no clear tendency to unify around the general subject of solids. Rather, many forces pulled toward specialization on a microstructural scale. Nevertheless, there was indeed a unifying force. This has been touched on already in some of the preceding chapters, and was also revealed by close inspection of the co-citation patterns. This force was the growing intellectual unity of the subject. Solid-state physics could become a social community only after its cognitive parts had drawn together in the minds of some physicists. The social institutions would follow hard on the heels of this new way of thinking and would bring it to the attention of the rest of the physics community.

The first institutions where the tendency toward unity appeared were conference proceedings, review articles, and textbooks: the proto-field had a proto-literature. At first, the essential elements of solid-state physics all came together only sporadically. The Fourth Solvay Conference in 1924 ("Electric Conductivity of Metals and Associated Problems") and a 1928 *Lehrbuch* article by Ewald are examples of places where x-ray crystallography, quantum mechanics, and the details of structural irregularities were all thrown together as a unit, even while the correct quantum theory of metals was still lacking. These were isolated cases, inadequate for most teaching purposes. A better start was made in 1936 and 1940 by the pair of textbooks that Mott co-authored, one on metals and alloys and the other on ionic crystals; taken together, these approximately spanned what would later be solid-state physics. During the 1930s, Slater too conceived a textbook to cover the field, but like Mott he found the field sprawled beyond a single book, and by the end of the decade he had not yet written a book on solids. Finally, in 1940, a single textbook apeared that in both title and contents heralded a separate field of physics: Seitz's *Modern Theory of Solids.*[17]

Seitz later remarked that his textbook descended from the lectures that Wigner had given in the autumn of 1932 at Princeton, where Seitz had attended as a student and taken the course notes (Chapter 3). Wigner had shown some interest not only in crystal structure and, especially, quantum theory, but also in the imperfections that were responsible for so many of the real properties of materials. Seitz began to draw things together in 1936 in an invited talk, "Applications of the Electron Theory of Solids," which he gave at the Symposium on Applied Physics held by the American Physical Society. This was expanded the following year in a series of review articles, "Modern Theory of Solids," co-authored with R. P. Johnson and published in the *Journal of Applied Physics.* The editor took the unusual step of adding a prefatory comment: he was "particularly glad" to present the series, he said, "because of the widespread interest of metallurgists" in the fundamental progress that physicists had been making in understanding the solid state. Seitz and Johnson began by stressing the newly achieved coherence of their subject:

> Until recently, the various theories of different solids were conspicuously lacking in unity. To interpret the distinctive properties of the three solids copper, diamond, and rocksalt, for example, one had to begin with three widely differing pictures of their internal constitution.

This problem had now been solved, they announced. A "unified theory" had been created through the coming of quantum mechanics, and in particular through band theory, which was the chief subject of the series.[18]

This same unifying spirit permeated the more complete textbook that Seitz published in 1940. He later remarked that he had become familiar with every significant paper, both experimental and theoretical, that dealt with the area he covered. To be sure, the textbook did not fully span the subject of solids. Seitz specifically excluded "plastic properties" and many other "structure-sensitive properties of solids," for which the theory was "not one that grows naturally out of modern quantum theory." However, he plainly foresaw steady progress toward including more and more details until a single theory would cover every feature of solids.[19]

Seitz was groping toward a still wider unification: theory with practical applications. That interest was shared by Mott, Slater, and other theorists who were beginning to see the theory of solids together with its applications as a whole, and also by physical metallurgists like Hume-Rothery who were approaching the same point from the side of applications. Concern for applications as well as theory would become one of the distinguishing marks of the new field. Seitz had developed his text while working both in academia and at the General Electric Labs, and by the time he published the book he was in Pittsburgh, where relations between the university and the steel industry were important. In fact, he was giving a series of evening lectures to a mixed audience of physicists and metallurgists, neither of whom had ever known much of the other's trade. Seitz's lectures used a strictly atomic-physics viewpoint, and he was surprised and gratified to find that the metallurgists found this as interesting as the physicists did. These lectures (which in 1943 became another book) at last bundled together in a single package quantum theory and structure-sensitive properties of practical importance.[20]

In its day, Seitz's text on the theory of solids was a unique creation. If in retrospect the book looks like an obvious harbinger of the coming of a field matching its title, that was far from obvious at the time. In 1940 nobody could have confidently predicted that a unified field of solid-state physics would ever become a dominant feature of the physics community. The specialties later assembled under that title were certainly growing, yet they might have continued to follow separate intellectual and institutional paths. How a coherent discipline of solid-state physics could at last arise, barely maintaining itself against centrifugal forces, will become clear if we look into the curious details of the creation of the first permanent institution bearing the name of solid state.

9.2 A Solid-State Physics Division

Around the end of 1943, a few physicists and metallurgists began writing letters to one another about the idea of a new division of the American Physical Society, one that would bring together physicists working on metals or perhaps on solids in general. There were already other divisions of the society in various stages of formation, so the idea did not at first seem controversial. But in fact the proposal for a new division became a strand in a tangled fight, one of the longest and fiercest internal battles in the history of physics institutions.

The idea of forming divisions within the society had originated around 1930 as a possible answer to demographic changes within the physics community. In 1910 there had been roughly 500 physicists in the society, whereas in 1930 there were some 2500 and the number was still rising rapidly. Particularly striking was the jump in the number of physicists in industry following the First World War. By 1930 there were perhaps a thousand of these—but less than half had joined the Physical Society, whose members were chiefly to be found in colleges and universities.

Physicists in industry did not share the academics' fascination with the higher reaches of quantum theory and the innermost workings of the nucleus, subjects that were becoming more and more dominant in the society's journal, the *Physical Review,* and its meetings. "We did not feel at home in the Physical Society Meetings," an applied physicist complained. "I once read a paper before the Physical Society and I am sure the Physical Society was not interested in the paper." Here status came into play, for academic scientists had traditionally looked down on those in industry, feeling they had abandoned "pure" research in return for mere monetary reward. The industrial researchers resented the suggestion that they had "prostituted" themselves. Neither did they share the career concerns of the university professors who ran the society. Now the physicists in industry were gradually becoming numerous enough to think of forming their own societies where they could find others really interested in a paper on applied physics or a conversation on industrial careers and, indeed, where they could find respect. Already in 1916 a group at the Eastman Kodak laboratories had founded the Optical Society of America, and in 1929 a group at Bell Labs followed with the Acoustical Society, while men interested in the flow of materials founded the Society of Rheology. Talk began to circulate that there should be a Society of Applied Physics, and perhaps also a Society of Applied Mathematics.[21]

The leaders of the American Physical Society, if they were not at the time in some industrial laboratory, in many cases had done industrial work at some point in their careers, or had received grants from industrial philanthropists, or at least drew their professorial salaries as payment for teaching students who were destined for industry. They recognized that in the long run the success of physics as a whole depended on close relations between industrial and academic workers. Fearing a break in these relations, they appointed a committee to study the matter. In 1930 the committee reported that the problem was becoming critical. Indeed, "numerous papers dealing with pure and applied physics are not even submitted [to the *Physical Review*] but are published in various chemical, engineering, photographic and geological journals." Leaders of the physics community responded promptly. Symposia on applied subjects were organized for Physical Society meetings, and the schismatic Optical, Acoustical, and Rheology societies, as well as the newborn American Association of Physics Teachers, were loosely connected back together with the Physical Society as members of an American Institute of Physics, which began to publish the *Journal of Applied Physics.*

The committee also recommended that the Physical Society make room for special sections, beginning with a division of applied physics and a division of mathematical physics to forestall the formation of separate societies in those areas. The formation of divisions required a change in the society's constitution, but that was a small impediment. The new article was approved by an almost unanimous vote at the end of 1931.[22]

By then conditions had drastically changed, for the Great Depression was bearing down. As business looked desperately for ways to contract, between 1931 and 1933 the number of physicists in industry dropped as rapidly as it had risen a few years earlier. The researchers scarcely had the strength to keep their jobs, let alone to form independent groups. The proposed divisions of the Physical Society were forgotten.

Within a few years, the world economy and especially industrial physics bounced back. As preparations began for the Second World War, the number of physicists in American industry, and their relative strength compared with the academic sector, reached an unprecedented height. The critical size that had been approached in 1930 was passed by 1940, and talk of forming new societies returned. Now, even more than in 1930, the idea circulated in specialties close to what would later be called solid-state physics.

Nothing was more central to the study of solids than x-ray crystallography. As we have seen, this had developed into a distinct research area, having not only its own characteristic problems and methods but also its own classic textbooks, its own occasional international conferences, and even its own journal. The matter might have stopped there, but these workers had a particular need that drew them together: as the number of crystalline powders investigated with x-ray electron diffraction grew into the thousands, the workers required detailed exchanges of information and standardized compilations of data. With the cooperation of chemists and engineers, around 1940 American crystal researchers formed a Committee on X-ray and Electron Diffraction under the National Research Council of the National Academy of Sciences, joining a number of other National Research Council committees that compiled data.

This was not enough for the American researchers, however. They wanted to hold regular meetings, and with the European war endangering the *Zeitschrift für Kristallographie* they also thought of creating their own journal. Spurred by Maurice Huggins, a chemist at Eastman Kodak research labs, a 1940 meeting was held devoted to applications of x-ray and electron diffraction. The following year Huggins founded the American Society for X-ray and Electron Diffraction (ASXRED) with 135 charter members.[23] Diffraction studies made a rather narrow focus for a society, and by 1946 an MIT mineralogist, Martin J. Buerger, had expanded a local Cambridge group into the broad Crystallographic Society of America. In 1950 the two societies achieved a fusion—a rarity in the history of scientific organizations—to become the American Crystallographic Association. Meanwhile, the Americans laid plans for a journal.

Much the same hopes circulated in the British Isles, where the X-ray Analysis Group was formed within the Institute of Physics in 1943. The prospect of several rival national journals disturbed P. P. Ewald, then at Queens University College in Belfast. He was also disturbed by his field's ambiguous status, suspended between physics, chemistry, and mineralogy. In 1944, in a talk at Oxford to a meeting of the X-ray Analysis Group, he suggested the creation of the International Union of Crystallography. This union could become a center for publishing standardized tables of crystal structure data and the like. Ewald spoke up frankly for separatism:

> specialization . . . is sometimes deplored as a disintegration of science. Should it not rather be looked upon as the formation of a new branch on the tree of science, which conveys the sap to a region hitherto not well provided for, a branch necessary for rounding off the shape of the tree as a whole?

W. L. Bragg and others took up the idea, and in 1946 they organized an international meeting in London. A large group came from the United States, and many came from other nations; even the Russians sent a contingent, arriving too late for the meeting but not too late to take part in forming the union. In 1948 the new union held its first congress at Harvard University, and meanwhile began publishing *Acta Crystallographica,* an international journal. In sum, by the end of the

1940s crystallography had become a well-organized and distinct social community and scientific field.[24]

Just what did this field contain? The definition written in 1949 for the constitution of the American Crystallographic Society was imperially general: the society would cover "the study of the arrangement of the atoms in matter, its causes, its nature and its consequences." Similarly, in his 1944 Oxford address, Ewald had offered a vision of his field expanding until it embraced "the whole system of research on the structure and properties of matter in the solid and related states," whether crystalline or not—in effect, what some would call solid-state physics. The overlap with physics was visible from the outset—for example, in the program of the 1940 meeting that preceded the creation of ASXRED. There were several talks on organic chemical compounds, a subject remote from the interests of physicists of the time, but also a talk on the theory of x-ray and electron diffraction (by Peter Debye), two talks on metals, and one on thin films.[25] Subsequently, the American Crystallographic Association became a member of the American Institute of Physics, and during its early years on one occasion ASXRED met jointly with the American Physical Society.

But on other occasions, ASXRED met with the Mineralogical Society of America and the Electron Microscope Society, and over the years the crystallographers developed close ties with engineering groups such as the American Society for Testing Materials. Links with chemistry, engineering, and even biochemistry proved to be as strong as links with the physics of solids. In short, the case of crystallography demonstrates how an area of research, in some respects fundamental to the physics of solids, could find an interdisciplinary niche and take on a permanent, separate identity of its own. How many such specialties would find niches, almost independent of the rest of the physics community?

In 1941, Huggins had informed the Physical Society that he was establishing a society of diffraction crystallographers. The Physical Society's council voted a motion to assist, but at the same time wanted to "point out the provision in the constitution which makes possible the formation of 'Divisions.'"[26] Huggins could not be diverted from forming ASXRED, but the idea of wooing breakaway groups by forming divisions within the Physical Society had returned from limbo.

A number of people began to suggest forming divisions to avert fragmentation of the physics community. Now, as in 1930, physicists were complaining that at Physical Society meetings it was hard to find common ground between industrial and academic researchers. The industrial men, for the most part young, shy in encounters with the famous professors, and following very different interests, could not be expected to initiate contact—no doubt sensing the low regard academics held for their career choice. Further, as one letter to the society remarked, it seemed that "modern physics has become largely a study of electrons and nuclei with less attention given to anything as large as a molecule"; the physicists who were interested in visible substances might well feel that engineering societies or the like would give a warmer welcome to their papers. The industrial researchers could never find a status comparable to that of academic physicists, unless they had some sort of organization of their own and a fitting name for what they did.

In late 1942 the Physical Society received proposals for a division of textile physics and also for an all-encompassing division of industrial physics. The society's council discussed the proposals but did not endorse them, probably because the one seemed much too narrow, the other much too broad. Meanwhile, in 1942 another group had broken away: the Electron Microscope Society. Electron microscopes were a genuinely new specialty, opening up a rapidly growing research area that interested people in many fields and also attracted manufacturers like the Radio Corporation of America. In January 1943, no doubt with an eye to the new society, the secretary of the Physical Society got in touch with a friend, the electron microscope expert Ladislaus Marton, who had recently moved from RCA to Stanford. The secretary drew Marton's attention to the possibility of a division within the Physical Society. A few months later, Marton submitted a petition signed by 40 members of the society, including such luminaries as Felix Bloch and Ernest O. Lawrence, to establish a division of electron and ion optics. This time, the council gave tentative approval. Some 700 members of the society expressed interest in joining such a division.[27]

Clearly the proposed division was only a beginning. Talking the matter over among themselves, the society's officers decided to encourage industrial physicists to form divisions in a variety of fields. Examples mentioned included biophysics, nuclear physics, and "physics of metals and semiconductors." In September 1943, the society announced in the pages of the *Journal of Applied Physics* that it would welcome proposals for divisions.[28]

The next specialty to make a formal proposal was chiefly concerned with solids. A small group of Physical Society members who met one another at a textile physics conference in September 1943 decided to petition once again for a division in their field. One of the group, Warren F. Busse of the General Aniline & Film Corporation's research laboratories, was then organizing the Symposium on the Physics of Rubber and Other High Polymers, to be held that November during a Physical Society meeting—one of the symposia that the society hoped would help keep industrial workers within its fold. Busse advised his colleagues to include more than textiles in their proposed division. Physical studies of plastics, paper, and paint were also on the rise, and rising still faster was the study of elastic polymers, driven by the government's desperate wartime effort to found a synthetic-rubber industry. Busse was soon backed up by petitions from the rubber-tire laboratories in Akron, Ohio. After a flurry of organizing activity, in April 1944 the society's council formally established the Division of High Polymer Physics, along with the Division of Electron and Ion Optics.[29]

Meanwhile, the organizing activity spread as people in many applied fields considered divisions. In 1944, the council received tentative proposals for divisions of biophysics, spectroscopy, chemical physics—and a division of physics of the solid state. This last, in particular, would bring into the open certain views, vehemently opposed, that the society's officers had not yet taken into account.[30]

At first, it all seemed fairly simple. The initiative came from Roman Smoluchowski, then in his early thirties. Formerly the head of the Physics Section of the Institute of Metals of the Warsaw Institute of Technology, in 1940 he made his way to Princeton and in 1941 joined the General Electric Research Laboratories. In the 10 years since he had begun his career, Smoluchowski had worked on topics ranging from fluorescence to the theory of Brillouin zones, and he was now deep into practical work on iron alloys.

Roman Smoluchowski in the GE Research Lab, Schenectady, New York, around 1945. (AIP Niels Bohr Library, courtesy of R. Smoluchowski)

Smoluchowski had noted the organization activity going on at the November 1943 rubber physics symposium. He talked the matter over with Saul Dushman, assistant director of the General Electric labs, and others at GE who had an interest in metals. They were all attracted by the idea of some sort of organization within the Physical Society to unite "metal-physicists."

Before taking any further steps, Smoluchowski carefully touched base with the one man whose views on the matter would be crucial: Seitz, already recognized as a central figure in American solid-state research. Seitz's feelings were mixed. As he recalled a few years later, he doubted that research in solids could be kept within the physics community. He knew that fields tended to split off from physics when they moved away from fundamental research and toward applications; such had been the fate of electrical engineering, for example. Seitz suspected that research on solids would sooner or later become a branch of physical chemistry, and that any physicists still working in the field would end up joining the American Chemical Society. In the meanwhile, however, Smoluchowski's proposed division could be helpful, even if it did divide physicists into categories more than Seitz would prefer. He wrote back, "I feel that this type of organization is a necessary evil." The term "metal-physicist" seemed wrong to him, but he could suggest no alternative.

With Seitz's hesitant approval in hand, Smoluchowski brought others into the act, in particular Thomas Read and Sidney Siegel, both of Westinghouse, and William Shockley of Bell Labs, then doing war work in Washington. Except for Seitz, who straddled the border between academia and industry, all those involved were industrial men. In January 1944, Smoluchowski's group of six (including Dushman) signed a circular letter to some 50 "metal-physicists," inviting comments on the idea of organizing within the Physical Society. The letter noted that

> after the war we will have more demand for physicists and more students studying physics than we ever had before. We can expect also a development of various branches of Pure and Applied Physics, and we would like them to remain branches

of *physics* rather than to become new "pure sciences" or new types of "engineering."

Doing this—keeping industrial researchers connected with academic physicists—would be, as one correspondent warned Smoluchowski, "an uphill fight."[31]

Certain respected figures within the Physical Society—for example, Wigner and Lee DuBridge—were having second thoughts about divisions. Most eminent and most determined of all these men was the Harvard quantum theorist John Van Vleck. Replying to Smoluchowski's circular letter, he wrote that he was "a rather violent opponent of what I like to call 'Balkanization' of the American Physical Society." He recalled the spirit of the meetings the society had held each spring during the 1930s in Washington, D.C., where a few score physicists gathered to hear papers on any topic related to physics, or if they were not interested in the topic, to sit on the sunny lawn of the National Bureau of Standards and talk with their friends. Bringing in a great many industrial researchers, Van Vleck suspected, would stifle this spirit. "The idea that various groups whose main interest is not always physics must be coddled, in order to make them members of the American Physical Society, has never appealed to me," he wrote, "as just mere numbers is not everything."

Smoluchowski replied by return mail. Continuing Van Vleck's political simile, he said that he did not want "Balkanization" either; he was concerned "to prevent 'an invasion and partition by powerful neighbors' (i.e., chemists and metallurgists)." Van Vleck shot back that he was less worried that the Physical Society would be invaded than that it would attempt its own imperialistic invasions. In the springtime when he sat on the lawn of the Bureau of Standards, he did not want to feel

Isidor Isaac Rabi, Edward U. Condon, Henry D. Watson, and William W. Watson at an American Physical Society meeting in Washington, D.C., 1930s. (AIP Niels Bohr Library)

Karl K. Darrow (1891–1982). (AIP Niels
Bohr Library, Francis Simon Collection)

crowded out by "too many hangers-on whose main interest is not that of pure physcis or science." If industrial men wanted status, the Harvard professor would prefer
that they sought it elsewhere.[32]

Van Vleck's views were shared by only a minority. Of the 50 metal researchers
to whom Smoluchowski had sent his letter, about 30 replied, and about half of them
wanted to set up some sort of permanent group, while another third wholeheartedly
favored starting up a new division. Accordingly, after touching base with Seitz and
other members of his group of six, in April 1944 Smoluchowski wrote a letter to
Karl K. Darrow to request the formation within the society of a "Division of Metal
Physics."[33]

Darrow was the society's secretary, and because it was the secretary who handled
daily business while presidents and council members came and went with each elec-
tion, Darrow was the most powerful officer in the society. A lean-faced and spare
man, Darrow had a sense of elegance and propriety, an ironic sense of humor, a gift
for clear words, and strong opinions. As might be expected of an employee of Bell
Labs (where he had begun as a physicist, but soon developed into a respected science writer), Darrow's opinion regarding divisions was wholly in favor. "One of the
things most greatly to be desired," he wrote in the *Journal of Applied Physics,* "is a
unification of the physicists called 'academic' and the physicists called 'industrial.'
It is important to avert the danger of a lack or loss of interest by either group in the
problems and the enterprises of the other." Darrow also had a more personal problem. It was difficult to carry out his secretarial duty of arranging all the sessions and
symposia at society meetings, for he did not know every field well enough to decide
among subjects and speakers. Divisions would take this task off his shoulders.[34]

Darrow was surely willing to propose a new division to the council, but Smoluchowski's letter was not the sort of formal petition he needed. To make the matter
more difficult, Darrow meanwhile received a letter from Seitz. Van Vleck and others had been bombarding Seitz with their views. "These are men whose opinions I

respect very much," Seitz wrote to Darrow, "and I am impressed by the fact that they feel as if divisionalization would be catastrophic." Seitz found that his own feelings were mixed. He advised that the question be set aside until after the war, when travel restrictions and pressure of work would be lifted and people would be able to get together to discuss the question thoroughly.

Seitz's misgivings sparked a carefully wrought reply. Darrow warned him that "there is a great tendency to form small new societies, too small to guarantee a continued successful existence." (He mentioned ASXRED as an example.) The force behind this worrisome trend, Darrow knew, was the rapid growth in the numbers of physicists. Already at the last peacetime Washington meetings, he reminded Seitz, the secretary had been forced to divide the talks into separate, concurrent sessions. This was de facto divisionalization of an awkward sort. Reflecting on Darrow's determined letter, Seitz at last made up his mind that a division had to be formed. However, it remained true that a time when everyone was frantically involved in war work was a poor time for a step that disturbed many. Darrow met in June with Smoluchowski, along with Léon Brillouin, for lunch at the Columbia University Faculty Club. In view of wartime pressures, they agreed to take only a tentative first step, the formation of a committee that would organize a symposium. The society's council duly agreed and appointed Smoluchowski's group of six for the task.[35]

Meanwhile, there had been some changes in the name of the proposed division. In the original circular letter of January 1944, Smoluchowski had spoken of "Physics of Metals"—a term familiar in both the United States and Britain—and many of his correspondents endorsed that succinct definition of their specialty. Others did not. Brillouin, for example, wanted "to include *all* problems of crystal physics, metal or non metal." After all, he explained, "the distinction between metals and other solids has no scientific basis . . . so let us speak of *Physics of Solids* in general." Conyers Herring of Bell Labs agreed that solids as a whole made a more clearly defined field, and added some practical information: scanning recent papers and abstracts, he found that work on solids made up somewhere between a sixth and a third of all physics research, and was about equally divided between metals and nonmetals. Smoluchowski was not convinced. "It would sound peculiar," he wrote Herring, "to call ourselves solid-physicists." More important, "cooperation from purely metallurgical quarters may be more active if we are 'all out for metals.'"

The debate expanded as Seitz offered his views. On the intellectual side, he noted that "one of the things we have gained during the past fifteen years is a unified viewpoint of metals and nonmetals." On the practical side, a "Division of the Physics of Solids" would attract people interested in ceramics, pigments, glass, and so forth, groups that "are very large in number, and are no less well organized than the metallurgists."[36] Darrow was firmest of all. Whenever Smoluchowski wrote to him about the proposed committee in the "Physics of Metals," Darrow wrote back about the proposed "Division of Solid-State Physics." This term, "solid-state," was beginning to circulate. Gregory Wannier, for example, thought of it spontaneously when he heard about the proposed division in early 1944, although he confessed that "solid state physics sounds kind of funny." Darrow may have heard the term circulating at Bell Labs, where it had appeared as early as 1940, perhaps in analogy to the German term *Festkörperphysik,* which had already seen occasional use for a

John H. Van Vleck at an American Physical Society meeting in Washington, D.C., 1930s. (AIP Niels Bohr Library, Goudsmit Collection)

decade. Faced with Seitz's views and Darrow's firmness, Smoluchowski gave in and the more general term was adopted.[37]

Considerably later, Clarence Zener suggested that the word *solids* in the division's name be formally restricted to crystalline solids, to the exclusion of plastics. Smoluchowski and Seitz happened to be visiting with Shockley at Bell Labs, and the three discussed the question. They again chose the most general definition. After all, they reflected, phenomena closely related to crystals showed up in glasses and the like. In principle, there was only a vague boundary between crystalline and amorphous materials, and they did not want to limit their division too strictly.[38]

The symposium organized by the Committee of Six, held at the society's January 1945 meeting, was accordingly titled "Symposium on the Solid State." Seitz served as keynote speaker, and many other distinguished men, even Van Vleck, delivered papers. The great lecture hall of the Pupin Laboratories at Columbia was filled to overflowing by hundreds of physicists, so that even those willing to stand could often scarcely find room. Time was set aside for a discussion of the proposal to form a division, and for the first time the strongly opposed views came out in public. Meanwhile, Smoluchowski's Committee of Six had begun to gather signatures on a formal petition to establish their division.[39]

Van Vleck struck back by addressing the most urgent problem, that of sessions at meetings, with an ingenious counterproposal. In a petition signed with 11 others, he asked the Physical Society to set up a committee on programs to encompass representatives of any and all specialties. The solid-state Committee of Six responded with its own letter to the society, drafted as usual by Smoluchowski. "The field for physics has grown enormously," he noted, creating a "difficulty in maintaining contact among younger physicists themselves and between them and the Society." To prevent "schisms" it was essential either to proceed with divisions or to elaborate the proposed program committee with elected subcommittees—which in practice would be little different from divisions.

In its May 1945 meeting, the Physical Society's council got into a long and difficult discussion over the entire question of divisions. Finally, the members agreed to appoint a program committee—which would include Darrow, Seitz, Van Vleck, and Wigner, among others—and defer further discussion until peacetime.[40] By the time the January 1946 meeting came around, the issue could no longer be put off. The proposal for a solid-state division remained pending. Unlike the earlier idea of a metal physics committee serving industrial researchers, a solid-state division, with a definition generously embracing all solid matter, would plainly reorganize a major portion of the physics community. A similar prospect was offered by the Division of Electron and Ion Optics, which now petitioned to change its title to the far more sweeping Division of Electron Physics.

Darrow permitted his opinions to impart some flavor to the ordinarily dry minutes of the council meeting. "During a long discussion," he recorded,

> it became evident that there is still a marked difference between those who want the Divisions restricted to marginal topics such as have not hitherto been fully represented in the programs of the Society, and those who have no such restrictions. . . . The crucial vote went in favor of the former class by the smallest possible margin.

In June the council confronted the issue again and adopted the classic response to an impasse: it asked the president to appoint a Committee on Policy Regarding Divisions. The president, Edward Condon (recently moved from Westinghouse labs to the National Bureau of Standards), deliberately appointed Seitz and others who favored new divisions, and when the committee reported in September, the members were unanimous. The council decided to approve a policy of encouraging divisions, although by what margin of votes, Darrow did not record. Smoluchowski, as soon as he heard the news of the council's action, dashed off a letter on behalf of the Committee of Six, renewing the petition for a "Division of Solid State."

But did the official policy really allow such a broadly defined division? A new committee was appointed to implement the policy, and it soon found itself, along with the council as a whole, "confronted with irreconcilable viewpoints," as Darrow put it. In the January 1947 council meeting, his minutes said, "the interminable question of Divisions came up again, and to the consternation of some of those present it was evident that the policy adopted at the September meeting still does not command the unanimous consent of the Council." Resolution of the question was once again put off until the next meeting, in May. There, Darrow recorded, "the Council turned to its favorite pastime of discussing the question of Divisions. . . . Many complaints and lamentations were uttered." Yet another committee was appointed. At last, in June 1947, the council came to a final decision in favor of broad divisions. As a result, a set of bylaws that had meanwhile been drawn up for the Division of Solid State Physics was accepted: this was the formal stamp of approval. Darrow and the solid-state committee had already begun to enroll Physical Society members in the newborn division. For the first time, then, there could be such a thing as a "solid-state physicist"—a creature with a unique institutional affiliation, a generally recognized name and status, and perhaps even a distinct field of study.[41]

Could it have turned out otherwise? Certainly there were strong pressures working in favor of cohesion, pressures ranging from the intellectual unity offered by the concept of a solid, to the desire of farsighted men to bridge the gap between pure and applied science. However, intense centrifugal pressures were also at work, not only in solids but in all of physics. Most physicists recognized that the increasing size of their community made some sort of subdivision inevitable, if only so they could hold meetings where they could find their colleagues in the crowd. Less obvious at the time but equally important, physics had grown in intellectual complexity and depth to the point where a single researcher could not expect to master more than one or two specialties. There were sound reasons for groups like the metal physicists to break away, either to follow the siren song of the metallurgical societies or to find their own interdisciplinary niche, as the crystallographers were doing. Even within the Physical Society, if the proposed division had been restricted to metals, then other groups might eventually have demanded divisions of semiconductors, amorphous solids, or whatever, following the example of the high-polymer group. The tendency toward subcommunities, tightly specialized in subject and applied in orientation, could have led to the "Balkanization" that Van Vleck rightly feared. Standing at the point where all these fiercely opposed forces collided, a few thoughtful people were able to sway the balance in favor of a community that was intellectually and socially open, yet internally coherent—a solid-state physics community.

9.3 Consolidation

At the end of the Second World War, and even in the early 1950s, solid-state physics still lacked most of the supports of a mature scientific field. It had no regular meetings, no designated professorships, no exclusive funding sources, no journals, no prizes. It remained a fragile entity, then, and few physicists could feel confident that it would not fragment into various specialties, each pulled in a different direction. To see how solid-state physics established a lasting cohesion, I will remain a while longer with the community in the United States. Things happened there first. That was partly because the American physics community, unlike others, had been able to keep up its institutional life during the war. More important, during the years 1940 to 1960, when solid-state physics built its institutional foundation, the United States had more scientists than any other nation—in fact, as we will see, nearly as many as the rest of the non-Communist world put together. American scientists had an influence even greater than their numbers alone would suggest, for during these years the Americans were the best funded of all scientists; they trained legions of students from abroad; and they benefited from the general (if temporary) predominance of their nation in world political, economic, and cultural life.

During the late 1940s there existed only one permanent institution bearing the name of solid-state physics, the Division of Solid-State Physics of the American Physical Society, and that did not amount to a great deal. Its only activity was to hold symposia at one or another society meeting, irregularly, in the same fashion as the High-Polymer Division, the Electron and Ion Optics Division, and the new Committee on Fluid Dynamics. The first sign of something different came at a meeting the Physical Society held in Cleveland in March 1949. This meeting was meant to be a revival of a prewar practice, a small March meeting for Midwestern physicists, sandwiched between the much larger January (New York) and April (Washington) meetings. The Solid-State Division organized a symposium of three half-day sessions of invited papers for Cleveland. The event attracted a number of contributed 10-minute papers on related topics, so that the final count showed 27 contributed papers on solid-state physics, accompanied by a mere 13 on nuclear physics and 9 on other subjects. In effect, then, this became a meeting of back-to-back sessions on the solid state, alongside two minor sessions on other topics. As expected, it was a small and local meeting, with only 257 registrants of whom perhaps 150 showed up to hear the solid-state papers. But it would be remembered as the start of something much larger—the first meeting, open to all physicists, devoted to solid-state physics in the comprehensive modern sense. In Seitz's recollection, that was the sort of thing that had been wanted from a division all along: those in the field wanted to "all be in the same hotel and meet in the same bar and talk over the problems."[42]

Nothing was done along these lines in 1950, but in 1951 the March meeting, held in Pittsburgh, again boasted a symposium on solid-state physics. The division also took the occasion to hold its annual business meeting. "Nuclear physics will feel rather lonesome at this meeting," Darrow remarked when he announced it in the society's *Bulletin.* The society wanted to build up the March gathering as another general meeting, for the January and April meetings were beginning to feel painfully overcrowded. But the nuclear physicists were accustomed to those other meetings and would not go to the Midwest. The trend continued at the next year's March meeting, where the Division of Solid-State Physics ambitiously arranged no fewer than five half-day sessions of invited papers. This brought in another 70 contributed papers in solid-state physics, joined by 24 in high-polymer physics, many of them dealing with solids; only 20 papers came from all the rest of physics.

By 1953, Darrow was accepting and even welcoming the trend. He noted in the *Bulletin* that a "preponderance of nuclear physics has become customary at our Washington meeting," and that the various divisions had perforce abandoned that meeting to concentrate on other occasions. Nuclear physics, a term that at the time still included the study of elementary particles at high energies, had become a distinct field of its own. The American government's buildup of fission and fusion bomb inventories was advancing at top speed, with the nuclear industry consuming about a tenth of all the electricity generated in the United States; particle accelerators, which in the reflected glow of the bombs seemed to be essential to national security, were funded with a generosity never before known in science; and an impatient enthusiasm for civilian uses of uranium fission was gathering momentum. Nuclear science dominated the Physical Society, so that nuclear physicists felt no need to set up a special division of their own.

Van Vleck might still sit beneath a tree on the lawn of the National Bureau of Standards each April, but instead of greeting 100 or so colleagues he would now be lost among some 2000, largely nuclear physicists. American physics was no longer like a small town where everyone knew one another. At the council session held in connection with the 1953 Washington meeting, Van Vleck announced that some physicists felt there were too many gatherings each year (although in fact there were no more than in the 1930s). He and others proposed that the March meetings be canceled and that the divisions, most particularly the Solid-State Division, be forced to meet with the rest of the society—for example, at the June or November meeting. This was a futile rear-guard effort, and the council took no action, allowing the Division of Solid-State Physics to adopt the March meeting as its principal meeting of the year. It remains so to this day, and in fact has become the nation's largest annual gathering of physicists.[43]

Meanwhile, solid-state physics took on another feature of existence as a separate field: an annual prize. In the summer of 1952, the Bell Telephone Laboratories gave the Physical Society $50,000 to endow a solid-state physics prize to be named after the labs' recently retired president, Oliver E. Buckley. Thus physicists would get an annual reminder of the triumphs of solid-state physics and, at the same time, of Bell Labs' interest in the field. The following January, a committee awarded the first of the Buckley prizes, amounting to $1000, to William Shockley of Bell Labs.[44]

Still missing was the most central of all features of a separate field, a journal of its own. Typically, the rise of a specialty is signaled by the creation of a journal. New societies and organizations, like the International Union of Crystallography, have often been founded with publications as a central purpose. However, solid-state physics was not typical of a struggling new specialty. Along with nuclear physics, it already held such a prominent place in the general physics journals that there was no intense pressure for a separate publication. What pressure there was, came from the same force that had led to separate meetings—the increasing weight of numbers in the physics community as a whole.

The American Physical Society's *Physical Review* was growing ever thicker. It seemed unfair to force individual physicists to bear the rising cost, when they could not possibly have time to read more than a small and dwindling fraction of the articles in each issue. After years of discussion, in 1955 the council suggested that the journal be broken into two sections. As announced in the *Bulletin,* one section might be devoted to "nuclear physics" and the other to "non-nuclear physics." But a table in the *Bulletin,* suggesting how the split would have been made for recent articles, showed that the proposed section *B* would be about one-half "Nuclear Physics" and one-quarter "High Energy," with the rest given to related "Theory," while the proposed section *A* would be one-half "Solid State," with the rest about equally divided among "Atoms, Molecules and Ions," "General Physics," and related "Theory." In effect, then, solid-state physics would dominate the first section about as much as nuclear physics would dominate the second. The *Bulletin* included a ballot to poll society members on the issue, and several hundred ballots were clipped and sent in. Nearly all of them favored the split, but nearly all of these came from solid-state physicists. In a formal vote, the council decided not to split the *Physical Review* after all. In this instance, then, although the urge to subdivide was as plain as in the case of meetings, the solid-state physicists had failed to create a new institution within the Physical Society.[45]

Nevertheless, it was "felt by many that the coming of age of solid-state science should be recognized by the publication of a journal devoted exclusively to this field." Such was the conclusion of Harvard professor Harvey Brooks, who in 1956 began editing the *Journal of the Physics and Chemistry of Solids.* Although, as the title indicated, it was never possible to define a boundary with chemistry, there was little doubt that this was a solid-state physics journal. It was an international one;

the publisher, Pergamon Press, was transatlantic, and some associate editors were as far away as Moscow (E. M. Lifshitz and L. D. Landau). The success of Pergamon's venture was followed up in 1959 by the Russian-language journal *Fizika Tverdogo Tela (Physics of Solids),* demonstrating that the same division into fields was proceeding in the Soviet Union. The East Germans followed two years later with *Physica Status Solidi.*[46]

Yet the *Physical Review* remained by a substantial margin the most frequently cited of all journals in regard to solid-state physics papers. Gradually such papers became segregated into their own set of pages within each issue. By 1964 the *Physical Review* had grown so large that neither the editor nor even the printer could handle the size of a monthly issue, and it was published in two sections, one primarily for nuclear and high-energy physics, and the other three-quarters solid state.[47] Finally, in 1970, the *Physical Review* split into completely separate journals, with *Phys. Rev. B,* devoted to solids, the largest. Meanwhile in 1968, the *Journal of Physics* of the Institute of Physics in Britain had also divided, with one part given over to solid-state physics, and in 1971 the *Zeitschrift für Physik* obeyed the same fissile urge. In sum, the field was demarcated, as both a social and an intellectual structure, in an unmistakable way: what got between the covers of these journals was almost by definition solid-state physics.

A final proof that the field had come of age during the 1950s was the appearance of a secondary literature. *Solid State Abstracts* began publishing in 1960, expanding out of an abstract journal that at first had handled only semiconductor electronics. The following year, the *Bulletin signalétique* began devoting a separate publication to the field. (We shall see what happened in *Physics Abstracts* around the same time.) Still earlier, in 1955, Seitz and David Turnbull had begun a sort of heavyweight review journal with the straightforward title *Solid State Physics.* The monographs within these volumes, Seitz recalled, were intended to serve the same function for solid-state physics as the classic *Handbuch* articles had once served for all of physics.

Seitz had launched the series of volumes in place of a new edition of his textbook, for he no longer felt he could review the entire field by himself, and meanwhile excellent books for graduate students were beginning to appear. The paradigm was Charles Kittel's *Introduction to Solid State Physics.* Introduced in 1953, it was so successful that three years later it went into a second edition some 200 pages thicker. Generally speaking, the second edition's contents were coextensive with solid-state physics as the field has been defined ever since.[48]

Although textbooks and other publications were essential to keep the new field together, communications were increasingly done in person. Already in 1948, the International Conference on the Physics of Metals drew over 300 people to Amsterdam, including 30 from the United States, who brought their Continental colleagues abreast of wartime developments. In this case, "metals" embraced everything from plasticity to ferromagnetism. The high point of the conference came when W. L. Bragg ran motion pictures of congregations of tiny bubbles, for the first time showing in a striking and visible analogy the formation of grains and the appearance of dislocations, concepts that applied across all of solid-state physics (Chapter 5). Three years later, the Ninth Solvay Conference was devoted to "L'Etat solide." In fact, this conference covered mainly crystal growth, dislocations, and the like, with only occasional references to quantum theory, but the next conference, in 1954, made up for that by covering electron theory from Fermi surfaces to superconductivity; taken together, the pair of conferences suggested a growing international acceptance for the existence of a field of solid-state physics.[49]

Such conferences not only recognized but helped to spread the concepts of the field. We have already seen—for example, in the existence of journals—that during the 1950s solid-state physics rapidly established itself as a field in the Soviet Union and Britain in particular. It seems that the British physicists both encouraged and were encouraged by their American colleagues, so that developments in the two places were almost part of a single movement, whereas the Soviet case suggests an autonomous rise in parallel with that in Britain and the United States. But in both cases, it must be left for future research to determine just how the consolidation of the field was furthered by local institutions and by foreign influences.

The separate existence of solid-state physics was not always easily recognized, and in several cases it had to be imported from America and Britain. In France, for example, not only the physics of solids but much of modern physics as it related to quantum theory had made little progress in the 1930s and had stopped almost entirely during the war. By 1945 many French physicists and particularly the younger ones, forced to admit that they were far behind, resolved to educate themselves. Just as a century earlier British and American scientists had found a stop in Paris important to their education, now young French scientists went to study in Britain and the United States, and some brought solid-state physics back with them. Others managed to import not only the teachings but the teachers themselves, especially at the summer school run by the University of Grenoble at Les Houches. The idea of this institution was itself imported, going back to the congenial summer school that had flourished at the University of Michigan in the 1930s. Solid-state physics was among the subjects taught at Les Houches—for example, in a lecture series delivered by Rudolf Peierls in 1953.[50]

French industry, increasingly interested in such matters as semiconductor electronics and the metallurgical problems of nuclear reactors, looked for students instructed in these specialties. In 1956 the Paris science faculty established as part of its curriculum an official "third cycle in solid state physics," teaching everything from quantum theory of solids to the properties of crystal defects; similar courses arose at the Universities of Grenoble and Strasbourg, which, next to Paris, became the most important centers of solid-state physics. The ultimate stamp of official existence came with a chair and a laboratory of Physique des Solides, awarded to André Guinier of the Paris science faculty at Orsay. By the early 1960s, there were perhaps a dozen French laboratories old and new with names like Magnetism and Physics of Solids and Physics of Metals. A similar development took place in Italy, although it labored under even greater difficulties. Beginning after the Second World War "with empty hands," as one physicist put it, pioneers at several universities gradually established solid-state groups. They had to bring in much of the initial knowledge from elsewhere, especially the United States.[51]

The process was easier in Germany, where there was a stronger prewar tradition to build on. In formal terms, the East Germans were in the lead. A conference on *Festkörper-physik* was held in Dresden in 1952, followed in later years by other conferences in the DDR; Paul Görtlich took a chair of solid-state physics in Jena in

1954; and, as mentioned earlier, in 1959 an entire journal devoted to the field was established.[52] Meanwhile, the West Germans, like the French, imported enthusiasm for solid-state physics from Britain and the United States—for example, in a 1949 color center conference called by Pohl. The first formally titled chair for *Festkörperphysik* in West Germany was taken in 1959 by Alfred Seeger, not at a university but at the Stuttgart Technische Hochschule, where Professor of Experimental Physics Heinz Pick had already established a major center for work on solids. By 1960 there were a dozen university and technical institute chairs in the Federal Republic that listed solid-state physics, if not in their title, then at least as one of the main fields they taught.[53]

The (West) German Physical Society followed, with some delay, the same path as the American one. The first signs of consolidation were *Physikertagung* symposia devoted to solids, held, for example, in 1953, 1957, and 1958. In 1959 several small specialized sections *(Fachausschüsse)* of the society were grouped under the rubric *Festkörperprobleme,* and in 1970 this became an official division of *Festkörperphysik.* To seal the process, in 1972 Siemens AG helped the society establish the Walter Schottky Prize for Solid State Physics Research.[54]

These developments (and similar examples that might be mentioned in Japan, Italy, and other countries we lack space to cover) did not take place automatically. Establishing each institution required a complex struggle with issues this chapter has barely touched on. Boundaries with chemistry and engineering had to be defined; the question of incorporating fields such as low-temperature work and metallurgy had to be considered; the needs of industry and perhaps national defense had to be invoked. Not all the initiatives came to anything, but enough succeeded to establish the field as a social entity.

Of course, such formal institutions as divisions of a society, prizes, conferences, journals, and even university chairs were only surface indications of the cohesion of a field. The body of actual research was always done by "schools"—that is, by groups of like-minded researchers. As mentioned in previous chapters, some of the prewar centers of solid-state research survived and expanded, such as Ioffe's at Leningrad. Others were built up almost from scratch, such as the one Seitz founded at the University of Illinois in Urbana and the metal theory group created by Ilya Lifshitz in Kharkov. Still others were founded as a union of divers earlier groups working in areas from low temperatures to crystallography, such as the solid-state school that coalesced in Cambridge after Mott arrived in 1953.[55]

These centers little resembled the prewar schools. Where those had been built impermanently on a single professor, and never devoted themselves specifically to a wide range of pure and applied solid-state physics problems, the postwar schools stood on a firm and broad foundation. A few significant examples will show how this worked.

The first institutional department to take the whole range of solid-state physics as its field, and even formally as its name, was at Bell Telephone Laboratories. As early as 1938, the farsighted director of research, Mervin Kelly, had established a little group in "Physics of the Solid State." By early 1945, Kelly was explaining that with the coming of quantum mechanics, "a unified approach to all of our solid state problems offers great promise," and in July he formed the Solid State Physics Department. Work in this and other, closely linked groups grew rapidly, bringing together dozens of researchers in such varied topics as magnetism, dielectrics, and semiconductors, researchers who produced everything from theoretical papers to highly purified crystals. Winning fame already in 1948 with the discovery of the transistor, the Solid State Physics Department quickly inspired the creation of like-minded groups at many other places.[56]

Slater, a part-time consultant for Bell, was among those inspired. He set up his Solid-State and Molecular Theory Group, which hoped to advance not only theory but also applications. Slater's group worked alongside a number of others in allied areas: the Chemistry Department crystallography group, which studied alloys, aided by MIT's pioneering Whirlwind digital computer; the new Lincoln Laboratory, devoted especially to military electronics, where research was done on topics like cyclotron resonance in semiconductors; and Francis Bitter's large-magnet laboratory, useful among other things for semiconductor and metallurgical work.

At Bristol, similarly, already by 1947 a section devoted to experimental work in the solid state was demarcated, including 15 physicists and two assistants, as well as theoretical sections in the same area with 11 more physicists. Alongside this were separate experimental and theoretical sections with Mott and five other physicists working on liquid helium. Obviously, this kind of institutional complex could easily muster the critical number of researchers and students necessary for a true solid-state school, such as had scarcely ever assembled in one place before the war.[57]

As the MIT and Bristol cases show, centers of solid-state work did not necessarily organize themselves into a single department, as at Bell. This was even clearer at General Electric. In 1956 some engineers created the Group on Solid State Applications and Measurements, which met a few times a year, but that was the limit of formal combination. In the 1950s and on into the 1980s, groups doing solid-state work at the GE research laboratories were spread throughout the organization, some in the metallurgy department, others grouped with electronics, others in physical chemistry, and so forth. Places like General Electric could mount a major program in solid state without using the name, simply by supporting the appropriate groups in various departments.[58]

Growth was not uniform. In 1960, even in the United States, there remained major university physics departments that, concentrating their forces on other subjects, almost entirely neglected solid-state physics. Into the 1980s, the field continued to grow, eventually triumphing nearly everywhere, although never without the necessity of struggle to retain its definition and cohesion. But that continuing story is beyond the scope of this book.

Solid-state groups could not have grown so large unless nourished by quantitites of money and manpower. To these subjects we now turn. Since the Americans set the pace during the postwar years, the United States will once again take up most of our attention, although other nations did not lag far behind.

9.4 Matériel

Of all the institutions that drew solid-state physics together, the most important one scarcely existed as an institution. Its official status was weak and variable, its membership poorly defined, its budget nonexistent. This was the Chowder & Marching

Society, otherwise known as the Solid State Physics Advisory Panel of the U.S. Navy's Office of Naval Research (ONR).

Like so may solid-state physics institutions, this one owed its existence and success above all to Seitz. A frequent commuter to Washington after the war, in 1948 he spent an entire month there reviewing the program of the Office of Naval Research, which during the early postwar years had become the central sustainer of scientific research in the universities of the United States. Seitz was still uncertain whether solid-state research would not inevitably migrate into physical chemistry and engineering, parting ways with physics; according to his later recollection, he toyed with the idea of personally taking time away from the field in order to build up a group in the booming new area of particle accelerators. But when he reviewed the research that ONR was funding, he found to his surprise that physicists remained at the center of solid-state work.[59]

Around this time an embryonic solid-state physics subpanel was set up under the navy's Research Development Board; shortly after the outbreak of the Korean War in 1950, this group, under Seitz's leadership, was made an official advisory panel of the ONR.[60] There was no budget, and the members had to pay for their travel and other expenses out of whatever resources they could command, which usually meant a government grant. Nevertheless, the group was quite active. Two, three, or even four times a year, members would take a train or a DC-3 airplane to some research site, especially isolated naval stations where solid-state work needed a lift, but also universities and other centers. In 1951, for example, the Chowder & Marching Society convened at MIT; at the remote naval test station in the desert at Inyokern, California; at the navy center in San Diego; and at the California Institute of Technology. Local solid-state scientists would present talks on their

Office of Naval Research Panel on Solid State Physics in China Lake, California, 9–10 April 1951. (AIP Niels Bohr Library, courtesy of Frederick Seitz)

research, intense discussions would follow, and then people would sit down over a beer for other kinds of work.[61]

The original intent was to help out the navy, and indeed everyone else, by carrying news of what was happening in solid-state physics from place to place. Essentially an ad hoc and self-appointed group, the panel had neither the right nor the wish to tell government officials how to spend their research money. However, the officials who had to make these difficult decisions were intelligent and responsible men, quick to seek advice from individuals such as they met on the panel; that was why Lawson McKenzie and other navy officials sympathetic to academia had sponsored the panel in the first place. Not just ONR listened to the members, for other government agencies began to show an interest in supporting fundamental science. For example, the Air Force Office of Scientific Research in 1952 established the Solid State Sciences Division, which began to pour out millions of dollars to universities, and by the latter 1950s this office was spending more than the ONR on solid-state physics.[62] To help everyone keep in touch, ONR officials invited their counterparts in other agencies to use the panel's services, so men from the science funding offices of the air force, the army, the Atomic Energy Commission (AEC), and the National Science Foundation began coming along to panel meetings.

A list of members of the Solid State Physics Panel in early 1957 shows officials from each of these agencies joining the navy representatives, along with a balanced list of 10 people from universities, 9 from industry, 7 from in-house government laboratories (most but not all navy), and 4 from government-contract laboratories (e.g., the AEC's Brookhaven National Laboratory). This membership list had been drawn up with a discriminating eye: it included one or at most two of the leading solid-state people from each leading American institution. Without particularly trying to do so, such a group could make sure that researchers in their field were as smoothly coordinated as a football team.[63]

On one occasion, the panel did decide to take a stand as a group. When they met at Oak Ridge National Laboratory in the spring of 1957, they talked to one another about a threatening development. Government support of research, which had grown wonderfully in the past decade under the leadership of the armed services, now seemed gravely endangered by sharp cutbacks that President Eisenhower was ordering for the military budget. The group decided to appoint a subpanel, chaired by Robert Sproull of Cornell, to plead the case for funding solid-state research, and above all fundamental research in the universities.

Sproull's report came to an ominous conclusion: "The stifling of fundamental research by lack of support inevitably threatens the development of new and improved weapons and leaves the country vulnerable to technological surprise." Completed in late September, this report landed on desks in the higher reaches of government in October 1957, just as *Sputnik* came circling overhead. The Russian satellite was the biggest technological surprise of the decade, and a frightening one for most Americans. Suddenly money began to pour forth with unheard-of largesse. The panel's report had arrived at exactly the right time to make sure that an ample share of the new riches would flow into university-based, fundamental, solid-state science.[64]

As will happen with any informal group, great success spelled an end of sorts for the panel. Pressure to add members and to take on a more formal status gradually mounted. When the Department of Defense created the high-level Advanced Research Projects Agency (ARPA) to help mobilize American science in reply to *Sputnik,* ARPA provided a secretariat for the Solid State Physics Panel. This still seemed too informal, so in 1962 the panel was officially transferred to the National Research Council of the National Academy of Sciences. By this time, the group comprised 28 members from universities, 19 from industrial and nonprofit institutions, and 18 from government, and its meetings had become large scientific conferences that sometimes led to published proceedings. The Solid State Physics Panel of the National Research Council was a Chowder & Marching Society no longer.[65]

The post-*Sputnik* burst of funding brought American solid-state research to the point where its identity as a field was literally set in concrete, in a series of buildings wholly dedicated to the study of solids. These new laboratories embodied the unifying tendency of the field—exactly the opposite of the separatist tendency often associated with specialization—the urge to bring traditionally distinct disciplines together literally under one roof. A model had been established early on at Los Alamos, where physicists, chemists, and metallurgists had been pushed together by the urgent pressures of the Manhattan Project, and this model had been perpetuated by Cyril Stanley Smith in his Institute for the Study of Metals in Chicago. The need for such interdisciplinary work was increasingly clear with new materials becoming indispensable for everything from microelectronics to jet turbines, from nuclear reactors to ballistic missiles. As Sproull's report had put it, "In a great variety of problems in the development of military equipment and weapons systems the limiting factor at the crucial point is the *material.*" Only an interdisciplinary laboratory could break through these barriers.[66]

In the late 1950s, a variety of such pleas for materials laboratories came before government officials and other members of the ONR Solid State Physics Panel. The idea ricocheted through a maze of offices and committees and was enthusiastically endorsed—for example, by the Department of Defense's top scientific advisory board (where the ubiquitous Seitz was a key member). In 1960 ARPA took on the task of funding the plan, with some help from the Atomic Energy Commission and other agencies; eventually, a dozen laboratories went up around the United States. The spirit behind this, the union of physicists, chemists, and metallurgists, was displayed perhaps most concretely at MIT, where the new Center for Materials Science and Engineering was built in the middle of campus, physically as well as intellectually accessible to each of the departments involved. There was an even higher level of integration when ARPA brought the directors of the various laboratories together each year for intensive discussions.

Materials science, as the new name implied, was not identical with solid-state physics. The search for funding, identity, and status led to constant border shifts among physicists, chemists, metallurgists, and engineers, so that even the chain of grand new laboratories did not guarantee a settled position to solid-state physics. The laboratories did guarantee that, under whatever name, the research would proceed more swiftly than ever.[67]

Up to this point, we have watched the emergence of some particular institutions, ranging from textbooks to the ARPA laboratories program. Next we shall take a look at the field as a whole. How many solid-state physicists were there, where did they work, how much money did they consume, how many papers did they produce in return? There is no direct way to count solid-state physicists prior to the time when the definition of the field coalesced. As will be seen, counting published papers suggests that in the early 1920s roughly one-tenth of physicists' effort went into what would later be called solid-state research, and that this rose to over one-quarter by the later 1950s. From the early 1950s on, good statistics on the people themselves were compiled for the United States, and with a little faith in extrapolations we can also estimate figures for the rest of the world.

After the Korean War broke out, the United States government set out to inventory its stock of scientists. The survey categorized solid-state physics separately, although only in part; such specialties as "fluorescent materials," which fell under the heading "electronics," and "photoelectric phenomena," under "optics," no doubt included some solid-state physicists. Taking all that into account, the survey can be used to estimate that in the United States there were some 300 to 400 Ph.D. solid-state physicists in 1951. (Including graduate students and other workers not holding a doctorate would about double the number.) This was roughly one-tenth of the total of all American Ph.D. physicists.[68]

Solid-state physics was thus by no means a dominating field. The unquestioned leader in 1951 was nuclear physics, with more than twice as many Ph.D.s as solid state; it was followed by electronics, which at that time was still concerned mainly with electrons in vacuum. These fields also seemed to hold the most promise for the future: graduate students flocked especially to nuclear physics and after that to electronics. In 1951 over 60% of nuclear physicists were under 35 years of age, compared with 40% of solid-state people, who were at about the average age of all physicists. The typical solid-state physicist was younger only than people in such stodgy fields as optics, not yet reinvigorated by the laser.

A 1960 survey produced a strikingly different picture. The number of American physicists had more than doubled, but solid-state physics had done still better. There were now roughly 2000 Ph.D. solid-state physicists in the United States, making up one-fifth of all physicists. This was twice the proportion of 1951 and had reached almost as high a proportion as nuclear physics. (By 1970, solid state would be well ahead.) Solid-state physics remained about average in its age structure, however, so we must conclude that the field had grown not only in the usual way through the graduate schools, but also by aggressively recruiting scientists from other fields of physics and even from chemistry and engineering.[69]

One characteristic of solid-state physics held steady. Both in 1951 and in 1960, roughly one-quarter of solid-state physicists were employed in higher education and one-half in industry (the other quarter were mostly in government laboratories). This was the reverse of the relation between academia and industry for physics as a whole. In industrial laboratories, then, solid-state physics was the largest field almost from its birth.[70]

Statistics on funding are less complete but equally striking. Partial surveys of various kinds show that funds for physics went up like a rocket. The Sproull report estimated that government-contract money for solid-state physics more than doubled in the few years between 1952 and 1957; during the same period at three leading universities, the annual funds for solid-state research from government contracts rose from a range of $50,000 to $100,000, to one of $150,000 to $400,000 apiece. In the next stage, between 1956 and 1960, a study for Congress found that total federal support for basic research in all fields of physics rose under the influence of *Sputnik* from $100 million to $250 million; the Department of Defense and the Atomic Energy Commission were the largest patrons. Federal government money was about half of the total available for solid-state and other physics research, and the nongovernment funds were rising at least as fast. For example, a U.S. Navy survey found that industry's expenditures for basic research increased more than fourfold between 1947 and 1957. Even allowing for moderate inflation, it seems that a physicist in the early 1960s used, very roughly speaking, twice as much money as one in the late 1940s and perhaps three times as much as one between the world wars. In this general increase, however, the field of solid-state physics seems to have lagged behind some other fields, receiving nothing like the money that was pouring into high-energy particle accelerators. At least some solid-state research could be done as cheaply in 1960 as in 1930.[71]

A detailed study was made in 1962 for "condensed matter"—the newly popular name included liquids and, like "materials science" in a different manner, reflected a persistent uncertainty as to whether "solid-state physics" was the best way to group subfields. The study estimated that in American universities, funds from all sources came to some $36,000 per condensed-matter Ph.D. per year. That was about three times a typical physics professor's salary, so that each professor represented an important investment in graduate students, technical assistants, and apparatus. In government laboratories, the average condensed-matter Ph.D. cost $83,000, and in industry $57,000, which was close to the weighted average.

If we apply the figure of roughly $50,000 per physicist to our estimate of about 2000 American solid-state physicists, we get an enterprise with a total budget for 1960 on the order of $100 million.[72]

In a still cruder way, these figures can be extended to other countries. Some studies in the early 1960s indicated that the total scientific and engineering personnel in the United States was about equaled by the total in the Soviet Union. The other major scientific countries—with Japan and Great Britain in the lead, followed at some distance by Germany and France, and some distance farther behind by others—together accounted for at least as many scientists and engineers as either of the two great powers. The funds spent per scientist in the various countries, adjusted for differences in costs of research, were found to be roughly comparable. The rates of increase also seem to have been of the same order; the United States was not alone in furthering research.[73]

This is compatible with the only world figures available for solid state physics itself: a study of the publications originating in various countries in 1961. The study found that the United States and the Soviet Union each accounted for about one-quarter of all solid-state physics papers; Great Britain and Japan, roughly comparable to each other, together accounted for another quarter; the remaining quarter came from West Germany and France (each publishing about one-twentieth of all papers), followed in descending order by the Netherlands, Czechoslovakia, Poland, Denmark, and other countries.[74]

If we assume that the number of publications crudely reflected the number of workers and their funding, and extrapolate from our American figures, then around

1960 the worldwide solid-state physics enterprise employed some 6,000 to 10,000 highly trained physicists, and spent several hundred million dollars. In size, then, the field was on a level with a single second-rank international corporation.

9.5 A Unified Way of Thinking

In this final section, we come to the social aspects of solid-state physics that were closest to the field's intellectual structure. Again, we will look into the rival forces pulling toward unity or separation. We shall find a compromise between these, a sort of confederation in which each part kept its own individual identity, not unlike the members of the Chowder & Marching Society. We shall find reason to suspect that this structure reflected deep characteristics of the field, distinguishing it in basic ways from other sorts of physics.

Solid-state physics was from the outset unusual as a field formed by bringing together widely differing specialties. Once formed, it faced a risk that the advance of specialization would tear it apart again. Seitz and Turnbull remarked in 1955 that because of the expansion of knowledge, "solid state physicists are finding that, in order to make significant contributions, it is necessary to concentrate their efforts in narrower fields than formerly."[75] We have already seen many of the mechanisms, including the *Solid State Physics* review volumes in which this statement appeared, that were designed to keep the subject unified and that by and large succeeded in their task. Yet at the same time, the specialties could retain or even create distinct social structures of their own.

Previous chapters have discussed the rise of communications networks such as a "many-body community," a "color center community," and a "magnetism community." A particularly interesting example is magnetism, which before the Second World War had seemed almost as strongly inclined as x-ray crystallography to form a specialty all by itself. After the war the specialty continued to grow, the workers assembling, for example, in a series of international conferences that began in 1950 with a meeting arranged by Néel in Grenoble, and in American conferences held annually from 1955 on. However, the subjects discussed at such meetings often came to include not molecules and nuclei but only solids; magnetism was increasingly seen as a part of solid-state physics, although a part that could stand confidently on its own feet.[76]

Even more striking than the continuing life of old specialties was the rise of powerful new ones, notably semiconductor physics. With the coming of the transistor, this mushroomed from a minor topic, scarcely worth calling a specialty, to a separate and major enterprise that nearly overshadowed other solid-state topics. For example, the General Electric laboratories in the 1950s created, not a solid-state physics division, such as existed at Bell Labs, but the vigorous Semiconductor Branch, organizationally distinct from the branch that studied metals. Similarly, in 1952 the Soviet Academy of Sciences took the semiconductor work that had always been a main theme of Ioffe's Leningrad institute, and set it up in the independent Semiconductor Institute. By the end of the 1950s, semiconductor physics possessed a great variety of institutions all its own. There were international conferences—

notably one held in Reading, England, in 1950, followed by a series beginning in Amsterdam in 1954—as well as textbooks, an annual series of volumes on *Progress in Semiconductors,* and so forth.

Yet it became clear that the subject belonged within a larger structure, the family of solid-state physics. For example, in West Germany during the 1950s the Physical Society had loosely grouped specialized sections devoted to various solids topics under the rubric "Semiconductor Problems," but in 1959, when not only Semiconductors but also other, still newer sections such as Metal Physics and Magnetism were growing rapidly, their subject was renamed *Festkörperprobleme.* In parallel with this, an annual series of volumes on semiconductors, begun in 1954 by Schottky, was changed in 1961 to a general solid-state series. The transformation of the international semiconductor abstracting journal into *Solid State Abstracts* in 1960 pointed in the same direction.[77]

It is not easy to see at a glance precisely how the various specialties came to be called solid-state physics. Fortunately, there were always certain people whose profession was to think hard and long, year in and year out, about exactly how physics should be subdivided and categorized. We will now inspect their conclusions.

Science Abstracts, for most of the first half of the twentieth century, grouped published articles under broad categories such as "Heat" and "Electricity." The judgments about subdivision were continually and vigorously criticized by scientists, who wanted to look up subjects of interest as easily as possible. The classification system was frequently reviewed and adjusted as the years passed, giving some clues as to how, when people were forced to make distinctions, they divided up physics.[78]

As late as 1939, the subjects we would group under solid-state physics were widely dispersed among the categories of the abstracts. For example, subjects like thermal conductivity and dielectric constants appeared as headings that were subdivided into solids, liquids, and gases. It was not, then, the state of matter that came in front, but the type of phenomenon and even the experimental technique. Theory took a back seat. A significantly different arrangement could be seen in the first complete postwar year (abstracts of 1946 papers published in 1947), when liquids, gases, and solids were sometimes set apart as main categories. Two major headings, "Structure of Solids" and "Elasticity. Strength. Rheology," were almost entirely concerned with solid-state research. However, both the theory of solids in general, and lattice dynamics in particular, still lay under the heading "Atomic and Molecular Structure," mixed in with such subjects as molecular spectra and nuclear theory. Crystallography, including subjects like crystal growth and the various physical properties of crystals, appeared as an entirely separate science on a par with physics.

Jumping ahead a decade to 1957, we find a modest regrouping in the Index of Subject Headings in *Physics Abstracts* (as the appropriate section of *Science Abstracts* was now named). Here "Theory of Solid State" possessed a category of its own, separated from molecules and nuclei. Semiconductors (an important category) and a number of related solid-state topics were now separated from electrolytes and the other traditional studies listed under "Electric Conduction." But other topics remained isolated—for example, studies of the radiowave spectra of solids, which were lumped with spectra of atoms. Crystallography, including such classic solid-state topics as color centers, stayed independent. In short, while in 1939 solids had appeared in the index on only the lowest level, always subsidiary to physical effects such as "Conduction" and "Polarization," by 1957 the idea of the solid state had worked its way into the middle levels of the system.

At last, in 1960, "Solid-State Physics" became a heading at the highest level under physics. The rubric covered almost everything having to do with solids, from magnets to rubber. Even crystallography had lost its independence and become a subheading. The editors gave a strictly intellectual reason for the change: "A feature of modern work on the solid state is the way in which formerly unrelated properties of solids are being shown to have common causes."[79]

The transformation could come even more abruptly. For example, the 1957 McGraw-Hill *American Institute of Physics Handbook* divided up physics in the traditional fashion among mechanics, optics, electricity, magnetism, and so on, with solid-state subjects scattered among all these categories. A year later the same publisher issued another comprehensive handbook of physics that inverted the arrangement, giving a major division to "the solid state," which included everything from crystallography to phase transformations. The difference in viewpoint was noted almost unconsciously as early as 1952 in a survey article by Slater. The physics of solids had a long history, he remarked, for even at the turn of the century, mechanics, heat, and so forth had each included "solid-state aspects." Proceeding to his survey of the modern situation, Slater said he would "go over various classical fields of solid-state physics—mechanics, heat," and so forth; now these had implicitly become aspects of the category "solids" instead of the other way around.[80]

This question of categorization in publications, lying as it does right at the boundary between intellectual and social processes, is so revealing of the evolution of the field that we studied it in some detail. Applying retrospective criteria to abstracts, Jerome Rowley aided by Kris Szymborski identified papers in *Physics Abstracts* that would now be considered solid-state physics, looking at the volumes for the three years 1921 to 1923 and also for 1955 to 1957.

First and most striking, we found an increase, as already mentioned: from 12% of all papers in the early 1920s, solid-state papers had risen to 29% in the late 1950s. More subtle but equally important was the shift in categorization. Consider those subheadings, at the lowest level in *Physics Abstracts,* that had at least 50 papers in them (for the sum of the years 1921 to 1923 or for 1957 alone). Among these substantial categories we looked at the headings that included at least three solid-state papers. In the early 1920s, solid-state physics was widely dispersed among even these categories: compared with headings under which more than half the papers were in solid-state physics, those in which solid state was present but only as a minority were three times more numerous. A study of *Physikalische Berichte* for the 1920s showed that in the other major abstracting journal, too, solid-state papers were spread thinly among various headings. The proportion had exactly reversed by the late 1950s. That is, of those substantial headings of *Physics Abstracts* under which solid-state physics figured at all, it made up the majority of papers three times more often than when it appeared in only a minority role. This last result should not be surprising, since we have been using postwar criteria for the field; the point is that those criteria would not have defined any coherent entity a generation earlier.

We have seen that solid-state physics had achieved a substantial degree of unity by the end of the 1950s: intellectual unity as a distinct field of science, and social unity as a community with its own set of permanent institutions. Now, it is commonly observed that members of a community tend to show similar ways of thinking. A mentality, an ideology, even a philosophy, may come to be generally shared. This would often be especially true for a community that only with some difficulty pulled itself together and maintained its identity, as was the case for solid-state physics. This poses our final question: to search for shared viewpoints.

Here, as everywhere else throughout this book, only more so, we can attempt nothing more than a first rough cut at the problem in hopes of stimulating other historians to do better. We will restrict ourselves to the Americans and British, and will inspect only the views of a handful of the most senior figures of the field. Even so, we will find no perfect unity of thought. As will presently be seen, it may even be characteristic of solid-state physicists that each retained an individual viewpoint. Nevertheless, there existed several important ideas with which most of the senior figures of the field would agree, and that together formed a consistent way of thinking about solid-state physics on the highest plane.

The first of these commonly held ideas about the field has already been pointed to by the fact that solids gradually moved from a category at the lowest level to one at the highest level. This reflected a deep shift in the approach to research, touched on repeatedly in previous chapters. Up through the early twentieth century, we recall, many physicists followed the program of studying solids in hopes of gaining a deeper understanding of the fundamental forces and units of matter. By probing crystals with light, x rays, electromagnetic and mechanical pressures, extremely low temperatures, and so forth, could one not eventually discover the basic nature of interatomic forces and even the very shape of atoms? As late as 1928, W. H. Bragg advocated the study of solids for that reason: "When we know their various designs we can begin to study those structural forces which hold atoms and molecules together." Yet in the end, the royal road to quantum mechanics lay not through solids but through gases, and particularly their spectra. From the point in 1927 when Sommerfeld began to walk around with a reference book on properties of solids, seeing how many he could calculate using the new quantum ideas, the study of solids became a matter of applying known fundamental theory.

Mott and Gurney, in the preface to their 1940 textbook, distinguished between two "aims of theoretical physics." One aim was to discover the basic properties of atoms, and the other was "to deduce from the properties of atoms the properties of matter as we know them"; it was this latter aim that their book would pursue. Widely accepted during the 1930s, the idea became basic to the definition of the field of solid-state physics. Seitz put it most succinctly in 1960: "The goal of the physics of solids is to understand the various properties in terms of atomic and nuclear theory." A number of authors declared that this was a chief "unifying influence" behind solid-state physics—the principle of pressing theory into every problem, until traditional empirical physics and even cookbook engineering would give way to rigorous fundamental understanding—with physicists sometimes coming almost like missionaries to the heathen metallurgists. As Peierls put it, quantum theory would be used "to bring order into an apparently disconnected list of observed facts."[81]

From the outset, this approach gave the solid-state physicists a galling problem of self-definition. Physics had often taken as its goal the search for fundamental laws, not their application. With quantum mechanics "solved," the next frontier for this program lay in the nucleus. The emerging field of solid-state physics therefore had to define its worth in relation to the aggressive new fields of nuclear and high-energy physics. Many of the reminiscences of solid-state physicists speak of this difficulty (and here it is the reminiscences that count, for we are not asking what actually happened in the 1930s, but what ideas were held later on among the solid-state physicists).

It was recalled that a problem was apparent in Göttingen already before the completion of quantum mechanics. Robert Pohl subsequently explained to British colleagues that he felt "our work was not much noticed at first for we did not relate it to Bohr's atomic model. At that time . . . only things related to Bohr's model were regarded as 'physics.'" Such feelings could only become more acute once quantum mechanics was completed. Felix Bloch, Peierls, and others who came out of Germany brought with them a strong recollection of Pauli's scornful attitude toward detailed elaboration of the electron theory of metals: "dirty physics."[82]

The conflict between the program of applying known theory to the details of solids and the program of searching for fundamental laws took on a practical side when people tried to recruit students and set up solid-state groups in the face of forces pushing for nuclear physics research. Philip Anderson recalled that his fellow graduate students at Harvard, just after the war, looked down on his decision to study solids; "it certainly wasn't until much later that I lost a certain defensiveness about the intellectual respectability of solid state theory." Slater recalled that at MIT "our department was often looked down on, to some extent, by those who felt that no physicist of any imagination would be in any field except nuclear and high-energy physics." Slater was able to create his group only because his superior, K. T. Compton, decided to diversify rather than to promote nothing but nuclear physics, a decision made possible by the Physics Department's large teaching load directed toward engineers and by the strong interest that MIT's industrial sponsors took in applications. Similarly, Seitz recalled that after the war, Illinois Physics Department head Wheeler Loomis had been urged to concentrate entirely on high-energy physics where "the great mysteries" awaited; Loomis hired Seitz, however, in order not to be a one-subject department. At Bristol, according to F.R.N. Nabarro's recollection there was some rivalry between Mott's solid-state group and the cosmic-ray group, which was "successfully exploring the foundations of physics. . . . The problems that Mott's group was tackling were less fundamental." At Cornell in the 1950s, Sproull had a stiff fight on his hands whenever he wanted to hire another solid-state physicist or simply get some space in a building for his group: the nuclear and high-energy people claimed the resources for their own work.[83]

Whether in terms of allocation of resources or of allocation of prestige, the matter remained as Peierls put it in 1955: "The quantum theory of solids has sometimes the reputation of being rather less respectable than other branches of modern theoretical physics." The field did not even seem worth explaining to the public. When Peierls published a popularization of physics, he gave only half a dozen pages to solid-state theory but entire chapters to nuclear physics and mesons. The only thing unusual about this was that he gave any space at all to solids. In contrast to the overwhelming flood of books, magazine and newspaper articles, even comic books and cartoon films that explained nuclear physics during the decade following the bombing of Hiroshima, solid-state physics remained unknown, almost as if too shy to boast of its emerging powers.[84]

Within the physics community itself, however, solid-state physicists gradually began to assert their right to stand on the same level as nuclear and high-energy physicists. They had two possible strategies: to assert that applications were just as worthy as fundamental research, or to assert that solid-state physics was itself "fundamental." Some people asserted both.

The first assertion—a confident claim that solid-state physics was especially close to applications of high practical value—was invariably a feature of programmatic statements about the field. "Results of the solid state sciences are the foundations and the frontiers of much of our present technological strength," proudly declared Smoluchowski in a National Research Council report, and many others said the same. That this idea was true is not the point here; the point is that it was one of the main ideas that gave solid-state physicists an image of the nature and worth of their field.

Considering what has been said in preceding pages about the role of industrial and military interests in the formation of solid-state physics, it is obvious that in the stress on practical applications, the social and intellectual characters of the field merged. Even in its organizations, from the Division of Solid State Physics of the American Physical Society to the Chowder & Marching Society, the most obvious characteristic of the field was its urge to combine fundamental academic science with applied work. Indeed, some pointed out that solid-state physics had "a special place with respect to science and technology in general" because it furnished "increasingly sophisticated tools" important to the advance of all research and engineering.[85]

Some were content to leave the matter at that. Slater, for example, agreed with the nuclear physicists that "everything outside the nucleus" had become "in a sense a finished subject." He simply added that unraveling solid-state problems could be just as "interesting"—that is, just as satisfying in a puzzle-solving fashion.

> It is a different sort of satisfaction from that of nuclear physics, where we may hope to uncover new particles and fundamental laws: but it is no less real a satisfaction, and the pleasure of investigation is coupled with the realization that the problems we are dealing with are fundamental to technology and the practical affairs of daily life.[86]

Others took a bolder stance. Was not solid-state physics, too, in its own way, fundamental? A 1972 National Research Council report said that one could call any research "basic" that was motivated solely by the search for deeper understanding of nature, and such understanding was what many solid-state physicists were after. The report also noted that in this field, more than in many others, the distinction between basic research and applications had become blurred. Another authoritative report on physics, from 1966, put the matter in the most outspoken form. Solid-state physics, the authors insisted, was "a fundamental branch of physics." In fact, "progress in this field could well turn out to be of greater significance to our knowledge of the world than further progress in elementary-particle physics."[87]

What did it mean to say that solid-state physics held fundamental significance for our knowledge of the world? Some sort of new idea was at work here, an idea seldom articulated outright, which implied a major shift in views about what was fundamental, what was knowledge, perhaps even what in truth was the world. Wigner, almost unwittingly, had pointed out the way to this new view already in 1936. The study of solids, he wrote, had recently turned into a second sort of physics, altogether different from the sort represented by nuclear physics. Although he believed that this second sort was "of less importance" because it did not lead toward "fundamental discoveries," he admitted that the study of solids had a unique attraction. It "deals in a scientific way with those subjects with which we must deal in our everyday experience. For example, we are never afraid when dropping a key that it will fly to pieces, as glass would." In short, solid-state physics would study the world as we actually meet it. Thirty years later, Mott gave a lecture in which he took the idea to its conclusion: research into the familiar properties of matter should be as esteemed, as much a proper part of physics, as research into the hidden and unfamiliar.[88]

As Wigner's example of the key and the glass suggests, the interest in everyday properties lies not in what unites things but in what makes them different. Where nuclear and high-energy physics would pursue the dream of working down layer by layer to the simplest forces and particles underneath all phenomena, solid-state physics would take the same urge in the reverse direction, using known laws to encompass ever more diverse phenomena. A fascination with diversity, in short, was another of those ideas that bound the field together.

The underlying feelings were well expressed by W. H. Bragg in his most widely read book:

> The universe is so rich in its variety, the earth and all that rests on it, the waters of the seas, the air and the clouds, all living things that move in earth or sea or air, our bodies and every different part of our bodies, the sun and moon and the stars . . .

That is all clichés, of course, but it is precisely the power of clichés that is under study here, the power of ideas held so widely and deeply that they did not need to be explained but only referred to. Slater perhaps put it best when he said that "atoms by themselves have only a few interesting properties." What attracted him and other solid-state physicists was the things that happened when atoms combined. A distinguishing characteristic of their field, they said, was its "richness," "wealth of phenomena . . . unending possibilities," "unique diversity . . . wide range of subjects," and the like.[89]

It was an ancient approach to nature, this fascination with the diversity that is the most obvious characteristic of our world. Over the long run, this approach held its own against the fascination, equally durable, with the search for hidden unities. To be sure, one of the main programs of solid-state physics was to reduce phenomena to the simple laws of atoms, but it always remained acutely aware that such a program could never be completed. As John Ziman put it, "Merely to *describe* all the creepy-crawly creatures to be seen in an electron micrograph of a thin film of stainless steel is beyond our analytical powers." Even with digital computers at hand, much lay beyond the range of a rigorous solution. A willingness to make approximations according to the case at hand, an appreciation for the limits of per-

Cyril Stanley Smith, 1975. (AIP Niels Bohr
Library, courtesy of C. S. Smith)

fect theory, became another one of those ideas, sharply contrasted with the hopes
of high-energy workers, by which solid-state physicists formed an image of their
field.[90]

These ideas had social corollaries. Above all, the insistence on approximate,
diverse, and everyday phenomena was inseparably linked with the interest in practical applications. Solid-state physics defined itself not only in terms of its contrast
with nuclear and high-energy physics, but also in terms of its affinities with metallurgy and other applied fields. Here, where the diversity of phenomena was overwhelming, analytic thinking always required help from the intuitive approach of
the craftsman. One of the most common clichés found in general writings about
solid-state physics was that the field's traditions extended back to ancient times, if
not indeed—a claim no other field of physics could match—back to the origins of
humanity itself among prehistoric flint-chippers.

Respect for craftsmanship, as part of the definition of solid-state research, suggested a number of ideas. Only one man thought the ideas through carefully, C. S.
Smith. He suggested that "the growth of solid-state physics marks . . . a basic change
in the attitude of physicists toward matter," for henceforth there would be "a concern with far more complex things than have been allowed in the domain of respectable physics in the past." The study of structures on the level of complexity comparable with our own bodies, he said, was philosophically not inferior to the high-
energy physicists' work on the structures of elementary particles or to the cosmol-
ogists' work on the shape of the universe as a whole. What distinguished the study
of solids, rather, was that because of its interest in the most complex and difficult
structures, it must turn to the qualities of thought long associated with craftsmanship and even art.

> A list of types of bricks used in the Hagia Sophia may help one to build an interesting brick wall, but it poorly suggests the great edifice from which they came.

> Science in the past has been almost synonymous with distrust of the senses, but it seems to me that we are now ready to make better use of the other properties of the human brain besides its capacity for logical rigor . . . we should seek a bridge to a more sensual study of whole systems.[91]

Not everyone agreed that solid-state physicists could learn much from artists. It was a more obvious aspect of diversity that was most commonly pointed out. The enormous range of types of solids and their behaviors, and the resulting dispersal of solid-state physics among a multitude of specialties, was reflected in a diverse solid-state community. Here, too, the sharpest contrast was with high-energy physics, notorious for its "big science" of monster accelerators and papers with dozens of co-authors. Solid-state physics, Bardeen explained, "is 'little physics,' requiring only modest outlays for equipment. . . . Students can design and carry out their own experiments, rather than just be a member of a large team." Of course, Bardeen's own work on the transistor and superconductivity showed how teamwork could pay off in solid-state physics, but those teams had still been small, temporary, and loosely coordinated. The variety of the phenomena under study thus had a counterpart in the individualism of the researcher, a sort of independence that many found appealing. Some felt, Seitz recalled, that solid-state physics "was a field where a person could still be himself and not a slave to a machine."[92]

That idea could have broad ramifications. Smoluchowski, for one, suggested that Americans gave too much emphasis to large-scale organized research, and should instead give special support to the "admittedly rare scientists who are capable of individual creative effort." A still broader view was expressed by Seitz when he suggested that the strength of American science as a whole must come from "diversity," from evolving much as a biological species evolves by making use of a profusion of individual variations. Seitz's nightmare was that "our species may be overwhelmed by homogeneity," but he put his trust in freedom and the range of different approaches that it allowed. C. S. Smith more explicitly related solid-state physics to views on all of society. "As in matter, so in the structure of ideas," he believed, local anomalies could become a nucleus for widespread changes of structure—a new crystal form, a new way of thinking, a social revolution. Like Seitz, he was attracted by the idea of a complex, diverse society, where new things could keep arising more or less as solid-state physics itself had grown since the 1930s.

Most explicit of all in relating the ideas of solid-state physics to political thought was Slater. He, too, pointed to the importance of a local nucleus for change; "The exceptional man produces ideas, the multitude follows them." Such a viewpoint, he believed, was fundamentally opposite to ideas adopted by certain thinkers from Hobbes to the Marxists, the view that society can be regarded as a predictable machine. Slater denied such reductionism because of

> the enormous complexity of all living things. . . . To one with the experience that I and my colleagues have, it would seem preposterous to believe that even with many years of further work we could predict by means of a computer what an individual person would do. The more one works with these things, the more one becomes impressed with the infinite variety of nature.[93]

U.S. Air Force officers visit the physicists of the University of Illinois Control Systems Laboratory, 1958. Frederick Seitz, Physics Department chairman, is second from left in the back row, next to Wheeler Loomis; Chalmers W. Sherwin (at Urbana, 1946–1964) is the last on right in the same row. Andrew Longacre (at Urbana, 1947–1957) is second from left in the front row.

We have come a long way from Wigner's comment on the practical difference between a key and a glass. Probably few solid-state physicists would have agreed to come the whole distance. Nevertheless, ideas regarding the importance of attention to the variety of the real world, of mingling theory with applications, of craftsmanship and art, of diversity and individuality, were all compatible with one another. They were compatible also with the intellectual form of solid-state physics and with its social institutions, which always worked to bring about a unity not of an imposed and artificial kind but of the kind sometimes found in art, where each part retains a separate identity while fitting into the whole.

Taken together, the ideas formed a familiar world view. Stress on the individual, especially the outstanding individual, and on diversity within the boundaries of an overall agreed-on system, is "conservative" in one of the older senses of the word.

Before calling solid-state physicists conservative, we must recall that it was always their intention to bring about tremendous, even revolutionary change in the life of people, and that in their own patient way they did that. This is akin to "conservatism" as it has evolved in countries like the United States, and particularly in industrial corporations. When the status quo is in fact a state of ceaseless technological innovation, conserving the status quo is compatible with a type of dynamism.[94]

The foregoing, incomplete and speculative, is meant not to answer questions about solid-state physics but to provoke others to take up the search. No structure of a solid, however complex, is remotely as intricate as the history of the field. This book is only an introduction to the subject, therefore, rather than a definitive history; no definitive history will ever be written, for alongside each intellectual and social interaction that we manage to understand there will be others calling for our investigation. In this regard, physicists and historians can share the same sort of quest that came to C. S. Smith in a recurrent dream. In his younger years at the American Brass Company, the metallurgist had often been in factories with great rolling mills, with dancing and clanging presses, with red-hot copper rod snaking back and forth as it was squeezed into wire. Long after, he would still dream "of wandering through complex assemblies of industrial buildings full of such machines, in search of something I never find."[95] If no final unitary solution can be expected, at each stage there will be something new and fascinating to understand.

Notes

In preparing this chapter, I have been helped by all members of the International Project in History of Solid State Physics, but I have a particular debt for useful comments to Conyers Herring, Paul Hoch, Lillian Hoddeson, Steven Keith, and Günter Küppers. I have also been helped by an unpublished paper by Yves Gingras, Université de Montréal. A large amount of statistical work was done by Jerome Rowley with help from Kris Szymborski.

1. B. L. Griffith, H. G. Small, J. A. Stouchill, and S. Dey, "The Structure of Scientific Literature, II: Toward a Macro- and Micro-structure of Science," *Science Studies* 4 (1974): 339-365, pp. 342–343. This and other works are summarized in Eugene Garfield, *Citation Indexing* (New York: Wiley, 1974); for 1973 clusters, see 122.

2. Co-citation maps and data for 1920 to 1929 kindly furnished by Henry Small et al. of the Institute for Scientific Information, Philadelphia; his work was performed under a grant from the National Science Foundation, Program in History and Philosophy of Science.

3. For example, Gerard Lemaine et al., eds., *Perspectives on the Emergence of Scientific Disciplines* (The Hague: Mouton; Chicago: Aldine, 1976); D. O. Edge and Michael J. Mulkay, *Astronomy Transformed: The Emergence of Radio Astronomy in Britain* (New York: Wiley, 1976); N. C. Mullins, "The Development of a Scientific Specialty: The Phage Group and the Origins of Molecular Biology," *Minerva* 10 (1972): 51–82; J. Lankford, "Amateurs and Astrophysics: A Neglected Aspect in the Development of a Scientific Specialty," *Social Studies of Science* 11 (1981):275–303.

4. For example, Ernest Rutherford, James Chadwick, and C. D. Ellis, *Radiations from Radioactive Substances* (Cambridge: Cambridge University Press, 1930); George Gamow, *Constitution of Atomic Nuclei and Radioactivity* (Oxford: Clarendon Press, 1931); Georg von Hevesy and F. A. Paneth, *Lehrbuch der Radioaktivität* (Leipzig: Barth, 1931); Marie Curie, *Radioactivité* (Paris: Hermann, 1935); Franco Rasetti, *Elements of Nuclear Physics* (New York: Prentice-Hall, 1936); Carl-Friedrich von Weizsäcker, *Die Atomkerne* (Leipzig: Akademische Verlagsgesellschaft, 1937); Instituts Solvay, Conseil de Physique [Seventh, 1933], *Structure et propriétés des noyaux* (Paris: Gauthier-Villars, 1934); E. Bretscher, ed., *Kernphysik: Vorträge gehalten am Physikalischen Institut der E.T.H. Zurich im Sommer 1936 . . .* (Berlin: Springer, 1936); H. A. Bethe and M. S. Livingston, *Reviews of Modern Physics* 9 (1937) (the series of review articles that were to nuclear physics what Sommerfeld and Bethe's was to quantum theory of solids); and others.

5. Most useful was the collection of bibliographies in the Niels Bohr Library, AIP, supplemented by the usual biographical sources. Full data were easiest to find for Americans, so they tend to dominate the sample: L. Apker, P. Anderson, J. Bardeen, H. Bethe, F. Bitter, F. Bloch, M. Born, R. Bozorth, W. Brattain, L. Brillouin, W. H. Bragg, W. L. Bragg, W. Brown, J. Burgers, L. Cooper, P. Debye, S. Dushman, P. Ewald, L. Giulotto, C. Herring, V. Johnson, A. Kip, C. Kittel, W. Kohn, L. Landau, A. Landé, M. von Laue, K. Lark-Horovitz, D. Lazarus, I. Langmuir, B. Matthias, W. Meissner, N. Mott, W. Pauli, R. Peierls, D. Pines, R. Pohl, J. Schrieffer, F. Seitz, W. Shockley, D.

Shoenberg, F. Simon, J. Slater, C. Slichter, R. Smoluchowski, A. Sommerfeld, J. Van Vleck, G. Wannier, E. Wigner, C. Zener.

6. For valuable insights into analogies between changes of state in social and physical systems, see C. S. Smith, *A Search for Structure: Selected Essays on Science, Art, and History* (Cambridge, Mass.: MIT Press, 1981), chaps. 13, 14.

7. Community: Besides other chapters of this volume, see Michael Eckert, "Propaganda in Science: Sommerfeld and the Spread of the Electron Theory of Metals," *Historical Studies in the Physical Sciences* 17, no. 2 (1987): 191–233; S. T. Keith and Paul K. Hoch, "Formation of a Research School: Theoretical Solid State Physics at Bristol 1930–54," *British Journal for the History of Science* 19 (1986): 19–44; F. Seitz, in "Biographical Notes," in *Beginnings,* 91; John C. Slater, *Solid-State and Molecular Theory: A Scientific Biography* (New York: Wiley, 1975), 5–7; and, especially, Krzysztof Szymborski, "The Physics of Imperfect Crystals: A Social History," *Historical Studies in the Physical Sciences* 14 (1984): 317–355. Paul R. Josephson, "The Leningrad Physico-Technical Institute and the Birth of Russian Physics" (Ph.D. diss., Massachusetts Institute of Technology, 1986).

8. National Research Council, Physics Survey Committee, *Physics: Survey and Outlook* [Pake Report]. *Reports of the Subfields . . .* (Washington, D.C.: National Academy of Sciences, 1966), 147.

9. For x rays, see the cogent remarks by Cyril S. Smith, "A Historical View of One Area of Applied Science—Metallurgy," in National Academy of Sciences, Panel on Applied Science and Technological Progress [Harvey Brooks, chair], *Applied Science and Technological Progress. A Report to the Committee on Science and Astronautics, U.S. House of Representatives,* Committee print (Washington, D.C.: Government Printing Office, 1967), 57-71, p. 66–67. For quantum mechanics, see Chapter 2. A. Smekal, "Aufbau der Zusammenhängende Materie," in *Handbuch der Physik* (Berlin: Springer, 1933), 24: pt. 2, 795–922. See, for example, W. G. Burgers, "How My Brother and I Became Interested in Dislocations," in *Beginnings,* 126. W. H. Bragg, quoted in Congrès International de Physique [London, 1934], *L'Etat solide de la matière* (Paris: Hermann, 1936), 12.

10. W. Voigt, *Lehrbuch der Kristallphysik* (Leipzig: Teubner, 1910); see John C. Slater, *Quantum Theory of Molecules and Solids,* vol. 3: *Insulators, Semiconductors and Metals* (New York: McGraw-Hill, 1967), 3.

11. France: Collège de France, *Annuaire* 14 (1914): 65; Michel Biezunski, "La Diffusion de la théorie de la relativité en France" (Ph.D. diss., University of Paris VII, 1981). Weissenberg: *Phys. Zs.* 1927–1930, information furnished by M. Eckert.

12. Smekal (n. 9). John C. Slater and Nathaniel H. Frank, *Introduction to Theoretical Physics* (New York: McGraw-Hill, 1933).

13. Instituts Solvay, Conseil de Physique, *Le Magnétisme* (Paris: Gauthier-Villars, 1932). *Leipziger Vorträge:* P. Debye, ed., *Magnetismus* (Leipzig: Hirzel, 1933). For the Strasbourg "Réunion sur le Magnétisme," see L. F. Bates, "A Link Between Past and Present in European Magnetism," *Contemporary Physics* 13 (1972): 601–614, p. 610, who cites *Le Magnétisme* (Paris: Institut International de Coopération Intellectuelle, 1940). Textbooks: R. Becker and W. Döring, *Ferromagnetismus* (Berlin: Springer, 1939); F. Bitter, *Introduction to Ferromagnetism* (New York: McGraw-Hill, 1937); Lothar Nordheim, "Quantentheorie des Magnetismus," in *Müller–Pouillets Lehrbuch der Physik,* 11th ed., ed. A. Euchen et al. (Braunschweig: Vieweg, 1934), 11: pt. 4, 798–876; E. C. Stoner, *Magnetism and Atomic Structure* (London: Methuen, 1926); Stoner, *Magnetism and Matter* (London: Methuen, 1934); J. H. Van Vleck, *The Theory of Electric and Magnetic Susceptibilities* (New York: Oxford University Press, 1932); Pierre Weiss and Gabriel Foëx, *Le Magnétisme* (Paris: Colin, 1926), and others.

14. The 1920s co-citation study used only physics journals, not physical chemistry or the *Zeitschrift für Kristallographie,* so it cannot tell where general crystallography, aside from x rays, stood in relation to physics. Frederick Wright's 1919 statement is quoted in G. Phair, *American Mineralogist* 54 (1969): 1233–1243, quoted in turn by C. Frondel, in *Crystallography in North America,* ed. Dan McLachlan, Jr., and Jenny P. Glusker (New York: American Crystallographic Association, 1983), 19. (Wright eventually joined the Mineralogical Society.) Russians (Fedorov, and Wulff following Voigt) quoted by A. V. Shubnikov, in *Fifty Years of X-ray Diffraction,* ed. P. P. Ewald (Utrecht: Oostehoek's Uitgeversmij, 1962), 494. See Paul Forman, "The Discovery of the

Diffraction of X-rays by Crystals: A Critique of the Myths," *Archives for the History of the Exact Sciences* 6 (1970): 38–71.

15. Ewald (n. 14). For the British x-ray crystallography community, see also John Law, "The Development of Specialties in Science: The Case of X-ray Protein Crystallography," in Lemaine et al. (n. 3), 123–152; W. H. Bragg and W. L. Bragg, *X-rays and Crystal Structure* (London: Bell, 1915); Max Born, *Dynamik der Kristallgitter* (Leipzig: Teubner, 1915); P. P. Ewald, *Kristalle und Röntgenstrahlen* (Berlin: Springer, 1923); Charles Mauguin, *La Structure des cristaux déterminées au moyen des rayons X* (Paris: Presses Universitaires de France, 1924); Ralph Wyckoff, *The Structure of Crystals* (New York: Chemical Catalog Co., 1924). Note also Arthur Schoenflies, *Theorie der Kristallstruktur* (Berlin: Bornträger, 1923), a new version of his classic *Kristallsysteme und Kristallstruktur* of 1891.

16. Carl Hermann, ed., *Internationale Tabellen zur Bestimmung von Kristallstruktur,* 2 vols. (Berlin: Gebr. Bornträger, 1935). For the *Zs. Kris.,* see P. P. Ewald, "Personal Reminiscences," *Acta Crystallographica* A24, pt. 1 (1968): 1–3; Ewald, in Ewald (n. 14), 493–497. For congresses, see Ewald, "Some Personal Experiences in the International Coordination of Crystal Diffraction," *Physics Today* 6, no. 12 (December 1953): 12–17; P. P. Ewald, interview with C. Weiner, 1968, AIP. International Congress of Physics [London, 1934], *The Solid State of Matter,* vol. 2 (London: Physical Society, 1935), also published as Congrès International (n. 9). Bragg and Bernal, quoted in Keith and Hoch (n. 7).

17. Instituts Solvay, Conseil de Physique, *Conductibilité électrique des métaux et problèmes connexes* (Paris: Gauthier-Villars, 1927). P. P. Ewald, in Ewald, Th. Pöschl, and L. Prandtl, *The Physics of Solids and Fluids,* trans. J. Dougall and W. M. Deans (London: Blackie, 1930), revised from *Müller–Pouillets Lehrbuch der Physik* (n. 13). N. F. Mott and H. Jones, *The Theory of the Properties of Metals and Alloys* (Oxford: Oxford University Press, 1936); N. F. Mott and R. W. Gurney, *Electronic Processes in Ionic Crystals* (Oxford: Oxford University Press, 1940). For J. C. Slater's plans, see his *Quantum Theory of Matter* (New York: McGraw-Hill, 1951), viii. F. Seitz, *The Modern Theory of Solids* (New York: McGraw-Hill, 1940).

18. For Wigner's course, see F. Seitz, "Eugene Wigner—A Tribute on His Seventieth Birthday," *Physics Today* 25, no. 10 (October 1972): 40–43; Chapter 3. F. Seitz and R. P. Johnson, "Modern Theory of Solids," *Journal of Applied Physics* 8 (1937): 84–97, 186–204, 246–260.

19. Seitz (n. 7), 89. Seitz (n. 17), 1; see, for example, p. 420 on band theory: "A large part of this discussion is qualitative and probably will remain so until computational technique has been developed much further."

20. F. Seitz, *The Physics of Metals* (New York: McGraw-Hill, 1943).

21. "We did not feel at home," L. A. Jones, quoted in Wheeler P. Davey, "Report of Conference on Applied Physics," *Review of Scientific Instruments* 7 (1936): 119. See S. R. Weart, "The Rise of 'Prostituted' Physics," *Nature* 262 (1976): 13–17; Weart, "The Physics Business in America, 1919–1940: A Statistical Reconnaissance," in *The Sciences in the American Context: New Perspectives,* ed. Nathan Reingold (Washington, D.C.: Smithsonian Institution Press, 1979), 295–358.

22. Proceedings and Minutes of the American Physical Society (APS), 1930–1931, records of the APS, Niels Bohr Library, AIP.

23. Dan McLachlan, "The Organizations of American Crystallographers," in McLachlan and Glusker (n. 14), 125–133, see also 153–55.

24. P. P. Ewald, "The Beginnings of the Union of Crystallography," in McLachlan and Glusker (n. 14), 134–135; Ewald, "International Status of Crystallography, Past and Future," *Nature* 154 (1944): 628–631; Ewald (n. 14), 700–705.

25. McLachlan and Glusker (n. 14), 127, 130. Ewald (1944) (n. 24).

26. Council Minutes of the APS, copy at APS, 241st meeting, May 1941, New York City.

27. "Current Trends in the American Physical Society," *Journal of Applied Physics* 14 (1943): 437–442, p. 439. Council Minutes of the APS, 253rd meeting, January 1943; 254th meeting, April 1943, APS. L. Marton, "APS Division of Electron Physics. The First 20 Years," *Physics Today* 17, no. 10 (1964): 44–50.

28. Council Minutes of the APS, 255th meeting, June 1943, APS. "Current Trends" (n. 27).

29. W. James Lyon, remarks, and other materials, APS, Division of High-Polymer Physics, Papers relating to the formation of the division, Niels Bohr Library, AIP. Council minutes of the APS, 257th meeting, November 1943; 260th meetng, April 1944, APS.

30. Council Minutes of the APS, 260th meeting, April 1944; 261st meeting, June 1944; 263rd meeting, December 1944, APS.

31. F. Seitz, "Solid," *Physics Today* 2, no. 6 (1949): 18–22. R. Smoluchowski to F. Seitz, 23 November 1943; Seitz to Smoluchowski, 29 November 1943; letter of Committee of Six, January 1944; Wheeler Davey to Smoluchowski, 28 January 1944, and other materials, APS, Division of Solid State Physics, records, Niels Bohr Library, AIP. Smoluchowski had also approached metallurgist Charles Barrett of the Carnegie Institute of Technology, but Barrett hesitated to join the group.

32. J. Van Vleck to R. Smoluchowski, 29 January 1944; Smoluchowski to Van Vleck, 3 February 1944; Van Vleck to Smoluchowski, 16 February 1944, APS, Division of Solid State Physics, records.

33. R. Smoluchowski to F. Seitz, 28 February 1944; R. Smoluchowski to K. K. Darrow, 25 April 1944, APS, Division of Solid State Physics, records.

34. K. K. Darrow, "Current Trends in the American Physical Society," *Journal of Applied Physics* 14 (1943): 437–438. K. K. Darrow to F. Seitz, 16 May 1944, copy in APS, Division of Solid State Physics, records.

35. Seitz to Darrow, 6 May 1944; Darrow to Seitz, 16 May 1944; Seitz to Darrow, 25 May 1944; Darrow, memorandum, 16 June 1944, and other materials, APS, Division of Solid State Physics, records. Council Minutes of the APS, 261st meeting, June 1944, APS.

36. L. Brillouin to R. Smoluchowski, 25 January 1944; C. Herring to R. Smoluchowski, 13 February 1944; Smoluchowski to Herring, 15 February 1944; F. Seitz to R. Smoluchowski, 3 March 1944, 22 May 1944, 14 June 1944, APS, Division of Solid State Physics, records.

37. D. Wooldridge, "Luminescence and Projection Television," Bell Labs Technical Memorandum, no. 9, January 1940, quoted in Wooldridge, interview with L. Hoddeson, 1976, p. 41, AIP. In early 1944, Bert Warren wrote to Seitz to suggest the name "solid state"; excerpts of his and Wannier's letters are in APS, Division of Solid State Physics, records. "Division of Solid-State Physics": K. K. Darrow to R. Smoluchowski, 22 March 1944, APS, Division of Solid State Physics, records. Darrow tended to favor the hyphen, but not even he was entirely consistent in using it.

38. C. Zener to R. Smoluchowski, 24 March 1947; Smoluchowski to Zener, 9 April 1947, APS, Division of Solid State Physics, records.

39. "Symposium on the Solid State," *Reviews of Modern Physics* 17, no. 1 (1945); Proceedings of the American Physical Society, *Phys. Rev.* 67 (1945): 199–200.

40. Committee of Six to Harvey Fletcher, 4 May 1945, APS, Division of Solid State Physics, records; Council Minutes of the APS, 266th meeting, May 1945, APS.

41. Council Minutes of the APS, 270th meeting, January 1946, through 279th meeting, June 1947, APS. R. Smoluchowski to council of the APS, 20 September 1946, and other materials, APS, Division of Solid State Physics, records. At this point, Zener had joined the Committee of Six to replace Dushman, who retired, citing pressure of other work.

42. *Bulletin of the American Physical Society* 24, no. 3 (March 1949); Proceedings of the American Physical Society, *Phys. Rev.* 75 (1949): 1624; David Lazarus, manuscript autobiography, Niels Bohr Library, AIP. F. Seitz, interview with S. Weart, October 1982, AIP. Earlier solid-state symposia were held at Montreal, June 1947, and New York, January 1948.

43. *Bulletin of the American Physical Society* 26, no. 2 (March 1951); 27, no. 2 (March 1952); 30, no. 2 (March 1955). "Preponderance": K. Darrow, "Proceedings of the American Physical Society, Minutes of the 1953 Spring Meeting Held at Washington D.C., April 30, May 1 and 2, 1953," *Phys. Rev.* 91 (1953): 428. Van Vleck: Council Minutes of the APS, 319th meeting, April 1953, APS.

44. *Physics Today* 6, no. 3 (1953): 21.

45. *Bulletin of the American Physical Society* 30, no. 1 (January 1955); Council Minutes of the APS, 330th meeting, January 1955; 332nd meeting, April 1955, APS.

46. Note also the *Journal of the Mechanics and Physics of Solids* begun by Pergamon in 1952. The Soviet journal was translated by the American Institute of Physics as *Soviet Physics—Solid State.*

47. Most cited: my counts from *Solid State Physics* and from several leading review monographs of the 1950s. Split: Council minutes of the APS, 385th meeting, January 1963, APS.

48. *Solid State Abstracts* (1960–). Seitz interview (n. 42). F. Seitz and D. Turnbull, *Solid State*

Physics: Advances in Research and Applications (New York: Academic Press, 1955–). Note also the annual volumes *Progress in Semiconductors,* ed. A. F. Gibson, R. E. Burgess, and P. Aigrain (London: Heywood, 1956–). C. Kittel, *Introduction to Solid State Physics* (New York: Wiley, 1953, 1956); see 2nd ed., vii. The fourth edition (1971) was 150 pages thicker. Some other textbooks and monographs of the period were the second edition, little changed from the first (1940) edition, of N. F. Mott and R. W. Gurney, *Electronic Processes in Ionic Crystals* (Oxford: Clarendon Press, 1948); W. Shockley, *Electrons and Holes in Semiconductors* (Princeton, N.J.: Van Nostrand, 1950); Slater (n. 17); A. H. Wilson, *Theory of Metals,* 2nd ed. (Cambridge: Cambridge University Press, 1953); R. E. Peierls, *Quantum Theory of Solids* (Oxford: Clarendon Press, 1955).

49. J. C. Slater, "The Physics of Metals" (Amsterdam Conference), *Physics Today* 2, no. 1 (January 1949): 6–13. Instituts Solvay, Conseil de Physique [Ninth], *L'Etat solide* (Brussels: Stoops, 1952); Instituts Solvay [Tenth], *Les Electrons dans les métaux* (Brussels: Stoops, 1955).

50. Peierls (n. 48).

51. Jacques Friedel, *French Bibliographic Digest. Solid State Physics* (New York: Cultural Center of the French Embassy, 1963). A copy was kindly furnished by Stephen Keith. Fausto Fumi et al., "The Origins of Solid State Physics in Italy: 1945–1960," in "The Origins of Solid State Physics in Italy, 1945–1960," ed. G. Giulani [Conference Proceedings, Pavia, 21–24 September 1987], *Italian Physical Society Proceedings,* vol. 13 (Bologna: Italian Physical Society, 1988).

52. Conferences: for example, Physikalische Gesellschaft in der Deutschen Demokratische Republik, *Festkörperphysik und Physik der Leuchtstoffe* [Erfurt, 1957] (Berlin: Akademie, 1958). Note also the chair in *Festkörperphysik* taken by F. Matossi in Freiburg i. Br. in 1957 (*Physikalische Blätter* 13 [1957]). All of this information was furnished by Gisela Torkar.

53. Information from Seeger, October 1982. Note also the chair for *Festkörperphysik* taken by Wilhelm von Meyeren at the TH Hanover. *Taschenbuch für das Wissenschaftlichen Leben* (Vademecum Deutschen Lehr- und Forschungstätten, vol. 3 (Bonn: Festland, 1961).

54. Deutsche Physikalische Gesellschaft, *Organisationsübersicht,* various years. For this and other information, I am grateful to Günter Küppers.

55. Urbana: see Chapter 8. Cavendish: James G. Crowther, *The Cavendish Laboratory 1874–1974* (New York: Science History, 1974), chap. 29. Leningrad and Kharkov: *Soviet Physics—Solid State* 9 (1968): 2402, 2404.

56. Lillian Hoddeson, "The Discovery of the Point-Contact Transistor," *Historical Studies in the Physical Sciences* 12 (1981): 41–76; M. J. Kelly, authorization for work, case no. 38139, 1 January 1945, Bell Labs Archives.

57. Slater (n. 7), 227–237. Shoemaker, in McLachlan and Glusker (n. 14), 51–53. Francis Bitter, *Selected Papers and Commentaries,* ed. T. Erber and C. M. Fowler (Cambridge, Mass.: MIT Press, 1969). Bristol: department organization roster, November 1947, copy from the university archives kindly furnished by Steven Keith.

58. G. E. organization tables and other information kindly furnished by George Wise.

59. Seitz (n. 31); Seitz interview (n. 42).

60. The Research Development Board panel was established in 1947, according to Lawson McKenzie to Rear Admiral L. C. Coates, 19 February 1963, Archives of the National Academy of Sciences, Washington, D.C. (assistance kindly provided by Janice F. Goldblum). The panel was established in 1949, according to National Academy of Sciences, *Annual Report 1962–1963,* 91. I have not found direct documentation.

61. Seitz interview (n. 42). Robert Sproull, interview with S. Weart, July 1983, AIP. 1951 schedule: Slater (n. 7).

62. U.S. Air Force, Office of Scientific Research, Historical Division, "A Summary View of the AFOSR Solid State Sciences Program" (Washington, D.C.: AFOSR Information Division, 1961), copy at AIP.

63. Sources in notes 60–62 and "Distribution List—Solid State Advisory Panel Meeting at Oak Ridge" (Spring 1957), copy at AIP, provided by R. Sproull.

64. U.S. Navy, Office of Naval Research, Solid State Advisory Panel, "Solid State Physics Research, Performance and Promise" [Sproull Report], October 1957, copy of this and of relevant correspondence at AIP, provided by R. Sproull. See John Finney, "Defense Scientists Fear Cut," *New York Times,* 22 September 1957.

65. Membership list: National Academy of Sciences—National Research Council, Division of

Physical Sciences, *Annual Report, 1963,* Archives of National Academy of Sciences, Washington, D.C.

66. See Smith (n. 6), 356. U.S. Navy (n. 64), 4, 67–68.

67. Defense Science Board: Seitz interview (n. 42). Sproull interview (n. 61). Slater (n. 7), 268–271.

68. U.S. Federal Security Agency, Office of Education, *Manpower Resources in Physics 1951,* National Science Register, Scientific Manpower Series, no. 3 (Washington, D.C.: Government Printing Office, 1952). To the 429 solid-state physicists (with or without Ph.D.) listed, we add all those in the two specialties mentioned and in "quantum theory—solids," plus another 20 to allow for solids experts in such fields as cryogenics and physical electronics. This gives a total of about 500. We multiply this by the ratio, 3500/6597, of the census total of Ph.D. physicists to all physicists surveyed, and by the ratio, 8.1/6.5, of the percentages representing Ph.D. solid-state physicists and all solid-state physicists in the survey categories.

69. U.S. National Science Foundation, *American Science Manpower 1960: A Report of the National Register of Scientific and Technical Personnel* (Washington, D.C.: Government Printing Office, 1962), esp. 83–84. For 1970 figures and age structure, see National Research Council, Physics Survey Committee, *Physics in Perspective* [Bromley Report], 2: pt. A, *The Core Subfields of Physics* (Washington, D.C.: National Academy of Sciences, 1972), 451; 2: pt. C, *Statistical Data,* 1514.

70. Federal Security Agency (n. 68); National Science Foundation (n. 69).

71. U.S. Navy (n. 64), 11–13; U.S. Congress, House Committee on Science and Astronautics, *The National Science Foundation. A General Review of Its First Fifteen Years* (Washington, D.C.: Government Printing Office, 1966), 59. U.S. Navy, Naval Research Advisory Committee, *A Report to the Secretary of the Navy on Basic Research in the Navy* [Bacher Report, PB 151925] (Washington, D.C.: Department of Commerce, Office of Technical Services, 1959), 59. The estimate of funds per physicist over time is based on a variety of anecdotal and other evidence and is highly tentative.

72. The Air Force estimated that its 1960 budget of $3.5 million was about 10% of the total federal effort or 5% of the total national effort, which is of the same order of magnitude. U.S. Air Force (n. 62), 9. National Research Council (n. 8), 144–146. I do not use the Pake Report's estimates for the number of condensed-matter physicists (4000–6000, nearly half of all Ph.D. physicists), which seems much larger than the number of solid-state physicists. U.S. Navy (n. 64), 10.

73. E. Zaleski et al., *Science Policy in the USSR* (Paris: Organization for Economic Cooperation and Development, 1969). Organization for Economic Cooperation and Development, *International Statistical Year for Research and Development. A Study of Resources Devoted to R&D in OECD Member Countries in 1963/64,* vol. 2: *Statistical Tables and Notes* (Paris: OECD, 1968). The straight R&D scientists and engineers and gross national expenditures totals show the United States well ahead of the sum of other OECD countries, but closer inspection shows that for basic research and the natural sciences the United States was not predominant: the U.S. totals included social sciences and many lower-level industrial personnel. See p. 31 for costs of research.

74. Study of *Physics Abstracts* in National Research Council (n. 69), 2: pt. A, 513.

75. Seitz and Turnbull (n. 48), 1: ix.

76. See especially R. Bozorth, Preface to Conference on Magnetism and Magnetic Materials [Sixth, New York, 1960], Proceedings, *Journal of Applied Physics,* suppl. 32 (1961): 1S. For international conferences, see Bates (n. 13), 612–613.

77. GE Labs: information from George Wise. Conference on Semi-conducting Materials, University of Reading, England, 1950, organizers R. W. Ditchburn and N. F. Mott: *Semi-conducting Materials,* ed. H. K. Henisch (London: Butterworth, 1951); International Conference on Semiconductor Physics [First, Amsterdam, 1954], Proceedings, in *Physica,* 20, no. 11 (1954). Gibson et al. (n. 48). Some examples: Abram F. Ioffé, *Physics of Semiconductors* (New York: Academic Press, 1960), from his *Poluprovodniki v sovremenoi fizike* (1954); Shockley (n. 48); R. A. Smith, *Semiconductors* (New York: Cambridge University Press, 1959); E. Spenke, *Elektronische Halbleiter* (Wiesbaden: Springer, 1955); J. P. Suchet, *Physique des semiconducteurs* (Paris: Dunod, 1961); the journal *Fizika i Tekhnika Poluprovodnikov* (*Soviet Physics—Semiconductors* [1967–]). For Germany, see n. 54.

78. *Physics Abstracts,* Section A of *Science Abstracts* (London: Institution of Electrical Engineers).

79. *Physics Abstracts* 63, no. 745 (1960), 1.

80. American Institute of Physics, *American Institute of Physics Handbook* (New York: McGraw-Hill, 1957). E. U. Condon and Hugh Odishaw, *Handbook of Physics* (New York: McGraw-Hill, 1958). John C. Slater, "The Solid State," *Physics Today* 5, no. 1 (January 1952), 10–15, p. 10.

81. William Bragg, *An Introduction to Crystal Analysis* (London: Bell, 1928), 3. Mott and Gurney (nn. 17, 48). F. Seitz, "Solid-state Physics," in *McGraw-Hill Encyclopedia of Science and Technology* (New York: McGraw-Hill, 1960), 12: 479–481. "Unifying influence": Kittel (1953) (n. 48), v. Similarly, Frederick C. Brown, *The Physics of Solids: Ionic Crystals, Lattice Vibrations, and Imperfections* (New York: Benjamin, 1967), 1–2. For understanding over empiricism as the driving force behind the AFOSR funding program in the 1950s, see, for example, U.S. Air Force (n. 62), 6. R. E. Peierls, *The Laws of Nature* (New York: Scribner, 1956), 203.

82. R. Pohl, interview with C. A. Hempstead, 1974, in *Beginnings,* 114. Also in *Beginnings,* see N. F. Mott, "Memories of Early Days in Solid State Physics," 61; F. Bloch, "Memories of Electrons in Crystals," 27; R. Peierls, "Reflections of Early Solid State Physics," 34. For Pauli, the memory is authentic; see Chapter 2.

83. For Anderson, see Jeremy Bernstein, *Three Degrees Above Zero: Bell Labs in the Information Age* (New York: Scribner, 1984), 127. Slater (n. 7), 218; Seitz interview (n. 42). F.R.N. Nabarro, "Recollections of the Early Days of Dislocation Physics," in *Beginnings,* 132–133. Sproull interview (n. 61); cf. Edwin Salpeter, interview with S. Weart, March 1978, pp. 47–48, AIP.

84. Peierls (n. 48), v; Peierls (1956) (n. 81). For the popularization of nuclear physics, see S. Weart, *Nuclear Fear: A History of Images* (Cambridge, Mass.: Harvard University Press, 1988).

85. National Research Council, Solid States Sciences Panel [R. Smoluchowski, chair], *Research in Solid-State Sciences: Opportunities and Relevance to National Needs* (Washington, D.C.: National Academy of Sciences, 1968), 1. "Special place": ibid., 16, repeated verbatim from the panel report by W. Kohn et al. used in National Research Council (n. 8).

86. "Finished subject": Slater (n. 17), vii. Similarly, J. Bardeen (citing B. Pippard): "Much of the future of solid-state physics will be in applying fairly well established principles in new combinations . . . ," and so on ("Solid-State Physics—Time to Level Off?" *Scientific Research,* 17 March 1969, pp. 26–28). "Different sort of satisfaction": Slater (n. 80), 15. The word *interesting* is frequent in Slater's books. For a particularly eloquent statement of the satisfactions of applied work, see Smith (n. 9), 70.

87. "The separation between basic discoveries and applications in condensed matter is less distinct than in some of the other subfields of physics. As in other branches of physics, basic work is motivated by a quest for knowledge and understanding . . . " (Physics of Condensed Matter Panel [George H. Vineyard, chair], in National Research Council [n. 69], 2: pt. A, 455, 460). "fundamental . . . of greater significance": W. Kohn et al., in National Research Council (n. 8), 68. See Brown (n. 81), 1–2.

88. E. Wigner, "On the Structure of Solid Bodies," *Scientific Monthly* 42 (1936): 40–46, p. 40. Mott (1956) paraphrased in Crowther (n. 55), 343.

89. William H. Bragg, *Concerning the Nature of Things* (New York: Harper, 1925; reprint, New York: Dover, 1932, 1960), 5. Slater (n. 17), 224. "Richness": Bardeen (n. 86). "Wealth . . . possibilities": W. Kohn et al., in National Research Council (n. 8), 153. "Unique diversity": Martin Blume and Roman Smoluchowski, *Comments on Solid State Physics* (London: Gordon and Breach, 1968), 1: no. 1, ii.

90. On unity and diversity as viewpoints, see S. Weart, "The Last Fifty Years—A Revolution?" *Physics Today* 34, no. 11 (November 1981): 37–49. John M. Ziman, "Solid State," *Physics Today* 21, no. 5 (May 1968): 53–58, p. 57.

91. C. S. Smith, "The Prehistory of Solid-State Physics," *Physics Today* 18, no. 12 (December 1965): 18–30; Smith (n. 6), chap. 14, esp. 367–369. Cf. F. Seitz, "The Contribution of Modern Physics to Metallurgy," *Journal of Applied Physics* 19 (1948); 973–987.

92. Bardeen (n. 86). Similarly, National Research Council (n. 69), 2: pt. A, 455, and similar reports. Seitz interview (n. 42), referring to Harvey Brooks and Harvard.

93. Slater (n. 7), 336 ff.

94. R. Smoluchowski, "Creative Research," *Physics Today* 2, no. 4 (April 1949): 16–17. F. Seitz, "Promises and Constraints in Science," in U.S. Office of Naval Research, *Research in the*

Service of National Purpose (Washington, D.C.: Government Printing Office, 1966), 48–65, pp. 49–50. Smith (n. 6), 375, 377–378. The views of W. Shockley regarding the importance of individual and racial differences in intelligent tests may also be relevant. Some may be tempted to seek a connection between the ideas of American solid-state physicists and the ideas prevalent in the capitalist industries with which they have been closely involved; I hope those so inclined will begin with a comparative investigation of the ideas held by Soviet solid-state physicists.

95. Smith (n. 6), 355.

Contributors

Gordon Baym, born in 1935, graduated from Brooklyn Technical High School, and went to Cornell University, receiving his A.B. in 1956, and then to Harvard University, where he received his Ph.D. in 1960 for work in theoretical physics. After two years of research at the Niels Bohr Institute in Copenhagen and a year at the University of California, Berkeley, he joined the faculty of the University of Illinois, Urbana, where he has been ever since. He is a member of the U.S. National Academy of Sciences. His principal research has been in the physics of condensed matter in systems ranging from liquid helium to neutron stars and high-energy nuclear collisions.

Christine Blondel was born in 1949, studied physics in Paris, got a diploma in astrophysics, and received the physics *agrégation* in 1973. Then, at the Centre Alexandre Koyré, she prepared a thesis on the history of Ampère's electrodynamics. After teaching physics for four years, she is now a researcher of the CNRS at the Centre de Recherche en Histoire des Sciences et des Techniques in the Cité des Sciences et de l'Industrie (La Villette), Paris. She works on the history of electricity in science and technology.

Ernest Braun, born in 1925, was trained as a physicist, receiving his Ph.D. at the University of Bristol. His research was in solid-state physics in industry. He was lecturer at several universities and professor of physics at the University of Aston in Birmingham (1967). He changed his direction to social aspects of science and technology, and founded and headed the Technology Policy Unit at the University of Aston in Birmingham, for postgraduate teaching and research into technological innovation, technology policy, and technology assessment. His interest is in the history of science in the context of technological innovation. He is the author of numerous articles and books on these topics.

Michael Eckert, born in 1949, received a university education in physics with a 1976 diploma; his 1979 doctoral dissertation was on theoretical physics. Until 1981 he taught and did research in theoretical physics and biophysics. From 1981 to 1984, he collaborated in the International Project in the History of Solid State Physics team at the Deutsches Museum, Munich. Since 1985 he has been collaborating in a research project on Scientific Transfer in Nuclear Physics and Electronics and on the emigration of theoretical physicists. He is presently employed by the Bayer der Schullsuch Verlag in Munich.

Paul Hoch, born in 1942, studied mathematical and theoretical physics at City College of New York and Brown University, receiving his Ph.D. in 1967. After postdoctoral research at the University of Toronto, he went in 1968 to study philosophy and the history of science at London University. In 1973 he became professor of humanities at Dawson College, Montreal; while working on the International Project in the History of Solid State Physics, he was at the University of Aston in Birmingham. He is currently principal fellow and head of the Science Policy & Innovation Unit at the University of Nottingham. His work has been chiefly on the scientific migrations from Germany and central Europe in the 1930s and science policy.

Lillian Hoddeson, born in 1940, graduated from the Bronx High School of Science in 1957 and attended Brandeis University and Barnard College. She received her Ph.D. in physics at Columbia University in 1966 with a dissertation in solid-state physics. After eight years of teaching physics at Barnard and Rutgers University, she turned to the history of physics, taking graduate courses at

Princeton University. Her research from 1974 to 1979 was on solid-state developments at Bell Laboratories leading to the transistor. Since 1978, as the Fermilab historian, she has been studying the history of large-scale research, including particle accelerators and particle physics. Since 1984 she has been the principal historian of the team that wrote *Critical Assembly,* the first technical history of the building of the atomic bomb. During her work on the International Project in the History of Solid State Physics she was based at the University of Illinois, Urbana, where she teaches in the Department of History and is a member of the Department of Physics.

Stephen Keith, born in 1951, studied physics at the University of Sussex and social aspects of science and technology at the University of Aston in Brimingham, where he received his Ph.D. in 1982; he remained there until 1984, working on the International Project in the History of Solid State Physics. At present, he is teaching physics at Southwark College in London and is a visiting fellow of the Royal Institution Centre for the History of Science and Technology. His research to date has focused on the institutional aspects of modern science.

Gisela Oittner-Torkar, born in 1939, was trained at Siemens as "engineer assistant" and until 1973 worked on electronic tubes. She then received a university education in physics with a 1978 diploma; her 1980 doctoral dissertation was on experimental physics. From 1981 to 1984, she collaborated in the International Project in the History of Solid State Physics team at the Deutsches Museum, Munich.

Pierre Quédec was born in 1940. His 1964 doctoral dissertation was on solid-state physics. Since 1967 he has been *maître de conférences* at Paris 7 University. He was head of the Physics Department at the French Centre d'Etudes et de Recherches Mathématiques et Physiques in Beirut from 1970 to 1976 and head of the French Scientific Documentation Center in São Paulo from 1978 to 1980.

Helmut Schubert, born in 1950, had a university education in physics, graduating in 1977. His 1981 doctoral dissertation at the Technische Universität of Munich was on experimental solid-state physics. He worked at the Deutsches Museum in Munich from 1981 to 1984 on the International Project in the History of Solid State Physics, and from 1985 to 1987 on a research project on Scientific Transfer in Nuclear Physics and Electronics. He is currently working in the Patentstelle der Fraunhofer Gesellschaft in Munich.

Krzysztof Szymborski, born in 1941, studied theoretical physics at Warsaw University in Poland, where he received his master's degree in 1964. He did experimental research in biophysics and solid-state physics before receiving his Ph.D. in the history of science from the Polish Academy of Sciences in 1979. He also worked as a free-lance science writer and journalist until 1981, when he joined the International Project in the History of Solid State Physics, working out of the University of Illinois, Urbana. He is now science librarian at Skidmore College, where he teaches the history of science.

Jürgen Teichmann was born in 1941 and received a university education in science with a 1967 diploma in experimental physics; his 1972 doctoral dissertation was in the history of physics. Since 1970 he has been employed by the Deutsches Museum of Munich; since 1986 he has been the director of the museum's Department of Sciences. He received the *Habilitation* degree *(Privatdozent)* in 1986 from Munich University for a work on Robert Pohl and the history of color centers (*Zur Geschichte der Festkörperphysik, Farbzentrenforschung bis 1940* [Stuttgart: Steiner, 1988]). His interests include the history of electricity, astronomy, the relation between science and technology, and questions of pedagogy and popular education involving the history of physics. He is also the author of a book about interrelations between physics and astronomy in history (*Wandel des Weltbildes* [Hamburg: Rowohlt, 1989]).

Spencer Weart, born in 1942, received a B.A. in physics at Cornell University in 1963 and a Ph.D. in physics and astrophysics at the University of Colorado, Boulder, in 1968. He then did postdoctoral work in solar physics at the California Institute of Technology. In 1971 he enrolled as a graduate student in the History Department of the University of California, Berkeley, and in 1974 he became director of the Center for History of Physics of the American Institute of Physics in New York City, conducting oral history, documentation, and education programs in the history of nuclear physics, astrophysics, and other areas of the history of science. His books include *Physics circa 1900: Personnel, Funding and Productivity of the Academic Establishments,* with Paul Forman and John L. Heilbron (Princeton, N.J.: Princeton University Press, 1975); *Leo Szilard: His Version of the Facts,* edited with Gertrud Weiss Szilard (Cambridge, Mass.: MIT Press, 1978); *Scientists in Power* (Cambridge, Mass.: Harvard University Press, 1979); and *Nuclear Fear: A History of Images* (Cambridge, Mass.: Harvard University Press, 1988).

Index